Handbook of Experimental Pharmacology

Volume 166

Editor-in-Chief

K. Starke, Freiburg i. Br.

Editorial Board

G.V.R. Born, London
M. Eichelbaum, Stuttgart
D. Ganten, Berlin
F. Hofmann, München
B. Kobilka, Stanford, CA
W. Rosenthal, Berlin
G. Rubanyi, Richmond, CA

Transcription Factors

Contributors

S. Akira, A.Z. Ansari, D.N. Arnosti, C. Asker, E. Assenat,
B. Baumann, D. Bohl, V.J.N. Bykov, N. Corbi, R.B. Dickson,
L. Drocourt, Z. Dvorák, G. Foos, S. Gerbal-Chaloin,
R. von Harsdorf, L. Hauck, C.A. Hauser, J.-M. Heard,
A. Hecht, F.J. Herrera, Y. Kaneda, C.M. Klinge,
M. Kretzschmar, V. Kumar, J.J. La Pres, A. Lasar, J. Lecanda,
V. Libri, Z. Lu, A.K. Mapp, R. Marienfeld, P. Maurel,
C. Mendez-Vidal, R. Morishita, T. Ogihara, J.M. Pascussi,
C. Passananti, D.P. Sarkar, G. Selivanova, J.-H. Sheen,
D.D. Shooltz, K. Takeda, I. Timokhina, N. Tomita,
S.J. Triezenberg, M.J. Vilarem, M.T. Wilhelm, K.G. Wiman,
T. Wirth, Z. Wu, T.R. Zacharewski

Editors
Manfred Gossen, Jörg Kaufmann
and Steven J. Triezenberg

Dr. Manfred Gossen
Max-Delbrück-Centrum
für Molekulare Medizin
Robert-Rössle-Strasse 10
13125 Berlin, Germany
e-mail: mgossen@mdc-berlin.de

Dr. Jörg Kaufmann
atugen AG
Robert-Rössle-Strasse 10
13125 Berlin, Germany
e-mail: kaufmann@atugen.com

Professor
Steven J. Triezenberg
Department of Biochemistry
and Molecular Biology
Associate Director, Graduate Program
in Cell and Molecular Biology
510 Biochemistry Building
Michigan State University
East Lansing, MI 48824-1319, USA
e-mail: triezenb@msu.edu

With 72 Figures and 15 Tables

ISSN 0171-2004
ISBN 3-540-21095-4 Springer Berlin Heidelberg New York

Library of Congress Cataloging-in-Publication Data
Transcription factors / contributors, S. Akira ... [et al.] ; editors, Manfred Gossen, Jörg Kaufmann, and Steven J. Triezenberg. p. ; cm. – (Handbook of experimental pharmacology ; v. 166)
Includes bibliographical references and index.
ISBN 3-540-21095-4 (alk. paper)
1. Transcription factors. I. Akira, S. II. Gossen, Manfred. III. Kaufmann, Jörg. IV. Triezenberg, Steven J. V. Series.
[DNLM: 1. Transcription Factors–genetics. 2. Transcription Factors–pharmacology. 3. Transcription, Genetic. QH 450.2 T77173 2005]
QP905.H3 vol. 166   [QP552.T68]   615.1s–dc22   [572.8'845]   2004045233

This work is subject to copyright. All rights are reserved, whether the whole or part of the material is concerned, specifically the rights of translation, reprinting, re-use of illustrations, recitation, broadcasting, reproduction on microfilm or in any other way, and storage in data banks. Duplication of this publication or parts thereof is permitted only under the provisions of the German Copyright Law of September 9, 1965, in its current version, and permission for use must always be obtained from Springer-Verlag. Violations are liable to Prosecution under the German Copyright Law.

Springer is a part of Springer Science+Business Media
springeronline.com

© Springer-Verlag Berlin Heidelberg 2004
Printed in Germany

The use of general descriptive names, registered names, etc. in this publication does not imply, even in the absence of a specific statement, that such names are exempt from the relevant protective laws and regulations and free for general use.

Product liability: The publishers cannot guarantee the accuracy of any information about dosage and application contained in this book. In every individual case the user must check such information by consulting the relevant literature.

Editor: Dr. R. Lange
Desk Editor: S. Dathe
Cover design: design & production GmbH, Heidelberg
Typesetting: Stürtz GmbH, Würzburg
Printed on acid-free paper   27/3150 hs – 5 4 3 2 1 0

Preface

Transcription factors are key players in the execution of the genomic program of any given cell. Their control over complex patterns of gene expression governs essentially every step in the development, growth and differentiation of an organism as well as its physiological responses to external stimuli.

An Introduction to Transcription Factors, the first part of this volume, describes the varied and combinatorial mechanisms involved in the two basic modes of transcriptional control, activation (the chapter by Herrera et al.) and repression (the chapter by Arnosti). Overlaying both modes is an emerging emphasis on epigenetic regulation that uses intricate patterns of chromatin modification. Although many of these mechanisms have been elucidated by studying prototypical genes or transcription factors, the recently developed tools of genomic biology allow us to unravel the complexity of global transcriptional programs with an ever-increasing speed and accuracy (the chapter by Zacharewski and La Pres).

Given the central role that transcription factors play as relays positioned between the genome and the intracellular and extracellular signals to which cells must respond, it is not surprising that these regulatory proteins play a major role in pathological processes. Part 2 of this volume, Transcription Factors in Pathophysiology, introduces a selection of specific transcription factors and their families. Each of these factors has a well-established function in basic cellular mechanisms that, upon dysregulation, can cause or contribute to human disease. Transcription factors with a direct involvement in signaling pathways and/or growth control are emphasized in the chapters by Takeda and Akira, Hecht, Timokhina et al., Asker et al., Hauser et al., Hauck and von Harsdorf, Sheen and Dickson, and Lasar et al. Many of the chapters focus on the role of transcription factors in cellular transformation and cancer (those by Hecht, Timokhina et al., Asker et al., Hauser et al., Hauck and von Harsdorf, and Sheen and Dickson). Other chapters highlight the contributions of transcription factors to inflammatory responses (Lasar et al.) or xenobiotic responses (Pascussi et al.). Moreover, the chapter by Kumar and Sarkar illustrates how a viral transcription factor interferes with the physiology of its host cell.

These intricate connections between transcription factors and disease lead to the suggestion that these regulatory proteins might be suitable targets for therapeutic interventions. The third part of this volume, Exogenous Control over Transcription Factor Activity, considers the classical pharmacological intervention of transcription factor activity (Klinge), the design of artificial transcription factors that might be employed in gene therapeutic regimens (chapters by Corbi et al., Bohl and Heard, and Mapp et al.) and novel therapeutic concepts based on the functional sequestration of transcription factors (Morishita et al.).

Despite the depth of our understanding of transcriptional regulatory mechanisms (Part 1) and the breadth of our understanding of their roles in affecting cellular and organismal health (Part 2), this last part of the volume illustrates that the clinical implementation of our knowledge of transcription factors is still in its infancy. We thank all of those colleagues who have brought us to where we now stand, and we hope that this book will help point the way for those who will continue to advance the ever more exciting field of transcription factor biology.

Manfred Gossen
Jörg Kaufmann
Steven J. Triezenberg

List of Contributors

(Addresses stated at the beginning of respective chapters)

Akira, S. 97
Ansari, A.Z. 535
Arnosti, D.N. 33
Asker, C. 209
Assenat, E. 409

Baumann, B. 325
Bohl, D. 509
Bykov, V.J.N. 209

Corbi, N. 491

Dickson, R.B. 309
Drocourt, L. 409
Dvorák, Z. 409

Foos, G. 259

Gerbal-Chaloin, S. 409

Harsdorf, R. von 277
Hauck, L. 277
Hauser, C.A. 259
Heard, J.-M. 509
Hecht, A. 123
Herrera, F.J. 3

Kaneda, Y. 439
Klinge, C.M. 455
Kretzschmar, M. 167
Kumar, V. 377

La Pres, J.J. 69
Lasar, A. 325

Lecanda, J. 167
Libri, V. 491
Lu, Z. 535

Mapp, A.K. 535
Marienfeld, R. 325
Maurel, P. 409
Mendez-Vidal, C. 209
Morishita, R. 439

Ogihara, T. 439

Pascussi, J.M. 409
Passananti, C. 491

Sarkar, D.P. 377
Selivanova, G. 209
Sheen, J.-H. 309
Shooltz, D.D. 3

Takeda, K. 97
Timokhina, I. 167
Tomita, N. 439
Triezenberg, S.J. 3

Vilarem, M.J. 409

Wilhelm, M.T. 209
Wiman, K.G. 209
Wirth, T. 325
Wu, Z. 535

Zacharewski, T.R. 69

List of Contents

Part 1. An Introduction to Transcription Factors

Mechanisms of Transcriptional Activation in Eukaryotes 3
 F. J. Herrera, D. D. Shooltz, S. J. Triezenberg

Multiple Mechanisms of Transcriptional Repression in Eukaryotes 33
 D. N. Arnosti

Genomic Approaches to the Study of Transcription Factors 69
 T. R. Zacharewski, J. J. La Pres

Part 2. Transcription Factors in Pathophysiology

Biological Roles of the STAT Family in Cytokine Signaling 97
 K. Takeda, S. Akira

Members of the T-Cell Factor Family of DNA-Binding Proteins
and Their Roles in Tumorigenesis. 123
 A. Hecht

Transcription Factors in the Control of Tumor Development
and Progression by TGF-β Signaling 167
 I. Timokhina, J. Lecanda, M. Kretzschmar

The p53 Transcription Factor as Therapeutic Target in Cancer 209
 C. Asker, V. J. N. Bykov, C. Mendez-Vidal, G. Selivanova,
 M. T. Wilhelm, K. G. Wiman

The Role of Ets Transcription Factors in Mediating Cellular
Transformation. 259
 G. Foos, C. A. Hauser

Function of the E2F Transcription Factor Family
During Normal and Pathological Growth. 277
 L. Hauck, R. von Harsdorf

c-Myc in Cellular Transformation and Cancer 309
    J.-H. Sheen, R.B. Dickson

NF-κB: Critical Regulator of Inflammation and the Immune Response 325
    A. Lasar, R. Marienfeld, T. Wirth, B. Baumann

Hepatitis B Virus X Protein: Structure–Function
Relationships and Role in Viral Pathogenesis 377
    V. Kumar, D.P. Sarkar

Regulation of Xenobiotic Detoxification by PXR, CAR, GR, VDR
and SHP Receptors: Consequences in Physiology 409
    J.M. Pascussi, Z. Dvorák, S. Gerbal-Chaloin, E. Assenat, L. Drocourt,
    P. Maurel, M.J. Vilarem

Part 3. Exogenous Control of Transcription Factor Activity

Potential of Transcription Factor Decoy Oligonucleotides
as Therapeutic Approach... 439
    R. Morishita, N. Tomita, Y. Kaneda, T. Ogihara

Selective Estrogen Receptor Modulators
as Therapeutic Agents in Breast Cancer Treatment 455
    C.M. Klinge

Artificial Zinc Finger Peptides:
A Promising Tool in Biotechnology and Medicine..................... 491
    N. Corbi, V. Libri, C. Passananti

Tetracycline-Controlled Transactivators and Their Potential Use
in Gene Therapy Applications....................................... 509
    D. Bohl, J.-M. Heard

Modulating Transcription with Artificial Regulators................... 535
    A.K. Mapp, A.Z. Ansari, Z. Wu, Z. Lu

Subject Index 573

Part I
An Introduction to Transcription Factors

Mechanisms of Transcriptional Activation in Eukaryotes

F. J. Herrera · D. D. Shooltz · S. J. Triezenberg (✉)

Department of Biochemistry and Molecular Biology, Michigan State University,
East Lansing, MI 48824-1319, USA
triezenb@msu.edu

1	Modular Design of Transcriptional Activator Proteins.	4
1.1	DNA-Binding Domains	5
1.2	Activation Domains	5
2	Activating the Activators	7
3	Activator Association with Promoters.	8
3.1	Core Promoter Architecture and Activator Specificity	8
3.2	Enhancers and Upstream Activating Sequence Elements.	9
4	Actions of Activators at Promoters	10
4.1	Stepwise Recruitment of Basal Transcription Machinery.	11
4.1.1	TFIID, TBP, and TAFs	11
4.1.2	TFIIA, TFIIB, TFIIF, and TFIIH.	12
4.2	Holoenzyme Recruitment	13
4.3	Associations with Mediator Complex.	14
4.4	Chromatin-Modifying Coactivators	15
4.4.1	Covalent Modifications of Histones.	15
4.4.2	ATP-Dependent Remodeling Complexes	18
4.4.3	The p160 SRC Coactivator Family	20
4.5	Leaving a Mark: Activator Effects After Initiation.	20
5	Orchestration: Modular Design and Combinatorial Control	21
References	. . . .	21

Abstract Eukaryotic cells respond to growth, developmental and environmental cues in large part by regulating the expression of specific sets of genes. Befitting the wide range of these signals and the proper gene regulatory response, mechanisms of transcriptional activation in eukaryotes are impressively diverse. These mechanisms are built on the modular design of *cis*-acting DNA regulatory sequences and of *trans*-acting regulatory proteins, coupled with flexibility and diversity in the protein:protein interactions linking activators to chromatin-modifying enzymes and general transcription factors. This review summarizes and illustrates these principles of modular design and combinatorial logic underlying transcriptional activation in eukaryotes.

Keywords Gene regulation · Transcription · RNA polymerase II · Chromatin · Enhancer

The genome of any organism contains not only the information specifying the primary sequence of the many proteins required for growth, development and defense, but also regulatory information for determining when a given gene is to be expressed. The link between *cis*-acting regulatory DNA sequences and the expression of the corresponding genes is made by *trans*-acting transcriptional activator and repressor proteins. These regulatory proteins must fulfill two functions. They must accurately recognize which genes they are to regulate, and once associated with their target gene they must either stimulate (for activator proteins) or inhibit (for repressor proteins) transcription by RNA polymerase II (RNA pol II). The mechanisms for accomplishing these two functions are built on principles of modular design and combinatorial logic. For any given gene, the *cis*-regulatory elements comprise a set of short DNA sequences that together dictate the appropriate expression of that gene. The *trans*-regulatory proteins that bind such elements bring particular domains that interact with either chromatin-modifying enzymes or components of the general transcription machinery. These protein:protein interactions are highly diverse, both in the array of activators and target proteins present in a particular cell and also in that any given activator might bind to several target proteins. Given that many pharmacological agents also exert their effects (directly or indirectly) by modulating gene expression, this review will summarize the basic elements of these design principles. Subsequent chapters in this volume will expand on the specific biological roles and mechanistic details of individual transcriptional regulators of pharmacological interest.

1
Modular Design of Transcriptional Activator Proteins

The dual functions of a transcriptional activator protein, *cis*-element recognition and transcriptional activation, are typically fulfilled by distinct regions of the protein's primary structure. For example, the DNA-binding domain of the yeast Gal4 protein (a prominent activator in the yeast *Saccharomyces cerevisiae*) resides within the amino-terminal 100 amino acids, whereas the major transcriptional activation domain resides within the carboxyl-terminal 120 amino acids. This modular design seems advantageous both for evolutionary and technological appropriation. In the latter sense, a fusion protein linking the DNA-binding domain of Gal4 with the transcriptional activation domain of the VP16 protein from herpes simplex virus (Sadowski et al. 1988) is widely used in both in vitro and in vivo investigations into the mechanisms of transcriptional activation. A second example, comprising a fusion of the DNA-binding domain of the tetracycline repressor with the VP16 activation domain, allows the regulation of DNA binding by the presence or absence of the tetracycline ligand (Gossen and Bujard

1992). This regulatable artificial activator can function in a wide range of eukaryotes ranging from plants to mammals (Weinmann et al. 1994; Gossen et al. 1995; Gossen and Bujard 2002).

1.1
DNA-Binding Domains

The DNA-binding domains of a large number of eukaryotic transcriptional activator proteins have been extensively characterized by genetic, biochemical, and structural approaches. Recent reviews catalog the known structures and specificities (Luscombe and Thornton 2002) and highlight the common themes in structure and recognition (Garvie and Wolberger 2001). In many cases of pharmacological interest, the binding activity or specificity of a DNA-binding domain may be modulated by ligand binding, by dimerization with other DNA-binding proteins, or by association with other factors (reviewed by Marmorstein and Fitzgerald 2003). Subsequent chapters in this volume will describe various transcription factors that illustrate the diverse structures of DNA-binding domains and the effects of ligand interactions on their activities.

The principles of protein:DNA interaction have now been established to a sufficient degree to permit the design of DNA-binding modules of engineered specificity (Falke and Juliano 2003; Beerli and Barbas 2002). This is particularly true for the zinc finger families of transcription factors (Corbi et al. 1997; Kamiuchi et al. 1998; Wolfe et al. 2000; Sera and Uranga 2002). This ability to tailor novel chimeric transcriptional activators for recognition of DNA sequences that might not serve as native regulatory elements has profound implications for potential pharmacological application (Zhang et al. 2000; Liu et al. 2001; Lin et al. 2003). These principles and their implications will be further developed in the chapters by Corbi et al. and by Mapp et al. in this volume.

1.2
Activation Domains

In contrast to DNA-binding domains and despite substantial research, relatively little is known about the structures of transcriptional activation domains (TADs). By analysis of primary sequence, TADs have been broadly classified based on the abundance of particular amino acids, resulting in acidic, glutamine rich, proline rich, and other classes (Mitchell and Tjian 1989; Triezenberg 1995). Despite these amino acid preferences, however, careful mutational analyses have frequently indicated that the most critical elements of activation domains are the patterns of hydrophobic and aromatic amino acids (Cress and Triezenberg 1991; Triezenberg 1995; Jackson et al. 1996; Candau et al. 1997; Almlof et al. 1998).

Less is known about the secondary and tertiary structures of activation domains. A number of biophysical analyses of various regulatory proteins have shown that activation domains are largely unstructured in solution under physiological conditions (O'Hare and Williams 1992; Schmitz et al. 1994; Dahlman-Wright et al. 1995; Shen et al. 1996a). However, key amino acids in an activation domain can become conformationally constrained upon interaction with a target protein, suggesting that the most promising candidates for structural studies will be binary complexes between activators and targets. For example, circular dichroism spectra indicate that the c-Myc transactivation domain is induced to form a helical structure upon binding to TATA-binding protein (TBP) (McEwan et al. 1996). The activation domains of VP16 and of the estrogen receptor also become conformationally constrained upon interaction with TBP (Shen et al. 1996b; Warnmark et al. 2001). Furthermore, the VP16 activation domain appears to become helical upon binding a TBP associated factor, human TAF9 (Uesugi et al. 1997). An amphipathic helix from the p53 activation domain fills a hydrophobic cleft in the Mdm2 oncoprotein (Kussie et al. 1996). An amphipathic helix structure is also seen in the interface between the activator CREB and its coactivator protein CBP (Radhakrishnan et al. 1997). These examples support the model that an activator target provides a folding template for an unstructured activation domain, which might allow activation domains to interact with a number of different target proteins.

In some cases the tables may be turned: that is, an activation domain may provide the folding template for a potential target. Nuclear hormone receptors have a conserved carboxyl-terminal activation domain, known as AF-2. Ligand binding leads to a conformational change in the activator that opens a hydrophobic groove for interaction with transcriptional co-activators. In this case, the unstructured LxxLL peptide motif present in several coactivators (Heery et al. 1997) folds into an amphipathic helix upon binding the AF-2 region of a hormone receptor (reviewed in Warnmark et al. 2003).

Although in many cases the activation domains seem to adopt α-helical structures, that rule is not universal. Mutational and biophysical analysis of the Gal4 activation domain suggested that it might form a β-strand instead (Van Hoy et al. 1993). The activation domain of E2F-2, upon interaction with the Rb tumor suppressor protein, assumes a combination of helical and β-strand conformations (Lee et al. 2002). Together, these observations suggest that activators and their target proteins bind to each other using a highly diverse set of interaction surfaces.

2
Activating the Activators

The specificity of gene regulation in a given cell at different times or in response to various external or internal signals demands that the panoply of transcription factors in that cell should not all be active all of the time. This obvious observation implies that the activity of activators must be regulated, and indeed this principle has been demonstrated in a surprisingly diverse array of mechanisms. Some activators are regulated by control of their cellular localization from cytoplasm to nucleus. For example, NF-κB, an activator of cytokine-induced gene expression, is retained in the cytoplasm by interaction with IκB. Kinases triggered to action by ligand-bound cytokine receptors phosphorylate IκB, which then releases NF-κB to traverse to the nucleus (reviewed by Ghosh and Karin 2002). The sterol regulatory element-binding protein (SREBP) proprotein is anchored in the endoplasmic reticulum (ER) by two *trans*-membrane domains. Proteolytic cleavage releases SREBP from the ER membrane and allows its translocation to the nucleus where it activates genes involved in sterol biosynthesis (reviewed by Rawson 2003). Some nuclear hormone receptors are localized in the cytoplasm until binding to the relevant ligand allows translocation to the nucleus to activate transcription of target genes (reviewed by McKenna and O'Malley 2002a).

The activities of transcriptional activators can also be regulated by covalent modifications including phosphorylation, methylation, acetylation, and ubiquitinylation. A particularly impressive example is the tumor suppressor protein p53, which is subjected to multiple covalent modifications that tightly regulate its function and stability (reviewed by Brooks and Gu 2003; see also the chapter by Asker et al., this volume). Ubiquitinylation of p53 by Mdm2 (an E3 ubiquitin ligase) and subsequent destruction in the proteasome is responsible for maintaining p53 protein at low levels. Phosphorylation of p53 in response to ionizing radiation stabilizes the p53 protein. Acetylation of p53 also stabilizes p53 and stimulates its transcriptional activity.

One modification that has attracted much interest lately is the addition of ubiquitin to activators by ubiquitin ligases. Although long known as a 'tag' for protein degradation by the proteasome, ubiquitin (and a related polypeptide termed SUMO) is increasingly recognized for nonproteolytic roles as well. Most intriguing has been the suggestion that ubiquitin is required de facto for the function of activators in yeast and mammalian cells (reviewed by Muratani and Tansey 2003). One clue is that the peptide motif that targets activators for degradation overlaps with the transcriptional activation domain in proteins including various nuclear hormone receptors, the tumor suppressor p53, the proto-oncoproteins c-*Myc* and c-*Fos*, and the VP16 protein (Salghetti et al. 2000). Mutations in specific E3 ubiquitin ligases can stabilize an activator protein but reduce its transcriptional activity (Salghetti et al. 2001). Moreover, ectopic addition of a single ubiquitin moiety to an activator can enhance its transcription activity in a nonproteolytic manner. The counterpart to these clues regarding ubiquitin is the evidence that components of the proteasome can also be found associated with transcriptional promoters in conjunction with gene regulation (Ferdous et al. 2002; Gonzalez et al. 2002; Ottosen et al. 2002). Although the explicit roles of ubiquitin and the proteasome in the mechanism of transcriptional activation remain uncertain, one model proposes that the addition of ubiquitin may promote interaction of activators with key target proteins. Subsequent ligation of additional ubiquitin groups may lead to the degradation of the activator by the proteasome, providing a rapid downregulation of the activating signals (reviewed by Lipford and Deshaies 2003).

3
Activator Association with Promoters

3.1
Core Promoter Architecture and Activator Specificity

The *cis*-acting regulatory sequences for genes transcribed by RNA pol II typically comprise a combination of core promoter elements (reviewed in Smale and Kadonaga 2003) which serve as the binding site for the basal transcriptional factors and define the transcriptional startpoint, and upstream activating or repressing sequences, which serve as the binding sites for transcriptional activators and repressors. The most obvious feature in many core promoters is the TATA box, located about 30 bp upstream of the transcription start site. Surrounding the start site itself may be found an initiator (Inr) element. A downstream promoter element (DPE) has been defined in *Drosophila* and in mammalian genes, but not in yeast (Kadonaga 2002). Furthermore, a TFIIB recognition element, flanking the TATA box, has been described in some organisms (Smale and Kadonaga 2003).

Differences among core promoters of various genes, with respect to the presence and strength of these core elements, have pronounced effects on how those promoters respond to particular transcriptional activators. This specificity may allow a particular enhancer to differentially regulate various target genes, and may allow a particular core promoter to selectively respond to different enhancers. For example, the Sp1 activation domain strongly activates core promoters containing TATA elements, Inr elements, or both, whereas the Gal4–VP16 activator is most effective at a core promoter with both TATA and Inr (Emami et al. 1995). In an 'enhancer trapping' study in *Drosophila*, many enhancers were able to drive expression from both TATA dependent and DPE dependent promoters, but some enhancers preferentially activated either one or the other class of core promoter (Butler and Kadonaga 2001). The mechanistic differences leading to enhancer-core promoter specificity may involve the recruitment of transcriptional cofactors that display core promoter specificity. For example, the transcriptional cofactor NC2 represses transcription from TATA dependent promoters, but activates transcription from DPE dependent promoters (Willy et al. 2000). Thus, either direct or indirect recruitment of transcriptional cofactors by an activator may differentially regulate transcription from different core promoters, leading to activator-promoter specificity.

3.2
Enhancers and Upstream Activating Sequence Elements

In addition to the core promoter elements, transcription of many genes depends on *cis*-acting regulatory elements termed enhancers or upstream activating sequences, which provide the binding sites for transcriptional activators. These DNA elements can vary in sequence and affinity for a particular DNA-binding domain, and may exist in promoter-proximal locations or hundreds to thousands of base pairs upstream or downstream of a promoter. The consensus sequences of various *cis*-acting regulatory elements and the proteins that recognize and bind to those elements have been cataloged in various databases now available at Internet websites, including the Eukaryotic Promoter Database at www.epd.isb-sib.ch (Praz et al. 2002) and the TRANSFAC database at www.gene-regulation.com (Matys et al. 2003).

Combinations of activator-binding sites may be clustered into more complex regulatory elements. In such cases, cooperative binding of activators leads to the formation of large DNA–protein structures termed enhanceosomes, resulting in synergistic effects on transcription. In the prototypical virus-inducible interferon (IFN)-β enhanceosome (reviewed in Merika and Thanos 2001), a cluster of three different activator-binding sites directs the expression of IFN-β in response to viral infection. However, none of the activator-binding sites act alone; only the combination of all three activator-binding sites recapitulates the logic necessary to drive proper expression and specificity of the *IFN-B* gene. In this model system, the enhancer represents not simply the sum of individual activator functions, but rather an integration of inputs from different sources interpreted by the particular combination of transcription factors present at the enhancer.

Taking this organizational theme one step further, many genes may have multiple enhancers, each of which is poised to respond to particular developmental, growth, or environmental signals. Each of these independent enhancers may be simultaneously signaling to the core promoter either to stimulate or to repress transcription, depending on the signal inputs received by the regulatory proteins that bind there. This diverse and sometimes conflicting information must be integrated and interpreted at the promoter to make a final decision on whether or not transcription is to proceed. This 'information display' model for genes with multiple enhancers has arisen from studies of developmentally related genes in *Drosophila* (Kulkarni and Arnosti 2003), but will probably be relevant for many hormonally or pharmacologically regulated genes in humans as well.

With the increasing availability of genomic sequences for prominent experimental organisms, computational analysis for identifying *cis*-acting regulatory sequences has become a growth industry. In some cases, these searches focus on a particular *cis* regulatory elements, such as the estrogen response elements in mammalian genomes (Bajic et al. 2003). Other programs are designed for broader application, searching for sites corresponding to any of the transcription factors in the TRANSFAC database and combining site searches to increase the likelihood of identifying legitimate regulatory regions rather than idiosyncratic consensus sequence matches (Kel et al. 2003; Grabe 2002; Boardman et al. 2003). These computational approaches yield results that still must be validated by direct evidence of the function of putative elements in gene regulation and their interaction with specific transcription factors. Although this is often done on a case-by-case basis, either by mutational analysis of the *cis* elements or by in vitro binding assays, more global assessments are also now possible. For several transcription factors in yeast (Ren et al. 2000) and in mammalian cells (Weinmann et al. 2001; Ren et al. 2002), genomic mapping of transcription factor-binding sites has been accomplished by combining chemical crosslinking and immunoprecipitation (so-called chromatin IP or ChIP assays) with DNA microarrays comprising intergenic or putative regulatory sequences. These 'ChIP on chip' assays and other genomic aspects of transcription factor function are described in greater detail in the chapter by Zacharewski and LaPres in this volume.

4
Actions of Activators at Promoters

Once localized to a promoter, a transcriptional activator can interact with a number of different targets, including RNA pol II, the basal transcription factors, the mediator complex, coactivators, and chromatin-remodeling machinery. A common theme in models of activation is recruitment, where a promoter-bound activator localizes either a component of the transcriptional machinery or a transcriptional cofactor. This model is supported by evidence of direct physical interactions of activators with basal transcription factors, and by activator bypass experiments (reviewed in Ptashne and Gann 1997). In the latter experiments, a component of the transcriptional machinery is fused directly to a DNA-binding domain, and this artificial recruitment serves to activate transcription. Alternatively, in a variation of recruitment, a transcriptional activator may modulate the activity of components of the transcriptional machinery, facilitating the assembly of the preinitiation complex.

4.1
Stepwise Recruitment of Basal Transcription Machinery

Transcription of protein-coding genes requires the assembly of a preinitiation complex (PIC) comprising RNA pol II, the general transcription factors (GTFs), and a number of associated factors. In the stepwise model of PIC assembly, the TBP-containing TFIID complex binds to a promoter, followed by TFIIA, TFIIB, TFIIF and RNA pol II, and TFIIE and TFIIH (Buratowski et al. 1989). TFIID is the only general transcription factor that can specifically bind a promoter in the absence of interactions with other GTFs, suggesting that it nucleates the assembly of the PIC. Any step of PIC assembly might be rate limiting, and the recruitment of GTFs by association with activators may facilitate assembly.

4.1.1
TFIID, TBP, and TAFs

The TFIID protein complex comprises TBP and several TBP-associated factors (TAFs) (Burley and Roeder 1996; Pugh 2000). TBP binds selectively to the TATA core promoter element, while the TAFs extend the footprint to include the Inr and DPE elements. Studies in vitro show that although TBP is sufficient for basal transcription, the TAFs are required for activated transcription (Dynlacht et al. 1991; Sauer et al. 1995; Martel et al. 2002).

TBP can bind directly to transcriptional activation domains, as demonstrated by in vitro binding assays using a wide range of activator proteins, and mutations in activators that weaken activation also weaken interaction with TBP (Ingles et al. 1991; Shen et al. 1996b). While these results might suggest that activators simply recruit TBP, other evidence suggests that the mechanism is more complicated. The acidic activator Gal4 binds TBP competitively with TATA DNA (Xie et al. 2000a) and TBP and Gal4 do not bind cooperatively to promoters (Xie et al. 2000b), suggesting that competition for the DNA-binding domain of TBP may be involved in activation mechanisms.

The conserved core region of TBP is approximately symmetrical, as is the TATA DNA element, and in vitro, purified TBP binds the TATA element with little orientational preference (Cox et al. 1997). However, the interfaces of TBP with the rest of the transcriptional machinery are not symmetrical: the orientation of TBP on DNA defines the orientation of the overall preinitiation complex. The VP16 activation domain can significantly increase the orientational and axial specificity of TBP on the TATA DNA element (Kays and Schepartz 2002). This represents a variation on recruitment that might prime the TBP–DNA complex for productive association with additional GTFs and possibly work against the formation of incorrectly oriented preinitiation complexes.

The TFIID complex associates with TATA box DNA more slowly than does isolated TBP, implying that the TAFs contain inhibitory functions (Kokubo et al. 1993). Some activators, including VP16 and the Zta protein of Epstein-Barr virus, can stimulate the assembly of a TFIID–TFIIA–DNA complex (Lieberman and Berk 1994; Kobayashi et al. 1995), but do not stimulate a ternary complex when TBP is used instead of TFIID, suggesting that activators may counteract inhibitory functions of the TAFs. This ability of activators to stimulate ternary complex assembly appears to be relevant for activation, as mutations in an activation domain that reduce transcriptional activation potential in vivo also diminish the in vitro D–A assembly function (Kobayashi et al. 1998). In yeast, one of the TAF inhibitory activities has been mapped to two amino-terminal domains in the largest TAF subunit (Kotani et al. 1998). These TAF amino-terminal domains, called TAND1 and TAND2, bind to TBP and competitively inhibit the interactions of TBP with DNA and TFIIA, respectively (Kokubo et al. 1998; Kotani et al. 1998). The VP16C activation domain can compete with TAND1 for binding to TBP (Nishikawa et al. 1997). These observations have led to a handoff model for activation (Kotani et al. 2000), where VP16C competes with TAND1, in turn loosening the TAND2–TBP connection, which allows TFIIA to associate with TBP. Then, in a step that is still not understood, TBP is handed-off from VP16C to DNA, leading to TFIID–TFIIA–DNA complex assembly.

Other TAFs have a direct affinity for certain transcriptional activators, suggesting a more direct role in recruitment or modulation of activity. For example, the glutamine-rich activators Sp1, NFAT, and CREB interact with the TAF4 protein from *Drosophila* or human cells (Gill et al. 1994; Chiang and Roeder 1995; Felinski and Quinn 1999; Rojo-Niersbach et al. 1999; Wolstein et al. 2000; Kim et al. 2001). The acidic VP16 and p53 activation domains can interact with TAF9 (Goodrich et al. 1993; Klemm et al. 1995; Uesugi et al. 1997). Many other examples also exist. These interactions cannot always be interpreted to imply an effect on TFIID, as a significant number of these TAF proteins are also present in protein complexes lacking TBP that nonetheless influence transcriptional activation (Grant et al. 1998; Wieczorek et al. 1998).

4.1.2
TFIIA, TFIIB, TFIIF, and TFIIH

TFIIA is a positive cofactor in PIC assembly, as it binds cooperatively with TFIID at TATA DNA elements. TFIIA also functions as an antirepressor, inhibiting the TBP–DNA destabilizing actions of Mot1 and NC2 (reviewed in Pugh 2000). The formation of the ternary TFIID–TFIIA–DNA complex is a rate-limiting step in PIC formation, and activators can enhance this step (Lieberman and Berk 1994; Kobayashi et al. 1995). Some evidence points to direct association of the VP16 AD with subunits of TFIIA (Kobayashi et al. 1995, 1998).

TFIIB also stabilizes the TBP-TATA complex, and serves as a docking site for other components of the PIC. Several activators including the VP16 TAD have been shown to bind TFIIB with a high affinity (Lin et al. 1991), and the TAD-TFIIB connection has been implicated in transcriptional activation (Roberts et al. 1993). Interaction with the VP16 TAD has been shown to alter the conformation of TFIIB, possibly priming it for incorporation in the PIC (Roberts and Green 1994; Hayashi et al. 1998) or altering TFIIB-DNA contacts (Evans et al. 2001). Although this evidence is intriguing in pointing to TFIIB as a potential activator target, other reports have failed to find evidence supporting this association (Goodrich et al. 1993; Shen et al. 1996b).

TFIIH, which contains both protein kinase and nucleic acid helicase activities, also appears as a target for activation domains. The activation domains of VP16, p53, and E2F1 can interact with TFIIH (Xiao et al. 1994; Pearson and Greenblatt 1997) and recruitment of TFIIH may stimulate promoter escape (Kumar et al. 1998), but no clear evidence exists that activators stimulate the enzymatic activities of TFIIH. TFIIF and TFIIE might also be targets for activation domains. The serum response factor (SRF) interacts with the RAP74 subunit of TFIIF (Joliot et al. 1995), and Fos-Jun dimers can interact with both TFIIF and TFIIE (Martin et al. 1996). Although the mechanistic implications of these interactions have not yet been fully developed, the dual role of TFIIF as both an initiation and elongation factor suggests the possibility that activators might modulate promoter escape or elongation in addition to assembly of the preinitiation complex.

4.2
Holoenzyme Recruitment

Many of the models described in preceding sections are predicated on the premise that transcriptional activation involves a sequential recruitment of the basal transcription factors and RNA Pol II to form the PIC at the target promoter. This premise was challenged, however, by the biochemical purification from yeast cells of an extraordinarily large protein complex comprising RNA Pol II stably associated with a subset of GTFs together with additional polypeptides (the Mediator proteins, described below; Kim et al. 1994). Certain transcriptional activators were shown capable of recruiting this 'holoenzyme' to a promoter in a manner sufficient to achieve transcriptional activation in vitro (Kim et al. 1994; Hengartner et al. 1995). The model arising from these observations is that rather than separately and sequentially recruiting each general transcription factor, activation might more simply involve recruitment of the distinct TFIID and holoenzyme complexes. Similar RNA Pol II holoenzyme complexes have been also purified from human cells (Koleske and Young 1994) suggesting that the recruitment of the holoenzyme by activators might be an evolutionarily conserved mechanism (reviewed by Hampsey and Reinberg 1999). Kinetic and thermodynamic questions arising from the two competing models have not been fully resolved. How could the stepwise assembly occur quickly enough (given diffusion parameters for each component) for efficient transcriptional activation? And yet, how can a complex the size of the holoenzyme be translocated to specific genes at specific times quickly enough to respond to transcriptional activation?

4.3
Associations with Mediator Complex

The mediator complex, first identified as a component of the yeast RNA pol II holoenzyme (Kim et al. 1994), is composed of approximately 20 subunits forming three major domains (Gal 11, Med9/10 and Srb modules) that wrap around the RNA Pol II (reviewed by Myers and Kornberg 2000; Boube et al. 2002). Homologs of the yeast mediator subunits and similar protein complexes have since been identified in a wide range of organisms. Mammalian protein complexes resembling yeast mediator were described independently by several laboratories using biochemical purifications of proteins stably bound to different activator proteins (reviewed by Malik and Roeder 2000; Rachez and Freedman 2001). These different purifications lead to very similar complexes (variously termed TRAP, DRIP, ARC, CRSP, SMCC and PC2) suggesting that different activators bind the same or highly related human mediator complexes. The slight differences in protein compositions of these human mediators might represent variations due to differences in the biochemical purification procedures used or might represent different forms of the mediator complex that associate with different activators.

Several activators are known to interact physically and functionally with the mediator complex and the particular mediator subunits involved in these interactions are being identified. For example, the p53 tumor suppressor protein interacts with the TRAP80 subunit of the human mediator complex whereas the thyroid hormone receptor and PPARγ2 interact with TRAP220 (Ito et al. 1999; Ge et al. 2002). The VP16 activation domain may associate either with TRAP80 or with ARC92 (Ito et al. 1999; Mittler et al. 2003). Interferon-stimulated transcription depends on an interaction of STAT3 with the DRIP150 mediator component (Lau et al. 2003). These and other examples indicate that the mediator complex can be considered as a modular interface connecting activators with RNA Pol II allowing the integration of different signals during transcriptional activation.

4.4
Chromatin-Modifying Coactivators

Transcription activation must overcome the physical barriers presented by the packaging of eukaryotic DNA into chromatin. It is now clear that histones are not only a static scaffold for the compaction of DNA but rather they participate actively in the regulation of gene expression. Transcriptional activators affect chromatin with the assistance of two general classes of coactivator proteins. Some of these coactivators are enzymes that covalently modify amino acids within the histones themselves. These modifications then either directly alter chromatin structure, or serve as recognition signals for binding additional proteins that modulate that structure (Iizuka and Smith 2003). Other coactivators use the energy of ATP to remodel chromatin by sliding or removing nucleosomes.

4.4.1
Covalent Modifications of Histones

Covalent modifications that have been identified in histone proteins include acetylation, methylation, phosphorylation, ubiquitinylation, and ADP-ribosylation (reviewed by Berger 2002). The panoply of such modifications on the various histones has been likened to a 'code' (Strahl and Allis 2000; Jenuwein and Allis 2001) that, when deciphered and integrated, signals whether and how strongly a given gene is to be expressed. This histone code might be considered in several levels. First, any given modification at a given position on a given histone might either be present or absent. This implies the existence of complementary enzymes that either put on or take off the modifying mark. Second, at any given nucleosome or at several nucleosomes in any given promoter, various sets of such signals might be present. Third, the signals might occur sequentially—that is, one modification might serve as a signal to permit or stimulate another, or conversely one modification might block or inhibit the deposition of another. In recent years much effort has been dedicated to correlate specific histone modifications with the levels of gene expression of particular loci. The mechanisms of 'reading' this histone code might involve the binding of proteins to a particular histone modification or to specific combinations of these modifications. For instance, bromodomains and chromodomains, present in many transcription factors, can bind acetylated and methylated histones respectively. The bromodomain of the histone acetyltransferase Gcn5 can itself bind to acetylated histone H3 peptides (Hudson et al. 2000), which might contribute to the stability of these complexes at promoters of actively transcribed genes (Hassan et al. 2002). The heterochromatin protein HP1, through its chromodomain, binds methylated H3-K9, a modification prominent in heterochromatin regions. In contrast, the chromodomain present in the polycomb protein binds to methylated H3-K27, demonstrating that different chromodomains can bind different methylated residues in the histone proteins (Fischle et al. 2003; Min et al. 2003).

Hyperacetylation of the amino-terminal tails of histones (especially of H3 and H4) is generally correlated with gene activation, and conversely hypoacetylation is associated with gene repression (reviewed by Roth et al. 2001). The acetylation of the ϵ-amino group by histone acetyltransferases (HATs) neutralizes the positive charge of the lysine side chain, which may affect the affinity of histones for DNA, altering the packaging of the chromatin, or may affect protein–histone interactions contributing to the recruitment of specific transcription factors to active promoters. An important connection between histone modification and transcriptional activation was made when genetic evidence for transcriptional coactivator proteins was linked to biochemical evidence for histone acetylation (Brownell et al. 1996; Marcus et al. 1996). The first of these enzymes identified was Gcn5, a HAT present in the yeast multiprotein complexes termed SAGA and ADA. Gcn5 is recruited to promoters by activators and contributes to the hyperacetylation of histones associated with the promoter region during transcriptional activation (Kuo et al. 1998; Larschan and Winston 2001). Homologs of the yeast Gcn5 in other organisms play similar roles in transcription regulation, implying conserved mechanisms of regulation through evolution (Candau et al. 1996; Kusch et al. 2003; Vlachonasios et al. 2003).

Several distinct families of evolutionarily conserved HATs have been identified in a wide range of organisms (Carrozza et al. 2003). These groups include the GNAT family (*G*cn5-related *N*-*a*cetyl*t*ransferase), the MYST family (*M*OZ, *Y*bf2/Sas3, *S*as2 and *T*ip60), the highly related p300 and CBP coactivators, and the basal transcription factors TAF1 and TFIIIC. A major challenge for future research is to define the roles for each of the families and for each member of those families. This characterization should include identifying the specific amino acids in various histones that are substrates for various HATs, identifying the genes affected by their enzymatic activity, and identifying the regulatory proteins that recruit each HAT to the respective promoters.

Intriguingly, many of these HAT enzymes can also acetylate substrates other than histones and might thereby alter (either positively or negatively) the function of the target protein. Known nonhistone substrates include activator proteins such as p53, c-Myb, E2F, GATA-1, and MyoD; the general transcription factors TFIIE and TFIIF; and the high-mobility-group chromatin-associated proteins (reviewed by Sterner and Berger 2000). Thus the relevant in vivo substrates of a particular HAT might not necessarily be restricted to the histones.

Methylation of histones occurs on both Arg and Lys residues, most prominently in histones H3 and H4 (reviewed by Kouzarides 2002). Individual Lys residues can accommodate one, two or three methyl groups, and each isomer can exert a distinct and separate downstream effect. In contrast to histone acetylation, which typically is associated with transcriptionally active genes, methylation of certain histone amino acids corresponds to active loci whereas methylation of other residues is associated with inactive genes or even heterochromatic regions (Hampsey and Reinberg 2003). Thus, methylation of Arg3 of histone H4 is associated with transcriptional activation, as is methylation of lysines 4, 36 and 79 of histone H3. In contrast, methylation of lysines 9 and 27 of histone H3 or of Lys20 of histone H4 are correlated with transcriptional repression or silencing. Interestingly, histone methylation is not associated solely with nucleosomes near the promoter, but extends throughout the coding region of the gene. For instance, trimethylation of H3-K4 by the Set1 enzyme is a mark of the early phase of transcriptional elongation (Ng et al. 2003) whereas Set2 methylation of H3-K36 seems to mark subsequent stages of elongation (Krogan et al. 2003).

Discrete enzymes are responsible for methylating different amino acids in the histones. The Arg methyltransferases best known for their transcriptional role are CARM1 and PRMT1, both especially prominent as coactivators for nuclear hormone signaling (Xu et al. 2003). The Lys-specific histone methyltransferases typically possess a conserved catalytic domain known as the SET domain originally found in the SUV39, E(Z) and trithorax proteins of *Drosophila* (Lachner and Jenuwein 2002). Curiously, whereas the enzymes that attach other covalent modifications all have complimentary removal enzymes (histone deacetylases or phosphatases, for example) no histone demethylase has yet been described (Bannister et al. 2002).

Phosphorylation of histones, most notably at Ser10 of H3, has been correlated with both mitotic condensation of chromosomes and activation of gene expression in yeast, insects and mammals (Cheung et al. 2000; Nowak and Corces 2000; Lo et al. 2001). For transcriptional activation, this modification stimulates the subsequent acetylation of H3 Lys14 (Lo et al. 2000). Several kinases responsible for the phosphorylation of H3 Ser10 have been identified. In yeast, SNF-1 phosphorylates H3-S10 during transcription activation of genes involved in the biosynthesis of inositol (Lo et al. 2001). In mammalian cells, the immediate early response to epidermal growth factor through the mitogen activated protein kinase pathway results in the phosphorylation of H3 at the c-*fos* and c-*jun* gene promoters by the MSK1 and MSK2 kinases (Cheung et al. 2000; Thomson et al. 2001; Soloaga et al. 2003). The Ikk-α protein is the kinase responsible for the phosphorylation of H3 at promoters of nuclear factor (NF)-κB-regulated genes during cytokine-induced gene expression (Anest et al. 2003; Yamamoto et al. 2003). The Aurora family of H3 kinases are responsible for mitotic phosphorylation (Pascreau et al. 2003). Histone phosphatases have also been identified, most notably the Glc7 protein of yeast, but their roles in chromatin modification and gene regulation are relatively poorly defined at this time.

Although modification of proteins by covalent addition of the ubiquitin polypeptide is typically associated with subsequent degradation by the proteasome, this mark on histone proteins seems to serve other roles. Histones H2A and H2B are most often the targets of this modification, although ubiquitinylation of H3 and H1 has also been described. In contrast to the proteolytic pathway, in which multiple ubiquitin moieties might be added, histone ubiquitinylation typically comprises addition of a single group. Monoubiquitinylation of histones might affect higher-order chromatin folding or might affect histone–protein interactions by creating binding sites for particular transcription factors (reviewed by Zhang 2003). In yeast, transcriptional activation by Gal4 depends (in part) on both ubiquitinylation of H2B (by the ubiquitin ligase Rad6) and on subsequent removal of the ubiquitin by a Ub-specific protease, Ubp8. This latter enzyme is a component of the SAGA complex which also contains the Gcn5 histone acetyltransferase, suggesting an intimate relationship between histone ubiquitinylation and acetylation. Failure either to add or remove ubiquitin results in diminished gene activation and in altered levels of gene-associated methylation of Lys4 and Lys36 of histone H3, both modifications linked to transcription activation (Henry et al. 2003).

4.4.2
ATP-Dependent Remodeling Complexes

The second general class of chromatin-modifying transcriptional coactivators is the ATP-dependent chromatin remodeling complexes. These multiprotein complexes use the energy from ATP to 'remodel' chromatin by mechanisms that include alterations of DNA—histone contacts. The ATPase subunits of these complexes belong to the SNF2-like family of ATPases and can be classified in different subfamilies according to the presence of other protein motifs such as bromodomains, chromodomains and SANT domains (reviewed by Eisen et al. 1995; Narlikar et al. 2002). These complexes have both common and distinctive biochemical characteristics and are involved not only in transcription activation but also in other cellular functions that require unwrapping of DNA such as DNA repair, homologous recombination and chromatin assembly (reviewed by Lusser and Kadonaga 2003).

ATPases of the SNF2 subfamily contain a bromodomain and in general are part of multiprotein complexes of approximately 10–12 subunits designated SWI/SNF and RSC in yeast and Brahma in *Drosophila*. The human genome encodes two SNF2 homologs, BRG1 and BRM, present in similar but distinct protein complexes. These complexes regulate different sets of genes as demonstrated by the variations in mutant phenotypes. In yeast, SWI/SNF seems to control approximate 5% of yeast genes whereas RSC seems to play a more global effect in gene regulation (Sudarsanam et al. 2000; Ng et al. 2002). In mice, null mutations of the *Brg1* gene result in death of homozygotic embryos during the peri-implantation stage, whereas $BRM^{-/-}$ mutant mice develop normally although cell proliferation seems to be misregulated (Reyes et al. 1998; Bultman et al. 2000). The recruitment of SWI/SNF to gene promoters by activators has been observed in vivo and in vitro, in biological systems ranging from yeast to human. Mammalian activators including nuclear receptors, erythroid Kruppel-like factor, C/EBPβ, c-Myc, MyoD, HSF-1 and viral activators such as EBNA2 and VP16 are known to recruit human SWI/SNF complexes during transcriptional activation (reviewed by Narlikar et al. 2002).

Members of the ISWI subfamily of ATPases contain a SANT domain (SWI3, ADA2, N-CoR, TFIIIB) and are present in complexes comprising two to four subunits. These complexes include NURF, CHRAC and ACF from Drosophila, and vertebrate complexes such as RSF, hACF, hCHRAC and NoRC. Biochemical characterizations indicate that several of these complexes are involved in the assembly of chromatin (reviewed by Lusser and Kadonaga 2003). Members of the CHD1 subfamily of ATPases are characterized by the presence of a chromodomain and are present in complexes including NurD and Mi-2. These complexes also contain HDAC activities and are known to be recruited by certain repressor proteins (Kim et al. 1999; Schultz et al. 2001). The yeast Ino80 and Swr1 ATPases comprise another subfamily. Ino80 is present in a large (~12 subunits) protein complex that also contains DNA helicase activity. This complex is involved in transcription of genes encoding enzymes for phospholipid biosynthesis and also of genes such as PHO5, GAL1, CYC1 and ICL1 involved in unrelated pathways (Ebbert et al. 1999). Besides its role in transcription, this complex seems to play a role in DNA damage repair (Shen et al. 2000). The Swr1 ATPase is present in a complex that is able to exchange histone H2A variants (Mizuguchi et al. 2003). Other ATPases known to have chromatin-remodeling activities include the Arabidopsis DDM1, involved in maintaining normal levels of DNA methylation (Jeddeloh et al. 1999); Rad54, involved in homologous recombination (Alexeev et al. 2003); and CSB (Cockayne Syndrome protein B) involved in DNA excision repair (Citterio et al. 2000).

The mechanisms of these remodeling activities are not yet fully understood but include local and stable alterations of the DNA–histone contacts leading to sliding of nucleosomes along the DNA or transfer of nucleosomes from one DNA to another in trans. These protein complexes can also alter the superhelicity of DNA in vitro (reviewed by Narlikar et al. 2002). Repositioning of nucleosomes may alleviate chromatin-mediated transcription repression, for example by exposing DNA-binding sites for additional activators or by exposing core promoter elements that might be critical for the binding of general transcription factors and the formation of the pre-initiation complex.

4.4.3
The p160 SRC Coactivator Family

One additional group of transcriptional coactivators, known as the p160 steroid receptor coactivator (SRC) family, is particularly well known for their roles in gene regulation stimulated by the nuclear hormone receptors (reviewed by McKenna and O'Malley 2002b; Xu and Li 2003). This family contains three highly related members known as SRC-1 (or NCoA-1), SRC-2 (also known as GRIP1, TIF2 and NCoA-2), and SRC-3 (or p/CIP, RAC3, ACTR, and TRAM-1). Although these proteins possess weak histone acetyltransferase activity, their major role is to link the ligand-bound nuclear receptors to the chromatin-modifying enzymes detailed above. The SRC proteins exist in various complexes that also contain the HATs p300, CBP or PCAF, or the histone methyltransferases CARM1 and PRMT1. The recruitment of the SWI/SNF chromatin remodeling complex to hormone-regulated genes depends (probably indirectly) on the SRC proteins. Moreover, the Mediator complex (also known as TRAP, DRIP or SMCC) can be recruited to target promoters by direct interactions with SRC-containing complexes and perhaps also by direct interactions with the receptors themselves.

4.5
Leaving a Mark: Activator Effects After Initiation

Although most studies of transcriptional activation have focused on recruitment and initiation, subsequent steps including promoter escape and elongation can also be stimulated by activator proteins. Several lines of evidence indicate that activators might also work in post-initiation steps. One well-characterized example corresponds to the human and *Drosophila* gene encoding heat shock protein (hsp) 70. A paused RNA polymerase resides near the 5' end of the uninduced *hsp*70 gene. In response to heat shock, not only does the transcriptional initiation rate increase, but the pausing time is dramatically reduced (Rougvie and Lis 1988; Brown et al. 1996). Transcriptional activators can also stimulate rates of transcriptional elongation. For example, the heat shock factor-1 (HSF-1) involved in the activation of the *hsp*70 gene and the viral activators VP16 and E1A can stimulate elongation by mechanisms that are apparently different from that of stimulation of initiation (Yankulov et al. 1994; Brown et al. 1998). The interaction of activators including HIV tat, c-Myc, and NF-κB with the elongation factor P-TEFb (a kinase that modifies the carboxyl-terminal tail of RNA pol II) further illustrates this link (Kanazawa et al. 2000; Barboric et al. 2001; Cho et al. 2001; Kanazawa et al. 2003).

Transcription and RNA processing have often been considered as separate and sequential events, but a more recent perspective views these as a single integrated pathway (reviewed by Proudfoot et al. 2002). Capping, splicing and polyadenylation are tightly coupled to RNA Pol II through the carboxyl-terminal domain of the largest subunit (reviewed by Maniatis and Reed 2002). Selection of splice sites in the nascent RNA can be influenced by promoter elements in a manner that is independent of the promoter strength, suggesting that activators might also regulate alternative splicing decisions (Cramer et al. 1997; Auboeuf et al. 2002; Nogues et al. 2002).

5
Orchestration: Modular Design and Combinatorial Control

The unifying themes of modular design and combinatorial control can help make sense of the panoply of activators, targets, and mechanisms described in the preceding sections. A core promoter can be viewed as a combination of independently acting core elements, present with varying strengths. When taken together, the combination forms a core promoter that selectively responds to activating signals. Likewise, the collection of activators present in an enhancer determines the strength and specificity of transcriptional response, as in the IFN-β enhancer. In this conceptual framework of modular design and combinatorial control, the weakest link at present is our understanding of the collection and superposition of activating modules present within transcriptional activation domains. Although many activator–target interactions have been qualitatively described, few thorough quantitative assessments of particular interactions and of their integrated effects have been completed. In the future, the combination of structural details and quantification of affinities for targets will allow the dissection and classification of activators based on mechanisms and mechanistic potential, helping us to understand and use the combinatorial control that bridges activators to a transcriptional response.

References

Alexeev A, Mazin A, Kowalczykowski SC (2003) Rad54 protein possesses chromatin-remodeling activity stimulated by the Rad51-ssDNA nucleoprotein filament. Nat Struct Biol 10:182–186

Almlof T, Wallberg AE, Gustafsson JA, Wright AP (1998) Role of important hydrophobic amino acids in the interaction between the glucocorticoid receptor τ 1-core activation domain and target factors. Biochemistry 37:9586–9594

Anest V, Hanson JL, Cogswell PC, Steinbrecher KA, Strahl BD, Baldwin AS (2003) A nucleosomal function for IκB kinase-α in NF-κB-dependent gene expression. Nature 423:659–663

Auboeuf D, Honig A, Berget SM, O'Malley BW (2002) Coordinate regulation of transcription and splicing by steroid receptor coregulators. Science 298:416–419

Bajic VB, Tan SL, Chong A, Tang S, Strom A, Gustafsson JA, Lin CY, Liu ET (2003) Dragon ERE Finder version 2: A tool for accurate detection and analysis of estrogen response elements in vertebrate genomes. Nucl Acids Res 31:3605–3607

Bannister AJ, Schneider R, Kouzarides T (2002) Histone methylation: dynamic or static? Cell 109:801–806

Barboric M, Nissen RM, Kanazawa S, Jabrane-Ferrat N, Peterlin BM (2001) NF-κB binds P-TEFb to stimulate transcriptional elongation by RNA polymerase II. Mol Cell 8:327–337

Beerli RR, Barbas CF, 3rd (2002) Engineering polydactyl zinc-finger transcription factors. Nat Biotechnol 20:135–141

Berger SL (2002) Histone modifications in transcriptional regulation. Curr Opin Genet Dev 12:142–148

Boardman PE, Oliver SG, Hubbard SJ (2003) SiteSeer: Visualisation and analysis of transcription factor binding sites in nucleotide sequences. Nucl Acids Res 31:3572–3575

Boube M, Joulia L, Cribbs DL, Bourbon HM (2002) Evidence for a mediator of RNA polymerase II transcriptional regulation conserved from yeast to man. Cell 110:143–151

Brooks CL, Gu W (2003) Ubiquitination, phosphorylation and acetylation: the molecular basis for p53 regulation. Curr Opin Cell Biol 15:164–171

Brown SA, Imbalzano AN, Kingston RE (1996) Activator-dependent regulation of transcriptional pausing on nucleosomal templates. Genes Dev 10:1479–1490

Brown SA, Weirich CS, Newton EM, Kingston RE (1998) Transcriptional activation domains stimulate initiation and elongation at different times and via different residues. EMBO J 17:3146–3154

Brownell JE, Zhou J, Ranalli T, Kobayashi R, Edmondson DG, Roth SY, Allis CD (1996) Tetrahymena histone acetyltransferase A: a homolog to yeast Gcn5p linking histone acetylation to gene activation. Cell 84:843–851

Bultman S, Gebuhr T, Yee D, La Mantia C, Nicholson J, Gilliam A, Randazzo F, Metzger D, Chambon P, Crabtree G, Magnuson T (2000) A Brg1 null mutation in the mouse reveals functional differences among mammalian SWI/SNF complexes. Mol Cell 6:1287–1295

Buratowski S, Hahn S, Guarente L, Sharp PA (1989) Five intermediate complexes in transcription initiation by RNA polymerase II. Cell 56:549–561

Burley SK, Roeder RG (1996) Biochemistry and structural biology of transcription factor IID (TFIID). Annu Rev Biochem 65:769–799

Butler JE, Kadonaga JT (2001) Enhancer-promoter specificity mediated by DPE or TATA core promoter motifs. Genes Dev 15:2515–2519

Candau R, Moore PA, Wang L, Barlev N, Ying CY, Rosen CA, Berger SL (1996) Identification of human proteins functionally conserved with the yeast putative adaptors ADA2 and GCN5. Mol Cell Biol 16:593–602

Candau R, Scolnick DM, Darpino P, Ying CY, Halazonetis TD, Berger SL (1997) Two tandem and independent sub-activation domains in the amino terminus of p53 require the adaptor complex for activity. Oncogene 15:807–816

Carrozza MJ, Utley RT, Workman JL, Cote J (2003) The diverse functions of histone acetyltransferase complexes. Trends Genet 19:321–329

Cheung P, Tanner KG, Cheung WL, Sassone-Corsi P, Denu JM, Allis CD (2000) Synergistic coupling of histone H3 phosphorylation and acetylation in response to epidermal growth factor stimulation. Mol Cell 5:905–915

Chiang CM, Roeder RG (1995) Cloning of an intrinsic human TFIID subunit that interacts with multiple transcriptional activators. Science 267:531–536

Cho EJ, Kobor MS, Kim M, Greenblatt J, Buratowski S (2001) Opposing effects of Ctk1 kinase and Fcp1 phosphatase at Ser 2 of the RNA polymerase II C-terminal domain. Genes Dev 15:3319–3329

Citterio E, Van Den Boom V, Schnitzler G, Kanaar R, Bonte E, Kingston RE, Hoeijmakers JH, Vermeulen W (2000) ATP-dependent chromatin remodeling by the Cockayne syndrome B DNA repair-transcription-coupling factor. Mol Cell Biol 20:7643–7653

Corbi N, Perez M, Maione R, Passananti C (1997) Synthesis of a new zinc finger peptide; comparison of its 'code' deduced and 'CASTing' derived binding sites. FEBS Lett 417:71–74

Cox JM, Hayward MM, Sanchez JF, Gegnas LD, van der Zee S, Dennis JH, Sigler PB, Schepartz A (1997) Bidirectional binding of the TATA box binding protein to the TATA box. Proc Natl Acad Sci USA 94:13475–13480

Cramer P, Pesce CG, Baralle FE, Kornblihtt AR (1997) Functional association between promoter structure and transcript alternative splicing. Proc Natl Acad Sci USA 94:11456–11460

Cress WD, Triezenberg SJ (1991) Critical structural elements of the VP16 transcriptional activation domain. Science 251:87–90

Dahlman-Wright K, Baumann H, McEwan IJ, Almlof T, Wright AP, Gustafsson JA, Hard T (1995) Structural characterization of a minimal functional transactivation domain from the human glucocorticoid receptor. Proc Natl Acad Sci USA 92:1699–1703

Dynlacht BD, Hoey T, Tjian R (1991) Isolation of coactivators associated with the TATA-binding protein that mediate transcriptional activation. Cell 66:563–576

Ebbert R, Birkmann A, Schuller HJ (1999) The product of the SNF2/SWI2 paralogue INO80 of Saccharomyces cerevisiae required for efficient expression of various yeast structural genes is part of a high-molecular-weight protein complex. Mol Microbiol 32:741–751

Eisen JA, Sweder KS, Hanawalt PC (1995) Evolution of the SNF2 family of proteins: subfamilies with distinct sequences and functions. Nucl Acids Res 23:2715–2723

Emami KH, Navarre WW, Smale ST (1995) Core promoter specificities of the Sp1 and VP16 transcriptional activation domains. Mol Cell Biol 15:5906–5916

Evans R, Fairley JA, Roberts SG (2001) Activator-mediated disruption of sequence-specific DNA contacts by the general transcription factor TFIIB. Genes Dev 15:2945–2949

Falke D, Juliano RL (2003) Selective gene regulation with designed transcription factors: implications for therapy. Curr Opin Mol Ther 5:161–166

Felinski EA, Quinn PG (1999) The CREB constitutive activation domain interacts with TATA-binding protein-associated factor 110 (TAF110) through specific hydrophobic residues in one of the three subdomains required for both activation and TAF110 binding. J Biol Chem 274:11672–11678

Ferdous A, Kodadek T, Johnston SA (2002) A nonproteolytic function of the 19S regulatory subunit of the 26S proteasome is required for efficient activated transcription by human RNA polymerase II. Biochemistry 41:12798–12805

Fischle W, Wang Y, Jacobs SA, Kim Y, Allis CD, Khorasanizadeh S (2003) Molecular basis for the discrimination of repressive methyl-lysine marks in histone H3 by Polycomb and HP1 chromodomains. Genes Dev 17:1870–1881

Garvie CW, Wolberger C (2001) Recognition of specific DNA sequences. Mol Cell 8:937–946

Ge K, Guermah M, Yuan CX, Ito M, Wallberg AE, Spiegelman BM, Roeder RG (2002) Transcription coactivator TRAP220 is required for PPAR gamma 2-stimulated adipogenesis. Nature 417:563–567

Ghosh S, Karin M (2002) Missing pieces in the NF-κB puzzle. Cell 109 Suppl: S81–96

Gill G, Pascal E, Tseng ZH, Tjian R (1994) A glutamine-rich hydrophobic patch in transcription factor Sp1 contacts the dTAFII110 component of the Drosophila TFIID complex and mediates transcriptional activation. Proc Natl Acad Sci USA 91:192–196

Gonzalez F, Delahodde A, Kodadek T, Johnston SA (2002) Recruitment of a 19S proteasome subcomplex to an activated promoter. Science 296:548–550

Goodrich JA, Hoey T, Thut CJ, Admon A, Tjian R (1993) Drosophila TAFII40 interacts with both a VP16 activation domain and the basal transcription factor TFIIB. Cell 75:519–530

Gossen M, Bujard H (1992) Tight control of gene expression in mammalian cells by tetracycline-responsive promoters. Proc Natl Acad Sci USA 89:5547–5551

Gossen M, Bujard H (2002) Studying gene function in eukaryotes by conditional gene inactivation. Annu Rev Genet 36:153–173

Gossen M, Freundlieb S, Bender G, Muller G, Hillen W, Bujard H (1995) Transcriptional activation by tetracyclines in mammalian cells. Science 268:1766–1769

Grabe N (2002) AliBaba2: context specific identification of transcription factor binding sites. In Silico Biol 2: S1–15

Grant PA, Schieltz D, Pray-Grant MG, Steger DJ, Reese JC, Yates JR, 3rd, Workman JL (1998) A subset of TAF(II)s are integral components of the SAGA complex required for nucleosome acetylation and transcriptional stimulation. Cell 94:45–53

Hampsey M, Reinberg D (1999) RNA polymerase II as a control panel for multiple coactivator complexes. Curr Opin Genet Dev 9:132–139

Hampsey M, Reinberg D (2003) Tails of intrigue: phosphorylation of RNA polymerase II mediates histone methylation. Cell 113:429–432

Hassan AH, Prochasson P, Neely KE, Galasinski SC, Chandy M, Carrozza MJ, Workman JL (2002) Function and selectivity of bromodomains in anchoring chromatin-modifying complexes to promoter nucleosomes. Cell 111:369–379

Hayashi F, Ishima R, Liu D, Tong KI, Kim S, Reinberg D, Bagby S, Ikura M (1998) Human general transcription factor TFIIB: conformational variability and interaction with VP16 activation domain. Biochemistry 37:7941–7951

Heery DM, Kalkhoven E, Hoare S, Parker MG (1997) A signature motif in transcriptional co-activators mediates binding to nuclear receptors. Nature 387:733–736

Hengartner CJ, Thompson CM, Zhang J, Chao DM, Liao SM, Koleske AJ, Okamura S, Young RA (1995) Association of an activator with an RNA polymerase II holoenzyme. Genes Dev 9:897–910

Henry KW, Wyce A, Lo WS, Duggan LJ, Emre NC, Kao CF, Pillus L, Shilatifard A, Osley MA, Berger SL (2003) Transcriptional activation via sequential histone H2B ubiquitylation and deubiquitylation, mediated by SAGA-associated Ubp8. Genes Dev 17:2648–2663

Hudson BP, Martinez-Yamout MA, Dyson HJ, Wright PE (2000) Solution structure and acetyl-lysine binding activity of the GCN5 bromodomain. J Mol Biol 304:355–370

Iizuka M, Smith MM (2003) Functional consequences of histone modifications. Curr Opin Genet Dev 13:154–160

Ingles CJ, Shales M, Cress WD, Triezenberg SJ, Greenblatt J (1991) Reduced binding of TFIID to transcriptionally compromised mutants of VP16. Nature 351:588–590

Ito M, Yuan CX, Malik S, Gu W, Fondell JD, Yamamura S, Fu ZY, Zhang X, Qin J, Roeder RG (1999) Identity between TRAP and SMCC complexes indicates novel pathways for the function of nuclear receptors and diverse mammalian activators. Mol Cell 3:361–370

Jackson BM, Drysdale CM, Natarajan K, Hinnebusch AG (1996) Identification of seven hydrophobic clusters in GCN4 making redundant contributions to transcriptional activation. Mol Cell Biol 16:5557–5571

Jeddeloh JA, Stokes TL, Richards EJ (1999) Maintenance of genomic methylation requires a SWI2/SNF2-like protein. Nat Genet 22:94–97

Jenuwein T, Allis CD (2001) Translating the histone code. Science 293:1074–1080

Joliot V, Demma M, Prywes R (1995) Interaction with RAP74 subunit of TFIIF is required for transcriptional activation by serum response factor. Nature 373:632–635

Kadonaga JT (2002) The DPE, a core promoter element for transcription by RNA polymerase II. Exp Mol Med 34:259–264

Kamiuchi T, Abe E, Imanishi M, Kaji T, Nagaoka M, Sugiura Y (1998) Artificial nine zinc-finger peptide with 30 base pair binding sites. Biochemistry 37:13827–13834

Kanazawa S, Okamoto T, Peterlin BM (2000) Tat competes with CIITA for the binding to P-TEFb and blocks the expression of MHC class II genes in HIV infection. Immunity 12: 61–70

Kanazawa S, Soucek L, Evan G, Okamoto T, Peterlin BM (2003) c-Myc recruits P-TEFb for transcription, cellular proliferation and apoptosis. Oncogene 22:5707–5711

Kays AR, Schepartz A (2002) Gal4-VP16 and Gal4-AH increase the orientational and axial specificity of TATA box recognition by TATA box binding protein. Biochemistry 41:3147–3155

Kel AE, Gossling E, Reuter I, Cheremushkin E, Kel-Margoulis OV, Wingender E (2003) MATCH: A tool for searching transcription factor binding sites in DNA sequences. Nucl Acids Res 31:3576–3579

Kim J, Sif S, Jones B, Jackson A, Koipally J, Heller E, Winandy S, Viel A, Sawyer A, Ikeda T, Kingston R, Georgopoulos K (1999) Ikaros DNA-binding proteins direct formation of chromatin remodeling complexes in lymphocytes. Immunity 10:345–355

Kim LJ, Seto AG, Nguyen TN, Goodrich JA (2001) Human Taf(II)130 is a coactivator for NFATp. Mol Cell Biol 21:3503–3513

Kim YJ, Bjorklund S, Li Y, Sayre MH, Kornberg RD (1994) A multiprotein mediator of transcriptional activation and its interaction with the C-terminal repeat domain of RNA polymerase II. Cell 77:599–608

Klemm RD, Goodrich JA, Zhou S, Tjian R (1995) Molecular cloning and expression of the 32-kDa subunit of human TFIID reveals interactions with VP16 and TFIIB that mediate transcriptional activation. Proc Natl Acad Sci USA 92:5788–5792

Kobayashi N, Boyer TG, Berk AJ (1995) A class of activation domains interacts directly with TFIIA and stimulates TFIIA-TFIID-promoter complex assembly. Mol Cell Biol 15:6465–6473

Kobayashi N, Horn PJ, Sullivan SM, Triezenberg SJ, Boyer TG, Berk AJ (1998) DA-complex assembly activity required for VP16C transcriptional activation. Mol Cell Biol 18:4023–4031

Kokubo T, Gong DW, Yamashita S, Horikoshi M, Roeder RG, Nakatani Y (1993) Drosophila 230-kD TFIID subunit, a functional homolog of the human cell cycle gene product, negatively regulates DNA binding of the TATA box-binding subunit of TFIID. Genes Dev 7:1033–1046

Kokubo T, Swanson MJ, Nishikawa JI, Hinnebusch AG, Nakatani Y (1998) The yeast TAF145 inhibitory domain and TFIIA competitively bind to TATA-binding protein. Mol Cell Biol 18:1003–1012

Koleske AJ, Young RA (1994) An RNA polymerase II holoenzyme responsive to activators. Nature 368:466–469

Kotani T, Banno K, Ikura M, Hinnebusch AG, Nakatani Y, Kawaichi M, Kokubo T (2000) A role of transcriptional activators as antirepressors for the autoinhibitory activity of TATA box binding of transcription factor IID. Proc Natl Acad Sci USA 97:7178–7183

Kotani T, Miyake T, Tsukihashi Y, Hinnebusch AG, Nakatani Y, Kawaichi M, Kokubo T (1998) Identification of highly conserved amino-terminal segments of dTAFII230 and yTAFII145 that are functionally interchangeable for inhibiting TBP-DNA interactions in vitro and in promoting yeast cell growth in vivo. J Biol Chem 273:32254–32264

Kouzarides T (2002) Histone methylation in transcriptional control. Curr Opin Genet Dev 12:198–209

Krogan NJ, Kim M, Tong A, Golshani A, Cagney G, Canadien V, Richards DP, Beattie BK, Emili A, Boone C, Shilatifard A, Buratowski S, Greenblatt J (2003) Methylation of histone H3 by Set2 in Saccharomyces cerevisiae is linked to transcriptional elongation by RNA polymerase II. Mol Cell Biol 23:4207–4218

Kulkarni MM, Arnosti DN (2003) Information display by transcriptional enhancers. Development 130:6569–6575

Kumar KP, Akoulitchev S, Reinberg D (1998) Promoter-proximal stalling results from the inability to recruit transcription factor IIH to the transcription complex and is a regulated event. Proc Natl Acad Sci USA 95:9767–9772

Kuo MH, Zhou J, Jambeck P, Churchill ME, Allis CD (1998) Histone acetyltransferase activity of yeast Gcn5p is required for the activation of target genes in vivo. Genes Dev 12:627–639

Kusch T, Guelman S, Abmayr SM, Workman JL (2003) Two Drosophila Ada2 homologues function in different multiprotein complexes. Mol Cell Biol 23:3305–3319

Kussie PH, Gorina S, Marechal V, Elenbaas B, Moreau J, Levine AJ, Pavletich NP (1996) Structure of the MDM2 oncoprotein bound to the p53 tumor suppressor transactivation domain. Science 274:948–953

Lachner M, Jenuwein T (2002) The many faces of histone lysine methylation. Curr Opin Cell Biol 14:286–298

Larschan E, Winston F (2001) The S. cerevisiae SAGA complex functions in vivo as a coactivator for transcriptional activation by Gal4. Genes Dev 15:1946–1956

Lau JF, Nusinzon I, Burakov D, Freedman LP, Horvath CM (2003) Role of metazoan mediator proteins in interferon-responsive transcription. Mol Cell Biol 23:620–628

Lee C, Chang JH, Lee HS, Cho Y (2002) Structural basis for the recognition of the E2F transactivation domain by the retinoblastoma tumor suppressor. Genes Dev 16:3199–3212

Lieberman PM, Berk AJ (1994) A mechanism for TAFs in transcriptional activation: activation domain enhancement of TFIID-TFIIA–promoter DNA complex formation. Genes Dev 8:995–1006

Lin Q, Barbas CF, 3rd, Schultz PG (2003) Small-molecule switches for zinc finger transcription factors. J Am Chem Soc 125:612–613

Lin YS, Ha I, Maldonado E, Reinberg D, Green MR (1991) Binding of general transcription factor TFIIB to an acidic activating region. Nature 353:569–571

Lipford JR, Deshaies RJ (2003) Diverse roles for ubiquitin-dependent proteolysis in transcriptional activation. Nat Cell Biol 5:845–850

Liu PQ, Rebar EJ, Zhang L, Liu Q, Jamieson AC, Liang Y, Qi H, Li PX, Chen B, Mendel MC, Zhong X, Lee YL, Eisenberg SP, Spratt SK, Case CC, Wolffe AP (2001) Regulation of an endogenous locus using a panel of designed zinc finger proteins targeted to accessible chromatin regions. Activation of vascular endothelial growth factor A. J Biol Chem 276:11323–11334

Lo WS, Duggan L, Tolga NC, Emre, Belotserkovskya R, Lane WS, Shiekhattar R, Berger SL (2001) Snf1–a histone kinase that works in concert with the histone acetyltransferase Gcn5 to regulate transcription. Science 293:1142–1146

Lo WS, Trievel RC, Rojas JR, Duggan L, Hsu JY, Allis CD, Marmorstein R, Berger SL (2000) Phosphorylation of serine 10 in histone H3 is functionally linked in vitro and in vivo to Gcn5-mediated acetylation at lysine 14. Mol Cell 5:917–926

Luscombe NM, Thornton JM (2002) Protein-DNA interactions: amino acid conservation and the effects of mutations on binding specificity. J Mol Biol 320:991–1009

Lusser A, Kadonaga JT (2003) Chromatin remodeling by ATP-dependent molecular machines. Bioessays 25:1192–1200

Malik S, Roeder RG (2000) Transcriptional regulation through Mediator-like coactivators in yeast and metazoan cells. Trends Biochem Sci 25:277–283

Maniatis T, Reed R (2002) An extensive network of coupling among gene expression machines. Nature 416:499–506

Marcus GA, Horiuchi J, Silverman N, Guarente L (1996) ADA5/SPT20 links the ADA and SPT genes, which are involved in yeast transcription. Mol Cell Biol 16:3197–3205

Marmorstein R, Fitzgerald MX (2003) Modulation of DNA-binding domains for sequence-specific DNA recognition. Gene 304:1–12

Martel LS, Brown HJ, Berk AJ (2002) Evidence that TAF-TATA box-binding protein interactions are required for activated transcription in mammalian cells. Mol Cell Biol 22:2788–2798

Martin ML, Lieberman PM, Curran T (1996) Fos-Jun dimerization promotes interaction of the basic region with TFIIE-34 and TFIIF. Mol Cell Biol 16:2110–2118

Matys V, Fricke E, Geffers R, Gossling E, Haubrock M, Hehl R, Hornischer K, Karas D, Kel AE, Kel-Margoulis OV, Kloos DU, Land S, Lewicki-Potapov B, Michael H, Munch R, Reuter I, Rotert S, Saxel H, Scheer M, Thiele S, Wingender E (2003) TRANSFAC: transcriptional regulation, from patterns to profiles. Nucl Acids Res 31:374–378

McEwan IJ, Dahlman-Wright K, Ford J, Wright AP (1996) Functional interaction of the c-Myc transactivation domain with the TATA binding protein: evidence for an induced fit model of transactivation domain folding. Biochemistry 35:9584–9593

McKenna NJ, O'Malley BW (2002a) Combinatorial control of gene expression by nuclear receptors and coregulators. Cell 108:465–474

McKenna NJ, O'Malley BW (2002b) Minireview: nuclear receptor coactivators—an update. Endocrinology 143:2461–2465

Merika M, Thanos D (2001) Enhanceosomes. Curr Opin Genet Dev 11:205–208

Min J, Zhang Y, Xu RM (2003) Structural basis for specific binding of Polycomb chromodomain to histone H3 methylated at Lys 27. Genes Dev 17:1823–1828

Mitchell PJ, Tjian R (1989) Transcriptional regulation in mammalian cells by sequence-specific DNA binding proteins. Science 245:371–378

Mittler G, Stuhler T, Santolin L, Uhlmann T, Kremmer E, Lottspeich F, Berti L, Meisternst M (2003) A novel docking site on Mediator is critical for activation by VP16 in mammalian cells. EMBO J 22:6494–6504

Mizuguchi G, Shen X, Landry J, Wu WH, Sen S, Wu C (2004) ATP-Driven Exchange of histone H2AZ variant catalyzed by SWR1 chromatin remodeling complex. Science 303:343–348

Muratani M, Tansey WP (2003) How the ubiquitin-proteasome system controls transcription. Nat Rev Mol Cell Biol 4:192–201

Myers LC, Kornberg RD (2000) Mediator of transcriptional regulation. Annu Rev Biochem 69:729–749

Narlikar GJ, Fan HY, Kingston RE (2002) Cooperation between complexes that regulate chromatin structure and transcription. Cell 108: 475–487

Ng HH, Robert F, Young RA, Struhl K (2002) Genome-wide location and regulated recruitment of the RSC nucleosome-remodeling complex. Genes Dev 16:806–819

Ng HH, Robert F, Young RA, Struhl K (2003) Targeted recruitment of Set1 histone methylase by elongating Pol II provides a localized mark and memory of recent transcriptional activity. Mol Cell 11:709–719

Nishikawa J, Kokubo T, Horikoshi M, Roeder RG, Nakatani Y (1997) Drosophila TAF(II)230 and the transcriptional activator VP16 bind competitively to the TATA box-binding domain of the TATA box-binding protein. Proc Natl Acad Sci USA 94:85–90

Nogues G, Kadener S, Cramer P, Bentley D, Kornblihtt AR (2002) Transcriptional activators differ in their abilities to control alternative splicing. J Biol Chem 277:43110–43114

Nowak SJ, Corces VG (2000) Phosphorylation of histone H3 correlates with transcriptionally active loci. Genes Dev 14:3003–3013

O'Hare P, Williams G (1992) Structural studies of the acidic transactivation domain of the Vmw65 protein of herpes simplex virus using 1H NMR. Biochemistry 31:4150–4156

Ottosen S, Herrera FJ, Triezenberg SJ (2002) Transcription. Proteasome parts at gene promoters. Science 296:479–481

Pascreau G, Arlot-Bonnemains Y, Prigent C (2003) Phosphorylation of histone and histone-like proteins by aurora kinases during mitosis. Prog Cell Cycle Res 5:369–374

Pearson A, Greenblatt J (1997) Modular organization of the E2F1 activation domain and its interaction with general transcription factors TBP and TFIIH. Oncogene 15: 2643–2658

Praz V, Perier R, Bonnard C, Bucher P (2002) The Eukaryotic Promoter Database, EPD: new entry types and links to gene expression data. Nucl Acids Res 30:322–324

Proudfoot NJ, Furger A, Dye MJ (2002) Integrating mRNA processing with transcription. Cell 108:501–512

Ptashne M, Gann A (1997) Transcriptional activation by recruitment. Nature 386:569–577

Pugh BF (2000) Control of gene expression through regulation of the TATA-binding protein. Gene 255:1–14

Rachez C, Freedman LP (2001) Mediator complexes and transcription. Curr Opin Cell Biol 13:274–280

Radhakrishnan I, Perez-Alvarado GC, Parker D, Dyson HJ, Montminy MR, Wright PE (1997) Solution structure of the KIX domain of CBP bound to the transactivation domain of CREB: a model for activator:coactivator interactions. Cell 91:741–752

Rawson RB (2003) The SREBP pathway–insights from Insigs and insects. Nat Rev Mol Cell Biol 4:631–640

Ren B, Cam H, Takahashi Y, Volkert T, Terragni J, Young RA, Dynlacht BD (2002) E2F integrates cell cycle progression with DNA repair, replication, and G(2)/M checkpoints. Genes Dev 16:245–256

Ren B, Robert F, Wyrick JJ, Aparicio O, Jennings EG, Simon I, Zeitlinger J, Schreiber J, Hannett N, Kanin E, Volkert TL, Wilson CJ, Bell SP, Young RA (2000) Genome-wide location and function of DNA binding proteins. Science 290:2306–2309

Reyes JC, Barra J, Muchardt C, Camus A, Babinet C, Yaniv M (1998) Altered control of cellular proliferation in the absence of mammalian brahma (SNF2α). EMBO J 17:6979–6991

Roberts SG, Green MR (1994) Activator-induced conformational change in general transcription factor TFIIB. Nature 371:717–720

Roberts SG, Ha I, Maldonado E, Reinberg D, Green MR (1993) Interaction between an acidic activator and transcription factor TFIIB is required for transcriptional activation. Nature 363:741–744

Rojo-Niersbach E, Furukawa T, Tanese N (1999) Genetic dissection of hTAF(II)130 defines a hydrophobic surface required for interaction with glutamine-rich activators. J Biol Chem 274:33778–33784

Roth SY, Denu JM, Allis CD (2001) Histone acetyltransferases. Annu Rev Biochem 70:81–120

Rougvie AE, Lis JT (1988) The RNA polymerase II molecule at the 5' end of the uninduced hsp70 gene of D. melanogaster is transcriptionally engaged. Cell 54:795–804

Sadowski I, Ma J, Triezenberg S, Ptashne M (1988) GAL4-VP16 is an unusually potent transcriptional activator. Nature 335:563–564

Salghetti SE, Caudy AA, Chenoweth JG, Tansey WP (2001) Regulation of transcriptional activation domain function by ubiquitin. Science 293:1651–1653

Salghetti SE, Muratani M, Wijnen H, Futcher B, Tansey WP (2000) Functional overlap of sequences that activate transcription and signal ubiquitin-mediated proteolysis. Proc Natl Acad Sci USA 97:3118–3123

Sauer F, Hansen SK, Tjian R (1995) Multiple TAFIIs directing synergistic activation of transcription. Science 270:1783–1788

Schmitz ML, dos Santos Silva MA, Altmann H, Czisch M, Holak TA, Baeuerle PA (1994) Structural and functional analysis of the NF-κ B p65 C terminus. An acidic and modular transactivation domain with the potential to adopt an α-helical conformation. J Biol Chem 269:25613–25620

Schultz DC, Friedman JR, Rauscher FJ, 3rd (2001) Targeting histone deacetylase complexes via KRAB-zinc finger proteins: the PHD and bromodomains of KAP-1 form a cooperative unit that recruits a novel isoform of the Mi-2 subunit of NuRD. Genes Dev 15:428–443

Sera T, Uranga C (2002) Rational design of artificial zinc-finger proteins using a nondegenerate recognition code table. Biochemistry 41:7074–7081

Shen F, Triezenberg SJ, Hensley P, Porter D, Knutson JR (1996a) Critical amino acids in the transcriptional activation domain of the herpesvirus protein VP16 are solvent-exposed in highly mobile protein segments. An intrinsic fluorescence study. J Biol Chem 271:4819–4826

Shen F, Triezenberg SJ, Hensley P, Porter D, Knutson JR (1996b) Transcriptional activation domain of the herpesvirus protein VP16 becomes conformationally constrained upon interaction with basal transcription factors. J Biol Chem 271:4827–4837

Shen X, Mizuguchi G, Hamiche A, Wu C (2000) A chromatin remodelling complex involved in transcription and DNA processing. Nature 406:541–544

Smale ST, Kadonaga JT (2003) The RNA polymerase II core promoter. Annu Rev Biochem 72:449–479

Soloaga A, Thomson S, Wiggin GR, Rampersaud N, Dyson MH, Hazzalin CA, Mahadevan LC, Arthur JS (2003) MSK2 and MSK1 mediate the mitogen- and stress-induced phosphorylation of histone H3 and HMG-14. EMBO J 22:2788–2797

Sterner DE, Berger SL (2000) Acetylation of histones and transcription-related factors. Microbiol Mol Biol Rev 64:435–459

Strahl BD, Allis CD (2000) The language of covalent histone modifications. Nature 403:41–45

Sudarsanam P, Iyer VR, Brown PO, Winston F (2000) Whole-genome expression analysis of snf/swi mutants of Saccharomyces cerevisiae. Proc Natl Acad Sci USA 97:3364–3369

Thomson S, Clayton AL, Mahadevan LC (2001) Independent dynamic regulation of histone phosphorylation and acetylation during immediate-early gene induction. Mol Cell 8:1231–1241

Triezenberg SJ (1995) Structure and function of transcriptional activation domains. Curr Opin Genet Dev 5:190–196

Uesugi M, Nyanguile O, Lu H, Levine AJ, Verdine GL (1997) Induced α helix in the VP16 activation domain upon binding to a human TAF. Science 277:1310–1313

Van Hoy M, Leuther KK, Kodadek T, Johnston SA (1993) The acidic activation domains of the GCN4 and GAL4 proteins are not α helical but form β sheets. Cell 72:587–594

Vlachonasios KE, Thomashow MF, Triezenberg SJ (2003) Disruption mutations of ADA2b and GCN5 transcriptional adaptor genes dramatically affect Arabidopsis growth, development, and gene expression. Plant Cell 15:626–638

Warnmark A, Treuter E, Wright AP, Gustafsson JA (2003) Activation functions 1 and 2 of nuclear receptors: molecular strategies for transcriptional activation. Mol Endocrinol 17:1901–1909

Warnmark A, Wikstrom A, Wright AP, Gustafsson JA, Hard T (2001) The N-terminal regions of estrogen receptor α and β are unstructured in vitro and show different TBP binding properties. J Biol Chem 276:45939–45944

Weinmann AS, Bartley SM, Zhang T, Zhang MQ, Farnham PJ (2001) Use of chromatin immunoprecipitation to clone novel E2F target promoters. Mol Cell Biol 21:6820–6832

Weinmann P, Gossen M, Hillen W, Bujard H, Gatz C (1994) A chimeric transactivator allows tetracycline-responsive gene expression in whole plants. Plant J 5:559–569

Wieczorek E, Brand M, Jacq X, Tora L (1998) Function of TAF(II)-containing complex without TBP in transcription by RNA polymerase II. Nature 393:187–191

Willy PJ, Kobayashi R, Kadonaga JT (2000) A basal transcription factor that activates or represses transcription. Science 290:982–985

Wolfe SA, Ramm EI, Pabo CO (2000) Combining structure-based design with phage display to create new Cys(2)His(2) zinc finger dimers. Structure Fold Des 8:739–750

Wolstein O, Silkov A, Revach M, Dikstein R (2000) Specific interaction of TAFII105 with OCA-B is involved in activation of octamer-dependent transcription. J Biol Chem 275:16459–16465

Xiao H, Pearson A, Coulombe B, Truant R, Zhang S, Regier JL, Triezenberg SJ, Reinberg D, Flores O, Ingles CJ, et al. (1994) Binding of basal transcription factor TFIIH to the acidic activation domains of VP16 and p53. Mol Cell Biol 14:7013–7024

Xie Y, Denison C, Yang SH, Fancy DA, Kodadek T (2000a) Biochemical characterization of the TATA-binding protein-Gal4 activation domain complex. J Biol Chem 275:31914–31920

Xie Y, Sun L, Kodadek T (2000b) TATA-binding protein and the Gal4 transactivator do not bind to promoters cooperatively. J Biol Chem 275:40797–40803

Xu J, Li Q (2003) Review of the in vivo functions of the p160 steroid receptor coactivator family. Mol Endocrinol 17:1681–1692

Xu W, Cho H, Evans RM (2003) Acetylation and methylation in nuclear receptor gene activation. Methods Enzymol 364:205–223

Yamamoto Y, Verma UN, Prajapati S, Kwak YT, Gaynor RB (2003) Histone H3 phosphorylation by IKK-α is critical for cytokine-induced gene expression. Nature 423:655–659

Yankulov K, Blau J, Purton T, Roberts S, Bentley DL (1994) Transcriptional elongation by RNA polymerase II is stimulated by transactivators. Cell 77:749–759

Zhang L, Spratt SK, Liu Q, Johnstone B, Qi H, Raschke EE, Jamieson AC, Rebar EJ, Wolffe AP, Case CC (2000) Synthetic zinc finger transcription factor action at an endogenous chromosomal site. Activation of the human erythropoietin gene. J Biol Chem 275:33850–33860

Zhang Y (2003) Transcriptional regulation by histone ubiquitination and deubiquitination. Genes Dev 17:2733–2740

Multiple Mechanisms of Transcriptional Repression in Eukaryotes

D. N. Arnosti

Department of Biochemistry and Molecular Biology, Michigan State University,
East Lansing, MI 48824-1319, USA
arnosti@msu.edu

1	What Is Transcriptional Repression?.	34
2	What Is a Transcriptional Repressor?	36
2.1	Functional Complexity.	36
2.2	Use of Phylogeny to Illuminate Function.	36
3	*Cis*-Elements Regulated by Transcriptional Repressors	37
3.1	Initial Studies of Transcriptional Switches and Terminology.	37
3.2	*Cis*-Element Design	38
4	Categorizing Transcriptional Repressors	39
4.1	Repressors and Corepressors	39
4.2	Transcriptional Repression Domains.	40
4.3	Types of Activities	40
5	Methods Used to Characterize Repressors.	41
5.1	Measuring Transcriptional Repression—Biochemical Approaches	41
5.2	Genetic Approaches	42
5.3	Molecular Biological Approaches.	43
5.4	Criteria for Identifying the Physiological Activity of a Repressor	44
6	Mechanisms of Transcriptional Repression.	45
6.1	Covalent Modification of Transcriptional Activators	45
6.2	Direct Competition.	47
6.3	Direct Interaction with Basal Machinery.	48
6.4	Masking of Activation Domains.	50
6.5	Interaction with Chromatin-Modifying Factors	50
6.5.1	Chromatin Remodeling and Repression	51
6.5.2	HDACs	52
6.5.3	Histone Methyltransferases	53
6.5.4	DNA Methylation.	54
6.6	Polycomb Proteins and Epigenetic States— Combinations of Repression Mechanisms	55
6.7	Context Influences the Activity of Putative Transcriptional Repressors	56
7	Multiple Activities of Transcriptional Repressors	57
7.1	Qualitative Effects	58
7.2	Quantitative Effects	59
7.3	Artifactual Repression?	59
	References	60

Abstract The selective transcription of eukaryotic genes is regulated by both positive and negative inputs from sequence-specific DNA-binding factors. These proteins provide the information essential for correct temporal and spatial control of transcription. The activities of transcriptional repressors have been characterized by a variety of methods, but in many cases the physiological relevance of proposed mechanisms has not been established. This chapter reviews pathways of repression, critically evaluates criteria by which repression mechanisms can be analyzed, and discusses recent progress in identifying the functional relevance of multiple repression activities of transcriptional repressors.

Keywords Transcription · Repression · Corepressor · Chromatin

1
What Is Transcriptional Repression?

A broad view of transcriptional repression includes cellular processes that interfere with the transcription of genes on a global or local level (Fig. 1). Processes that degrade, sequester, covalently modify, or remove from the

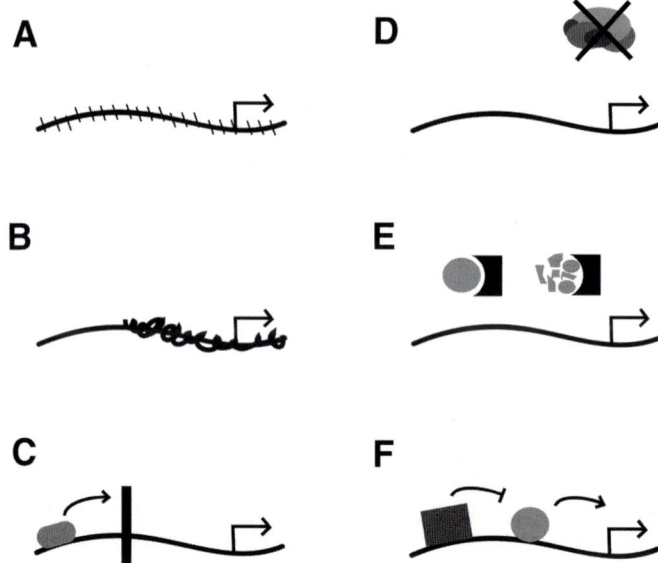

Fig. 1A–F General models of transcriptional repression. **A** Genome- or chromosome-wide effects of chromatin structure (*hatch marks*). **B** Directed assembly of localized chromatin structures under the control of silencers. **C** Local inhibition of transcriptional enhancers (*gray oval*) by boundary elements or insulators (*black bar*). **D** Covalent modifications to components of the general transcriptional machinery. **E** Sequestration, modification, or degradation of positively acting sequence-specific transcriptional activators (*gray circle*) by inhibitory proteins or proteolytic machinery (*black shape*). **F** Gene-specific repression by DNA-binding proteins (*dark gray rectangle*) and their associated corepressors through interactions with activators, basal machinery, and chromatin

nucleus positively acting transcription factors can cause transcriptional repression of genes by eliminating an activating signal. Similarly, global gene regulation can be effected by modification of the general transcriptional machinery, i.e., RNA polymerases and basal factors used at most promoters. In these cases, negatively acting factors need not directly interact with genes to inhibit transcription. Many forms of transcriptional repression, however, involve the activity of DNA-binding proteins that do directly contact genes. The binding of repressors can be relatively nonsequence-specific. For example, large-scale modifications of chromatin influence expression of genes on a chromosomal scale. These effects are seen with specialized chromatin structures present at telomeres and centromeres, in loci controlled by mating type loci in yeast, and with sex-linked dosage compensation systems (Akhtar 2003; Huang 2002; Henikoff 2000). In other cases, genes are negatively regulated by boundary elements, which modulate interactions between regulatory regions and transcriptional start sites (Gerasimova and Corces 2001).

These pathways of repression are of documented importance, but it is clear that many repressors act primarily at a local level, targeting specific genes by means of sequence-specific DNA-binding domains. This chapter is focused on DNA-binding repressors and their cofactors because of their importance in controlling gene switches critical for a variety of biological processes. In addition, many such candidate transcription factors, identified in sequencing projects by virtue of their characteristic DNA-binding motifs, remain to be functionally characterized. This chapter describes approaches currently used to characterize repressors and lays out some of the limitations of these techniques. The discussion centers on repression of protein coding genes, which are transcribed by RNA polymerase II. The regulation of RNA polymerase I and III genes are discussed in recent reviews (Grummt 2003; White 2004).

Consistent with the variety of inhibitory processes involved in repression, the effects of repression are also heterogeneous. A single gene can be transiently repressed by the action of factors binding directly to that gene, or whole classes of genes can be regulated by modification of general transcription machinery or chromatin. Repression can be limited to the time when negative regulatory factors are found at a gene, or repression can leave an 'imprint' so that transcription is blocked for the life of an organism. In general, more is understood mechanistically about how a repressed state is established; less well studied but equally important is how repression is reversed.

What Is a Transcriptional Repressor?

Functional Complexity

Some proteins appear to have dedicated roles as transcriptional repressors or activators; that is, in any context in which they have been studied, they have only one activity. In many other cases, however, regulatory proteins can work in a context-dependent manner, either facilitating or inhibiting transcription depending on the regulatory element, signals from signal transduction cascades, responding basal promoter, or levels of cofactors (Barolo and Posakony 2002; Rogatsky et al. 2002). Such physiological diversity is complicated by the heterogeneous assays used to measure factor activity. In many cases, the description of a particular factor as a repressor comes from limited analysis that does not necessarily correlate with true activities. By testing repressors in dechromatinized systems, overexpressing them at high levels, and tethering them directly to a basal promoter as a fusion protein, potential repressors may be induced to exhibit novel activities. Thus, the panoply of mechanisms for repressors comprehensively reviewed in recent articles (Roberts 2000; Gaston and Jayaraman 2003) must be taken as a picture of possible mechanisms involved in gene regulation, rather than as proof that a particular protein actually uses such mechanisms to effect gene control.

2.2
Use of Phylogeny to Illuminate Function

The human genome project and other genomic efforts have identified thousands of presumptive transcriptional regulatory factors, mainly based on sequence conservation especially in the DNA-binding domain. In many cases, homologous factors are able to mediate similar activities in a heterologous organism, indicating functional conservation (Malicki et al. 1992; Halder et al. 1995). However, sequence conservation alone does not guarantee that functional activity is conserved. In some cases, obvious transcription factor homologs, often identified by means of highly conserved DNA-binding domains, have functionally distinct activities in divergent species. For example, the SMAD pathway inhibitor TGIF plays a negative role in vertebrates, but appears to activate in *Drosophila* (Hyman et al. 2003). Evolutionary studies have revealed in some cases particular changes in proteins that correlate with change in function from a dedicated activator to a facultative repressor (Galant and Carroll 2002; Ronshaugen et al. 2002). The strict conservation of residues critical for recruiting corepressors provides compelling evidence for function, but in most cases the rapid evolution of regions outside the

DNA-binding domains makes it difficult to predict whether homologs retain function as positive or negative acting factors.

3
Cis-Elements Regulated by Transcriptional Repressors

3.1
Initial Studies of Transcriptional Switches and Terminology

In prokaryotes, work with the paradigmatic Lac repressor led to an initial view that most genes were controlled by repression. Later studies demonstrated the importance of both positive and negative elements of gene switches, including the *lac* operon (Muller-Hill 1996). In eukaryotes, early studies of transcriptional regulation were based on viral gene regulation. A number of well-characterized viral regulatory elements function as dedicated transcriptional activation elements, leading to a focus on transcriptional activation. The term 'transcriptional enhancer' was coined to describe a *cis*-regulatory element that functioned to activate transcription in an orientation and distance independent manner (Banerji et al. 1981). Early biochemical studies of eukaryotic regulatory factors also focused on transcriptional activation, because technically it is easier to monitor enhanced, rather than reduced, transcriptional initiation. However, later studies of cellular enhancers determined that these elements are often more complex, mediating repression as well as activation functions (Davidson 2001). In most cases, repressor-binding sites are found within regions that also contain binding sites for activators, producing a *cis*-regulatory element with multiple potential activities. In some cases, dedicated negatively acting elements termed 'silencers' have been identified. This term has been applied both to elements that work in a nondirectional manner and to elements that mediate directional gene silencing in yeast (Busturia et al. 1997; Henikoff 2000). In recent work, some authors have chosen to adopt the general term '*cis*-regulatory element' rather than 'enhancer,' to more accurately reflect possible positive and negative aspects of control (Carroll et al. 2001). The related term 'promoter' is firmly embedded in the literature, however. Although the word connotes a strictly positive effect, it can refer to the entire set of *cis*-regulatory elements that are required for expression of a gene, including positive and negative elements. Alternatively, the term 'promoter' may refer to just the minimal region surrounding the initiation site containing sequences necessary for binding of the basal machinery (Arnosti 2003; Smale and Kadonaga 2003).

3.2
Cis-Element Design

Eukaryotic *cis*-regulatory elements rarely consist of single binding sites for a transcription factor. Instead, they generally comprise short segments of DNA to which a number of proteins bind in a sequence-specific manner (Fig. 2). Such elements form nucleoprotein complexes that can have varying degrees of higher order structure (Merika and Thanos 2001; Struhl 2001; Kulkarni and Arnosti 2003). In some cases, the distribution of binding sites for activators and repressors represents a carefully selected design that reflects highly cooperative interactions between the factors, while in others, binding site locations are relatively flexible, permitting binding sites to drift over the course of evolutionary time (Ludwig et al. 2000). Depending on the range of action, the repression can be local, limited to the enhancer in which the repressor binds, or long-range, which results in inhibition of multiple enhancers. Repressor sites may be located close to the transcriptional initiation site, a fact that may be important if repressors compete with components of the basal machinery for binding to DNA. In other cases, repressors

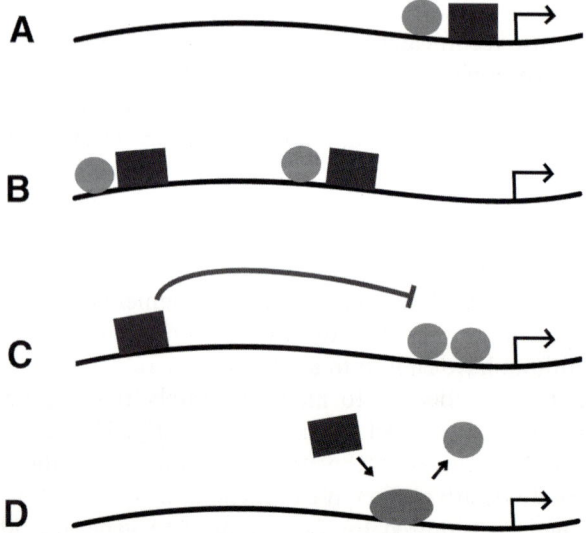

Fig. 2A–D Design of *cis-regulatory* elements. A Binding sites for sequence-specific activators and repressors (*circles* and *rectangles*, respectively) may be closely linked to the basal promoter. B Alternatively, sites may be distributed at some distance from the basal promoter, 5′ or 3′ of the start of transcription, in modules termed enhancers. C Dedicated negatively acting elements may be located at some distance from the transcriptional start site; such repression may involve chromatin remodeling or insulator function. D Some DNA-binding proteins (*gray oval*) can act as either activators or repressors, depending on association with a coactivator (*circle*) or corepressor (*rectangle*)

are located many kilobases from the transcriptional start site, and most likely function by interfering with neighboring activators. A conserved architecture has been identified for some *cis*-regulatory elements responding to a number of signal transduction pathways, such as the Wnt and Hedgehog pathways. *Cis*-elements targeted by these pathways can contain binding sites for bifunctional proteins that act as default repressors, which are converted to activators in response to signaling activity. As these activators are apparently often rather weak, the *cis*-elements contain binding sites for constitutively active activators that boost signal output and provide an additional level of cell type specific control (Barolo and Posakony 2002).

4
Categorizing Transcriptional Repressors

4.1
Repressors and Corepressors

Transcriptional repressors have been categorized by a number of criteria, including the presence or absence of DNA-binding activity, the types of amino acids present in 'repression domains,' and functional aspects of repression. Proteins that are able to recognize DNA in a sequence-specific manner are termed repressors, while those that are recruited to a gene via protein–protein interactions with a repressor are commonly termed corepressors. Confusingly, in some systems, non-DNA-binding factors, such as the yeast Gal80 protein, are also termed repressors. Some authors have termed Gal80 an 'inhibitor' to differentiate its mechanism of action, masking of the Gal4 activation domain, from autonomous repression activity (Ansari et al. 1998). The distinction between repressor and corepressor does underscore the importance in many cases of a particular protein's role in targeting repression to specific loci, but the nomenclature is to some extent arbitrary, because some proteins can interact with genes either by directly contacting the DNA or via protein–protein interactions. For example, nuclear hormone receptors are able to mediate activation or repression via specific DNA-binding sites, but also inhibit gene expression through 'composite' elements that bind to AP-1 and NF-κB transcriptional activators (Nissen and Yamamoto 2000). The CTIP corepressor facilitates COUP-TF repression as a corepressor, and also binds to separate DNA elements independently to repress (Senawong et al. 2003). As a further indication of the similarity of repressors and corepressors, many corepressors such as C-terminal-binding protein (CtBP), Groucho/TLE and NCoR exhibit repression activity when fused to a heterologous DNA-binding domain (Wen et al. 2000; Flores-Saaib and Courey 2000; Ryu et al. 2001).

4.2
Transcriptional Repression Domains

By deletion analysis, portions of repressor proteins have been identified as critical and sufficient for repression, leading to their identification as 'repression domains' (often without regard to the notion of a domain in structural terms as an autonomously folding portion of a protein). Because of their diversity, repression domains have not been successfully classified by the overall amino acid content, although some repressors have been found to contain alanine-rich segments (Hanna-Rose and Hansen 1996). In many cases, repression domains contain structural motifs that interact with distinct classes of corepressors. These motifs can be short and degenerate, but they are often conserved between homologous genes, facilitating their identification. For example, members of the Hairy/E(spl) family, from insects to vertebrates, all contain a WRPW-like motif at or near the C terminus important for contacting Groucho/TLE corepressors, and repressors that interact with the CtBP cofactor contain PXDLS related motifs (Chen and Courey 2000; Davis and Turner 2001; Turner and Crossley 2001).

4.3
Types of Activities

DNA-binding repressors can be functionally categorized as 'passive' (direct competition for a factor-binding site) and 'direct' or 'active' (all other mechanisms) (Jaynes and O'Farrell 1991; Gaston and Jayaraman 2003). An alternate classification distinguishes 'short-range' (able to interfere with positively-acting elements within about 100 bp) from 'long-range' (active over more than 1 kbp) repressors (Gray and Levine 1996; Courey and Jia 2001). The ability to repress locally or over a larger distance can be exploited to provide unique forms of gene regulation; for example, short-range repression is indispensable for selective repression of an individual module of a complex regulatory locus. These distinctions have some utility in the interpretation of particular repression assays, but repressors can in some cases fall into more than one category, depending on the gene context (Nibu et al. 2001). As molecular mechanisms of repression are better understood, these simple categories will likely be superseded.

5
Methods Used to Characterize Repressors

5.1
Measuring Transcriptional Repression—Biochemical Approaches

Assays used to study transcriptional repression include techniques from biochemistry, genetics, and molecular biology, although in practice these categories overlap to a great extent (Table 1). Biochemical assays for tran-

Table 1 Experimental approaches to characterizing transcriptional regulatory proteins

Approach	Advantages	Disadvantages
Biochemical		
In vitro transcription	Obtain mechanistic details of transcription mechanisms	Repression may be artifactual
	Determine effects of molecular modifications to transcription factors	May lack chromatin
		Difficult to replicate long-range or multi-enhancer regulation
Genetic		
Mutational analysis, gene array studies	Endogenous genes studied	Difficult to differentiate direct from indirect effects
	Can study even large, complex genes	(footprinting or chromatin immunoprecipitation used to establish direct effects)
	Native chromatin environment.	
Molecular biological		
Transient transfections	Rapid	DNA not integrated in chromosome
	In vivo setting with endogenous cofactors often present	Chromatin structure not well defined
		Factors often overexpressed at nonphysiological levels
Stable transfections	Reporter genes integrated in chromosome	Establishment of cell lines time-consuming
	Regulated expression of transcriptional effectors with IPTG, Tet, Dex and other small molecule inducers	Gene often integrates as multiple, tandem copies
		Difficult to analyze large regulatory regions
Transgenic organisms	Reporter genes integrated in chromosome	Can be tedious, especially with vertebrate systems
	Present in all cell types, developmental regulation recapitulated	Can be difficult to analyze large regulatory regions
	By using transgenesis to misexpress repressor, effects on endogenous target genes can be observed	Gene expression patterns and levels can be influenced by position effects i.e., site of integration on chromosome

scriptional repression rely on reconstituted in vitro systems, where fractionated or highly purified transcription factors are allowed to interact with a genetic regulatory region, usually in the absence of chromatin. Transcription is measured in the presence or absence of a DNA-binding repressor by quantification of RNA produced in vitro. This method has the advantage of providing a well-characterized setting for repression, and it has been possible to identify putative biochemical mechanisms of repression, e.g., to measure directly by DNA footprinting the effects of a repressor on the binding of basal machinery (Ross et al. 1999). With in vitro assays, it is critical to differentiate nonspecific repressive effects from specific ones, because in vitro transcription is readily inhibited by addition of salt, denaturants, or large molar excesses of DNA-binding proteins. In addition, previous in vitro studies have suffered from the limitation that they have been generally carried out in the absence of chromatin, and recent work indicates that chromatin is critical for the activity of many repressors (see below). Long-distance repression by elements located far from the transcriptional initiation site has also been difficult to simulate in vitro. Thus, the results from in vitro assays must be tested using a complementary approach that more closely approximates an in vivo situation.

5.2
Genetic Approaches

Genetic studies have focused on gene expression in mutants with defects in genes for repressors or corepressors. Genetic approaches have been extensively utilized in model organisms such as yeast or *Drosophila* to characterize the activity of the transcriptional apparatus, including repressors. In many cases, the physiological roles of transcriptional repressors and their placement in genetic networks were established before details of their molecular modes of action were elaborated. In vertebrates, by contrast, many proteins have been identified first as putative transcription factors through biochemical or molecular biological approaches, and in many cases their activities in a physiological setting are still poorly understood.

Genetic screens allow for an unbiased search for putative repressor targets or cofactors. In addition, genetic characterization of repression has the advantage that the genes regulated by repressors are in their natural chromatinized state. In all cases, however, a challenge faced in genetic characterizations is interpreting and differentiating direct from indirect effects. In one case, a genetic screen for cofactors of the Knirps repressor instead identified cell cycle regulatory genes; mutations in these genes slow the cell cycle sufficiently to permit expression of a related gene, *knirps-related*, which rescued the original phenotype (Ruden and Jackle 1995). Even in more directed experiments, where responses of target genes are measured to the presence or absence of a putative repressor, similar questions arise. For example, in gene array experiments that measure the effects of a repressor, cellular steady state levels of mRNAs are measured in the presence or absence of the repressor (DeRisi et al. 1997; Holstege et al. 1998). mRNA species whose levels increase in the absence of a transcriptional regulator may be true, direct transcriptional targets of a repressor, but the regulation may also be indirect, effected by means of an activator that is itself controlled by the repressor. In conditional knockout experiments, indirect effects can be minimized by measuring gene response within a short period of time after inactivation of the repressor. For a temperature-sensitive allele, the temperature is shifted up or down and gene responses are measured within a short period of time, minimizing the possibility of additional layers of regulation (Moqtaderi et al. 1996). The use of temperature-sensitive alleles is largely restricted to microbial systems, unless a temperature-sensitive allele is fortuitously available. In some cases, direct and indirect effects can be distinguished by carrying out the experiment in the presence of cycloheximide to inhibit translation, but blocking translation can itself introduce a variety of nonspecific effects. Direct effects can also be established if the repressor in question can be shown to bind to the target gene in vivo. One widely used method involves chromatin immunoprecipitation (ChIP) assays (Kuo and Allis 1999). Cells are treated with formaldehyde to crosslink proteins to a gene, then chromatin is prepared from the cells, after which antibodies to the protein being investigated are used to precipitate the gene fragments in question. The fragments are characterized by PCR or by hybridization to a gene array. These assays can establish the presence of a protein somewhere within approximately 300 bp of a given regulatory sequence. More precise binding information can come from in vitro DNAse I footprinting or in vivo footprinting (Dai et al. 2000).

5.3
Molecular Biological Approaches

Molecular biological characterization of repressors includes cell-based assays, where target genes are transfected into cells in the presence or absence of a repressor, and whole organism assays, where the target gene and/or the repressor in question is reintroduced into the organism by transgenesis, either in the intact state or modified state. Cell culture studies allow rapid characterization of particular repressors and target genes, but suffer from a number of drawbacks. In the case of transiently transfected reporter genes, the chromatin state of the reporter gene may differ from that found for chromosomally located genes, which may influence transcriptional regulation. Transient expression of repressors usually involves overexpression at non-physiological levels, which might also introduce artifacts, just as with in vitro assays. These limitations can be overcome by analyzing repression of endogenous genes in cell culture, or by stably integrating reporter genes. Expression of repressors can be titrated in regulated systems, using for example the Tet system (see the chapter by Bohl and Heard, this volume).

Many putative transcriptional repressors have been characterized solely when overexpressed in transfections of heterologous cells, in the context of multimerized binding sites for the factor introduced adjacent to a basal promoter element. Unless proper controls are carried out to verify that the repressive effects are gene specific, i.e., dependent on the binding sites, and also are not simply effects of steric hindrance of a bulky protein blocking access to the promoter, the labeling of factors as repressors based solely on such assays must be viewed with caution. In some cases, the activity of repressors has been analyzed in parallel in cell culture and in whole organisms; here, the similarities of the activities suggests that at least some of the repressor function can be recapitulated on transiently introduced target genes (Ryu and Arnosti 2003). In other cases, activities of repressor proteins or domains have been identified in transfection assays that have subsequently not been confirmed from in vivo analysis (Tolkunova et al. 1998).

To capture some of the features of repression that transient assays lack, more labor-intensive approaches have been used, involving stable transformation of intact organisms to introduce reporter genes or modified versions of repressors. Such analysis is readily done in yeast, as well as model organisms such as *Drosophila* or *Arabidopsis*. Chromosomal integration of reporter genes is usually preferable, to avoid artifacts arising from high copy number plasmids. Reporter gene expression is also affected by the site of chromosomal integration, an effect that can be reduced in some cases by surrounding the reporter gene with boundary elements. Such analyses are less common in vertebrate systems, reflecting the greater difficulty and expense of working with transgenic mice and fish. However, only through such approaches are the contributions of transcriptional repression to developmental timing and patterning apparent, as well as more subtle responses of regulatory elements to repressors.

5.4
Criteria for Identifying the Physiological Activity of a Repressor

In general, transfection assays have been most successful in identifying *potential* activities of factors, such as their ability to repress, possible cofactors, and characterization of structural features of the protein, whereas more physiological assays are required to determine how the repressor functions on endogenous regulatory elements. The best understood function of transcriptional repressors on true target genes comes from studies in which three criteria have been satisfied: (1) the repressor has been shown genetically to affect expression of the target gene; (2) the repressor has been shown to interact directly with the gene in vitro by footprinting or gel-shift assays, or in vivo by ChIP; and (3) the specific mutation of the repressor binding sites disrupts repression.

6
Mechanisms of Transcriptional Repression

6.1
Covalent Modification of Transcriptional Activators

As discussed above, modification, degradation or sequestration of transcriptional activators can lead to repression of genes. Examples of such modifications are indicated in Table 2. In many cases, the modification can occur when the factor is not binding to a gene and/or is not located within the nucleus, or alternatively, modification can also occur to DNA-bound transcriptional activators. The proteins that carry out such modifications generally

Table 2 Modifications of transcriptional regulators that contribute to transcriptional repression

Modification	Effect	Example
Deacetylation	Reduction in DNA-binding affinity	E2F1 deacetylation mediated by Rb associated HDACs Martinez-Balbas et al. 2000
Glycosylation	Blocking coactivator contact	Sp1 glycosylation by O-GlcNAc transferase Yang et al. 2001
Ubiquitylation	Loss of nuclear localization or protein degradation	p53 nuclear export and turnover Brooks and Gu 2003
Sumoylation	Unknown	Sumolyation of glucocorticoid nuclear hormone receptor inhibits activation Verger et al. 2003
Arginine methylation	Prevention of activator binding by coactivator	CBP coactivator methylation by CARM-1 prevents binding to CREB transcription factor Xu et al. 2001
Heterodimerization with non-DNA-binding partner	Inhibition of DNA binding by basic helix loop helix factors	Id interaction with E2A family members E12/E47 Ruzinova and Benezra 2003
Binding of cytoplasmic inhibitor	Prevention of nuclear import	I-κB binding to NF-κB (rel family) proteins Karin and Ben-Neriah 2000
Phosphorylation	Inhibition of cofactor binding	Interaction with E2F transcription factors inhibited by Rb corepressor phosphorylation Taya 1997
Allosteric inhibition of DNA-binding domain	Inhibition of DNA binding	Ets-1 autoinhibitory domain Pufall and Graves 2002

are not DNA-binding proteins themselves, but are targeted to activators via protein–protein interactions. Constitutive pathways that remove covalent modifications can provide a level of negative regulation. The SMAD proteins, downstream effectors of transforming growth factor (TGF)-β pathway signaling, are phosphorylated in response to pathway activation, and dephosphorylation returns them to an inactive state (Inman et al. 2002).

Major pathways for DNA-bound repressors and corepressors are: (1) direct competition between a repressor and an activator for the same or overlapping binding sites; (2) interactions with the basal machinery, including RNA polymerase II and associated factors, to inhibit initiation or elongation; (3) direct interaction between (co)-repressors and activators, resulting in inhibition of activator function, sometimes termed 'masking'; and (4) chromatin alterations, involving the change of DNA/nucleosomal structure, or covalent modification of chromatin or DNA (Fig. 3). Additional, less understood mechanisms have also been suggested, including intranuclear targeting of repressed genes to heterochromatic sites, 'repulsion' of coactivators, or possible direct proteolysis by transcriptional repressors (He et al. 1995; Brown et al. 1997; Senger et al. 2000).

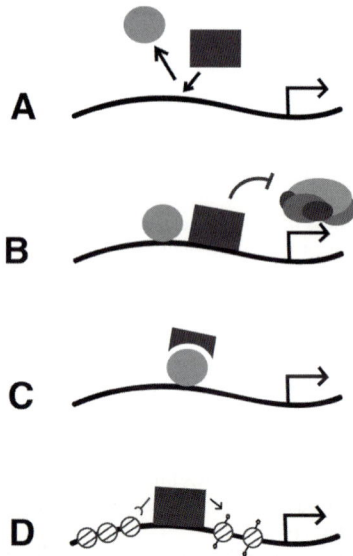

Fig. 3A–D Modes of repression by DNA-associated transcriptional repressors and corepressors. **A** Direct displacement of an activator (*circle*) by the competitive binding of a DNA-binding repressor (*rectangle*). **B** Direct interactions between the transcriptional repressor and the basal transcriptional machinery. **C** 'Masking' of a transcriptional activation domain by a corepressor. **D** Chromatin remodeling, either by nucleosome repositioning or covalent modification of histones. Additional levels of regulation not shown include controlled proteolysis of transcriptional activators, subnuclear targeting, and covalent modifications of activators (see Table 2)

6.2
Direct Competition

Some transcriptional repressors have been documented to inhibit gene expression by displacing activators from identical or overlapping sites. In most instances this activity has been noted under conditions when the repressor has been overexpressed, usually in cell culture assays, or in vitro, where arbitrarily large amounts of repressor can be added. For instance, the chicken ovalbumin upstream promoter transcription factors (COUP-TFI and II, also known as Ear 2 and 3) are widely conserved orphan nuclear hormone receptors that have binding specificity similar to that of RAR, VDR, and SF-1 receptors (Pereira et al. 2000). Support for a physiological role as transcriptional repressors comes from the complementary expression patterns between COUP-TF factors and putative target genes (Shibata et al. 2003a). When COUP TF factors are overexpressed in transfection assays, activation by SF-1 of the vertebrate *CYP17* promoter is inhibited (Shibata et al. 2003b). In vitro, COUP-TFII can bind to and displace SF-1 from the DNA (Bakke and Lund 1995). Similar effects are noted with overexpression experiments involving other nuclear hormone receptors, such as RXR, RAR, HNF-4, and the *Drosophila* EcR (Zelhof et al. 1995). However, from these experiments, it has not been possible to establish whether competition at the level of DNA binding is a significant factor under physiological conditions. Other work suggests that COUP TF factors operate via noncompetitive modes. For example in repressing the *MHC I* promoter, COUP TFII antagonizes the activity of NF-κB from nonoverlapping sites, suggesting a mode of repression that is independent of competition (Zhao et al. 2003). Further substantiating the notion of active repression, portions of COUP TFII can mediate repression when bound to the heterologous Gal4 DNA-binding domain, apparently by recruiting the histone deacetylases (HDACs) or the N-CoR corepressor (Smirnov et al. 2000).

A second example of competition comes from studies of the *Drosophila* Knirps transcriptional repressor. This protein binds to regulatory elements of the *Krüppel* and *even-skipped* (*eve*) genes, countering the activity of the Bicoid activator (Small et al. 1991). In cell culture studies, overexpression of Knirps inhibited activation by Bicoid from small element derived from the *Krüppel* promoter, and Knirps blocked Bicoid from binding to a 16-bp target sequence derived from this element in vitro, suggesting a repression mechanism that involved competition (Hoch et al. 1992). However, later work showed that Knirps contains active repression domains that interact with the CtBP corepressor and other factors, enabling it to act from nonoverlapping sequences (Arnosti et al. 1996b; Nibu et al. 1998; Keller et al. 2000). In the embryo, the Knirps DNA-binding domain alone is unable to mediate repression of *eve*, indicating that competition does not play a significant role (Struffi et al. 2004). Finally, a recent study of the Krüppel repressor indicated that competition can be observed in vivo if the binding sites are directly overlapping Dorsal and Twist activator sites, but bioinformatics analysis of native regulatory elements indicates that Kruppel tends to be located ~15 bp from Bicoid activator sites, suggesting that competition is not a preferred mode of regulation (Nibu et al. 2003; Makeev et al. 2003).

6.3
Direct Interaction with Basal Machinery

Transcriptional repressors have also been suggested to act by direct interactions with the basal transcriptional machinery. Targets within the basal machinery that have been identified in functional studies include TFIIA, TFIIB, TFIID, RNA polymerase, and holoenzyme subunits Srb10/11. Comprehensive summaries of potential targets are reviewed elsewhere (Gaston and Jayaraman 2003), and are addressed here to present the limits of our knowledge. In vertebrates, the transcriptional effector of the conserved Notch signaling pathway is the DNA-binding CBF-1 protein (also known as RBP-Jκ, CSL, and LMP-2), with which the intracellular portion of Notch interacts in the nucleus to activate target genes. In the absence of Notch signaling, CBF-1 acts as a repressor. To identify the mechanism of repression, the adenovirus pI promoter was studied; this simple promoter contains a CBF-1 site positioned between a single Sp1 activator site and the TATA box (Olave et al. 1998). One strength of this study is that a direct comparison was made between the activity of CBF in vivo (by transfection of reporter genes) and in vitro on a variety of reporter genes. CBF-1 repressed only when positioned between the Sp1 site and the TATA box, and in a highly purified transcription system, repression activity correlated with CBF-1 contacts on TFIIA and TFIID (specifically the TAFII110 subunit). Preassembling the activator complex, or moving the repressor site 5' of the activator, prevented repression. This specific promoter appears to take advantage of direct protein–protein interactions between the repressor and basal machinery, but in general it appears that CBF-1 proteins use other mechanisms to repress cellular promoters. The *Drosophila* CBF-1 homolog, Su(H), represses in the absence of Notch signaling, and genetic and biochemical studies clearly indicate that repression is dependent on the activity of the CtBP and Groucho cofactors that interact with HDAC. Vertebrate CBF-1 has also been found to interact with HDAC-recruiting corepressors, namely SMRT and NCoR (Lai 2002). Furthermore, the binding sites of Su(H) are effective at mediating repression from a variety of locations, not just between an activator and the TATA box, suggesting that the very stringent positioning seen with the pI promoter is an exception (Furriols and Bray 2001). Whether CBF-1 interactions with basal machinery plays a role on other promoters is unclear, but from the extreme level of context dependence identified previously, it is unlikely that the same protein–protein interactions are generally utilized.

Cell cycle and cancer studies have focused on the retinoblastoma (Rb) tumor suppressor and proteins functioning in this pathway, which are thought to play key roles in cell cycle regulation (see the chapter by Hauck and von Harsdorf, this volume). A number of cell cycle regulated genes are controlled by the E2F family of DNA-binding transcriptional activators and repressed by Rb family corepressors. Biochemical studies with purified transcriptional factors suggest that repression of E2F regulated promoters by the Rb may in some cases involve the prevention or displacement of contacts between the E2F activator and the basal transcriptional factors TBP and TFIIH (Pearson and Greenblatt 1997). Notably, however, E2F cannot activate in vitro in the absence of TBP associated factors (TAFs). Further in vitro studies indicated that Rb may also prevent the recruitment of TFIIA, perhaps by direct interactions with TAF1 (also known as TAFII250) (Ross et al. 1999). In the nonchromatinized in vitro E2F-activated system, Rb was suggested to act at the level of preventing the assembly of the preinitiation complex, but in a chromatinized in vitro system, in which Rb repressed not E2F but Sp1, repression was found to function at a step following recruitment of the preinitiation complex, apparently independently of HDAC activity (Ross et al. 2001). As discussed below, chromatin modifications have been linked in a variety of ways to Rb-mediated repression. It remains unknown how much, if at all, direct interactions with the basal machinery contribute to overall repression by Rb.

In the cases mentioned above, detailed biochemical information has been produced about direct interactions with the basal transcriptional machinery, but the in vivo relevance of such contacts still remains unresolved. In the case of the Tup1 corepressor in yeast, strong evidence links repression to activity of the basal machinery, but there is no agreement whether this is a direct effect. The Tup1 corepressor is recruited to a variety of promoters by contacting different classes of DNA-binding proteins (Smith and Johnson 2000). Genetic suppression screens linked the activity of the Tup1 corepressor to the activity of the Srb10/11 cyclin/cyclin dependent kinase subunits of the RNA Polymerase II holoenzyme, suggesting that this repressor might function by affecting the phosphorylation of the C terminus of RNA polymerase. One line of inquiry suggests that the Tup1 protein directly contacts the Srb10/11 proteins; alternatively, Tup1 has also been suggested to mediate nucleosome repositioning on affected genes (Li and Reese 2001; Schuller and Lehming 2003). In this latter view, the loss of Srb 10/11 function might reduce polymerase sensitivity to the changes in chromatin, providing an indirect route for suppression of Tup1 activity. Tup1 also appears to work by recruiting HDACs (Deckert and Struhl 2001; Davie et al. 2003). Thus, despite the excellent genetic and molecular biological tools available in yeast, there is still considerable uncertainty about how Tup1 acts at different promoters.

Chromatin immunoprecipitation assays were used to investigate the activity of glucocorticoid receptor (GR) in repressing genes activated by

NF-κB proteins. In the case of the interleukin-8 and intracellular adhesion molecule type 1 genes, neither activator binding nor polymerase loading was prevented on promoters repressed by GR. Instead, repression is associated with failure of RNA polymerase II to become phosphorylated in the serine 2 position of the C-terminal domain (CTD) tail, a modification that is normally associated with promoter clearance (Nissen and Yamamoto 2000). In this case, it was not determined whether GR directly interacts with a CTD kinase protein to block its activity, but it was possible to rule out activator exclusion or prevention of preinitiation complex assembly as possible mechanisms.

6.4
Masking of Activation Domains

As opposed to displacing activators from the DNA, some inhibitory proteins bind to and block an activation domain. As discussed above, Rb binds to E2F proteins, thereby interfering with their interactions with the basal transcriptional machinery, although this is clearly only one facet of Rb activity. In yeast, repression of the *GAL1* promoter is effected in part by the Gal80 protein binding to and obstructing the activation domain of the Gal4 activator protein, preventing the recruitment of HAT complexes (Carrozza et al. 2002). Interestingly, Gal80 is not entirely displaced from the Gal4 activator when the *GAL1* promoter is activated, but rather is bound in an alternative conformation (Leuther and Johnston 1992; Sil et al. 1999). A similar 'masking' role has been ascribed to the Mdm2 repressor in inhibition of the p53 tumor suppressor protein, although the binding of this protein also leads to destabilization of the activator (reviewed in the chapter by Asker et al., this volume). In contrast to the examples of these corepressors, there are few documented cases where DNA bound transcriptional repressors directly contact the activation domain of an adjacent activator to block its action. Many repressors have been found to inhibit a wide variety of transcriptional activators in a nonselective manner, suggesting that they do so without having to make specific activator–repressor contacts. Such promiscuity has been demonstrated for a number of repressors, such as short-range repressors in *Drosophila* (Arnosti et al. 1996a).

6.5
Interaction with Chromatin-Modifying Factors

Earlier studies had established firm correlations between certain DNA and chromatin modifications and transcriptional activity of genes. Repressed genes were generally found to exhibit lower levels of histone acetylation and increased DNA methylation (generally 5-methyl C), and subsequent studies have gone on in many cases to show that a wide variety of DNA-binding transcriptional repressors and corepressors interact with DNA and chromatin modifying activities. The best understood activities include: (1) chromatin remodeling complexes; (2) HDACs; (3) histone methyl transferases; and (4) DNA methyl transferases.

6.5.1
Chromatin Remodeling and Repression

As discussed in the chapter by Herrera et al. in this volume, chromatin remodeling complexes are ATP-driven multiprotein machines that can change the physical interaction between DNA and core nucleosomes (Peterson 2002; Narlikar et al. 2002). Genomic analyses have revealed that in most cases, these proteins are involved in gene activation, but a subset of cellular target genes in yeast are upregulated when snf/swi or chd1 activity is removed from the cell (Tran et al. 2000; Sudarsanam et al. 2000). In animal cells, the activity of the Brg1 and Brm chromatin remodeling complexes have been linked to the Rb corepressor protein (Zhang and Dean 2001). Genetic evidence from *Drosophila* places Brm components in the Rb pathway, and Rb-mediated cell cycle arrest depends on the presence of these complexes, but it is not clear whether this involved a direct interaction of the complexes with Rb (Strobeck et al. 2000; Brumby et al. 2002). Recent reports have suggested that Brg/Brm dependence of Rb activity might be indirect, mediated through activity of the p21 promoter, thus influencing the phosphorylation status of Rb (Kang et al. 2004). Other studies have found Brg/Brm complexes bound directly to Rb regulated promoters (Strobeck et al. 2000; Wang et al. 2002).

Evidence for a direct role for Swi/Snf proteins in repression comes from chromatin immunoprecipitation studies that demonstrate that these proteins are present at the repressed *ser3* promoter in yeast, and that a physically distinct chromatin structure was present (reviewed in Martens and Winston 2003). It is not clear at this point how remodeling actually induces loss of gene expression. One idea suggests that remodeling of chromatin structure might permit assembly of a stable repressor complex that would involve the activity of other proteins, such as HDACs. An additional clue is provided by the phenotype of the *ddm1* mutant of *Arabidopsis*; this mutant was originally identified by a loss of DNA methylation, but turned out to encode a DNA remodeling enzyme (Bourc'his and Bestor 2002). Similar findings have subsequently been made in the mouse, suggesting that one general pathway for chromatin remodeling-mediated repression is by induction of a structure that provides access for DNA methylases, discussed below (Bourc'his and Bestor 2002).

6.5.2
HDACs

As discussed in the chapter by Herrera et al. (this volume), acetylation of lysine residues of the core histone proteins can facilitate transcription of genes, possibly by reducing the affinity of histone proteins for DNA or by allowing the association of activating proteins via acetyl-lysine binding bromodomains. Removal of these acetyl groups by HDACs therefore has an inhibitory effect on gene expression. There are three classes of HDAC, all of which have been linked to transcriptional repression: class I enzymes, homologous to the yeast Rpd3 protein, are found in the nucleus; class II enzymes, homologous to yeast Hda1 protein, are found both in cytoplasm and nucleus; the structurally distinct class III enzymes are homologous to yeast Sir2, an NAD-dependent deacetylase that functions in maintenance of stable repressed chromatin structure at telomeres and silent mating type loci (Thiagalingam et al. 2003). In a few cases, DNA-binding transcriptional repressors have been found to contact these enzymes directly, while in a number of other examples, these enzymes are recruited by corepressors, including CtBP, N-CoR, SMRT, Sin3, Rb, Groucho/TLE, CoREST, and others (Chen and Courey 2000; Ahringer 2000; Jepsen and Rosenfeld 2002; Chinnadurai 2002; Lunyak et al. 2002). HDACs have been found generally in large protein complexes, including the Sin3 complex, and in the NuRD complex, which includes chromatin remodeling proteins. Because class I and II enzymes are sensitive to the inhibitor trichostatin A (TSA), many studies have sought to identify a potential role for HDACs in repression by comparing transcriptional signals with and without TSA. Care must be taken when interpreting results of such experiments, however, because this drug can cause nonspecific activation of gene expression. Therefore, it is important to verify that increases in reporter gene activity after treatment with TSA are specific to a reporter that is bound by the repressor.

Important targets of HDACs are presumably histone tail residues targeted by histone acetyltransferases, including histone H3 lysines 9 and 14, and H4 lysine 5 (discussed in the chapter by Herrera et al. in this volume). Although numerous repressors appear to recruit histone deacetylases, we have an incomplete understanding of the mechanisms by which HDACs interfere with transcription. Loss of histone acetylation has been proposed to interfere with transcriptional activation by several means, including the induction of a more compact chromatin structure due to restoration of DNA contacts with positively charged lysine tails, which would disfavor binding by activators or general transcription factors. This simple view is contradicted by findings that chromatin that has been deacetylated and is in an inactive conformation can still be accessed by components of the general transcription machinery, suggesting that deacetylation might have more subtle effects (Sekinger and Gross 2001; Saurin et al. 2001). An alternative consideration is that acetyl-lysine residues provide recognition sites for bromodomain containing transcription factors, and binding of specific factors might be destabilized by deacetylation. Finally, although HDACs have demonstrated activity on histones, it is possible that important targets may include other proteins whose functional properties are modified by acetylation, such as the transcriptional activators E2F and p53 (Martinez-Balbas et al. 2000; Brooks and Gu 2003).

The extent of histone deacetylation at repressed genes has been mapped by ChIP assays. In some cases, deacetylation is restricted to a single nucleosome, as in the case of the Rb-repressed *cyclinE* gene and the Rpd3-repressed genes in yeast (within the limits of measurement for this technique) (Morrison et al. 2002; Deckert and Struhl 2002). In other cases, silenced regions affected by the Sir2 deacetylase in the *Saccharomyces cerevisiae* mating-type locus were found to be deacetylated over a range of kilobases (Braunstein et al. 1993, 1996). The limit of deacetylation is not necessarily a function of the deacetylase itself, as Rpd3 is associated with local deacetylation as well as long-range repression, but may rather reflect the assembly of repressor proteins that recruit this deacetylase.

6.5.3
Histone Methyltransferases

Lysine residues of histone proteins H3 (K4, 9, 27, 36, 79) and H4 (K20) have been identified as targets of cellular methylases. Arginine residues of histone tails are also methylated, but this modification appears to be associated exclusively with gene activation (Zhang and Reinberg 2001). As with histone acetylation, histone methylation patterns are specific for widespread areas, reflecting chromatin accessibility, and are also modified at the level of individual genes, influencing their individual expression. At a bulk level, methylation patterns can predict overall gene activity; for example, genomic surveys of the region around the silenced mating type locus in the yeast *Schizosaccharomyces pombe* revealed a strong correlation of H3 K4 methylation in active regions, and H3 K9 within inactive regions (Noma et al. 2001). The methylation associated with inactive genes is carried out by a specific class of methyltransferase first identified through genetic studies of heterochromatin function in *Drosophila*. The *Su(var)3-9* gene and its vertebrate homologues (SUV39H1 and 2) encode proteins containing a conserved SET domain with histone methyltransferase activity (reviewed in Kouzarides 2002; Sims et al. 2003). SUV39 methyltransferases are recruited by corepressors such as KAP1, CtBP, and Rb. Some 73 putative genes in the human genome encode SET domains, only a few of which have been characterized. SUV39 proteins are specific for histone H3 lysine 9, providing a specific binding motif recognized by the so-called chromodomain of the HP-1 protein. HP-1 is found to be enriched on heterochromatin, and also on specific repressed genes located in euchromatic regions (Eissenberg and Elgin 2000). HP-1 is thought to contribute directly to repression by recruiting HDACs. The SET1 family protein EZH has been implicated in methylation of histone H3 lysine 27, a modification involved in binding by the Polycomb protein (discussed below).

Unlike histone acetylation, histone methylation does not appear to be readily reversible. No verified histone demethylases are known to function in transcription, but it has been proposed that histone methylation is reversed by unknown oxidative activities, by proteolysis of methylated tails, or by turnover of methylated histone subunits. Consistent with the relative stability of methylation, previous models suggested that HP-1 binding established a condensed, inactive locus. However, recent work demonstrates that this protein is not tightly bound, but is rather in dynamic equilibrium with associated loci, and that HP-1 at repressed E2F regulated promoters, for instance, is readily dissociated (Cheutin et al. 2003; Young and Longmore 2004).

6.5.4
DNA Methylation

In many, but not all, multicellular organisms, DNA methylation plays an important role in repression of gene expression (Attwood et al. 2002; Robertson 2002). The 5-methyl cytosine modification at CpG dinucleotides has been associated not with the rapid, reversible repression found at some promoters, but rather with stable repression. Thus, DNA methylation has been implicated in the repression of transposable elements and establishment of long-term transcriptionally repressed states, although recent research indicates that it may be important for more flexible regulation as well (Klose and Bird 2003). DNA methylation is catalyzed by DNA methyltransferases that transfer an activated methyl group from S-adenosyl methionine, either in de novo reactions (establishment of a new site of methylation) or in maintenance methylase reactions (for hemimethylated sites produced after DNA replication). As with histone modifications, a correlation with gene inactivity and DNA methylation was established long before the mechanism of action was known. The treatment of cells with 5-azacytidine, a nonmethylatable substrate, was sufficient to reactivate genes that had been silenced, suggesting that this modification played a causative role, but it was the identification of methyl-C-binding proteins that provided a link to repression pathways (Ballestar and Wolffe 2001; Jaenisch and Bird 2003). Methyl-C-binding proteins have been found to recruit HDAC, which may propagate the repressed state through chromatin effects. In addition to this influence of DNA methylation on chromatin structure, there are also clear indications that chromatin structure regulates DNA methylation. As indicated above, chromatin remodeling complexes have been found to regulate DNA methylation in plants, and the dim-5 histone methyltransferase has been shown to be important for DNA methylation in *Neurospora* (Tamaru and Selker 2001). Methyl-C-binding protein 2 has also been found to interact with histone methyltransferases, providing a cooperative link between these two forms of methlylation (Fuks et al. 2003).

6.6
Polycomb Proteins and Epigenetic States— Combinations of Repression Mechanisms

The differentiation of cell types during development often involves the stable repression of genes. Such repression can involve epigenetic mechanisms, that is, stably inherited modifications to a gene that affects its expression but does not affect its DNA sequence (Jaenisch and Bird 2003). Most recent evidence suggests that such epigenetic marks include covalent modification of DNA and chromatin by methylases and deacetylases. The role of Polycomb group (PcG) proteins has been extensively analyzed in this regard. *PcG* genes were originally identified in screens of *Drosophila* for genes that affect the developmental fate of the adult structures (homeotic transformations), and vertebrate homologs have been found to be involved in various forms of cancer (reviewed in Simon and Tamkun 2002; Orlando 2003). In *PcG* mutants in *Drosophila*, initial repression of homeotic genes is observed, but later in development these genes are ectopically expressed. In *Drosophila*, PcG proteins have been found to function in complexes that can be localized to ~100 sites on polytene chromosomes of the salivary gland. In vertebrates and the fly, at least two distinct complexes containing PcG proteins have been identified. E(z) and Esc proteins (or their vertebrate homologs) are included in related, developmentally dynamic complexes of 600 kDa and greater that also possess HDAC and histone methyl transferase activities (Furuyama et al. 2003). A distinct 2-MDa PRC-1 complex includes the Polycomb protein itself (Shao et al. 1999). The PRC-1 complex has been demonstrated to antagonize chromatin remodeling activity by SWI/SNF complexes in vitro, although it is not known if a similar activity is used in vivo (King et al. 2002).

The Polycomb protein itself possesses a chromodomain, a motif also found in HP1. Whereas HP1 binds preferentially to methylated H3 lysine 9, Polycomb specificity is directed toward trimethylated histone H3 lysine 27(Min et al. 2003). These observations have led to a model for Polycomb-mediated repression. Transient repression of target genes early in embryogenesis by means of repressors such as Krüppel and Hunchback leads to the recruitment of the E(z) complex proteins, which deacetylate the local region and add stable histone methylation signatures. Methylation of histone H3 lysine 27 provides a histone code that leads to association of the PRC-1 complex proteins, thereby blocking the activity of positively acting histone remodeling complexes. In addition, Polycomb proteins can associate with elements of the basal machinery, suggesting an additional level of targeting the general transcriptional machinery (Saurin et al. 2001). Polycomb proteins dissociate and reassociate with target genes during mitosis, and can be removed from repressed genes for brief periods of time, suggesting that a chromatin 'mark' provides the stable memory for continuous association of polycomb gene products (Beuchle et al. 2001). Additional specificity comes from interactions of PcG proteins with sequence specific-binding proteins such as Pleiohomeotic, Zeste and Trithoraxlike (GAGA factor). The importance of some of the activities associated with PcG proteins is not clear, but genetic studies suggest that chromatin modifying activities are necessary for complete and stable repression of target genes. In some cases, subsets of the PcG proteins, rather than one of these subcomplexes, may be utilized to effect transient repression at promoters. Rb has been found to interact with the vertebrate Polycomb homolog PC2, and the PcG protein Osa is found to work through SWI/SNF remodeling complexes to effect repression in *Drosophila* (Collins and Treisman 2000; Dahiya et al. 2001).

6.7
Context Influences the Activity of Putative Transcriptional Repressors

Relevant to the study of novel transcription factors for putative repression activity, it is worthwhile to examine factors that influence repressor activity, for in the absence of the proper context, a factor may not exhibit its physiological activity. Some of the factors known to affect the function of a repressor include: (1) the nature of the binding site that the protein contacts; (2) the nature of flanking binding sites that permit different factors to bind adjacent to the repressor; (3) the physiological state of the cell with respect to hormonal signaling, signal transduction pathways, and cofactor expression.

Relatively modest differences in DNA-binding sites can induce alternative corepressor complexes to form. For instance, the two DNA-binding moieties of the POU family Pit-1 factor can bind differently to cognate sites so that in one configuration it acts as a constitutive activator, while in the other as a repressor or activator (Scully et al. 2000). Pit-1 activates the growth hormone gene in rat pituitary somatotrope cells and the prolactin gene in lactotrope cells. The growth hormone promoter is not active in lactotrophs, despite the presence of the Pit-1 at the promoter, because the conformation adopted by Pit-1 at the binding site in this promoter allows Pit-1 to recruit the NCoR corepressor, in cooperation with additional factors. Substitution of the Pit-1 binding site with an alternative Pit-1 binding sequence allowed ectopic expression of the gene in lactotropes.

Flanking DNA-binding sites can also be critical for repressor activity. The rel-family Dorsal protein activates a number of genes in the *Drosophila* embryo in cooperation with Twist transcriptional activators (Szymanski and

Levine 1995). On regulatory elements of the *zen* and *dpp* genes, however, Dorsal functions in a negative mode by recruiting the Groucho corepressor. This Dorsal repressor function is dependent on the binding of Cut and Dri proteins to sites immediately flanking Dorsal (Valentine et al. 1998). The positioning of Cut and Dri binding sites relative to Dorsal is intolerant of any changes, unlike the flexibility seen with disposition of Dorsal and Twist sites for activation (Cai et al. 1996). Flanking binding sites are also important for selection of particular E2F family members that function primarily as repressors or activators (Giangrande et al. 2003).

Modification of cellular factors through hormonal or signal transduction cascades are also critical for determining the potential activities of transcriptional repressors. Nuclear hormone receptors exchange corepressor complexes for coactivator complexes upon binding of ligand. For example, unliganded thyroid hormone receptor or retinoic acid receptors interact with the NCoR/SMRT corepressors to mediate active repression. Upon binding of ligand, these cofactors are replaced by coactivators (reviewed in Aranda and Pascual 2001; Jones and Shi 2003). In addition to the modulation of repressor activity by hormone binding, different availability of transcriptional cofactors can also modify activity. Tamoxifen and related drugs that bind to the estrogen receptor demonstrate tissue-specific effects; depending on the types of ER cofactors present, the drug acts as an antagonist in some tissues and as an agonist in others (Smith et al. 1997).

Signal transduction pathways also converge on numerous DNA-binding factors that act facultatively as repressors. For example, the Notch pathway effector Su(H)/CBF-1 protein complexes with HDAC complexes, as described above, in the absence of Notch signaling, or with a portion of Notch itself as coactivator when the pathway is activated. In this instance, as with the Wnt, TGF-β, and Hedgehog pathways, default repression activities have been suggested to provide especially tight gene regulation to amplify the differences between active and inactive states (Barolo and Posakony 2002). These sorts of contextual variables indicate that characterization of a novel protein as a repressor based only on its ability to depress gene expression from a simplified reporter gene in transfection assays is inadequate. Nonetheless, many fruitful studies of transcriptional repressors have initiated with such modest beginnings.

7
Multiple Activities of Transcriptional Repressors

As suggested by the discussion of the Polycomb protein complexes, transcriptional repression can involve multiple enzymatic activities to provide multiple levels of genetic programming, including establishment and maintenance of repression. Even at the level of a single transcriptional repressor or corepressor with temporally limited effects, multiple repression activities are commonly identified. The *Drosophila* Knirps repressor has two repression activities, one dependent on the CtBP corepressor and the other CtBP-independent (Keller et al. 2000). The vertebrate nuclear hormone RIP140 corepressor was recently reported to possess no fewer than four autonomous repression domains as defined in cotransfection assays (Christian et al. 2004). Corepressors such as NCoR and CtBP have each been found to associate with multiple activities. Three nonexclusive explanations for the utilization of multiple repression pathways and cofactors by individual repressors are: (1) qualitatively distinct repression—different repression activities are deployed to interfere with different activators or basal promoters; (2) quantitative effects—multiple activities are simultaneously deployed to provide sufficient inhibitory activity; (3) artifactual repression—some of the multiple activities ascribed to repressors do not play a meaningful role in physiological contexts.

7.1
Qualitative Effects

Activator or promoter specificity has been documented for individual activities of a number of metazoan repressors. Rb family corepressors repress early S phase promoters such as *cyclin E* via HDAC-dependent pathways, while promoters repressed into G_2 phase are repressed by other mechanisms, including the activity of histone methyltransferases, SWI/SNF remodeling complexes, and Polycomb protein (Strobeck et al. 2000; Dahiya et al. 2001; Nielsen et al. 2001). Similarly, the vertebrate REST/NRSF protein represses genes in non-neuronal tissues using HDAC specific mechanisms and other mechanisms, in a promoter specific manner (Lunyak et al. 2002). The *Drosophila* Brinker transcriptional repressor is an effector of the TGF-β pathway, and represses a variety of genes during development (Affolter et al. 2001). Brinker binds directly to the Groucho and CtBP corepressors, both of which are known to recruit HDAC to effect long- and short-range repression, respectively. A genetic assay of the cofactor requirement for Brinker activity on endogenous genes revealed that this protein requires either CtBP or Groucho for repression of genes at different points in development, and for some genes in the blastoderm embryo, neither cofactor (Hasson et al. 2001).

In all three of these cases, it is not known which particular attributes of the target genes contribute to the context-specific requirement for individual corepressors. In at least some cases, however, multiple repression activities are suggested to be directed at distinct activators. The zinc finger/homeodomain Zeb transcriptional repressor regulates lymphocyte and muscle differentiation. This protein was found to repress the MEF2C activator in a

CtBP-independent manner, and the Ets and Myb activators in a CtBP-dependent manner (Postigo and Dean 1999).

7.2
Quantitative Effects

As noted above, different repressor activities have been linked to distinct genes, but the molecular features of these target genes that dictate such specificity is largely unknown. In addition to possible activator specificity, repressors might use multiple repressor activities to achieve quantitatively adequate levels of activity. Knirps repression of distinct enhancers directing expression of the *even-skipped* gene in *Drosophila* was used as a test case for comparison of CtBP corepressor-dependent and -independent activities. *eve* is expressed in the early embryo in a series of seven transverse stripes under the control of stripe enhancers located 5' and 3' of the coding sequence. In the absence of CtBP, Knirps is unable to repress the *eve* stripe 4/6 enhancer, but can still repress the *eve* 3/7 enhancer by means of its CtBP-independent activity (Struffi et al. 2004). The structure of the stripe 3/7 enhancer suggests that it is more sensitive to Knirps due to the larger number and higher affinity of the binding sites (Clyde et al. 2003). The lower sensitivity *eve* stripe 4/6 enhancer can respond to the CtBP-independent activity of Knirps if levels of this protein are elevated. This result suggests that the activators driving the *eve* stripe 4/6 enhancer are not intrinsically resistant to the CtBP-independent activity, and that the requirement for CtBP-dependent activity is a quantitative effect. Consistent with this quantitative hypothesis, when tested independently, the CtBP-dependent and -independent domains of Knirps are sufficient to mediate repression of a sensitive reporter gene, but only when combined together are they able to mediate repression of a more robust reporter (Sutrias-Grau and Arnosti 2004).

7.3
Artifactual Repression?

Given the heterogeneity of assays used to measure repression, it is likely that some of the observed repression activities do not represent those used in physiological settings. Caution should be exercised when interpreting results of repression assays. The identification of closely overlapping repressor and activator binding sites within the *eve* stripe enhancers led to the suggestion that direct competition might be the basis of repression (Stanojevic et al. 1991). Indeed, evidence from transfection assays suggested that Knirps could compete with the Bicoid activator at directly overlapping binding sites in DNA (Hoch et al. 1992). However, subsequent work indicated that repressor sites could be moved some distance from activators without ablating repression, and overexpression of the Knirps DNA-binding domain alone did not have a repressive effect (Arnosti et al. 1996b; Struffi et al. 2004). Thus, in this particular case, the purported competition for DNA binding is probably not physiologically relevant.

As discussed above, in vitro experiments can be susceptible to artifacts caused by addition of large amounts of DNA-binding proteins, which might compete for binding to the basal promoter element. Even in cases where specific contacts have been mapped between repressors and components of the basal transcriptional machinery, it is necessary to determine whether mutations that disrupt the interaction surfaces will also disrupt repression. Pharmacological treatments, such as the use of the HDAC inhibitor TSA have been used extensively in transfection assays to determine whether a particular repressor utilizes HDAC activity. This drug can nonspecifically stimulate a variety of promoters, and so suitable controls (i.e., testing for TSA responses in the absence of the repressor) are required.

In many cases, putative corepressors have been identified by yeast two-hybrid screening or by co-immunoprecipitation assays. In the latter case, when two proteins are demonstrated to co-immunoprecipitate after having been coexpressed in cell culture, it is important to establish whether the co-immunoprecipitation reflects direct protein contacts. In some cases, it has not been possible to distinguish whether protein association is merely mediated by fragments of DNA. Strong support for initially characterized repressor–corepressor interactions comes from the mapping of contact surfaces in the factors, which when mutated, lead to loss of repression. In other cases, subsequent genetic analysis has strongly supported these findings, in that mutation of the putative cofactor in question impaired the repression activity of the partner.

In summary, modern genetic and genomic approaches have greatly expanded our abilities to identify and characterize transcriptional repressors. The complexity of these regulatory proteins, and the contextual factors that affect their activities, demand that the investigator use well planned and complementary approaches to fully understand their biological functions.

References

Affolter M, Marty T, Vigano MA, Jazwinska A (2001) Nuclear interpretation of Dpp signaling in Drosophila. EMBO J 13:3298–3305

Ahringer J (2000) NuRD and SIN3 histone deacetylase complexes in development. Trends Genet 16:351–356

Akhtar A (2003) Dosage compensation: an intertwined world of RNA and chromatin remodelling. Curr Opin Genet Dev 13:161–169

Ansari AZ, Reece RJ, Ptashne M (1998) A transcriptional activating region with two contrasting modes of protein interaction. Proc Natl Acad Sci USA 95:13543–13548

Aranda A, Pascual A (2001) Nuclear hormone receptors and gene expression. Physiol Rev 81:1269–1304

Arnosti DN (2003) Analysis and function of transcriptional regulatory elements: insights from Drosophila. Annu Rev Entomol 48: 579–602

Arnosti DN, Barolo S, Levine M, Small S (1996a) The eve stripe 2 enhancer employs multiple modes of transcriptional synergy. Development 1:205–214

Arnosti DN, Gray S, Barolo S, Zhou J, Levine M (1996b) The gap protein knirps mediates both quenching and direct repression in the Drosophila embryo. EMBO J 14:3659–3666

Attwood JT, Yung RL, Richardson BC (2002) DNA methylation and the regulation of gene transcription. Cell Mol Life Sci 59:241–257

Bakke M, Lund J (1995) Mutually exclusive interactions of two nuclear orphan receptors determine activity of a cyclic adenosine 3',5'-monophosphate-responsive sequence in the bovine CYP17 gene. Mol Endocrinol 9:327–339

Ballestar E, Wolffe AP (2001) Methyl-CpG-binding proteins. Targeting specific gene repression. Eur J Biochem 268:1–6

Banerji J, Rusconi S, Schaffner W (1981) Expression of a beta-globin gene is enhanced by remote SV40 DNA sequences. Cell 2:299–308

Barolo S, Posakony JW (2002) Three habits of highly effective signaling pathways: principles of transcriptional control by developmental cell signaling. Genes Dev 16:1167–1181

Beuchle D, Struhl G, Muller J (2001) Polycomb group proteins and heritable silencing of Drosophila Hox genes. Development 6:993–1004

Bourc'his D, Bestor TH (2002) Helicase homologues maintain cytosine methylation in plants and mammals. BioEssays 24:297–299

Braunstein M, Rose AB, Holmes SG, Allis CD, Broach JR (1993) Transcriptional silencing in yeast is associated with reduced nucleosome acetylation. Genes Dev 7:592–604

Braunstein M, Sobel RE, Allis CD, Turner BM, Broach JR (1996) Efficient transcriptional silencing in Saccharomyces cerevisiae requires a heterochromatin histone acetylation pattern. Mol Cell Biol 16:4349–4356

Brooks CL, Gu W (2003) Ubiquitination, phosphorylation and acetylation: the molecular basis for p53 regulation. Curr Opin Cell Biol 15:164–171

Brown KE, Guest SS, Smale ST, Hahm K, Merkenschlager M, Fisher AG (1997) Association of transcriptionally silent genes with Ikaros complexes at centromeric heterochromatin. Cell 91:845–854

Brumby AM, Zraly CB, Horsfield JA, Secombe J, Saint R, Dingwall AK, Richardson H (2002) Drosophila cyclin E interacts with components of the Brahma complex. EMBO J 21:3377–3389

Busturia A, Wightman CD, Sakonju S (1997) A silencer is required for maintenance of transcriptional repression throughout Drosophila development. Development 21:4343–4350

Cai HN, Arnosti DN, Levine M (1996) Long-range repression in the Drosophila embryo. Proc Natl Acad Sci USA 18:9309–9314

Carroll SB, Grenier JK, Weatherbee SD, Grenier J, Wetherbee S (2001) From DNA to Diversity: Molecular Genetics and the Evolution of Animal Design. Blackwell Scientific, Malden

Carrozza MJ, John S, Sil AK, Hopper JE, Workman JL (2002) Gal80 confers specificity on HAT complex interactions with activators. J Biol Chem 277:24648–24652

Chen G, Courey AJ (2000) Groucho/TLE family proteins and transcriptional repression. Gene 1/2:1–16

Cheutin T, McNairn AJ, Jenuwein T, Gilbert DM, Singh PB, Misteli T (2003) Maintenance of stable heterochromatin domains by dynamic HP1 binding. Science 299:721–725

Chinnadurai G (2002) CtBP, an unconventional transcriptional corepressor in development and oncogenesis. Mol Cell 2:213–224

Christian M, Tullet JM, Parker MG (2004) Characterisation of four autonomous repression domains in the corepressor RIP140. J Biol Chem (in press)

Clyde DE, Corado MS, Wu X, Pare A, Papatsenko D, Small S (2003) A self-organizing system of repressor gradients establishes segmental complexity in Drosophila. Nature 426:849–853

Collins RT, Treisman JE (2000) Osa-containing Brahma chromatin remodeling complexes are required for the repression of wingless target genes. Genes Dev 24:3140–3152

Courey AJ, Jia S (2001) Transcriptional repression: the long and the short of it. Genes Dev 15:2786–2796

Dahiya A, Wong S, Gonzalo S, Gavin M, Dean DC (2001) Linking the Rb and polycomb pathways. Mol Cell 8:557–569

Dai SM, Chen HH, Chang C, Riggs AD, Flanagan SD (2000) Ligation-mediated PCR for quantitative in vivo footprinting. Nat Biotechnol 18:1108–1111

Davidson EH (2001) Genomic Regulatory Systems: Development and Evolution

Davie JK, Edmondson DG, Coco CB, Dent SY (2003) Tup1-Ssn6 interacts with multiple class I histone deacetylases in vivo. J Biol Chem 278:50158–50162

Davis RL, Turner DL (2001) Vertebrate hairy and Enhancer of split related proteins: transcriptional repressors regulating cellular differentiation and embryonic patterning. Oncogene 58:8342–8357

Deckert J, Struhl K (2001) Histone acetylation at promoters is differentially affected by specific activators and repressors. Mol Cell Biol 8:2726–2735

Deckert J, Struhl K (2002) Targeted recruitment of Rpd3 histone deacetylase represses transcription by inhibiting recruitment of Swi/Snf, SAGA, and TATA binding protein. Mol Cell Biol 18:6458–6470

DeRisi JL, Iyer VR, Brown PO (1997) Exploring the metabolic and genetic control of gene expression on a genomic scale. Science 278:680–686

Eissenberg JC, Elgin SC (2000) The HP1 protein family: getting a grip on chromatin. Curr Opin Genet Dev 10:204–210

Flores-Saaib RD, Courey AJ (2000) Analysis of Groucho-histone interactions suggests mechanistic similarities between Groucho- and Tup1-mediated repression. Nucl Acids Res 21:4189–4196

Fuks F, Hurd PJ, Wolf D, Nan X, Bird AP, Kouzarides T (2003) The methyl-CpG-binding protein MeCP2 links DNA methylation to histone methylation. J Biol Chem 278:4035–4040

Furriols M, Bray S (2001) A model Notch response element detects Suppressor of Hairless-dependent molecular switch. Curr Biol 11:60–64

Furuyama T, Tie F, Harte PJ (2003) Polycomb group proteins ESC and E(Z) are present in multiple distinct complexes that undergo dynamic changes during development. Genesis 35:114–124

Galant R, Carroll SB (2002) Evolution of a transcriptional repression domain in an insect Hox protein. Nature 6874:910–913

Gaston K, Jayaraman PS (2003) Transcriptional repression in eukaryotes: repressors and repression mechanisms. Cell Mol Life Sci 60:721–741

Gerasimova TI, Corces VG (2001) Chromatin insulators and boundaries: effects on transcription and nuclear organization. Annu Rev Genet 35:193–208

Giangrande PH, Hallstrom TC, Tunyaplin C, Calame K, Nevins JR (2003) Identification of E-box factor TFE3 as a functional partner for the E2F3 transcription factor. Mol Cell Biol 23:3707–3720

Gray S, Levine M (1996) Transcriptional repression in development. Curr Opin Cell Biol 3:358–364
Grummt I (2003) Life on a planet of its own: regulation of RNA polymerase I transcription in the nucleolus. Genes Dev 17:1691–1702
Halder G, Callaerts P, Gehring WJ (1995) Induction of ectopic eyes by targeted expression of the eyeless gene in Drosophila. Science 5205:1788–1792
Hanna-Rose W, Hansen U (1996) Active repression mechanisms of eukaryotic transcription repressors. Trends Genet 12:229–234
Hasson P, Muller B, Basler K, Paroush Z (2001) Brinker requires two corepressors for maximal and versatile repression in Dpp signalling. EMBO J 20:5725–5736
He GP, Muise A, Li AW, Ro HS (1995) A eukaryotic transcriptional repressor with carboxypeptidase activity. Nature 378:92–96
Henikoff S (2000) Heterochromatin function in complex genomes. Biochim Biophys Acta 1470:O1–O8
Hoch M, Gerwin N, Taubert H, Jackle H (1992) Competition for overlapping sites in the regulatory region of the Drosophila gene Kruppel. Science 5053:94–97
Holstege FC, Jennings EG, Wyrick JJ, Lee TI, Hengartner CJ, Green MR, Golub TR, Lander ES, Young RA (1998) Dissecting the regulatory circuitry of a eukaryotic genome. Cell 5:717–728
Huang Y (2002) Transcriptional silencing in Saccharomyces cerevisiae and Schizosaccharomyces pombe. Nucl Acids Res 30:1465–1482
Hyman CA, Bartholin L, Newfeld SJ, Wotton D (2003) Drosophila TGIF proteins are transcriptional activators. Mol Cell Biol 23:9262–9274
Inman GJ, Nicolas FJ, Hill CS (2002) Nucleocytoplasmic shuttling of Smads 2, 3, and 4 permits sensing of TGF-beta receptor activity. Mol Cell 10:283–294
Jaenisch R, Bird A (2003) Epigenetic regulation of gene expression: how the genome integrates intrinsic and environmental signals. Nat Genet 33 Suppl:245–254
Jaynes JB, O'Farrell PH (1991) Active repression of transcription by the engrailed homeodomain protein. EMBO J 6:1427–1433
Jepsen K, Rosenfeld MG (2002) Biological roles and mechanistic actions of co-repressor complexes. J Cell Sci 4:689–698
Jones PL, Shi YB (2003) N-CoR-HDAC corepressor complexes: roles in transcriptional regulation by nuclear hormone receptors. Curr Top Microbiol Immunol 274:237–268
Kang H, Cui K, Zhao K (2004) BRG1 controls the activity of the retinoblastoma protein via regulation of p21CIP1/WAF1/SDI. Mol Cell Biol 24:1188–1199
Karin M, Ben-Neriah Y (2000) Phosphorylation meets ubiquitination: the control of NF-κB activity. Annu Rev Immunol 18:621–663
Keller SA, Mao Y, Struffi P, Margulies C, Yurk CE, Anderson AR, Amey RL, Moore S, Ebels JM, Foley K, Corado M, Arnosti DN (2000) dCtBP-dependent and -independent repression activities of the Drosophila Knirps protein. Mol Cell Biol 19:7247–7258
King IF, Francis NJ, Kingston RE (2002) Native and recombinant polycomb group complexes establish a selective block to template accessibility to repress transcription in vitro. Mol Cell Biol 22:7919–7928
Klose R, Bird A (2003) Molecular biology. MeCP2 repression goes nonglobal. Science 302:793–795
Kouzarides T (2002) Histone methylation in transcriptional control. Curr Opin Genet Dev 12:198–209
Kulkarni MM, Arnosti DN (2003) Information display by transcriptional enhancers. Development 130:6569–6575

Kuo MH, Allis CD (1999) In vivo cross-linking and immunoprecipitation for studying dynamic Protein:DNA associations in a chromatin environment. Methods 19:425–433

Lai EC (2002) Keeping a good pathway down: transcriptional repression of Notch pathway target genes by CSL proteins. EMBO Rep 9:840–845

Leuther KK, Johnston SA (1992) Nondissociation of GAL4 and GAL80 in vivo after galactose induction. Science 256:1333–1335

Li B, Reese JC (2001) Ssn6-Tup1 regulates RNR3 by positioning nucleosomes and affecting the chromatin structure at the upstream repression sequence. J Biol Chem 276:33788–33797

Ludwig MZ, Bergman C, Patel NH, Kreitman M (2000) Evidence for stabilizing selection in a eukaryotic enhancer element. Nature 6769:564–567

Lunyak VV, Burgess R, Prefontaine GG, Nelson C, Sze SH, Chenoweth J, Schwartz P, Pevzner PA, Glass C, Mandel G, Rosenfeld MG (2002) Corepressor-dependent silencing of chromosomal regions encoding neuronal genes. Science 5599:1747–1752

Makeev VJ, Lifanov AP, Nazina AG, Papatsenko DA (2003) Distance preferences in the arrangement of binding motifs and hierarchical levels in organization of transcription regulatory information. Nucl Acids Res 31: 6016–6026

Malicki J, Cianetti LC, Peschle C, McGinnis W (1992) A human HOX4B regulatory element provides head-specific expression in Drosophila embryos. Nature 358:345–347

Martens JA, Winston F (2003) Recent advances in understanding chromatin remodeling by Swi/Snf complexes. Curr Opin Genet Dev 13:136–142

Martinez-Balbas MA, Bauer UM, Nielsen SJ, Brehm A, Kouzarides T (2000) Regulation of E2F1 activity by acetylation. EMBO J 19:662–671

Merika M, Thanos D (2001) Enhanceosomes. Curr Opin Genet Dev 11:205–208

Min J, Zhang Y, Xu RM (2003) Structural basis for specific binding of Polycomb chromodomain to histone H3 methylated at Lys 27. Genes Dev 17:1823–1828

Moqtaderi Z, Bai Y, Poon D, Weil PA, Struhl K (1996) TBP-associated factors are not generally required for transcriptional activation in yeast. Nature 6596:188–191

Morrison AJ, Sardet C, Herrera RE (2002) Retinoblastoma protein transcriptional repression through histone deacetylation of a single nucleosome. Mol Cell Biol 22:856–865

Muller-Hill B (1996) The Lac Operon: A Short History of a Genetic Paradigm.

Narlikar GJ, Fan HY, Kingston RE (2002) Cooperation between complexes that regulate chromatin structure and transcription. Cell 4:475–487

Nibu Y, Senger K, Levine M (2003) CtBP-independent repression in the Drosophila embryo. Mol Cell Biol 23:3990–3999

Nibu Y, Zhang H, Bajor E, Barolo S, Small S, Levine M (1998) dCtBP mediates transcriptional repression by Knirps, Kruppel and Snail in the Drosophila embryo. EMBO J 23:7009–7020

Nibu Y, Zhang H, Levine M (2001) Local action of long-range repressors in the Drosophila embryo. EMBO J 9:2246–2253

Nielsen SJ, Schneider R, Bauer UM, Bannister AJ, Morrison A, O'Carroll D, Firestein R, Cleary M, Jenuwein T, Herrera RE, Kouzarides T (2001) Rb targets histone H3 methylation and HP1 to promoters. Nature 412:561–565

Nissen RM, Yamamoto KR (2000) The glucocorticoid receptor inhibits NFkappaB by interfering with serine-2 phosphorylation of the RNA polymerase II carboxy-terminal domain. Genes Dev 18:2314–2329

Noma K, Allis CD, Grewal SI (2001) Transitions in distinct histone H3 methylation patterns at the heterochromatin domain boundaries. Science 293:1150–1155

Olave I, Reinberg D, Vales LD (1998) The mammalian transcriptional repressor RBP (CBF1) targets TFIID and TFIIA to prevent activated transcription. Genes Dev 11:1621–1637

Orlando V (2003) Polycomb, epigenomes, and control of cell identity. Cell 112:599–606

Pearson A, Greenblatt J (1997) Modular organization of the E2F1 activation domain and its interaction with general transcription factors TBP and TFIIH. Oncogene 15:2643–2658

Pereira FA, Tsai MJ, Tsai SY (2000) COUP-TF orphan nuclear receptors in development and differentiation. Cell Mol Life Sci 57:1388–1398

Peterson CL (2002) Chromatin remodeling: nucleosomes bulging at the seams. Curr Biol 12:R245–R247

Postigo AA, Dean DC (1999) Independent repressor domains in ZEB regulate muscle and T-cell differentiation. Mol Cell Biol 12:7961–7971

Pufall MA, Graves BJ (2002) Autoinhibitory domains: modular effectors of cellular regulation. Annu Rev Cell Dev Biol 18:421–462

Roberts SG (2000) Mechanisms of action of transcription activation and repression domains. Cell Mol Life Sci 57:1149–1160

Robertson KD (2002) DNA methylation and chromatin—unraveling the tangled web. Oncogene 21:5361–5379

Rogatsky I, Luecke HF, Leitman DC, Yamamoto KR (2002) Alternate surfaces of transcriptional coregulator GRIP1 function in different glucocorticoid receptor activation and repression contexts. Proc Natl Acad Sci USA 26:16701–16706

Ronshaugen M, McGinnis N, McGinnis W (2002) Hox protein mutation and macroevolution of the insect body plan. Nature 6874:914–917

Ross JF, Liu X, Dynlacht BD (1999) Mechanism of transcriptional repression of E2F by the retinoblastoma tumor suppressor protein. Mol Cell 3:195–205

Ross JF, Naar A, Cam H, Gregory R, Dynlacht BD (2001) Active repression and E2F inhibition by pRB are biochemically distinguishable. Genes Dev 15:392–397

Ruden DM, Jackle H (1995) Mitotic delay dependent survival identifies components of cell cycle control in the Drosophila blastoderm. Development 1:63–73

Ruzinova MB, Benezra R (2003) Id proteins in development, cell cycle and cancer. Trends Cell Biol 13:410–418

Ryu JR, Arnosti DN (2003) Functional similarity of Knirps CtBP-dependent and CtBP-independent transcriptional repressor activities. Nucl Acids Res 31:4654–4662

Ryu JR, Olson LK, Arnosti DN (2001) Cell-type specificity of short-range transcriptional repressors. Proc Natl Acad Sci USA 23:12960–12965

Saurin AJ, Shao Z, Erdjument-Bromage H, Tempst P, Kingston RE (2001) A Drosophila Polycomb group complex includes Zeste and dTAFII proteins. Nature 6847:655–660

Schuller J, Lehming N (2003) The cyclin in the RNA polymerase holoenzyme is a target for the transcriptional repressor Tup1p in Saccharomyces cerevisiae. J Mol Microbiol Biotechnol 5:199–205

Scully KM, Jacobson EM, Jepsen K, Lunyak V, Viadiu H, Carriere C, Rose DW, Hooshmand F, Aggarwal AK, Rosenfeld MG (2000) Allosteric effects of Pit-1 DNA sites on long-term repression in cell type specification. Science 5494:1127–1131

Sekinger EA, Gross DS (2001) Silenced chromatin is permissive to activator binding and PIC recruitment. Cell 105:403–414

Senawong T, Peterson VJ, Avram D, Shepherd DM, Frye RA, Minucci S, Leid M (2003) Involvement of the histone deacetylase SIRT1 in chicken ovalbumin upstream promoter transcription factor (COUP-TF)-interacting protein 2-mediated transcriptional repression. J Biol Chem 278:43041–43050

Senger K, Merika M, Agalioti T, Yie J, Escalante CR, Chen G, Aggarwal AK, Thanos D (2000) Gene repression by coactivator repulsion. Mol Cell 6:931–937

Shao Z, Raible F, Mollaaghababa R, Guyon JR, Wu CT, Bender W, Kingston RE (1999) Stabilization of chromatin structure by PRC1, a Polycomb complex. Cell 1: 37–46

Shibata H, Kobayashi S, Kurihara I, Saito I, Saruta T (2003a) Nuclear receptors and co-regulators in adrenal tumors. Horm Res 59 Suppl 1:85–93

Shibata H, Kurihara I, Kobayashi S, Yokota K, Suda N, Saito I, Saruta T (2003b) Regulation of differential COUP-TF-coregulator interactions in adrenal cortical steroidogenesis. J. Steroid Biochem. Mol Biol 85:449–456

Sil AK, Alam S, Xin P, Ma L, Morgan M, Lebo CM, Woods MP, Hopper JE (1999) The Gal3p-Gal80p-Gal4p transcription switch of yeast: Gal3p destabilizes the Gal80p-Gal4p complex in response to galactose and ATP. Mol Cell Biol 19:7828–7840

Simon JA, Tamkun JW (2002) Programming off and on states in chromatin: mechanisms of Polycomb and trithorax group complexes. Curr Opin Genet Dev 12:210–218

Sims RJr, Nishioka K, Reinberg D (2003) Histone lysine methylation: a signature for chromatin function. Trends Genet 19:629–639

Smale ST, Kadonaga JT (2003) The RNA polymerase II core promoter. Annu Rev Biochem 72:449–479

Small S, Kraut R, Hoey T, Warrior R, Levine M (1991) Transcriptional regulation of a pair-rule stripe in Drosophila. Genes Dev 5:827–839

Smirnov DA, Hou S, Ricciardi RP (2000) Association of histone deacetylase with COUP-TF in tumorigenic Ad12-transformed cells and its potential role in shut-off of MHC class I transcription. Virology 268:319–328

Smith CL, Nawaz Z, O'Malley BW (1997) Coactivator and corepressor regulation of the agonist/antagonist activity of the mixed antiestrogen, 4-hydroxytamoxifen. Mol Endocrinol 11:657–666

Smith RL, Johnson AD (2000) Turning genes off by Ssn6-Tup1: a conserved system of transcriptional repression in eukaryotes. Trends Biochem Sci 25:325–330

Stanojevic D, Small S, Levine M (1991) Regulation of a segmentation stripe by overlapping activators and repressors in the Drosophila embryo. Science 5036:1385–1387

Strobeck MW, Knudsen KE, Fribourg AF, DeCristofaro MF, Weissman BE, Imbalzano AN, Knudsen ES (2000) BRG-1 is required for RB-mediated cell cycle arrest. Proc Natl Acad Sci USA 97:7748–7753

Struhl K (2001) Gene regulation. A paradigm for precision. Science 5532:1054–1055

Struffi P, Corado M, Kulkarni M, Arnosti DN (2004) Quantitative contributions of CtBP-dependent and -independent repression activities of Knirps. Development 131:2419–2429

Sudarsanam P, Iyer VR, Brown PO, Winston F (2000) Whole-genome expression analysis of snf/swi mutants of Saccharomyces cerevisiae. Proc Natl Acad Sci USA 97:3364–3369

Sutrias-Grau M, Arnosti DN (2004) CtBP contributes quantitatively to Knirps repression activity in an NAD binding dependent manner. Mol Cell Biol (in press)

Szymanski P, Levine M (1995) Multiple modes of dorsal-bHLH transcriptional synergy in the Drosophila embryo. EMBO J 10:2229–2238

Tamaru H, Selker EU (2001) A histone H3 methyltransferase controls DNA methylation in Neurospora crassa. Nature 414:277–283

Taya Y (1997) RB kinases and RB-binding proteins: new points of view. Trends Biochem Sci 22:14–17

Thiagalingam S, Cheng KH, Lee HJ, Mineva N, Thiagalingam A, Ponte JF (2003) Histone deacetylases: unique players in shaping the epigenetic histone code. Ann NY Acad Sci 983:84–100

Tolkunova EN, Fujioka M, Kobayashi M, Deka D, Jaynes JB (1998) Two distinct types of repression domain in engrailed: one interacts with the groucho corepressor and is preferentially active on integrated target genes. Mol Cell Biol 5:2804–2814

Tran HG, Steger DJ, Iyer VR, Johnson AD (2000) The chromo domain protein chd1p from budding yeast is an ATP-dependent chromatin-modifying factor. EMBO J 19:2323–2331

Turner J, Crossley M (2001) The CtBP family: enigmatic and enzymatic transcriptional co-repressors. BioEssays 8:683–690

Valentine SA, Chen G, Shandala T, Fernandez J, Mische S, Saint R, Courey AJ (1998) Dorsal-mediated repression requires the formation of a multiprotein repression complex at the ventral silencer. Mol Cell Biol 11:6584–6594

Wang S, Zhang B, Faller DV (2002) Prohibitin requires Brg-1 and Brm for the repression of E2F and cell growth. EMBO J 21:3019–3028

Wen YD, Perissi V, Staszewski LM, Yang WM, Krones A, Glass CK, Rosenfeld MG, Seto E (2000) The histone deacetylase-3 complex contains nuclear receptor corepressors. Proc Natl Acad Sci USA 13:7202–7207

White RJ (2004) RNA polymerase III transcription–a battleground for tumour suppressors and oncogenes. Eur J Cancer 40:21–27

Xu W, Chen H, Du K, Asahara H, Tini M, Emerson BM, Montminy M, Evans RM (2001) A transcriptional switch mediated by cofactor methylation. Science 294:2507–2511

Young AP, Longmore GD (2004) Differences in stability of repressor complexes at promoters underlie distinct roles for Rb family members. Oncogene 23:814–823

Zelhof AC, Yao TP, Chen JD, Evans RM, McKeown M (1995) Seven-up inhibits ultraspiracle-based signaling pathways in vitro and in vivo. Mol Cell Biol 15:6736–6745

Zhang HS, Dean DC (2001) Rb-mediated chromatin structure regulation and transcriptional repression. Oncogene 20:3134–3138

Zhang Y, Reinberg D (2001) Transcription regulation by histone methylation: interplay between different covalent modifications of the core histone tails. Genes Dev 18:2343–2360

Zhao B, Hou S, Ricciardi RP (2003) Chromatin repression by COUP-TFII and HDAC dominates activation by NF-kappaB in regulating major histocompatibility complex class I transcription in adenovirus tumorigenic cells. Virology 306: 68–76

Genomic Approaches to the Study of Transcription Factors

T. R. Zacharewski · J. J. La Pres (✉)

Michigan State University, 402 Biochemistry Building, East Lansing, MI 48824-1319, USA
lapres@msu.edu

1	Introduction	70
2	Background	70
2.1	Transcription Factors	70
2.2	Genomics	72
2.3	Relevance to Pharmacology	73
3	Genomics and TF Research	74
3.1	Model System	74
3.2	Experimental Design	78
3.3	The Platform and Its Use	79
3.4	Microarray Image and Data Analysis	81
4	In Silico Approaches to TF Target Gene Identification	84
5	The Future	87
	References	88

Abstract Transcription factors (TFs) are responsible for regulating the temporal and tissue specific expression of the genome. Identifying the complete battery of genes that are targeted by individual TFs, how these target genes interact within a cellular network, and the role these networks play in various disease states is essential to pharmacology and toxicology research. Genomic tools have been and are being developed to address these issues on a global scale. The most widely used tool among them is the microarray. cDNA and oligonucleotide arrays have become standard tools to study TFs and identify regulatory networks. Understanding microarray data requires careful attention to experimental design, platforms, model systems and data analysis. Microarray data analysis has also become increasingly complex with the use of various error models and clustering algorithms. Finally, new in silico approaches have been utilized to further characterize TF signaling networks. These approaches utilize computer based modeling, chromatin immunoprecipitation and microarray experiments. The application of these approaches and some concerns with the various approaches and their application to TF research are discussed.

Keywords Microarray · Genomics · Experimental design · Gene networks · Response elements · GeneChips

1
Introduction

Genomics is defined as the discipline of mapping, sequencing and analyzing genomes. Recent advances in sequencing have rapidly increased our ability to sequence and map genomes (Lander and Weinberg 2000). Analyzing these sequences for coding regions and assigning functions to these identified genes has now become an active area of research which has taken different forms. Most prominent are the tools used for expression profiling, including oligonucleotide and cDNA microarrays (for a historical review see Jordan 2002). Expression profiling can offer insights into cellular perturbation, such as the addition of ligands or gene disruption, with subsequent consequences at the level of transcription. Though these changes in gene expression do not necessarily translate into physiological outcomes, they can be used to identify possible target genes for cellular transcription factors and help develop signaling networks. Many recent advances in computational analysis of genome sequences have identified transcription factor binding sites and possible regulatory networks. These bioinformatic methods, when coupled with the functional data generated using expression profiling, transgenic technologies, and high-throughput biochemical screens, provide powerful tools for whole genome analysis, creating transcription factor regulatory networks and assigning functions to all genes in an organism.

2
Background

2.1
Transcription Factors

Transcription factors (TFs) are proteins responsible for directing the expression of genes in spatially and temporally distinct patterns in response to external stimuli. Transcription factors come from numerous different classes and families. These include the general TFs of eukaryotes that function in the process of transcription (Green 2000). Most notable among the general TFs is TFIID, which recognizes and binds the TATA box and recruits other holoenzyme components (Struhl 1997). General TFs play a critical role in transcription but are unable to generate the specificity of expression required for normal development and homeostasis. Specificity and differential gene expression is achieved by a diverse group of TFs that bind specific sequences in regulatory regions of responsive genes and direct the RNA polymerase holoenzyme to begin transcription. These factors are the focus of this review.

Diverse TFs including POU domain containing proteins, steroid receptors, basic-leucine zipper and PAS proteins (Gu et al. 2000; Phillips and Luisi 2000; Aranda and Pascual 2001; Don and Stelzer 2002) to name a few, respond to unique cellular signals and differentially direct the transcription of specific genes. These signals, initiated by growth factor-stimulated phosphorylation, acetylation and in some cases, direct ligand binding activate TFs and facilitate their binding to unique regulatory sequences, referred to as response elements (Dolwick et al. 1993; Jeong et al. 2002; Guren et al. 2003; Lu et al. 2003; Mottet et al. 2003). These response elements can be found in regulatory regions of the genome and direct the tissue and temporal specific expression of the target gene.

Transcription factors vary in structure and in the genes they regulate. However, most share common structural features (Warren 2002). These include a DNA-binding domain that is responsible for recognizing the specific response elements and a regulatory domain that interacts with the basic transcriptional machinery (basic TFs mentioned above) to stimulate or inhibit transcription. Some TFs are activated by ligands and contain distinct ligand-binding domains, such as the nuclear receptor superfamily (Aranda and Pascual 2001). Those that do not require ligands for activation have residues or motifs that are post-translationally regulated by phosphorylation, hydroxylation or acetylation. These are important distinctions when analyzing signal transduction using genomic technology. Studies examining ligand activated TF, such as nuclear receptors, can be performed in the presence or absence of ligand or with structural analogs that exhibit agonist or antagonist properties. Studies of nonliganded TFs require specific cellular signals or mutant cell lines that do not express the TF of interest to identify a complete set of target genes using genomic technology. These types of manipulations, however, may influence other signals within the cell and complicate the interpretation of the expression analysis. These types of issues highlight the need for careful study design and statistical support.

Strategies for the identification of target genes for individual TFs have significantly evolved over the last two decades. Initially, responsive genes were identified using low throughput differential expression techniques that suffered from high false positive rates. Regulatory genomic sequence for these genes were then isolated and systematically analyzed for critical motifs to identify sequence motifs capable of transcriptional control. This 'bottom-up' approach was time consuming and did not always consider the complex interactions that may be required for gene expression. Alternatively, a 'top-down' approach that identifies global changes in gene expression using DNA microarrays is currently leading the way. When genomic screens are used in combination with chromatin immunoprecipitations, transgenic or knock-out cell lines or mice strains, RNA interference (RNAi) and complex computational models, target genes for specific TFs can be identified and their role in complex gene regulatory networks can be described.

2.2
Genomics

In 1984, leading researchers were brought together by the U.S. Department of Energy and the International Commission for Protection Against Environmental Mutagens and Carcinogens to determine if increases in mutations could be detected in survivors of the Hiroshima and Nagasaki nuclear detonations or their children (Cook-Deegan 1989). It was concluded that the technology did not exist to offer this type of analysis; however, the subsequent meeting report served as the impetus for the Human Genome Initiative (No author listed 1991). One objective of the Human Genome Initiative was the creation of new technology to increase sequencing throughput and to analyze identified genes, which facilitated the development of cDNA and the oligonucleotide microarrays (Collins et al. 2003).

The scientific basis for the cDNA and oligonucleotide microarrays is that labeled polynucleotide probes can bind complementary nucleic acid sequences attached to a solid support (Southern 1975). The specificity of hybridization is dictated by the inherent information within the DNA molecule itself. Using this same principle, thousands of hybridizations can be performed simultaneously to capture a snapshot of gene expression.

cDNA microarrays and oligonucleotide arrays differ in construction and in experimental considerations. However, both have proven useful in analyzing the expression patterns of thousands of genes simultaneously (Holloway et al. 2002). cDNA microarrays consist of thousands of cDNA samples that have been systematically spotted onto a solid support such as a chemically modified glass slide (Schena et al. 1995). These spotted cDNAs represent known genes or expressed sequence tags (ESTs) identified in sequencing projects. EST clones are created by reverse transcribing mRNA isolated from various tissues or cell culture and cloning into a suitable sequencing vector. The nature of their construction dictates that these genes are expressed and usually contain only a portion of the open reading frame (usually biased towards the 3' end of the gene), thus giving them the name 'expressed sequence tags.' Affymetrix GeneChips are high density arrays that group oligonucleotides into probe sets to simultaneously interrogate the expression levels of thousands of specific genes (Fodor et al. 1993). The synthesis of the oligonucleotides is done by photolithography onto a glass surface. The Affymetrix GeneChips, unlike cDNA microarrays, can be used for only one sample at a time and are considerably more expensive. However, the GeneChip offers the advantage of ready-made microarrays, substantial annotation and a common platform for comparison among different researchers.

Though microarrays are the most common type of technology currently available for expression profiling other alternatives (each with inherent limitations) are also available. These other methods for expression profiling include total gene expression analysis (TOGA), representational differential analysis (RDA), amplified differential gene expression (ADGE) and serial analysis of gene expression (SAGE) (for a review see Green et al. 2001). Of these alternative technologies, SAGE is the most widely used. SAGE is an open platform system that requires no a priori knowledge of the sequence of genes of interest and therefore, in theory, places no limitation on which genes can be monitored. In contrast, cDNA microarrays and GeneChips require cDNA sequence information for each gene represented on the microarray and therefore, expression information is limited to genes represented on the array. SAGE generates 'tags' for each expressed gene in a model system that can be sequenced and counted (Jiang et al. 2002; Menssen and Hermeking 2002; Seth et al. 2002; Wu et al. 2002). The tag count correlates with the level of expression for its corresponding gene. SAGE data are warehoused into a searchable database that can be used to compare results from different tissues and treatments (http://www.ncbi.nlm.nih.gov/SAGE/). In addition, this data can be used to generate an expression profile for any desired gene that is referred to as a 'virtual Northern.' SAGE is resource intensive and therefore is not readily replicated. Moreover, the ability to measure low levels of expression is dependent on the number of tags that are sequenced and therefore, sensitivity is based on available resources and statistical evaluation to determine the extent of coverage.

Genomic technology is uniquely suited for the study the role of TFs in altering the expression pattern of genes. Integration of genomic technology with both standard (e.g., knock-out, ligand activation, overexpression) and emerging (e.g., chromatin immunoprecipitation, RNAi) technologies can yield answers to complex scientific questions concerning TF function, gene expression networks and target genes.

2.3
Relevance to Pharmacology

Genomic technology has changed the face of drug discovery and pharmacological and toxicological testing. The identification of complete lists of genes influenced by drug targets, high-throughput test compound production and multifaceted testing, haplotype mapping for multidimensional diseases and the formulation of novel therapeutics such as RNAi have now become standard for pharmaceutical research and development. Transcription factors and their target genes are increasingly the focus of this area of research, due to their direct role in regulating various disease states and their indirect roles in cellular homeostasis. The formulation of a genomic approach to the study of TFs as a target for pharmacology research requires several important areas of research. First, the characterization of the differential regulation of these factors at the protein and signaling level is critical to understanding of TFs in cellular homeostasis and disease. Second, if loss or overstimulation of a given TF is important in the pathology of disease, then the genes that are regulated by that TF must play a role in the process. Only by identifying all of the target genes for a disease-causing TF in the pertinent tissue will a complete understanding of the process be reached. Genomic tools are uniquely suited to identifying these targets and their role in regulatory networks. Finally, the influence of these target genes on the disease state can be addressed using, genomic tools, such as high thorough-put RNAi screens and transient transfection analysis. The future of drug discovery, pharmacology and toxicology testing is dependent upon genomic research and the new tools that can be applied to these problems on a global scale.

3
Genomics and TF Research

Microarrays are the most widely used technology to study the genome-wide activities of transcription factors. Microarrays are defined as an array of thousands of bio-molecules (i.e., cDNAs, oligonucleotides or proteins) immobilized on a solid support such as glass or membrane. The three basic components of a microarray experiment include the model system under investigation, the microarray, and the analysis of microarray data. Model systems are limited to organisms with complementary arrays but can include in vitro and in vivo models. Inherent in the model system are the treatments that are to be tested. For example, MCF7 cells are a widely used cell line in the study of breast cancer and have been used to study the mechanism of action of estrogenic compounds. The second part of a microarray experiment is the construction and use of the microarray which includes the selection of the type of platform to be used (i.e., cDNA microarray or Affymetrix GeneChip). In addition, this portion includes the hybridization conditions and image analysis of the microarray. The third portion of a microarray experiment, and the most challenging, is data analysis. This involves the normalization, identification of significant changes in gene expression resulting from treatment, and other higher order forms of analysis (e.g., cluster analysis and graphical representations) to place changes in gene expression into biological context.

3.1
Model System

In the study of a particular TF, the experimental system chosen is critical. In the case of a ligand-activated TF, a major consideration is that the TF is expressed and is functional in the cell line or tissue under investigation (Omoto and Hayashi 2002). In most cases the cells or tissue is a target for the ligand in question. In the case of nonligand activated TFs, the consideration of the system may be more complex. Ectopic activation of endogenous

TFs in a given cell or tissue may be problematic if activation triggers other activities. For example, the activation of CREB through the upregulation of cAMP has other cellular consequences that may confound data analysis and interpretation (Kvietikova et al. 1995; Kako and Ishida 1998).

Many loci have been targeted for deletion in mice or in cell lines that may be useful in determining the role of the TF in a given response (Lee et al. 2003; Poguet et al. 2003). These resources can be used to identify target genes for the given TF when compared to intact models and have the advantage of providing a completely null background for comparative purposes (Lindberg et al. 2002). In addition, animals with TF gene disruptions can be used to determine the effects of the given TF at various developmental time points, while null cells can be used to identify target genes in any tissue or cell type. However, many TFs are critical for normal development and loci disruptions may be embryonic lethal, thus limiting the utility of these mice to early stages of development (Lichtlen et al. 2001). For example, the null alleles of the hypoxia inducible factors (HIF) 1α and HIF2α are embryonic lethal when homozygous. Therefore, these animals cannot be used to generate useful genomic data for later stages of development. These problems can be addressed with the creation of conditional null animals. These animals are engineered to contain a loxP site within the genomic sequence of the gene of interest. The null genotype is created only after exposure to Cre-recombinase, which can be done at any given developmental time point or in a specific tissue. This overcomes the problem of embryonic lethality and offers a powerful tool for the study of critical TFs. However, the availability of these conditional nulls is not widespread.

Overexpression by stably integrating an expression vector into a certain cell type has also been used to investigate TF function (Shtutman et al. 2002; Mackay et al. 2003). Overexpression increases the level of a given TF and its signal in relation to the normal cell and facilitates comparisons to wild type cells in order to identify target genes. This system does not require complex genetic engineering or special materials but is limited by several drawbacks. Overexpression may affect other signaling systems, potentially resulting in misleading results. In addition, the integration site of the TF during the construction of the cell line may disrupt other signaling proteins that could also confound analysis and interpretation. These drawbacks can be minimized by the selection of several clonal populations of the cell line which would be analyzed independently for consistency. Finally, overexpression is physiologically artifactual and may disrupt the natural homeostasis within the cell itself as well as the balance between various signaling cascades.

RNA interference can be used to manipulate the levels of expression of the TF of interest; it was initially described in plants, and subsequently in lower eukaryotes before being adapted for use in mammalian cells (Baulcombe 1996; Fire et al. 1998; Elbashir et al. 2001; Paul et al. 2002; Yu et al. 2002). RNA interference uses short interfering RNAs (siRNAs) to induce the degradation of particular mRNA sequences thus leading to lower levels of expression of the target gene product. siRNAs can be designed to target any mRNA, are readily accessible and are not limited by the availability of resources and expertise required to create knock-out animals. Several different siRNAs for a family of TFs can be generated to determine the role of each family member in a particular cell type (Zhou et al. 2003). In addition, RNAi can be coupled with standard cell line tools to create stable cell lines that express the siRNA cassette. The major disadvantage of this approach is that RNA disruption is usually not complete. Although substantial reductions can be achieved, the complete 'knock-down' of mRNA or protein levels is rarely observed.

The molecular manipulation of an in vivo or in vitro system may offer suitable models to study TFs on a genome scale. However, these systems may have problems in terms of specificity and artifactual results. For example, it is difficult to determine if the genes identified in these screens are influenced by the TF of interest directly or indirectly. One possible mechanism to address this shortcoming is the chromatin immunoprecipitation (ChIP) assay. The basic process for a ChIP assay is to induce the TF to bind to its response element in the regulatory region of a target gene. The cells are then treated with a cross-linking reagent that covalently links the TF protein to the genomic DNA. The protein:DNA complexes are then isolated and sheared by sonication. The sheared TF:DNA complex of interest is then immunoprecipitated by using an antibody specific for the TF. In traditional ChIP assays, the DNA is released by reversing the crosslinking reaction and analyzed by PCR using primers for a specific gene of interest (Fig. 1). Consequently, a priori knowledge of which genes are responsive is required, as well as information regarding the location of functional response elements to which the TF may bind.

Recent advances by various genome sequencing projects has made it possible to run ChIPs on a genome scale (Fig. 1) (Ren et al. 2000). These 'ChIP on Chip' assays use the precipitated genomic DNA as a probe for a microarray spotted with regulatory regions of multiple genes (Ren et al. 2000; Lieb 2003). This higher throughput unsupervised assay facilitates genome wide analysis of TF:response element interactions and offers a tool to identify functional response elements and to determine whether changes seen in standard microarray experiments are a direct effect of the TF action.

Two complementary technologies that provide additional information regarding TF function are tissue arrays and atomic force microscopy (AFM). Tissue arrays are tissue sections arrayed onto slides for the purpose of in situ hybridization and immunohistochemistry (Dolled-Filhart et al. 2003; Freier et al. 2003; Tolgay Ocal et al. 2003; Tzankov et al. 2003). These arrays can be used to screen multiple normal, disease and treated tissues in a side by side format in order to obtain comparative, longitudinal or dose–response information for a particular gene or gene product. These arrays are

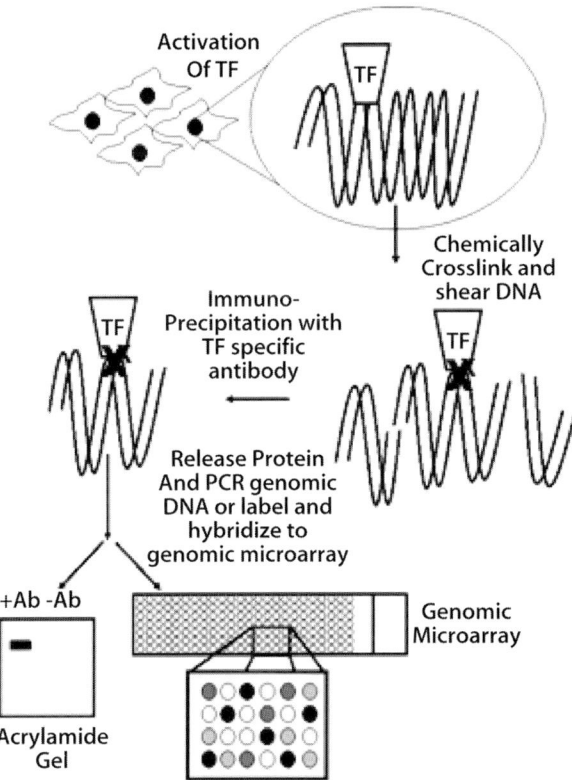

Fig. 1 Chromatin immunoprecipitation assays. The cells are induced to activate the TF of interest. The TF is crosslinked to the genomic DNA, DNA is sheared and the TF:DNA complex is purified by immmunoprecipitation. Proteins are released and the DNA can be analyzed by PCR with promoter specific primers. Alternatively, the DNA can be labeled and hybridized to a microarray of genomic DNA fragments or putative regulatory regions

particularly useful in the analysis of TFs that are thought to play a role in tumorigenesis. Though not technically 'genomic' they do offer a high-throughput method for TF distribution, expression and may help assign function in different disease states. AFM has advanced to the point that is capable of visualizing TFs in contact with DNA. Using other markers on the chromosomal segment it is possible to use this method to analyze gene start sites, TF complex formation and other genomic events (Allison et al. 1997; Hoyt et al. 2001).

Microarrays and the alternative strategies described above can be applied to the study of TFs and the identification of target genes for the TF of interest. Each has advantages and disadvantages. The research question must be critically considered and the methodology that best suits these aims should be used. For example, microarray platforms cannot be easily employed in research using obscure organisms as the gene sets are not available. In other cases, the long-term infrastructure costs necessary to produce custom cDNA microarrays may be prohibitive and Affymetrix GeneChips may suit the research agenda in the short term. The researcher must realize that the platform chosen will directly affect the size, quality and substance of the TF target genes that can be identified.

3.2
Experimental Design

Experimental design is an important factor in any study and is predicated on the biological question being asked. In microarray studies, two main designs have been used (Fig. 2). Each of these procedures has the goal of making a direct comparison between the treated and control samples to identify changes in gene expression. The more popular universal reference design (Fig. 2, left), involves comparing all perturbed systems to a reference sample, usually an untreated control. Each array is run twice, exchanging the dye used between the two samples. This dye-swap is necessary to control for the differences between dyes with respect to their spectral properties and incorporation efficiencies. Although relatively intuitive, the reference design has limited statistical power and data collection is biased to the reference sample. Alternatively, the loop model permits a more systematic analysis of all of the samples within a microarray experiment (Fig. 2, right) (Kerr 2003). In this model, each sample is labeled an equal number of times and the dye-swap is inherent in the design. The most striking difference between this design and the reference design is that expression ratios are not calculated until all of the analysis has been performed. The loop design allows the reference sample to be compared to any other sample even though they were never competitively hybridized to the same microarray. This increases the efficiency of the design and creates more statistical power for comparisons.

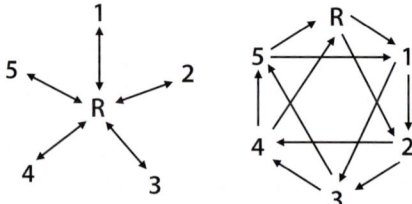

Fig. 2 Two basic designs for microarray experiments. In the standard reference design (*left*) signals from the various experimental conditions (*1–5*) are compared to a reference sample (*R*). In the loop design (*right*) each experimental condition is compared to four other groups, but not necessarily to the reference

Whatever experimental design is selected, existing standards for microarray experiments should be followed. Minimum information about a microarray experiment (MIAME) standards have been proposed and several leading journals require compliance with the guidelines for publication (Brazma 2001). These standards detail the minimal information needed for the accurate description of experimental design and analysis of microarray experiments. The MIAME standards are designed to facilitate replication of the study and the comparison of microarray data between laboratories.

3.3
The Platform and Its Use

Two main platforms are usually considered for microarray studies, although other platforms exist (Fig. 3, right). Affymetrix oligonucleotide DNA Chips, also referred to as simply GeneChips, have hundreds of thousands of in situ synthesized tiled oligonucleotides. Although expensive, Affymetric GeneChips provide a readily available and reproducible platform that can also identify alternatively spliced variants and with some chips, single nucleotide polymorphisms. However, Affymetrix chips analyze only one sample at a time and therefore, at least two GeneChips are required to identify treatment-related effects on gene expression. This means that twice as many chips are needed for any given experiment when compared to two-colored cDNA microarrays (discussed below). GeneChips also require proprietary hybridization and scanning equipment.

The Affymetrix platform is available for several prominent experimental organisms. The GeneChip is capable of representing thousands of gene of interest for multiple species including human, rat, mouse, Arabidopsis, *Caenorhabditis elegans*, *Drosophila* and yeast expressed genomes. For example, the human U133 GeneChip contains 45,000 probe sets representing 39,000 transcripts from 33,000 human genes.

cDNA or long oligonucleotide microarrays have up to 20,000 features spotted onto a glass surface; in theory this could represent as many genes. In contrast to the Affymetrix DNA chip, on which probes are synthesized in situ using photolithography, the cDNA and long oligonucleotide microarrays are constructed using PCR amplified cDNAs or synthesized oligonucleotides, respectively, that are robotically spotted onto a solid substrate. The spotted cDNA PCR products typically range in size from 200 to 2,000 bp depending on the source. Spotted cDNA PCR products and long oligonucleotides are usually biased to the $3'$ end of the gene reflecting the bias in EST sequencing project procedures and because sequence similarity within the $3'$ untranslated region is low thus minimizing cross-hybridization.

cDNA microarray assays involve a competition between two samples labeled with either cy3 or cy5 fluorescent dyes (Fig. 3, left). Advances in fluorescent dyes may extend this to as many as four colors, although use of these

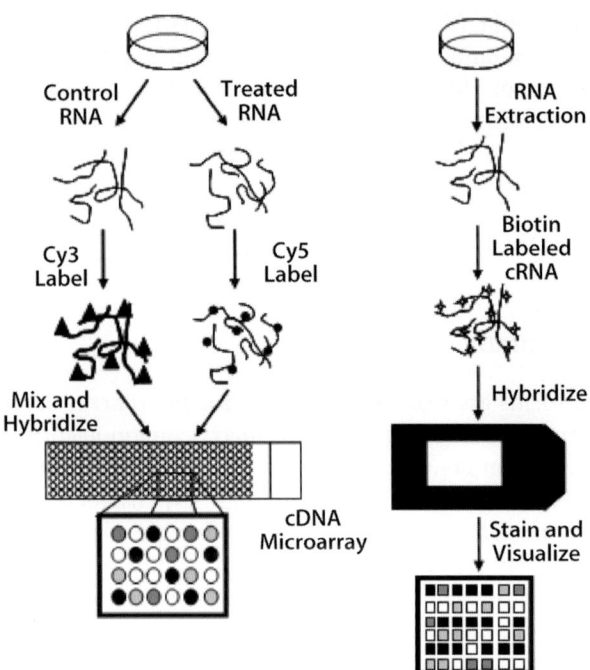

Fig. 3 Microarray platforms and general protocol. The custom cDNA (*left panel*) and the Affymetrix GeneChip (*right panel*) are the two main microarray platforms. Both protocols start with RNA isolated from cells or tissues. The cDNA microarray utilizes two fluorescent probes to control for differences in loading and RNA and probe preparations and to maximize resources. Following labeling and hybridization the cDNA microarray is scanned and expression values are calculated. Differences in expression are determined by signal intensities in the two channels. Those genes expressed disproportionally in one channel (*black* or *white circles*) will appear as those colors, while genes unaffected by treatment will appear *gray*. The two channels can also be merged to generate pseudo-color representation (*bottom*). The Affymetrix Gene Chip uses a unique biotin labeling protocol. Each GeneChip is hybridized with a single probe. Data are analyzed using proprietary algorithms to generate expression values. Expression patterns are observed as increasing intensities within the array image (*black* being highest, *white* being lowest)

alternative dyes is limited primarily by lack of infrastructure and analysis paradigms (Hessner et al. 2003). Two-color cDNA and long oligonucleotide microarrays have a significant cost and flexibility advantage over Affymetrix GeneChips but still require dedicated infrastructure including a robotic spotter and scanner. In addition, liquid handling robots, PCR machines and storage facilities to maintain the clones set and amplified products properly are significant issues that are avoided with GeneChips. As cDNA and oligonucleotide sets increase in size, computer assisted tracking of clones becomes necessary. Several companies offer laboratory information management software for handling this type of information. Once all of these components are in place, the cDNA and oligonucleotide microarrays are relatively inexpensive to produce and can be tailored to a particular interest. For example, if a researcher is interested in all of the genes that have been shown to be influenced by the E2F transcription factor, clones can be purchased and incorporated into an existing clone set or spotted to create a customized array.

Standard microarray studies aimed at investigating the function of a TF are initiated by the activation of the TF in a model system in comparison to the nonperturbed state of the model. For example, studies investigating the estrogen receptor would compare changes in gene expression in estrogen responsive cells following treatment with 17β-estradiol or vehicle, as the control. Total RNA is independently extracted from the treated and control cells and in some instances further purified to obtain mRNA. Fluorescent probes are generated in reverse transcription reactions using isolated RNA and different types of modified nucleotides. In the Affymetrix system, the standard labeling protocol involves the production of cDNA from purified mRNA. Following cDNA synthesis, cRNA is synthesized in the presence of a biotin labeled nucleotide. This labeled cRNA is then sheared and hybridized to the microarray. Finally, the labeled microarray is exposed to fluorescently labeled strepavidin beads. In contrast, the cDNA microarrays can be labeled by either of two protocols. The one-step or direct labeling method incorporates fluorescent modified dUTP (i.e., Cy3-dUTP or Cy5-dUTP) into the cDNA during the reverse transcription reaction. This method offers convenience and speed but introduces a dye bias as the Cy5 dye is incorporated less efficiently into the cDNA than the Cy3 dye. This bias can be minimized by including a 'dye swap' microarray to compensate for variations due to dye. The two-step or indirect method involves incorporation of an amino-modified dUTP during the reverse transcription reaction. The incorporated modified nucleotide is then chemically coupled to the fluorescent dyes in the second step. The dye swap is not required for the indirect incorporation because the same modified nucleotide is used in the labeling; however, dye swaps are still incorporated into the experimental design to compensate for spectral differences. This comparison is done on the same chip for the microarray and between two separate chips with the Affymetrix platform. The hybridization and washing procedures are standard and very similar to Northern Blots. Once hybridized, these microarrays must be scanned and analyzed.

3.4
Microarray Image and Data Analysis

The process of analyzing microarray data is complicated by the inherent sources of systemic variation. In addition, the type of image, and the statistical and higher order analyses that are performed can alter the final results; for this reason these factors are critical to the integrity of the conclusions that can be drawn regarding changes in gene expression levels (Smyth et al. 2003). Numerous commercial and freeware packages are available to facilitate analysis and visualization of microarray data. The process involves the following steps:

1. Segmentation of the image using software capable of locating the spots on the microarray
2. Computer aided background correction to normalize the spots against local averages of background fluorescence (Yang et al. 2000)
3. A second normalization to correct for dye variability, RNA preparation, hybridization and experimental differences (Claverie 1999; Callow et al. 2000; Dudoit et al. 2000)
4. Statistical analysis of spot intensities and the ratios generated in the experiment (Dudoit et al. 2000)
5. Graphical representation of the data (Eisen et al. 1998)

Normalization is necessary in microarray experiments to control for dye biases, differences between samples in the amount and quality of the starting RNA, and differences inherent in the system that change expression levels (Quackenbush 2002). This is done to ensure that meaningful comparisons can be made between the expression of the various genes on the microarray. Several possible normalization methods are used, but the most common approaches are global mean normalization and locally weighted linear regression (LOWESS) (Yang et al. 2002b; Chen et al. 2003). Global mean normalization utilizes the mean intensity across the entire microarray as a normalization factor for the expression level of individual spots. This type of normalization has many different variations, but all rely on a linear normalization across the surface of the array. The main advantage to this type of normalization is the ease of calculation and its general applicability. Disadvantages to global normalization include major influences of outliers and a requirement for a representative gene set and may mask global effects on transcription that might arise from an experimental condition. The LOWESS process involves using locally calculated linear regression to minimize changes seen due to spotting pin differences, hybridization inconsistencies across the surface of the microarray as well as inherent differences seen from dye bias and RNA quality (Yang et al. 2002b; Chen et al. 2003). This method is more complicated mathematically but it offers the advantage of correcting for differential labeling across the surface of the microarray.

Several other methods can also be used to normalize microarray data. Reference models, both internal and external, can be applied. Internal references rely on a gene or group of genes that do not change over treatment or time within the experiment. The expression of the entire data set is normalized to these internal reference genes. Obviously, this requires the prior identification of gene(s) that do not change over the course of the experiment. In addition, the expression level of the genes within the reference set is a concern. However, this method has the advantage of being applicable to incomplete or specialized gene sets. External reference models rely on spotting genes not found in experimental model. For example, varying concentrations of bacterial genes can be spotted on a mammalian gene microarray. Bacterial DNA can then be spiked into the labeling reaction and intensities from these bacterial genes used to normalize the microarray data. Taking advantage of various spotted DNA concentrations, this method offers the advantage of determining the linearity of labeling and detection. However, the level of expression for these bacterial genes is still a concern.

Various approaches are available to identify differentially expressed genes once normalized values of the filtered expression intensities have been obtained (Yang et al. 2002a; Cui and Churchill 2003; Park et al. 2003). These include the use of arbitrary cut-offs, common statistical tests such as the t-test or analysis of variance (ANOVA), more sophisticated statistical algorithms such as general linear mixed models or empirical Bayes models. However, when more than 20,000 pair-wise comparisons are considered on a single microarray it is necessary to control for various statistical anomalies such as family-wise type I errors. This is usually done with a combination of ANOVA, and adjustment of the *P* values for large data sets using corrections such as Bonferoni or false discovery rates. Other methods, such as nonparametric Bayesian models, are now being developed to address the statistical concerns of microarray data. However, in most circumstances, *t*-tests are sufficient if the experiment includes proper controls (Slonim 2002).

A subset of the data analysis field is the production of minimal data sets that can provide clues to biological function. In the case of transcription factors, the identification of a more complete set of target genes is the main focus. To accomplish this type of minimization, higher order 'clustering' algorithms are used (Datta 2003). Currently, every commercially available microarray data analysis software package contains a series of clustering programs. The most widely used method for clustering microarray data is the hierarchical cluster (HC) which draws comparisons between different genes across each treatment group. For example, in a ligand activated transcription factor genomic screen, the target cells are treated with ligand over a time course. The expression patterns are then normalized and clustered along the various time points. Genes that cluster will exhibit comparable expression behavior over the time course of study. This type of analysis can offer useful clues for novel TF target genes. If known TF target genes are located in a cluster with genes that had not been previously identified as a TF targets, these latter genes can be tested for direct activation by a 'guilt by association' mentality. HCs are primarily unsupervised forms of clustering and can be performed in two main ways. First is the agglomerative method (bottom-up). Here the expression data are analyzed and placed into a cluster and at each successive round the two closest clusters are merged. This process is continued until a criterion for each cluster is met; several algorithms are utilized to perform this type of analysis. The second form of unsupervised clustering is divisive (top-down). Divisive clustering places all of the microarray data into a single cluster and then at each successive round the clusters are split into two. This process continues until the cluster meets a specific criterion. These methods are simple, computationally and logically (in a tree form that most researchers recognize). However, they cannot be modified during the process and early mistakes can be magnified as the process continues.

Other clustering methods have also been applied to microarray data. These processes require researcher input to specify the optimum number of clusters to assign. This type of portioning cluster analysis allows the researcher to define the optimum number of clusters; however, the decision is not always simple and the methodology can be computationally expensive. The most widely used of these methods include self-organizing maps and K-means. Recent advances have also applied partitioning around mediods (PAM) and fuzzy-C means. All of these methods have the main purpose of organizing a complex data set to facilitate hypothesis generation and target gene identification. The choice of clustering algorithm can significantly influence representation of the expression data (Waring et al. 2001).

4
In Silico Approaches to TF Target Gene Identification

The wealth of genomic data for a number of species has made it possible to identify sets of target gene for a given TF using computational approaches. The in silico approach to identifying response elements (REs) is now becoming a discipline of its own. Initially, the idea was to search promoters of genes for consensus RE sequences based on expression data from microarrays. For example, microarrays can be used to identify a battery of genes whose expression is altered upon exposure to estrogen. Estrogen is a ligand activated transcription factor with a known consensus response element sequence [estrogen response element (ERE), 5'-AGGTCA-3']. The regulatory region of the genes identified in the microarray experiments can be isolated from current databases and analyzed for consensus EREs. Presumably, those genes that contain an ERE and are altered upon exposure to estrogen are activated directly by the hormone. Mathematically, this approach is flawed because the ERE consensus sequence of 6 base pairs (bp) will occur randomly approximately every 4,000 bp. This problem becomes even greater when one considers TFs with a core RE of less than 6 bp. In addition, the verification of the functionality of the putative REs requires experimental testing. However, this type of analysis for a set of critical genes is straightforward to perform and can provide further information regarding the role of the TFs in the network.

The next generation of in silico approaches to identify TF–RE interactions uses microarray cluster analysis to identify groups of genes that are regulated in a similar fashion over a treatment or dose schedule. The promoter sequences of these genes are then analyzed by pattern recognition software that identifies common DNA sequence motifs (Savoie et al. 2003; Pilpel et al. 2001; Zhu et al. 2002). This differs from the ERE example described above because no a priori information is necessary and all conserved putative regulatory sequences will be identified for the coordinately regulated gene cluster. In addition, comparative promoter analysis can be performed to identify conserved regulatory motifs across species (Yeh and Lim 2000; Nykanen et al. 2001; Oh et al. 2001; Xu and Blaser 2001).

Elegant regulatory motif and network elucidation studies have been completed in yeast for a small number of transcription factors. *Saccharomyces cerevisiae* offers the advantage of a sequenced genome, well-defined regulatory regions, ease of maintenance, a complete array of mutants and a rich history of genetic data, much of which has involved the function of TFs. The goal of this research was to map the networks of regulatory signals and TFs that control the expression of a complete genome. This includes the identification of response elements, their respective TFs and the relationships between various signals.

The basis of this type of analysis is represented in Fig. 4, starting with an environmental stimulus that activates the TF of interest. Once activated, the TF binds to its response elements within the regulatory regions of primary target genes. These genes can code for: (a) other TFs that drive the expression of secondary target genes; (b) signaling proteins that can influence other TFs in the cascade; and (c) functional proteins that have specific roles to perform in response to the environmental stimulus. For example, the HIF1α can activate a group of TFs, signaling molecules and the glycolytic enzymes responsible for cell survival in a low oxygen environment (Semenza et al. 1996; Minet et al. 2000; Oikawa et al. 2001; Christensen et al. 2002; Laderoute et al. 2002; Minchenko et al. 2002). Direct comparison of expression profiles and clustering of entire gene sets can be used to identify those genes involved in primary, secondary or later responses. It is assumed that those genes that are involved primary responses will show the earliest time course of regulation (Fig. 4). Expression of these genes can remain elevated if their response is necessary, or they can return to normal levels. The genes involved in secondary response will increase during the mid-phase of the exposure and the later phase response will go up in succession (Ashburner 1973).

Computational biologists take the data from expression profiles and cluster the genes according to their expression pattern after a given stimulus;

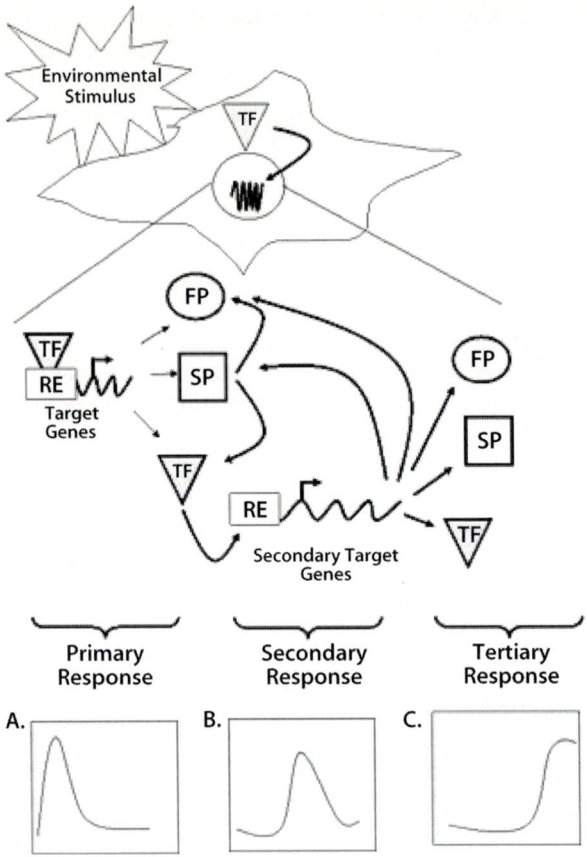

Fig. 4 Graphical representation of TF signaling network. The TF is stimulated, binds to its corresponding response element (*RE*) and drives the expression of three different classes of proteins: other TFs, signaling proteins (*SP*), and functional proteins (*FP*). These primary proteins can then influence the expression of other secondary genes. Representative expression profiles are also given for primary (*A*), secondary (*B*) and tertiary (*C*) response genes

they then begin to formulate possible interactions and networks (Ideker et al. 2001; Brazhnik et al. 2002). When a network computational model can be generated that can be used to predict the behavior of the network, it can be empirically tested (Ideker et al. 2002; Masuda and Church 2003; Savoie et al. 2003; Tegner et al. 2003). The initial testing is performed by obtaining the regulatory regions of these genes and analyzing them for conserved sequences among each group (i.e., primary, secondary and tertiary response genes). It is assumed that those genes that behave in a similar fashion over a treatment schedule will share regulatory TFs. Once patterns have been identified, it is possible to identify putative TFs responsible for interacting with these response elements. Presumably, those TFs whose expression is altered in the previous responses (i.e., primary or secondary) will interact with response elements identified from genes whose expression patterns change in later stages (i.e., tertiary). Determining where these response element:TF pairs fit into a signaling network will connect external stimuli to complex biological responses.

Obviously, this model is overly simplified. Gene regulatory networks will also involve multiple levels of complex regulation such as feedback inhibition, feed forward activation, self-regulation, protein–protein interactions and be influenced by changes in endogenous metabolite levels. Researchers will not be capable of distinguishing these types of regulation without further information provided by ChIP on chip assays, proteomics and metabonomics (reviewed by Banerjee and Zhang 2002). As mentioned above, this assay involves the application of chromatin immunoprecipitations that are widely used by TF researchers and couples it to arrays of regulatory regions. This method has proved successful in yeast where the regulatory regions are easily identified. Lee et al. have used this type of analysis to characterize 106 TFs from yeast (Lee et al. 2002). Directly relating the TFs that bind the promoter regions of these same TFs the authors were capable of describing various forms of regulation (i.e., self regulation, a TF that binds its own promoter). Similar research is going on in mammalian systems (Elkon et al. 2003). The success of this research in mammalian systems is dependent upon the identification and characterization of regulatory regions for every expressed gene. This can be a large hurdle given that regulatory regions can be 10 kb away from the start codon and in some cases, downstream of the start methionine.

5
The Future

In order to assess the role of TFs in complex biological processes such as development, homeostasis and disease, more complete knowledge of expression profiles under a wide range of stimuli is required. This will require a comprehensive understanding of the human genome including an inclusive index of all open reading frames, non-protein-coding genes, transcriptional regulatory elements, and determinants of chromosome structure and function. New public research consortia are being organized to test, compare and develop more comprehensive high-throughput methods to exhaustively identify and verify functional sequence elements within the human genome. For example, the National Human Genome Research Institute recently launched the Encyclopedia of DNA Elements (ENCODE) Project with the goal of developing and using high-throughput strategies to comprehensively identifying all functional TF response elements in the human genome (http://www.genome.gov/Pages/Research/ENCODE). The integration of this genomic features index with gene, protein and metabolite expression data as well as other metadata (e.g., high-throughput in situ hybridization, immunohistochemistry, histopathology) will provide the knowledge base needed to create interaction networks. These networks can then serve as a scaffold to create more deterministic computational models that will have greater power to be able to predict complex biological responses. These models, when combined with chemical genomic approaches, will not only identify new targets for therapeutic strategies—many of which are likely to be TFs or modulators of TFs—but could also be used to computationally screen drugs in development for toxicity (Schieber 2003; Willson 2003). With mounting public pressure on pharmaceutical manufactures and government regulators to provide more effective and safer drugs at lower cost, future drug development that exploits TFs as targets will require multidisciplinary efforts with input from biomedical researchers, statisticians, informaticians, computational modelers, and physicians.

References

Allison DP, Kerper PS, Doktycz MJ, Thundat T, Modrich P, Larimer FW Johnson DK, Hoyt PR, Mucenski ML, Warmack RJ (1997) Mapping individual cosmid DNAs by direct AFM imaging. Genomics 41:379–384

Aranda A and Pascual A (2001) Nuclear hormone receptors and gene expression. Physiol Rev 81:1269–1304

Ashburner M (1973) Sequential gene activation by ecdysone in polytene chromosomes of drosophila melangaster. I. Dependence upon ecdysone concentration. Dev. Biol. 35:47–61

Banerjee N, Zhang MQ (2002) Functional genomics as applied to mapping transcription regulatory networks. Curr Opin Microbiol 5:313–317

Baulcombe DC (1996) RNA as a target and an initiator of post-transcriptional gene silencing in transgenic plants. Plant Mol Biol 32:79–88

Brazhnik P, de la Fuente A, Mendes P (2002) Gene networks: how to put the function in genomics. Trends Biotechnol 20:467–472

Brazma A, Hingamp P, Quackenbush J, Sherlock G, Spellman P, Stoeckert C, Aach J, Ansorge W, Ball CA, Causton HC, Gaasterland T, Glenisson P, Holstege FCP, Kim IF, Markowitz V, Matese JC, Parkinson H, Robinson A, Sarkans U, Schulze-Kremer S, Stewart J, Taylor R, Vilo J, Vingron M (2001) Minimum information about a microarray experiment (MIAME)—toward standards for microarray data. Nat Genet 29:365–371

Callow MJ, Dudoit S, Gong EL, Speed TP, Rubin EM (2000) Microarray expression profiling identifies genes with altered expression in HDL-deficient mice. Genome Res 10:2022–2029

Chen YJ, Kodell R, Sistare F, Thompson KL, Morris S, Chen JJ (2003) Normalization methods for analysis of microarray gene-expression data. J Biopharm Stat 13:57–74

Christensen RA, Fujikawa K, Madore R, Oettgen P, Varticovski L (2002) NERF2, a member of the Ets family of transcription factors, is increased in response to hypoxia and angiopoietin-1: a potential mechanism for Tie2 regulation during hypoxia. J Cell Biochem 85:505–515

Claverie JM (1999) Computational methods for the identification of differential and coordinated gene expression. Hum Mol Genet 8:1821–1832

Collins FS, Morgan M, Patrinos A (2003) The Human Genome Project: lessons from large-scale biology. Science 300:286–290

Cook-Deegan RM (1989) The Alta summit, December 1984. Genomics 5:661–663

Cui X, Churchill GA (2003) Statistical tests for differential expression in cDNA microarray experiments. Genome Biol 4:210

Datta S (2003) Comparisons and validation of statistical clustering techniques for microarray gene expression data. Bioinformatics 19:459–466

Dolled-Filhart M, Camp RL, Kowalski DP, Smith BL, Rimm DL (2003) Tissue microarray analysis of signal transducers and activators of transcription 3 (Stat3) and phosphostat3 (Tyr705) in node-negative Breast cancer shows nuclear localization is associated with a better prognosis. Clin Cancer Res 9:594–600

Dolwick KM, Swanson HI, Bradfield CA (1993) In vitro analysis of Ah receptor domains involved in ligand-activated DNA recognition. Proc Natl Acad Sci USA 90:8566–8570

Don J, Stelzer G (2002) The expanding family of CREB/CREM transcription factors that are involved with spermatogenesis. Mol Cell Endocrinol 187:115–124

Dudoit S, Yang YH, Callow MJ, Speed TP (2000) Statistical methods for identifying differentially expressed genes in replicated cDNA microarray experiments. University of California at Berkeley-Technical Reports 578: http://www.stat.berkeley.edu/tech-reports/

Eisen MB, Spellman PT, Brown PO, Botstein D (1998) Cluster analysis and display of genome-wide expression patterns. Proc Natl Acad Sci USA 95:14863–14868.

Elbashir SM, Harborth J, Lendeckel W, Yalcin A, Weber K, Tuschl T (2001) Duplexes of 21-nucleotide RNAs mediate RNA interference in cultured mammalian cells. Nature 411:494–498

Elkon RCL, Sharan R, Shamir R, Shiloh Y (2003) Genome-wide in silico identification of transcriptional regulators controlling cell cycle in human cells. Genome Res 13:773–780

Fire A, Xu S, Montgomery MK, Kostas SA, Driver SE, Mello CC (1998) Potent and specific genetic interference by double-stranded RNA in Caenorhabditis elegans. Nature 391:806–811

Fodor SP, Rava RP, Huang XC, Pease AC, Holmes CP, Adams CL (1993) Multiplexed biochemical assays with biological chips. Nature 364:555–556

Freier K, Joos S, Flechtenmacher C, Devens F, Benner A, Bosch FX, Lichter P, Hofele C (2003) Tissue microarray analysis reveals site-specific prevalence of oncogene amplifications in head and neck squamous cell carcinoma. Cancer Res 63:1179–1182

Green CD, Simons JF, Taillon BE, Lewin DA (2001) Open systems: panoramic views of gene expression. J Immunol Methods 250:67–79

Green MR (2000) TBP-associated factors (TAFIIs): multiple, selective transcriptional mediators in common complexes. Trends Biochem Sci 25:59–63

Gu Y-Z, Hogenesch J, Bradfield C (2000) The PAS superfamily: Sensors of environmental and developmental signals. Annu Rev Pharmacol Toxicol 40: 519–561

Guren TK, Abrahamsen H, Thoresen GH, Susa M, Andersson Y, Christoffersen T (2003) EGF receptor-mediated, c-Src-dependent, activation of Stat5b is downregulated in mitogenically responsive hepatocytes. J Cell Physiol 196:113–123

Hessner MJ, Wang X, Khan S, Meyer L, Schlicht M, Tackes J, Datta MW, Jacob HJ, Ghosh S (2003) Use of a three-color cDNA microarray platform to measure and control support-bound probe for improved data quality and reproducibility. Nucl Acids Res 31: e60

Holloway AJ, van Laar RK, Tothill RW, Bowtell DD (2002) Options available–from start to finish–for obtaining data from DNA microarrays II. Nat Genet 32 Suppl: 481–489

Hoyt PR, Doktycz MJ, Warmack RJ, Allison DP (2001) Spin-column isolation of DNA-protein interactions from complex protein mixtures for AFM imaging. Ultramicroscopy 86:139–143

Ideker T, Thorsson V, Ranish JA, Christmas R, Buhler J, Eng JK, Bumgarner R, Goodlett DR, Aebersold R, Hood L (2001) Integrated genomic and proteomic analyses of a systematically perturbed metabolic network. Science 292:929–934

Ideker T, Ozier O, Schwikowski B, Siegel AF (2002) Discovering regulatory and signalling circuits in molecular interaction networks. Bioinformatics 18 Suppl 1: S233–S240

Jeong JW, Bae MK, Ahn MY, Kim SH, Sohn TK, Bae MH, Yoo MA, Song EJ, Lee KJ, Kim KW (2002) Regulation and destabilization of HIF-1alpha by ARD1-mediated acetylation. Cell 111:709–720

Jiang C, Lu H, Vincent KA, Shankara S, Belanger AJ, Cheng SH, Akita GY, Kelly RA, Goldberg MA, Gregory RJ (2002) Gene expression profiles in human cardiac cells subjected to hypoxia or expressing a hybrid form of HIF-1 alpha. Physiol Genomics 8:23–32

Jordan B (2002) Historical background and anticipated developments. Ann N Y Acad Sci 975:24–32

Kako K, Ishida N (1998) The role of transcription factors in circadian gene expression. Neuroscience Research 31:257–264

Kerr MK (2003) Experimental design to make the most of microarray studies. Methods Mol Biol 224:137–147

Kvietikova I Wenger RH Marti HH, Gassmann M (1995) The transcription factors ATF-1 and CREB-1 bind constitutively to the hypoxia-inducible factor-1 (HIF-1) DNA recognition site. Nucl Acids Res 23:4542–4550

Laderoute KR, Calaoagan JM, Gustafson-Brown C, Knapp AM, Li GC, Mendonca HL, Ryan HE, Wang Z, Johnson RS (2002) The response of c-jun/AP-1 to chronic hypoxia is hypoxia-inducible factor 1 alpha dependent. Mol Cell Biol 22:2515–2523

Lander ES, Weinberg RA (2000) Genomics: journey to the center of biology. Science 287:1777–1782

Lee JM, Calkins MJ, Chan K, Kan YW, Johnson JA (2003) Identification of the NF-E2-related factor-2-dependent genes conferring protection against oxidative stress in primary cortical astrocytes using oligonucleotide microarray analysis. J Biol Chem 278:12029–12038

Lee TI, Rinaldi NJ, Robert F, Odom DT, Bar-Joseph Z, Gerber GK, Hannett NM, Harbison CT, Thompson CM, Simon I, Zeitlinger J, Jennings EG, Murray HL, Gordon DB, Ren B, Wyrick JJ, Tagne JB, Volkert TL, Fraenkel E, Gifford DK, Young RA (2002) Transcriptional regulatory networks in Saccharomyces cerevisiae. Science 298:799–804

Lichtlen P, Wang Y, Belser T, Georgiev O, Certa U, Sack R, Schaffner W (2001) Target gene search for the metal-responsive transcription factor MTF-1. Nucl Acids Res 29:1514–1523

Lieb JD (2003) Genome-wide mapping of protein-DNA interactions by chromatin immunoprecipitation and DNA microarray hybridization. Methods Mol Biol 224:99–109

Lindberg MK, Weihua Z, Andersson N, Moverare S, Gao H, Vidal O, Erlandsson M, Windahl S, Andersson G, Lubahn DB, Carlsten H, Dahlman-Wright K, Gustafsson JA, Ohlsson C (2002) Estrogen receptor specificity for the effects of estrogen in ovariectomized mice. J Endocrinol 174:167–178

Lu Q, Hutchins AE, Doyle CM, Lundblad JR, Kwok RP (2003) Acetylation of cAMP-responsive element-binding protein (CREB) by CREB- binding protein enhances CREB-dependent transcription. J Biol Chem 278:15727–15734

Mackay A, Jones C, Dexter T, Silva RL, Bulmer K, Jones A, Simpson P, Harris RA, Jat PS, Neville AM, Reis LF, Lakhani SR, O'Hare MJ (2003) cDNA microarray analysis of genes associated with ERBB2 (HER2/neu) overexpression in human mammary luminal epithelial cells. Oncogene 22:2680–2688

Masuda N, Church GM (2003) Regulatory network of acid resistance genes in Escherichia coli. Mol Microbiol 48:699–712

Menssen A, Hermeking H (2002) Characterization of the c-MYC-regulated transcriptome by SAGE: identification and analysis of c-MYC target genes. Proc Natl Acad Sci USA 99:6274–6279

Minchenko A, Leshchinsky I, Opentanova I, Sang N, Srinivas V, Armstead V, Caro J (2002) Hypoxia-inducible factor-1-mediated expression of the 6-phosphofructo-2-kinase/fructose-2,6-bisphosphatase-3 (PFKFB3) gene. Its possible role in the Warburg effect. J Biol Chem 277:6183–6187

Minet E, Arnould T, Michel G, Roland I, Mottet D, Raes M, Remacle J, Michiels C (2000) ERK activation upon hypoxia: involvement in HIF-1 activation. FEBS Lett 468:53–58

Mottet D, Dumont V, Deccache Y, Demazy C, Ninane N, Raes M, Michiels C (2003) Regulation of HIF-1alpha protein level during hypoxic conditions by the PI3 K/AKT/GSK3beta pathway in HepG2 cells. J Biol Chem 278:31277–31285

No author listed (1991) The human genome initiative. FASEB J 5:4–78

Nykanen P, Alastalo TP, Ahlskog J, Horelli-Kuitunen N, Pirkkala L, Sistonen L (2001) Genomic organization and promoter analysis of the human heat shock factor 2 gene. Cell Stress Chaperones 6:377–385

Oh Y, Lee S, Yoon J, Han KS, Baek K (2001) Promoter analysis of the Drosophila melangaster gene encoding transcription elongation factor TFIIs. Biochim Biophys Acta 1518:276–281

Oikawa M, Abe M, Kurosawa H, Hida W, Shirato K, Sato Y (2001) Hypoxia induces transcription factor ETS-1 via the activity of hypoxia- inducible factor-1. Biochem Biophys Res Commun 289:39–43

Omoto Y, Hayashi S (2002) A study of estrogen signaling using DNA microarray in human breast cancer. Breast Cancer 9:308–311

Park T, Yi SG, Lee S, Lee SY, Yoo DH, Ahn JI, Lee YS (2003) Statistical tests for identifying differentially expressed genes in time-course microarray experiments. Bioinformatics 19:694–703

Paul CP, Good PD, Winer I, Engelke DR (2002) Effective expression of small interfering RNA in human cells. Nat Biotechnol 20:505–508

Phillips K, Luisi B (2000) The virtuoso of versatility: POU proteins that flex to fit. J Mol Biol 302:1023–1039

Pilpel Y, Sudarsanam P, Church GM (2001) Identifying regulatory networks by combinatorial analysis of promoter elements. Nat Genet 29:153–159

Poguet AL, Legrand C, Feng X, Yen PM, Meltzer P, Samarut J, Flamant F (2003) Microarray analysis of knockout mice identifies cyclin D2 as a possible mediator for the action of thyroid hormone during the postnatal development of the cerebellum. Dev Biol 254:188–199

Quackenbush J (2002) Microarray data normalization and transformation. Nat Genet 32 Suppl: 496–501

Ren B, Robert F, Wyrick JJ, Aparicio O, Jennings EG, Simon I, Zeitlinger J, Schreiber J, Hannett N, Kanin E, Volkert TL, Wilson CJ, Bell SP, Young RA (2000) Genome-wide location and function of DNA binding proteins. Science 290:2306-2309

Savoie CJ, Aburatani S, Watanabe S, Eguchi Y, Muta S, Imoto S, Miyano S, Kuhara S, Tashiro K (2003) Use of gene networks from full genome microarray libraries to identify functionally relevant drug-affected genes and gene regulation cascades. DNA Res 10:19-25

Schena M, Shalon D, Davis RW, Brown PO (1995) Quantitative monitoring of gene expression patterns with a complementary DNA microarray. Science 270:467-470.

Schieber S (2003) The small-molecule approach to biology. Chem Engin News March 3:51-61

Semenza GL, Jiang BH, Leung SW, Passantino R, Concordet JP, Maire P, Giallongo A (1996) Hypoxia response elements in the aldolase A, enolase 1, and lactate dehydrogenase A gene promoters contain essential binding sites for hypoxia-inducible factor 1. J Biol Chem 271:32529-32537

Seth P, Krop I, Porter D, Polyak K (2002) Novel estrogen and tamoxifen induced genes identified by SAGE (Serial Analysis of Gene Expression). Oncogene 21:836-843

Shtutman M, Zhurinsky J, Oren M, Levina E, Ben-Ze'ev A (2002) PML is a target gene of beta-catenin and plakoglobin, and coactivates beta-catenin-mediated transcription. Cancer Res 62:5947-5954

Slonim DK (2002) From patterns to pathways: gene expression data analysis comes of age. Nat Genet 32 Suppl: 502-508

Smyth GK, Yang YH, Speed T (2003) Statistical issues in cDNA microarray data analysis. Methods Mol Biol 224:111-136

Southern EM (1975) Detection of specific sequences among DNA fragments separated by gel electrophoresis. J Mol Biol 98:503-517

Struhl K (1997) Selective roles for TATA-binding-protein-associated factors in vivo. Genes Funct 1:5-9

Tegner J, Yeung MK, Hasty J, Collins JJ (2003) Reverse engineering gene networks: Integrating genetic perturbations with dynamical modeling. Proc Natl Acad Sci USA 100:5944-5949

Tolgay Ocal I, Dolled-Filhart M, D'Aquila TG, Camp RL, Rimm DL (2003) Tissue microarray-based studies of patients with lymph node negative breast carcinoma show that met expression is associated with worse outcome but is not correlated with epidermal growth factor family receptors. Cancer 97:1841-1848

Tzankov A, Zimpfer A, Lugli A, Krugmann J, Went P, Schraml P, Maurer R, Ascani S, Pileri S, Geley S, Dirnhofer S (2003) High-throughput tissue microarray analysis of G1-cyclin alterations in classical Hodgkin's lymphoma indicates overexpression of cyclin E1. J Pathol 199:201-207

Waring JF, Jolly RA, Ciurlionis R, Lum PY, Praestgaard JT, Morfitt DC, Buratto B, Roberts C, Schadt E, Ulrich RG (2001) Clustering of hepatotoxins based on mechanism of toxicity using gene expression profiles. Toxicol Appl Pharmacol 175:28-42

Warren AJ (2002) Eukaryotic transcription factors. Curr Opin Struct Biol 12:107-114

Willson T (2003) Chemical genomics of orphan nuclear receptors. Ernst Schering Res Found Workshop: 29-42

Wu CG, Forgues M, Siddique S, Farnsworth J, Valerie K, Wang XW (2002) SAGE transcript profiles of normal primary human hepatocytes expressing oncogenic hepatitis B virus X protein. FASEB J 16:1665-1667

Xu Q, Blaser MJ (2001) Promooters of the CATG-specific methyltransferase gene hypIM differ between iceA1 and iceA2 Helicobacter pylori strains. J Bacteriol 183:3875-3884

Yang IV, Chen E, Hasseman JP, Liang W, Frank BC, Wang S, Sharov V, Saeed AI, White J, Li J, Lee NH, Yeatman TJ, Quackenbush J (2002a) Within the fold: assessing differential expression measures and reproducibility in microarray assays. Genome Biol 3: 0062.1–0062.12

Yang YH, Buckley M, Dudoit S, Speed TP (2000) Comparison of methods for image analysis on cDNA microarray data. University of California at Berkeley-Technical Reports 584: http://www.stat.berkeley.edu/tech-reports/

Yang YH, Dudoit S, Luu P, Lin DM, Peng V, Ngai J, Speed TP (2002b) Normalization for cDNA microarray data: a robust composite method addressing single and multiple slide systematic variation. Nucl Acids Res 30: e15

Yeh K, Lim RW (2000) Genomic organization and promoter analysis of the murine id3 gene. Gene 254:163–171

Yu JY, DeRuiter SL, Turner DL (2002) RNA interference by expression of short-interfering RNAs and hairpin RNAs in mammalian cells. Proc Natl Acad Sci USA 99:6047–6052

Zhou A, Scoggin S, Gaynor RB, Williams NS (2003) Identification of NF-kappa B-regulated genes induced by TNFalpha utilizing expression profiling and RNA interference. Oncogene 22:2054–2064

Zhu Z, Pilpel Y, Church GM (2002) Computational identification of transcription factor binding sites via a transcription-factor-centric clustering (TFCC) algorithm. J Mol Biol 318:71–81

Part II
Transcription Factors in Pathophysiology

Part B

Biological Roles of the STAT Family in Cytokine Signaling

K. Takeda · S. Akira (✉)

Department of Host Defense, Research Institute for Microbial Diseases, Osaka University, 3-1 Yamada-oka, Suita, 565-0871 Osaka, Japan
sakira@biken.osaka-u.ac.jp

1	Introduction	98
2	The JAK–STAT Pathway	98
3	Structure of JAK and STAT Proteins	99
4	Biological Roles of the JAK Family Revealed by Knockout Mice	100
5	Biological Roles of the STAT Family Revealed by Knockout Mice	101
5.1	Stat1 Knockout Mice	102
5.2	Stat2 Knockout Mice	102
5.3	Stat3 Knockout Mice	103
5.4	Stat4 Knockout Mice	103
5.5	Stat5a and Stat5b Knockout Mice	103
5.6	Stat6 Knockout Mice	105
6	Tissue-Specific Roles of Stat3 Revealed by Conditional Gene Targeting	106
6.1	Deletion of Stat3 in T cells	107
6.2	Deletion of Stat3 in B Cells	107
6.3	Deletion of Stat3 in Myeloid Lineage Cells	107
6.4	Deletion of Stat3 in Hematopoietic Cells	108
6.5	Deletion of Stat3 in the Mammary Gland	109
6.6	Deletion of Stat3 in the Skin	109
6.7	Deletion of Stat3 in Thymic Epithelium	110
6.8	Deletion of Stat3 in the Liver	110
6.9	Deletion of Stat3 in Neurons	110
7	Negative Regulation of STAT Activity	111
7.1	SOCS Family	111
7.2	Other Regulatory Mechanisms of STAT Activity	112
8	Concluding Remarks	113
References		114

Abstract The JAK–STAT pathway, originally identified in the interferon signaling pathway, has been shown to play an essential role in cytokine-mediated biological responses mainly through the analysis of knockout mice. The JAK family consists of four members (Jak1–3 and Tyk2), and the STAT family consists of seven members (Stat1–4, Stat5a, Stat5b, and Stat6). The JAK family of proteins plays an essential role mediated by cytokine receptors, whereas STAT proteins are responsible for their specific cytokines. In addition, there is a considerable body of evidence demonstrating that activity of the

JAK-STAT pathway is regulated by several mechanisms, including the SOCS family of proteins induced by activation of the JAK–STAT pathway. Gene targeting of the SOCS family revealed that individual SOCS proteins negatively regulate some cytokine-mediated signaling pathways.

Keywords Signal transduction · Knockout mice · STAT · JAK · SOCS

1
Introduction

Growth, differentiation, and activation of cells in the immune system are finely regulated by soluble factors called cytokines. Cytokines exert multiple biological functions through interaction with their specific receptors. The binding of cytokines to these receptors induces homo- or hetero-dimerization of the receptors and triggers activation of intracellular signaling cascades (Kishimoto et al. 1994). Accumulating evidence indicates that the latent transcription factors, STAT (signal transducers and activators of transcription), play an essential role in cytokine signaling pathways. STAT proteins were first identified in interferon (IFN) signaling pathways. Subsequently, several STAT family members have been identified, and the STAT family now consists of seven members (Stat1–4, Stat5a, Stat5b, and Stat6). The biological roles of the STAT family of transcription factors have been revealed mainly through the analysis of gene targeted mice. In this section, we will focus on the function of molecules in JAK–STAT pathways in vivo.

2
The JAK–STAT Pathway

The binding of cytokines induces dimerization of its receptors. This dimerization subsequently activates an intracellular signaling pathway by phosphorylation of Janus kinases (JAK), which are constitutively associated with the cytokine receptor (Ihle et al. 1995; Taniguchi et al. 1995). Phosphorylation of JAK in turn induces phosphorylation of the tyrosine residue in the cytoplasmic region of the receptor, and activates several substrates including STAT. STAT proteins are then recruited to the phosphorylated tyrosine residue of the receptor, and activated by JAK with tyrosine phosphorylation. The activated STATs then dissociate from the receptor, form homophilic or heterophilic dimers, and rapidly translocate into the nucleus, where they induce expression of target genes (Fig. 1) (Schindler et al. 1995; Ihle 1996). Thus, STAT proteins facilitate a direct connection between receptors on the plasma membrane and the nucleus. Four members of mammalian JAK kinase family and seven members of STAT family proteins have been identified to date.

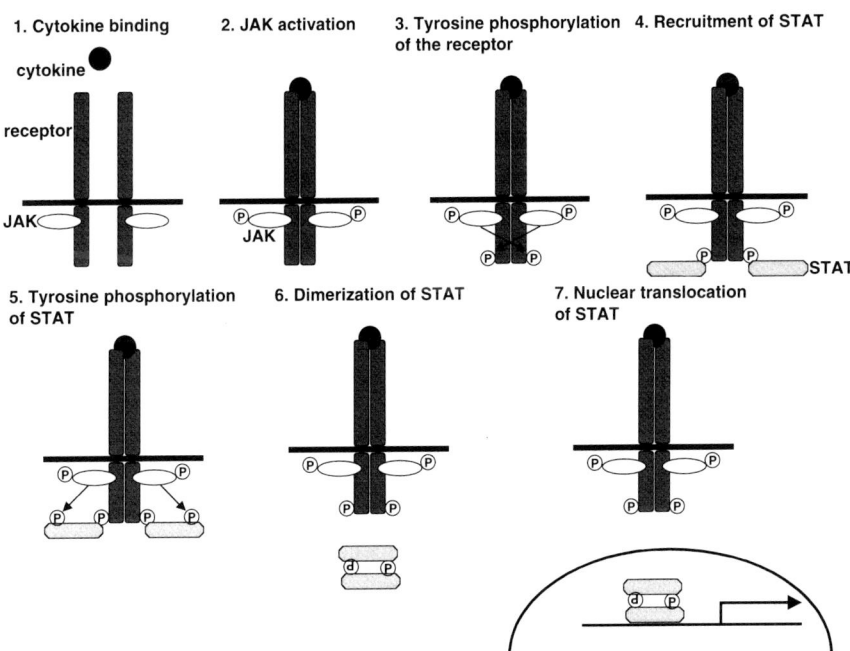

Fig. 1 Cytokine signaling pathways. Cytokine induces dimerization of the cytokine receptor, which triggers activation of receptor-associated JAK kinases. Activated JAK kinases phosphorylate the tyrosine residue of the cytokine receptor. STAT proteins residing in the cytoplasm are recruited to the phosphorylated tyrosine of the receptor, and tyrosine phosphorylated by JAK kinases. Once phosphorylated, STAT proteins dissociate from the receptor, form homophilic or heterophilic dimers, and rapidly translocate into the nucleus, to induce expression of the target genes

3
Structure of JAK and STAT Proteins

JAK proteins have seven regions (JH1–JH7) that possess a high degree of similarity between family members (Fig. 2A). JH1 is active as a tyrosine kinase, while JH2 represents a pseudokinase domain, which is required for the catalytic activity of JH1 (Yeh et al. 2000). JH3–JH7, that are located at the N-terminus of JAK proteins, are implicated by their association with the cytokine receptor (Leonard and O'Shea 1998).

STAT proteins have several conserved domains, which are characteristic of both a signal transducing molecule and a transcription factor (Fig. 2B). As a transcription factor, STAT proteins possess a DNA-binding domain in the central region, and a transactivation domain in the C terminus. Both the tyrosine residue that is essential for STAT activation in the transactivation domain, and the SH2 domain that is essential for the binding to phosphorylated tyrosine residues, precede the transactivation domain. In addition, there is a

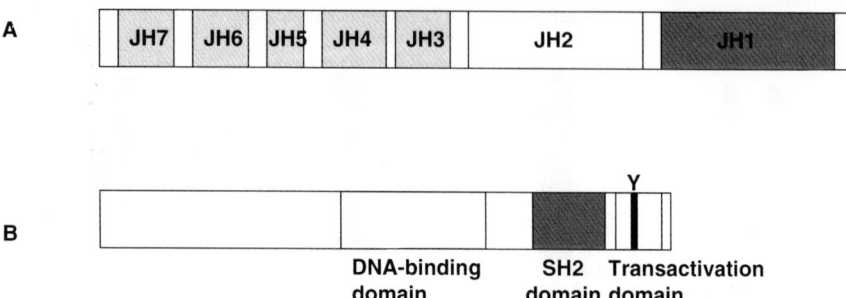

Fig. 2A, B Schematic structures of JAK and STAT proteins. **A** Structure of JAK. JAK kinases have JH1–JH7 domains, which are based on sequence similarity between JAK family members. JH1 is the kinase domain, and JH2 is the pseudokinase domain. *JH*, JAK homology. **B** Structure of STAT. There is a DNA-binding domain in the middle portion of STAT protein, and transactivation domain in the C terminus. The SH2 domain is located just N terminal of the transactivation domain. The tyrosine residue, which is critical to STAT activation, is seen in the C terminus

coiled-coil domain in the N-terminal region, which probably mediates protein–protein interaction. These domains have been visualized by three-dimensional analysis of Stat1 and Stat3 (Chen et al. 1998; Becker et al. 1998).

4
Biological Roles of the JAK Family Revealed by Knockout Mice

For each JAK family member, the generation of knockout mice has revealed that JAK kinases play a crucial role in the biological responses triggered by cytokine receptors (Table 1).

Jak1 is associated with several cytokine receptors, such as the interleukin (IL)-2 receptor family (IL-2R, IL-7R, IL-9R and IL-15R), the IL-4 receptor family (IL-4R, IL-13R), the gp130 receptor family (IL-6R, IL-11R, LIF-R, OSM-R CT-1R, CNTF-R, and leptin-R) and class II cytokine receptors (IFN-α/βR, IFN-γR, IL-10R). Indeed, Jak1 knockout mice exhibit postnatal mortality, probably due to defective responses by all class II cytokine receptors, cytokine receptors that utilize the γc subunit for signaling, and the family of cytokine receptors that depend on the gp130 subunit for signaling (Rodig et al. 1998).

Jak2 knockout mice suffered from embryonic mortality due to impaired definitive erythropoiesis, a phenotype very similar to erythropoietin (Epo) knockout mice (Neubauer et al. 1998; Parganas et al. 1998). Hence, Jak2 is essential for the Epo receptor-mediated signaling pathway. In addition, cells from Jak2 knockout embryos were unresponsive to IFN-γ, Tpo, IL-3 and granulocyte-macrophage colony stimulating factor (GM-CSF), but normally responded to G-CSF, IL-6, or the IL-2 and IL-4 family of cytokines.

Table 1 Phenotype of JAK knockout mice

JAK proteins	Phenotype
Jak1	Perinatal mortality, impaired response to cytokines that use the IL-2 receptor family, the IL-4 receptor family, the gp130 receptor family, and type II cytokine receptor family
Jak2	Embryonic mortality, impaired erythropoiesis probably due to impaired signaling via the Epo receptor
Jak3	Severe combined immunodeficiency disease phenotype due to impaired signaling via the γc receptor
Tyk2	Highly susceptible to microbial pathogens due to an impaired response to IFN-α/β and IL-12

Unlike other JAK family members, Jak3 is preferentially expressed in hematopoietic cells, and associated with the γc chain of the cytokine receptor. Jak3 knockout mice showed severely impaired lymphopoiesis, similar to that observed in mice deficient in γc (Nosaka et al. 1995; Park et al. 1995; Thomis et al. 1995).

Tyk2 knockout mice showed defects in their response to IFN-α/β (Karaghiosoff et al. 2000; Shimoda et al. 2000). In addition, Tyk2 knockout mice had an impaired response to IL-12 (Shimoda et al. 2002). Furthermore, it has been shown that Tyk2 knockout mice are resistant to lipopolysaccharide (LPS)-induced endotoxin shock, and defective in LPS induction of IFN-β and IFN-α4 mRNA in macrophages (Karaghiosoff et al. 2003). Thus, Tyk2 is likely to be involved in the signaling pathways of LPS as well as the cytokines such as IL-12 and IFN-α/β.

5
Biological Roles of the STAT Family Revealed by Knockout Mice

Biological functions of all STAT family proteins have now been elucidated through the generation of knockout mice. Knockout mice of each member

Table 2 Phenotype of STAT knockout mice

STAT protein	Phenotype
Stat1	Impaired response to IFN-α/β, IFN-γ
Stat2	Impaired response to IFN-α/β
Stat3	Embryonic mortality
Stat4	Impaired response to IL-12
Stat5a	Impaired response to prolactin
Stat5b	Impaired response to growth hormone, IL-2, and IL-15
Stat6	Impaired response to IL-4 and IL-13

of the STAT family displayed impaired cytokine-mediated function (Table 2).

5.1
Stat1 Knockout Mice

Stat1 is activated by IFN-α/β and IFN-γ. Stat1 knockout mice exhibit a defective response to IFN-α/β and IFN-γ. Interferons play an important role in macrophage activation and host defense against microbial pathogens. Accordingly, Stat1 knockout mice are vulnerable to infection with viral and microbial pathogens, including vesicular stomatitis virus, mouse hepatitis virus, and *Listeria monocytogenes*. Thus, Stat1 is essential for IFN-mediated functions in vivo (Meraz et al. 1996; Durbin et al. 1996). Patients with a partial deficiency of IFN-γ receptors and heterozygous germline *STAT1* mutation show increased susceptibility to mycobacterial disease, demonstrating the importance of Stat1 in human IFN-γ mediated anti-mycobacterial responses (Dupuis et al. 2001). However, recent evidence indicates there is a Stat1-independent pathway in IFN-γ signaling. A microarray analysis has shown that expression of several genes, such as macrophage inflammatory proteins (MIP-1α and MIP-1β) and arginase, are induced by IFN-γ in bone marrow-derived macrophages from Stat1 knockout mice (Gil et al. 2001; Rammana et al. 2002). The expression of the immediate-early genes, such as c-*myc* and c-*jun*, was transiently and rapidly induced in serum starved embryonic fibroblasts from Stat1 knockout mice, but not in wild type cells (Rammana et al. 2000). Although Stat1 is very important for antiviral responses, Stat1 knockout mice show more resistance to murine cytomegalovirus and Sindbis virus than IFN-γ receptor and IFN-α/β receptor knockout mice, indicating the presence of Stat1-independent antiviral responses (Gil et al. 2001).

5.2
Stat2 Knockout Mice

An important role of Stat2 in IFN-α/β signaling has been demonstrated in Stat2 knockout mice (Park et al. 2000). Stat2 knockout mice displayed a defective response to IFN-α/β, and high susceptibility to infection by vesicular stomatitis virus (Park et al. 2000). Embryonic fibroblasts from Stat2 knockout mice exhibited a more significant defect in their response to IFN-α/β than macrophages, indicating tissue-specific differences in the response to IFN-α/β.

5.3
Stat3 Knockout Mice

Stat3 knockout mice suffered from embryonic mortality (Takeda et al. 1997). Stat3 knockout embryos showed rapid degeneration between 6.5 days postcoitum (dpc) and 7.5 dpc. In accordance with the phenotype of Stat3 knockout mice, Stat3 mRNA was detected in both maternal and extraembryonic tissues during the early postimplantation stages of murine development (Duncan et al. 1997). Furthermore, analysis of Stat3 DNA-binding activity in E4.5–E9.5 decidual swellings, indicated that Stat3 is active during the early period of development. Together these data demonstrate that Stat3 is essential during early embryogenesis.

5.4
Stat4 Knockout Mice

The essential role of Stat4 in IL-12 signaling has been demonstrated in Stat4 knockout mice (Kaplan et al. 1996; Thierfelder et al. 1996). Stat4 knockout mice exhibited impaired IL-12 mediated increases in IFN-γ production, cellular proliferation, and NK cell cytotoxic activity of lymphocytes. Furthermore, IL-12 induced development of Th1 cells was abolished in Stat4 knockout mice. Adoptive transfer of T cells into immunodeficient mice induced Th1 cell-mediated immune disorders, such as enterocolitis. Transfer of T cells from Stat4 knockout mice induced a significantly milder form of colitis than in the wild type, indicating that a Stat4-mediated pathway is involved in the pathogenesis of Th1-mediated colitis (Simpson et al. 1998).

However, in Stat4 knockout mice, an additional gene deletion of Stat6, which is essential for IL-4 signaling, increases Th1 development (Kaplan et al. 1998a). Furthermore, in a model of the classic Th1 cell-mediated immune response showing the delayed-type hypersensitivity reaction, Stat4 and Stat4/Stat6 double knockout mice produce significant amounts of IFN-γ from CD4 T cells. Thus, a potent Th1-inducing stimulus can induce a Th1 response even in the absence of Stat4 indicating the presence of a Stat4-independent pathway in Th1 cell development.

In addition to IL-12, IFN-α/β directly activates Stat4, and induces production of IFN-γ particularly in CD8 T cells (Nguyen et al. 2002). Thus, Stat4 is probably implicated in antiviral responses induced by IFN-α/β.

5.5
Stat5a and Stat5b Knockout Mice

Stat5a and Stat5b are closely related proteins with 96% similarity in their amino acid sequence, and both are tyrosine phosphorylated in response to a variety of cytokines, including prolactin, growth hormone, Epo, thrompbopoietin (Tpo), GM-CSF, IL-2, IL-3, IL-5, IL-7, IL-9, and IL-15. Stat5a and Stat5b knockout mice have been generated and mice lacking both Stat5a and Stat5b have also been generated. The first characterization of Stat5a knockout mice focused on its function in the mammary gland. Stat5a knockout mice displayed defective lobuloalveolar outgrowth during pregnancy and defective lactation, features that are both associated with defective prolactin responses (Liu et al. 1997). Although Stat5b knockout females did not show significant impairment in mammopoiesis, female mice lacking both Stat5a and Stat5b were infertile and exhibited a more severe phenotype in mammopoiesis, which was not observed in Stat5a and Stat5b single knockout mice (Teglund et al. 1998). These findings suggest that Stat5a and Stat5b have compensatory functions for each other. Although Stat5b knockout mice did not exhibit a significant phenotype in mammary gland development, Stat5b knockout males exhibited loss of their sexually dimorphic pattern due to an impaired growth hormone response (Udy et al. 1998; Teglund et al. 1998).

Although Stat5 proteins have been shown to be tyrosine phosphorylated in response to Epo and Tpo, Stat5a/Stat5b double knockout mice did not exhibit any reduction in the number of erythrocytes and platelets (Teglund et al. 1998). However, another report has indicated that Stat5a/Stat5b double knockout embryos are anemic due to increased apoptosis of erythroid progenitor cells, probably through a defective response to Epo (Socolovsky et al. 1999).

Bone marrow cells from Stat5a/Stat5b double knockout mice display a strong increase in apoptotic cell death during GM-CSF dependent functional maturation (Kieslinger et al. 2000). In addition, the recruitment of myeloid cells to inflammatory sites was significantly impaired in Stat5a/Stat5b double knockout mice. These findings indicate that Stat5 is required for both GM-CSF dependent survival and proliferation of differentiating myeloid progenitor cells during inflammation.

The role of Stat5 proteins in lymphocytes has also been characterized. T cells from Stat5a knockout mice showed a partially impaired IL-2 induced proliferation due to the partially impaired IL-2 induced expression of the IL-2 receptor α chain (Nakajima et al. 1997). Stat5b knockout mice exhibited a more severely impaired IL-2 induced proliferation of T cells than Stat5a knockout mice (Imada et al. 1998). Furthermore, IL-2 induced proliferation was profoundly reduced in T cells from Stat5a/Stat5b double knockout mice (Moriggl et al. 1999). Thus, in T cells Stat5a and Stat5b are likely to have compensatory roles in IL-2 mediated signaling. Stat5b knockout mice also had defective NK cell cytotoxic activity due to impaired IL-15 mediated signaling (Imada et al. 1998). Stat5a/Stat5b double knockout mice showed a significantly reduced number of peripheral blood B cells, which was attributed to impaired IL-7 responses in pre- and pro-B cells (Sexl et al. 2000).

Growth and differentiation of mast cells are controlled by IL-3 and stem cell factor (SCF), which both activate Stat5. Stat5a/Stat5b double knockout mice showed impaired mast cell development in vivo, which was probably due to defective responses to IL-3 and SCF (Shelburne et al. 2003).

5.6
Stat6 Knockout Mice

Stat6 knockout mice showed an impaired response to IL-4, a Th2-type cytokine and Th2 inducing cytokine (Takeda et al. 1996a; Shimoda et al. 1996; Kaplan et al. 1996). In addition, Stat6 knockout mice exhibited an impaired response to another Th2-type cytokine, IL-13, which shares the IL-4 receptor α chain in the receptor complex (Takeda et al. 1996b). Thus, Stat6 knockout mice were defective in the signaling pathways of IL-4 and IL-13. Accordingly, Stat6 knockout mice showed an impaired Th2-type response in several models of infection such as *Nippostrongylus brasiliensis*, *Leishmania mexicana* and *Schistosoma mansoni* (Takeda et al. 1996; Kaplan et al. 1998b; Hayashi et al. 1999; Stamm et al. 1999). Stat6 is also involved in Th2-mediated allergic diseases including asthma (Kuppermann et al. 1998; Akimoto et al. 1998; Miyata et al. 1999). Ovalbumin-treated Stat6 knockout mice did not exhibit any airway inflammatory change, such as eosinophilia in bronchoalveolar lavage, a peribronchial pathological change, or increased airway reactivity, indicating that a Stat6-mediated pathway plays a critical role in Th2-mediated airway inflammation. Asthma-like responses are induced by infection with viruses such as the respiratory syncytial virus (RSV). RSV infection caused airway hyperreactivity, which is associated with increased IL-13 production. In Stat6 knockout mice, RSV-induced airway hyperreactivity was not observed, indicating that Stat6 is also essential for virus-induced airway inflammation (Tekkanat et al. 2001). Balb/c mice are sensitive to infection with Ectromelia virus (EV), which causes a disease known as mousepox. Similarly, Balb/c mice are prone to mount a Th2-type response against infection by the helminth parasite *Taenia crassiceps* that causes murine cysticercosis. Stat6 knockout Balb/c mice exhibited a strong induction of Th1-type responses to effectively control infection by both EV and *Taenia crassiceps* (Mahalingam et al. 2001; Rodriguez-Sosa et al. 2002). Thus, Stat6 is involved in infection-mediated Th2 dominant pathogenesis.

Stat6 activity is regulated by a cell-type specific cleavage of Stat6 in mast cells (Suzuki et al. 2002; Sherman et al. 2002). Proteolytic processing of Stat6 leads to production of the protein lacking the C-terminal transactivation domain and acts as a dominant-negative molecule to Stat6-mediated transcription in mast cells. Thus, the C-terminal cleaved Stat6 acts as a lineage-specific negative regulator of Stat6-dependent signaling in mast cells.

In a murine model of septic peritonitis induced by cecal ligation and puncture, the Th1-type response is required for clearance of bacteria, whereas the Th2-type response modulates systemic organ failure, indicating that a balanced Th1/Th2 response is critical in host defense during sepsis. Both Stat6 and Stat4 knockout mice are resistant to septic peritonitis by enhancing local bacterial clearance and modulating systemic organ damage, respectively (Matsukawa et al. 2001). In contrast, Stat4 and Stat6 knockout mice display a high susceptibility to the septic model induced by administration of LPS (Lentsch et al. 2001). Although there is controversy between these two septic model studies, together they indicate that regulation of Stat4/Stat6 activity has a considerable effect on septic condition.

6
Tissue-Specific Roles of Stat3 Revealed by Conditional Gene Targeting

As described above, Stat3 knockout mice died during early embryogenesis, while knockout mice of other STAT family members were all viable. To avoid early embryonic mortality and analyze the role of Stat3 in adult tissues, conditional gene targeting was performed with the Cre–loxP recombination system (Rajewsky et al. 1997). Using mice lacking Stat3 in a tissue- or cell-specific manner, the role of Stat3 has been characterized in various aspects of biological responses (Table 3).

Table 3 Phenotype of tissue-specific Stat3 knockout mice

Tissue	Promoter used for Cre expression	Phenotype
T cells	Lck	Impaired IL-6-mediated prevention of apoptosis
		Impaired IL-2-mediated expression of IL-2Rα
B cells	CD19	Compromised Ig production
Myeloid cells	LysM	Enhanced inflammatory responses,
		leading to development of chronic enterocolitis
Granulocytes	Mx	Enhanced G-CSF-dependent proliferation
		of granulocytes
Skin	K5	Impaired wound healing and hair cycle
Thymic epithelium	K5	Thymic hypoplasia in adults
Mammary gland	BLG	Delayed involution of mammary gland
Liver	Mx	Impaired expression of acute phase protein genes
	Albumin	Reduced growth during liver regeneration
Neurons	NFL	Impaired survival of motoneurons after injury
	Bal	Increased apoptosis of sensory neurons

6.1
Deletion of Stat3 in T cells

Conditional gene targeting of Stat3 was first achieved in T cells using transgenic mice expressing Cre recombinase under the control of the Lck promoter (Takeda et al. 1998). These Stat3 mutant mice displayed normal T cell development. However, Stat3-deficient T cells showed a defective proliferative response to stimulation by IL-6/anti-CD3 antibody or IL-6/concanavalin A. Although IL-6 did not induce cell cycle progression by itself, it did prevent apoptosis of normal T cells, but not in Stat3-deficient T cells. Thus, Stat3 deficiency in T cells resulted in an impaired IL-6 mediated inhibition of apoptosis leading to an impaired growth response in vitro.

Stat3-deficient T cells also exhibited some impairment of IL-2 induced cell proliferation, accompanied by defective IL-2 induced expression of the IL-2 receptor α chain (Akaishi et al. 1998). Thus, Stat3 is likely to be indirectly involved in T-cell proliferation through the prevention of apoptosis and expression of cytokine receptors.

6.2
Deletion of Stat3 in B Cells

B cell-specific deletion of Stat3 was generated using mice expressing Cre recombinase under the control of the CD19 promoter (our unpublished data). These Stat3 mutant mice showed no impairment in B-cell development, including B1 cells. However, the serum concentration of immunoglobulins (Igs) was compromised in the Stat3 mutant mice. In Stat3 mutant mice, the concentrations of IgM and IgG2a were normal, but IgG1 levels were significantly reduced, and IgG2b and IgG3 levels partially reduced. Given that IL-6 induces the final differentiation of B cells into Ig-producing plasma cells, Stat3-deficient B cells may have a defect in a differentiation program.

6.3
Deletion of Stat3 in Myeloid Lineage Cells

Myeloid cell-specific Stat3 knockout mice have been created using mice expressing Cre recombinase controlled by the lysozyme M promoter (Takeda et al. 1999). Lysozyme M is exclusively expressed in cells of myeloid lineages, such as macrophages, neutrophils, and dendritic cells. These Stat3 mutant mice show a high level of sensitivity to LPS-induced endotoxin shock. The Stat3 mutant mice also showed enhanced Th1 activity and developed chronic enterocolitis with age. These phenotypes were reminiscent of that observed in IL-10 knockout mice (Kühn et al. 1993; Berg et al. 1996). IL-10 suppresses the activity of myeloid cells through a Stat3-dependent signaling pathway (Riley et al. 1999). Consistent with these findings, LPS induced a high level of inflammatory cytokines, such as tumor necrosis factor (TNF)-α and IL-12 in Stat3-deficient macrophages. In normal macrophages that were treated with IL-10, LPS-induced cytokine production was substantially suppressed. In contrast, Stat3-deficient macrophages did not show IL-10 mediated suppression of LPS-induced cytokine production. Thus, Stat3 plays an essential role in IL-10 mediated suppression of myeloid cell activity, thereby preventing chronic inflammatory diseases such as enterocolitis.

The mechanism by which abnormal activity of Stat3-deficient myeloid cells induce chronic enterocolitis in vivo has been analyzed using Stat3 mutant mice with additional gene deletions of Stat1, TNF-α, IL-12p40, RAG2, and TLR4 (Kobayashi et al. 2003). Although Stat1 was constitutively tyrosine phosphorylated in Stat3 mutant mouse macrophages, Stat1/Stat3 double mutant mice developed chronic enterocolitis, and produced increased levels of inflammatory cytokines such as TNF-α and IL-12p40 from macrophages. TNF-α/Stat3 double mutant mice also developed severe chronic enterocolitis with enhanced Th1 activity. However, IL-12p40/Stat3 double mutant mice showed neither an inflammatory change in the colon nor enhanced Th1 activity. Similarly, RAG2/Stat3 double mutant mice with inhibited T-cell development did not develop chronic enterocolitis. These findings established a sequential mechanism that induces chronic inflammation in vivo. Abnormally activated Stat3 mutant macrophages overproduced inflammatory cytokines. Among these cytokines, IL-12 induced an amplified Th1 response and induced Th1-dependent chronic enterocolitis. TLR4 is a member of the family of Toll-like receptors that plays an essential role in the activation of innate immune cells, including macrophages, through recognition of microbial components (Takeda et al. 2003). LPS, a TLR4 ligand, triggers the overproduction of IL-12 in Stat3 mutant macrophages. TLR4/Stat3 double mutant mice showed no LPS-induced IL-12p40 production in macrophages or reduced Th1 activity. Furthermore, the incidence of chronic enterocolitis was delayed and greatly reduced. These findings indicate that TLR4-mediated recognition of microbial components, probably from bacteria residing in the large intestine, triggers abnormal activation of Stat3 mutant macrophages, and leads to the development of chronic enterocolitis.

6.4
Deletion of Stat3 in Hematopoietic Cells

Stat3 was deleted in hematopoietic progenitor cells using mice bearing Cre recombinase driven by the IFN-inducible Mx promoter, to permit analysis of the role of Stat3 in G-CSF mediated function (McLemore et al. 2001). These Stat3 mutant mice displayed no impairment of terminal differentiation, proliferation, and function of granulocytes. Rather, Stat3 mutant mice showed enhanced G-CSF dependent proliferation of granulocytes, indicating that Stat3 is a negative regulator of G-CSF induced proliferation and homeostasis of granulocytes. This study revealed an unexpected role of Stat3 in G-CSF mediated signaling.

6.5
Deletion of Stat3 in the Mammary Gland

In sharp contrast with Stat5a, which is activated during pregnancy and lactation, and essential for development of the mammary gland, Stat3 is activated in the mammary gland during involution (Philp et al. 1996). To analyze the role of Stat3 in the mammary gland, it was disrupted in mice expressing Cre recombinase driven by the β-lactoglobulin promoter (Chapman et al. 1998). Stat3 mutant mice showed delayed involution after weaning, due to decreased apoptosis of mammary gland epithelial cells. Insulin-like growth factor (IGF)-1 is responsible for the survival of mammary gland epithelial cells. Mammary gland epithelial cells synthesize and secrete high levels of IGFBP-5 during involution, and is proposed to induce epithelial apoptosis by sequestering IGF-1. In the Stat3 mutant tissue, induction of IGFBP-5 was substantially reduced, which may result in reduced levels of apoptosis. Thus, Stat3 plays an important role in the induction of apoptosis in mammary gland epithelial cells. Mice lacking leukemia inhibitory factor (LIF), which activates Stat3, showed delayed involution of the mammary gland (Kritikou et al. 2003). This study indicates that LIF is one of factors responsible for Stat3 activation and induction of apoptosis in mammary gland.

6.6
Deletion of Stat3 in the Skin

Utilization of mice expressing Cre recombinase under the control of the keratin 5 (K5) promoter, enabled us to delete Stat3 in keratinocytes (Sano et al. 1999). These Stat3 mutant mice were born with no apparent abnormality in development of the epidermis and hair follicles. However, mutant mice show a compromised process in the second hair cycle and spontaneous development of skin ulcers and alopecia with age. In addition, the mutant mice show significantly delayed wound healing. The proliferation and differentiation of keratinocytes are regulated by several growth factors and cytokines, including epidermal growth factor, transforming growth factor-α, keratinocyte growth factor, and IL-6, all of which have an ability to induce phosphorylation of Stat3. Factor-induced in vitro migration of Stat3-deficient keratinocytes is severely impaired, despite the normal proliferative response. These findings indicate that Stat3 is involved in skin remodeling, including hair cycles and wound healing, through growth factor-dependent regulation of keratinocyte migration.

6.7
Deletion of Stat3 in Thymic Epithelium

Keratin 5 is expressed in thymic as well as stratified epithelium. Therefore, deletion of Stat3 was also achieved in thymic epithelial cells of the mutant mice (Sano et al. 2001). Adult mutant mice had substantially hypoplastic thymi with altered cortical thymic epithelial architecture and a reduced number of thymocytes. Although thymocytes isolated from Stat3 mutant mice exhibited both a normal growth response and sensitivity to apoptotic stimuli such as dexamethasone or γ-irradiation, the thymus was highly susceptible to in vivo treatment with dexamethasone or γ-irradiation, even in young mutant mice lacking significant symptoms. Thus, Stat3 in thymic epithelium has been shown to be essential for the maintenance of thymic architecture required for the survival of thymocytes.

6.8
Deletion of Stat3 in the Liver

Stat3 was disrupted in the liver using Mx–Cre transgenic mice, in which Cre recombinase is expressed in response to type I interferon (Alonzi et al. 2001). The Stat3 mutant mice showed defective IL-6 induced expression of acute phase proteins in the liver. In addition, LPS-induced expression of several acute phase proteins was also compromised in the Stat3 mutant mice. These findings demonstrate that Stat3 is critical for the induction of acute phase proteins during inflammation of the liver.

The role of Stat3 during liver regeneration was analyzed in adult mice lacking Stat3 in the liver using an albumin promoter-driven Cre-loxP recombination system (Li et al. 2002). Although IL-6 knockout mice displayed severe impairment of liver regeneration after hepatectomy, Stat3 mutant mice showed no such impairment. However, the liver of Stat3 mutant mice exhibited a defective DNA synthetic response after hepatectomy. Thus, Stat3 is likely to be involved in the mitogenic response during liver regeneration.

6.9
Deletion of Stat3 in Neurons

Using mice in which Cre recombinase is expressed with the neurofilament light chain promoter, the expression of Stat3 was suppressed in neurons (Schweizer et al. 2002). The Stat3 mutant mice displayed no apparent change in the development of motoneurons during the embryonic period. However, adult Stat3 mutant mice showed a dramatic loss of motoneurons after facial nerve transection. Thus, Stat3 has been shown to play an important role in motor neuron survival after nerve lesions in postnatal mice but not during embryonic development.

The role of Stat3 in sensory neurons was also analyzed using a Cre-balancer strain of mouse that produces mosaic mice that differentially influence gene expression in separate tissues (Alonzi et al. 2001). Stat3 mutant mice displayed enhanced sensory neuronal death and severely impaired ciliary neurotrophic factor (CNTF)- or LIF-mediated survival of sensory neurons.

7
Negative Regulation of STAT Activity

Cytokine-induced phosphorylation of STAT proteins is only transient, suggesting that some negative regulatory mechanisms act on STAT activity. In this context, several mechanisms by which STAT activity is negatively regulated have been elucidated. Among these mechanisms, the inhibition of STAT activation by SOCS proteins is the most striking and intensively investigated.

7.1
SOCS Family

The first SOCS protein was discovered concurrently by three independent groups as a suppresser of cytokine signaling-1 (SOCS1), JAK-binding protein (JAB), and STAT-induced STAT inhibitor-1 (SSI-1) (Starr et al. 1997; Endo et al. 1997; Naka et al. 1997). Before the identification of SOCS-1/JAB/SSI-1, an original member of this family was identified as CIS1 that is induced by IL-2, IL-3, and Epo, and inhibits cytokine-induced Stat5 activation (Yoshimura et al. 1995). Subsequently, several proteins that are structurally related to CIS1 and SOCS1/JAB/SSI-1 have been identified, and the SOCS family of proteins now consist of eight members (CIS1, SOCS1/JAB/SSI-1, CIS2–7/SOCS2–7) (Masuhara et al. 1997). While there is no amino acid similarity at the N terminus among SOCS family members, all of these members harbor an SH2 domain and a conserved C terminus designated as the SOCS-box or CIS-homology domain. SOCS protein expression is induced by cytokines, SOCS protein in turn binds to the cytokine receptor or Jak kinases to inhibit activation of STAT proteins. In addition to this negative regulation of the cytokine signaling, SOCS proteins mediate a ubiquitin–protease cascade through the interaction of the SOCS box with elongins B and C (Zhang et al. 1999).

The physiological roles of SOCS proteins have been elucidated through the generation of knockout mice. SOCS1 knockout mice died within 3 weeks of birth suffering from massive liver degeneration that was probably due to the high level of IFN-γ and IL-4 activity (Naka et al. 1998; Starr et al. 1998). Indeed, the introduction of other gene deletions such as IFN-γ, Stat1, and Stat6 in SOCS1 knockout mice rescued the liver from degeneration and prevented early mortality (Alexander et al. 1999; Marine et al. 1999; Naka et al. 2001). In addition to the critical role of SOCS1 in IFN-γ and IL-4 signaling, SOCS1 is involved in the negative regulation of signaling pathways of prolactin, insulin, and TNF-α (Lindeman et al. 2001; Kawazoe et al. 2001; Morita et al. 2000). Most recently, SOCS1 has been shown to negatively regulate the LPS signaling pathway (Kinjo et al. 2002; Nakagawa et al. 2002). Deletion of the SOCS2 gene resulted in mice with gigantism, demonstrating that SOCS2 is a negative regulator of growth hormone and IGF-1 (Metcalf et al. 2000). SOCS3 knockout mice exhibiting exaggerated erythrocytosis died during embryogenesis, indicating that SOCS3 plays a role in the regulation of fetal erythropoiesis (Marine et al. 1999). Given that Epo is essential for erythropoiesis, SOCS3 is likely to negatively regulate Epo signaling. In addition, SOCS3 regulates LIF-mediated signaling and trophoblast differentiation (Takahashi et al. 2003). Most recently, SOCS3 has been shown to inhibit the signaling pathway of IL-6, but not of IL-10, both of which activate Stat3 (Lang et al. 203, Crocker et al. 2003; Yasukawa et al. 2003). SOCS6 knockout mice exhibited mildly retarded growth, and SOCS6 interacted with IRS4, indicating the possible involvement of SOCS6 in insulin- or IGF-1 mediated signaling pathways (Krebs et al. 2002).

The SOCS family of proteins also plays a critical role in the IL-12 mediated Stat4 component of the Th1 response and the IL-4 mediated Stat6 component of the Th2 response. IL-4 induced Stat6 activation was severely reduced in Th1 cells, whereas IL-12 induced Stat4 activation was markedly suppressed in Th2 cells. In Th1 cells, SOCS5 has been shown to be preferentially expressed and bind to the IL-4R α chain. T cell specific SOCS5 Tg mice showed impaired IL-4 dependent Stat6 activation and Th2 differentiation, indicating that SOCS5 is a negative regulator of IL-4 signaling in Th1 cells (Seki et al. 2002). SOCS3 is selectively expressed in Th2 cells. Furthermore, SOCS3 expression is enhanced in patients with atopic dermatitis and asthma. T-cell specific SOCS3 Tg mice exhibited enhanced allergic airway inflammation which was accompanied by increased Th2-mediated responses, whereas SOCS3$^{+/-}$ mice showed defective Th2-mediated responses (Seki et al. 2003). These results indicate that SOCS3 is likely to be a critical inhibitor of IL-12 induced Stat4 activation. Thus, SOCS5 and SOCS3 are critically involved in Th1 and Th2 responses, respectively.

7.2
Other Regulatory Mechanisms of STAT Activity

In addition to the SOCS family, several mechanisms that negatively regulate STAT activity have been proposed. The degradation of STAT proteins and receptors regulate the activity of cytokine signaling pathways (Kim et al. 1996; Strous et al. 1996). Naturally occurring C-terminal truncated forms of Stat3 (called Stat3β) and Stat5 have been reported to act as dominant negative inhibitors (Wang et al. 1996; Azam et al. 1997; Caldenhoven et al. 1996). By gene targeting, specific ablation of Stat3β in mice impaired their recovery from endotoxin shock, demonstrating that Stat3β plays an important role in the regulation of inflammatory responses (Yoo et al. 2002).

The protein inhibitor of the activated STAT (PIAS) family of proteins has been shown to negatively regulate the transcriptional activity of STAT proteins in the nucleus. So far, four members of the PIAS family have been identified: PIAS1, PIAS3, PIASy, and PIASx (Shuai 2000). PIAS1 and PIAS3 inhibit Stat1- and Stat3-mediated transcription, respectively, by blocking their DNA-binding activity (Chung et al. 1997; Liu et al. 1998). PIASy and PIASx inhibit transcriptional activity of Stat1 and Stat4, respectively, without affecting the DNA-binding activity (Liu et al. 2001; Arola et al. 2003). At least two members of the PIAS family, PIAS1 and PIASx, act as specific SUMO E3-like ligases that inhibit the activity of several transcription factors, such as LEF1, p53 and c-Jun (Schmidt et al. 2002; Sachdev et al. 2001). Thus, the PIAS family of proteins has additional functions that are independent of STAT activity. Physiological roles of PIAS proteins remain to be elucidated through the generation of knockout mice.

A considerable body of evidence indicates that tyrosine phosphatases, such as SHP-1, SHP-2, and PTP1B, are negative regulators of cytokine signaling pathways. SH2 domain-containing protein tyrosine phosphatase (SHP-1) induces the dephosphorylation of JAK kinases through direct binding with JAKs (Haque et al. 1998; Migone et al. 1998; David et al. 1995; Jiao et al. 1996). In familial polycythemia a point mutation of the Epo receptor, that causes disruption of the SHP-1-binding site, has been identified (Kralovics et al. 1997). In these mutant cells, Epo treatment resulted in prolonged phosphorylation of Jak2, indicating that SHP-1 is physiologically involved in the negative regulation of JAK activity (Klingmuller et al. 1995). SHP-2 regulates gp130- and IFN-mediated signaling pathways (Kim et al. 1998; You et al. 1999). PTP1B knockout mice were highly sensitive to insulin and resistant to obesity. The mice also showed leptin hypersensitivity and enhanced leptin-induced hypothalamic Stat3 phosphorylation that can probably be attributed to reduced Jak1 dephosphorylation (Zabolotny et al. 2002).

8
Concluding Remarks

The JAK-STAT pathway was identified in cytokine signaling pathways. Subsequent studies, especially through the generation of knockout mice, have revealed that the JAK and STAT family of proteins have an essential role in cytokine-mediated biological responses. Furthermore, constitutive activation of STAT proteins, particularly Stat3, is associated with the progression of tumors (Bromberg et al. 1999; Bowman et al. 2000; Yoshikawa et al. 2001).

In contrast, IFN-γ mediated activation of Stat1 is required for prevention of tumors (Kaplan et al. 1998; Shankaran et al. 2001). Thus, STAT activation plays a critical role in human diseases, such as tumors. Further understanding of the mechanisms that underlie the induction of distinct types of gene expression by individual STAT proteins will be helpful in future applications of STAT protein regulators for the clinical treatment of several human diseases.

Acknowledgements We thank M. Hashimoto for excellent secretarial assistance.

References

Akaishi H, Takeda K, Kaisho T, Shineha R, Satomi S, Takeda J,Akira S. (1998) Defective IL-2 mediated IL-2 receptor α chain expression in Stat3-deficient T lymphocytes. Int Immunol 10:1747–1751

Akimoto T, Numata F, Tamura M, Takata Y, Higashida N, Takashi T, Takeda K, Akira S (1998) Abrogation of bronchial eosinophilic inflammation and airway hyperreactivity in STAT6-deficient mice. J Exp Med 187:1537–1542

Alexander WS, Starr R, Fenner JF, Scott CL, Handman E, Sprigg NS, Corbin JE, Cornish AL, Darwiche R, OwczarekCM, Kay TWH, Nicola NA, Hertzog PJ, Metcalf D, Hilton DJ (1999) SOCS1 is a critical inhibitor of interferon gamma signaling and prevents the potentially fatal neonatal actions of this cytokine. Cell 98:597–608

Alonzi T, Maritano D, Gorgoni B, Rizzuto G, Libert C, Poli, V (2001) Essential role of STAT3 in the control of the acute-phase response as revealedby inducible gene inactivation [correction of activation] in the liver. Mol Cell Biol 21: 1621–1632

Alonzi T, Middleton G, Wyatt S, Buchman V, Betz UA, Müller W, Musiani P, Poli V, Davies A M (2001) Role of STAT3 and PI 3-kinase/Akt in mediating the survival actions of cytokines on sensory neurons. Mol Cell Neurosci 18:270–282

Arora T, Liu B, He H, Kim J, Murphy TL, Murphy KM, Modlin RL, Shuai K (2003) PIASx Is a Transcriptional co-repressor of signal transducer and activator of transcription 4. J Biol Chem 278:21327–21330

Azam M, Lee C, Strehlow I, Schindler C (1997) Functionally distinct isoforms of STAT5 are generated by protein processing. Immunity 6:691–701

Becker S, Groner B, Muller CW (1998) Three-dimensional structure of the Stat3β homodimer bound to DNA. Nature 394:145–151

Berg DJ, Davidson N, Kuhn R, Muller W, Menon S, Holland G, Thompson-Snipes L, Leach MW, Rennick D (1996). Enterocolitis and colon cancer in interleukin-10-deficient mice are associated with aberrant cytokine production and CD4(+) TH1-like responses. J Clin Invest 98:1010–1020

Bowman T, Garcia R, Turkson J, Jove R (2000) STATs in oncogenesis. Oncogene 19:2474–2488

Bromberg JF, Wrzeszczynska MH, Devgan G, Zhao Y, Pestell RG, Albanese C, Darnell JE Jr. (1999) Stat3 as a oncogene. Cell 98:295–303

Caldenhoven E, van Dijk TB, Solari R, Armstrong J, Raaijmakers JA, Lammers JW, Koenderman L, de Groot RP (1996). STAT3β, a splice variant of transcription factor STAT3, is a dominant negative regulator of transcription. J Biol Chem 271:13221–13227

Chapman RS, Lourenco P, Tonner E, Flint DJ, Selbert S, Takeda K, Akira S, Clarke AR, Watson CJ (1999) Suppression of epithelial apoptosis and delayed mammary gland involution in mice with a conditional knockout of Stat3. Genes Dev 13:2604–2616

Chen X, Vinkemeier U, Zhao Y, Jeruzalmi D, Darnell JE Jr, Kuriyan J (1998) Crystal structure of a tyrosine phosphorylated STAT-1 dimer bound to DNA. Cell 93:827–839

Chung CD, Liao J, Liu B, Rao X, Jay P, Berta P, Shuai K. (1997) Specific inhibition of Stat3 signal transduction by PIAS3. Science 278:1803–1805

Croker BA, Krebs DL, Zhang JG, Wormald S, Willson TA, Stanley EG, Robb L, Greenhalgh CJ, Forster I, Clausen BE, Nicola NA, Metcalf D, Hilton DJ, Roberts AW, Alexander WS. (2003) SOCS3 negatively regulates IL-6 signaling in vivo. Nat Immunol 4, 540–545

David M, Chen HE, Goelz S, Larner AC, Neel BG. (1995) Differential regulation of the alpha/beta interferon-stimulated Jak/Stat pathway by the SH2 domain-containing tyrosine phosphatase SHPTP1. Mol Cell Biol 15:7050–7058

Duncan SA, Zhong Z, Wen Z, Darnell J E Jr (1997) Requirement of the transcription factor GATA4 for heart tube formation and ventral morphogenesis. Dev Dyn 208:190–198

Dupuis S, Dargemont C, Fieschi C, Thomassin N, Rosenzweig S, Harris J, Holland SM, Schreiber RD, Casanova JL (2001) Impairment of mycobacterial but not viral immunity by a germline human STAT1 mutation. Science 293:300–303

Durbin JE, Hackenmiller R, Simon MC, Levy DE (1996) Targeted disruption of the mouse Stat1 gene results in compromised innate immunity to viral disease. Cell 84:443–450

Endo TA, Masuhara M, Yokouchi M, Suzuki R, Sakamoto H, Mitsui K, Matsumoto A, Tanimura S, Ohtsubo M, Misawa H, Miyazaki T, Leonor N, Taniguchi T, Fujita T, Kanakura Y, Komiya S, Yoshimura A (1997) A new protein containing an SH2 domain that inhibits JAK kinases. Nature 387:921–924

Gil MP, Bohn E, O'Guin AK, Ramana CV, Levine B, Stark GR, Virgin HW, Schreiber RD (2001) Biologic consequences of Stat1-independent IFN signaling. Proc Natl Acad Sci USA 98: 6680–6685

Haque SJ, Harbor P, Tabrizi M, Yi T, Williams BR (1998) Protein-tyrosine phosphatase Shp-1 is a negative regulator of IL-4- and IL-13-dependent signal transduction. J Biol Chem 273:33893–33896

Hayashi N, Matsui K, Tsutsui H, Osada Y, Mohamed RT, Nakano H, Kashiwamura S, Hyodo Y, Takeda K, Akira S, Hada T, Higashino K, Kojima S, Nakanishi K (1999) Kuppfer cells from Schistosoma mansoni-infected mice participate in the prompt type 2 differentiation of hepatic T cells in response to worm antigens. J Immunol 163:6702–6711

Ihle JN, Witthuhn BA, Quelle FW, Yamamoto K, Silvennoinen O (1995) Signaling through the hematopoietic cytokine receptors. Annu Rev Immunol 13:369–398

Ihle JN (1996) STATs: signal transducers and activators of transcription. Cell 84:331–334.

Imada K, Bloom ET, Nakajima H, Horvath-Arcidiacono JA, Udy GB, Davey HW, Leonard WJ (1998) Stat5b is essential for natural killer cell-mediated proliferation and cytotoxic activity. J Exp Med 188:2067–2074

Jiao H, Berrada K, Yang W, Tabrizi M, Platanias LC, Yi T (1996) Direct association with and dephosphorylation of Jak2 kinase by the SH2-domain-containing protein tyrosine phosphatase SHP-1. Mol Cell Biol 16:6985–6992

Kaplan MH, Schindler U, Smiley ST, Grusby MJ (1996) Stat6 is required for mediating responses to IL-4 and for the development of Th2 cells. Immunity 4: 313–319

Kaplan MH, Sun YL, Hoey T Grusby MJ (1996) Impaired IL-12 responses and enhanced development of Th2 cells in Stat4-deficient mice. Nature 382:174–177

Kaplan MH, Wurster AL, Grusby MJ (1998) A signal transducer and activator of transcription (Stat)4-independent pathway for the development of T helper type 1 cells. J Exp Med 188: 1191–1196

Kaplan MH, Whitfield JR, Boros DL, Grusby MJ (1998) Th2 cells are required for the Schistosoma mansoni egg-induced granulomatous response. J Immunol 160:1850–1856

Kaplan DH, Shankaran V, Dighe AS, Stockert E, Aguet M, Old LJ, Schreiber RD (1998) Demonstration of an interferon γ-dependent tumor surveillance system in immunocompetent mice. Proc Natl Acad Sci USA 95:7556–7561

Karaghiosoff M, Neubauer H, Lassnig C, Kovarik P, Schindler H, Pircher H, McCoy B, Bogdan C, Decker T, Brem G, Pfeffer K Muller M (2000) Partial impairment of cytokine responses in Tyk2-deficient mice. Immunity 13:549–560

Karaghiosoff M, Steinborn R, Kovarik P, Kriegshauser G, Baccarini M, Donabauer B, Reichart U, Kolbe T, Bogdan C, Leanderson T, Levy D, Decker T, Muller M (2003) Central role for type I interferons and Tyk2 in lipopolysaccharide-induced endotoxin shock. Nat Immunol 4:471–477

Kawazoe Y, Naka T, Fujimoto M, Kohzaki H, Morita Y, Narazaki M, Okumura K, Saitoh H, Nakagawa R, Uchiyama Y, Akira S, Kishimoto T. (2001) Signal transducer and activator of transcription (STAT)-induced STAT inhibitor 1 (SSI-1)/suppressor of cytokine signaling 1 (SOCS1) inhibits insulin signal transduction pathway through modulating insulin receptor substrate 1 (IRS-1) phosphorylation. J Exp Med 193:263–269

Kieslinger M, Woldman I, Moriggl R, Hofmann J, Marine JC, Ihle JN, Beug H, Decker T (2000) Anti-apoptotic activity of Stat5 required during terminal stages of myeloid differentiation. Genes Dev 14:232–244

Kim TK, Maniatis T (1996) Regulation of interferon-γ-activated STAT1 by the ubiquitin-proteasome pathway. Science 273:1717–1719

Kim H, Hawley TS, Hawley RG, Baumann H (1998) Protein tyrosine phosphatase 2 (SHP-2) moderates signaling by gp130 but is not required for the induction of acute-phase plasma protein genes in hepatic cells. Mol Cell Biol 18:1525–1533

Kinjyo I, Hanada T, Inagaki-Ohara K, Mori H, Aki D, Ohishi M, Yoshida H, Kubo M, Yoshimura A (2002) SOCS1/JAB is a negative regulator of LPS-induced macrophage activation. Immunity 17:583–591

Kishimoto T, Taga T, Akira S (1994) Cytokine signal transduction. Cell 76:253–262

Klingmuller U, Lorenz U, Cantley LC, Neel BG, Lodish HF (1995) Specific recruitment of SH-PTP1 to the erythropoietin receptor causes inactivation of JAK2 and termination of proliferativesignals. Cell 80:729–738

Kobayashi M, Kweon M, Kuwata H, Kiyono H, Takeda K, Akira S (2003) Toll-like receptor-dependent IL-12p40 production causes chronic enterocolitis in myeloid cell-specific Stat3-deficient mice. J Clin Invest 111:1297–1308

Kralovics R, Indrak K, Stopka T, Berman BW, Prchal JF, Prchal JT (1997) Two new EPO receptor mutations: truncated EPO receptors are most frequently associated with primary familial and congenital polycythemias. Blood 90:2057–2061

Krebs DL, Uren RT, Metcalf D, Rakar S, Zhang JG, Starr R, De Souza DP, Hanzinikolas K, Eyles J, Connolly LM, Simpson RJ, Nicola NA, Nicholson SE, Baca M, Hilton DJ, Alexander WS (2002) SOCS-6 binds to insulin receptor substrate 4, and mice lacking the SOCS-6 gene exhibit mild growth retardation. Mol Cell Biol 22:4567–4578

Kritikou EA, Sharkey A, Abell K, Came PJ, Anderson E, Clarkson RW, Watson CJ (2003) A dual, non-redundant, role for LIF as a regulator of development and STAT3-mediated cell death in mammary gland. Development 130:3459–3468

Kuhn R, Lohler J, Rennick D, Rajewsky K, Muller W (1993) Interleukin-10-deficient mice develop chronic enterocolitis. Cell 75:263–274

Lang R, Pauleau AL, Parganas E, Takahashi Y, Mages J, Ihle JN, Rutschman R, Murray PJ (2003) SOCS3 regulates the plasticity of gp130 signaling. Nat Immunol 4:546–550

Lentsch AB, Kato A, Davis B, Wang W, Chao C, Edwards MJ (2001) STAT4 and STAT6 regulate systemic inflammation and protect against lethal endotoxemia. J Clin Invest 108:1475–1482

Leonard W, O'Shea JJ (1998) JAKS and STATS: Biological implications. Annu Rev Immunol 16:293–322

Li W, Liang X, Kellendonk C, Poli V, Taub R. (2002) STAT3 contributes to the mitogenic response of hepatocytes during liver regeneration. J Biol Chem 277:28411–28417

Lindeman GJ, Wittlin S, Lada H, Naylor MJ, Santamaria M, Zhang JG, Starr R, Hilton DJ, Alexander WS, Ormandy CJ, Visvader (2001) SOCS1 deficiency results in accelerated mammary gland development and rescues lactation in prolactin receptor-deficient mice. Genes Dev 15:1631–1636

Liu B, Liao J, Rao X, Kushner SA, Chung CD, Chang DD, Shuai K (1998) Inhibition of Stat1-mediated gene activation by PIAS1. Proc Natl Acad Sci USA 95:10626–10631

Liu B, Gross M, ten Hoeve J, Shuai K (2001) A transcriptional corepressor of Stat1 with an essential LXXLL signature motif. Proc Natl Acad Sci USA 98:3203–3207

Liu X, Robinson GW, Wagner KU, Garrett L, Wynshaw-Boris A, Hennighausen L (1997) Stat5a is mandatory for adult mammary gland development and lactogenesis. Genes Dev 11:179–186

Mahalingam S, Karupiah G, Takeda K, Akira S, Matthaei KI, Foster PS (2001) Enhanced resistance in STAT6-deficient mice to infection with ectromelia virus. Proc Natl Acad Sci USA 98:6812–6817

Marine J-C, Topham DJ, McKay C, Wang D, Parganas E, Stravopodis D, Yoshimura A, Ihle JN (1999) SOCS1 deficiency causes a lymphocytedependent perinatal lethality. Cell 98:609–616

Masuhara M, Sakamoto H, Matsumoto A, Suzuki R, Yasukawa H, Mitsui K, Wakioka T, Tanimura S, Sasaki A, Misawa H, Yokouchi M, Ohtsubo M, Yoshimura A (1997) Cloning and characterization of novel CIS family genes. Biochem Biophys Res Commun 239:439–446

McLemore ML, Grewal S, Liu F, Archambault A, Poursine-Laurent J, Haug J, Link DC (2001) STAT-3 activation is required for normal G-CSF-dependent proliferation and granulocytic differentiation. Immunity 14: 193–204

Meraz MA, White JM, Sheehan KC, Bach EA, Rodig SJ, Dighe AS, Kaplan DH, Riley JK, Greenlund AC, Campbell D, Carver-Moore K, DuBois RN, Clark R, Aguet M, Schreiber RD (1996) Targeted disruption of the Stat1 gene in mice reveals unexpected physiological specificity in the JAK-STAT pathway. Cell 84:431–442

Metcalf D, Greenhalgh CJ, Viney E, Willson TA, Starr R, Nicola NA, Hilton DJ, Alexander WS. (2000) Gigantism in mice lacking suppressor of cytokine signalling-2. Nature 405:1069–1073

Migone TS, Cacalano NA, Taylor N, Yi T, Waldmann TA, Johnston JA. (1998) Recruitment of SH2-containing protein tyrosine phosphatase SHP-1 to the interleukin 2 receptor; loss of SHP-1 expression in human T-lymphotropic virus type I-transformed T cells. Proc Natl Acad Sci USA 95:3845–3850

Moriggl R, Topham DJ, Teglund S, Sexl V, McKay C, Wang D, Hoffmeyer A, van Deursen J, Sangster MY, Bunting KD, Grosveld GC, Ihle JN. (1999) Stat5 is required for IL-2 induced cell cycle progression of peripheral T cells. Immunity 10:249–259

Morita Y, Naka T, Kawazoe Y, Fujimoto M, Narazaki M, Nakagawa R, Fukuyama H, Nagata S, Kishimoto T (2000) Signals transducers and activators of transcription (STAT)-induced STAT inhibitor-1 (SSI-1)/suppressor of cytokine signaling-1 (SOCS-1) suppresses tumor necrosis factor alpha-induced cell death in fibroblasts. Proc Natl Acad Sci USA 97:5405-5410

Naka T, Narazaki M, Hirata M, Matsumoto T, Minamoto S, Aono A, Nishimoto N, Kajita T, Taga T, Yoshizaki K, Akira S, Kishimoto T (1997) Structure and function of a new STAT-induced STAT inhibitor. Nature 387:924-929

Naka T, Matsumoto T, Narazaki M, Fujimoto M, Morita Y, Ohsawa Y, Saito H, Nagasawa T, Uchiyama Y, Kishimoto T (1998) Accelerated apoptosis of lymphocytes by augmented induction of Bax in SSI-1 (STAT-induced STAT inhibitor-1) deficient mice. Proc Natl Acad Sci USA 95:15577-15582

Naka T, Tsutsui H, Fujimoto M, Kawazoe Y, Kohzaki H, Morita Y, Nakagawa R, Narazaki M, Adachi K, Yoshimoto T, Nakanishi K, Kishimoto T (2001) SOCS-1/SSI-1-deficient NKT cells participate in severe hepatitis through dysregulated cross-talk inhibition of IFN-gamma and IL-4 signaling in vivo. Immunity 14:535-545

Nakagawa R, Naka T, Tsutsui H, Fujimoto M, Kimura A, Abe T, Seki E, Sato S, Takeuchi O, Takeda K, Akira S, Yamanishi K, Kawase I, Nakanishi K, Kishimoto T (2002) SOCS-1 participates in negative regulation of LPS responses. Immunity 17:677-687

Nakajima H, Liu X-W, Wynshaw-Boris A, Rosenthal LA, Imada K, Finbloom DS, Hennighausen L, Leonard WJ (1997) An indirect effect of Stat5a in IL-2 induced proliferation: a critical role for Stat5a in IL-2 mediated IL-2 receptor α chain induction. Immunity 7:691-701

Neubauer H, Cumano A, Mueller M, Wu H, Huffstadt U, Pfeffer K (1998) Jak2 deficiency defines an essential developmental checkpoint in definitive hematopoiesis. Cell 93:397-409

Nguyen KB, Watford WT, Salomon R, Hofmann SR, Pien GC, Morinobu A, Gadina M, O'Shea JJ, Biron CA (2002) Critical role for STAT4 activation by type 1 interferons in the interferon-γ response to viral infection. Science 297:2063-2066

Nosaka T, Van Deursen JM, Tripp RA, Thierfelder WE, Witthuhn BA, McMickle AP, Doherty PC, Grosveld GC, Ihle JN (1995) Defective lymphoid development in mice lacking Jak3. Science 270:800-802

Parganas E, Wang D, Stravopidis D, Topham D, Marine JC, Teglund S, Vanin EF, Bodner S, Colamonici OR, van Deursen JM, Gorsveld G, Ihle JN (1998) Jak2 is essential for signaling through a variety of cytokine receptors. Cell 93:385-395

Park SY, Saijo K, Takahashi T, Osawa M, Arase H, Hirayama N, Miyake K, Nakauchi H, ShirasawaT Saito T (1995) Developmental defects of lymphoid cells in Jak3 kinase-deficient mice. Immunity 3:771-782

Philp JA, Burdon TG, Watson CJ (1996) Differential activation of STATs 3 and 5 during mammary gland development. FEBS Lett 396:77-80

Rajewsky K, Gu H, Kuhn R, Betz UA, Muller W, Roes J, Schwenk F (1996) Conditional gene targeting. J Clin Invest 98:600-603

Ramana CV, Grammatikakis N, Chernov M, Nguyen H, Goh KC, Williams BR, Stark GR (2000) Regulation of *c-myc* expression by IFN-γ through Stat1-dependent and -independent pathways. EMBO J 19:263-272

Ramana CV, Gil MP, Schreiber RD, Stark GR (2002) Stat1-dependent and -independent pathways in IFN-γ-dependent signaling. Trends Immunol 23:96-101

Riley JK, Takeda K, Akira S, Schreiber RD (1999) Interleukin-10 receptor signaling through the JAK-STAT pathway. Requirement for two distinct receptor-derived signals for anti-inflammatory action. J Biol Chem 274:16513-16521

Rodig SJ, Meraz MA, White JM, Lampe PA, Riley JK, Arthur CD, King KL, Sheehan KC, Yin L, Pennica D, Johnson EM Jr, Schreiber RD (1998) Disruption of the Jak1 gene demonstrates obligatory and nonredundant roles of the Jaks in cytokine-induced biologic responses. Cell 93: 373–383

Rodriguez-Sosa M, David JR, Bojalil R, Satoskar AR, Terrazas LI (2002) Cutting edge: susceptibility to the larval stage of the helminth parasite Taenia crassiceps is mediated by Th2 response induced via STAT6 signaling. J Immunol 168:3135–3139

Sachdev S, Bruhn L, Sieber H, Pichler A, Melchior F, Grosschedl (2001) PIASy, a nuclear matrix-associated SUMO E3 ligase, represses LEF1 activity by sequestration into nuclear bodies. Genes Dev 15:3088–3103

Sano S, Itami S, Takeda K, Tarutani M, Yamaguchi Y, Miura H, Yoshikawa K, Akira S, Takeda J (1999) Keratinocyte-specific ablation of Stat3 exhibits impaired skin remodeling, but does not affect skin morphogenesis. EMBO J 18:4657–4668

Sano S, Takahama Y, Sugawara T, Kosaka H, Itami S, Yoshikawa K, Miyazaki J, van Ewijk W, Takeda J (2001) Stat3 in thymic epithelial cells is essential for postnatal maintenance of thymic architecture and thymocyte survival. Immunity 15:261–273

Schindler C, Darnell JE Jr (1995) Transcriptional responses to polypeptide ligands: the JAK-STAT pathway. Annu Rev Biochem 65:621–651

Schmidt D, Muller S (2002) Members of the PIAS family act as SUMO ligases for c-Jun and p53 and repress p53 activity. Proc Natl Acad Sci USA 99:2872–2877

Schweizer U, Gunnersen J, Karch C, Wiese S, Holtmann B, Takeda K, Akira S, Sendtner M. (2002) J Cell Biol 156:287–297

Seki Y, Hayashi K, Seki N, Matsumoto A, Ransom J, Naka T, Kishimoto T, Yoshimura A, Kubo M (2002) Selective expression of suppressor of cytokine signaling-5 (SOCS5) in type 1 helper T cell negatively regulates IL-4 dependent STAT6 activation. Proc Natl Acad Sci USA 99:13003–13008

Seki YI, Inoue H, Nagata N, Hayashi K, Fukuyama S, Matsumoto K, Komine O, Hamano S, Himeno K, Inagaki-Ohara K, Cacalano N, O'Garra A, Oshida T, Saito H, Johnston JA, Yoshimura A, Kubo M (2003) Suppressor of cytokine signaling-3 (SOCS3) regulates onset and maintenance of type 2 helper T cell mediated allergic responses. Nat Med 9:1047–1054

Sexl V, Piekorz R, Moriggl R, Rohrer J, Brown MP, Bunting KD, Rothammer K, Roussel MF, Ihle JN (2000) Stat5a/b contribute to interleukin 7-induced B-cell precursor expansion, but abl- and bcr/abl-induced transformation are independent of Stat5. Blood 96:2277–2283

Shankaran V, Ikeda H, Bruce AT, White JM, Swanson PE, Old LJ, Schreiber RD (2001) IFNγ and lymphocytes prevent primary tumour development and shape tumour immunogenicity. Nature 410:1107–1111

Sherman MA, Powell DR, Brown MA (2002) IL-4 induces the proteolytic processing of mast cell STAT6. J Immunol 169:3811–3818

Shimoda K, Kato K, Aoki K, Matsuda T, Miyamoto A, Shibamori M, Yamashita M, Numata A, Takase K, Kobayashi S, Shibata S, Asano Y, Gondo H, Sekiguchi K, Nakayama K, Nakayama T, Okamura T, Okamura S, Niho Y (2000) Tyk2 plays a restricted role in IFN alpha signaling, although it is required for IL-12-mediated T cell function. Immunity 13:561–571

Shimoda K, Tsutsui H, Aoki K, Kato K, Matsuda T, Numata A, Takase K, Yamamoto T, Nukina H, Hoshino T, Asano Y, Gondo H, Okamura T, Okamura S, Nakayama K, Nakanishi K, Niho Y, Harada M (2002) Partial impairment of interleukin-12 (IL-12) and IL-18 signaling in Tyk2-deficient mice. Blood 99:2094–2099

Shuai K (2000) Modulation of STAT signaling by STAT-interacting proteins. Oncogene 19:2638–2644

Shelburne CP, McCoy ME, Piekorz R, Sexl V, Roh KH, Jacobs-Helber SM, Gillespie SR, Bailey DP, Mirmonsef P, Mann MN, Kashyap M, Wright HV, Chong HJ, Bouton LA, Barnstein B, Ramirez CD, Bunting KD, Sawyer ST, Lantz CS, Ryan JJ (2003) Stat5 expression is critical for mast cell development and survival. Blood 102:1290–1297

Simpson SJ, Shah S, Comiskey M, de Jong YP, Wang B, Mizoguchi E, Bhan AK, Terhorst C (1998) T cell-mediated pathology in two models of experimental colitis depends predominantly on the interleukin 12/ signal transducer and activator of transcription (Stat)-4 pathway, but is not conditional on interferon γ expression by T cells. J Exp Med 187:1225–1234

Socolovsky M, Fallon AE, Wang S, Brugnara C, Lodish HF (1999) Fetal anemia and apoptosis of red cell progenitors in Stat5a −/− 5b −/− mice: A direct role for Stat5 in Bcl-XL induction. Cell 98:181–191

Stamm LM, Raisanen-Sokolowski A, Okano M, Russell ME, David JR, Satoskar AR (1999) Mice with STAT6-targeted gene disruption develop a Th1 response and control cutaneous Leishmaniasis. J Immunol 161:6180–6188

Starr R, Willson TA, Viney EM, Murray LJ, Rayner JR, Jenkins BJ, Gonda TJ, Alexander WS, Metcalf D, Nicola NA, Hilton DJ (1997) A family of cytokine inducible inhibitors of signalling. Nature 387:917–921

Starr R, Metcalf D, Elefanty AG, Brysha M, Willson TA, Nicola NA, Hilton DJ, Alexander WS (1998) Liver degeneration and lymphoid deficiencies in mice lacking suppressor of cytokine signaling-1. Proc Natl Acad Sci USA 95:14395–14399

Strous GJ, van Kerkhof P, Govers R, Ciechanover A, Schwartz AL (1996) The ubiquitin conjugation system is required for ligand-induced endocytosis and degradation of the growth hormone receptor. EMBO J 15:3806–3812

Suzuki K, Nakajima H, Kagami S, Suto A, Ikeda K, Hirose K, Hiwasa T, Takeda K, Saito Y, Akira S, Iwamoto I (2002) Proteolytic processing of Stat6 signaling in mast cells as a negative regulatory mechanism. J Exp Med 196:27–38

Takahashi Y, Carpino N, Cross JC, Torres M, Parganas E, Ihle JN (2003) SOCS3: an essential regulator of LIF receptor signaling in trophoblast giant cell differentiation. EMBO J 22:372–384

Takeda K, Tanaka T, Shi W, Matsumoto M, Minami M, Kashiwamura S, Nakanishi K, Yoshida N, Kishimoto T, Akira S (1996) Essential role of Stat6 in IL-4 signalling. Nature 380:627–630

Takeda K, Kamanaka K, Tanaka T, Kishimoto T, Akira S (1996) Impaired IL-13-mediated functions of macrophages in STAT6-deficient mice. J Immunol 157:3220–3222

Takeda K, Noguchi K, Shi W, Tanaka T, Matsumoto M, Yoshida N, Kishimoto T, Akira, S (1997) Targeted disruption of the mouse *Stat3* gene leads to early embryonic lethality. Proc Natl Acad Sci USA 94:3801–3804

Takeda K, Kaisho T, Yoshida N, Takeda J, Kishimoto T, Akira S (1998) Stat3 activation is responsible for IL-6-dependent T cell proliferation through preventing apoptosis: Generation and characterization of T cell-specific Stat3-deficient mice. J Immunol 161:4652–4660

Takeda K, Clausen B, Kaisho T, Tsujimura T, Terada N, Förster I, Akira S (1999) Enhanced Th1 activity and development of chronic enterocolitis in mice devoid of Stat3 in macrophages and neutrophils. Immunity 10:39–49

Takeda K, Kaisho T, Akira S (2003) Toll-like receptors. Annu Rev Immunol 21:335–376

Taniguchi T (1995) Cytokine signal through nonreceptor protein tyrosine kinases. Science 268:251–255

Teglund S, McKay C, Schuetz E, van Deursen JM, Stravopodis D, Wang D, Brown M, Bodner S, Grosveld G, Ihle JN (1998) Stat5a and Stat5b proteins have essential roles and nonessential, or redundant, roles in cytokine responses. Cell 93:841–850

Tekkanat KK, Maassab HF, Cho DS, Lai JJ, John A, Berlin A, Kaplan MH, Lukacs NW (2001) IL-13-induced airway hyperreactivity during respiratory syncytial virus infection is STAT6 dependent. J Immunol 166:3542–3548

Thierfelder WE, van Deursen JM, Yamamoto K, Tripp RA, Sarawar SR, Carson RT, Sangster MY, Vignali DA, Doherty PC, Grosveld G Ihle JN (1996) Requirement for Stat4 in interleukin-12 mediated response of natural killer and T cells. Nature 382:171–174

Thomis DC, Gurniak CB, Tivol E, Sharpe AH, Berg LJ (1995) Defects in B lymphocyte maturation and T lymphocyte activation in mice lacking Jak3. Science 270:794–797

Udy GB, Towers RP, Snell RG, Wilkins RJ, Park SH, Ram PA, Waxman DJ, Davey DW (1997) Requirement of Stat5b for sexual dimorphism of body growth rates and liver gene expression. Proc Natl Acad Sci USA 94:7239–7244

Wang D, Stravopodis D, Teglund S, Kitazawa J, Ihle JN (1996) Naturally occurring dominant negative variants of Stat5. Mol Cell Biol 16:6141–6148

Yasukawa H, Ohishi M, Mori H, Murakami M, Chinen T, Aki D, Hanada T, Takeda K, Akira S, Hoshijima M, Hirano T, Chien KR, Yoshimura A (2003) IL-6 induces an anti-inflammatory response in the absence of SOCS3 in macrophages. Nat Immunol 4:551–556

Yeh TC, Dondi E, Uze G, Pellegrini S (2000) A dual role for the kinase-like domain of the tyrosine kinase Tyk2 in interferon-alpha signaling. Proc Natl Acad Sci USA 97:8991–8996

Yoo JY, Huso DL, Nathans D, Desiderio S (2002) Specific ablation of Stat3beta distorts the pattern of Stat3-responsive gene expression and impairs recovery from endotoxic shock. Cell 108:331–344

Yoshikawa H, Matsubara K, Qian GS, Jackson P, Groopman JD, Manning JE, Harris CC, Herman JG (2001) SOCS-1, a negative regulator of the JAK/STAT pathway, is silenced by methylation in human hepatocellular carcinoma and shows growth-suppression activity. Nat Genet 28:29–35

Yoshimura A, Ohkubo T, Kiguchi T, Jenkins NA, Gilbert DJ, Copeland NG, Hara T, Miyajima A (1995) A novel cytokineinducible gene CIS encodes an SH2 containing protein that binds to tyrosinephosphorylated interleukin 3 and erythropoietin receptors. EMBO J 14:2816–2826

You M, Yu DH, Feng GS (1999) Shp-2 tyrosine phosphatase functions as a negative regulator of the interferonstimulated Jak/STAT pathway. Mol Cell Biol 19:2416–2424

Zabolotny JM, Bence-Hanulec KK, Stricker-Krongrad A, Haj F, Wang Y, Minokoshi Y, Kim YB, Elmquist JK, Tartaglia LA, Kahn BB, Neel BG (2002) PTP1B regulates leptin signal transduction in vivo. Dev Cell 2:489–495

Zhang JG, Farley A, Nicholson SE, Willson TA, Zugaro LM, Simpson RJ, Moritz RL, Cary D, Richardson R, Hausmann G, Kile BJ, Kent SB, Alexander WS, Metcalf D, Hilton DJ, Nicola NA, Baca M (1999) The conserved SOCS box motif in suppressors of cytokine signaling binds elongins B and C and may couple bound proteins to proteasomal degradation. Proc Natl Acad Sci USA 96:2071–2076

Members of the T-Cell Factor Family of DNA-Binding Proteins and Their Roles in Tumorigenesis

A. Hecht

Institut für Molekulare Medizin und Zellforschung, Universität Freiburg,
Stefan-Meier-Str. 17, 79104 Freiburg, Germany
andreas.hecht@mol-med.uni-freiburg.de

1	Introduction	124
2	Structure of TCF Proteins	126
2.1	Protein Domains	126
2.2	Domain Functions	128
2.2.1	DNA-Binding and Promoter Recognition	128
2.2.2	Context-Dependent Activation and Repression Domains	129
2.3	Protein–Protein Interactions	130
2.3.1	β-Catenin/Armadillo/Plakoglobin	130
2.3.2	Groucho-Related Genes/TLE	132
2.3.3	C-Terminal-Binding Protein	133
2.3.4	HBP1	133
2.3.5	Runx2	134
2.3.6	CREB-Binding Protein/p300	135
2.3.7	I-mfa	135
2.3.8	Smads	136
2.3.9	Microphthalmia-Associated Transcription Factor	136
2.3.10	ALY	137
2.3.11	PIASy	137
2.3.12	Notch-1	138
2.3.13	Nemo-Like Kinase	138
2.3.14	Other Factors	139
2.4	Post-translational Modifications	139
2.4.1	Phosphorylation	139
2.4.2	Acetylation	141
2.4.3	Sumoylation	143
3	TCF Genes: Structure and Regulation	145
4	TCF Function in Gene Regulation	150
4.1	Dual Repressor/Activator Function in Wnt/β-Catenin Signaling	150
4.2	Wnt/β-Catenin Independent Gene Activation	152
5	TCFs in Tumorigenesis	153
5.1	Deregulation of Wnt/β-Catenin Signaling and the Development of Cancer	153
5.2	Mutations in the TCF4 Gene and Its Deregulation in Cancer Cells	154
5.3	A Role for LEF1 in Tumor Progression?	155
5.4	TCF1 as a Tumor Suppressor Gene	156
6	Summary	157
	References	158

Abstract Uncontrolled Wnt/β-catenin signaling activity contributes to the etiology of a growing number of human cancers. Regardless of whether the underlying genetic lesions activate a proto-oncogene component or rather inactivate a tumor suppressor component of the Wnt/β-catenin signal transduction cascade, the critical outcome is the inappropriate activation of genes which control cellular proliferation. The culprits in these processes are heterodimeric transcription factor complexes formed by a transactivating component, β-catenin, and a DNA-binding subunit, a member of the T-cell factor (TCF) family of proteins. Often, TCFs are simply perceived as carriers for β-catenin, but actually they carry out more complex functions in gene regulation. In Wnt signaling processes they behave as binary switches, both repressing and activating target genes. Additionally, outside Wnt/β-catenin signaling, they act as architectural transcription factors. T-cell factor function and activity is modulated by numerous protein interaction partners in addition to β-catenin. Various post-translational modifications affect the subcellular distribution of TCFs and their interactions with DNA and proteins. A multitude of TCF isoforms with different gene regulatory potential is generated by dual promoter usage and alternative splicing. TCF genes are subject to mutation and deregulation in cancer cells and TCF family members possess qualities of tumor suppressors as well as tumor promoters. And yet, despite all the progress made, the full range of the functional diversity and the regulatory complexity of TCF proteins is just beginning to emerge.

Keywords Adenomatous polyposis coli · β-Catenin · Lymphoid enhancer factor · T-cell factor · Wnt signaling

1
Introduction

The development of multicellular organisms and the maintenance of tissue homeostasis in the adult require an exquisitely well-tuned balance between cell proliferation and differentiation. The principle decision as to whether a cell enters the division cycle or activates a particular differentiation program is controlled through the combinatorial action of a surprisingly small number of growth factor families and signal transduction pathways (Ghazi and VijayRaghavan 2000). Manipulation of these signaling systems by molecular genetic approaches and analyses of naturally occurring mutations, for example in tumor cells, revealed that both too much and too little signal activity can have deleterious consequences (for reviews see Bienz and Clevers 2000; Derynck et al. 2001; Polakis 2000). Disruption of a signal transduction cascade can halt development or lead to malformation and dysfunction of an organ, whereas excess signaling activity frequently leads to hyperproliferation and thus initiates a series of events which may culminate in tumor formation. Signaling by the family of Wnt growth factors provides an extremely well-suited example to illustrate the importance of carefully controlled signaling activity for organogenesis and the avoidance of disease (Bienz and Clevers 2000; Polakis 2000). Wnt growth factors can stimulate several different intracellular signal transduction pathways (Miller et al. 1999), the best understood of which is the canonical Wnt/β-catenin pathway. In this context, Wnt factors modulate the activity of the so-called destruction complex which in turn controls the activity of the multifunctional protein β-catenin (reviewed by Huelsken and Birchmeier 2001; Miller et al. 1999). As a component of the Wnt pathway β-catenin acts a transcriptional activator, but it also fulfills additional tasks outside of Wnt signaling (e.g., it acts as a molecular adaptor connecting components of the cytoskeleton to members of the cadherin family of cell–cell adhesion molecules) (Hecht and Kemler 2000; Huelsken and Birchmeier 2001; Miller et al. 1999). Within the destruction complex, the tumor suppressor proteins adenomatous polyposis coli (APC) and axin (or the related conductin/axin2/axil) facilitate the phosphorylation of β-catenin by the sequential action of casein kinase I (CK I) and glycogen synthase kinase 3β (GSK3β) which primes β-catenin for ubiquitination and subsequent degradation by the proteasome (Amit et al. 2002; Huelsken and Birchmeier 2001; Liu et al. 2002; Miller et al. 1999). Consequently, in the absence of a Wnt signal, β-catenin levels are kept low and Wnt target genes remain transcriptionally silent. Only upon Wnt-mediated inhibition of the destruction complex is hypophosphorylated β-catenin able to enter the nucleus where it interacts with members of the T-cell factor (TCF) family of DNA-binding proteins (Hecht and Kemler 2000). In mammals, there are four TCF family members: TCF1, LEF1 (lymphoid enhancer factor-1), TCF3 and TCF4. The interaction with one of these guides β-catenin to specific Wnt target genes and thereby allows it to enhance their transcription. Wnt/β-catenin target genes include cell cycle regulators such as cyclinD1 and the c-*Myc* proto-oncogene as well as several growth factors (He et al. 1998; Kratochwil et al. 2002; Levy et al. 2002; Shtutman et al. 1999; Tetsu and McCormick 1999). It is thus readily conceivable that deregulated, excessive Wnt signaling activity—as caused by mutational inactivation of the APC and axin tumor suppressors or by hyperactivation of β-catenin carrying oncogenic lesions—pave the way towards colorectal and numerous other tumors (Bienz and Clevers 2000; Polakis 2000). On the other hand, blocking Wnt signals by inactivation of the β-catenin or the TCF genes in the mouse genome by homologous recombination results in a wide spectrum of developmental defects (Huelsken and Birchmeier 2001).

As ultimate nuclear effectors of the Wnt/β-catenin signal transduction cascade TCF family members carry out important and crucial functions in both cell cycle control as well as cell fate decisions and differentiation. Although the initial findings which uncovered the interaction between β-catenin and TCF proteins suggested that they cooperate with β-catenin in the stimulatory branch of the Wnt/β-catenin pathway, many further studies established a more complicated and diverse picture of TCF functions (Hecht and Kemler 2000). In fact, as implied by their names, the two family members TCF1 and LEF1 also participate in Wnt- and β-catenin independent gene regulatory processes and this was known long before it was realized that TCFs interact with β-catenin (reviewed by van de Wetering et al.

2002a). Even in the context of Wnt signaling TCF proteins are not only required for the activation of Wnt target genes, but also for their transcriptional repression in the absence of Wnt stimuli (Hecht and Kemler 2000; Huelsken and Birchmeier 2001). T-cell factor family members act in a cell-context and promotor-specific manner in which they fulfill partially redundant but also specific tasks (Atcha et al. 2003; Galceran et al. 1999; Hecht and Stemmler 2003; Held et al. 2003). In addition, they are part of both positive and negative regulatory feedback loops which dampen or enhance Wnt signals (Gallagher et al. 2002; Hovanes et al. 2001; Roose et al. 1998). The complexity and the diversity of their functions in Wnt signaling and transcriptional control and the need to control these activities precisely are reflected at several organizational levels: the structure of TCF genes and the regulation of their expression, the structure and diversity of the transcripts and polypeptides derived from these genes, the large and growing number of proteins interacting with TCFs, and the numerous reversible post-translational modifications which influence TCF activity. This review summarizes the multifaceted features of TCF genes and proteins and describes the various and sometimes opposing roles of TCFs in human cancer.

2
Structure of TCF Proteins

2.1
Protein Domains

The principle structure of a prototypic TCF dedicated to Wnt/β-catenin signal transduction is depicted schematically in Fig. 1. The N-terminal part contains a short amino acid stretch that mediates the interaction with β-catenin and related proteins (see Sect. 2.3.1). The adjacent section harbors binding sites for transcriptional coregulators, and has additional modulating effects on transactivation (Brantjes et al. 2001; Cavallo et al. 1998; Gradl et al. 2002; Levanon et al. 1998; Pukrop et al. 2001; Roose et al. 1998). This part abuts the high-mobility-group (HMG) domain which is responsible for sequence-specific DNA binding of TCFs (Grosschedl et al. 1994). The C-terminal parts of TCFs which are highly variable in length, provide contacts for coactivator and corepressor proteins and again alter the transactivation properties of TCFs (Atcha et al. 2003; Brannon et al. 1999; Hecht and Stemmler 2003; Valenta et al. 2003). Although a similar overall organization of TCF proteins cannot be denied, TCFs possess a high degree of structural variability which arises from alternative splicing and the usage of different promoters (see Sect. 3). Not surprisingly, there are reports suggesting that different TCF family members and their isoforms fulfill specific, noninterchangeable functions in transcriptional control processes (Atcha et al. 2003; Hecht and Stemmler 2003; Held et al. 2003). A major challenge for the future will be to

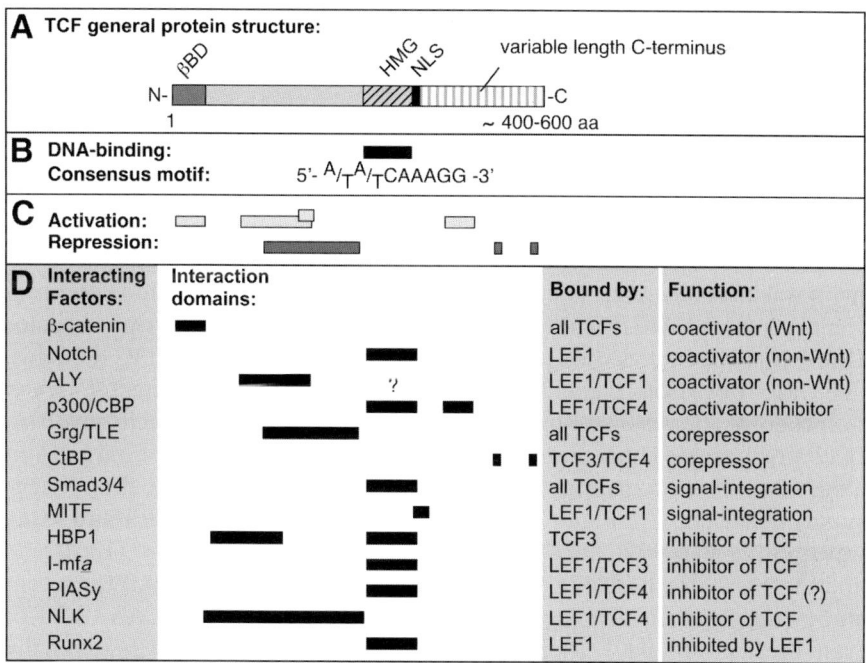

Fig. 1A–D General structure of TCF proteins. A TCF proteins are composed of an N-terminal domain of approximately 300 amino acids, an 80-amino acid HMG domain (*hatched bar*), an adjacent nuclear localization signal (*NLS*; *black box*), and a C-terminal tail which is highly variable in length (*striped bar*). The binding domain for β-catenin (*βBD*) is a short amino acid sequence motif close to the very N terminus. B The HMG domain of TCF mediates sequence-specific DNA binding to the consensus recognition motif indicated. C Transcription regulatory domains of TCFs. Several domains located at the very N terminus and within the central part of the N terminus contribute to context-dependent transcriptional activation by TCF proteins. The C-terminal tail regions of certain TCF isoforms (TCF1E and TCF4E) harbor an additional promoter-specific activation domain. Transcriptional repression mediated by TCFs depends mainly on a region within the N-terminal half of the proteins. TCF3 and TCF4 variants with long C-terminal tails contain additional repressing elements composed of two short peptide motifs at their C termini. Most of these functional domains correlate with interaction sites for coactivators and corepressors of TCFs. D Interaction partners of TCFs. Listed are known interaction partners of TCFs for which the binding sites in TCFs have been mapped and for which the functional importance of the interaction is known. The location of the interaction domains is indicated by *black bars*. In addition, the interacting TCF is denoted. The *question marks* indicate that it is not known whether p300/CBP interact with the HMG domain of vertebrate LEF1, and that there are conflicting results as to the inhibitory role of PIASy. For further information see text

explore fully the range of regulatory tasks carried out by the various TCF gene products and to assign these tasks to specific TCF proteins and their isoforms. The known functions of the TCF domains which have been identified and characterized to date are described below.

2.2
Domain Functions

2.2.1
DNA-Binding and Promoter Recognition

The TCF proteins belong to a group of DNA-binding transcription factors which as a signature motif contain a single copy of the so-called HMG domain (reviewed by Grosschedl et al. 1994) (Fig. 1). This 80-amino acid domain was first recognized to be present in the abundant nonhistone chromosomal proteins HMG-1 and -2 and the RNA polymerase I transcription factor UBF (Grosschedl et al. 1994). Factors with multiple copies of HMG domains, like HMG-1 or HMG-2, exhibit no sequence-specific DNA binding but appear to recognize altered DNA structures such as kinked DNA or cruciform DNA. In contrast, single HMG domain proteins like TCFs additionally bind DNA in a sequence-specific manner (Grosschedl et al. 1994). Originally, the consensus sequence for TCF-binding elements (TBE) recognized by LEF1 and TCF1 was a rather loosely defined 8-bp motif 5′–A/T A/T CAAAGG–3′ (Hurlstone and Clevers 2002). Based on a binding site selection assay with TCF1, a recent study suggested an extended recognition sequence 5′–AGATCAAAGGG–3′ (van Beest et al. 2000). It should be pointed out though, that functionally characterized TBEs can deviate considerably from the both the extended and the short consensus sequences (He et al. 1998; Hecht and Stemmler 2003; Hovanes et al. 2001; McKendry et al. 1997; Saito et al. 2002; Tetsu and McCormick 1999). This is a complication which one has to keep in mind when searching for TBEs in putative target genes. The occurrence of a consensus motif may be a good indicator, but this does not exclude the existence of additional nonconsensus motifs.

The three-dimensional structure of the HMG domain of LEF1 has been determined by nuclear magnetic resonance (Love et al. 1995). The overall solution structure of the LEF1 HMG domain compares well to other HMG domains. In all examples known, three α-helices, arranged in a twisted L-form, shape a concave surface which accommodates a highly bent DNA molecule. In the case of LEF1 the axis of the bound DNA forms an angle of 130° (Love et al. 1995). Based on this property, TCF proteins were initially termed architectural transcription factors (Grosschedl et al. 1994; see Sect. 4.2). The strong DNA bending induced by TCFs appears to result at least partly from the intercalation of hydrophobic amino acid side chains into the DNA helix (Love et al. 1995). As another unusual feature of the LEF1:DNA complex, the HMG domain makes contacts with a widened minor grove of the DNA whereas most other sequence-specific DNA-binding domains interact with the major grove (Love et al. 1995). In addition to the α-helices a stretch of basic residues located at the 3′-end of the HMG domain also contacts the bound DNA molecule. The E-isoforms of TCF proteins harbor another block of positively charged amino acids, the so-called CRARF domain in their extended C termini (Hovanes et al. 2000; Hurlstone and Clevers 2002). The CRARF domain, named after a characteristic stretch of cysteine and arginine residues, is evolutionary conserved and highly similar sequences are present in the mammalian TCF1E and TCF4E isoforms, *Drosophila* dTCF/Pangolin, the *Caenorhabditis elegans* TCF protein Pop1, as well as other transcription factors not related to TCFs (Atcha et al. 2003; Hurlstone and Clevers 2002). The presence or absence of the CRARF domain has been shown to influence the DNA-binding properties of TCF4E (Hecht and Stemmler 2003). Even though the HMG domains in TCF proteins are nearly identical in amino acid sequence (Hurlstone and Clevers 2002) and TCF proteins are thought to interact with the same DNA sequences, displaying only minor differences in affinity (Pukrop et al. 2001), this raises the possibility that regions outside of the HMG domain contribute to target gene recognition of TCF proteins.

The repertoire of promoters occupied by TCFs could be further extended by protein–protein interactions. The TCF4E isoform, for example, preferentially binds multimerized TCF-binding sites which are frequently found in Wnt target genes (Hecht and Stemmler 2003). In contrast, its affinity for a single consensus motif derived from the T-cell receptor (TCR) α-chain enhancer, a non-Wnt TCF target gene, was markedly reduced (Hecht and Stemmler 2003). The differences in DNA binding of TCF4 depended on the presence of the extended C-terminal E-tail in TCF4E. Although TCF proteins bind DNA as monomers these observations allow the speculation that homophilic interactions among TCF proteins occur and that these interactions affect their DNA-binding properties. This may have an impact on the ability of particular TCFs to form active transcription factor complexes when bound to tandem arrays of TBEs. In addition, TCF proteins are known to interact with other sequence-specific DNA-binding factors, for example the helix-loop-helix leucine-zipper protein microphthalmia-associated transcription factor (MITF) or the Smad proteins (Labbe et al. 2000; Nishita et al. 2000; Saito et al. 2002; Yasumoto et al. 2002). At least under experimental conditions the interaction with Smads allowed the formation of a TCF:promoter complex even upon mutational inactivation of the cognate TCF-binding sites in the *Xenopus* Twin promoter (Nishita et al. 2000). Hence, heterodimerization with Smads or MITF could enable TCFs to recognize certain promoters even if these lack a cognate TBE, a mechanism that could alter the spectrum of potential TCF target genes considerably.

2.2.2
Context-Dependent Activation and Repression Domains

DNA-bound TCF proteins are functionally neutral on their own. Depending on the particular circumstances, however, TCF proteins can exert both activating and repressing functions. Transcriptional activation and repression can be assigned to defined sequence elements within TCFs (Fig. 1). For Wnt/β-catenin regulated TCF target genes, the very N terminus is of critical importance (Hecht and Kemler 2000; Miller et al. 1999). At a subset of Wnt-regulated promoters, the CRARF-domain specifically present at the C termini of TCF1E and TCF4E additionally functions as obligatory but context-dependent transcriptional activation domain (Atcha et al. 2003; Hecht and Stemmler 2003). In mammals, the CRARF domain allows for selective activation of the β-catenin/TCF-responsive LEF1 promoter by TCF1E and the Wnt-responsive Cdx1 promoter by TCF4E (Atcha et al. 2003; Hecht and Stemmler 2003). A further contribution to promoter activation can be made by amino acid sequences encoded by an alternatively spliced exon upstream of the HMG domain (Gradl et al. 2002; Pukrop et al. 2001). The N-terminal half of LEF1 harbors yet another transactivation domain, which functions specifically in the context of the TCRα enhancer but not at Wnt-inducible promoters (Bruhn et al. 1997; Hsu et al. 1998). In the case of sequences derived from the alternatively spliced exon, the molecular basis for their activation potential is not known but the other activation domains coincide with interaction domains for well-known binding partners of TCFs: β-catenin, ALY and p300 (Fig. 1; see Sect. 2.3) suggesting that these domains function by mediating protein–protein interactions and cofactor recruitment. A similar scenario is probably true for the repression domains in TCFs which were mapped to a region upstream of the HMG domain and to two short sequence motifs at the C termini of the TCF3E and TCF4E isoforms (Fig. 1). While the former interacts with Grg/TLE corepressors (Brantjes et al. 2001; Cavallo et al. 1998; Levanon et al. 1998; Roose et al. 1998), the latter interact with CtBP (see Sects. 2.3.2 and 2.3.3) (Brannon et al. 1999; Valenta et al. 2003). Which activation or repression domains are used is probably determined by the exact promoter architecture and the molecular framework provided by the various transcription factors and their cofactors which assemble into higher order structures at a given regulatory element.

2.3
Protein–Protein Interactions

2.3.1
β-Catenin/Armadillo/Plakoglobin

The event which propelled TCF proteins into center stage was the discovery that LEF1 and XTCF3 interacted with the multifunctional protein β-catenin (Bienz and Clevers 2000; Hecht and Kemler 2000; Miller et al. 1999; Polakis 2000). The reason why this sparked such enormous interest is that the formation of this heterodimeric protein complex closed a gap in the signal transduction pathway of Wnt growth factors and provided an explanation for how the signal could be relayed into the nucleus. As further studies showed, β-catenin appears to function as a mobile transcriptional activation domain which is guided to its target genes by exploiting the sequence-specific DNA-binding capacity of TCFs (Hecht and Kemler 2000 and references therein).

By now it is clearly established that the two highly related proteins β-catenin, plakoglobin (also known as γ-catenin) and their *Drosophila* ortholog, armadillo, as well as more distant relatives in diverse species including polyps and nematodes all bind to an N-terminal domain in TCFs spanning approximately amino acids 10–55 (Fig. 1). The crystal structures of the β-catenin-binding domains (βBD) of TCF3 and TCF4 in complex with a β-catenin fragment have been solved (Graham et al. 2000, 2001; Poy et al. 2001). The structural hallmark of β-catenin, armadillo, and the other family members are 12 so-called armadillo (Arm) repeat motifs which make up the central three-quarters of the proteins (Peifer et al. 1994). In their sum, the Arm repeats form an elongated superhelix of helices with a positively charged groove (Huber et al. 1997) that provides the contact points for several critical interaction partners of β-catenin and its relatives. Among these interaction partners are TCFs, but also cadherins, APC and axin which appear to interact competitively with the same or closely overlapping sections of the Arm repeat region of β-catenin (Graham et al. 2000, 2001; Huber and Weis 2001; Poy et al. 2001; von Kries et al. 2000). Slight structural differences of βBDs from different TCFs appear to exist, but the overall structure of the βBD in TCFs is similar to the corresponding β-catenin-interacting domains in E-cadherin and APC (Graham et al. 2000, 2001; Huber and Weis 2001; Poy et al. 2001). In all cases, the Arm repeat region and the βBD are arranged in an antiparallel fashion. Main contacts with β-catenin are made by an extended, negatively charged part of the βBD in TCFs. Additional interactions are provided by residues located within an α-helical segment at the C-terminal end of the βBD. A certain conformational flexibility of the βBD in TCF4 indicates that TCF4 docks to β-catenin in a stepwise process in which the contacts are gradually improved (Graham et al. 2001). Although these structural analyses present a considerable advancement for the understanding not only of the β-catenin–TCF interactions, they also led to a certain frustration. The observed, but perhaps unexpected, similarity between βBDs in different interaction partners of β-catenin will make it very difficult, if not impossible, to design and develop small molecule inhibitors which specifically and selectively disrupt the β-catenin/TCF complex but not the complexes between β-catenin and cadherins or β-catenin and its destruction complex (Daniels et al. 2001).

2.3.2
Groucho-Related Genes/TLE

TCFs have the potential to act as binary switches which can impose either a repressed or an activated state on a promoter. In both cases, though, functional properties of the targeted promoter are determined by specific cofactors associated with TCFs. The primary binding partners employed in promoter inactivation are transcriptional corepressors which are encoded by a group of genes called 'Groucho-related genes' in the mouse or 'transducin-like-enhancer-of-split' in humans, short Grg/TLE (reviewed by Fisher and Caudy 1998). Grg/TLE proteins are required in many developmental programs where they act as DNA nonbinding repressor proteins which are brought to their target promoters through interactions with a wide range of diverse transcription factors (Fisher and Caudy 1998). Grg/TLE proteins possess a similar domain structure: an N-terminal dimerization and repression domain with glutamine and glycine/proline-rich stretches is followed by a more variable domain which contains the nuclear localization sequence and protein kinase target sites plus a serine/proline-rich sequence element. The C-terminal part is characterized by the presence of so-called WD40 repeats which are involved in protein–protein interactions. Alternative splicing generates short and long sequence variants of Grg/TLE proteins which can lack some of these domains and which therefore are endowed with different functional properties. In some cases, for example the Runt-domain transcriptional regulators AML1, AML2, and AML3, Grg/TLE proteins were shown to interact with a characteristic 'WRPW' or 'WRPY' amino acid motif present in their binding partners (Levanon et al. 1998). TCF proteins, however, do not contain such an amino acid sequence. Still, TCFs were shown to interact with Grg/TLE biochemically and genetically (Brantjes et al. 2001; Cavallo et al. 1998; Levanon et al. 1998; Roose et al. 1998). There is no selectivity among TCFs and Grg/TLE family members, i.e., all TCFs interact with all Grg/TLE proteins (Brantjes et al. 2001). The interactions are mediated by a region in TCFs located immediately upstream of the HMG domain and by the Gln-rich N-terminal domain in Grg/TLE (Levanon et al. 1998; Roose et al. 1998) (Fig. 1). In conjunction with TCFs only Grg/TLE variants retaining the Gly/Pro-rich domain function as transcriptional corepressors (Brantjes et al. 2001; Lepourcelet and Shivdasani 2002; Roose et al. 1998). Their activity probably requires an additional biochemical interaction with histone deacetylase (HDAC) (Chen et al. 1999). A role for Grg/TLE in repressing genes regulated by Wnt/β-catenin and TCF activity was firmly established by genetic studies in *Drosophila melanogaster* and *Xenopus laevis*, as well as biochemical analyses in mammalian tissue culture cells (Cavallo et al. 1998; Levanon et al. 1998; Roose et al. 1998). The abundance of Grg/TLE proteins, their extensive coexpression with various TCFs and the functional redundancy among Grg/TLE family members (Brantjes et al. 2001), leave it as an important but open question how Wnt signaling can rapidly and efficiently overcome Grg/TLE-based promoter repression and catalyze transcription initiation.

2.3.3
C-Terminal-Binding Protein

Whereas all TCFs are endowed with the ability to interact with the Grg/TLE corepressors, TCF3 and TCF4 splice variants with long C-terminal extensions additionally interact with the C-terminal-binding protein (CtBP) (Brannon et al. 1999; Valenta et al. 2003). CtBP is an evolutionarily conserved nuclear phosphoprotein (reviewed by Chinnadurai 2002). In mammals, two highly related genes encoding CtBP1 and CtBP2 exist and both proteins were shown to bind to TCF4 (Valenta et al. 2003). Together with Groucho and its relatives, CtBP represents the major transcriptional corepressor activity with important roles in developmental processes in vertebrates and invertebrates. The mechanism of repression is not yet clear but both HDAC-dependent and independent means could be used. CtBP binds to a characteristic amino acid sequence motif 'PLDLS' and closely related variants thereof. Two copies of this motif are present in TCF3 (PLSLT and PLSLV) and TCF4 (PLSLS and PLSLV) (Fig. 1). Functional assays in *Xenopus* embryos indicated that in conjunction with XTCF3 Grg/TLE repression functionally dominates over CtBP activity at least during early embryonic development (Brannon et al. 1999). Recent studies in human tissue culture cells showed, however, that CtBP mediates a HDAC-dependent repression of the axin 2/conductin gene (Valenta et al. 2003). TCF proteins hence may utilize the Grg/TLE and CtBP corepressors in a highly context-dependent—albeit sometimes redundant—manner.

2.3.4
HBP1

HBP1 has been described as an Rb-binding protein which inhibits cell cycle progression by reducing expression levels of certain growth promoting genes including N-*myc* and c-*myc* (Sampson et al. 2001). Like TCFs, HBP1 contains an HMG box DNA-binding domain but transcriptional repression by HBP1 can occur through DNA binding-dependent as well as -independent mechanisms (Sampson et al. 2001 and references therein). Overexpression of HBP1 abrogates the induction of Wnt target genes by β-catenin, Wnt, or the Wnt-surrogate LiCl. Activation of both the artificial reporter construct TOPFLASH and endogenous target genes, CyclinD1 and c-*myc*, was impaired. In the case of the Wnt target genes the inhibitory effect of HBP1 did not require the HBP1 HMG domain but a separate repression domain which mediates a physical interaction with TCF4 (Sampson et al.

2001). Two interaction domains in TCF4 were identified, one located between amino acid residues 53 and 171 and a second one which coincides with the HMG domain (Fig. 1). Interaction of the HBP1 repression domain and the TCF4 HMG domain appears to be more critical for the observed inhibitory effects of HBP1 because it prevents DNA binding of TCF4 and TCF4/β-catenin complexes. Mechanistically, therefore, HBP1 appears to function by interfering with promoter occupation by TCF complexes.

2.3.5
Runx2

Runx2, also known as Cbfa-1, AML-3 or PEBP2αA is a sequence-specific DNA-binding protein with important functions in the development of osteoblasts and chondrocytes (Kahler and Westendorf 2003 and references therein). Whereas the related factor Runx1 was already known to cooperate with LEF1 in the formation of a multicomponent transcription factor complex at the TCRα enhancer (Giese et al. 1995), a more recent report indicates that both Runx1 and Runx2 can physically interact with LEF1 through their respective DNA-binding domains, the Runt-homology domain and the HMG domain (Kahler and Westendorf 2003) (Fig. 1). Interestingly, though, a Runx1/LEF1 interaction was not observed in the earlier studies. Moreover, and in contrast to the functional synergism between LEF1 and Runx1 which was observed at the TCRα enhancer, the study by Kahler and Westendorf describes that LEF1 overexpression inhibits Runx2-dependent activation of the osteocalcin promoter (Kahler and Westendorf 2003). How this is achieved mechanistically is unclear. Aside from the HMG domain which mediates the Runx interaction, also the extreme LEF1 C terminus and the β-catenin-binding domain but not the Grg/TLE interaction domains are required for inhibition. Overexpression of β-catenin in the presence of LEF1 contributes to the repressive effect but in the absence of exogenous LEF1, β-catenin stimulates osteocalcin promoter activity (Kahler and Westendorf 2003). Various explanations for these somewhat contradictory observations are possible— ranging from an obstruction of Runx2 DNA binding by LEF1 or the β-catenin/LEF1 complex, to an artifact based on overexpression of the factors involved. For example, excess LEF1 or β-catenin might sequester a limiting cofactor from Runx2 and cause repression of the osteocalcin promoter without invoking physical interactions between LEF1 and Runx2. Alternatively, β-catenin and LEF1 may indirectly affect the osteocalcin promoter activity by upregulating a transcriptional repressor. While the described interaction between LEF1 and Runx2 is interesting because it opens up new possibilities of how the functional properties of TCF factors can be diversified, it has to be considered as very preliminary and requires a fair amount of follow up investigation.

2.3.6
CREB-Binding Protein/p300

The CREB-binding protein (CBP) and the related p300 are bimodal transcriptional coactivators which can modulate transcription processes by acetylating histones and nonhistone transcription factors or by bridging transcriptional activators to components of the basal transcription machinery (Goodman and Smolik 2000). In *D. melanogaster*, dCBP has been reported to be a negative regulator of Wnt signaling, whereas in vertebrates, p300 and CBP are essential coactivators of β-catenin which are necessary for the activation of Wnt targets (Hecht et al. 2000; Miyagishi et al. 2000; Sun et al. 2000; Takemaru and Moon 2000; Waltzer and Bienz 1998). As described in a later section (see Sect. 2.4.2), LEF1 is thought to be acetylated by dCBP which was claimed to interact with the LEF1 HMG domain (Waltzer and Bienz 1998) (Fig. 1). In *C. elegans*, CBP targets the TCF family member Pop1 but in this case acetylation is required for Pop1 nuclear retention and activity (Gay et al. 2003). In a third report, p300 was shown to interact with a region containing the CRARF domain at the C terminus of the TCF4E isoform but acetylation of TCF4E was not investigated (Hecht and Stemmler 2003). Whether other TCF family members also interact with CBP or p300 and which domains in TCFs are capable of interacting with p300/CBP is not yet fully investigated. Nonetheless, it appears that p300 and CBP are among the interaction partners of TCF proteins, but binding of TCFs to p300/CBP may be mediated by multiple TCF domains. Moreover, the functional consequences of these interactions appear to vary between species and perhaps also between TCFs.

2.3.7
I-mf*a*

The I-mf proteins are inhibitory factors which prevent nuclear localization and DNA binding of myogenic regulatory proteins (Snider et al. 2001 and references therein). Inhibition is mediated by a cystein-rich regulatory domain in I-mf proteins. In mice and humans, three I-mf isoforms (I-mf*a*, I-mf*b* and I-mf*c*) with divergent C-terminal sequences are known. Overexpression specifically of the I-mf*a* isoform prevented proper formation of the anterior–posterior body axis in *X. laevis* embryos, a phenotype typically obtained after inhibition of β-catenin/TCF activity at early developmental stages (Snider et al. 2001). Epistasis experiments established that the Wnt/β-catenin signaling pathway was blocked at the level of XTCF3. Biochemical studies subsequently revealed a direct physical interaction between I-mf and the HMG domain of XTCF3 (Fig. 1). As a consequence of this interaction DNA binding by TCF3 was impaired which explains the inhibition of the Wnt/β-catenin signaling cascade in the presence of I-mf. More recently, also mammalian LEF1 was found in a complex with I-mfa (Kusano and Raab-Traub 2002). However, the domain(s) in LEF1 interacting with I-mfa were not defined. A severe complication for assessing the functional importance of the I-mfa-TCF interaction is posed by the finding that I-mfa and the related human HIC bind to the GSK-binding domain of axin, thereby allowing for the increased accumulation of cytosolic β-catenin (Kusano and Raab-Traub 2002). How simultaneously raising the levels of β-catenin and blocking TCF activity is meaningfully incorporated into cellular signaling processes remains to be seen.

2.3.8
Smads

Certain developmental processes such as embryonic axis determination in *X. laevis* are controlled in a combinatorial fashion by Wnt growth factors and members of the transforming growth factor-β (TGFβ) superfamily (Nishita et al. 2000). Nuclear effectors of TGFβ signaling are the constituents of the Smad family of sequence-specific DNA-binding proteins (Derynck et al. 2001). In *X. laevis*, a common target of Wnt and TGFβ growth factors is the Twin gene promoter (Nishita et al. 2000). A rather elegant way of nature to integrate the activities of both signal cascades is built on the physical interaction of LEF1 with the Smad family members Smad3 and Smad4 (Labbe et al. 2000; Nishita et al. 2000). Two short sequence motifs within the LEF1 HMG domain confer the interaction with the Smad MH1 and MH2 domains, respectively (Labbe et al. 2000) (Fig. 1). The LEF1–Smad interaction generates heterodimeric transcription factor complexes with combined DNA sequence specificities which explains why not all TCF target genes are simultaneously TGFβ- and Wnt-responsive (Nishita et al. 2000). Thus, certain Wnt target gene promoters lacking Smad-binding elements are TGFβ insensitive while other regulatory elements with binding sites for TCFs and Smad can be activated by both signaling cascades.

2.3.9
Microphthalmia-Associated Transcription Factor

The gene coding for MITF is upregulated by Wnt/β-catenin signaling during differentiation of neural crest cells (Dorsky et al. 2000; Takeda et al. 2000). The MITF protein is a transcriptional activator with a combined basic helix–loop–helix/leucine zipper (bHLH/LZ) DNA-binding domain (reviewed by Goding 2000). MITF and LEF1 synergize to activate the promoter of melanocyte-specific dopachrome tautomerase gene (Yasumoto et al. 2002). A potential positive autoregulatory loop has also been proposed whereby MITF together with LEF1 would activate its own gene (Saito et al. 2002). Underlying the functional cooperation between MITF and LEF1 appears to be a physical interaction of the proteins. The combined results of glutathione S-transferase (GST) pull-down assays in vitro and functional tests of LEF1 deletion mutants in vivo indicate that a region C-terminal to the LEF1 HMG domain is required for the MITF/LEF1 interaction but it cannot be ruled out that other domains in LEF1 also play a role (Yasumoto et al. 2002) (Fig. 1). Although the functional cooperation with MITF seems to be a unique feature of LEF1, and TCF1 behaves differently in activation assays, we have recently observed that TCF1 also interacts with MITF in coimmunoprecipitation experiments (K. Bruser and A. Hecht, unpublished results). Because of the distinct functional properties exhibited by LEF1 and TCF1 when coexpressed with MITF this raises the possibility that the balance between LEF1 and TCF1 levels influence the activities of certain promoters that are coregulated by MITF and TCF proteins.

2.3.10
ALY

Amino acids 99–217 of LEF1 constitute a transcriptional activation domain which functions specifically in the context of the TCR α enhancer (Grosschedl et al. 1994) (Fig. 1). A yeast two-hybrid screen carried out by Bruhn and colleagues identified ALY (Ally of AML-1 and LEF1) as a novel LEF1-interacting factor that bound the context-dependent transactivation domain and to a minor extent the HMG domain (Bruhn et al. 1997). A weak interaction between ALY and TCF1 has also been reported. Whether ALY also interacts with other TCFs is not known, but this appears unlikely because TCFs are less similar to each other in the ALY-binding region than in other regions. ALY functions as a coactivator of LEF1 at the TCR α enhancer, but it neither synergizes with β-catenin nor functions at a Wnt-responsive LEF1 target promoter (Bruhn et al. 1997; Hsu et al. 1998). Therefore, the cooperation of LEF1 and ALY appears to be restricted to non-Wnt controlled transcriptional processes. The simultaneous interaction of ALY with LEF1 and AML, another transcription factor binding to the TCR α enhancer, and the ability of ALY to stimulate DNA-binding of LEF1 and AML suggest that ALY increases TCR α enhancer activity by facilitating the formation of higher order transcription factor complexes (see Sect. 4.2).

2.3.11
PIASy

PIASy belongs to a group of SUMO-E3 ligases. It was initially identified as an interaction partner for LEF1, but more recently also for TCF4 (Sachdev et al. 2001; Yamamoto et al. 2003). The binding domain for PIASy in LEF1 colocalizes with the HMG domain whereas additional sequences immediately upstream of the HMG domain appear to be required for the PIASy/TCF4 interaction or to allow PIASy to exert its effects on TCF4 (Fig. 1). As described below (see Sect. 2.4.3), the interaction with PIASy and sumoylation change both the subnuclear distribution and the transactivation properties of LEF1 and TCF4.

2.3.12
Notch-1

Notch proteins are cell surface receptors involved in juxtacrine signaling processes (Ross and Kadesch 2001, and references therein). Upon ligand stimulation Notch proteins are proteolytically cleaved and their intracellular domains (ICD) translocate into the nucleus where they act as transcriptional activation domain for the CSL (RBP-J/CBF-1/LAG-1/Supressor-of-hairless) DNA-binding proteins. Overexpression of Notch-1 and Notch-2 ICDs stimulates LEF-1 dependent activation of specific reporter gene constructs carrying synthetic or natural TCF-responsive promoters (Ross and Kadesch 2001). Results from mammalian two-hybrid analyses and GST pull-down assays demonstrated a weak interaction between the transactivation domain of the Notch-1 ICD and the HMG domain of LEF1 (Fig. 1). As the affinity of the Notch-1 ICD for CSL appears to be considerably higher, the ICD/CSL interaction therefore may prevail over ICD/LEF1 complex formation under competitive conditions (Ross and Kadesch 2001). At higher levels of ICD availability the ICD–LEF1 complex may form additionally and allow the activation of a wider range of target genes. Although genetic data from *D. melanogaster* back up the results obtained in mammalian tissue culture cells and in vitro (Lawrence et al. 2001), currently it is not clear whether the interaction between Notch-1 and LEF1 is of physiological importance.

2.3.13
Nemo-Like Kinase

The Nemo-like kinase (NLK) is a Ser/Thr-specific kinase related to the *C. elegans* Lit-1 gene product (Ishitani et al. 1999; Rocheleau et al. 1999). NLK and Lit-1 are probably downstream components of a mitogen-activated protein kinase cascade. Both enzymes are negative regulators of TCF activity in the nematode and in vertebrates, albeit there are some differences concerning the molecular mechanisms whereby Lit-1 and NLK act (Ishitani et al. 1999, 2003; Meneghini et al. 1999; Rocheleau et al. 1999) (see Sect. 2.4.1). NLK was shown biochemically to directly interact with a broad domain of LEF1 located between the β-catenin-binding domain and the HMG domain (Ishitani et al. 2003). As described below, Lit-1 and NLK phosphorylate TCFs and change their subcellular localization and their DNA-binding capacities.

2.3.14
Other Factors

Aside from the factors mentioned so far several other proteins have been shown to somehow form complexes with TCFs. Among these are the multi-HMG domain factor UBF2, HDAC1, CK Iε, CK II, GSK3β, and the estrogen receptors (Billin et al. 2000; El-Tanani et al. 2001; Gradl et al. 2002; Grueneberg et al. 2003; Lee et al. 2001). Currently it is not known, though, whether the observed interactions are direct or whether they are indirectly mediated by bridging factors. Moreover, the TCF domains required for the association with these factors have not been determined. Despite these uncertainties, some of these proteins are of critical importance for the biological activities of TCFs as will be described below.

2.4
Post-translational Modifications

2.4.1
Phosphorylation

Protein phosphorylation is one of the most widespread post-translational modifications and—not surprisingly—several different kinases are known to target TCFs thereby changing their biochemical properties and consequently their biological functions. Studies in invertebrates and amphibia are pioneering with respect to the identification of these kinases and their effects on TCF proteins. In early embryonic development of *C. elegans* the TCF family member Pop1 is involved in the specification of different cell lineages by generating anterior–posterior asymmetries between two sister cells arising from a common precursor blastomere (Rocheleau et al. 1999). The generation of these asymmetries requires the suppression of Pop1 activity in the anterior sister cell which is achieved by Lit-1, a mitogen-activated protein kinase-related to the Nemo and NLKs in *D. melanogaster* and vertebrates (Meneghini et al. 1999; Rocheleau et al. 1999). At the molecular level it has been shown that Lit-1 phosphorylates Pop1 and that the action of Lit-1 correlates with a relocation of Pop-1 from the nucleus to the cytoplasm (Rocheleau et al. 1999). In vertebrates, NLK phosphorylates the TCF family members mLEF1, XTCF3 and hTCF4 (Ishitani et al. 1999). In the case of hTCF4, phosphorylation by NLK reduces its DNA-binding affinity (Ishitani et al. 1999). Consequently, the activation of target genes in tissue culture cells and in *Xenopus* embryos is impaired (Ishitani et al. 1999, 2003). The amino acid residues which are modified by NLK have been mapped to Thr_{155} and Ser_{166} in mLEF1, and to the corresponding residues Thr_{178} and Thr_{189} in hTCF4 (Ishitani et al. 2003) (Fig. 2). Substitution of these residues by alanine or valine allow LEF1 and TCF4 to evade the inhibitory influence

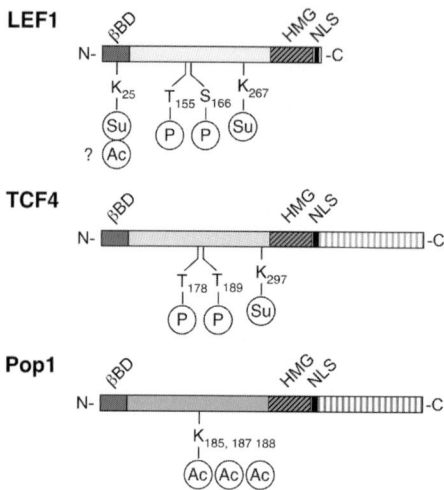

Fig. 2 Post-translational modifications of TCFs. Schematically shown are LEF1, TCF4 (E-isoform) and Pop1, a TCF family member in *C. elegans*. Amino acid residues (in *one-letter code*) that are modified post-translationally are shown and their positions indicated. Acetylation of K_{25} in LEF1 has been reported in *D. melanogaster*, but these results have not yet been confirmed in other organisms. Therefore, acetylation of LEF1-K_{25} is labeled with a *question mark*. *Ac*, acetylation; *Su*, Sumoylation; *P*, Phosphorylation; *βBD*, β-catenin-binding domain; *HMG*, HMG domain; *NLS*, nuclear localization signal

of NLK. Despite the overall high degree of similarity between the inhibition of TCF proteins in *C. elegans* and in vertebrate cells there are some differences, though. Whereas in *C. elegans* Lit-1 interacts with Wrm-1 (a relative of β-catenin), it turned out that NLK interacts directly with LEF1 and TCF4 and only indirectly with β-catenin upon complex formation between TCFs and β-catenin (Ishitani et al. 2003; Rocheleau et al. 1999). Thus, while the players appear to be conserved, the details of their mode of action seem to have changed.

Aside from NLK there are three additional kinases which have been implicated in regulating TCF activity. As a component of the β-catenin destruction complex GSK3β plays a central role in the transmission of Wnt signals. In extracts from *Xenopus* embryos GSK3β was additionally found to be associated with XTCF3 and recombinant GSK3β phosphorylates XTCF3 in vitro (Lee et al. 2001). This phosphorylation results in decreased complexation of XTCF3 and β-catenin. Because TCFs compete with APC for binding to β-catenin, a reduced binding capacity of XTCF3 would render β-catenin more readily susceptible to being shuttled into the destruction complex by APC. Interestingly, the effect of GSK3β-mediated phosphorylation of XTCF3 may be counteracted by another kinase, namely CK Iε. CK Iε, which like GSK3β has multiple functions in Wnt signal transduction and other cellular processes, also associates with XTCF3 (Lee et al. 2001). In contrast to GSK3β, however, phosphorylation by CK Iϵ furthers complex formation of β-catenin and XTCF3 and thereby protects β-catenin from degradation. Which sites in XTCF3 are targeted by GSK3β and CK Iϵ is not known but the two kinases do not appear to compete for the same binding sites or target residues (Lee et al. 2001). The observation that phosphorylation of XTCF3 interferes with its binding to β-catenin has also been made by Pukrop and colleagues, although their study did not identify the kinase responsible (Pukrop et al. 2001). Interestingly, these authors showed that phosphatase treatment of TCFs, while enhancing their association with β-catenin, simultaneously reduced the affinity of mLEF1, XTCF3 and XTCF4 for their DNA recognition motives. This suggests that the DNA-binding properties of TCF proteins are also positively regulated by phosphorylation and not just subject to negative control as exemplified by the effect of NLK. Experiments in vitro suggest that CK II could be the relevant enzyme with targets TCFs and stimulates their DNA-binding activities but this awaits further confirmation (Pukrop et al. 2001). Still, based on these findings, an unusual regulatory principle by opposing kinase activities emerges. Whether or not there are still more pairs of kinases which antagonistically regulate TCF functions and which phosphatases are involved in these processes remains to be seen. Phosphorylation nonetheless appears to be an important and versatile regulatory tool which serves to control major functional properties of TCF proteins: their intracellular distribution, their DNA-binding properties and the interaction with their transcriptional coactivator β-catenin.

2.4.2
Acetylation

A link between protein modification through acetylation and transcriptional control was first revealed by the observation that transcriptionally active chromosomal regions are enriched for hyperacetylated histones. Today, numerous other nonhistone proteins are known to be acetylated and a variety of protein functions such as protein-stability, protein-protein interactions, protein-DNA binding and subcellular localization can be influenced by acetylation (reviewed by Kouzarides 2000). Two groups of antagonistically acting enzymatic activities control protein acetylation. Acetylases transfer acetyl groups from acetyl-coenzyme A to the ϵ-NH$_2$ groups of lysine side chains, whereas deacetylases catalyze the removal of the acetyl moiety (Kouzarides 2000).

Acetylation has been linked to various aspects of TCF function albeit at present most of the information relating to acetylation of TCFs has been obtained from studies with invertebrates. There are some observations which indicate that perhaps not all findings can be extended to the mammalian system but currently it cannot be strictly ruled out that acetylation and acetylases play a similar role in mammalian cells. Therefore, the two examples for acetylation of TCF family members in invertebrates are listed below. The first case was reported in D. *melanogaster*, where the acetylase dCBP was found to interact with the HMG box of Pangolin/dTCF (Waltzer and Bienz 1998). Genetic analyses suggested an inhibitory role of dCBP in Wnt signaling processes. A potential molecular explanation for this effect was provided by the finding that dCBP acetylates lysine K_{25} in the β-catenin-binding site of dTCF (Fig. 2), thereby weakening the interaction between β-catenin/armadillo and dTCF (Waltzer and Bienz 1998). To date, however, there is no evidence for a similar function of CBP or p300 in vertebrates. Rather, both CBP and p300 serve as important coactivators of β-catenin and TCFs in Wnt signaling (Hecht and Stemmler 2003; Hecht et al. 2000; Miyagishi et al. 2000; Sun et al. 2000; Takemaru and Moon 2000), and functional analyses of a K_{25} mutant form of LEF1 in mammalian tissue culture or *Xenopus* embryogenesis did not reproduce the hyperactivity phenotype seen in the fly (A. Hecht and K. Vleminckx, unpublished results).

The second, more recent report about acetylation of a TCF protein is from C. *elegans*. Here, the TCF family member Pop1 is acetylated by CBP/p300 at residues K_{185}, K_{187} and K_{188}, but apparently not at the N-terminus (Gay et al. 2003) (Fig. 2). This again is in conflict with the findings in the *Drosophila* system. In addition, acetylation of Pop1 has a positive regulatory function because it is required for its nuclear localization and retention (Gay et al. 2003). Unfortunately, it is doubtful whether the findings of Gay and coauthors can be extrapolated to vertebrate TCFs, because the main Pop1 acetylation target K_{185} is not conserved in the mammalian orthologs of Pop1.

While it thus remains largely unclear to what extent TCF proteins themselves are targets for acetylation in vertebrates, TCFs nonetheless utilize acetylases and deacetylases to execute their regulatory functions in mammalian cells. A novel activation domain present specifically in the 'E' isoforms of TCF1 and TCF4 has been shown to interact with p300 in vitro and in cellular extracts (Hecht and Stemmler 2003). It is likely that this interaction is important for the context-dependent activation of certain promoters by TCF1 and TCF4. The mechanism underlying p300 activity in this context is unknown but it could involve acetylation of TCFs. This is suggested by the observation that in TCF4 each of the two binding sites for the transcriptional repressor CtBP is flanked by a lysine residue (Chinnadurai 2002). In the case of E1A, which also binds both p300 and CtBP, a similarly configured lysine near the CtBP-binding site is reversibly acetylated by p300/CBP and this acetylation modulates the interaction between E1A and CtBP. By extrapolation, acetylation of the C terminus of certain TCF splice variants might be required for promoter activation by weakening the interaction with the CtBP repressor proteins.

In addition to this stimulatory interaction between TCF4 and the p300 acetylase, TCFs can also interact with a deacetylase to repress transcription. LEF1 has been found in a complex with HDAC1 (Billin et al. 2000). The interaction may occur either directly or indirectly through Groucho/TLE repressors which are known to bind HDAC1. Inhibition of HDAC1 by Trichostatin enhances Wnt/β-catenin target gene expression and alters the acetylation state near Wnt/β-catenin regulated promoters (Billin et al. 2000). This indicates that the association of LEF1 and HDAC1 contributes to the transcriptional repression of target genes in the absence of a Wnt stimulus through a chromatin-based mechanism. However, chromatin may not be the only target of LEF1-associated HDAC1. In a recent report, HDAC1 overexpression counteracted LEF1-dependent nuclear accumulation and retention of β-catenin (Henderson et al. 2002). This is clearly reminiscent of the role of acetylation/deacetylation of Pop1 in subcellular localization of *C. elegans*. Unfortunately, however, it was not analyzed whether it was only β-catenin which failed to accumulate in the nucleus in the presence of excess HDAC1, or whether the subcellular distribution of LEF1 was affected as well. This would be expected if the Pop1 paradigm was applicable to LEF1 and mammalian cells.

2.4.3
Sumoylation

The reversible addition of SUMO (small ubiquitin-related modifier) represents another post-translational modification of proteins which controls a variety of cellular processes (reviewed by Muller et al. 2001). SUMO is covalently linked to the ϵ-amino group of lysine residues by a so-called SUMO E3 ligase (Muller et al. 2001). In contrast to ubiquitiylation, sumoylation does not earmark its substrates for degradation by the proteasome, and SUMO appears to be added in single units per lysine rather than multi-SUMO chains. LEF1 was the first TCF family member which was found to be sumoylated (Sachdev et al. 2001). Subsequently it was shown that PIASy acts as an E3 ligase for LEF1 and TCF4 in reconstituted sumoylation systems in vitro (Sachdev et al. 2001; Yamamoto et al. 2003). A search for consensus sumoylation target motifs ΨKXD/E (Ψ: hydrophobic residue, X: any amino acid) within the LEF1 and TCF4 amino acid sequences turned up two potential target sites in LEF1 (K_{25} and K_{267}) and six potential sites in TCF4 ($K_{22, 297, 317, 318, 407, 516}$) (Sachdev et al. 2001; Yamamoto et al. 2003) (Fig. 2). Mutation of the two lysines in LEF1 indeed abolished sumoylation, whereas in TCF4 only K_{297} could be shown to be sumoylated. As in other cases, sumoylation of TCFs is reversible and is catalyzed by a SUMO-specific protease, SENP2/axam. Axam has previously been linked to Wnt signaling because of its interaction with axin and its role in the control of β-catenin degradation (Rui et al. 2002).

Coexpression of PIASy with LEF1 or TCF4 affects the TCFs in several ways (Sachdev et al. 2001; Yamamoto et al. 2003). First, in agreement with its role as E3 ligase PIASy enhances sumoylation of LEF1 and TCF4. Second, upon overexpression of PIASy, the diffuse nuclear distribution of LEF1 and TCF4 is altered and both proteins colocalize with SUMO and PIASy to the so-called promyelocytic leukemia (PML) bodies. The precise function of PML bodies is unknown but recruitment of diverse protein factors to PML bodies and their modification by SUMO conjugation may either induce or repress the activity of transcription factors and other regulatory proteins. Finally, the transcriptional activation capacity of LEF1 and TCF4 changed in the presence of PIASy. However, while one laboratory reported that PIASy reduced both Wnt/β-catenin and non-Wnt related activities of LEF1, another group reported that PIASy synergized with β-catenin to stimulate LEF1 and TCF4-mediated target gene expression (Sachdev et al. 2001; Yamamoto et al. 2003). Interestingly, neither the changes in transcriptional activation/repression nor the recruitment to PML bodies seem to require sumoylation, because LEF1 and TCF4 mutant proteins lacking the sumoylated lysine residues behaved like their wild-type counterparts. Consistent with this, sumoylation had no influence on DNA binding of LEF1 and TCF4 and their interaction with β-catenin remained unaffected. The potential sumoylation of LEF1 K_{25} is nonetheless of particular interest because K_{25} has been suggested to be subject to inhibitory acetylation (see Sect. 2.4.2). These two mutually exclusive modifications could be used as a regulatory tool to impose alternative functional states upon acetylated versus sumoylated LEF1.

How PIASy affects the transcriptional activities of LEF1 and TCF4 and how their localization to PML bodies relates to this is another open question. The predominant localization of PIASy to nuclear bodies implies that it exerts its effect on TCFs in this subnuclear compartment. Indeed, Sachdev and coauthors suggested that the sequestration of LEF1 to PML bodies accounts for the repressive effect of PIASy coexpression (Sachdev et al. 2001). However, because of the contradicting results of Yamamoto and coworkers such a simple scenario seems to be an unlikely explanation. Formally it is possible that PIASy displaces either coactivators or corepressors from TCF proteins regardless of their subnuclear distribution. Perhaps even minor differences in the experimental set up change the way in which TCF proteins respond to PIASy, in particular under conditions of overexpression. The physiological role of the interaction between TCF proteins and PIASy as well as the significance of their sumoylation and of their altered subnuclear distribution therefore still need to be determined.

3
TCF Genes: Structure and Regulation

TCFs constitute a highly conserved family of proteins and the corresponding genes are represented in *D. melanogaster*, *Hydra magnipapillata*, *C. elegans* and all vertebrates (Hobmayer et al. 2000; Hurlstone and Clevers 2002; Miller et al. 1999). Due to the completion of the human genome sequencing projects the entire or nearly entire sequences of the LEF1 (human Chr. 4), TCF1 (human Chr. 5), TCF3 (human Chr. 2) and TCF4 (human Chr. 17) genes can be found in the databases. Generally, these genes are rather large, covering 100 kb and more of sequence (Duval et al. 2000; Hovanes et al. 2000; Van de Wetering et al. 1996) (Fig. 3). Exon/intron boundaries, as far as they have been identified and confirmed, appear to be conserved between family members (Hovanes et al. 2000). The minimum number of exons known is 12 in case of the human TCF3 gene, but this is probably an underestimate. Extensive analyses of transcripts from other TCF genes has revealed the presence of up to at least 17 exons which undergo extensive alternative splicing (Duval et al. 2000; Hovanes et al. 2000; Mayer et al. 1995; Shiina et al. 2003; Van de Wetering et al. 1996). Because of this, the cDNA species which were initially reported contain only a subset of exons, and exons which were not represented in the first cDNA isolates naturally could not be assigned in the genomic databases. Careful analyses and laborious cloning efforts with cDNA from different tissues and cell types are needed to reveal the full spectrum of alternatively spliced TCF transcripts and their coding capacity. Currently available data indicate, however, that not all theoretically possible exon combinations exist. In case of the TCF4 gene only exons 4, 13, 14, 15 and 16 appear to be alternatively spliced (Duval et al. 2000; Shiina et al. 2003) (Fig. 4). Additional sequence variation is generated by the use of alternative splice acceptor sequences in exons 7, 9, 16 and 17. Similar observations have been made for the TCF1 and LEF1 genes (Hovanes et al. 2000; Van de Wetering et al. 1996). These variations mainly affect the amino acid sequences located downstream of the HMG domain. Roughly, long (E-), medium- and short (B-) isoforms can be distinguished which differ with respect to the length of the C-terminal domains (Fig. 4). The inclusion or omission of certain amino acid motifs, such as the CRARF domain specifically present in the TCF1E and TCF4E isoforms or the CtBP-binding motifs found only in the TCF3E- and TCF4E isoforms confer distinct functional properties to these variants (see above).

However, the N-termini of TCFs are also subject to sequence variability. In *Xenopus*, an alternatively spliced exon has been linked to different gene regulatory properties of LEF1 and TCF4 variants which either carry or lack this sequence element (Gradl et al. 2002; Pukrop et al. 2001). Recent reports also describe TCF forms consisting of only N-terminal portions excluding the HMG domain and the NLS (Kennell et al. 2003). Finally, not even the

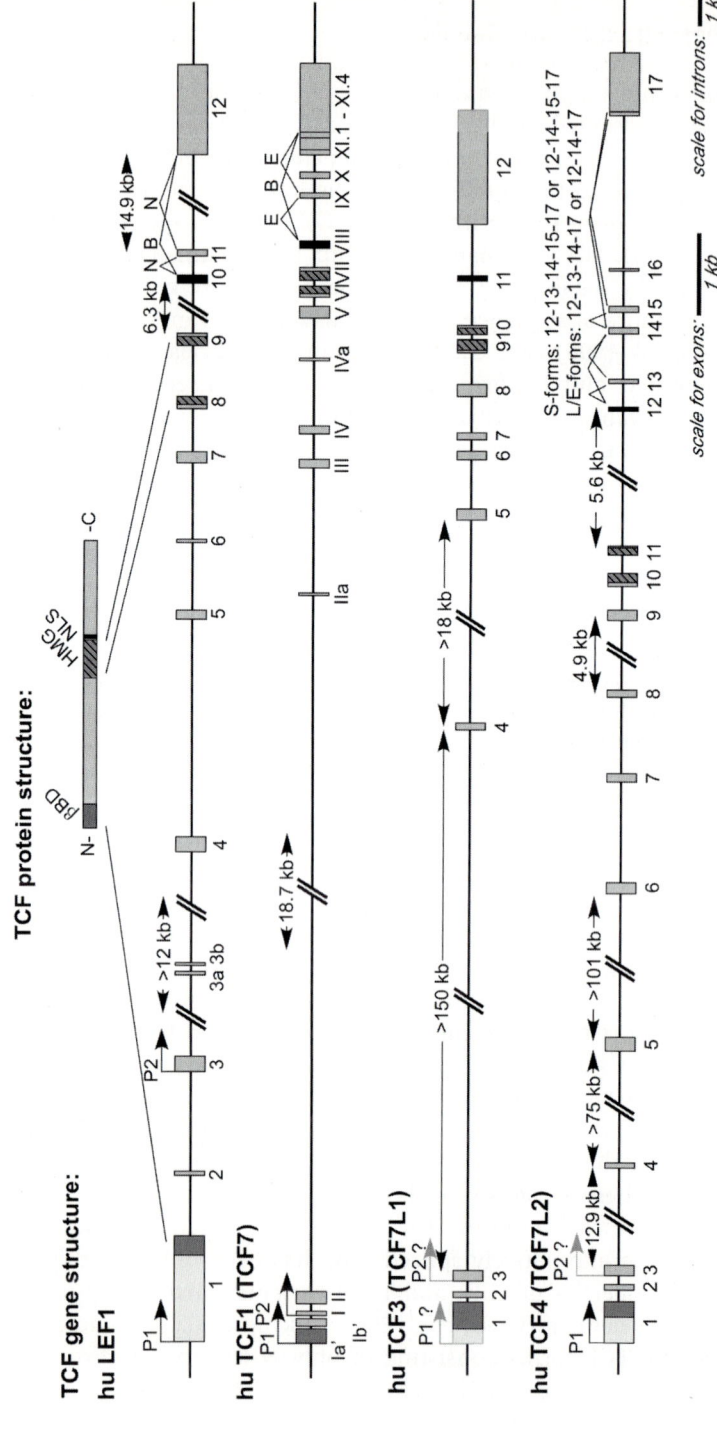

βBD is an obligatory component of TCFs. Both, in the case of the TCF1 and the LEF1 genes, two separate promoters P1 and P2 located upstream of exon 1 and upstream of exon 3 have been described (Hovanes et al. 2000; Roose et al. 1999). Transcripts initiated at promoter 2 encode polypeptides lacking the βBD. Interestingly, the LEF1 and the TCF1 genes are targets of Wnt/β-catenin signaling. Importantly, in the case of the LEF1 gene, the P1 and the P2 promoters are differentially regulated by Wnt. P1 promoter activity is induced, whereas at the P2 promoter seems unresponsive. Accordingly, Wnt signaling initiates a positive feed-forward loop at the LEF1 gene whereas stimulation of the TCF1 gene by Wnt signaling leads mainly to the production of dominant negative TCF1 isoforms (Roose et al. 1999). Taken together, dual promoter usage, alternative splicing and splice acceptor utilization generate a panoply of TCF isoforms with considerable variability as to the inclusion or omission of protein domains. This will certainly have an impact on protein–protein interactions and patterns of post-translational modifications (see above). Very likely, the regulatory potential of TCFs is much greater than currently known, and control of TCF activity is more complex than anticipated.

Aside from the Wnt responsiveness of the LEF1 and TCF1 genes, respectively, little is known about the mechanisms controlling TCF gene expression. One notable exception is the induction of LEF1 expression by bone morphogenetic protein 4 (BMP4) in the presumptive dental mesenchyme of

Fig. 3 Structure of human genes coding for LEF1, TCF1, TCF3 and TCF4. The TCF1 gene is also called TCF7, and the genes for TCF3 and TCF4 are deposited as TCF7-like 1 (*TCF7L1*) and TCF7-like 2 (*TCF7L2*) in the databases. Exons are depicted by *gray boxes*. Numbering of exons is according to Hovanes et al. 1997, Duval et al. 2000 and the Ensembl genomic database (http://www.ensembl.org). For comparison, the protein structure of TCFs is shown at the *top* of the figure. The same coloring and patterns have been used for the βBD, the HMG domain, the NLS and the corresponding exons coding for them. Different scales were used for exons and introns as indicated. The existence of two different promoters P1 and P2 was demonstrated for the TCF1 and LEF1 genes. By analogy, two promoters are thought to be present at the TCF3 and TCF4 genes, but experimental evidence for the TCF4 promoter P2 and the TCF3 promoters P1 and P2 is missing (as implied by the *question marks* and *gray coloring*). To indicate that the 5′-end and the 3′-end of the TCF3 gene are unknown, gray border lines were used for exons 1 and 12. Certain patterns of alternative splicing are shown for the LEF1, TCF1 and TCF4 genes. The isoforms generated are LEF1-N, LEF1-B, TCF1-B, TCF1-E, TCF4-S, TCF4-L, also known as TCF4E. Alternative splice acceptors in exons XI of the TCF1 gene and exon 17 of the TCF4 gene are indicated by *vertical lines*. (Data were compiled from Hovanes et al. 1997 and 2000, Duval et al. 2000, and Shiina et al. 2003. Additional information can be found at http://www.ensembl.org. The Ensembl gene numbers are: LEF1, ENSG00000138795; TCF1, ENSG00000081059; TCF3, ENSG00000152284; and TCF4, ENSG00000148737)

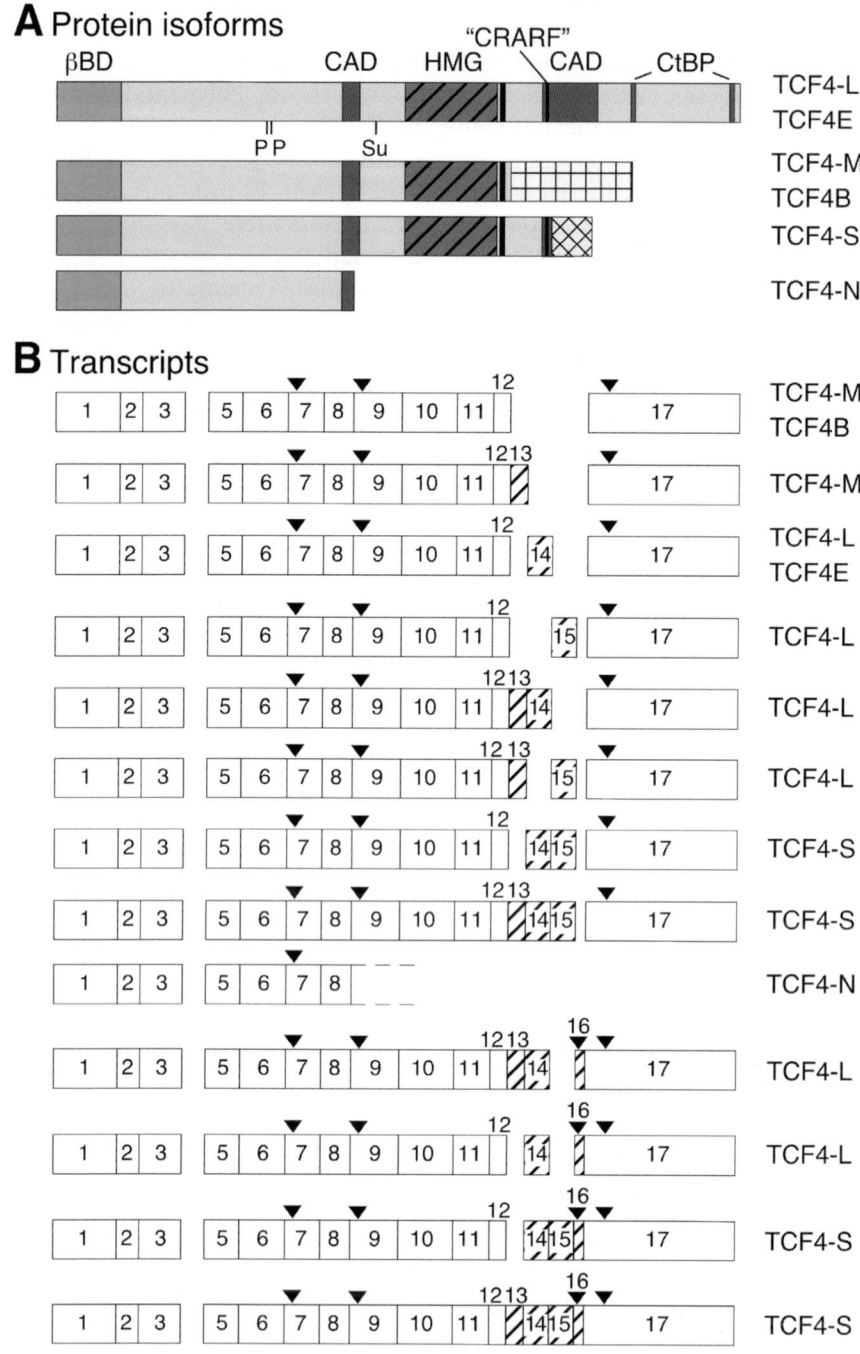

developing murine embryos (Kratochwil et al. 1996). However, it has not been determined whether BMP4 induction of LEF1 expression occurs directly or indirectly and which regulatory DNA elements in the LEF1 gene are involved in BMP4 induction. Still Wnt and BMP inducibility can serve to highlight that TCF gene expression is tissue specific and under tight control which fits well with the diverse functions carried out by TCFs in various tissues. For obvious reasons, a complete survey of embryonic and post-embryonic expression sites in humans is not possible. Given the high degree of structural and functional conservation between murine and human TCFs it is probably justified to refer to the murine system for TCF expression patterns during embryonic development. TCF family members display partially overlapping but highly restricted, nonubiquituos expression with considerable temporal dynamics. (Cho and Dressler 1998; Galceran et al. 1999; Korinek et al. 1998b; Lee et al. 1999; Oosterwegel et al. 1993). In view of the many alternatively spliced isoforms with differential functional properties it would be interesting to know whether all these sequence variants are expressed simultaneously with the same cell-type specificity. The differential Wnt inducibility of the LEF1 promoters suggests that maybe not, but a definitive answer to this question cannot yet be given. In addition, our current knowledge about TCF expression patterns is likely to be incomplete. TCF

Fig. 4A, B Protein isoforms and alternative splice products of the TCF4 gene. **A** Schematic representation of four different types of translation products from TCF4 transcripts. The main difference is the presence of C-terminal domains of variable length and sequence composition (indicated by differently *hatched boxes* in TCF4-M and -S). Accordingly, the C-terminal context-dependent activation domain (*CAD*) and the CtBP-binding motifs are either present or absent from TCF4-L, TCF4-M or TCF4-S isoforms. Because parts of the CAD are encoded by two different exons (14 and 17), TCF4-S isoforms exist which can harbor only the N-terminal part of the CAD with the CRARF amino acid motif. One particular TCF4-L variant is also known as TCF4E, and a TCF4-M variant as TCF4B. The newly identified TCF4-N variant terminates upstream of the HMG domain. Other structural features of the TCF4 protein are the βBD, phosphorylation and sumoylation sites (*P, Su*) and the HMG DNA-binding domain (*hatched box*). **B** Exon composition of transcripts of the TCF4 gene found in colorectal cancer cells, renal cancer cells and samples from normal kidney. Only a subset of alternative splicing products is shown. Exons affected are depicted as *hatched boxes.* Sequence composition of TCF4 transcripts can vary further, depending on the incorporation or omission of the alternatively spliced exon 4, and the use of alternative splice acceptors in exons 7, 9, 16 and 17 (*black triangles*). The presence or absence of exons 13, 14, or 15 alters the reading frame used to translate exon 17. Thus, the C-terminal amino acid sequence of the corresponding TCF4 isoforms changes. Transcript collections of similar complexity are also known for the TCF1 and LEF1 genes (van de Wetering et al. 1996; Hovanes et al. 2000). The 3′-end of the mRNA coding for TCF4-N was not determined. (Data are from Duval et al. 2000, Shiina et al. 2003, and Kennell et al. 2003)

genes were initially thought to be expressed almost exclusively during embryonic development, but new expression domains, also in adult tissues are still being discovered (Merrill et al. 2001). Postnatal expression of TCFs can also be postulated based on the growing list of human cancers involving the Wnt/β-catenin pathway (Polakis 2000). Thus, one or more of the TCFs has to be expressed in adult epithelial cells of the entire gastrointestinal tract, in mammary gland, in hepatocytes, pancreatic, kidney and prostate cells, melanocytes, endometrial and ovarian cells, epidermal cells and most likely a few more.

4
TCF Function in Gene Regulation

4.1
Dual Repressor/Activator Function in Wnt/β-Catenin Signaling

The initial finding that β-catenin interacts with TCF proteins immediately led to the proposal that TCFs act as positive regulators of Wnt signaling by directing the transcriptional activator β-catenin to target gene promoters. Based on recent reports of functional differences among TCF family members and TCF isoforms, the enormous structural diversity generated by alternative splicing, the largely unexplored relevance of post-translational modifications and the still growing number of interaction partners one can expect that more and more sophisticated scenarios of TCF function in gene regulation will emerge. However, current knowledge suggests that these new models will be variations of a common principle which is depicted in Fig. 5. The hallmarks of this model are that TCFs can bind to specific promoter elements of their target genes, but TCFs on their own have neither a stimulating nor a repressing influence on transcription (Hecht and Kemler 2000; Hurlstone and Clevers 2002). Only when complexed with additional cofactors can TCF proteins establish a promoter environment that favors or prevents the expression of their target genes. In the context of Wnt signaling, the relevant corepressor proteins are CtBP or Grg/TLE factors in conjunction with a HDAC (Brannon et al. 1999; Brantjes et al. 2001; Cavallo et al. 1998; Chen et al. 1999; Levanon et al. 1998; Roose et al. 1998; Valenta et al. 2003). Upon Wnt stimulation, β-catenin interacts with TCFs to alleviate promoter repression and activate transcription (Hecht and Kemler 2000; Miller et al. 1999). For this, β-catenin uses a number of additional coactivators many of which target chromatin, such as the p300/CBP acetylases, or Brg1 and Pontin52 which are components of chromatin remodeling complexes (Barker et al. 2001; Bauer et al. 2000; Hecht and Kemler 2000; Hecht et al. 2000). To involve these cofactors constitutes a plausible antagonism considering the role of chromatin structure in gene repression by Grg/TLE and

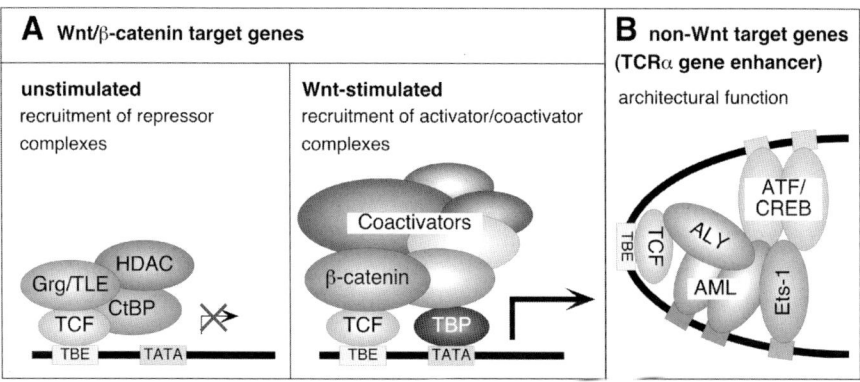

Fig. 5A, B Models for transcriptional control by TCFs. A TCFs play a dual role in the regulation of Wnt/β-catenin target genes. In unstimulated cells, promoter-bound TCFs are thought to be complexed with repressor protein complexes containing Grg/TLE gene products, HDAC and conditionally CtBP. Accordingly, promoter activity is suppressed. In Wnt-stimulated cells, an activating protein complex is formed in which TCFs serve to recruit β-catenin and associated coactivators. Promoter repression is relieved and transcription is induced. The underlying theme for both conditions is that TCFs function as chromosomal docking sites for non-DNA-binding transcriptional regulators. B TCFs additionally act as architectural transcription factors at *cis*-regulatory elements which are not Wnt responsive. As an example, a model for the situation at the TCRα gene enhancer is shown. The DNA-bending capacity of TCFs induces a DNA topology which is favorable for the formation of a multiprotein complex which is additionally stabilized by numerous protein–protein interactions among several DNA-binding transcription factors and the coactivator ALY (adapted from Giese et al. 1995)

HDAC. While the functional dualism of TCFs and the Wnt/β-catenin induced switch in promoter activity represent an attractive model for the control of Wnt/β-catenin target genes, it still contains several aspects which deserve to be investigated. For example the fate of Grg/TLE and CtBP upon Wnt stimulation is unclear. Do these repressors stay bound to TCFs or are they replaced? Is the same TCF protein used as docking site for the repressors and the activator β-catenin, or is there an exchange of TCFs? If all TCFs recognize the same DNA sequences, what determines which TCF or TCF isoform is allowed to bind to a promoter? Aside from Wnt growth factors also other extracellular signals can activate β-catenin and stimulate TCF activity (Monga et al. 2002; Novak et al. 1998; Papkoff and Aikawa 1998). Does this affect the same range of target genes that would be activated by Wnt signals? The answers to some of these question may be largely of academic interest, however, in particular the latter issue is highly relevant to the role of TCFs in tumorigenesis.

4.2
Wnt/β-Catenin Independent Gene Activation

Their names are clear reminders that TCFs play important roles outside of Wnt/β-catenin signaling. Unfortunately, these functions are somewhat forgotten and nowadays TCFs are best known for their roles as nuclear effectors in Wnt signaling. Nonetheless, TCF1 and LEF1 were first identified as cell type-specific nuclear proteins with DNA-binding sites in *cis*-regulatory elements of lymphocyte-specific genes such as CD3ε, Ly49a and the TCR genes (van de Wetering et al. 2002a). A TBE is also found in the enhancer of the human immunodeficiency virus-1 (HIV-1) (van de Wetering et al. 2002a). The description of LEF1 and TCF1 as architectural transcription factors is based largely on the observation that LEF1-induced DNA bending contributes to the formation of multifactorial transcription factor complexes as depicted for the TCRα element in Fig. 5 (Giese et al. 1995). When bound to their recognition sequences on an extended DNA molecule the various transcription factors involved in TCRα enhancer function would be well separated from each other. The conformational change induced by LEF1, however, allows these factors to interact with each other and to synergistically form a functional enhancer complex. This complex additionally includes the coactivator ALY which simultaneously interacts with LEF1 and AML. Although TCFs exert both Wnt-independent and Wnt-dependent functions in lymphocytes, there appears to be a clear distinction between those target genes that are Wnt/β-catenin responsive and those that are not. For example the TCRα enhancer cannot be stimulated by β-catenin (Hsu et al. 1998) and the presence of a TCF variant lacking the βBD permits its activation (van de Wetering et al. 2002a and references therein). In contrast, typical Wnt-inducible target genes are refractory to ALY overexpression and require the presence of a TCF isoform that contains the βBD. At first glance, the TCRα and the HIV-1 enhancers and the emphasis which is put on the architectural function of TCFs at these elements suggests that TCFs can function as transcription factors in at least two fundamentally different ways, either as architectural or as cofactor-dependent regulators. However, upon closer examination this distinction may disappear. There is a clear functional analogy between β-catenin and ALY as context-dependent transcriptional coactivators of TCFs. Moreover, while it has been shown that DNA bending alone is not sufficient to activate transcription at the HIV-1 enhancer (van de Wetering et al. 2002a), currently it is not known whether or not DNA topology plays a role in Wnt/β-catenin target gene control. Finally, it is very unlikely that β-catenin/TCF complexes are the sole regulatory factors at Wnt target genes. Interactions between TCFs and other DNA-binding transcription factors such as Smads, MITF and others (see above) clearly indicate that regulation of Wnt/β-catenin target genes as well relies on combinatorial control and functional cooperation of multiple distinct transcription factors very similar to the situation at the HIV-1 or TCRα enhancers. Thus, the differences may be the players involved, but not the principles.

5
TCFs in Tumorigenesis

5.1
Deregulation of Wnt/β-Catenin Signaling and the Development of Cancer

An active Wnt/β-catenin signaling cascade has been found in many different human cancers including among others pilomatricomas (a rare form of tumor derived from epidermal cells), melanomas, hepatocellular carcinomas and, very prominently, tumors of the gastrointestinal tract, most notably colorectal cancer (for recent reviews see Bienz and Clevers 2000; Polakis 2000). The widespread and recurrent involvement of the Wnt/β-catenin signaling in tumorigenesis fits well with the observation that the Wnt/β-catenin signaling pathway plays an essential role in maintaining multipotent stem cells in an undifferentiated, proliferative and self-renewing state (Huelsken et al. 2001; Korinek et al. 1998a; Merrill et al. 2001; Reya et al. 2003; van de Wetering et al. 2002b; Zechner et al. 2003). In particular stem cells, however, with their high proliferative capacity and their life-long persistence in an organism are prone to acquire a critical number of mutations which ultimately lead to an escape from growth control mechanisms and to cellular transformation. Moreover, among the many Wnt target genes are several groups which may critically contribute to the processes of tumorigenesis and carcinogenesis. These genes include proliferation stimulating genes (*CyclinD1*, c-*myc*, c-*jun*), growth factor genes (interleukin-8, fibroblast growth factor-4, vascular-endothelial growth factor) and genes coding for extracellular matrix degrading enzymes (matrix-metalloproteinases-7 and -26) (for a comprehensive listing and references see http://www.stanford.edu/~rnusse/pathways/targets.html).

Critical for liberating the tumorigenic potential of the Wnt/β-catenin signaling pathway is the activation of β-catenin as transcriptional activator. This can be achieved both by loss-of-function and gain-of-function type mutations in several distinct pathway components (Bienz and Clevers 2000; Polakis 2000). Most frequently observed in colon cancer are biallelic alterations of the APC gene with at least one allele displaying truncating mutations in the so-called mutation cluster region (Bienz and Clevers 2000; Polakis 2000). Disabling mutations in the axin gene have also been found in hepatocellular carcinoma and in colorectal cancer (Satoh et al. 2000; Shimizu et al. 2002). Inactivating mutations in the two tumor suppressor genes APC and axin alleviate negative control of β-catenin by the destruction complex and allow β-catenin accumulation and activation. Searches for oncogenes cooperating in lymphomagenesis revealed that this effect can also be achieved by overexpression of GBP/FRAT, an inhibitor of the destruction complex which acts on GSK3β (Berns et al. 1999; Jonkers et al. 1997). GBP/FRAT is thus classified as a proto-oncogene, just like the Wnt1 gene (Bienz and Clevers 2000; Polakis 2000). The β-catenin gene itself, too, possesses characteristics of a proto-oncogene and is subject to mutation. In the majority of cases the mutations lead to amino acid alterations immediately at or in the vicinity of residues that are targets for phosphorylation by CK Iϵ and GSK3β (Polakis 2000). Again, these mutations allow β-catenin to escape from degradation and enable its function as transcriptional activator. Experiments showing that β-catenin derivatives lacking transactivating properties lose their transforming capacity, and the finding that TCF fusion proteins with heterologous transactivation domains acquire the ability to transform cell lines in the absence of β-catenin demonstrated that transactivation is essential for oncogenic transformation by β-catenin/TCF complexes (Aoki et al. 1999, 2002; Kolligs et al. 1999). As interaction partners of β-catenin and nuclear effectors of Wnt signals TCFs are willingly or unwillingly tied into pathological events underlying tumor formation involving the Wnt/β-catenin signal transduction pathway. But how exactly do TCFs contribute to tumorigenesis and is their role only to bring β-catenin to some critical target genes?

5.2
Mutations in the TCF4 Gene and Its Deregulation in Cancer Cells

The TCF4 gene is most commonly analyzed in conjunction with aberrant Wnt/β-catenin activity in cancer cells. It is subject to mutation as well as deregulation in colorectal cancer, renal cancer and hepatocellular cancer cells (Cui et al. 2003; Duval et al. 2000; Jiang et al. 2002; Ruckert et al. 2002; Shiina et al. 2003). The coding region of exon 17 of the TCF4 gene contains a stretch of nine consecutive adenines. A one-base pair deletion in this polyadenine run is frequently observed in cells with defects in the DNA mismatch repair system, which display high microsatellite instability (Duval et al. 2000; Ruckert et al. 2002). Changing the length of the polyadenine run introduces a frame shift and premature translation termination signal. As a consequence, the TCF4 mutant proteins can no longer interact with the CtBP transcriptional corepressor. Additional nonsense mutations and deletions as well as changes in the pattern of alternative splicing events were reported in colorectal and renal cancer cells (Duval et al. 2000; Ruckert et al. 2002; Shiina et al. 2003). In contrast, other studies found no mutations in the TCF4 gene in hepatocellular carcinomas, but significantly upregulated mRNA levels (Cui et al. 2003; Jiang et al. 2002). The significance of these various observations is difficult to evaluate. The vast majority of mutations affected only one of the two TCF4 alleles leaving one copy of the gene intact

(Duval et al. 2000). Moreover, TCF4 gene mutations were observed only in cellular backgrounds with additional mutations in the APC or β-catenin genes. The coexistence of defects either in the β-catenin or the APC genes and of monoallelic TCF4 mutations, therefore, seems to preclude a causative role of the TCF4 gene mutations. This conclusion was also reached by Ruckert and colleagues who functionally analyzed TCF4 mutants derived from colorectal carcinomas with high microsatellite instability (Ruckert et al. 2002). Because analyses of the transactivation properties of the TCF4 mutants revealed no changes compared to the wild-type counterpart, TCF4 mutations appear to be passenger mutations with no impact on tumorigenesis (Ruckert et al. 2002). Changes in the production of certain splice variants and TCF4 overexpression, however, may have some relevance for tumor development and progression. A correlation was found between a more abundant expression of TCF4 l-isoforms in renal cancer and a worse prognosis (Shiina et al. 2003). l-Isoform prevalence coincided with an inhibition of apoptosis (Shiina et al. 2003) although the molecular mechanisms linking these two findings need to be clarified. Interestingly, the l-isoforms retain the CtBP-binding sites. Hence, the ability of TCF4 to interact with CtBP and its impact on tumor development is either highly cell type-specific or is irrelevant. The observation that TCF4 represses axin2 expression through CtBP also implies that loss of CtBP binding is of minor importance (Valenta et al. 2003) because alleviating axin2-repression would initiate a negative feedback loop dampening β-catenin/TCF activity (Lustig et al. 2002). Overexpression of TCF4 suggests a more obvious way in which TCF4 could contribute to tumor development. High levels of both LEF1 and TCF3 can stabilize β-catenin probably because the TCFs compete with APC for binding to β-catenin and thereby impede access of the destruction complex to β-catenin (Henderson et al. 2002; Lee et al. 2001). Increased levels of TCF4 accordingly could promote the uptake and/or the retention of β-catenin into the nucleus and aggravate the phenotypic consequences of APC or axin mutations.

5.3
A Role for LEF1 in Tumor Progression?

After it was first reported that β-catenin/TCF-mediated transcription was constitutively activated in colorectal cancer cells subsequent research focused on TCF4 because this is the TCF family member normally expressed postnatally in gastrointestinal tissue (Korinek et al. 1997). However, colorectal cancer cell lines additionally express LEF1, which is upregulated as a result of aberrant activation of Wnt/β-catenin signaling (Hovanes et al. 2001) (see Sect. 3). Thus, activation of Wnt/β-catenin signaling in cancer cells shifts the spectrum of TCFs expressed which may have an impact on tumor progression. The synergism between TGFβ and Wnt-signaling pathways at certain promoters notwithstanding, TGFβ has an inhibitory effect on the growth of certain tumor cell types and the TGFβ receptor 2 gene as well as the Smad2 and Smad4 genes are known as tumor suppressors in colon cancer (for a review see Derynck et al. 2001). The c-*myc* proto-oncogene is a key regulator of cell cycle progression and a common target of Wnt/β-catenin and TGFβ signaling. Importantly, c-*myc* expression is regulated in an opposing manner by the two growth factor families. While examining different possibilities for the failure of TGFβ to downregulate c-*myc* expression under certain circumstances, it was recently described that active TGFβ signaling could dissociate β-catenin/TCF4 complexes at c-*myc* regulatory elements, whereas LEF1/β-catenin complexes were refractory to TGFβ inhibition (Sasaki et al. 2003). Through upregulating LEF1 gene expression, uncontrolled Wnt/β-catenin signaling activity thus appears to promote further loss of growth restraints in tumor cells.

5.4
TCF1 as a Tumor Suppressor Gene

Postnatally, expression of the TCF1 gene seems to be largely confined to lymphocytes and is absent from intestinal epithelial cells (Roose et al. 1999). Colorectal cancer cell lines, however, were found to frequently express the TCF1 gene (Mayer et al. 1997; Roose et al. 1999). It is thought that this results from the activation of a β-catenin/TCF responsive enhancer located upstream of the TCF1 gene promoter P1 (Roose et al. 1999). The most abundant TCF1 isoforms in colorectal cancer cells lack the βBD but retain the ability to bind to DNA. Therefore, upregulation of TCF1 expression may constitute a negative feedback loop of the Wnt/β-catenin signaling system because TCF1 without βBD would displace transactivating β-catenin/TCF complexes from promoter regions (Roose et al. 1999). TCF1 thus seems to possess qualities of a tumor suppressor gene. In support of this view, loss of TCF1 appears to act synergistically with APC mutations in the development of adenoacanthomas of the intestine and of the mammary gland. This is suggested by higher frequency and earlier onset of neoplastic growth in compound mutant animals with deficiencies of both the APC and the TCF1 genes (Gallagher et al. 2002; Roose et al. 1999). Due to the nature of the TCF1 knock out allele, however, both dominant negative as well as potentially activating isoforms are eliminated. Thus, the theory of TCF1 acting as a dominant negative factor may be correct, but alternative models cannot be ruled out entirely. This is particularly true because the upstream enhancer is expected to preferentially act on the more closely situated promoter P1 which would drive expression of TCF1 variants including the βBD. Moreover, in the immune system, a β-catenin-dependent activating function of TCF1 is well documented (Held et al. 1999, 2003; Kunz and Held 2001; Staal et al. 2001). TCF1, therefore, cannot generally act as an inhibitory molecule in Wnt signaling. Unfortunately, Roose and colleagues did not analyze whether the Wnt-responsive TCF1 gene enhancer selectively activates one of the two TCF1 gene promoters and how the production of TCF1 splice forms is regulated (Roose et al. 1999). In fact, Mayer and colleagues presented a contrasting view as to the abundance of TCF1 isoforms lacking the βBD and even suggested a role for TCF1 in late stages of tumor progression (Mayer et al. 1997). Thus, a more sophisticated view of the role of TCF1 which also takes into account the functional differences between TCF1 isoforms might more accurately describe how TCF1 functions as a tumor suppressor. For example, inappropriate activation of β-catenin/TCF4 complexes in colorectal cancer cells appears to shift gene expression towards a proliferation-stimulating program und suppresses cellular differentiation (van de Wetering et al. 2002b). Gene-specific activities of TCF1 isoforms could antagonize TCF4 in these processes not only by blocking gene expression but also by specifically activating genes which stimulate cellular differentiation or which induce growth inhibitory or apoptotic stimuli (Atcha et al. 2003; Held et al. 2003; Kunz and Held 2001).

6
Summary

Much progress has been made since the initial findings that TCFs are DNA-binding proteins with the ability to act as architectural transcription factors and to serve as carriers for β-catenin. Their dual task as repressors and activators in Wnt/β-catenin signaling is well established. Recent investigations have revealed a multitude of new interaction partners for TCFs. Phosphorylation, acetylation, and sumoylation affect the subcellular distribution and the protein functions of TCFs. Alternative splicing produces a multitude of different TCF isoforms. The full extent of the structural diversity of TCF isoforms generated thereby is not yet fully known but already functional differences among TCF variants have begun to emerge. TCFs contribute to tumorigenesis in different, sometimes antagonistic, ways both as helpers of the oncoprotein β-catenin and as tumor suppressors. And yet, our understanding of TCFs and their roles in healthy cells and under pathological conditions appears to be far from complete. The mechanisms that control the spatial and temporal expression patterns of TCF genes are unknown, just as it is unclear whether and how alternative splicing events are regulated. The many translation products of alternatively spliced TCF transcripts might differ in their ability to interact with TCF-binding partners. There are likely to be even more interaction partners of TCFs than are currently known. Post-translational modifications of TCFs can be expected to affect these interactions and possibly other functions of TCFs as well. A comprehensive description of the determinants which govern promoter-specific activities of TCFs and the events which accompany the switch from a transcriptionally repressed to an activated promoter is lacking. While there may be a rough idea about the roles of TCFs in tumorigenesis, their precise functions and the significance of TCF mutations in tumor cells and the deregulation of their expression still need to be uncovered. For example, if both LEF1 and TCF1 are upregulated in tumor cells, then what tips the balance between tumor progression versus tumor suppression? There appears to be only one thing for sure: TCFs are far more complex than one would have anticipated.

Acknowledgements There is already an enormous body of information about Wnt/β-catenin signaling and TCF proteins, and more and more data are accumulating. Unfortunately, not all of this information could be incorporated into this chapter. Therefore, I apologize to all scientists whose work could not be included due to space limitations.

References

Amit S, Hatzubai A, Birman Y, Andersen JS, Ben-Shushan E, Mann M, Ben-Neriah Y, Alkalay I (2002) Axin-mediated CKI phosphorylation of β-catenin at Ser 45: a molecular switch for the Wnt pathway. Genes Dev 16:1066–1076

Aoki M, Hecht A, Kruse U, Kemler R, Vogt PK (1999) Nuclear endpoint of Wnt signaling: neoplastic transformation induced by transactivating lymphoid-enhancing factor 1. Proc Natl Acad Sci USA 96:139–144

Aoki M, Sobek V, Maslyar DJ, Hecht A, Vogt PK (2002) Oncogenic transformation by β-catenin: deletion analysis and characterization of selected target genes. Oncogene 21:6983–6991

Atcha FA, Munguia JE, Li TW, Hovanes K, Waterman ML (2003) A new β-catenin-dependent activation domain in T cell factor. J Biol Chem 278:16169–6175

Barker N, Hurlstone A, Musisi H, Miles A, Bienz M, Clevers H (2001) The chromatin remodelling factor Brg-1 interacts with β-catenin to promote target gene activation. EMBO J 20:4935–4943

Bauer A, Chauvet S, Huber O, Usseglio F, Rothbacher U, Aragnol D, Kemler R, Pradel J (2000) Pontin52 and reptin52 function as antagonistic regulators of β-catenin signalling activity. EMBO J 19:6121–6130

Berns A, Mikkers H, Krimpenfort P, Allen J, Scheijen B, Jonkers J (1999) Identification and characterization of collaborating oncogenes in compound mutant mice. Cancer Res 59:1773s–1777s

Bienz M, Clevers H (2000) Linking colorectal cancer to Wnt signaling. Cell 103:311–320

Billin AN, Thirlwell H, Ayer DE (2000) β-catenin-histone deacetylase interactions regulate the transition of LEF1 from a transcriptional repressor to an activator. Mol Cell Biol 20:6882–6890

Brannon M, Brown JD, Bates R, Kimelman D, Moon RT (1999) XCtBP is a XTcf-3 co-repressor with roles throughout Xenopus development. Development 126:3159–3170

Brantjes H, Roose J, van De Wetering M, Clevers H (2001) All Tcf HMG box transcription factors interact with Groucho-related co-repressors. Nucleic Acids Res 29:1410–1419

Bruhn L, Munnerlyn A, Grosschedl R (1997) ALY, a context-dependent coactivator of LEF-1 and AML-1, is required for TCRalpha enhancer function. Genes Dev 11:640–653

Cavallo RA, Cox RT, Moline MM, Roose J, Polevoy GA, Clevers H, Peifer M, Bejsovec A (1998) Drosophila Tcf and Groucho interact to repress Wingless signalling activity. Nature 395:604–608

Chen G, Fernandez J, Mische S, Courey AJ (1999) A functional interaction between the histone deacetylase Rpd3 and the corepressor groucho in Drosophila development. Genes Dev 13:2218–2230

Chinnadurai G (2002) CtBP, an unconventional transcriptional corepressor in development and oncogenesis. Mol Cell 9:213–224

Cho EA, Dressler GR (1998) TCF-4 binds β-catenin and is expressed in distinct regions of the embryonic brain and limbs. Mech Dev 77:9–18

Cui J, Zhou X, Liu Y, Tang Z, Romeih M (2003) Wnt signaling in hepatocellular carcinoma: analysis of mutation and expression of β-catenin, T-cell factor-4 and glycogen synthase kinase 3-β genes. J Gastroenterol Hepatol 18:280–287

Daniels DL, Eklof Spink K, Weis WI (2001) β-catenin: molecular plasticity and drug design. Trends Biochem Sci 26:672–678

Derynck R, Akhurst RJ, Balmain A (2001) TGF-β signaling in tumor suppression and cancer progression. Nat Genet 29:117–129

Dorsky RI, Raible DW, Moon RT (2000) Direct regulation of nacre, a zebrafish MITF homolog required for pigment cell formation, by the Wnt pathway. Genes Dev 14:158–162

Duval A, Rolland S, Tubacher E, Bui H, Thomas G, Hamelin R (2000) The human T-cell transcription factor-4 gene: structure, extensive characterization of alternative splicings, and mutational analysis in colorectal cancer cell lines. Cancer Res 60:3872–3879

El-Tanani M, Fernig DG, Barraclough R, Green C, Rudland P (2001) Differential modulation of transcriptional activity of estrogen receptors by direct protein-protein interactions with the T cell factor family of transcription factors. J Biol Chem 276:41675–41682

Fisher AL, Caudy M (1998) Groucho proteins: transcriptional corepressors for specific subsets of DNA-binding transcription factors in vertebrates and invertebrates. Genes Dev 12:1931–1940

Galceran J, Farinas I, Depew MJ, Clevers H, Grosschedl R (1999) Wnt3a-/—like phenotype and limb deficiency in Lef1(-/-)Tcf1(-/-) mice. Genes Dev 13:709–717

Gallagher RC, Hay T, Meniel V, Naughton C, Anderson TJ, Shibata H, Ito M, Clevers H, Noda T, Sansom OJ, Mason JO, Clarke AR (2002) Inactivation of Apc perturbs mammary development, but only directly results in acanthoma in the context of Tcf-1 deficiency. Oncogene 21:6446–6457

Gay F, Calvo D, Lo MC, Ceron J, Maduro M, Lin R, Shi Y (2003) Acetylation regulates subcellular localization of the Wnt signaling nuclear effector POP-1. Genes Dev 17:717–722

Ghazi A, VijayRaghavan KV (2000) Developmental biology. Control by combinatorial codes. Nature 408:419–420

Giese K, Kingsley C, Kirshner JR, Grosschedl R (1995) Assembly and function of a TCR alpha enhancer complex is dependent on LEF-1-induced DNA bending and multiple protein-protein interactions. Genes Dev 9:995–1008

Goding CR (2000) Mitf from neural crest to melanoma: signal transduction and transcription in the melanocyte lineage. Genes Dev 14:1712–1728

Goodman RH, Smolik S (2000) CBP/p300 in cell growth, transformation, and development. Genes Dev 14:1553–1577

Gradl D, Konig A, Wedlich D (2002) Functional diversity of Xenopus lymphoid enhancer factor/T-cell factor transcription factors relies on combinations of activating and repressing elements. J Biol Chem 277:14159–14171

Graham TA, Ferkey DM, Mao F, Kimelman D, Xu W (2001) Tcf4 can specifically recognize β-catenin using alternative conformations. Nat Struct Biol 8:1048–1052

Graham TA, Weaver C, Mao F, Kimelman D, Xu W (2000) Crystal structure of a β-catenin/Tcf complex. Cell 103:885–896

Grosschedl R, Giese K, Pagel J (1994) HMG domain proteins: architectural elements in the assembly of nucleoprotein structures. Trends Genet 10:94–100

Grueneberg DA, Pablo L, Hu KQ, August P, Weng Z, Papkoff J (2003) A functional screen in human cells identifies UBF2 as an RNA polymerase II transcription factor that enhances the β-catenin signaling pathway. Mol Cell Biol 23:3936–3950

He TC, Sparks AB, Rago C, Hermeking H, Zawel L, da Costa LT, Morin PJ, Vogelstein B, Kinzler KW (1998) Identification of c-MYC as a target of the APC pathway. Science 281:1509–1512

Hecht A, Kemler R (2000) Curbing the nuclear activities of β-catenin. Control over Wnt target gene expression. EMBO Rep 1:24–28

Hecht A, Stemmler MP (2003) Identification of a promoter-specific transcriptional activation domain at the C terminus of the Wnt effector protein T-cell factor 4. J Biol Chem 278:3776–3785

Hecht A, Vleminckx K, Stemmler MP, van Roy F, Kemler R (2000) The p300/CBP acetyltransferases function as transcriptional coactivators of β-catenin in vertebrates. EMBO J 19:1839–1850

Held W, Clevers H, Grosschedl R (2003) Redundant functions of TCF-1 and LEF-1 during T and NK cell development, but unique role of TCF-1 for Ly49 NK cell receptor acquisition. Eur J Immunol 33:1393–1398

Held W, Kunz B, Lowin-Kropf B, van de Wetering M, Clevers H (1999) Clonal acquisition of the Ly49A NK cell receptor is dependent on the *trans*-acting factor TCF-1. Immunity 11:433–442

Henderson BR, Galea M, Schuechner S, Leung L (2002) Lymphoid enhancer factor-1 blocks adenomatous polyposis coli-mediated nuclear export and degradation of β-catenin. Regulation by histone deacetylase 1. J Biol Chem 277:24258–24264

Hobmayer B, Rentzsch F, Kuhn K, Happel CM, von Laue CC, Snyder P, Rothbacher U, Holstein TW (2000) WNT signalling molecules act in axis formation in the diploblastic metazoan Hydra. Nature 407:186–189

Hovanes K, Li TW, Munguia JE, Truong T, Milovanovic T, Lawrence Marsh J, Holcombe RF, Waterman ML (2001) β-catenin-sensitive isoforms of lymphoid enhancer factor-1 are selectively expressed in colon cancer. Nat Genet 28:53–57

Hovanes K, Li TW, Waterman ML (2000) The human LEF-1 gene contains a promoter preferentially active in lymphocytes and encodes multiple isoforms derived from alternative splicing. Nucleic Acids Res 28:1994–2003

Hsu SC, Galceran J, Grosschedl R (1998) Modulation of transcriptional regulation by LEF-1 in response to Wnt-1 signaling and association with β-catenin. Mol Cell Biol 18:4807–4818

Huber AH, Nelson WJ, Weis WI (1997) Three-dimensional structure of the armadillo repeat region of β-catenin. Cell 90:871–882

Huber AH, Weis WI (2001) The structure of the β-catenin/E-cadherin complex and the molecular basis of diverse ligand recognition by β-catenin. Cell 105:391–402

Huelsken J, Birchmeier W (2001) New aspects of Wnt signaling pathways in higher vertebrates. Curr Opin Genet Dev 11:547–553

Huelsken J, Vogel R, Erdmann B, Cotsarelis G, Birchmeier W (2001) β-Catenin controls hair follicle morphogenesis and stem cell differentiation in the skin. Cell 105:533–545

Hurlstone A, Clevers H (2002) T-cell factors: turn-ons and turn-offs. EMBO J 21:2303–2311

Ishitani T, Ninomiya-Tsuji J, Matsumoto K (2003) Regulation of lymphoid enhancer factor 1/T-cell factor by mitogen-activated protein kinase-related Nemo-like kinase-dependent phosphorylation in Wnt/β-catenin signaling. Mol Cell Biol 23:1379–1389

Ishitani T, Ninomiya-Tsuji J, Nagai S, Nishita M, Meneghini M, Barker N, Waterman M, Bowerman B, Clevers H, Shibuya H, Matsumoto K (1999) The TAK1-NLK-MAPK-related pathway antagonizes signalling between β-catenin and transcription factor TCF. Nature 399:798–802

Jiang Y, Zhou XD, Liu YK, Wu X, Huang XW (2002) Association of hTcf-4 gene expression and mutation with clinicopathological characteristics of hepatocellular carcinoma. World J Gastroenterol 8:804–807

Jonkers J, Korswagen HC, Acton D, Breuer M, Berns A (1997) Activation of a novel proto-oncogene, Frat1, contributes to progression of mouse T-cell lymphomas. EMBO J 16:441–450

Kahler RA, Westendorf JJ (2003) Lymphoid enhancer factor-1 and β-catenin inhibit Runx2-dependent transcriptional activation of the osteocalcin promoter. J Biol Chem 278:11937–11944

Kennell JA, O'Leary EE, Gummow BM, Hammer GD, MacDougald OA (2003) T-cell factor 4N (TCF-4N), a novel isoform of mouse TCF-4, synergizes with β-catenin to coactivate C/EBPalpha and steroidogenic factor 1 transcription factors. Mol Cell Biol 23:5366–5375

Kolligs FT, Hu G, Dang CV, Fearon ER (1999) Neoplastic transformation of RK3E by mutant β-catenin requires deregulation of Tcf/Lef transcription but not activation of c-myc expression. Mol Cell Biol 19:5696–5706

Korinek V, Barker N, Moerer P, van Donselaar E, Huls G, Peters PJ, Clevers H (1998a) Depletion of epithelial stem-cell compartments in the small intestine of mice lacking Tcf-4. Nat Genet 19:379–383

Korinek V, Barker N, Morin PJ, van Wichen D, de Weger R, Kinzler KW, Vogelstein B, Clevers H (1997) Constitutive transcriptional activation by a β-catenin-Tcf complex in APC-/- colon carcinoma. Science 275:1784–1787

Korinek V, Barker N, Willert K, Molenaar M, Roose J, Wagenaar G, Markman M, Lamers W, Destree O, Clevers H (1998b) Two members of the Tcf family implicated in Wnt/β-catenin signaling during embryogenesis in the mouse. Mol Cell Biol 18:1248–1256

Kouzarides T (2000) Acetylation: a regulatory modification to rival phosphorylation? EMBO J 19:1176–1179

Kratochwil K, Dull M, Farinas I, Galceran J, Grosschedl R (1996) Lef1 expression is activated by BMP-4 and regulates inductive tissue interactions in tooth and hair development. Genes Dev 10:1382–1394

Kratochwil K, Galceran J, Tontsch S, Roth W, Grosschedl R (2002) FGF4, a direct target of LEF1 and Wnt signaling, can rescue the arrest of tooth organogenesis in Lef1(-/-) mice. Genes Dev 16:3173–3185

Kunz B, Held W (2001) Positive and negative roles of the *trans*-acting T cell factor-1 for the acquisition of distinct Ly-49 MHC class I receptors by NK cells. J Immunol 166:6181–6187

Kusano S, Raab-Traub N (2002) I-mfa domain proteins interact with Axin and affect its regulation of the Wnt and c-Jun N-terminal kinase signaling pathways. Mol Cell Biol 22:6393–6405

Labbe E, Letamendia A, Attisano L (2000) Association of Smads with lymphoid enhancer binding factor 1/T cell-specific factor mediates cooperative signaling by the transforming growth factor-β and wnt pathways. Proc Natl Acad Sci USA 97:8358–8363

Lawrence N, Langdon T, Brennan K, Arias AM (2001) Notch signaling targets the Wingless responsiveness of a Ubx visceral mesoderm enhancer in Drosophila. Curr Biol 11:375–385

Lee E, Salic A, Kirschner MW (2001) Physiological regulation of β-catenin stability by Tcf3 and CK1epsilon. J Cell Biol 154:983–993

Lee YJ, Swencki B, Shoichet S, Shivdasani RA (1999) A possible role for the high mobility group box transcription factor Tcf-4 in vertebrate gut epithelial cell differentiation. J Biol Chem 274:1566–1572

Lepourcelet M, Shivdasani RA (2002) Characterization of a novel mammalian Groucho isoform and its role in transcriptional regulation. J Biol Chem 277:47732–47740

Levanon D, Goldstein RE, Bernstein Y, Tang H, Goldenberg D, Stifani S, Paroush Z, Groner Y (1998) Transcriptional repression by AML1 and LEF-1 is mediated by the TLE/Groucho corepressors. Proc Natl Acad Sci USA 95:11590–11595

Levy L, Neuveut C, Renard CA, Charneau P, Branchereau S, Gauthier F, Van Nhieu JT, Cherqui D, Petit-Bertron AF, Mathieu D, Buendia MA (2002) Transcriptional activation of interleukin-8 by β-catenin-Tcf4. J Biol Chem 277:42386–42393

Liu C, Li Y, Semenov M, Han C, Baeg GH, Tan Y, Zhang Z, Lin X, He X (2002) Control of β-catenin phosphorylation/degradation by a dual-kinase mechanism. Cell 108:837–847

Love JJ, Li X, Case DA, Giese K, Grosschedl R, Wright PE (1995) Structural basis for DNA bending by the architectural transcription factor LEF-1. Nature 376:791–795

Lustig B, Jerchow B, Sachs M, Weiler S, Pietsch T, Karsten U, van de Wetering M, Clevers H, Schlag PM, Birchmeier W, Behrens J (2002) Negative feedback loop of Wnt signaling through upregulation of conductin/axin2 in colorectal and liver tumors. Mol Cell Biol 22:1184–1193

Mayer K, Hieronymus T, Castrop J, Clevers H, Ballhausen WG (1997) Ectopic activation of lymphoid high mobility group-box transcription factor TCF-1 and overexpression in colorectal cancer cells. Int J Cancer 72:625–630

Mayer K, Wolff E, Clevers H, Ballhausen WG (1995) The human high mobility group (HMG)-box transcription factor TCF-1: novel isoforms due to alternative splicing and usage of a new exon IXA. Biochim Biophys Acta 1263:169–172

McKendry R, Hsu SC, Harland RM, Grosschedl R (1997) LEF-1/TCF proteins mediate wnt-inducible transcription from the Xenopus nodal-related 3 promoter. Dev Biol 192:420–431

Meneghini MD, Ishitani T, Carter JC, Hisamoto N, Ninomiya-Tsuji J, Thorpe CJ, Hamill DR, Matsumoto K, Bowerman B (1999) MAP kinase and Wnt pathways converge to downregulate an HMG-domain repressor in Caenorhabditis elegans. Nature 399:793–797

Merrill BJ, Gat U, DasGupta R, Fuchs E (2001) Tcf3 and Lef1 regulate lineage differentiation of multipotent stem cells in skin. Genes Dev 15:1688–1705

Miller JR, Hocking AM, Brown JD, Moon RT (1999) Mechanism and function of signal transduction by the Wnt/β-catenin and Wnt/Ca2+ pathways. Oncogene 18:7860–7872

Miyagishi M, Fujii R, Hatta M, Yoshida E, Araya N, Nagafuchi A, Ishihara S, Nakajima T, Fukamizu A (2000) Regulation of Lef-mediated transcription and p53-dependent pathway by associating β-catenin with CBP/p300. J Biol Chem 275:35170–35175

Monga SP, Mars WM, Pediaditakis P, Bell A, Mule K, Bowen WC, Wang X, Zarnegar R, Michalopoulos GK (2002) Hepatocyte growth factor induces Wnt-independent nuclear translocation of β-catenin after Met-β-catenin dissociation in hepatocytes. Cancer Res 62:2064–2071

Muller S, Hoege C, Pyrowolakis G, Jentsch S (2001) SUMO, ubiquitin's mysterious cousin. Nat Rev Mol Cell Biol 2:202–210

Nishita M, Hashimoto MK, Ogata S, Laurent MN, Ueno N, Shibuya H, Cho KW (2000) Interaction between Wnt and TGF-β signalling pathways during formation of Spemann's organizer. Nature 403:781–785

Novak A, Hsu SC, Leung-Hagesteijn C, Radeva G, Papkoff J, Montesano R, Roskelley C, Grosschedl R, Dedhar S (1998) Cell adhesion and the integrin-linked kinase regulate the LEF-1 and β-catenin signaling pathways. Proc Natl Acad Sci USA 95:4374–4379

Oosterwegel M, van de Wetering M, Timmerman J, Kruisbeek A, Destree O, Meijlink F, Clevers H (1993) Differential expression of the HMG box factors TCF-1 and LEF-1 during murine embryogenesis. Development 118:439–448

Papkoff J, Aikawa M (1998) WNT-1 and HGF regulate GSK3 β activity and β-catenin signaling in mammary epithelial cells. Biochem Biophys Res Commun 247:851–858

Peifer M, Berg S, Reynolds AB (1994) A repeating amino acid motif shared by proteins with diverse cellular roles. Cell 76:789–791

Polakis P (2000) Wnt signaling and cancer. Genes Dev 14:1837–1851

Poy F, Lepourcelet M, Shivdasani RA, Eck MJ (2001) Structure of a human Tcf4-β-catenin complex. Nat Struct Biol 8:1053–1057

Pukrop T, Gradl D, Henningfeld KA, Knochel W, Wedlich D, Kuhl M (2001) Identification of two regulatory elements within the high mobility group box transcription factor XTCF-4. J Biol Chem 276:8968–8978

Reya T, Duncan AW, Ailles L, Domen J, Scherer DC, Willert K, Hintz L, Nusse R, Weissman IL (2003) A role for Wnt signalling in self-renewal of haematopoietic stem cells. Nature 423:409–414

Rocheleau CE, Yasuda J, Shin TH, Lin R, Sawa H, Okano H, Priess JR, Davis RJ, Mello CC (1999) WRM-1 activates the LIT-1 protein kinase to transduce anterior/posterior polarity signals in C. elegans. Cell 97:717–726

Roose J, Huls G, van Beest M, Moerer P, van der Horn K, Goldschmeding R, Logtenberg T, Clevers H (1999) Synergy between tumor suppressor APC and the β-catenin-Tcf4 target Tcf1. Science 285:1923–1926

Roose J, Molenaar M, Peterson J, Hurenkamp J, Brantjes H, Moerer P, van de Wetering M, Destree O, Clevers H (1998) The Xenopus Wnt effector XTcf-3 interacts with Groucho-related transcriptional repressors. Nature 395:608–612

Ross DA, Kadesch T (2001) The notch intracellular domain can function as a coactivator for LEF-1. Mol Cell Biol 21:7537–7544

Ruckert S, Hiendlmeyer E, Brueckl WM, Oswald U, Beyser K, Dietmaier W, Haynl A, Koch C, Ruschoff J, Brabletz T, Kirchner T, Jung A (2002) T-cell factor-4 frameshift mutations occur frequently in human microsatellite instability-high colorectal carcinomas but do not contribute to carcinogenesis. Cancer Res 62:3009–3013

Rui HL, Fan E, Zhou HM, Xu Z, Zhang Y, Lin SC (2002) SUMO-1 modification of the C-terminal KVEKVD of Axin is required for JNK activation but has no effect on Wnt signaling. J Biol Chem 277:42981–42986

Sachdev S, Bruhn L, Sieber H, Pichler A, Melchior F, Grosschedl R (2001) PIASy, a nuclear matrix-associated SUMO E3 ligase, represses LEF1 activity by sequestration into nuclear bodies. Genes Dev 15:3088–3103

Saito H, Yasumoto K, Takeda K, Takahashi K, Fukuzaki A, Orikasa S, Shibahara S (2002) Melanocyte-specific microphthalmia-associated transcription factor isoform activates its own gene promoter through physical interaction with lymphoid-enhancing factor 1. J Biol Chem 277:28787–28794

Sampson EM, Haque ZK, Ku MC, Tevosian SG, Albanese C, Pestell RG, Paulson KE, Yee AS (2001) Negative regulation of the Wnt-β-catenin pathway by the transcriptional repressor HBP1. EMBO J 20:4500–4511

Sasaki T, Suzuki H, Yagi K, Furuhashi M, Yao R, Susa S, Noda T, Arai Y, Miyazono K, Kato M (2003) Lymphoid enhancer factor 1 makes cells resistant to transforming growth factor β-induced repression of c-myc. Cancer Res 63:801–806

Satoh S, Daigo Y, Furukawa Y, Kato T, Miwa N, Nishiwaki T, Kawasoe T, Ishiguro H, Fujita M, Tokino T, Sasaki Y, Imaoka S, Murata M, Shimano T, Yamaoka Y, Nakamura Y (2000) AXIN1 mutations in hepatocellular carcinomas, and growth suppression in cancer cells by virus-mediated transfer of AXIN1. Nat Genet 24:245–250

Shiina H, Igawa M, Breault J, Ribeiro-Filho L, Pookot D, Urakami S, Terashima M, Deguchi M, Yamanaka M, Shirai M, Kaneuchi M, Kane CJ, Dahiya R (2003) The human T-cell factor-4 gene splicing isoforms, Wnt signal pathway, and apoptosis in renal cell carcinoma. Clin Cancer Res 9:2121–2132

Shimizu Y, Ikeda S, Fujimori M, Kodama S, Nakahara M, Okajima M, Asahara T (2002) Frequent alterations in the Wnt signaling pathway in colorectal cancer with microsatellite instability. Genes Chromosomes Cancer 33:73–81

Shtutman M, Zhurinsky J, Simcha I, Albanese C, D'Amico M, Pestell R, Ben-Ze'ev A (1999) The cyclin D1 gene is a target of the β-catenin/LEF-1 pathway. Proc Natl Acad Sci USA 96:5522–5527

Snider L, Thirlwell H, Miller JR, Moon RT, Groudine M, Tapscott SJ (2001) Inhibition of Tcf3 binding by I-mfa domain proteins. Mol Cell Biol 21:1866–1873

Staal FJ, Meeldijk J, Moerer P, Jay P, van de Weerdt BC, Vainio S, Nolan GP, Clevers H (2001) Wnt signaling is required for thymocyte development and activates Tcf-1 mediated transcription. Eur J Immunol 31:285–293

Sun Y, Kolligs FT, Hottiger MO, Mosavin R, Fearon ER, Nabel GJ (2000) Regulation of β-catenin transformation by the p300 transcriptional coactivator. Proc Natl Acad Sci USA 97:12613–12618

Takeda K, Yasumoto K, Takada R, Takada S, Watanabe K, Udono T, Saito H, Takahashi K, Shibahara S (2000) Induction of melanocyte-specific microphthalmia-associated transcription factor by Wnt-3a. J Biol Chem 275:14013–14016

Takemaru KI, Moon RT (2000) The transcriptional coactivator CBP interacts with β-catenin to activate gene expression. J Cell Biol 149:249–254

Tetsu O, McCormick F (1999) β-catenin regulates expression of cyclin D1 in colon carcinoma cells. Nature 398:422–426

Valenta T, Lukas J, Korinek V (2003) HMG box transcription factor TCF-4's interaction with CtBP1 controls the expression of the Wnt target Axin2/Conductin in human embryonic kidney cells. Nucleic Acids Res 31:2369–2380

van Beest M, Dooijes D, van De Wetering M, Kjaerulff S, Bonvin A, Nielsen O, Clevers H (2000) Sequence-specific high mobility group box factors recognize 10–12-base pair minor groove motifs. J Biol Chem 275:27266–27273

Van de Wetering M, Castrop J, Korinek V, Clevers H (1996) Extensive alternative splicing and dual promoter usage generate Tcf-1 protein isoforms with differential transcription control properties. Mol Cell Biol 16:745–752

van de Wetering M, de Lau W, Clevers H (2002a) WNT signaling and lymphocyte development. Cell 109 Suppl: S13–S9

van de Wetering M, Sancho E, Verweij C, de Lau W, Oving I, Hurlstone A, van der Horn K, Batlle E, Coudreuse D, Haramis AP, Tjon-Pon-Fong M, Moerer P, van den Born M,

Soete G, Pals S, Eilers M, Medema R, Clevers H (2002b) The β-catenin/TCF-4 complex imposes a crypt progenitor phenotype on colorectal cancer cells. Cell 111:241–250

von Kries JP, Winbeck G, Asbrand C, Schwarz-Romond T, Sochnikova N, Dell'Oro A, Behrens J, Birchmeier W (2000) Hot spots in β-catenin for interactions with LEF-1, conductin and APC. Nat Struct Biol 7:800–807

Waltzer L, Bienz M (1998) Drosophila CBP represses the transcription factor TCF to antagonize Wingless signalling. Nature 395:521–525

Yamamoto H, Ihara M, Matsuura Y, Kikuchi A (2003) Sumoylation is involved in β-catenin-dependent activation of Tcf-4. EMBO J 22:2047–2059

Yasumoto K, Takeda K, Saito H, Watanabe K, Takahashi K, Shibahara S (2002) Microphthalmia-associated transcription factor interacts with LEF-1, a mediator of Wnt signaling. EMBO J 21:2703–2714.

Zechner D, Fujita Y, Hulsken J, Muller T, Walther I, Taketo MM, Crenshaw EB, 3rd, Birchmeier W, Birchmeier C (2003) β-Catenin signals regulate cell growth and the balance between progenitor cell expansion and differentiation in the nervous system. Dev Biol 258:406–418

Transcription Factors in the Control of Tumor Development and Progression by TGF-β Signaling

I. Timokhina · J. Lecanda · M. Kretzschmar (✉)

Derald H. Ruttenberg Cancer Center, Mount Sinai School of Medicine,
One Gustave Levy Place, Box 1130, New York, NY 10029, USA
marcus.kretzschmar@mssm.edu

1	Introduction: The TGF-β Signaling System	168
2	SMADs: Transcriptional Mediators of TGF-β Signals	170
3	SMAD Interacting Proteins in the Nucleus	172
4	SMAD-Independent TGF-β Signaling Mechanisms	176
5	TGF-β Signaling in Normal Epithelial and Carcinoma Cells: Roles in Tumor Suppression and Promotion	177
5.1	Apoptosis and Survival	179
5.2	Cell Cycle Arrest	181
5.3	Epithelial-to-Mesenchymal Transition	182
5.4	Metastasis	185
6	TGF-β Responses as Targets of Tumor-Associated Alterations	187
6.1	Inactivating Mutations in TGF-β Receptors and SMADs	187
6.2	Other Mechanisms of Altered TGF-β Responsiveness	189
7	TGF-β Signaling in Vascular Endothelial Cells	192
	References	194

Abstract Transforming growth factor-β (TGF-β) is a cytokine that controls homeostasis and/or remodeling in many tissues, including most if not all epithelial tissues. These functions are mediated by the regulation of diverse cellular processes, such as cell proliferation, cell death or survival, cell migration and invasion, and cell differentiation. In human cancers TGF-β functions as a tumor suppressor in early stages, but converts into a tumor promoter in later stages of the disease. Understanding the molecular basis for this switch in TGF-β function is highly relevant, if TGF-β is to be exploited as a target in cancer therapy. Since most of the cellular TGF-β responses characterized to date are mediated by the transcriptional control of direct target genes of TGF-β signaling, this inevitably entails the analysis of the transcriptional mediators of TGF-β function in normal cells as well as tumor cells of different malignant potential. This review will discuss some of the important transcriptional mediators involved, with emphasis on the mammary epithelium as an example of the complex role of TGF-β in regulating normal and tumorigenic tissues.

Keywords Transforming growth factor-β · Receptor signaling · Smad · Tumor suppressor · Tumor promoter

1
Introduction: The TGF-β Signaling System

Transforming growth factor-β (TGF-β) is the prototypic member of a superfamily of peptide growth factors that includes the activins, TGF-βs, nodal, bone morphogenic proteins (BMPs), growth and differentiation factors (GDFs), and the müllerian inhibitory substance (MIS) (Fig. 1). Collectively these factors play important roles in the development, maintenance and function of most tissues and organs in species ranging from worms and flies to mammals (Massague et al. 2000). Characterization of knockout and

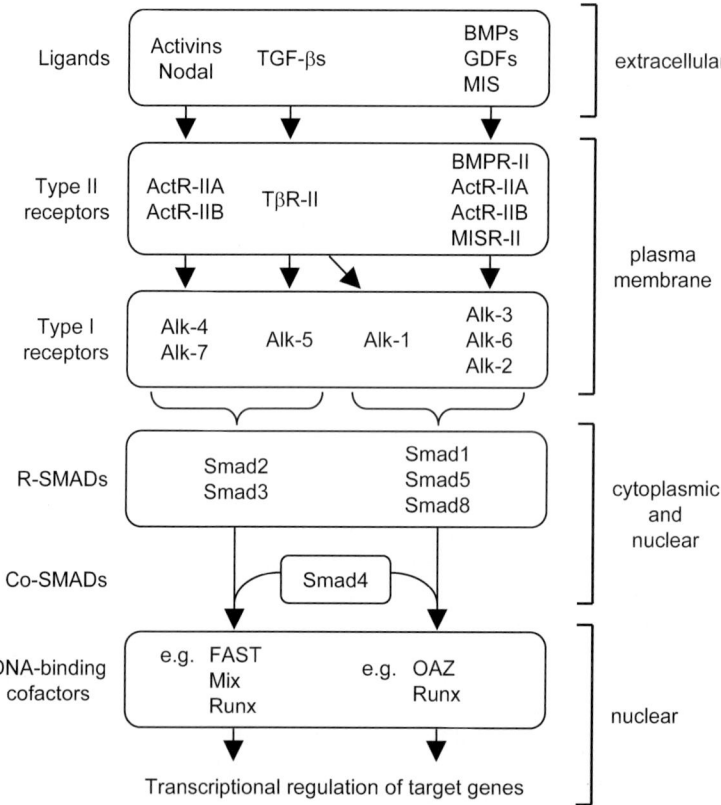

Fig. 1 Important players and interactions of the TGF-β/SMAD signaling system. For simplicity SMAD-independent signaling pathways are not depicted here. Alternative nomenclature includes TβR-I for Alk-5, BMPR-IA for Alk-3 and BMPR-IB for Alk-6

knockin transgenic mice, as well as the analysis of naturally occurring genetic mutations affecting superfamily members or their regulatory proteins in humans or other mammals, have provided a wealth of information on the specific roles of many of these proteins in vivo (Chang et al. 2002).

Structurally TGF-β growth factors are characterized by the so-called 'cysteine-knot,' a tight structure formed by extended β-strands that are interlocked by three conserved disulfide bonds (Sun and Davies 1995). Furthermore, the active forms of these factors are homodimers that are stabilized by hydrophobic interactions and in many cases by an intermolecular disulfide bridge.

TGF-β growth factors initiate their biological responses by inducing the formation of a cell surface receptor complex consisting of a type I and a type II receptor serine/threonine kinase (RS/TK) (Attisano and Wrana 2002; Massague et al. 2000). Both type I and type II RS/TKs represent families of related proteins, with seven identified type I receptors (Alk-1–Alk-7) and five identified type II receptors in mammalian cells (Fig. 1). All of these receptors consist of an amino-terminal extracellular ligand-binding domain, a transmembrane domain, and a carboxy-terminal intracellular domain which includes the serine/threonine kinase activity. An active receptor complex is formed by the ligand-induced association of the dimeric ligand, two molecules of a type I receptor and two molecules of a compatible type II receptor. Within this complex the constitutively active type II receptor kinase transphosphorylates the type I receptor at a characteristic serine- and glycine-rich domain, called the GS domain, that is located just amino terminal of the kinase domain in all type I receptors (Massague et al. 2000). This phosphorylation event activates the kinase activity of the type I receptor, thereby enabling the type I receptor to initiate intracellular signaling pathways and hence biological responses.

These signals are transmitted in large part by the direct phosphorylation of specific SMAD proteins by the activated type I receptors (Kretzschmar et al. 1997; Macias-Silva et al. 1996). There are two major branches of SMAD signaling activity, the 'TGF-β/activin/nodal branch' which includes Smad2 and 3 and is activated by the Alk-4, -5 and -7 receptors, and the 'BMP/GDF/MIS branch' which includes Smad1, 5, and 8 and is activated by the Alk-1, -2, -3 and -6 receptors. Receptor-mediated phosphorylation of these SMADs leads to their association with the common interaction partner Smad4, translocation into the nucleus, and the direct transcriptional regulation of target genes in cooperation with other DNA-binding cofactors (Fig. 1) (Attisano and Wrana 2002; Massague et al. 2000).

In addition to the SMAD pathways TGF-β growth factors also signal through SMAD-independent mechanisms, involving a variety of different signaling molecules (see Section 4 of this review and the chapter by Takeda and Akira, this volume).

2
SMADs: Transcriptional Mediators of TGF-β Signals

The discovery of the founding member of the SMAD family, dMAD, by genetic screens in the fruit fly *Drosophila* (Sekelsky et al. 1995) was rapidly followed by the identification of a total of eight SMAD proteins in mammals (Kretzschmar and Massagué 1998). Based on structural and functional criteria, the SMAD proteins can be divided into three subgroups. Subgroup one (receptor-activated SMADs or R-SMADs) comprises SMADs that are directly phosphorylated and activated by type I TGF-β receptor kinases (i.e., Smads 1–3, 5, and 8). The second subgroup (Co-SMADs) includes SMADs that are not substrates of the receptor kinases but participate in signaling by associating with receptor-activated SMADs (i.e., Smad4). Subgroup three (inhibitory SMADs or I-SMADs) consists of proteins that inhibit SMAD signaling activity by various mechanisms (i.e., Smads 6 and 7) (Shi and Massague 2003). Representatives of each subgroup of SMADs are found in species ranging from worms, flys, and frogs to humans.

Structurally SMAD proteins consist of three major domains, the highly conserved amino-terminal (MAD-homology 1) MH1 and carboxy-terminal MH2 domains which are separated by a more divergent linker domain.

The MH1 domains of all R-SMADs (with the exception of the major splice form of Smad2) and the Co-SMAD Smad4 possess sequence-specific DNA-binding activity. The major splice form of Smad2 contains a unique 30-residue insert in the MH1 domain that disrupts the structure of the conserved DNA-binding motif. However, an alternatively spliced isoform of Smad2 has been detected that lacks this insert and therefore has its DNA-binding activity restored (Yagi et al. 1999). I-SMADs display only weak homology to the R-SMADs in their amino-terminal domains and they do not bind DNA.

The DNA-binding motif of Smad3 was characterized by a crystal structure of the Smad3 MH1 domain bound to an oligonucleotide containing a Smad3-binding site (Shi et al. 1998). Interestingly, the β hairpin that makes specific contacts to the DNA as well as the surrounding residues are highly conserved among Smad4 and the R-SMADs, suggesting that all these SMADs have a very similar DNA-binding specificity. A basic SMAD-binding element, consisting of the four base pair sequence 5'-AGAC-3', was initially identified as the optimal binding site for Smad3 and Smad4 (Dennler et al. 1998; Yingling et al. 1997; Zawel et al. 1998). Many SMAD binding sites found in natural promoters contain an additional C at the 5' end, making it a 5'-CAGAC-3' sequence. In addition to the 5'-(C)AGAC-3' sequence SMADs have also been reported to bind to G/C-rich sequences. This preference for G/C-rich sequences might be mediated by a highly positively charged helix (helix H2) of the MH1 domain which is located next to the DNA-binding β hairpin in the crystal structure (Shi and Massague 2003). Importantly, as the DNA-binding specificity of the various SMADs is so similar, SMAD-specific target gene selection must involve additional mechanisms. This idea has been validated by the detailed analysis of a number of different SMAD target genes: selection of these genes depends on the specific interaction between SMADs and other DNA-binding transcription factors in the nucleus (see Section 3 of this review and the chapter by Zacharewski and La Pres, this volume).

Besides the DNA-binding activity the MH1 domain of SMADs also fulfills other functions. It negatively regulates the activity of the MH2 domain, it plays a role in nuclear import, and it mediates protein–protein interactions with other transcription factors in the nucleus.

The MH2 domain is regarded as an important effector domain of the SMADs, since it mediates both homomeric and heteromeric SMAD interactions, interactions with the nuclear pore complex for nucleocytoplasmic shuttling, and interactions with a plethora of transcription factors for transcriptional regulation of target genes. In addition, the MH2 domain is also responsible for interactions with the TGF-β type I receptors and in the case of the R-SMADs it contains the phosphorylation sites that are targeted by the receptor kinase activity (see below). Thus, the MH2 domain contributes essential protein–protein interactions in all stages of the SMAD signaling pathway and in the case of the R-SMADs contains the critical phosphorylation sites that initiate the pathway.

Resolution of the crystal structure of the MH2 domains of Smad2 and Smad4 have shed light onto the structural basis for some of the important protein–protein interactions (Shi et al. 1997; Wu et al. 2000). The Smad4 MH2 domain forms a homotrimer in crystals as well as in solution which is formed via a protein–protein interface with key residues that are highly conserved among the SMADs. A loop structure (the 'L3 loop') protrudes from each subunit of the trimeric complex, providing sites for heteromeric SMAD interactions as well as for interactions with the TGF-β type I receptor (Shi et al. 1997). Based on sequence similarities it is predicted that this overall structure is conserved in all the R-SMADs. Cancer-associated mutations in the SMAD interface residues destabilize the homotrimer whereas mutations in the Smad4 L3 loop prevent interaction with receptor-activated Smad2, thereby leading to inactivation of the pathway (see Sect. 6.1).

The binding affinity between R-SMADs and the type I receptor depends on a positively charged surface patch that is adjacent to the L3 loop (Wu et al. 2000). This patch is missing in the Co-SMAD Smad4 which does not bind to the receptor. The specificity of the R-SMAD–type I receptor interaction is determined primarily by the sequence of the L3 loop on the SMAD side and a loop structure (the 'L45 loop') on the receptor side (Chen et al. 1998b; Feng and Derynck 1997; Lo et al. 1998). Matching sets of L3 and L45 loops therefore specify which receptor interacts with which R-SMAD.

All R-SMADs contain a conserved SSXS motif at their carboxy-terminal end which serves as a phosphorylation site for activated TGF-β type I receptor kinases (Kretzschmar et al. 1997; Macias-Silva et al. 1996). Phosphorylation at this motif triggers R-SMAD association with the Co-SMAD Smad4, nuclear translocation and transcriptional activity of the SMAD complex. Structural studies revealed that this phosphorylation event creates a functional interface which mediates specific interactions with the MH2 domain of other SMAD molecules, including Smad4 (Wu et al. 2001). The phosphoserine binding surface on the MH2 domain overlaps with the surface that is required for docking of the R-SMADs to the activated type I receptors. Hence, phosphorylation of the SSXS motif leads to a dissociation of the R-SMAD from the receptor and to association with Smad4, facilitating the movement of the SMAD complex into the nucleus.

Even though the linker domains are the most divergent regions of the SMADs, they contain several regulatory phosphorylation sites as well as a PY protein interaction motif. A cluster of several mitogen activated protein (MAP) kinase phosphorylation sites is conserved in the R-SMADs and mediates cross-talk with other growth factor signaling pathways, such as the Ras/MEK/Erk MAP kinase pathway activated via receptor tyrosine kinases (Kretzschmar et al. 1997, 1999). Phosphorylation at these sites leads to inhibition of nuclear accumulation of R-SMADs, thereby affecting their ability to regulate transcription. This mechanism plays important roles in regulating cell competence to activin signals during mesoderm formation (Grimm and Gurdon 2002) and in the inhibition of BMP signals during neural induction (Pera et al. 2003; Sater et al. 2003). The linker domains also contain separate phosphorylation sites for calcium–calmodulin-dependent protein kinase II (Wicks et al. 2000). Phosphorylation of Smad2 at these sites similarly leads to an inhibition of SMAD nuclear accumulation.

The PY motif serves as a binding site for Smurfs, E3 ubiquitin ligases which link the SMADs to the proteasome-dependent protein degradation mechanism. Smurf1 targets Smad1 and Smad5 for destruction in the cytoplasm, possibly regulating the basal levels of Smad1/5 activity in unstimulated cells (Zhu et al. 1999). Smurf2 targets Smad1 and Smad2 but not Smad3 for proteasome-dependent degradation, thereby regulating the competence of cells to various TGF-β family growth factors (Lin et al. 2000; Zhang et al. 2001). Smad3 degradation is mediated by yet another ubiquitin ligase complex involving ROC1 (Fukuchi et al. 2001).

3
SMAD Interacting Proteins in the Nucleus

Structural and biochemical studies demonstrated that the DNA-binding specificities among various SMAD complexes, e.g., those activated by TGF-β, activin or BMPs, are very similar, and that their affinities are relatively low (Shi et al. 1998; Zawel et al. 1998). Selectivity in target gene regulation by

SMAD complexes is therefore in many cases achieved by interactions of SMADs with other transcription factors that bind to DNA via their own sequence-specific recognition motifs. These DNA-binding cofactors include many cell-type and/or cell-state restricted factors, but also more ubiquitously expressed factors.

As an example, receptor-activated Smad2 forms a complex with the winged-helix protein FAST-1 on the activin-response element of the *mix-2* gene promoter (Chen et al. 1996), and with the paired-like homeodomain proteins milk and mixer on the activin-inducible promoter of the *goosecoid* gene (Germain et al. 2000). Similarly, Smad1 associates with the zinc finger transcription factor OAZ on the BMP-inducible *vent-2* gene promoter (Hata et al. 2000). These SMAD-interacting factors are expressed in a cell-restricted fashion. In several cases SMAD-binding sites are located in close proximity to the binding sites of the auxiliary factors (Chen et al. 1996; Hata et al. 2000), whereas in other cases SMADs are recruited to the promoter exclusively through the DNA-binding capacity of the auxiliary factor (Germain et al. 2000).

Once a SMAD complex has been formed on the promoter SMADs associate with coactivators or corepressors which in turn recruit chromatin modifying enzymes. TGF-β-activated SMAD complexes interact with the corepressors TGIF, Ski, or SnoN (Akiyoshi et al. 1999; Luo et al. 1999; Stroschein et al. 1999; Sun et al. 1999; Wotton et al. 1999) which then recruit histone deacetylases to promote a transcriptionally repressed state of the chromatin via deacetylation of nucleosomal histones (Nomura et al. 1999; Wotton et al. 1999). The same SMAD complexes can alternatively interact with coactivators, such as p300 and CBP, that possess histone acetyl transferase (HAT) activity and thereby promote a transcriptionally active state of the chromatin (Feng et al. 1998; Janknecht et al. 1998; Nishihara et al. 1998; Pouponnot et al. 1998; Shen et al. 1998). Additional SMAD-interacting coactivators have been described, including SMIF, MSG1 and ARC105 (Bai et al. 2002; Kato et al. 2002; Yahata et al. 2000). SMIF enhances TGF-β/SMAD signaling via its direct interaction with the Co-SMAD Smad4 (Bai et al. 2002), while ARC105, a component of the Mediator complex, specifically interacts with Smad2/3-containing but not Smad1-containing transcription complexes and plays an essential role in TGF-β/activin/nodal signaling (Kato et al. 2002). MSG1 enhances TGF-β/SMAD signaling by stabilizing the interactions between SMADs and p300/CBP (Yahata et al. 2000).

The observation that the function of activated SMAD complexes in many cases depends on the interaction with cell type-specific, SMAD-specific, DNA-binding transcription factors is highly relevant for our understanding of the generation of temporal and spatial specificity in the signals by TGF-β family growth factors. The same growth factor can induce a different set of genes and thereby elicit a different cellular response depending on the cell-specific presence or absence of interacting transcription factors. This is important during developmental processes (Germain et al. 2000), and most likely also in physiological and pathophysiological processes in the adult, including wound healing, neoangiogenesis, and tumor progression. Identifying the players that mediate the cell type- and cell state-specific responses to TGF-β-related factors is an important challenge for this field of research.

A more comprehensive listing of SMAD-interacting factors has been provided recently (Moustakas et al. 2001). Here only a few selected examples will be discussed to illustrate the variety of the different interactions. Additional SMAD-interacting transcription factors are discussed in the subsequent chapters of this review.

FAST-1. Xenopus *f*orkhead *a*ctivin *s*ignal *t*ransducer 1 (FAST-1), a member of the winged-helix family of DNA-binding proteins, was the first transcription factor partner for SMADs to be identified (Chen et al. 1996). Studies of the Smad3/Smad4/FAST-1 transcription factor complex formed on the *Mix-2* gene promoter provided the paradigm for the mechanisms of target gene selection by SMADs that involves cooperative interactions between SMADs and other DNA-binding transcription factors (Chen et al. 1997; Liu et al. 1997).

OAZ. The 30-zinc finger protein OAZ interacts with BMP-activated SMADs to regulate the transcription of the *Xenopus* gene *Xvent-2* (Hata et al. 2000). OAZ uses different zinc fingers to contact SMADs and to bind to a specific DNA element in the *Xvent-2* promoter. Only a complex comprising both OAZ and SMADs is capable of mediating *Xvent-2* activation in response to BMP signaling (Hata et al. 2000). Hence, this study extended the paradigm of target gene selection to the BMP-regulated SMADs. OAZ is also expressed in a cell-type and cell state-restricted fashion, contributing to the tight regulation of *Xvent-2* expression during development.

p53. A recent study described TGF-β-inducible, direct physical interactions between the tumor suppressor protein p53 and the TGF-β mediators Smad2 and Smad3 (Cordenonsi et al. 2003). Furthermore, the TGF-β-induced cytostatic response as well as the transcriptional activation of the cyclin-dependent kinase (CDK) inhibitor $p21^{Cip1}$ were impaired in p53 deficient cells. The promoter of the *Mix-2* gene requires both p53 and separate SMAD-binding sites for full TGF-β induction. Hence, this study uncovered an unexpected direct link between two of the major tumor suppressor pathways in mammalian cells and suggests that a subset of TGF-β target genes, some of which are critical for the cytostatic response, require cooperation between p53 and SMADs at the transcriptional level (Cordenonsi et al. 2003).

bHLHZIP Proteins. The basic helix-loop-helix leucine-zipper (bHLHZIP) family members TFE3, TFEB and Max are all able to associate with Smad3 and Smad4 (Grinberg and Kerppola 2003). These interactions are mediated by the leucine-zipper and the MH1 domains of the respective proteins, and based on the distant relationship between TFE3 and Max within the family it is likely that SMADs may also interact with additional bHLHZIP family members. bHLHZIP transcription factors bind to DNA via their specific recognition sequence, the E-box. Direct interactions between TFE3 and Smad3 mediate synergistic activation of the *PAI-1* and *Smad7* genes in response to TGF-β (Hua et al. 1998, 2000). In both gene promoters SMAD-binding elements are separated from an E-box by a spacer of three nucleotides, suggesting that such an arrangement of sites may be characteristic of a subset of TGF-β target genes. Functionally, the interactions between bHLHZIP family members and SMADs may have different consequences. For example, while the interaction between TFE3 and SMADs facilitates synergistic activation of *PAI-1*, interaction between Max and SMADs inhibits TGF-β activation of the same gene (Grinberg and Kerppola 2003). These distinct outcomes are likely explained by the differential recruitment of coactivators such as p300/CBP or corepressors such as mSin3 and N-CoR by the related TFE3 and Max proteins.

ZEB Proteins. The two-handed zinc finger proteins ZEB-1 and ZEB-2 both interact with the MH2 domain of activated SMAD proteins in the TGF-β and BMP pathways (Postigo 2003). Their interaction with SMADs has opposite effects on TGF-β-induced transcriptional activation, with ZEB-1 acting synergistically and ZEB-2 acting antagonistically. These opposite effects are mediated through differential recruitment of coactivators (p300, P/CAF) and corepressors (carboxy-terminal-binding protein; CtBP) to the SMAD complex. Hence, the ZEB proteins provide another example (similar to the bHLHZIP proteins) where the direct interaction of related family members with SMADs has opposite consequences for TGF-β-mediated transcriptional activation of target genes.

AML/Runx Proteins. The three Runt domain transcription factors, AML-1/Runx1, AML-3/Runx2 and AML-2/Runx3, are essential components of the signaling responses to both TGF-β and BMPs in several important biological processes, including the regulation of hematopoietic cells (Runx1 and 3), osteogenesis (Runx2), and apoptosis in gastric epithelial cells (Runx3) (Ito and Miyazono 2003). Each of the R-SMADs interacts directly with each of the three Runx factors when overexpressed (Hanai et al. 1999). In addition, interactions of endogenous Smad1 and Runx2 in mesenchymal cells, as well as of endogenous Smad3 and Runx2 in osteoblast-like cells have been observed in response to BMP or TGF-β treatment, respectively, indicating that these interactions are physiologically relevant (Alliston et al. 2001; Zhang et al. 2000). TGF-β-induced interaction of Smad3 with Runx2 occurred at the Runx2-binding element of a target gene promoter, leading to repression of the transcriptional activity of Runx2 (Alliston et al. 2001). This repression occurred in mesenchymal but not in epithelial cells and was dependent on the promoter sequence. Hence, in contrast with other described examples this interaction between Smad3 and Runx2 results in transcriptional repression rather than activation of the target promoter.

4
SMAD-Independent TGF-β Signaling Mechanisms

While the SMAD pathways have been recognized as major signaling pathways activated by TGF-β growth factors, there is also strong evidence for the importance of various other signaling cascades in mediating specific TGF-β growth factor responses. This notion is consistent with the observation that some cell lines which are deficient in the co-SMAD Smad4 retain certain TGF-β responsiveness (Dai et al. 1999; Sirard et al. 2000). More directly, several laboratories have reported the rapid activation of MAP-kinases, including Erk, c-Jun N-terminal kinase (JNK) and p38, by TGF-β in various cellular systems. TGF-β-induced activation of JNK appears to occur independently of SMADs (Engel et al. 1999; Itoh et al. 2003) and leads to the rapid phosphorylation of the transcription factors c-Jun and ATF-2 (Engel et al. 1999; Hocevar et al. 1999; Sano et al. 1999a). These JNK-activated transcription factors mediate certain TGF-β-dependent gene responses independently of activated SMAD complexes, as observed for the transcriptional upregulation of the *fibronectin* gene (Hocevar et al. 1999). This is consistent with the finding that TGF-β induction of fibronectin synthesis is not affected by the lack of Smad2 or Smad3 in knockout fibroblasts (Piek et al. 2001).

The relevance of MAP kinase activation for TGF-β responsiveness greatly varies between different cellular systems. For instance, the inhibition of the MEK/Erk MAPK pathway leads to a block of TGF-β induction of epithelial-to-mesenchymal transition (EMT) in kidney cells (MDCK) (Peinado et al. 2003) but has no such effect on EMT induction in mammary epithelial cells (NMuMG) (Yu et al. 2002). In contrast, EMT induction in the latter is blocked when the p38 MAP kinase pathway is inhibited (Yu et al. 2002).

The biochemical link between the Alk-5 receptor and the MAP kinases has not been elucidated yet. However, the MAP-kinase kinase kinase (MAPKKK) TAK1 (TGF-β-activated kinase 1) functions upstream of JNK and p38 in a MAP kinase cascade and is involved in TGF-β signaling in mammalian cells (Yamaguchi et al. 1995) and in BMP signaling in early embryonic development in *Xenopus* (Shibuya et al. 1998; Yamaguchi et al. 1999). How TAK1 is functionally connected to TGF-β or BMP receptors is also not clearly understood. Possible links include XIAP (X-chromosome linked inhibitor of apoptosis protein) which has been suggested to serve as an adaptor between the BMP receptors and TAK1 (Shibuya et al. 1996; Yamaguchi et al. 1999) and HPK1 (hematopoietic progenitor kinase 1) which has been suggested to function upstream of TAK1 in TGF-β-mediated activation of JNK in 293T cells (Zhou et al. 1999). More recently, the adaptor protein disabled-2 (Dab2) was identified as a TGF-β receptor interacting protein that is required for the activation of SMADs by the receptor complex (Hocevar et al. 2001). This raises the possibility that Dab2, which possesses a phosphotyrosine-binding domain, or a similar adaptor molecule is responsible for linking the type I receptor to SMADs and/or to other downstream signaling mediators (Hocevar et al. 2001). In addition to MAP kinases, TGF-β also rapidly activates protein phosphatase-2A (PP2A), PI3 kinase, the GTPase RhoA, and Akt kinase in certain experimental settings, and these signaling pathways contribute to the effects of TGF-β on cell cycle arrest, EMT, and cell survival (see Sects. 5.1–5.3).

At the transcriptional level the SMAD-independent pathways may affect the expression of a separate set of target genes via the activation of downstream transcription factors such as AP-1. However, they may also interact with the SMAD pathways in a synergistic or antagonistic manner on an overlapping set of target genes. Examples for such a mechanism are provided by common target genes such as *collagenase I* and *endothelin-1* that contain binding sites for both SMADs and AP-1 and are regulated by these factors in a cooperative manner (Qing et al. 2000; Rodriguez-Pascual et al. 2003).

Taken together, these results establish that TGF-β family receptors utilize a variety of signaling mechanisms besides the SMAD pathways to mediate their biological responses and that at least some of these pathways converge with the SMAD pathways at the transcriptional level.

5
TGF-β Signaling in Normal Epithelial and Carcinoma Cells: Roles in Tumor Suppression and Promotion

In normal epithelia TGF-β plays a key role in the control of tissue homeostasis and/or remodeling by regulating diverse cellular processes such as proliferation, adhesion, migration, differentiation, and cell death or survival. An excellent example is the mammary gland where TGF-β regulates the development, differentiation, and involution of the mammary epithelium during puberty, estrous cycles, pregnancy, and weaning (Barcellos-Hoff and Ewan 2000). The three isoforms of TGF-β are expressed in the epithelium during all phases of mammary development (Robinson et al. 1991) and exogenous delivery of active TGF-β to the mammary epithelium causes potent and reversible inhibition of mammary growth and morphogenesis (Jhappan et al. 1993; Pierce et al. 1993; Silberstein and Daniel 1987).

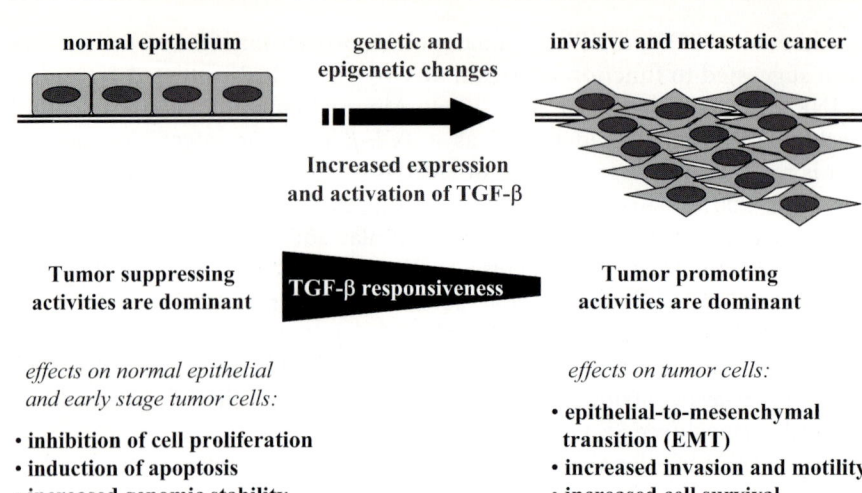

Fig. 2 Changes in TGF-β responsiveness and function that occur during tumor progression. The tumor suppressive activities of TGF-β in normal epithelial cells and early stage tumor cells are contrasted with the tumor promoting activities of TGF-β in invasive and metastatic cancer. *ECM*, extracellular matrix. (Modified from Wakefield and Roberts 2002)

During tumor formation and progression the TGF-β responsiveness of the transformed epithelial cells changes (Fig. 2). While the tumor suppressor functions of TGF-β, including the inhibition of cell proliferation and induction of apoptotic cell death, are compromised, other cellular responses that contribute to tumor progression become more dominant (Derynck et al. 2001; Wakefield and Roberts 2002). These tumor promoting functions include direct effects on the tumor cells, leading to EMT transition, increased invasion and motility, increased cell survival, and enhanced metastatic ability. TGF-β also has important effects on blood vessel forming endothelial cells, immune cells, and stromal cells. Here we will restrict our discussion to TGF-β effects on normal and transformed epithelial cells as well as endothelial cells, and refer to other review articles for discussion of TGF-β effects on the immune system and the tumor stroma.

5.1
Apoptosis and Survival

The key role of TGF-β in tissue homeostasis and remodeling is in large part mediated by the context-dependent control of cell death and survival. Pro-apoptotic signaling is considered as an important component of the tumor suppressor activity of TGF-β, suggesting that this function may be lost during breast cancer progression, similar to the cell cycle arrest response (Kretzschmar 2000).

TGF-β-induced apoptosis is observed in various cell types and tissues, including the mammary and prostate epithelium, the vascular endothelium, hepatocytes in the liver, and B- and T-cell populations in the immune system. In the mammary gland tissue-specific transgene expression of TGF-β1 causes elevated epithelial cell apoptosis during various stages of mammary development (Kordon et al. 1995). Furthermore, transplantation experiments provided direct evidence that TGF-β induces epithelial cell apoptosis during the involution of the mammary epithelium following weaning (Nguyen and Pollard 2000).

Even though TGF-β-induced apoptosis has been studied in several cell types, including hematopoietic and epithelial cells, our understanding of the involved signaling pathways and mediators is still fragmentary. Several molecules have recently been identified as potential effectors of TGF-β-induced apoptosis. These include the septin-like mitochondrial protein ARTS in rat prostate epithelial cells (Larisch et al. 2000), the TGF-β type II receptor-binding protein DAXX in B-lymphoma cells and mouse hepatocytes (Perlman et al. 2001), the lipid phosphatase SHIP in hematopoietic cell types (Valderrama-Carvajal et al. 2002), the inhibitory SMAD, Smad7, in prostate carcinoma cells (Landstrom et al. 2000), and the death-associated protein kinase (DAP-kinase) in hepatoma cells (Jang et al. 2002). Interestingly, SHIP and DAP-kinase are direct transcriptional targets of the Smad2/3 mediators (Jang et al. 2002; Valderrama-Carvajal et al. 2002), consistent with TGF-β-induced apoptosis being an Alk-5 receptor and SMAD-dependent process in these cell types (Patil et al. 2000; Yamamura et al. 2000).

Reports on the role of Smad7, an inhibitory SMAD that is directly upregulated by the Alk-5/Smad2/3 pathway and that blocks signaling by the Alk-5 receptor in a negative feedback loop, have been conflicting. Smad7 has been alternately described as an inhibitor or an effector of TGF-β-induced apoptosis in lymphocytes and prostate carcinoma cells (Landstrom et al. 2000; Patil et al. 2000). Interestingly, in cultured mammary epithelial cells it has been difficult to observe TGF-β induction of apoptosis (Shin et al. 2001; Yang et al. 2002b). However, one study has proposed an Alk-5 mediated,

SMAD-independent mechanism of TGF-β-induced apoptosis that depends on the activation of the p38 MAP-kinase pathway (Yu et al. 2002). In contrast, an in vivo analysis implicated the Smad3 pathway in TGF-β-induced mammary epithelial cell apoptosis (Yang et al. 2002b). Thus, while a number of molecules and mechanisms have been implicated in TGF-β-induced apoptosis, a coherent picture of the mechanisms of this process has yet to emerge.

Complicating matters further is a report that TGF-β can also promote cell survival instead of inducing apoptosis, as observed in a mouse mammary epithelial cell line (Shin et al. 2001). This pro-survival effect is mediated by activation of Akt kinase which leads to the phosphorylation and consequent inhibition of the forkhead transcription factor FKHRL1 (Shin et al. 2001). Taken together these results suggest that the outcome of TGF-β action on cell death or survival may differ depending on the cell type and the cell context. This notion is supported by the analysis of epithelial cell death in the mammary glands of TGF-β1 heterozygote mice. In these mice decreased mammary epithelial cell apoptosis is observed during estrus, while elevated mammary epithelial cell apoptosis is observed during proestrus and in early pregnancy, suggesting that depending on the stage of mammary gland development TGF-β may act either as a survival factor or a death-inducing factor (Ewan et al. 2002).

In gastric epithelial cells a putative transcriptional interaction partner of the SMAD proteins in the apoptotic response has recently been identified (Li et al. 2002). This factor, Runx3, belongs to the family of Runt domain transcription factors, which comprises three related members, Runx1, 2 and 3 (Ito and Miyazono 2003). Each of the receptor-activated SMADs can interact directly with each of the three Runx transcription factors (Hanai et al. 1999). (see Section 3). The three Runx factors play important roles in various biological processes that are also regulated by TGF-β growth factors (Ito and Miyazono 2003). Importantly, Runx3 was identified as a tumor suppressor gene in gastric epithelial cells which is inactivated by hemizygous deletion and promoter hypermethylation in about 40% of early stage carcinomas and in nearly 90% of advanced stage carcinomas (Li et al. 2002). Homozygous deletion of Runx3 in normal gastric epithelial cells leads to a complete loss of sensitivity to TGF-β-induced apoptosis, implicating Runx3 as an essential mediator of TGF-β-induced apoptosis in gastric epithelial cells (Li et al. 2002). Hence, Runx3 may act as a tumor suppressor by mediating the pro-apoptotic function of TGF-β, possibly by directing the transcriptional expression of pro-apoptotic genes in a direct interaction with the SMAD proteins (Balmain 2002; Li et al. 2002).

5.2
Cell Cycle Arrest

TGF-β induces growth inhibition of epithelial cells by arresting them in the G_1 phase of the cell cycle, in some circumstances leading to terminal differentiation or induction of apoptosis. G_1 arrest is achieved by several mechanisms which may act in a complimentary fashion in the same cell and include the transcriptional upregulation of the CDK inhibitors p15^{Ink4B} and p21$^{cip1/waf1}$, the transcriptional downregulation of the CDK-activating phosphatase Cdc25A and the proto-oncogene c-*myc*, and the post-transcriptional mobilization of inactive pools of the CDK2 inhibitor p27^{Kip1}. Together these events lead to a hypophosphorylated, activated state of the pRB tumor suppressor protein, the consequent inhibition of target gene activation by the pRB controlled transcription factors E2F-1 to -3, and hence to cell cycle arrest in the G_1 phase (Massague et al. 2000). The cytostatic gene expression program is activated by TGF-β signaling via the type I receptor Alk-5 and the Smad2/3/4 signaling pathway that leads to direct transcriptional regulation of some of the critical cell cycle regulators.

c-Myc is a central molecule in the G_0/G_1 to S phase transition as it directly controls the transcription of other regulatory genes involved in this process, such as cyclins D1 and D2, the cyclin D partner CDK4, the CDK phosphatase cdc25A, and the transcription factors E2F-1 to -3 (Pelengaris et al. 2002). C-Myc also controls the expression of the CDK inhibitors p15^{Ink4b} and p21^{Cip1} by direct transcriptional repression (Herold et al. 2002; Seoane et al. 2001, 2002; Staller et al. 2001; van de Wetering et al. 2002). In quiescent or cell cycle arrested cells c-*myc* is in a transcriptionally repressed state, but is strongly induced during the transition from G_0/G_1 to S phase and is subsequently expressed at elevated levels in actively proliferating cells (Pelengaris et al. 2002). A promoter region upstream of the major transcriptional initiation site (P2) mediates the rapid inhibition of c-*myc* expression in response to TGF-β activated Alk-5 signaling in proliferating cells (Chen et al. 2001a).

This cell cycle-sensitive proximal promoter region contains an E2F-binding site that is critical for c-*myc* regulation and is occupied in vivo by E2F-1, -2, or -4 (Albert et al. 2001; Weinmann et al. 2001). Many E2F responsive genes are transcriptionally silenced in quiescent or cell cycle arrested cells by complexes containing E2F-4 and the pocket protein p130 which induce a transcriptionally repressed chromatin conformation (Rayman et al. 2002). The E2F-binding site, in conjunction with surrounding sequences, is indeed responsible for the TGF-β/Alk-5 responsiveness of the c-*myc* promoter (Chen et al. 2002; Yagi et al. 2002). TGF-β/Alk-5 signaling induces the formation of a transcriptional complex consisting of Smad3/4, E2F-4/5 and p107 proteins on this regulatory site, leading to rapid repression of c-*myc* (Chen et al. 2002; Yagi et al. 2002). Within this TGF-β responsive element the Smad3/4 complex binds to the DNA sequence that is directly adjacent to the E2F-binding site, and mutation of either recognition sequence causes a loss of TGF-β responsiveness of the c-*myc* promoter (Chen et al. 2001a, 2002; Yagi et al. 2002). The c-*myc* promoter therefore provides another example of the concept that SMAD transcription complexes must interact with specific DNA-binding cofactors to select and regulate their target genes.

Another important cytostatic gene response to TGF-β in epithelial cells is the activation of the CDK inhibitor $p15^{Ink4b}$. This transcriptional response is mediated by a SMAD complex that may also involve SP1 and perhaps other unidentified transcription factors (Feng et al. 2000; Seoane et al. 2001). In addition, $p15^{Ink4b}$ expression is directly repressed by c-Myc through its interaction with the transcription factor Miz-1 on the initiator region of the $p15^{Ink4b}$ promoter (Feng et al. 2000; Seoane et al. 2001; Staller et al. 2001) and this repression is relieved by TGF-β-induced inhibition of c-*myc* expression. Hence, the cytostatic response to TGF-β in epithelial cells entails the SMAD-mediated transcriptional repression of one gene, c-*myc*, and transcriptional activation of another gene, $p15^{Ink4b}$, in the same cell. This provides an excellent example of how the quality of TGF-β-induced gene responses depends on the promoter-specific interaction of the same SMAD complex with distinct DNA-binding transcription factors and the consequent recruitment of either transcriptional coactivators or corepressors.

In addition to the SMAD-mediated cytostatic response TGF-β has also been suggested to induce cell cycle arrest by a second, independent mechanism (Petritsch et al. 2000). This mechanism involves the inhibition of p70(s6 k), a serine/threonine kinase that is essential for G_1 to S phase progression by directly controlling the translation initiation machinery required for translation of essential mRNAs (Thomas and Hall 1997). In a mouse mammary epithelial cell line the inhibition of SMAD signaling was not sufficient to release the cells from TGF-β-induced G_1 arrest, suggesting that additional pathways may be involved in growth inhibition by TGF-β (Petritsch et al. 2000). Upon cell stimulation with TGF-β the Bα subunit of the PP2A directly associates with the TGF-β receptor Alk-5, triggering an interaction between PP2A and p70(s6 k) that results in the dephosphorylation and inactivation of p70(s6 k). While either p70(s6 k) activity or SMAD-mediated transcription was sufficient to induce cell cycle arrest, inhibition of both pathways was required to release the epithelial cells from TGF-β-induced cell cycle arrest. Hence, TGF-β controls cell cycle progression at both the transcriptional and translational levels.

5.3
Epithelial-to-Mesenchymal Transition

EMT is a process in which epithelial cells (1) downregulate epithelial-specific proteins, such as tight-junction, adherens-junction, and desmosome proteins (e.g., E-cadherin, ZO-1, desmoplakin); (2) induce various mesenchymal proteins (e.g., vimentin, fibronectin); (3) remodel the actin cytoskeleton; (4) acquire a fibroblastoid morphology; and (5) gain migratory and invasive properties that facilitate the digestion and movement through the extracellular matrix (Hay 1995; Thiery 2002). Thus, EMT is a mechanism that allows epithelial cells to dissociate from one location and to migrate to different locations where they again differentiate into specialized cell types. Such processes give rise to individual migratory cells that are crucial for developmental processes such as mesoderm formation during gastrulation and the emigration of neural crest cells from the neural tube (Hay 1995). Similar epithelial plasticity also contributes to tubulogenesis and branching in the mammary gland (Affolter et al. 2003). Furthermore, EMT-like cellular changes are thought to play an important role in the progression of epithelial cancers (Thiery 2002). Epithelial cells from benign tumors (adenomas) display altered epithelial polarity, and more de-differentiated malignant tumors are characterized by the acquisition of invasive and migratory properties of the tumor cells. These changes thereby constitute an initial step in the formation of metastasis. Interestingly, TGF-β is one of several growth factors capable of inducing EMT in vitro and in vivo and elevated levels of TGF-β are observed in human carcinomas that present tumor cells of fibroblastoid morphology and highly malignant and invasive phenotype (Grunert et al. 2003; Thiery 2002).

TGF-β induction of EMT has been studied in a variety of different cell types and assay systems, including normal and transformed epithelial cells in both two- and three-dimensional culture systems (Grunert et al. 2003). Perhaps the most detailed analyses were done using mammary ductal epithelial cell lines, such as NMuMG and EpRas. In these cell lines TGF-β induces EMT-associated changes within 24 h of treatment, including changes in cell morphology and the expression of epithelial and mesenchymal marker genes (Miettinen et al. 1994; Oft et al. 1996). A complete EMT in EpRas cells, however, requires prolonged (>4–6 days) TGF-β treatment (Oft et al. 1996).

A number of observations indicate that both SMAD-dependent and SMAD-independent signaling from the TGF-β receptor Alk-5 contributes to TGF-β induction of EMT. For example, overexpression of Smad2/4 or Smad3/4 in the presence of constitutively active Alk-5 receptor leads to the reorganization of the actin cytoskeleton and decreased and delocalized E-cadherin expression in NMuMG cells (Piek et al. 1999). Inhibition of SMAD signaling in the same cells by overexpression of a dominant-negative Smad3 mutant or the I-SMAD Smad7 causes a block of TGF-β induction of growth arrest and SMAD-mediated transcriptional responses but not of EMT (Bhowmick et al. 2001). Several signaling proteins have been identified that contribute to efficient progression of TGF-β-mediated EMT. Foremost among them are hyperactivated Ras and several downstream effector molecules. Analysis of the nontransformed mammary epithelial cell line EpH4 and the oncogenic Ras-expressing derivative cell line EpRas revealed that only in the presence of activated Ras signaling TGF-β is capable of inducing EMT (Oft et al. 1996). Hence, a cooperation between Ras and TGF-β signaling appears to be critical for EMT induction. This is consistent with studies of EMT in vivo, which showed that ligands for receptor tyrosine kinases (e.g., FGFs) cooperate with TGF-β family members in inducing EMT involving processes (Grunert et al. 2003).

Effectors of Ras function, including PI3-kinase, RhoA, Raf1 kinase, and the MEK/Erk MAP-kinase pathway have all been implicated in EMT induction as well. In particular, the Raf/MEK/Erk MAP-kinase pathway appears to be critical for the synergism between Ras and TGF-β in EMT induction (Ellenrieder et al. 2001; Grande et al. 2002; Janda et al. 2002). In some cell systems TGF-β itself can activate these Ras effector pathways in a SMAD-independent manner, thereby facilitating TGF-β induction of EMT. For example, in keratinocytes the ability of TGF-β to activate the MEK/Erk pathway and to cause a loss of E-cadherin-mediated adherens junctions, cell-cell separation and a fibroblastoid morphology is blocked by pharmacological inhibitors of the MEK/Erk pathway (Zavadil et al. 2001). Similarly, in NMuMG cells the inhibition of PI3-kinase or RhoA by dominant-negative constructs or pharmacological inhibitors blocked the ability of TGF-β to activate this pathway and to induce EMT (Bakin et al. 2000; Bhowmick et al. 2001). Finally, SMAD-independent activation of the p38 MAP-kinase pathway has also been described as a necessary but not sufficient event in the induction of EMT by TGF-β in NMuMG cells (Bakin et al. 2002; Yu et al. 2002). Together, the involvement of these various pathways in TGF-β induction of EMT illustrates the complexity of this process and indicates multiple points of control that serve to limit EMT to very specific events in normal physiology. Furthermore, the described studies imply that TGF-β induction of EMT involves both the activation of a transcriptional program by the SMADs as well as direct transcription-independent effects on the cytoskeleton and cell-cell junctions.

The molecular details of how these various pathways are activated by the TGF-β receptor and how they interact with each other in the induction of EMT require further characterization. Part of these complex interactions is most likely a direct cooperation between SMADs and other transcription factors that are activated by SMAD-independent pathways. Candidates include the transcription factors AP-1 and ATF-2 which are downstream effectors of the MAP-kinase pathways and directly interact with SMAD proteins on specific target gene promoters (Hanafusa et al. 1999; Sano et al. 1999b; Wong et al. 1999; Zhang et al. 1998). Furthermore, the β-catenin–Lef-1 transcription complex also interacts with SMAD proteins thereby connecting the Wnt and TGF-β signaling systems at the transcriptional level (Labbe et al. 2000). Both of these growth factors are inducers of EMT processes (Thiery 2002).

The identity of the most critical target genes of TGF-β signaling in the induction of EMT is still an open question, even though a gene expression profiling study in keratinocytes has revealed important leads (Zavadil et al. 2001). The downregulation of E-cadherin and the resulting loss of adherens junctions has attracted great interest since E-cadherin functions as a tumor suppressor and the loss of E-cadherin expression is a frequent event in many types of carcinoma. Several transcriptional repressors of *E-cadherin* have been identified, including members of the Snail family of zinc finger factors (Snail and Slug) (Batlle et al. 2000; Cano et al. 2000), members of the basic helix-loop-helix (bHLH) family (E47/12) (Perez-Moreno et al. 2001), and the zinc finger factor ZEB2/SIP1 (Comijn et al. 2001). All of these factors are able to bind specific E-boxes located in the *E-cadherin* promoter region and when overexpressed they inhibit *E-cadherin* transcription and the induction of EMT. Some of these factors may play an important role in TGF-β induction of EMT, since their expression levels are regulated upon TGF-β treatment of cells. Slug is rapidly and transiently induced by TGF-β treatment of keratinocytes (Zavadil et al. 2001), while Snail is upregulated in a SMAD-independent manner in canine kidney cells (MDCK) after long-term TGF-β treatment (Peinado et al. 2003). SIP1 is also upregulated in MDCK cells after extended TGF-β treatment (Comijn et al. 2001). Whether the regulation of these *E-cadherin* repressors is required or even sufficient for TGF-β induction of EMT remains to be determined.

5.4
Metastasis

There is clear in vivo evidence from both mice and humans that TGF-β signaling plays an important role in the progression of advanced tumors to the metastatic state, despite its function as a tumor suppressor in early tumorigenesis. Elevated levels of TGF-β are observed in the advancing edges of primary tumors and in lymph node metastases of breast cancer patients, as well as in most tumor cell lines derived from invasive breast carcinomas (Dalal et al. 1993; Gorsch et al. 1992). Immunostaining of surgical specimen also showed that TGF-β expression is significantly related to venous invasion and tumor staging in pancreatic cancer, and is associated with liver metastasis (Teraoka et al. 2001). High levels of circulating TGF-β in the plasma have been associated with poor prognosis and may serve as a predictor of metastasis in a variety of cancers, including colorectal, gastric, breast bladder, and prostate carcinoma (Ghellal et al. 2000; Saito et al. 2000; Shariat et al. 2001; Tsushima et al. 2001). Finally, patients with familial colon carcinoma with mismatch repair defects that show a loss of functional TGF-β type II receptor have a significantly better survival rate than those patients that have retained TGF-β signaling (Watanabe et al. 2001).

In mice the transgenic expression of TGF-β in the skin attenuated the formation of benign skin tumors (papillomas) but strongly enhanced their malignant progression to spindle cell carcinomas (Cui et al. 1996). In addition, inducible transgene expression of TGF-β in chemically induced papillomas leads to rapid metastasis, in contrast to its tumor suppressive effects when induced earlier in the tumorigenic process (Weeks et al. 2001). Overexpression of dominant-active Smad2 promoted metastasis formation in cell lines derived from Ras-transformed squamous carcinomas (Oft et al. 2002). In contrast, the expression of a dominant-negative TGF-β type II receptor to inhibit TGF-β signaling in various tumor cell lines suppressed their metastatic potential (McEarchern et al. 2001; Oft et al. 1998; Yin et al. 1999). The neutralization of endogenous TGF-β via the transgenic expression of a soluble TGF-β antagonist suppressed metastasis in tail vein metastasis assays and from endogenous mammary tumors of MMTV-neu mice (Yang et al. 2002a). Interestingly, the metastasis in these mice was suppressed without leading to an enhancement of primary tumorigenesis (Yang et al. 2002a). In a similar approach the systemic administration of a TGF-β antagonist did not alter primary mammary tumor latency in MMTV-Polyomavirus middle T antigen transgenic mice, but reduced tumor cell motility, intravasation, and lung metastases (Muraoka et al. 2002). In some of these studies the metastatic potential could be correlated with the induction of EMT and the associated gain of tumor cell invasiveness (Cui et al. 1996; Oft et al. 1998, 2002;). Hence, these results support the notion that cell autonomous TGF-β signaling may be a general requirement for the development of a highly metastatic tumor cell phenotype.

In addition to such a general role in enhancing the metastatic cell phenotype more recent studies have revealed an additional role of TGF-β signaling in organ site tropism of the metastasizing tumor cells. One of the major sites of metastatic burden in several types of human cancer, including breast and prostate cancer, is the skeletal bone. More than 80% of breast cancer patients with advanced disease show bone metastases, leading to bone destruction with all the accompanying symptoms (Coleman and Rubens 1987). The inhibition of TGF-β signaling by a dominant-negative TGF-β type II receptor expressed in a breast carcinoma cell line (MDA-MB-231) resulted in a reduction of the cell's ability to metastasize to the bone (Yin et al. 1999). This effect could be reversed by the concomitant expression of a constitutively active TGF-β type I receptor Alk-5. As a possible effector of TGF-β-induced bone metastasis the parathyroid hormone-related protein (PTHrP) was identified. TGF-β stimulates PTHrP expression in breast tumor cells by the cooperative action of two rapidly activated signaling pathways, the SMAD pathway and the p38 MAP-kinase pathway (Kakonen et al. 2002), possibly leading to transcriptional synergy on the PTHrP promoter between SMADs and p38 activated transcription factors such as ATF2. Elevated expression of PTHrP leads to increased bone metastasis and osteoclastic bone resorption, which in turn results in the release of active TGF-β. Hence, PTHrP and TGF-β may be involved in a paracrine loop that fosters and accelerates the formation of osteolytic bone metastasis (Kakonen et al. 2002; Yin et al. 1999).

A recent study investigated the molecular basis for bone metastasis by performing gene expression profiling on human breast cancer cell line subpopulations with differential ability to metastsize to the bone (Kang et al. 2003). Several genes were identified that are specifically overexpressed in those subpopulations which have high bone metastatic activity. Interestingly, two of these genes, *interleukin-11* and *CTGF*, are in addition induced by TGF-β treatment, suggesting that the upregulation of these genes in response to TGF-β plays an important role in this cytokine's bone metastasis enhancing function (Kang et al. 2003). Interleukin-11 is a potent inducer of the differentiation of progenitor cells into osteoclasts, the cells that directly mediate bone resorption in osteolytic bone metastases (Mundy 2002). Connective tissue growth factor (CTGF) is an inducer of osteoblast and cartilage cell proliferation and differentiation, as well as of neovascularization (Takigawa et al. 2003). As demonstrated by chromatin immunoprecipitation assays both of these genes rapidly bind Smad2/3 and Smad4 to their promoter regions in response to TGF-β treatment, indicating that they are direct transcriptional targets of the TGF-β activated SMAD complexes (Kang et al. 2003). An analysis of putative SMAD interaction partners required for the transcriptional activation of these genes holds the promise of identifying novel targets for blocking the bone metastatic function of TGF-β.

6
TGF-β Responses as Targets of Tumor-Associated Alterations

The tumor suppressor function of TGF-β observed in normal epithelial cells and early stage tumor cells is compromised in many types of human cancer, including breast cancer. These tumor suppressive effects, mediated by the inhibition of cell proliferation and the induction of apoptosis, can be lost by different molecular mechanisms. These include genetic events such as inactivating mutations in components of the signaling system as well as epigenetic events such as the modulation of SMAD signaling responses by aberrant activities of various regulatory molecules.

6.1
Inactivating Mutations in TGF-β Receptors and SMADs

Disruption of TGF-β/SMAD signaling by mutations in components of the signaling pathway has been observed in a variety of human cancers (Kretzschmar 2000; Massague et al. 2000). The TGF-β type II receptor

(TβR-II) is inactivated by mutation in a high percentage of certain cancers with microsatellite instability, i.e., colon cancers (~30%), gastric cancers (30-80%), and non small-cell lung cancers (~75%) (Gebert et al. 2000; Kim et al. 2000; Pinto et al. 2003) [for further references see (Kretzschmar 2000)]. In comparison, mutations in *TβR-II* are relatively rare in cancers of the pancreas, liver, and pituitary gland, and in myelodisplastic syndrome or endometrial cancers with microsatellite instability (Kretzschmar 2000). *TβR-II* mutations are also found at a lower frequency (~15%) in microsatellite stable colon cancers (Grady et al. 1999). Missense mutations in the kinase domain of *TβR-II* have been observed in several cases of recurrent breast cancer (Lucke et al. 2001). Overall, inactivation of *TβR-II* by mutation appears to be mostly selected for in gastrointestinal cancers (Markowitz and Roberts 1996) and certain other types of cancer with microsatellite instability.

Mutations in the TGF-β type I receptor Alk-5 are much less frequent; however, deletions were found in a small percentage of pancreatic and biliary adenocarcinomas (Goggins et al. 1998) and in one case of anaplastic large cell lymphoma (Schiemann et al. 1999). Furthermore, a polymorphism resulting in a small deletion has been observed in some colorectal and cervical carcinomas and homozygous carriers of this polymorphism may be at enhanced risk for cancer development (Chen et al. 1999; Pasche et al. 1999). A missense mutation (S387Y) has been reported in two of 31 primary breast carcinomas (6%) and five of 12 lymph node metastases (42%), indicating that this mutation may represent an important event in the malignant progression of breast cancer (Chen et al. 1998a). A somatic intragenic four base pair deletion and another missense mutation (A230T) have also been detected in head and neck cancer metastases (Chen et al. 2001b). These mutations lead to truncated or highly unstable receptor proteins with diminished signaling capacity. Besides inactivating mutations normal TGF-β receptor function may also be compromised by reduced expression of *TβR-I* and *TβR-II* genes (Kretzschmar 2000).

Inactivating mutations in *SMAD* genes are found in a number of human cancers, with the highest frequency in pancreatic and colon carcinomas. *Smad4/DPC4* was originally isolated as a tumor suppressor gene on chromosome 18q21 that is deleted or mutated in nearly half of all human pancreatic carcinomas (Hahn et al. 1996), and less frequently in other cancers (Kretzschmar 2000). *Smad2*, which is also located on 18q21, is mutated in a small number of colon, head and neck, and lung carcinomas, but appears to be unaffected in other types of carcinoma as well as leukemias and lymphomas. Inactivating mutations in *Smad3* have not been observed in a large number of tumors investigated, including those from gastrointestinal, breast, lung, ovarian, and pancreatic cancers. Using tissue microarrays expression of Smad2, receptor-phosphorylated Smad2, and Smad4 was demonstrated in a majority (92%) of human breast carcinomas, similar to breast cancer cell lines (Xie et al. 2002). Interestingly, among patients with stage II breast cancer the loss of phospho-Smad2 in the tumor strongly correlated with shorter overall survival (Xie et al. 2002).

Why inactivating mutations in *TβR-II* and *Smad4* are frequent in gastrointestinal, pancreatic, and lung cancers, but are rare in breast and other cancers is currently unclear. It has been suggested, however, that an intact TGF-β signaling pathway may be required for the malignant progression of certain carcinomas and that mutational inactivation of the pathway would therefore not confer a selective advantage (Derynck et al. 2001).

6.2
Other Mechanisms of Altered TGF-β Responsiveness

Numerous molecules have been identified that display altered activities in human cancer and that functionally or physically interact with the TGF-β/SMAD signaling pathway and thereby modulate or interfere with functions such as growth inhibition (Shi and Massague 2003). A few examples will be discussed to illustrate the variety of mechanisms that impact on TGF-β function.

Ras. Ras signaling is hyperactivated in a large percentage of human cancers due to oncogenic mutations or amplification and overexpression of Ras-activating growth factor receptors, such as the HER-2/Neu receptor tyrosine kinase in breast cancer (Malumbres and Barbacid 2003). TGF-β can override the proliferative effects of Ras-activating mitogens in normal epithelial cells; however, cells harboring oncogenic Ras mutations often show a loss of TGF-β anti-proliferative responses. Ras signaling can interfere with the anti-proliferative response at various levels, including the upregulation of the activities of CDKs which drive G_1 to S phase progression of the cell cycle (Marshall 1999). Oncogenic Ras can also modulate TGF-β/SMAD signaling by inducing direct phosphorylation of Smad2 and Smad3 in their linker domains via Erk MAP kinases (Kretzschmar et al. 1999). Mutation of these MAP kinase sites in Smad3 yields a Ras-resistant form that can partially rescue the growth inhibitory response to TGF-β in Ras-transformed cells (Calonge and Massague 1999; Kretzschmar et al. 1999). MAP kinase mediated phosphorylation negatively affects the nuclear accumulation of the SMADs and thereby also regulates their transcriptional activity (Kretzschmar et al. 1999). This mechanism plays important roles in various aspects of vertebrate development (Grimm and Gurdon 2002; Pera et al. 2003; Sater et al. 2003). The potency of MAP-kinase in inhibiting SMAD signaling seems to be highly dependent on cell-type and/or cell context and may be controlled by yet unidentified rate-limiting factors (Massagué 2003). The significance in human cancer remains to be explored.

Inhibitory SMADs. Several molecules have been identified that can interfere with SMAD signaling by competitive protein–protein interactions either in the cytoplasm or the nucleus. The two I-SMADs, Smad6 and Smad7, inhibit the formation of transcriptionally active SMAD complexes by ligand-induced association with the receptor complex (Hayashi et al. 1997; Imamura et al. 1997; Nakao et al. 1997) or Smad4 (Hata et al. 1998). Association with the receptor complex prevents the interaction between activated TβR-I and its SMAD substrates, thereby blocking the TGF-β-induced phosphorylation and activation of Smad2 and Smad3. Smad7 also recruits the E3 ubiquitin ligases Smurf1 and Smurf2 to the receptor complex, leading to increased proteasome-mediated turnover of the TGF-β type I receptor (Ebisawa et al. 2001; Kavsak et al. 2000). One physiological function of these I-SMADs is to provide a negative feedback regulation of TGF-β family signaling. Northern blot and in situ hybridization analyses indicated that Smad6 and 7 are overexpressed in pancreatic cancer tissues and cell lines, suggesting that elevated expression of inhibitory SMADs may contribute to the loss of growth inhibitory TGF-β responses in pancreatic carcinomas (Kleeff et al. 1999). Smad7 has also been proposed as a prognostic marker in patients with colorectal cancer (Boulay et al. 2003). In this study elevated expression levels of Smad7 due to gene duplication correlated in a dose-dependent manner with decreased patient survival, whereas loss of Smad7 expression due to deletion indicated a favorable outcome. Furthermore, the oncogenic receptor tyrosine kinase HER2/Neu can activate transcription of Smad7 in breast, endometrial and ovarian cancer cell lines, suggesting this as another mechanism of interference with the TGF-β/SMAD pathway by this oncogenic receptor (Dowdy et al. 2003).

Transcriptional Corepressors Ski and Sno-N. Another group of proteins that interfere with TGF-β/SMAD signaling by competitive protein–protein interactions are the Ski transcription factors. Two closely related members of this family, the proto-oncoproteins Ski and SnoN, directly associate with Smad2, 3, or 4 in the nucleus, recruit histone deacetylases, and thereby repress the ability of SMAD complexes to activate transcription (see the chapter by Zacharewski and La Pres, this volume). The recruitment of Ski and SnoN is likely to compete with binding of the transcriptional coactivators p300 and CBP that possess HAT activity. The acetylation state of core histones plays a critical role in transcriptional regulation (Kouzarides 1999) and thus the balance between SMAD corepressors and coactivators might determine the transcriptional activity of nuclear SMAD complexes. In addition to the recruitment of corepressors, binding of Skis to the SMAD complex may also disrupt the interface between R-SMADs and Smad4 by competitive displacement of the R-SMAD–Smad4 interaction (Wu et al. 2002). Skis contain independent binding sites for interaction with the MH2 domains of R-SMADs and Smad4, respectively, and the function of either one of these binding sites is sufficient for the inhibition of SMAD transcriptional activity. Only mutation of both binding sites abrogates this repressive effect (Wu et al. 2002).

Structural analysis of a complex between Ski and Smad4 fragments indicated that the interactions of the Smad4 MH2 domain with an R-SMAD and with Ski are mutually exclusive, thus explaining the competitive displacement (Wu et al. 2002).

Overexpression of c-*Ski* is thought to be sufficient for oncogenic activation (Colmenares et al. 1991) and for inhibition of the growth suppressive function of TGF-β (Xu et al. 2000). Binding to the SMADs is essential for the transforming ability of Skis, indicating the importance of TGF-β/SMAD signaling inhibition for the oncogenic activity of these proteins (He et al. 2003). Elevated expression of c-*Ski* was detected in several tumor cell lines derived from neuroblastoma, melanoma and prostate cancer (Fumagalli et al. 1993; Nomura et al. 1989) as well as in melanoma tissues (Reed et al. 2001). In addition, the subcellular localization of Ski appears to shift from nuclear in intraepidermal melanoma cells to nuclear and cytoplasmic in invasive and metastatic melanoma cells (Medrano 2003; Reed et al. 2001).

Leukemic Fusion Gene Product Evi-1. Evi-1 is a nuclear zinc finger protein that is involved in leukemic transformation of hematopoietic cells subsequent to chromosomal translocations that lead to expression of an AML1/Evi-1 fusion product under the control of the AML1 promoter (Nucifora 1997). The biological functions of Evi-1 or the AML1/Evi-1 fusion protein are not well defined, but one way of contributing to a transformed phenotype, e.g., in chronic myelogenous leukemia, may be through its ability to inhibit TGF-β/SMAD signaling and growth inhibition. Evi-1 associates specifically with the Smad3 protein, thereby preventing DNA binding and transcriptional activity of Smad3 containing complexes (Kurokawa et al. 1998b). Alternatively, Evi-1 recruits CtBP as a corepressor to the SMAD transcriptional complex, thereby possibly leading to transcriptional repression via histone deacetylation of genes that would otherwise be activated by the SMAD complexes (Izutsu et al. 2001). Evi-1 or the AML1/Evi-1 fusion protein can suppress TGF-β-induced growth inhibition when expressed in lung epithelial or myeloid cells (Kurokawa et al. 1998a, 1998b). It has been suggested that Evi-1 might also be involved in solid tumors as overexpression has been observed in ovarian cancer samples (Brooks et al. 1996) and Evi-1 causes transformation when exogenously expressed in fibroblasts (Kilbey et al. 1999; Kurokawa et al. 1995).

Interference with TGF-β/SMAD Responses Downstream of SMADs: p27Kip1. The TGF-β growth-inhibitory response in tumor cells can also be affected by aberrant expression or inactivation of cell cycle regulators that function downstream of or independently of the TGF-β/SMAD pathway. Such alterations are frequently found in human breast tumors and they include elevated expression of cyclinD1, cyclinE, MDM 2, or c-*myc*, decreased expression of $p16^{INK4A}$ or $p27^{Kip1}$, and mutations in pRB or p53 (Dillon et al. 1998).

One of these molecules, the CDK2 inhibitor p27^{Kip1}, has recently received special attention. Three reports demonstrated that p27^{Kip1} is mislocalized from the nucleus to the cytoplasm in about 40% of p27^{Kip1} positive primary breast carcinomas and that this tightly correlates with an activated state of the Akt kinase (Liang et al. 2002; Shin et al. 2002; Viglietto et al. 2002). Mechanistic studies showed that Akt directly phosphorylates p27^{Kip1} at a threonine residue located in its nuclear localization signal, thereby preventing nuclear entry and inhibition of CDK2 activity. Hyperactivation of Akt kinase in the tumors can be due to oncogenic activation of receptor tyrosine kinases, such as HER2 or EGF receptor, the mutational inactivation of the tumor suppressor PTEN, or activation of Akt itself. The analysis of 128 breast cancer cases furthermore indicated that the cytoplasmic mislocalization of p27^{Kip1} correlates with poor patient survival (Liang et al. 2002). Interestingly, hyperactivation of Akt kinase and consequent retention of p27^{Kip1} in the cytoplasm render cells unresponsive to the growth inhibitory function of TGF-β or interleukin-6 by preventing efficient inhibition of CDK2 activity (Liang et al. 2002). Hence, both reduced expression levels as well as altered subcellular localization of p27^{Kip1} contribute to breast cancer progression, at least in part by causing resistance to TGF-β growth inhibition, thereby providing a possible prognostic marker as well as a valid target for cancer therapy.

In conclusion, a variety of different molecular mechanisms have been identified that may contribute to the development of resistance to the tumor suppressive effects of TGF-β in tumor cells. Future work needs to address the relative significance of these various mechanisms for the etiology of different cancers and elucidate their functional interplay with other oncogenic events.

7
TGF-β Signaling in Vascular Endothelial Cells

In the vascular system TGF-β regulates the process of angiogenesis (Carmeliet 2000), which generally involves the activation, remodeling, and expansion of a preexisting endothelial vessel network. Two members of the TGF-β type I receptor family, Alk-1 and Alk-5, are indispensable for embryonic angiogenesis, as evidenced in mice lacking either one of them (Larsson et al. 2001; Oh et al. 2000; Urness et al. 2000). Homozygous *Alk-1* embryos (Alk1$^{-/-}$) die at around day 10.5 to 11.5 of gestation due to severe vascular abnormalities. Related vascular defects are also observed in the zebrafish mutant *violet beauregarde* (*vbg*) which encodes the ortholog of human Alk-1 (Roman et al. 2002). Homozygous *Alk-5* mouse embryos (Alk5$^{-/-}$) also die at midgestation, due to severe defects in vascular development of the yolk sac and placenta and an absence of circulating red blood cells (Larsson et al. 2001).

Interestingly, gene targeting in mice clearly implicated several additional members of the TGF-β signaling system in vasculogenesis, angiogenesis, and vascular maturation (Goumans and Mummery 2000). The vascular phenotype of Alk-1 knockout mice is reminiscent of mice lacking endoglin, Smad5, TGF-β1, or TGF-β type II receptor (TβR-II). Their vascular phenotypes are highly overlapping (but not identical), suggesting that these molecules might cooperate in the regulation of vascular development.

Inactivating mutations and resulting haplo-insufficiency of Alk-1 or endoglin are the cause of hereditary hemorrhagic telangiectasia (also called Osler–Weber–Rendu syndrome) (Johnson et al. 1996; McAllister et al. 1994), an autosomal dominant disorder characterized by vascular defects. These defects include lesions at various locations of the vascular system, primarily in mucous membranes, skin, and the linings of stomach, intestines, lungs and brain (Guttmacher et al. 1995). The highly similar phenotypes of hereditary hemorrhagic telangiectasia caused by mutations in Alk-1 or in endoglin clearly suggest that these two molecules are functionally linked in vascular endothelial cells.

Interestingly, there is significant evidence suggesting that endoglin plays an important role in tumor angiogenesis (Duff et al. 2003; Fonsatti et al. 2001). For example, elevated expression levels of endoglin in tumors correlate with proliferation of tumor endothelial cells (Miller et al. 1999) and the microvessel density in breast cancers (Kumar et al. 1999). The microvessel density in primary breast tumors, a measure of the extent of new blood vessel growth, has been suggested as a prognostic indicator for the development of metastatic disease (Weidner et al. 1991). Consistent with this, both high intratumoral microvessel density determined using endoglin antibody and elevated plasma levels of endoglin in colorectal cancer patients correlate with poor prognosis (Li et al. 2003). Because of the functional linkage between endoglin and Alk-1 in endothelial cells these results also argue for an involvement of Alk-1 receptor function in tumor angiogenesis.

The signaling pathways and transcriptional targets of TGF-β signaling via the Alk-1 receptor have remained largely uncharacterized. It has been shown, however, that a constitutively-activated Alk-1 receptor activates the 'BMP branch' of SMAD signaling, i.e., Smads 1, 5 and 8, but not the 'TGF-β branch,' i.e. Smads 2 and 3 (Chen and Massagué 1999; Macias-Silva et al. 1998). This has been confirmed in an analysis of TGF-β signaling via the endogenous Alk-1 receptor (Goumans et al. 2002). One cellular function of Alk-1 is the stimulation of endothelial cell migration (Goumans et al. 2002). This is at least in part mediated by the transcriptional upregulation of the Id1 protein, a member of the Id family of dominant inhibitors of bHLH transcription factors (Sikder et al. 2003). Inhibition of *Id1* expression by antisense oligonucleotides blocked the stimulation of endothelial cell migration by TGF-β (Goumans et al. 2002). Even though it has not been demonstrated it is likely that the transcriptional activation of the *Id1* gene by Alk-1 signaling is a Smad1/4 mediated process, as activation of the *Id1* promoter by BMP signaling depends on binding of these SMADs to a combination of different sequence elements (Korchynskyi and ten Dijke 2002; Lopez-Rovira et al. 2002). Besides *Id1* a number of other candidate target genes of Alk-1 signaling in endothelial cells has been reported, but as these genes were identified by expression profiling of cells that expressed constitutively-active Alk-1 receptor it is unclear whether these genes are direct or indirect targets of Alk-1 signaling pathways (Ota et al. 2002). The identification of direct Alk-1 target genes in endothelial cells should yield valuable insights into the mechanisms of TGF-β regulation of angiogenic processes, including tumor angiogenesis.

References

Affolter M, Bellusci S, Itoh N, Shilo B, Thiery JP, Werb Z (2003) Tube or not tube: remodeling epithelial tissues by branching morphogenesis. Dev. Cell 4:11–18

Akiyoshi S, Inoue H, Hanai J-i, Kusanagi K, Nemoto N, Miyazono K, Kawabata M (1999) c-Ski acts as a transcriptional co-repressor in transforming growth factor-β signaling through interaction with Smads. J Biol Chem 274:35269–35277

Albert T, Wells J, Funk JO, Pullner A, Raschke EE, Stelzer G, Meisterernst M, Farnham PJ, Eick D (2001) The chromatin structure of the dual c-myc promoter P1/P2 is regulated by separate elements. J Biol Chem 276:20482–20490

Alliston T, Choy L, Ducy P, Karsenty G, Derynck R (2001) TGF-beta-induced repression of CBFA1 by Smad3 decreases cbfa1 and osteocalcin expression and inhibits osteoblast differentiation. EMBO J 20:2254–2272

Attisano L, Wrana JL (2002) Signal transduction by the TGF-beta superfamily. Science 296:1646–1647

Bai RY, Koester C, Ouyang T, Hahn SA, Hammerschmidt M, Peschel C, Duyster J (2002) SMIF, a Smad4-interacting protein that functions as a co-activator in TGFbeta signalling. Nat Cell Biol 4:181–190

Bakin AV, Rinehart C, Tomlinson AK, Arteaga CL (2002) p38 mitogen-activated protein kinase is required for TGFbeta-mediated fibroblastic transdifferentiation and cell migration. J Cell Sci 115:3193–3206

Bakin AV, Tomlinson AK, Bhowmick NA, Moses HL, Arteaga CL (2000) Phosphatidylinositol 3-kinase function is required for transforming growth factor beta-mediated epithelial to mesenchymal transition and cell migration. J Biol Chem 275:36803–36810

Balmain A (2002) Cancer: new-age tumour suppressors. Nature 417:235–237

Barcellos-Hoff MH, Ewan KB (2000) Transforming growth factor-beta and breast cancer: Mammary gland development. Breast Cancer Res 2:92–99

Batlle E, Sancho E, Franci C, Dominguez D, Monfar M, Baulida J, Garcia De Herreros A (2000) The transcription factor snail is a repressor of E-cadherin gene expression in epithelial tumour cells. Nat Cell Biol 2:84–89

Bhowmick NA, Ghiassi M, Bakin A, Aakre M, Lundquist CA, Engel ME, Arteaga CL, Moses HL (2001) Transforming growth factor-beta1 mediates epithelial to mesenchy-

mal transdifferentiation through a RhoA-dependent mechanism. Mol Biol Cell 12:27–36
Boulay JL, Mild G, Lowy A, Reuter J, Lagrange M, Terracciano L, Laffer U, Herrmann R, Rochlitz C (2003) SMAD7 is a prognostic marker in patients with colorectal cancer. Int J Cancer 104:446–449
Brooks DJ, Woodward S, Thompson FH, Dos Santos B, Russell M, Yang JM, Guan XY, Trent J, Alberts DS, Taetle R (1996) Expression of the zinc finger gene EVI-1 in ovarian and other cancers. Br J Cancer 74:1518–1525
Calonge MJ, Massague J (1999) Smad4/DPC4 silencing and hyperactive ras jointly disrupt transforming growth factor-β antiproliferative responses in colon cancer cells. J Biol Chem 274:33637–33643
Cano A, Perez-Moreno MA, Rodrigo I, Locascio A, Blanco MJ, del Barrio MG, Portillo F, Nieto MA (2000) The transcription factor snail controls epithelial-mesenchymal transitions by repressing E-cadherin expression. Nat Cell Biol 2:76–83
Carmeliet P (2000) Mechanisms of angiogenesis and arteriogenesis. Nat Med 6:389–395
Chang H, Brown CW, Matzuk MM (2002) Genetic analysis of the mammalian transforming growth factor-beta superfamily. Endocrinol Rev 23:787–823
Chen CR, Kang Y, Massague J (2001a) Defective repression of c-myc in breast cancer cells: A loss at the core of the transforming growth factor beta growth arrest program. Proc Natl Acad Sci USA 98:992–999
Chen CR, Kang Y, Siegel PM, Massagué J (2002) E2F4/5 and p107 as Smad cofactors linking the TGFbeta receptor to c-myc repression. Cell 110:19–32
Chen T, Carter D, Garrigue-Antar L, Reiss M (1998a) Transforming growth factor β type I receptor kinase mutant associated with metastatic breast cancer. Cancer Res. 58:4805–4810
Chen T, de Vries EG, Hollema H, Yegen HA, Vellucci VF, Strickler HD, Hildesheim A, Reiss M (1999) Structural alterations of transforming growth factor-β receptor genes in human cervical carcinoma. Int J Cancer 82:43–51
Chen T, Yan W, Wells RG, Rimm DL, McNiff J, Leffell D, Reiss M (2001b) Novel inactivating mutations of transforming growth factor-beta type I receptor gene in head-and-neck cancer metastases. Int J Cancer 93:653–661
Chen X, Rubock MJ, Whitman M (1996) A transcriptional partner of MAD proteins in TGF-β signalling. Nature 383:691–696
Chen X, Weisberg E, Fridmacher V, Watanabe M, Naco G, Whitman M (1997) Smad4 and FAST-1 in the assembly of activin-responsive factor. Nature 389:85–89
Chen YG, Hata A, Lo RS, Wotton D, Shi Y, Pavletich N, Massague J (1998b) Determinants of specificity in TGF-β signal transduction. Genes Dev 12:2144–2152
Chen YG, Massagué J (1999) Smad1 recognition and activation by the ALK1 group of transforming growth factor-β family receptors. J Biol Chem 274:3672–3677
Coleman RE, Rubens RD (1987) The clinical course of bone metastases from breast cancer. Br J Cancer 55:61–66
Colmenares C, Sutrave P, Hughes SH, Stavnezer E (1991) Activation of the c-ski oncogene by overexpression. J Virol 65:4929–4935
Comijn J, Berx G, Vermassen P, Verschueren K, van Grunsven L, Bruyneel E, Mareel M, Huylebroeck D, van Roy F (2001) The two-handed E box binding zinc finger protein SIP1 downregulates E-cadherin and induces invasion. Mol Cell 7:1267–1278
Cordenonsi M, Dupont S, Maretto S, Insinga A, Imbriano C, Piccolo S (2003) Links between tumor suppressors: p53 is required for TGF-beta gene responses by cooperating with Smads. Cell 113:301–314

Cui W, Fowlis DJ, Bryson S, Duffie E, Ireland H, Balmain A, Akhurst RJ (1996) TGFβ1 inhibits the formation of benign skin tumors, but enhances progression to invasive spindle carcinomas in transgenic mice. Cell 86:531–542

Dai JL, Schutte M, Bansal RK, Wilentz RE, Sugar AY, Kern SE (1999) Transforming growth factor-beta responsiveness in DPC4/SMAD4-null cancer cells. Mol Carcinog 26:37–43

Dalal BI, Keown PA, Greenberg AH (1993) Immunocytochemical localization of secreted transforming growth factor-β1 to the advancing edges of primary tumors and to lymph node metastases of human mammary carcinoma. Am J Pathol 143:381–389

Dennler S, Itoh S, Vivien D, ten Dijke P, Huet S, Gauthier JM (1998) Direct binding of Smad3 and Smad4 to critical TGF beta-inducible elements in the promoter of human plasminogen activator inhibitor-type 1 gene. EMBO J 17:3091–3100

Derynck R, Akhurst RJ, Balmain A (2001) TGF-beta signaling in tumor suppression and cancer progression. Nat Genet 29:117–129

Dillon DA, Howe CL, Bosari S, Costa J (1998) The molecular biology of breast cancer: accelerating clinical applications. Crit Rev Oncog 9:125–140

Dowdy SC, Mariani A, Janknecht R (2003) HER2/Neu and TAK1 mediated up-regulation of the TGF-β inhibitor Smad7 via the ETS protein ER81. J Biol Chem 278:44377–44384

Duff SE, Li C, Garland JM, Kumar S (2003) CD105 is important for angiogenesis: evidence and potential applications. FASEB J 17:984–992

Ebisawa T, Fukuchi M, Murakami G, Chiba T, Tanaka K, Imamura T, Miyazono K (2001) Smurf1 interacts with transforming growth factor-beta type I receptor through Smad7 and induces receptor degradation. J Biol Chem 276:12477–12480

Ellenrieder V, Hendler SF, Boeck W, Seufferlein T, Menke A, Ruhland C, Adler G, Gress T (2001) Transforming growth factor beta1 treatment leads to an epithelial-mesenchymal transdifferentiation of pancreatic cancer cells requiring extracellular signal-regulated kinase 2 activation. Cancer Res 61:4222–4228

Engel ME, McDonnell MA, Law BK, Moses HL (1999) Interdependent SMAD and JNK signaling in transforming growth factor-β-mediated transcription. J Biol Chem 274:37413–37420

Ewan KB, Shyamala G, Ravani SA, Tang Y, Akhurst R, Wakefield L, Barcellos-Hoff MH (2002) Latent transforming growth factor-beta activation in mammary gland: regulation by ovarian hormones affects ductal and alveolar proliferation. Am J Pathol 160:2081–2093

Feng XH, Derynck R (1997) A kinase subdomain of transforming growth factor-β (TGF-β) type I receptor determines the TGF-ß intracellular signaling activity. EMBO J 16:3912–3922

Feng XH, Lin X, Derynck R (2000) Smad2, Smad3 and Smad4 cooperate with Sp1 to induce p15(Ink4B) transcription in response to TGF-beta. EMBO J 19:5178–93

Feng XH, Zhang Y, Wu RY, Derynck R (1998) The tumor suppressor Smad4/DPC4 and transcriptional adaptor CBP/p300 are coactivators for smad3 in TGF-β-induced transcriptional activation. Genes Dev 12:2153–2163

Fonsatti E, Del Vecchio L, Altomonte M, Sigalotti L, Nicotra MR, Coral S, Natali PG, Maio M (2001) Endoglin: An accessory component of the TGF-beta-binding receptor-complex with diagnostic, prognostic, and bioimmunotherapeutic potential in human malignancies. J Cell Physiol 188:1–7

Fukuchi M, Imamura T, Chiba T, Ebisawa T, Kawabata M, Tanaka K, Miyazono K (2001) Ligand-dependent degradation of Smad3 by a ubiquitin ligase complex of ROC1 and associated proteins. Mol Biol Cell 12:1431–1443

Fumagalli S, Doneda L, Nomura N, Larizza L (1993) Expression of the c-ski proto-oncogene in human melanoma cell lines. Melanoma Res 3:23–27

Gebert J, Sun M, Ridder R, Hinz U, Lehnert T, Moller P, Schackert HK, Herfarth C, von Knebel Doeberitz M (2000) Molecular profiling of sporadic colorectal tumors by microsatellite analysis. Int J Oncol 16:169–179

Germain S, Howell M, Esslemont GM, Hill CS (2000) Homeodomain and winged-helix transcription factors recruit activated Smads to distinct promoter elements via a common Smad interaction motif. Genes Dev 14:435–451

Ghellal A, Li C, Hayes M, Byrne G, Bundred N, Kumar S (2000) Prognostic significance of TGF beta 1 and TGF beta 3 in human breast carcinoma. Anticancer Res 20:4413–4418

Goggins M, Shekher M, Turnacioglu K, Yeo CJ, Hruban RH, Kern SE (1998) Genetic alterations of the transforming growth factor β receptor genes in pancreatic and biliary adenocarcinomas. Cancer Res 58:5329–5332

Gorsch SM, Memoli VA, Stukel TA, Gold LI, Arrick BA (1992) Immunohistochemical staining for transforming growth factor beta 1 associates with disease progression in human breast cancer. Cancer Res 52:6949–6952

Goumans MJ, Mummery C (2000) Functional analysis of the TGFbeta receptor/Smad pathway through gene ablation in mice. Int J Dev Biol 44:253–265

Goumans MJ, Valdimarsdottir G, Itoh S, Rosendahl A, Sideras P, ten Dijke P (2002) Balancing the activation state of the endothelium via two distinct TGF-beta type I receptors. EMBO J 21:1743–1753.

Grady WM, Myeroff LL, Swinler SE, Rajput A, Thiagalingam S, Lutterbaugh JD, Neumann A, Brattain MG, Chang J, Kim SJ, Kinzler KW, Vogelstein B, Willson JK, Markowitz S (1999) Mutational inactivation of transforming growth factor β receptor type II in microsatellite stable colon cancers. Cancer Res 59:320–324

Grande M, Franzen A, Karlsson JO, Ericson LE, Heldin NE, Nilsson M (2002) Transforming growth factor-beta and epidermal growth factor synergistically stimulate epithelial to mesenchymal transition (EMT) through a MEK-dependent mechanism in primary cultured pig thyrocytes. J Cell Sci 115:4227–4236

Grimm OH, Gurdon JB (2002) Nuclear exclusion of Smad2 is a mechanism leading to loss of competence. Nat Cell Biol 4:519–522

Grinberg AV, Kerppola T (2003) Both Max and TFE3 cooperate with Smad proteins to bind the plasminogen activator inhibitor-1 promoter, but they have opposite effects on transcriptional activity. J Biol Chem 278:11227–11236

Grunert S, Jechlinger M, Beug H (2003) Diverse cellular and molecular mechanisms contribute to epithelial plasticity and metastasis. Nat Rev Mol Cell Biol 4:657–665

Guttmacher AE, Marchuk DA, White RIJ (1995) Hereditary hemorrhagic telangiectasia. N Engl J Med 333:918–924

Hahn SA, Schutte M, Hoque ATMS, Moskaluk CA, da Costa LT, Rozenblum E, Weinstein CL, Fischer A, Yeo CJ, Hruban RH, Kern SE (1996) *DPC4*, a candidate tumor suppressor gene at human chromosome 18q21.1. Science 271:350–353

Hanafusa H, Ninomiya-Tsuji J, Masuyama N, Nishita M, Fujisawa J, Shibuya H, Matsumoto K, Nishida E (1999) Involvement of the p38 mitogen-activated protein kinase pathway in transforming growth factor-beta-induced gene expression. J Biol Chem 274:27161–27167

Hanai J, Chen LF, Kanno T, Ohtani-Fujita N, Kim WY, Guo WH, Imamura T, Ishidou Y, Fukuchi M, Shi MJ, Stavnezer J, Kawabata M, Miyazono K, Ito Y (1999) Interaction and functional cooperation of PEBP2/CBF with Smads. Synergistic induction of the immunoglobulin germline Calpha promoter. J Biol Chem 274:31577–31582

Hata A, Lagna G, Massagué J, Hemmati-Brivanlou A (1998) Smad6 inhibits BMP/Smad1 signaling by specifically competing with the Smad4 tumor suppressor. Genes Dev 12:186–197

Hata A, Seoane J, Lagna G, Montalvo E, Hemmati-Brivanlou A, Massague J (2000) OAZ uses distinct DNA- and protein-binding zinc fingers in separate BMP-Smad and Olf signaling pathways. Cell 100:229–240

Hay ED (1995) An overview of epithelio-mesenchymal transformation. Acta Anat. (Basel) 154:8–20

Hayashi H, Abdollah S, Qiu Y, Cai J, Xu YY, Grinnell BW, Richardson MA, Topper JN, Gimbrone MA, Wrana JL, Falb D (1997) The MAD-related protein Smad7 associates with the TGFβ receptor and functions as an antagonist of TGFβ signaling. Cell 89:1165–1173

He J, Tegen SB, Krawitz AR, Martin GS, Luo K (2003) The transforming activity of Ski and SnoN is dependent on their ability to repress the activity of Smad proteins. J Biol Chem 278:30540–30547

Herold S, Wanzel M, Beuger V, Frohme C, Beul D, Hillukkala T, Syvaoja J, Saluz HP, Haenel F, Eilers M (2002) Negative regulation of the mammalian UV response by Myc through association with Miz-1. Mol Cell 10:509–521

Hocevar BA, Brown TL, Howe PH (1999) TGF-β induces fibronectin synthesis through a c-Jun N-terminal kinase-dependent, Smad4-independent pathway. EMBO J 18:1345–1356

Hocevar BA, Smine A, Xu XX, Howe PH (2001) The adaptor molecule Disabled-2 links the transforming growth factor beta receptors to the Smad pathway. Embo J 20:2789–2801

Hua X, Liu X, Ansari DO, Lodish HF (1998) Synergistic cooperation of TFE3 and smad proteins in TGF-beta-induced transcription of the plasminogen activator inhibitor-1 gene. Genes Dev 12:3084–3095

Hua X, Miller ZA, Benchabane H, Wrana JL, Lodish HF (2000) Synergism between transcription factors TFE3 and Smad3 in transforming growth factor-beta-induced transcription of the Smad7 gene. J Biol Chem 275:33205–33208

Imamura T, Takase M, Nishihara A, Oeda E, Hanai J, Kawabata M, Miyazono K (1997) Smad6 inhibits signalling by the TGF-ß superfamily. Nature 389:622–626

Ito Y, Miyazono K (2003) RUNX transcription factors as key targets of TGF-beta superfamily signaling. Curr Opin Genet Dev 13:43–47

Itoh S, Thorikay M, Kowanetz M, Moustakas A, Itoh F, Heldin CH, ten Dijke P (2003) Elucidation of Smad requirement in transforming growth factor-beta type I receptor-induced responses. J Biol Chem 278:3751–3761

Izutsu K, Kurokawa M, Imai Y, Maki K, Mitani K, Hirai H (2001) The corepressor CtBP interacts with Evi-1 to repress transforming growth factor beta signaling. Blood 97:2815–2822

Janda E, Lehmann K, Killisch I, Jechlinger M, Herzig M, Downward J, Beug H, Grunert S (2002) Ras and TGFβ cooperatively regulate epithelial cell plasticity and metastasis: dissection of Ras signaling pathways. J. Cell Biol 156:299–313

Jang CW, Chen CH, Chen CC, Chen JY, Su YH, Chen RH (2002) TGF-beta induces apoptosis through Smad-mediated expression of DAP-kinase. Nat Cell Biol 4:51–58

Janknecht R, Wells NJ, Hunter T (1998) TGF-β-stimulated cooperation of smad proteins with the coactivators CBP/p300. Genes Dev 12:2114–2119

Jhappan C, Geiser AG, Kordon EC, Bagheri D, Henninghausen L, Roberts AB, Smith GH, Merlino G (1993) Targeting expression of a transforming growth factor β1 transgene to the pregnant mammary gland inhibits alveolar development and lactation. EMBO J 12:1835–1845

Johnson DW, Berg JN, Baldwin MA, Gallione CJ, Marondel I, Yoon SJ, Stenzel TT, Speer M, Pericak-Vance MA, Diamond A, Guttmacher AE, Jackson CE, Attisano L, Kucherlapati R, Porteous ME, Marchuk DA (1996) Mutations in the activin receptor-like kinase 1 gene in hereditary haemorrhagic telangiectasia type 2. Nature Genet 13:189–195

Kakonen SM, Selander KS, Chirgwin JM, Yin JJ, Burns S, Rankin WA, Grubbs BG, Dallas M, Cui Y, Guise TA (2002) Transforming growth factor-beta stimulates parathyroid hormone-related protein and osteolytic metastases via Smad and mitogen-activated protein kinase signaling pathways. J Biol Chem 277:24571–24578

Kang Y, Siegel PM, Shu W, Drobnjak M, Kakonen SM, Cordon-Cardo C, Guise TA, Massague J (2003) A multigenic program mediating breast cancer metastasis to bone. Cancer Cell 3:537–549

Kato Y, Habas R, Katsuyama Y, Naar AM, He X (2002) A component of the ARC/Mediator complex required for TGF beta/Nodal signalling. Nature 418:641–646

Kavsak P, Rasmussen RK, Causing CG, Bonni S, Zhu H, Thomsen GH, Wrana JL (2000) Smad7 binds to Smurf2 to form an E3 ubiquitin ligase that targets the TGF beta receptor for degradation. Mol Cell 6:1365–1375

Kilbey A, Stephens V, Bartholomew C (1999) Loss of cell cycle control by deregulation of cyclin-dependent kinase 2 kinase activity in Evi-1 transformed fibroblasts. Cell Growth Differ. 10:601–610

Kim WS, Park C, Hong SK, Park BK, Kim HS, Park K (2000) Microsatellite instability(MSI) in non-small cell lung cancer(NSCLC) is highly associated with transforming growth factor-beta type II receptor(TGF-beta RII) frameshift mutation. Anticancer Res 20:1499–502

Kleeff J, Ishiwata T, Maruyama H, Friess H, Truong P, Buchler MW, Falb D, Korc M (1999) The TGF-β signaling inhibitor Smad7 enhances tumorigenicity in pancreatic cancer. Oncogene 18:5363–5372

Korchynskyi O, ten Dijke P (2002) Identification and functional characterization of distinct critically important bone morphogenetic protein-specific response elements in the Id1 promoter. J Biol Chem 277:4883–4891

Kordon EC, McKnight RA, Jhappan C, Hennighausen L, Merlino G, Smith GH (1995) Ectopic TGF beta 1 expression in the secretory mammary epithelium induces early senescence of the epithelial stem cell population. Dev Biol 168:47–61

Kouzarides T (1999) Histone acetylases and deacetylases in cell proliferation. Curr Opin Genet Dev 9:40–48

Kretzschmar M (2000) Transforming growth factor-beta and breast cancer: Transforming growth factor-beta/SMAD signaling defects and cancer. Breast Cancer Res 2:107–115

Kretzschmar M, Doody J, Timokhina I, Massagué J (1999) A mechanism of repression of TGFβ/Smad signaling by oncogenic Ras. Genes Dev 13:804–816

Kretzschmar M, Liu F, Hata A, Doody J, Massagué J (1997) The TGF-β mediator Smad1 is directly phosphorylated and functionally activated by the BMP receptor kinase. Genes Dev 11:984–995

Kretzschmar M, Massagué J (1998) SMADs: mediators and regulators of TGFβ family signalling. Curr Opin Genet Dev 8:103–111

Kumar S, Ghellal A, Li C, Byrne G, Haboubi N, Wang JM, Bundred N (1999) Breast carcinoma: vascular density determined using CD105 antibody correlates with tumor prognosis. Cancer Res. 59:856–861

Kurokawa M, Mitani K, Imai Y, Ogawa S, Yazaki Y, Hirai H (1998a) The t(3;21) fusion product, AML1/Evi-1, interacts with Smad3 and blocks transforming growth factor-β-mediated growth inhibition of myeloid cells. Blood 92:4003-4012

Kurokawa M, Mitani K, Irie K, Matsuyama T, Takahashi T, Chiba S, Yazaki Y, Matsumoto K, Hirai H (1998b) The oncoprotein Evi-1 represses TGF-β signalling by inhibiting Smad3. Nature 394:92-96

Kurokawa M, Ogawa S, Tanaka T, Mitani K, Yazaki Y, Witte ON, Hirai H (1995) The AML1/Evi-1 fusion protein in the t(3;21) translocation exhibits transforming activity on Rat1 fibroblasts with dependence on the Evi-1 sequence. Oncogene 11:833-840

Labbe E, Letamendia A, Attisano L (2000) Association of Smads with lymphoid enhancer binding factor 1/T cell-specific factor mediates cooperative signaling by the transforming growth factor-β and wnt pathways. Proc Natl Acad Sci USA 97:8358-8363

Landstrom M, Heldin NE, Bu S, Hermansson A, Itoh S, ten Dijke P, Heldin CH (2000) Smad7 mediates apoptosis induced by transforming growth factor beta in prostatic carcinoma cells. Curr Biol 10:535-538

Larisch S, Yi Y, Lotan R, Kerner H, Eimerl S, Tony Parks W, Gottfried Y, Birkey Reffey S, de Caestecker MP, Danielpour D, Book-Melamed N, Timberg R, Duckett CS, Lechleider RJ, Steller H, Orly J, Kim SJ, Roberts AB (2000) A novel mitochondrial septin-like protein, ARTS, mediates apoptosis dependent on its P-loop motif. Nat Cell Biol 2:915-921

Larsson J, Goumans MJ, Sjostrand LJ, van Rooijen MA, Ward D, Leveen P, Xu X, ten Dijke P, Mummery CL, Karlsson S (2001) Abnormal angiogenesis but intact hematopoietic potential in TGF-beta type I receptor-deficient mice. EMBO J 20:1663-1673

Li C, Gardy R, Seon BK, Duff SE, Abdalla S, Renehan A, O'Dwyer ST, Haboubi N, Kumar S (2003) Both high intratumoral microvessel density determined using CD105 antibody and elevated plasma levels of CD105 in colorectal cancer patients correlate with poor prognosis. Br J Cancer 88:1424-1431

Li QL, Ito K, Sakakura C, Fukamachi H, Inoue K, Chi XZ, Lee KY, Nomura S, Lee CW, Han SB, Kim HM, Kim WJ, Yamamoto H, Yamashita N, Yano T, Ikeda T, Itohara S, Inazawa J, Abe T, Hagiwara A, Yamagishi H, Ooe A, Kaneda A, Sugimura T, Ushijima T, Bae SC, Ito Y (2002) Causal relationship between the loss of RUNX3 expression and gastric cancer. Cell 109:113-124

Liang J, Zubovitz J, Petrocelli T, Kotchetkov R, Connor MK, Han K, Lee JH, Ciarallo S, Catzavelos C, Beniston R, Franssen E, Slingerland JM (2002) PKB/Akt phosphorylates p27, impairs nuclear import of p27 and opposes p27-mediated G1 arrest. Nat Med 8:1153-1160

Lin X, Liang M, Feng XH (2000) Smurf2 is a ubiquitin E3 ligase mediating proteasome-dependent degradation of Smad2 in transforming growth factor-beta signaling. J Biol Chem 275:36818-36822

Liu F, Pouponnot C, Massagué J (1997) Dual role of the Smad4/DPC4 tumor suppressor in TGFß-inducible transcriptional responses. Genes Dev 11:3157-3167

Lo RS, Chen YG, Shi YG, Pavletich N, Massagué J (1998) The L3 loop: a structural motif determining specific interactions between SMAD proteins and TGF-β receptors. EMBO J 17:996-1005

Lopez-Rovira T, Chalaux E, Massague J, Rosa JL, Ventura F (2002) Direct binding of Smad1 and Smad4 to two distinct motifs mediates bone morphogenetic protein-specific transcriptional activation of Id1 gene. J Biol Chem 277:3176-3185

Lucke CD, Philpott A, Metcalfe JC, Thompson AM, Hughes-Davies L, Kemp PR, Hesketh R (2001) Inhibiting mutations in the transforming growth factor beta type 2 receptor in recurrent human breast cancer. Cancer Res 61:482-485

Luo K, Stroschein SL, Wang W, Chen D, Martens E, Zhou S, Zhou Q (1999) The Ski oncoprotein interacts with the Smad proteins to repress TGFβ signaling. Genes Dev. 13:2196–2206

Macias-Silva M, Abdollah S, Hoodless PA, Pirone R, Attisano L, Wrana JL (1996) MADR2 is a substrate of the TGFß receptor and phosphorylation is required for nuclear accumulation and signaling. Cell 87:1215–1224

Macias-Silva M, Hoodless PA, Tang SJ, Buchwald M, Wrana JL (1998) Specific activation of Smad1 signaling pathways by the BMP7 type I receptor, ALK2. J Biol Chem 273:25628–25636

Malumbres M, Barbacid M (2003) Ras oncogenes: the first 30 years. Nat Rev Cancer 3:459–465

Markowitz SD, Roberts AB (1996) Tumor suppressor activity of the TGF-β pathway in human cancers. Cytokine Growth Factor Rev 7:93–102

Marshall C (1999) How do small GTPase signal transduction pathways regulate cell cycle entry? Curr Opin Cell Biol 11:732 736

Massague J, Blain SW, Lo RS (2000) TGFbeta signaling in growth control, cancer, and heritable disorders. Cell 103:295–309

Massague J (2003) Integration of Smad and MAPK pathways: a link and a linker revisited. Genes Dev 17:2993–2997

McAllister KA, Grogg KM, Johnson DW, Gallione CJ, Baldwin MA, Jackson MK, Helmbold EA, Markel DS, McKinnon WC, Murrell J, McCormick MK, Pericak-Vance MA, Heutik P, Oostra BA, Haitjema T, Westerman CJJ, Porteus ME, Guttmacher AE, Letarte M, Marchuck DA (1994) Endoglin, a TGF-β binding protein of endothelial cells, is the gene for hereditary haemorrhagic telangiectasia type 1. Nature Genet 8:345–351

McEarchern JA, Kobie JJ, Mack V, Wu RS, Meade-Tollin L, Arteaga CL, Dumont N, Besselsen D, Seftor E, Hendrix MJ, Katsanis E, Akporiaye E (2001) Invasion and metastasis of a mammary tumor involves TGF-beta signaling. Int J Cancer 91:76–82

Medrano EE (2003) Repression of TGF-beta signaling by the oncogenic protein SKI in human melanomas: consequences for proliferation, survival, and metastasis. Oncogene 22:3123–3129

Miettinen PJ, Ebner R, Lopez AR, Derynck R (1994) TGF-β induced transdifferentiation of mammary epithelial cells to mesenchymal cells: involvement of type I receptors. J Cell Biol 127:2021–2036

Miller DW, Graulich W, Karges B, Stahl S, Ernst M, Ramaswamy A, Sedlacek HH, Muller R, Adamkiewicz J (1999) Elevated expression of endoglin, a component of the TGF-β-receptor complex, correlates with proliferation of tumor endothelial cells. Int J Cancer 81:568–572

Moustakas A, Souchelnytskyi S, Heldin CH (2001) Smad regulation in TGF-beta signal transduction. J Cell Sci 114:4359–4369

Mundy GR (2002) Metastasis: Metastasis to bone: causes, consequences and therapeutic opportunities. Nat Rev Cancer 2:584–593

Muraoka RS, Dumont N, Ritter CA, Dugger TC, Brantley DM, Chen J, Easterly E, Roebuck LR, Ryan S, Gotwals PJ, Koteliansky V, Arteaga CL (2002) Blockade of TGF-beta inhibits mammary tumor cell viability, migration, and metastases. J Clin Invest 109:1551–1559

Nakao A, Afrakhte M, Morén A, Nakayama T, Christian JL, Heuchel R, Itoh S, Kawabata M, Heldin NE, Heldin CH, ten Dijke P (1997) Identification of Smad7, a TGFβ-inducible antagonist of TGF-β signaling. Nature 389:631–635

Nguyen AV, Pollard JW (2000) Transforming growth factor $\beta 3$ induces cell death during the first stage of mammary gland involution. Development 127:3107–3118

Nishihara A, Hanai JI, Okamoto N, Yanagisawa J, Kato S, Miyazono K, Kawabata M (1998) Role of p300, a transcriptional coactivator, in signalling of TGF-β. Genes Cells 3:613–623

Nomura N, Sasamoto S, Ishii S, Date T, Matsui M, Ishizaki R (1989) Isolation of human cDNA clones of ski and the ski-related gene, sno. Nucleic Acids Res 17:5489–5500

Nomura T, Khan MM, Kaul SC, Dong HD, Wadhwa R, Colmenares C, Kohno I, Ishii S (1999) Ski is a component of the histone deacetylase complex required for transcriptional repression by Mad and thyroid hormone receptor. Genes Dev 13:412–423

Nucifora G (1997) The EVI1 gene in myeloid leukemia. Leukemia 11:2022–2031

Oft M, Akhurst RJ, Balmain A (2002) Metastasis is driven by sequential elevation of H-ras and Smad2 levels. Nat Cell Biol 4:487–494

Oft M, Heider KH, Beug H (1998) TGFβ signaling is necessary for carcinoma cell invasiveness and metastasis. Curr Biol 8:1243–1252

Oft M, Peli J, Rudaz C, Schwarz H, Beug H, Reichmann E (1996) TGF-β1 and Ha-Ras collaborate in modulating the phenotypic plasticity and invasiveness of epithelial tumor cells. Genes Dev 10:2462–2477

Oh SP, Seki T, Goss KA, Imamura T, Yi Y, Donahoe PK, Li L, Miyazono K, ten Dijke P, Kim S, Li E (2000) Activin receptor-like kinase 1 modulates transforming growth factor-β1 signaling in the regulation of angiogenesis. Proc Natl Acad Sci USA 97:2626–2631

Ota T, Fujii M, Sugizaki T, Ishii M, Miyazawa K, Aburatani H, Miyazono K (2002) Targets of transcriptional regulation by two distinct type I receptors for transforming growth factor-beta in human umbilical vein endothelial cells. J Cell Physiol 193:299–318

Pasche B, Kolachana P, Nafa K, Satagopan J, Chen YG, Lo RS, Brener D, Yang D, Kirstein L, Oddoux C, Ostrer H, Vineis P, Varesco L, Jhanwar S, Luzzatto L, Massague J, Offit K (1999) TβR-I(6A) is a candidate tumor susceptibility allele. Cancer Res 59:5678–5682

Patil S, Wildey GM, Brown TL, Choy L, Derynck R, Howe PH (2000) Smad7 is induced by CD40 and protects WEHI 231 B-lymphocytes from transforming growth factor-beta - induced growth inhibition and apoptosis. J Biol Chem 275:38363–38370

Peinado H, Quintanilla M, Cano A (2003) Transforming growth factor beta-1 induces snail transcription factor in epithelial cell lines: mechanisms for epithelial mesenchymal transitions. J Biol Chem 278:21113–21123

Pelengaris S, Khan M, Evan G (2002) c-MYC: more than just a matter of life and death. Nat Rev Cancer 2:764–76

Pera EM, Ikeda A, Eivers E, De Robertis EM (2003) Integration of IGF, FGF, and anti-BMP signals via Smad1 phosphorylation in neural induction. Genes Dev 17:3023–3028

Perez-Moreno MA, Locascio A, Rodrigo I, Dhondt G, Portillo F, Nieto MA, Cano A (2001) A new role for E12/E47 in the repression of E-cadherin expression and epithelial-mesenchymal transitions. J Biol Chem 276:27424–27431

Perlman R, Schiemann WP, Brooks MW, Lodish HF, Weinberg RA (2001) TGF-β-induced apoptosis is mediated by the adapter protein Daxx that facilitates JNK activation. Nat Cell Biol 3:708–714

Petritsch C, Beug H, Balmain A, Oft M (2000) TGF-beta inhibits p70 S6 kinase via protein phosphatase 2A to induce G(1) arrest. Genes Dev 14:3093–3101

Piek E, Ju WJ, Heyer J, Escalante-Alcalde D, Stewart CL, Weinstein M, Deng C, Kucherlapati R, Bottinger EP, Roberts AB (2001) Functional characterization of transforming growth factor beta signaling in Smad2- and Smad3-deficient fibroblasts. J Biol Chem 276:19945–19953

Piek E, Moustakas A, Kurisaki A, Heldin CH, ten Dijke P (1999) TGF-(beta) type I receptor/ALK-5 and Smad proteins mediate epithelial to mesenchymal transdifferentiation in NMuMG breast epithelial cells. J Cell Sci 112:4557–4568

Pierce DF, Johnson MD, Matsui Y, Robinson SD, Gold LI, Purchio AF, Daniel CW, Hogan BLM, Moses HL (1993) Inhibition of mammary duct development but not alveolar outgrowth during pregnancy in transgenic mice expressing active TGF-β1. Genes Dev 7:2308–2317

Pinto M, Oliveira C, Cirnes L, Carlos Machado J, Ramires M, Nogueira A, Carneiro F, Seruca R (2003) Promoter methylation of TGFbeta receptor I and mutation of TGFbeta receptor II are frequent events in MSI sporadic gastric carcinomas. J Pathol 200:32–38

Postigo AA (2003) Opposing functions of ZEB proteins in the regulation of the TGFbeta/BMP signaling pathway. EMBO J 22:2443–2452

Pouponnot C, Jayaraman L, Massague J (1998) Physical and functional interaction of SMADs and p300/CBP. J Biol Chem 273:22865–22868

Qing J, Zhang Y, Derynck R (2000) Structural and functional characterization of the transforming growth factor-beta -induced Smad3/c-Jun transcriptional cooperativity. J Biol Chem 275:38802–38812

Rayman JB, Takahashi Y, Indjeian VB, Dannenberg JH, Catchpole S, Watson RJ, te Riele H, Dynlacht BD (2002) E2F mediates cell cycle-dependent transcriptional repression in vivo by recruitment of an HDAC1/mSin3B corepressor complex. Genes Dev 16:933–947

Reed JA, Bales E, Xu W, Okan NA, Bandyopadhyay D, Medrano EE (2001) Cytoplasmic localization of the oncogenic protein Ski in human cutaneous melanomas in vivo: functional implications for transforming growth factor beta signaling. Cancer Res 61:8074–8078

Robinson SD, Silberstein GB, Roberts AB, Flanders KC, Daniel CW (1991) Regulated expression and growth inhibitory effects of transforming growth factor-beta isoforms in mouse mammary gland development. Development 113:867–878

Rodriguez-Pascual F, Redondo-Horcajo M, Lamas S (2003) Functional cooperation between Smad proteins and activator protein-1 regulates transforming growth factor-beta-mediated induction of endothelin-1 expression. Circ Res 92:1288–1295

Roman BL, Pham VN, Lawson ND, Kulik M, Childs S, Lekven AC, Garrity DM, Moon RT, Fishman MC, Lechleider RJ, Weinstein BM (2002) Disruption of acvrl1 increases endothelial cell number in zebrafish cranial vessels. Development 129:3009–3019

Saito H, Tsujitani S, Oka S, Kondo A, Ikeguchi M, Maeta M, Kaibara N (2000) An elevated serum level of transforming growth factor-beta 1 (TGF-beta 1) significantly correlated with lymph node metastasis and poor prognosis in patients with gastric carcinoma. Anticancer Res. 20:4489–4493

Sano Y, Harada J, Tashiro S, Gotoh-Mandeville R, Maekawa T, Ishii S (1999a) ATF-2 is a common nuclear target of Smad and TAK1 pathways in transforming growth factor-beta signaling. J Biol Chem 274:8949–8957

Sano Y, Harada J, Tashiro S, Gotoh-Mandeville R, Maekawa T, Ishii S (1999b) ATF-2 is a common nuclear target of Smad and TAK1 pathways in transforming growth factor-beta signaling. J Biol Chem 274:8949–8957

Sater AK, El-Hodiri HM, Goswami M, Alexander TB, Al-Sheikh O, Etkin LD, Akif Uzman J (2003) Evidence for antagonism of BMP-4 signals by MAP kinase during Xenopus axis determination and neural specification. Differentiation 71:434–444

Schiemann WP, Pfeifer WM, Levi E, Kadin ME, Lodish HF (1999) A deletion in the gene for transforming growth factor β type I receptor abolishes growth regulation by transforming growth factor β in a cutaneous T-cell lymphoma. Blood 94:2854–2861

Sekelsky JJ, Newfeld SJ, Raftery LA, Chartoff EH, Gelbart WM (1995) Genetic characterization and cloning of *Mothers against dpp*, a gene required for *decapentaplegic* function in Drosophila melanogaster. Genetics 139:1347–1358

Seoane J, Le HV, Massague J (2002) Myc suppression of the p21(Cip1) Cdk inhibitor influences the outcome of the p53 response to DNA damage. Nature 419:729–734

Seoane J, Pouponnot C, Staller P, Schader M, Eilers M, Massague J (2001) TGFbeta influences Myc, Miz-1 and Smad to control the CDK inhibitor p15INK4b. Nat Cell Biol 3:400–408

Shariat SF, Shalev M, Menesses-Diaz A, Kim IY, Kattan MW, Wheeler TM, Slawin KM (2001) Preoperative plasma levels of transforming growth factor beta(1) (TGF-beta(1)) strongly predict progression in patients undergoing radical prostatectomy. J Clin Oncol 19:2856–2864

Shen X, Hu PP, Liberati NT, Datto MB, Frederick JP, Wang XF (1998) TGF-β-induced phosphorylation of Smad3 regulates its interaction with coactivator p300/CREB-binding protein. Mol Biol Cell 9:3309–3319

Shi Y, Hata A, Lo RS, Massagué J, Pavletich NP (1997) A structural basis for mutational inactivation of the tumour suppressor Smad4. Nature 388:87–93

Shi Y, Massague J (2003) Mechanisms of TGF-beta signaling from cell membrane to the nucleus. Cell 113:685–700

Shi Y, Wang YF, Jayaraman L, Yang H, Massague J, Pavletich NP (1998) Crystal structure of a Smad MH1 domain bound to DNA: insights on DNA binding in TGF-β signaling. Cell 94:585–594

Shibuya H, Iwata H, Masuyama N, Gotoh Y, Yamaguchi K, Irie K, Matsumoto K, Nishida E, Ueno N (1998) Role of TAK1 and TAB1 in BMP signaling in early Xenopus development. EMBO J 17:1019–1028

Shibuya H, Yamaguchi K, Shirakabe K, Tonegawa A, Gotoh Y, Ueno N, Irie K, Nishida E, Matsumoto K (1996) TAB1: an activator of the TAK1 MAPKKK in TGF-β signal transduction. Science 272:1179–1182

Shin I, Bakin AV, Rodeck U, Brunet A, Arteaga CL (2001) Transforming growth factor beta enhances epithelial cell survival via Akt-dependent regulation of FKHRL1. Mol Biol Cell 12:3328–3339

Shin I, Yakes FM, Rojo F, Shin NY, Bakin AV, Baselga J, Arteaga CL (2002) PKB/Akt mediates cell-cycle progression by phosphorylation of p27(Kip1) at threonine 157 and modulation of its cellular localization. Nat Med 8:1145–1152

Sikder HA, Devlin MK, Dunlap S, Ryu B, Alani RM (2003) Id proteins in cell growth and tumorigenesis. Cancer Cell 3:525–530

Silberstein GB, Daniel CW (1987) Reversible inhibition of mammary gland growth by transforming growth factor-β. Science 237:291–293

Sirard C, Kim S, Mirtsos C, Tadich P, Hoodless PA, Itie A, Maxson R, Wrana JL, Mak TW (2000) Targeted disruption in murine cells reveals variable requirement for Smad4 in transforming growth factor beta-related signaling. J Biol Chem 275:2063–2070

Staller P, Peukert K, Kiermaier A, Seoane J, Lukas J, Karsunky H, Moroy T, Bartek J, Massague J, Hanel F, Eilers M (2001) Repression of p15INK4b expression by Myc through association with Miz-1. Nat Cell Biol 3:392–399

Stroschein SL, Wang W, Zhou S, Zhou Q, Luo K (1999) Negative feedback regulation of TGF-β signaling by the SnoN oncoprotein. Science 286:771–774

Sun PD, Davies D (1995) The cystine-knot growth factor superfamily. Ann Rev Biophys Biomol Struct 24:269–291

Sun Y, Liu X, Eaton EN, Lane WS, Lodish HF, Weinberg RA (1999) Interaction of the Ski oncoprotein with Smad3 regulates TGF-β signaling. Mol Cell 4:499–509

Takigawa M, Nakanishi T, Kubota S, Nishida T (2003) Role of CTGF/HCS24/ecogenin in skeletal growth control. J Cell Physiol 194:256–266

Teraoka H, Sawada T, Yamashita Y, Nakata B, Ohira M, Ishikawa T, Nishino H, Hirakawa K (2001) TGF-beta1 promotes liver metastasis of pancreatic cancer by modulating the capacity of cellular invasion. Int J Oncol 19:709–715

Thiery JP (2002) Epithelial-mesenchymal transitions in tumour progression. Nat Rev Cancer 2:442–454

Thomas G, Hall MN (1997) TOR signalling and control of cell growth. Curr Opin Cell Biol 9:782–787

Tsushima H, Ito N, Tamura S, Matsuda Y, Inada M, Yabuuchi I, Imai Y, Nagashima R, Misawa H, Takeda H, Matsuzawa Y, Kawata S (2001) Circulating transforming growth factor beta 1 as a predictor of liver metastasis after resection in colorectal cancer. Clin Cancer Res 7:1258–1262

Urness LD, Sorensen LK, Li DY (2000) Arteriovenous malformations in mice lacking activin receptor-like kinase-1. Nat Genet 26:328–331

Valderrama-Carvajal H, Cocolakis E, Lacerte A, Lee EH, Krystal G, Ali S, Lebrun JJ (2002) Activin/TGF-beta induce apoptosis through Smad-dependent expression of the lipid phosphatase SHIP. Nat Cell Biol 4:963–969

van de Wetering M, Sancho E, Verweij C, de Lau W, Oving I, Hurlstone A, van der Horn K, Batlle E, Coudreuse D, Haramis AP, Tjon-Pon-Fong M, Moerer P, van den Born M, Soete G, Pals S, Eilers M, Medema R, Clevers H (2002) The beta-catenin/TCF-4 complex imposes a crypt progenitor phenotype on colorectal cancer cells. Cell 111:241–250

Viglietto G, Motti ML, Bruni P, Melillo RM, D'Alessio A, Califano D, Vinci F, Chiappetta G, Tsichlis P, Bellacosa A, Fusco A, Santoro M (2002) Cytoplasmic relocalization and inhibition of the cyclin-dependent kinase inhibitor p27(Kip1) by PKB/Akt-mediated phosphorylation in breast cancer. Nat Med 8:1136–1144

Wakefield LM, Roberts AB (2002) TGF-beta signaling: positive and negative effects on tumorigenesis. Curr Opin Genet Dev 12:22–29

Watanabe T, Wu TT, Catalano PJ, Ueki T, Satriano R, Haller DG, Benson ABr, Hamilton SR (2001) Molecular predictors of survival after adjuvant chemotherapy for colon cancer. N Engl J Med 344:1196–1206

Weeks BH, He W, Olson KL, Wang XJ (2001) Inducible expression of transforming growth factor beta1 in papillomas causes rapid metastasis. Cancer Res 61:7435–7443

Weidner N, Semple JP, Welch WR, Folkman J (1991) Tumor angiogenesis and metastasis–correlation in invasive breast carcinoma. N Engl J Med 324:1–8

Weinmann AS, Bartley SM, Zhang T, Zhang MQ, Farnham PJ (2001) Use of chromatin immunoprecipitation to clone novel E2F target promoters. Mol Cell Biol 21:6820–6832

Wicks SJ, Lui S, Abdel-Wahab N, Mason RM, Chantry A (2000) Inactivation of smad-transforming growth factor beta signaling by Ca^{2+}-calmodulin-dependent protein kinase II. Mol Cell Biol 20:8103–8111

Wong C, Rougier-Chapman EM, Frederick JP, Datto MB, Liberati NT, Li JM, Wang XF (1999) Smad3-Smad4 and AP-1 complexes synergize in transcriptional activation of the c-Jun promoter by transforming growth factor beta. Mol Cell Biol 19:1821–1830

Wotton D, Lo RS, Lee S, Massague J (1999) A Smad transcriptional corepressor. Cell 97:29–39

Wu G, Chen YG, Ozdamar B, Gyuricza CA, Chong PA, Wrana JL, Massague J, Shi Y (2000) Structural basis of Smad2 recognition by the Smad anchor for receptor activation. Science 287:92–97

Wu JW, Hu M, Chai J, Seoane J, Huse M, Li C, Rigotti DJ, Kyin S, Muir TW, Fairman R, Massague J, Shi Y (2001) Crystal structure of a phosphorylated Smad2. Recognition of phosphoserine by the MH2 domain and insights on Smad function in TGF- beta signaling. Mol Cell 8:1277–1289

Wu JW, Krawitz AR, Chai J, Li W, Zhang F, Luo K, Shi Y (2002) Structural mechanism of Smad4 recognition by the nuclear oncoprotein Ski: insights on Ski-mediated repression of TGF-beta signaling. Cell 111:357–367

Xie W, Mertens JC, Reiss DJ, Rimm DL, Camp RL, Haffty BG, Reiss M (2002) Alterations of Smad signaling in human breast carcinoma are associated with poor outcome: a tissue microarray study. Cancer Res 62:497–505

Xu W, Angelis K, Danielpour D, Haddad MM, Bischof O, Campisi J, Stavnezer E, Medrano EE (2000) Ski acts as a co-repressor with Smad2 and Smad3 to regulate the response to type beta transforming growth factor. Proc Natl Acad Sci USA 97:5924–5929

Yagi K, Furuhashi M, Aoki H, Goto D, Kuwano H, Sugamura K, Miyazono K, Kato M (2002) c-myc is a downstream target of the Smad pathway. J Biol Chem 277:854–861

Yagi K, Goto D, Hamamoto T, Takenoshita S, Kato M, Miyazono K (1999) Alternatively spliced variant of Smad2 lacking exon 3. Comparison with wild-type Smad2 and Smad3. J Biol Chem 274:703–709

Yahata T, de Caestecker MP, Lechleider RJ, Andriole S, Roberts AB, Isselbacher KJ, Shioda T (2000) The MSG1 non-DNA-binding transactivator binds to the p300/CBP coactivators, enhancing their functional link to the Smad transcription factors. J Biol Chem 275:8825–8834

Yamaguchi K, Nagai S, Ninomiya-Tsuji J, Nishita M, Tamai K, Irie K, Ueno N, Nishida E, Shibuya H, Matsumoto K (1999) XIAP, a cellular member of the inhibitor of apoptosis protein family, links the receptors to TAB1-TAK1 in the BMP signaling pathway. EMBO J 18:179–187

Yamaguchi K, Shirakabe K, Shibuya H, Irie K, Oishi I, Ueno N, Taniguchi T, Nishida E, Matsumoto K (1995) Identification of a member of the MAPKKK family as a potential mediator of TGF-β signal transduction. Science 270:2008–2011

Yamamura Y, Hua X, Bergelson S, Lodish HF (2000) Critical role of Smads and AP-1 complex in transforming growth factor-beta-dependent apoptosis. J Biol Chem 275:36295–36302

Yang YA, Dukhanina O, Tang B, Mamura M, Letterio JJ, MacGregor J, Patel SC, Khozin S, Liu ZY, Green J, Anver MR, Merlino G, Wakefield LM (2002a) Lifetime exposure to a soluble TGF-beta antagonist protects mice against metastasis without adverse side effects. J Clin Invest 109:1607–1615

Yang YA, Tang B, Robinson G, Hennighausen L, Brodie SG, Deng CX, Wakefield LM (2002b) Smad3 in the mammary epithelium has a nonredundant role in the induction of apoptosis, but not in the regulation of proliferation or differentiation by transforming growth factor-beta. Cell Growth Differ 13:123–130

Yin JJ, Selander K, Chirgwin JM, Dallas M, Grubbs BG, Wieser R, Massague J, Mundy GR, Guise TA (1999) TGF-β signaling blockade inhibits PTHrP secretion by breast cancer cells and bone metastases development. J Clin Invest 103:197–206

Yingling JM, Datto MB, Wong C, Frederick JP, Liberati NT, Wang X-F (1997) Tumor suppressor Smad4 is a transforming growth factor ß-inducible DNA binding protein. Mol Cell Biol 17:7019–28

Yu L, Hebert MC, Zhang YE (2002) TGF-beta receptor-activated p38 MAP kinase mediates Smad-independent TGF-beta responses. EMBO J 21:3749–3759

Zavadil J, Bitzer M, Liang D, Yang YC, Massimi A, Kneitz S, Piek E, Bottinger EP (2001) Genetic programs of epithelial cell plasticity directed by transforming growth factor-beta. Proc Natl Acad Sci USA 98:6686–6691

Zawel L, Dai JL, Buckhaults P, Zhou S, Kinzler KW, Vogelstein B, Kern SE (1998) Human Smad3 and Smad4 are sequence-specific transcription activators. Mol Cell 1:611–617

Zhang Y, Chang C, Gehling DJ, Hemmati-Brivanlou A, Derynck R (2001) Regulation of Smad degradation and activity by Smurf2, an E3 ubiquitin ligase. Proc Natl Acad Sci USA 98:974–979

Zhang Y, Feng XII, Derynck R (1998) Smad3 and Smad4 cooperate with c-Jun/c-Fos to mediate TGF-β-induced transcription. Nature 394:909–913

Zhang YW, Yasui N, Ito K, Huang G, Fujii M, Hanai J, Nogami H, Ochi T, Miyazono K, Ito Y (2000) A RUNX2/PEBP2alpha A/CBFA1 mutation displaying impaired transactivation and Smad interaction in cleidocranial dysplasia. Proc Natl Acad Sci USA 97:10549–10554

Zhou G, Lee SC, Yao Z, Tan TH (1999) Hematopoietic progenitor kinase 1 is a component of transforming growth factor β-induced c-Jun N-terminal kinase signaling cascade. J Biol Chem 274:13133–13138

Zhu H, Kavsak P, Abdollah S, Wrana JL, Thomsen GH (1999) A SMAD ubiquitin ligase targets the BMP pathway and affects embryonic pattern formation. Nature 400:687–693

The p53 Transcription Factor as Therapeutic Target in Cancer

C. Asker · V. J. N. Bykov · C. Mendez-Vidal · G. Selivanova · M. T. Wilhelm · K. G. Wiman (✉)

Department of Oncology–Pathology, Karolinska Institute,
Cancer Center Karolinska (CCK), 17176 Stockholm, Sweden
Klas.Wiman@mtc.ki.se

1	Introduction	210
2	p53 Is a Transcription Factor	211
2.1	The p53 Tumor Suppressor Protein	213
2.2	The Cell Cycle Arrest Response	214
2.3	The Apoptotic Response	214
2.4	p53 at Mitochondria: Transactivation-Independent Induction of Apoptosis	216
2.5	Choice of Response by p53	216
2.6	Other p53 Functions	218
3	Mechanisms of p53 Regulation	219
4	p53 Mutations	222
5	In Vivo Models of p53	226
5.1	Tissue-Specific Inactivation of p53	227
5.2	Gain-of-Function of Mutant p53	228
5.3	Why Do p53-Deficient Mice Mainly Develop Lymphomas?	228
5.4	Mouse Models for Testing Carcinogens	229
5.5	p53 and Aging	230
6	p53 and Cancer Therapy	231
6.1	Clinical Relevance of p53 Status	233
7	p53 and Novel Therapeutic Strategies	234
7.1	Gene Therapy and Smart Viruses	234
7.2	Inhibition of p53	235
7.3	Mutant p53 Reactivation	236
7.4	Inhibition of MDM2	238
8	Future Perspectives	239
	References	239

Abstract The p53 tumor suppressor protein is a transcription factor that responds to cellular stress. p53 regulates growth arrest, apoptosis, senescense, angiogenesis and DNA repair, mainly by activating and/or repressing transcription of specific target genes. Its central role in tumor development is supported by the frequent mutation or loss of the p53 gene in human tumors of different tissue origin. Furthermore, animal models have underscored p53's key role in tumorigenesis and pointed to p53-induced apoptosis as a major mechanism for tumor suppression in vivo. Due to its role in the evolution of tumors and pivotal function in regulation of apoptosis, the p53 pathway is an important target for conventional and novel cancer therapy.

Keywords p53 · Cancer therapy · Apoptosis · Tumor target

1
Introduction

The development of a malignant tumor is a multistep process that involves a series of genetic alterations that disrupt normal regulation of cell growth and survival and endow the tumor cell with unlimited replicative potential, the ability to attract blood vessels, and eventually the capacity for tissue invasion and metastasis (reviewed by Hanahan and Weinberg 2000). A common theme in tumorigenesis is the activation of oncogenes and the inactivation of tumor suppressor genes. The p53 protein was identified as a 53-kDa cellular protein tightly bound to the large T oncoprotein of the SV40 DNA tumor virus (Chang et al. 1979; Kress et al. 1979; Lane and Crawford 1979; Linzer and Levine 1979; Melero et al. 1979). p53 was first classified as an oncogene due to its overexpression in many cancer cells. However, later studies revealed p53 gene rearrangements in Friend virus-induced mouse erythroleukemia (Mowat et al. 1985), that the wild-type protein could suppress transformation (Finlay et al. 1989; Hinds et al. 1990), and that the p53 gene was mutated at high frequency in common human tumors (Baker et al. 1989; Nigro et al. 1989). This established p53 as an important tumor suppressor gene.

Since then, p53 has been the focus of extensive research. Its central role in tumor suppression is supported by the finding of a high number of human tumors in which the p53 gene is lost or inactivated through mutations in the gene itself or by defects in some of the upstream or downstream molecules in the p53 signaling pathway (Hollstein et al. 1996). Furthermore, p53 is inactivated by the viral E6 protein in human papilloma virus (HPV)-induced anogenital tumors (Vousden and Farrell 1994), which account for up to 10% of all human cancers. In addition, genetically engineered p53-null mice show dramatically increased susceptibility to spontaneous development of tumors at young age (Donehower et al. 1992). Thus, data from both experimental systems and human tumors argue convincingly that intact p53 provides critical protection against tumor development.

p53 is member of a gene family that also includes the p63 and p73 genes (Yang et al. 2002). Whereas p53 clearly is an important tumor suppressor, p63 is mainly involved in development, and p73 has a role in DNA damage responses, apoptosis, neurogenesis, and inflammation. Neither p63 nor p73 seem to be inactivated in any significant fraction of human tumors.

2
p53 Is a Transcription Factor

Under normal conditions p53 levels are very low in most cells. However, stress conditions such as DNA damage, hypoxia or oncogenic stress cause p53 protein stabilization and p53-dependent cellular responses including cell cycle arrest, senescence, differentiation and apoptosis or DNA repair after genotoxic damage (Bargonetti and Manfredi 2002) (Fig. 1). p53 directly transactivates genes containing p53 binding sites in the vicinity of their regulatory regions. The p53 consensus binding site consists of two copies of the ten-base pair motif 5′-PuPuPuC(A/T)(T/A)GPyPyPy-3′ separated by 0–13 base pairs (el-Deiry et al. 1992). One copy of the motif is insufficient for

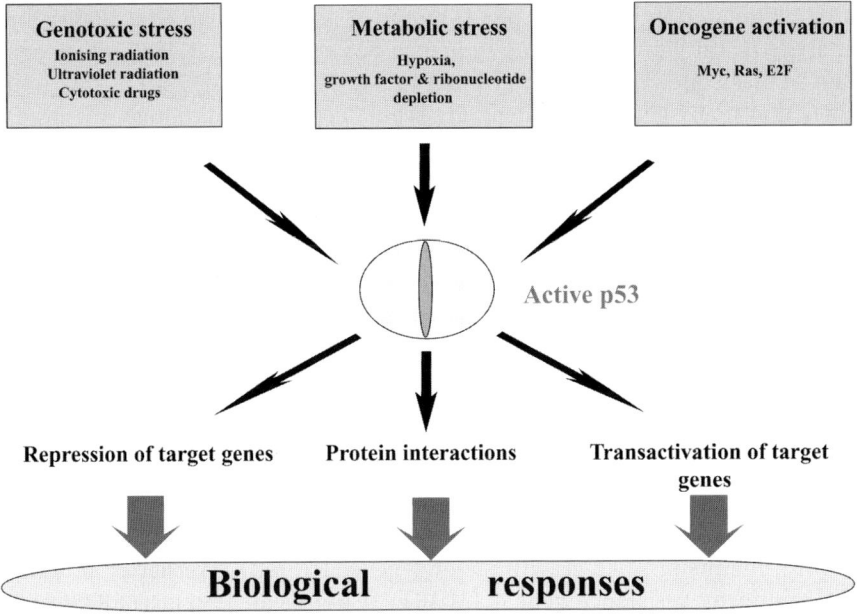

Fig. 1 The p53 protein responds to diverse stress conditions, such as DNA damage, hypoxia, and oncogene activation. This leads to increased stability of the p53 protein via inhibition of its interaction with the MDM2 protein and/or inhibition of MDM2 itself. DNA damage induces several kinases (ATM, ATR, CHk1 and Chk2) that phosphorylate p53 and MDM2, thus preventing MDM2-mediated degradation of p53 in the proteasome. Oncogene activation causes increased expression of the ARF protein that inhibits MDM2, resulting in accumulation of p53. Increased levels of activated p53 induces transcriptional transactivation of a whole set of p53 target genes (p21, GADD45, Bax, Fas, Noxa, PUMA, Wig-1) and repression of other target genes (Bcl-2, MAP4, IGF1R, hTERT). Accumulated p53 can possibly also act through protein–protein interactions, for instance at mitochondria. These activities of p53 will trigger a multifaceted biological response that includes cell cycle arrest, DNA repair, apoptosis, senescence and/or differentiation

binding and slight alterations of the motif results in loss of affinity for p53. Most tumor-derived mutants forms of p53 carrying missense point mutations in the core domain fail to bind to the consensus motif and consequently fail to transactivate target genes and trigger a p53-dependent biological response.

The growing list of p53 target genes contains genes involved in cell cycle arrest, apoptosis, DNA repair, and angiogenesis, as well as genes with unknown function (el-Deiry 1998; Vousden and Lu 2002). The p21, Gadd45, and 14-3-3sigma genes are effectors of p53-dependent cell cycle arrest, whereas transactivation of pro-apoptotic genes within the mitochondrial and death receptor pathways, including Bax, Noxa, PUMA, Fas and KILLER/DR5, promotes p53-dependent apoptosis. It is presumably the concerted action of a whole set of p53-induced genes that triggers a specific biological response. DNA microarray studies have indicated that several hundred genes are upregulated by wild-type p53 (Kannan et al. 2001; Zhao et al. 2000), and analysis of the human genome sequence has revealed that 4852 genes contain at least one consensus p53 binding site (Wang et al. 2001). This indicates that many genes are potentially regulated by p53, although not necessarily in all cells in response to all types of stress signaling. Rather, different stress conditions are likely to trigger transactivation of specific subsets of p53 target genes in different cell types.

One p53 target gene, *MDM2*, has a critical role as regulator of p53. The p53-induced MDM2 protein (Wu et al. 1993) binds p53 and inhibits p53's transactivation domain, and targets p53 for proteasome-mediated destruction in the cytoplasm (Haupt et al. 1997; Kubbutat et al. 1997). This negative feed back loop ensures low levels of p53 in the absence of cellular stress and tight regulation of the p53 response. In addition, p53 induces several proteins that modulate MDM2 in various ways, including PTEN, WISP1 and cyclin G (Okamoto et al. 2002; Vousden and Lu 2002).

Besides transcriptional transactivation of its target genes, p53 can also repress transcription of certain genes, including Bcl-2 (Miyashita et al. 1994a, 1994b), the insulin-like growth factor 1 gene (IGF1-R) (Werner et al. 1996), the multidrug resistance gene 1 (MDR1) (Chin et al. 1992), hTERT/telomerase (Xu et al. 2000), the microtubule-associated protein 4 gene (MAP4) (Murphy et al. 1996) and Op18/ Stathmin (Johnsen et al. 2000). This could occur through interaction with histone deacetylases via the corepressor Sin3a. Both wild-type p53 and Sin3a can be found bound to the promoter of the p53-repressed gene MAP4 when associated with deacetylated histones in vivo (Murphy et al. 1999).

2.1
The p53 Tumor Suppressor Protein

The 393-amino acid human p53 protein can be divided into several regions with distinct functions (Fig. 2). The acidic N-terminal domain (residues 1–50) confers p53 transcriptional activity by recruiting components of the basal transcriptional machinery (Fields and Jang 1990; O'Rourke et al. 1990). This domain is also important for regulation of p53 stability via interaction with the MDM2 protein (Lin et al. 1994) (see below). The induction of target genes is mediated by the specific binding of p53 to DNA through the core domain (residues 102–292) (Bargonetti et al. 1993; Cho et al. 1994). The majority of missense mutations seen in human tumors occur in this domain (Pavletich et al. 1993). The DNA-binding domain is separated from the N terminal domain by a proline-rich region (residues 61–94) typical of proteins interacting with signal transduction molecules that contain an SH3-binding domain (Walker and Levine 1996). The proline-rich domain is necessary to achieve full apoptotic p53 function in response to DNA damage caused by chemotherapeutic agents (Baptiste et al. 2002).

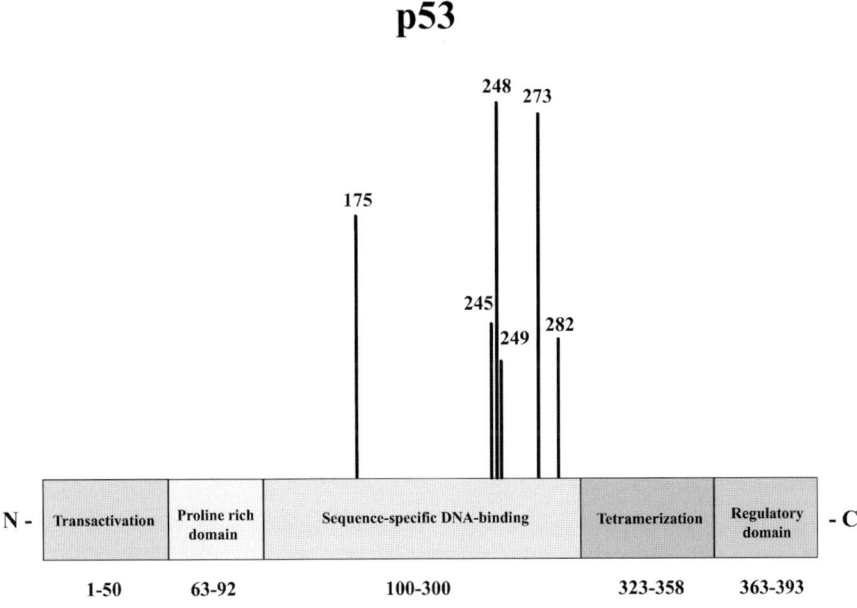

Fig. 2 Schematic representation of the p53 protein. Functional domains are indicated. *Numbers* refer to amino acid residues in human p53. The six most frequently mutated residues in human tumors are indicated by the *bars*. These six so-called hot spot mutations account for 28% of all p53 mutations. More than 95% of all p53 mutations are located in the core domain (residues 100–300), whereas less than 2% occur in the N-terminal region of p53 (1–99), and around 3% are located in the C-terminal region (300–393)

p53 binds DNA most efficiently as a tetramer. Four p53 molecules interact via the C-terminal tetramerization domain (residues 324–355). This domain is composed of basic residues and has important regulatory properties (Clore et al. 1995; Jeffrey et al. 1995). The very C terminus (residues 363–393) has been implicated in regulation of specific DNA binding by the core domain (Hupp et al. 1992; Jayaraman and Prives 1995).

2.2
The Cell Cycle Arrest Response

A key effector of p53-dependent cell cycle arrest is the p21 gene that encodes an inhibitor of cyclin-dependent kinases (CDKs) (el-Deiry et al. 1993) that are required for cell cycle progression. p21 Induces arrest in the G_1 phase of the cell cycle by inhibition of cyclin E-CDK2, and G_2 cell cycle arrest by inhibiting cyclin B/CDK1 activity (Agarwal et al. 1995; Bunz et al. 1998). Other activities, such as inhibition of DNA replication and DNA repair have also been attributed to p21 through its interaction with the proliferating cell nuclear antigen (PCNA) (Li et al. 1994; Luo et al. 1995; Rousseau et al. 1999). Another p53 target that regulates cell cycle progression is the 14-3-3 sigma gene, whose overexpression can block cells in G_2 (Hermeking et al. 1997). Cytoplasmic 14-3-3 sigma was shown to bind and inhibit the translocation of cyclin B/CDK1 to the nucleus, a process required for the initiation of mitosis (Chan et al. 1999). In addition, 14-3-3 sigma can cooperate with p21 to trigger cell cycle arrest at the G_2/M transition (Chan et al. 2000; Stewart et al. 1995). p53 also induces G_2 arrest through upregulation of GADD45 (growth arrest and DNA-damage inducible gene 45), a p53 target gene whose protein product binds and inhibits cyclin B/CDK1 (Zhan et al. 1998). Furthermore, the MCG10 protein containing KH motifs has also been implicated in p53-induced G_2 arrest and apoptosis (Zhu and Chen 2000). Interestingly, the RNA-binding capacity of this protein has been linked to its anti-proliferative functions. Finally, the p53 target gene Reprimo regulates G_2 cell cycle progression by inhibiting the activity of CDK1, although the exact molecular mechanism is unclear (Ohki et al. 2000). For a more complete list of p53 targets implicated in the G_2/M transition checkpoint see Taylor and Stark (2001).

2.3
The Apoptotic Response

The activation of p53 can also result in the irreversible inhibition of cell growth by induction of apoptosis. The importance of this response has been further accentuated in experiments performed using mice carrying Myc-driven lymphomas. This study suggested that apoptosis is the only p53 function selected against during lymphoma development and therefore, the only relevant function for tumor suppression in vivo (Schmitt et al. 2002a). The ability of p53 to trigger cell cycle arrest depends on its ability to transcriptionally upregulate particular target genes. Some studies have also proposed a transcription independent mechanism for p53-induced apoptosis (Bennett et al. 1998; Ding et al. 2000; Marchenko et al. 2000). Missense mutations which abrogate transcriptional activation introduced in both p53 alleles in mouse embryonic stem cells inhibited p53-dependent apoptosis (Chao et al. 2000; Jimenez et al. 2000). The number of p53 target genes that have been implicated in mediating p53-dependent apoptosis is rather large and the relative importance of each is difficult to assess. In any case, these observations indicate a considerable role of p53-mediated transactivation in the induction of apoptosis.

Some of the p53-regulated apoptotic effectors are part of the death receptor pathway, such as the tumor necrosis factor receptor family members Fas (Owen-Schaub et al. 1995) and KILLER/DR5 (Wu et al. 1997), or the death domain-containing protein Pidd (Lin et al. 2000). Other proapoptotic p53 targets belong to the mitochondrial pathway. The Bax gene was the first proapoptotic gene identified (Miyashita and Reed 1995). Bax is a member of the Bcl-2 family that, in certain cell types and after the appropriate apoptotic stimulus, can translocate from the cytoplasm to the mitochondria, facilitating cytochrome c release and the induction of apoptosis (Knudson and Korsmeyer 1997). The translocation of the Bax protein to mitochondria was shown to be mediated by Peg3/Pw1, another p53 target gene (Deng and Wu 2000). Bax accounts for at least part of the induced apoptosis as thymocytes derived from Bax-deficient mice had an intact apoptotic response to DNA damage (Knudson et al. 1995). However, a double knockout of Bax and the closely related Bak gene showed that p53-dependent apoptosis in thymocytes was completely abrogated (Lindsten et al. 2000).

Other effectors of p53-induced apoptosis through the mitochondrial pathways are Apaf-1 (Robles et al. 2001) and the mitochondrial proteins Noxa (Oda et al. 2000a), PUMA (Nakano and Vousden 2001; Yu et al. 2001) and p53AIP1 (Oda et al. 2000b). Noxa and PUMA are BH3 domain-containing proteins. Noxa induces Apaf-1 and caspase 9 activation whereas PUMA can interact with Bcl-2 family members and trigger cytochrome c release, which in turn activates downstream caspases. p53AIP1 appears to affect mitochondrial membrane potential, leading to cell death. Ultimately, p53 can also inhibit survival signaling through the upregulation of genes like PTEN (Stambolic et al. 2001) and IGF-BP3 (Buckbinder et al. 1995) in order to induce apoptosis.

Other p53 target genes defining novel apoptotic pathways have been described. For instance, the plasma membrane protein PERP represents a novel type of p53 target with sequence similarities to the PMP-22/gas3 tetraspan membrane protein (Attardi et al. 2000). PERP is expressed at high levels in apoptotic cells compared with G_1-arrested cells and its expression was found to cause cell death in fibroblasts. Likewise, the p53 target p53RDL1, located at the plasma membrane (Tanikawa et al. 2003), contains a death domain at the C-terminal end. Binding of p53RDL1 to its ligand Netrin blocks p53-induced apoptosis and cleavage of p53RDL1 by p53 restores its apoptotic function.

Given the complexity of p53 inducible apoptotic genes and the fact that no single gene appears to be crucial in p53-mediated apoptosis, it seems likely that p53-induced apoptosis is the result of the cumulative effect of a set of p53 target genes that cooperate to achieve a full response under specific stress conditions (Vogelstein et al. 2000).

2.4
p53 at Mitochondria: Transactivation-Independent Induction of Apoptosis

In addition to triggering apoptosis via transcriptional regulation of target genes, p53 may induce apoptosis by translocating to mitochondria. p53 has been shown to complex with BclXL and Bcl2, resulting in increased outer mitochondrial membrane permeability and cytochrome c release. Tumor-derived transactivation-deficient mutants of p53 failed to bind BclXL and trigger cytochrome *c* release (Mihara et al. 2003).

2.5
Choice of Response by p53

The factors that determine whether p53 activation will induce cell cycle arrest or apoptosis or other biological responses probably include cell type, presence of survival factors, and oncogenic transformation. Analysis of tumor-derived p53 mutants showed that the cellular response to p53 activation depends, at least in part, on which group of p53-responsive genes become transcriptionally activated (Friedlander et al. 1996; Ludwig et al. 1996). These mutants retained the ability to activate expression of G_1 cell cycle arrest genes but they had a reduced ability to bind p53-responsive DNA sequences in the promoters of proapoptotic genes. Moreover, these specific defects in transcriptional activation correlated with an impaired apoptotic function. p53 may have different affinities for binding sites in different promoters. Mutants with minor conformational changes are able to bind to high-affinity sites in the promoters of cell cycle arrest genes but are impaired in the binding of low-affinity sites in the promoters of apoptotic genes (Vousden and Lu 2002). This is in agreement with an earlier model proposing that low levels of p53 protein could induce cell cycle arrest whereas high levels would induce apoptosis (Chen et al. 1996).

As mentioned before, the functional outcome of p53-dependent repression has been related to the specific induction of p53-mediated apoptosis. The dissociation between transcriptional repression and apoptosis from p53-dependent transactivation has been correlated to the specific interaction of different pools of p53 with coactivator or corepressor molecules upon different stress signals (Koumenis et al. 2001). According to this model, DNA damage can induce the interaction of p53 with transcriptional activators like p300 but also with transcriptional repressors such as Sin3a. In contrast, stresses lacking a DNA damage response (i.e., hypoxia) mainly induce the interaction with corepressors. Apoptosis and transcriptional downregulation induced by p53 has also been correlated with the formation of a trimeric complex formed by p53, MDM2 and the retinoblastoma protein (pRB) in cells with DNA damage (Hsieh et al. 1999). In this complex pRB prevented MDM2 from inhibiting p53-mediated transrepression but not transactivation. The p53 in the trimeric complex was able to induce apoptosis. According to this model, pRB might protect p53 from MDM2-targeted degradation by occupying the site on MDM2 that confers sensitivity to degradation. However, this argues against a previous report showing that inactivation of pRB leads to p53-mediated induction of apoptosis (Almasan et al. 1995).

More recent studies have shown that the interaction of p53 with other proteins can also modulate the promoter choice. The interaction between p53 and the p300/CBP coactivator permits the acetylation of histones surrounding the p53-binding site, thus facilitating the access of the transcriptional machinery (Espinosa and Emerson 2001). Apart from this interaction that generally contributes to gene activation, p53 can also interact with other coactivators, for example, JMY or the ASPP family members to specifically induce the activation of proapoptotic target genes (Samuels-Lev et al. 2001; Shikama et al. 1999). During the cellular stress response, JMY interacts with p300 and the p300/JMY complex is recruited to activated p53. This augments p53-dependent transcription and apoptosis by inducing p53-dependent activation of Bax. The ASPP family members enhance the DNA binding and transactivation function of p53 on the promoters of proapoptotic genes in vivo. Moreover, the expression of ASPP was found frequently downregulated in human breast carcinomas expressing wild-type p53 but not mutant p53. It is likely that the availability of these cofactors and the regulation of their interaction with p53 can determine the outcome of the p53 response during cellular stress conditions.

The choice of p53 response is probably also to some extent determined by the p53 family members p63 and p73. The combined loss of p63 and p73 resulted in failure of cells containing functional p53 to undergo apoptosis in response to DNA damage (Flores et al. 2002). This study hypothesized that since p53 can not bind the promoters of proapoptotic genes in cells lacking p63 and p73, these proteins must localize at those sites to allow p53 binding.

Finally, phosphorylation of p53 at Ser46 has been linked to transcriptional activation of the apoptosis-inducing gene p53AIP1 (Oda et al. 2000b). This modification is in turn controlled by the p53-regulated genes p53DINP1, that activates phosphorylation of Ser46 (Okamura et al. 2001), and Wip1 which can dephosphorylate this site (Fiscella et al. 1997).

2.6
Other p53 Functions

Besides the two major responses discussed above, p53 also participates in other pathways that represent barriers against tumor development. Several lines of evidence have indicated a role of p53 in DNA excision repair and DNA double-strand break repair. It has been suggested that the C-terminal domain of p53 can interact with ends of double strand breaks generated by DNA damage (Bakalkin et al. 1995). It can also recognize insertion/deletion mismatches in DNA, possibly leading to the recruitment of other proteins to these sites and thus provide a signal for DNA damage (Lee et al. 1995). In addition, p53 can interact with proteins involved in DNA repair such as BRCA1 (Ouchi et al. 1998; Zhang et al. 1998) or DNA polymerase β (β-pol) (Seo et al. 2002). BRCA1 is a potent coactivator for p53 and can mediate the stabilization of wild-type p53 protein through the transcriptional regulation of p14ARF (Somasundaram et al. 1999; Xu et al. 2001). More recently, it has been suggested that BRCA1 selectively coactivates the p53 transcription factor towards genes that direct DNA repair and cell cycle arrest (MacLachlan et al. 2002). Examples of such genes are GADD45 (Harkin et al. 1999; Kastan et al. 1992; MacLachlan et al. 2000), and p21 (Somasundaram et al. 1997) that interacts with the PCNA also involved in DNA repair. Another gene involved in DNA repair processes that is directly regulated by p53 is p53R2. This gene encodes a nuclear ribonucleotide reductase induced by ultraviolet (UV) light, gamma irradiation and adriamycin in a wild-type p53-dependent manner (Tanaka et al. 2000).

Furthermore, p53 may participate in the induction of cellular senescence, which limits the proliferative capacity of most primary cells in culture. For example, certain mutations in p53 were sufficient to extend the lifespan of 'pre-aged' human diploid fibroblasts cultured until a near-senescent phenotype, but had no effect on 'young' fibroblasts (Bond et al. 1994). Moreover, the p53-regulated CDK inhibitor p21 was identified during the screening of cDNAs coding for inhibitors of DNA synthesis from senescent human diploid fibroblasts (Noda et al. 1994). More recently, p27, another CDK inhibitor, has been shown to cooperate with p21 to restrict the lifespan of differentiating ovarian cells (Jirawatnotai et al. 2003). However, it is not completely clear whether p21 is essential for p53-induced senescence (Pantoja and Serrano 1999).

Other functions attributed to p53 include inhibition of angiogenesis and metastasis. p53 upregulates a number of genes that normally function as potent angiogenesis inhibitors. Thrombospondin-1 -(Dameron et al. 1994) and the brain-specific p53-target gene BAI1 (Nishimori et al. 1997) are some examples. Moreover, p53 has been shown to transactivate molecular targets, such as the tumor suppressors Maspin (Zou et al. 2000) and KAI1 (Mashimo et al. 1998) or the plasminogen activator inhibitor 1 (Kunz et al. 1995) with a role in restraining tumor invasion and/or metastasis. Likewise, downregulation of the metalloproteinases MMP-1 and MMP-13 by p53 have been correlated with inhibition of cancer metastasis (Sun et al. 1999, 2000).

Yet the function of many p53 target proteins remains elusive. For example, the Wig-1 protein was found to be upregulated by p53 in a mouse T lymphoma line undergoing G_1 cell cycle arrest and apoptosis after expression of wild-type p53 from a temperature-sensitive construct (Varmeh-Ziaie et al. 1997). The *wig-1* gene contains two functional p53-binding motifs in its promoter (Wilhelm et al. 2002) and encodes a nuclear zinc finger protein able to inhibit cell growth in a colony formation assay (Hellborg et al. 2001). Interestingly, Wig-1 binds double-stranded RNA with high affinity via its most N-terminal zinc finger (Mendez-Vidal et al. 2002).

3
Mechanisms of p53 Regulation

p53 is a key regulator of the cellular response to diverse stress conditions. Regulation of p53 activity is very complicated, which is not surprising considering the diversity of biological responses induced by p53. A wide range of signals trigger a variety of post-translational modifications of p53, along with a whole network of protein–protein interactions. In the absence of stress conditions the p53 protein is kept inactive in the cell, mainly due to the high rate of p53 degradation through ubiquitin-dependent proteolysis. It is currently believed that various stress signals converge on the interaction between p53 and its negative regulator MDM2. DNA damage, oncogene activation, hypoxia, metabolic changes, cytokines, ribonucleotide depletion, telomere shortening and other types of stress conditions induce p53 stabilization and accumulation via inhibition of MDM2-mediated degradation of p53 (reviewed in Woods and Vousden 2001). The role of MDM2 as a key regulator of p53 has been strongly supported by the in vivo finding that the lethal phenotype of MDM2-null mice could be rescued in p53-null background (Montes de Oca Luna et al. 1995). MDM2 functions as an E3 ubiquitin ligase that attaches ubiquitin to several C-terminal lysine residues and thus targets p53 for proteasomal degradation (reviewed in Prives 1998). In addition, binding of MDM2 to the p53 transactivation domain inhibits its ability to activate gene expression.

DNA damage-induced accumulation of p53 is achieved through post-translational modifications of p53 and also MDM2. In response to gamma irradiation the PI-3 family kinase ATM (ataxia-telangiectasia mutated) phosphorylates p53, along with MDM2 and the checkpoint kinase Chk2, which in turn phosphorylates p53. ATM-dependent phosphorylation of Ser15 enhances the binding to p300 and the transcriptional transactivation function of p53, whereas Chk2-mediated Ser20 phosphorylation inhibits MDM2 binding and thus results in a rapid increase in p53 levels due to protein stabilization (Chehab et al. 2000; Dumaz and Meek 1999; Khosravi et al. 1999; Sakaguchi et al. 1998; Shieh et al. 2000). MDM2 protein expression is activated by p53 itself, creating a second feedback loop which terminates the p53 response (Wu et al. 1993). In addition to Ser15 and Ser20, five other serines and two threonines in the N terminus and two serines in the C terminus are targeted by DNA damage-induced kinases, including phosphorylation of Ser46 by p38MAPK, Thr81 by JNK, Ser315 by cdc2/CycB, and Ser 392 by CKII (for a comprehensive review, see Appella and Anderson 2001). DNA damage induces acetylation of C-terminal lysine residues by p300/CBP (Sakaguchi et al. 1998). Translocation of both p53 and p300 into promyelocytic leukemia (PML) nuclear bodies probably facilitates acetylation (Pearson et al. 2000). Another pathway of p53 stabilization occurs via direct deubiquitination of p53 by (herpesvirus-associated ubiquitin-specific protease) induced by DNA damage (Li et al. 2002a). Sumoylation of p53 at Lys 386 by SUMO-1 is associated with relocalization of p53 into nuclear bodies by PML, and has been shown to enhance p53-dependent transactivation in a promoter-dependent manner and to affect cell survival (Fogal et al. 2000; Gostissa et al. 1999; Muller et al. 2000; Rodriguez et al. 1999). It appears that modulation of p53 function by sumoylation might have different effects on p53 activity, depending on the cellular environment (Buschmann et al. 2001; Kwek et al. 2001; Schmidt and Muller 2002).

Disruption of the p16/pRb pathway leading to increased E2F1 activity, oncogenic Ras and Myc, and β-catenin dysregulation due to inactivation of its upstream regulator APC (adenomatous polyposis coli), or tumor-derived mutant β-catenin all trigger p53 activation and induction of apoptosis (Bates et al. 1998; Damalas et al. 1999, 2001; Palmero et al. 1998; Zindy et al. 1998). Oncogenic signaling leads to p53 stabilization via several mechanisms. Association between MDM2 and p53 could be counteracted by p14ARF (p19ARF in mice), a nucleolar protein that binds MDM2 and inhibits its E3-ligase activity (Honda and Yasuda 1999; Pomerantz et al. 1998). Activated p53 represses the expression of p14ARF, thus forming a feedback loop that controls p53 itself (Robertson and Jones 1998; Stott et al. 1998). Recent in vivo findings indicating that deficiency in p19ARF has no major impact on p53-mediated apoptosis induced by activated oncogenes points at the existence of an alternative, p14ARF-independent pathway of oncogene signaling to p53 (Tolbert et al. 2002). Oncogenes like Myc and deregulated E2F1 can induce production of ROS (reactive oxygen species) (Tanaka et al. 2002; Vafa et al. 2002). DNA damage due to oncogene-induced ROS results in activation of cell cycle checkpoints which in turn signal to p53. E2F1 stimulated phosphorylation of p53 at Ser15, Ser20 and Ser392 and induced p53 activity and apoptosis in a p14ARF-independent manner, whereas p19ARF was shown to play a significant role in promoting apoptosis in response to Myc in mouse epidermal tissues (Russell et al. 2002). Recent studies in human fibroblasts have demonstrated that activated Myc and E2F1 induce p53 and cell cycle arrest and/or apoptosis independently of p14ARF (Lindstrom and Wiman 2003).

Along with stabilization of p53, post-translational modifications directly affect p53 transcriptional activity. Phosphorylation at Ser315 affects selectivity of p53 binding to different DNA sequences and stimulates its transcriptional function (Blaydes et al. 2000; Wang and Prives 1995), whereas Ser392 phosphorylation activates DNA binding and facilitates p53 tetramerization (Hupp et al. 1992; Sakaguchi et al. 1997). Acetylation of p53 facilitates coactivator recruitment to promoters and histone acetylation (Barlev et al. 2001). p53 partner proteins can significantly modulate p53 function by altering promoter selectivity of p53 and thus the p53-dependent biological response. Several proteins, including Ref1, HMG1, and 14-3-3sigma, have been shown to interact with p53 and activate its specific DNA binding (for a review see Prives and Hall 1999). The novel tumor suppressor p33ING1 is a component of the p53 signaling pathway which cooperates with p53 in growth suppression by modulating p53-dependent transcriptional activation (Garkavtsev et al. 1998). An alternatively spliced product of p33ING1 gene, p33ING2, is induced by DNA damage and strongly enhances the transcriptional transactivation function of p53, probably by facilitating acetylation of p53 at Lys382 (Nagashima et al. 2001). Interestingly, BRCA1 selectively coactivates p53-mediated transcription of genes that operate in DNA repair and cell cycle arrest, but not those that control apoptosis (MacLachlan et al. 2002). ASPP family proteins specifically enhance the apoptotic activity of p53 without apparent effect on its ability to induce cell cycle arrest. Using chromatin immunoprecipitation, it has been demonstrated that ASPP1 and 2 facilitate the binding of p53 to the Bax and PIG3 promoters in vivo, but not to the p21 promoter (Samuels-Lev et al. 2001). Similarly, p53 family members p63 and p73 play an important role in the DNA damage-induced apoptotic response by directing p53 to the promoters of proapoptotic genes (Flores et al. 2002). In contrast, alternatively spliced forms of p63 and p73 that lack N-terminal transactivation domains (ΔN) can block p53 function (for a review see Irwin and Kaelin 2001).

The pattern of p53-induced genes in a given cell is affected by the pathway of stress signaling. For example, homeodomain-interacting protein kinase-2 (HIPK-2) mediates the p53 response to UV irradiation by phosphorylating p53 at Ser46, which in turn facilitates CBP-mediated acetylation of p53 at Lys382. Together these modifications promote expression of p53AIP-1 and induction of apoptosis (D'Orazi et al. 2002; Hofmann et al. 2002). Dephosphorylated hypoxia-induced factor HIF-1a has been shown to bind to p53 and modulate p53-dependent apoptosis (An et al. 1998; Suzuki et al.

2001). Poly-(ADP-ribose) polymerase PARP1 can also differentially modulate p53 activity (Valenzuela et al. 2002).

A set of negative regulators neutralizes p53 activity in the absence of cellular stress. For example, activation of the PI3K/Akt survival signaling cascade results in enhanced rate of p53 turnover due to phosphorylation of MDM2 by Akt kinase, leading to nuclear translocation and stabilization of MDM2 (Ashcroft et al. 2002; Gottlieb et al. 2002; Mayo and Donner 2001). Deacetylation of Lys382 by hSir2 and HDAC1 reduces p53's transcriptional activity and some of its biological functions, including its ability to suppress angiogenesis (Kim et al. 2001; Luo et al. 2001; Vaziri et al. 2001). Dephosphorylation of Ser33/Ser46 by PPMD1 might play an essential role in downregulation of p53 activity, as overexpression of PPMD1 in some tumors prevents p53 activation (Bulavin et al. 2002).

4
p53 Mutations

p53 is the most frequently mutated gene in human tumors (Hollstein et al. 1991; see www.iarc.fr/P53 and Fig. 2) and most mutations described in p53 are missense mutations (see www.iarc.fr/P53/). In contrast, other tumor suppressor genes such as pRb (Lohmann 1999), APC (Miyaki et al. 1994; Miyoshi et al. 1992) and BRCA1 (Futreal et al. 1994; Merajver et al. 1995) are inactivated preferentially by frameshift and nonsense mutations. The mutation spectra of tumor suppressor genes in tumors are the result of a selection process during tumor progression. It has been unclear whether the mere loss of function of wild-type p53 or a gain of novel oncogenic properties of mutant p53 are selected during tumorigenesis. The hetero-tetramerization between wild-type and mutant p53 molecules results in the reduction or complete loss of wild-type p53 DNA binding activity (Nicholls et al. 2002). In addition, binding of mutant p53 to the p53 family member protein p73 compromises p73's proapoptotic activity (Irwin et al. 2000; Stiewe and Putzer 2000; Strano et al. 2000). Such dominant negative effects have only been observed in the presence of excess of mutant p53 over wild-type p53, resulting in loss of wild-type p53 function (Blagosklonny 2000).

Dominant positive or gain-of-function activity of mutant p53 would endow the protein with novel properties that favor tumor growth (Cadwell and Zambetti 2001; Roemer 1999). Some studies have shown that mutant p53 can stimulate cell proliferation (Dittmer et al. 1993), promote resistance to cytostatics (Murphy et al. 2000; Pugacheva et al. 2002) and induce gene amplification (El-Hizawi et al. 2002). The intact transactivation domain appears essential for mutant p53 to exhibit its oncogenic characteristics (Matas et al. 2001; Pugacheva et al. 2002). Remarkably, p53 mutations found in human tumors are clustered in the DNA-binding domain (95% of all mutations)

(Olivier et al. 2002; see www.iarc.fr/P53/; Fig. 2). Six mutational 'hotspots' have been identified in this region at codons 175, 245, 248, 249, 273 and 282 (see www.iarc.fr/P53/ and Fig. 2). The exact mechanism behind the gain-of-function of mutant p53 is unknown. High levels of mutant p53 in cells might result in a gross exaggeration of some residual transcriptional transactivation as well as transcription-independent activities of the wild-type protein (Blagosklonny 2000). Furthermore, there are some indications that specific interactions of mutant p53 with components of the nuclear matrix may play a role in promoting new functions (Deppert et al. 2000). Promiscuous DNA binding and illegitimate transactivation of growth-promoting genes like *myc* or other genes that might contribute to the malignant phenotype, e.g., *MDR1*, could also be responsible for the gain-of-function activity of mutant p53 (Sampath et al. 2001).

Based on the crystal structure of the p53 core domain, two classes of p53 mutations can be discerned. Type I mutations (DNA contact mutants) affect amino acids participating in DNA contacts. These include, for instance, mutations at codons 248 and 273 (Cho et al. 1994). Such mutants do not usually show any gross protein misfolding and the protein maintains wild-type conformation (Legros et al. 1994). However, mutations at position 248 can cause structural distortions of p53 (Wong et al. 1999). The type II mutations (structural mutants) are characterized by changes in the amino acids crucial for the general fold of the protein (Cho et al. 1994). The most important residues for maintaining wild-type p53 conformation are 94–117, 175–185 and 282–294 (Chen et al. 2001). A more recent classification of p53 mutations describes five classes of p53 mutations by taking into consideration not only the site of mutation but also the thermodynamic stability and the DNA-binding properties of the mutated protein (Bullock and Fersht 2001; Wong et al. 1999).

Exposure to genotoxic agents induces damage in cellular DNA. Such damage often appears as a DNA lesion, resulting either from a chemical reaction between an active molecule and DNA bases (formation of DNA adducts) or by a reaction between nucleotides themselves which can be activated, for instance, by UV radiation (formation of photoproducts) (Anonymous 1994, "DNA Adducts: Identification and Biological Significance"). Other types of DNA modification such as deamination of 5-methylcytosine and formation of DNA strand breaks can also occur (Anonymous 1994, "DNA Adducts: Identification and Biological Significance"). Most of the DNA lesions will be repaired and only a minor portion will lead to changes in the genetic information. Such genetic changes are in many cases unique for a given agent, due to specific interactions between the agent and DNA. This implies that agents do not interact with DNA bases at random but rather modify certain DNA bases or even short DNA sequences (Denissenko et al. 1996; Tretyakova et al. 2002). Efficiency of the sequence-specific DNA repair (Tornaletti and Pfeifer 1994) and subsequent clonal selection of mutant cells are also important contributors to the formation of the pattern of mutation hotspots along p53 (Pfeifer and Denissenko 1998).

If not repaired, modifications of the primary DNA structure can result changes of the genetic information upon DNA replication. Tumor-derived p53 sequences from tissues exposed to genotoxic agents may sometimes reveal the natural history of a tumor (Hussain and Harris 1999; Pfeifer and Denissenko 1998). For UV radiation such genetic fingerprints are C→T transitions and CC→TT tandem mutations in skin tumors (Brash et al. 1991, 1996; Ziegler et al. 1993). In nonmelanoma skin tumors 92% of p53 mutations are located at dipyrimidine sites (Daya-Grosjean et al. 1995). This is consistent with high levels of dipyrimidine lesions induced in DNA in human skin even after very mild erythemal exposure to solar-simulated UV radiation (Bykov et al. 1998). The abundant exposure of skin to solar radiation results in the formation of multiple clones of keratinocytes with mutated p53 (Jonason et al. 1996; Ponten et al. 1997). Such patches of keratinocytes are covering our skin on frequently sun-exposed areas, with the number of lesions sometimes exceeding 30 patches per cm^2 (Jonason et al. 1996; Ponten et al. 1997; Ren et al. 1996). However, only a few of these clones of keratinocytes, if any, give rise to malignant tumors.

Induction of the G→T transversions by polycyclic aromatic hydrocarbons (products of incomplete combustion with mass exposure associated with smoking) was shown in experiments in vitro (Eisenstadt et al. 1982) and in vivo (Ruggeri et al. 1993). Subsequent epidemiological studies have demonstrated that the formation of G→T transversion 'hotspots' at codons 157, 248 and 273 in p53 in lung tumors is linked to smoking (Hainaut and Pfeifer 2001). Importantly, the frequency of G→T transversions in lung tumors of smokers was significantly higher than that of nonsmokers (Hainaut and Pfeifer 2001).

The high frequency of hepatocellular carcinomas in Africa, Asia and North America is caused by food contaminated with molds producing aflatoxin B1. This toxin is known to modify guanine residues in DNA, resulting in the high prevalence of G→T transversions in codon 249 of p53 in hepatocellular carcinomas of exposed individuals (Bressac et al. 1991; Hsu et al. 1991).

Loss of p53 tumor suppressor function may occur at different stages of tumorigenesis (Guimaraes and Hainaut 2002). It is plausible that the timing of p53 inactivation is influenced by the natural history of a given tumor. Tumors linked to strong exogenous carcinogenic exposure, such as UV-related skin basal cell carcinoma (Montesano et al. 1997; Ponten et al. 1997), tobacco-associated lung cancer (Hussain et al. 2001) and alcohol–tobacco-associated esophageal cancer (Montesano et al. 1997; Parenti et al. 1995) have a tendency to lose p53 function at earlier stages of tumor development. Moreover, p53 mutations are often found also in nontumorous tissue that has been subjected to excessive exogenous carcinogenic exposure, for instance as patches of keratinocytes harboring mutant p53 (Jonason et al. 1996). Interestingly, mutations in p53 were found in synoviocytes in rheumatoid arthritis, probably as a result of exposure to ROS during the course of a chronic inflammation (Firestein et al. 1997; Han et al. 1999). It is noteworthy that a recent report has linked rheumatism with a higher risk for developing lymphomas (Ekstrom et al. 2003).

Tumors caused by endogenous factors usually acquire p53 mutations at a later stage of tumor progression, such as sporadic colorectal cancers (Fearon and Vogelstein 1990; Kahlenberg et al. 1996), pancreatic cancer (Apple et al. 1999), endometrial cancer (Ayhan et al. 1998) and perhaps breast cancer (Megha et al. 2002; Norberg et al. 2001), although there are some contradicting data (Borresen-Dale 2003).

Around 50% of human tumors carry intact wild-type p53. However, accumulated data suggest that a majority of human tumors have defects in the p53 pathway. This may occur through alterations upstream or downstream of p53 itself. Alterations in INK4a/ARF locus are commonly observed in many human tumors (Sharpless and DePinho 1999). This may result in failure to induce p53 in response to oncogenic signaling and thus compromise the p53 pathway (Asker et al. 1999). Increased expression of the p53 antagonist MDM2 could be an alternative mechanism of inactivation of the p53 pathway. In non-Hodgkin's lymphomas inactivation of p14ARF was frequently detected in tumor cells harboring wild-type p53 (Pinyol et al. 2000). Overexpression of MDM2 has been observed in many types of tumors including leukemias (Bueso-Ramos et al. 1993; Faderl et al. 2000) and sarcomas (Cordon-Cardo et al. 1994). Burkitt lymphoma lines carrying wild-type p53 were shown to either overexpress MDM2 or lack p14ARF due to homozygous deletion (Lindstrom et al. 2001). Nonetheless, double inactivation of p53 and p14ARF seems to have the most sinister effect on the survival of patients (Pinyol et al. 2000). This suggests that p14ARF also has p53-independent effects and that mutant p53 itself somehow provides a selective advantage, consistent with the proposed gain-of-function activity (see above). Moreover, ATM mutations in patients with chronic lymphocytic leukemia contributed to the impaired induction of p21 in response to radiation in tumor cells with wild-type p53 and to the subsequent resistance to radiation therapy (Pettitt et al. 2001). Lack of expression of the Chk2 serine/threonine kinase, although extremely rare, can also inactivate the p53 pathway (Tort et al. 2002). However, more clinical data are need in order to estimate the role of these different alterations in the p53 tumor suppressor pathway and particularly to what extent they compromise the response to anticancer therapy.

5
In Vivo Models of p53

Although many aspects of p53 can be studied in cultured cells, a deeper understanding of p53's role in tumorigenesis and other processes can only be obtained by in vivo studies in mice. Numerous mouse models have been developed to study p53 and tumor development (reviewed in Attardi and Jacks 1999). The first evidence suggesting that p53 has tumor suppressor activity came from studies of erythroleukemic cell lines derived from spleens of mice infected with Friend leukemia virus (Ben-David and Bernstein 1991; Mowat et al. 1985). The p53 gene was found to be frequently rearranged, deleted or to contain proviral insertions, suggesting that inactivation of p53 is an important step in the development of virus-induced erythroleukemia in mice. A more direct approach to study p53 function was to generate transgenic mice that expressed either the p53Pro193 or p53Val135 mutant alleles under the control of their own promoters (Lavigueur et al. 1989). Both strains of mice showed 20–30% increased tumor susceptibility and mainly developed lung adenocarcinomas, osteosarcomas, and lymphoid tumors. This demonstrated that expression of mutant p53 resulted in tumor formation, probably due to the dominant negative nature of mutant p53 proteins that form hetero-oligomeres with wild-type p53 and thereby inhibit its normal function.

More conclusive evidence came from studies of p53-null mice generated by gene targeting technology (Donehower et al. 1992). Mice lacking p53 are viable, indicating that p53 is dispensable during embryonic development. However, these mice rapidly develop tumors at high frequency; 74% developed tumors by 6 months of age and all mice developed tumors or died by 10 months of age, with an average tumor latency of 4.5 months. The most frequently observed tumors were malignant lymphomas but hemiangosarcomas, undifferentiated sarcomas, osteosarcoma and occasional mammary adenocarcinomas were also found. The variety of tumors in p53-null mice indicates that p53 is involved in tumorigenesis in many tissues and cell types. It has also been shown that the genetic background affects the onset of tumors in p53-null mice (Donehower et al. 1995). 129/Sv p53-null mice developed tumors sooner on average than mixed inbred C57BL/6×129/Sv. This suggests that strain-specific modifier genes can modulate tumorigenesis in the context of p53 deficiency.

The p53 heterozygous ($p53^{+/-}$) mice were also found to be tumor prone (Harvey et al. 1993). However, they developed a slightly different spectrum of tumors than the homozygous mice and with longer latency. By the age of 18 months 50% of the heterozygous mice had developed tumors. Osteosarcomas and soft tissue sarcomas occurred in about 58% of the mice, but also lymphomas (32%) and carcinomas (10%) were observed. The majority of tumors (55%) in the heterozygous mice showed loss of the remaining wild-type allele. Nevertheless, the fact that some tumors retained the wild-type p53 allele indicates that a decrease in p53 gene dosage can be sufficient for tumor development. The $p53^{+/-}$ mice represent a mouse model for the Li-Fraumeni syndrome. Li-Fraumeni syndrome patients are predisposed to cancer due to a germline lesion in one p53 allele and 50% of the patients develop a variety of tumors before the age of 30 years (Malkin et al. 1990). Like the $p53^{+/-}$ mice, the Li-Fraumeni patients develop osteosarcomas and soft tissue sarcomas with approximately the same frequency and latency time. Still, mice and men differ in that Li-Fraumeni patients often develop breast carcinomas and brain tumors, which are rarely seen in mice.

5.1
Tissue-Specific Inactivation of p53

The reasons for the observed differences in tumor spectra between humans and mice are probably complex. However, they may to some extent simply reflect the nature of p53 deficient mice. The lack of p53 in every cell could result in a more severe phenotype, and tumor types that have been associated with p53 inactivation in humans would perhaps never have time to appear in p53-null mice. One way to get around this problem is to make conditional knockout mice. By using the Cre/loxP system it is possible to achieve tissue-specific deletion of p53 and thus study the effect of p53 loss in almost any tissue. This technique was used by Berns and colleagues to address the role of Brca2 and p53 in mammary tumorigenesis by generating mice that carry conditional Brca2 and p53 alleles (Jonkers et al. 2001). One-third of the cases of hereditary breast cancer carry a germline mutation in the Brca2 gene, and it has been shown that p53 is frequently mutated in Brca2-associated malignancies (Crook et al. 1998; Lee et al. 1999; Ramus et al. 1999). Brca2-deficient mice have been generated but they show embryonic lethality (Ludwig et al. 1997; Suzuki et al. 1997). Mice carrying the conditional Brca2 and p53 alleles were crossed with a K14-cre transgenic mice, which allows for epithelium-specific expression of Cre recombinase and subsequent deletion of Brca2 and p53 in epithelial cell types. All K14-cre females carrying conditional Brca2 and p53 alleles developed mammary and/or skin tumors. Mice carrying two conditional Brca2 alleles and one conditional p53 allele also developed mammary tumors. Notably, all of them had lost the remaining wild-type p53 allele. This indicates that inactivation of p53 is involved in the development of breast cancer in mice as well as in humans. Moreover, this mouse represents a suitable model for human breast cancer, and could be used for screening for novel potent therapeutic drugs to be used in the clinic.

5.2
Gain-of-Function of Mutant p53

The majority of p53 mutations in human tumors are missense mutations resulting in the expression of mutant p53 proteins (Hussain and Harris 1998). As already discussed above, this suggests that mutant p53 proteins have acquired novel functions that promote tumor growth. Also, there is evidence that Li-Fraumeni syndrome families with missense mutations in the core DNA-binding domain of p53 have a significantly higher incidence of cancer than families with a p53 mutation that lead to a truncated protein or no protein at all (Birch et al. 1998). In order to test whether mutant p53 has gain-of-function activity or only acts in a dominant negative fashion, a p53Val135 transgenic mouse was generated. These mice were then crossed onto a wild-type, $p53^{+/-}$ or $p53^{-/-}$ background (Harvey et al. 1995). The p53Val135 transgene accelerated tumor development in the wild-type and $p53^{+/-}$ mice but did not change the onset of tumorigenesis in the $p53^{-/-}$ mice. This suggests that the p53Val135 allele exerts a tumorigenic effect only when a wild-type copy of p53 is present, supporting a dominant negative rather than gain-of-function activity. However, it is also possible that the $p53^{-/-}$ mice die before the full effect of a gain-of function mutant can be detected.

The same question was addressed using a p53His172 mutant generated by homologous recombination (Liu et al. 2000). This mutation corresponds to the p53His175 hot spot mutation in human tumors and is associated with extremely poor prognosis (Goh et al. 1995). The advantage of using homologous recombination is that the mutant will not be overexpressed as in previous transgenic experiments, but rather expressed at levels similar to those in human tumors. The p53His172/$^+$ mice differed from $p53^{+/-}$ mice in tumor spectrum. They developed more carcinomas but the number of sarcomas was not significantly different. Interestingly, the osteosarcomas and carcinomas that developed in these mutant mice frequently metastasized which is rarely seen in tumors from $p53^{+/-}$ mice. Of 11 different tumors from p53His172/$^+$ mice, only one had lost the remaining p53 allele. This is in contrast to the $p53^{+/-}$ mice where more than half of the tumors loose the wild-type p53 allele. The different tumor spectrum and increased metastatic potential of the tumors in the p53His172/$^+$ mice could argue that mutant p53 protein has a gain-of-function activity. However, since the p53His172 mutant was not tested on a p53-null background, a dominant negative mechanism of action cannot be excluded.

5.3
Why Do p53-Deficient Mice Mainly Develop Lymphomas?

Normal thymocytes undergo apoptosis in response to DNA damage. However, p53-null thymocytes fail to do so, due to the absence of a p53-dependent

DNA double-strand break checkpoint (Clarke et al. 1993). This checkpoint could be monitoring the V(D)J recombination in lymphocytes, resulting in elimination of cells with illegitimate recombination. Failure to undergo apoptosis could result in potential oncogenic cells, thus facilitating lymphomagenesis. To test this theory, p53 deficient mice have been crossed with mice defective in V(D)J recombination in order to investigate whether inhibition of recombination diminishes lymphomagenesis (Nacht and Jacks 1998). RAG1 and RAG2 proteins initiate V(D)J recombination by introducing double-strand breaks precisely adjacent to recombination targeting signals that flank segments of immunoglobulin and T-cell receptor coding sequences. Both $Rag1^{-/-}/p53^{-/-}$ and $Rag2^{-/-}p53^{-/-}$ mice develop lymphomas at high frequency. There was no evidence for normal recombination in these animals, indicating that V(D)J recombination is not essential for lymphomagenesis in p53-deficent animals. Additionally, this suggests that aberrant rearrangements are not the trigger for tumorigenesis in the absence of p53.

p53 is involved in several cellular processes, including cell cycle arrest, apoptosis, senescence, DNA repair and angiogenesis (Tokino and Nakamura 2000). Which of these functions of p53 are really critical for p53-dependent tumor suppression? This question was addressed in a recent study using Emμ-Myc mice (Schmitt et al. 2002a). These mice overexpress the c-*myc* oncogene in the B-cell lineage and develop pre-B or B-cell lymphomas that frequently acquire p53 mutations on a $p53^{+/-}$ background (Eischen et al. 1999; Schmitt et al. 1999). Co-expression of a Bcl-2 transgene or a dominant-negative caspase 9 (C9DN) transgene that disrupt p53-dependent apoptosis resulted in lymphomas that retained the wild-type p53 allele. The lymphomas arising in the presence of Bcl-2 or C9DN and retaining wild-type p53 were as aggressive as those with p53 mutations, but they preserved key cell cycle checkpoints and were genetically stable. This implies that apoptosis is the only p53 effector function that limits tumor development in this setting. The cell cycle defect and aneuploidy associated with p53 mutations in the Emμ-Myc-derived lymphomas (Eischen et al. 1999; Elson et al. 1995; Gao et al. 2000; Hsu et al. 1995; McCormack et al. 1998; Schmitt et al. 1999, 2002a;) are by-products of p53 loss and do not contribute to lymphomagenesis.

5.4
Mouse Models for Testing Carcinogens

The high rate and early onset of spontaneous tumors in p53-null mice precludes testing the effect of various carcinogens. In contrast, p53 heterozygous mice with their relatively late onset of tumorigenesis are suitable for such studies. In one study, all three genotypes ($p53^{+/+}$, $p53^{+/-}$, and $p53^{-/-}$) were exposed to dimethylnitrosamine (DMN) (Harvey et al. 1993). This carcinogen is a particularly potent inducers of liver hemangiosarcomas and lung tumors (Clapp et al. 1968). As expected, the $p53^{-/-}$ mice developed spontaneous tumors too rapidly to correctly monitor the effect of DMN. The p53$^{+/-}$ mice developed multiple liver hemangiosarcomas and were significantly more susceptible to the carcinogen than wild-type mice, indicating that p53$^{+/-}$ mice can be a sensitive tool for some carcinogenesis studies.

Treatment of mice with dimethylbenzanthracene (DMBA) followed by 12-O-tetradecanoyl-phorbol-13-acetate (TPA) has been extensively used as a carcinogen model of skin cancer induction. This model allows studies of multistage carcinogenesis and the discrete steps of initiation, promotion and progression during conversion of benign papillomas to malignant skin carcinomas. Skin tumors initiated by DMBA and promoted by TPA have been examined at both the papilloma stage and carcinoma stage for p53 inactivation (Ruggeri et al. 1991, 1993). p53 was found to be mutated only in carcinoma samples, indicating that p53 loss is important for tumor progression but not initiation of these tumors. This has been further supported in a study where p53$^{+/-}$ mice were treated with DMBA (Kemp et al. 1993). The number, size, and growth rate of benign papillomas were not increased in the p53$^{+/-}$ mice compared to wild-type mice. However, the papillomas in the heterozygous mice underwent a more rapid malignant progression associated with the loss of the remaining wild-type allele. This confirms that loss of p53 is important for progression to a more malignant state rather than tumor initiation in chemically induced skin cancer.

5.5
p53 and Aging

Activation of p53 appears critical in cellular replicative senescence. When cultured primary cells senesce, the levels of activated p53 increase (Atadja et al. 1995; Bond et al. 1996). This is further supported by the fact that viral oncogenes delay senescence by inactivating p53 (Shay et al. 1991). The importance of p53 as a mediator of senescence has also been shown in genetically modified mice. Late generation telomerase-deficient mice have shortened telomeres, decreased longevity and early senescence-related phenotypes, and cells taken from these mice all have high levels of activated p53. Moreover, telomerase-deficient mice showed decreased senescence in crosses onto a p53-deficient background (Rudolph et al. 1999). Ku-80 deficient mice have defects in the double-stranded DNA break repair machinery and embryonic fibroblasts from these mice exhibit premature senescence, which is rescued by p53 deficiency (Lim et al. 2000). The involvement of p53 in organismal longevity was further supported by the serendipitous generation of a mouse that expresses a truncated form of p53, called the 'm allele'. This allele produces a C-terminal p53 fragment that probably results in constitutive activation of the remaining wild-type allele (Tyner et al. 2002). The m mouse displays an enhanced tumor resistance and a shortened lifespan accompanied by signs of early aging such as reduction in adipose tissue, hair growth and wound healing and also reduced cellularity and mass in several organs suggesting reduced stem cell proliferation. In contrast to the m mice, mice carrying one extra copy of p53 (i.e., three copies instead of two) show no premature aging (Garcia-Cao et al. 2002). These 'super p53' mice were generated by introducing large genomic DNA segments containing the entire p53 gene, reproducing the normal expression and regulation of the endogenous gene. Super p53 mice show an enhanced p53-mediated DNA damage response when treated with carcinogens and were resistant to tumor formation, proving that cancer resistance can be enhanced in a whole organism by a simple genetic modification without side effects such as premature aging.

To what extent do mouse models recapitulate human cancer and what are the criteria for an ideal mouse model of human cancer? Mouse tumors should be similar to their human counterparts with respect to histological features, and physiological and systemic effects. It is important that the same genes and/or pathways are affected in tumor initiation and progression and that gene expression profiles are similar in mouse and human tissues. Therapy responses in mice should mimic those in human patients and the model should be suitable for testing of new therapies, allowing predictions regarding clinical efficacy. However, cancer risk in both humans and mice are affected by environment and genetic background and eventually the major questions have to be answered in humans. Nevertheless, the mouse is a unique and powerful tool to study the complex influences of genetic background on tumor susceptibility.

6
p53 and Cancer Therapy

By virtue of its function as a sensor of DNA damage and integrator of several apoptotic pathways, p53 has been studied extensively in relation to sensitivity to chemotherapy and ionizing radiation in the past decade. Most of the accumulated data which support a role of p53 as a chemosensitizer rely on studies of in vitro cultured cell lines or murine cancer models. In an attempt to define drug sensitivity in terms of genotype, a large set of cell lines with defined p53 status have been screened by the US National Cancer Insitute (NCI) (Weinstein et al. 1997). While this screen cannot be representative of treatment of tumors in humans, it provides a profile for drugs and potential drug targets. Most clinically relevant chemotherapeutic agents screened appeared to achieve the best response in a wild-type p53 context, whereas a small group of compounds including colchicine derivatives and microtubule inhibitors (taxanes) were more effective on tumor cells expressing mutant p53 or did not show any preference that depended on p53 status (Weinstein et al. 1997). Stabilization of p53 by different chemotherapeutic agents also engage differential responses in downstream target genes (Weinstein et al.

1997). Paradoxically, p53-dependent and independent activation of p21 causes a growth arrest response which can promote cell survival in certain genetic backgrounds after exposure to several different chemotherapeutic agents and gamma irradiation (Bunz et al. 1998) (Mahyar-Roemer and Roemer 2001; Javelaud and Besancon 2002; Li et al. 2002b). This may constitute a mechanism for decreased drug sensitivity and should be considered when combinations of drugs with different activity are used (Blagosklonny 2002). In addition, the concentration of the drug of choice exposed to cells can influence the death mechanism; low dose exposure of 5-fluorouracil (5-FU) in colorectal cell lines may rather induce mitotic catastrophe and necrotic cell death while high dose engages apoptotic pathways. This response pattern was independent of p53 status but cells expressing wild-type p53 required lower drug concentrations (Yoshikawa et al. 2001). Microarray profiling showed induction of several p53-induced target genes upon treatment with high but not low doses of camptothecin in synchronized colorectal HCT116 cells carrying wild-type p53 (Zhou et al. 2002). Thus, several levels of complexity may exist in the p53-mediated response to genotoxic agents. The interpretation of this combined evidence is that efficiency of a given chemotherapeutic agent may depend on dose, tissue origin, genotype as well as p53 status (Weller 1998).

While data from in vitro systems are conflicting, disruption of the p53 pathway clearly serves as a resistance mechanism to DNA-damaging chemotherapeutic agents and gamma irradiation in animal models (for review see Schmitt et al. 2002a). Studies in the Emμ-Myc transgenic lymphoma model in mice have underscored p53's ability to induce apoptosis as its principal mode of tumor suppression after DNA damage (Schmitt and Lowe 2002). In addition, Emμ-Myc lymphomas derived from crosses with p53 or INK4a-null mice were more resistant to treatment by alkylating agents (cyclophosphamide) and radiation therapy compared to those from control animals (Schmitt et al. 1999). When p53's apoptotic response pathway was blocked by Bcl-2, ablation of either p53 or INK4a/p19ARF, but not p19ARF alone worsened the response to cyclophosphamide in mice reconstituted with these tumors (Schmitt et al. 2002b). The better prognosis for mice bearing wild-type p53 or p16 loci was related to the capability of inducing a p16 or p53-dependent prolonged growth arrest or senescence program in vivo upon exposure to the drug (Schmitt et al. 2002a). Thus, although derived in artificial systems, these data suggest that senescence and apoptosis may be two alternative responses to cyclophosphamide. A similar p53-dependent senescent response has been reported in human breast cancer cells following exposure to doxorubicin (Elmore et al. 2002).

6.1
Clinical Relevance of p53 Status

p53 mutations are frequent in many types of tumors including breast, ovarian, colon and lung cancer (Ferreira et al. 1999; Hollstein et al. 1991; Olivier et al. 2002). However, an accurate evaluation of the role of p53 in clinical outcome has been hampered by the fact that the techniques used for assessment of p53 status in many clinical studies, i.e., immunohistochemistry or incomplete sequencing, are not reliable (Bergh 1999; Borresen-Dale 2003). Furthermore, the p53 status in a primary tumor may not reflect an intact p53 pathway nor be representative of the genotype of metastases in advanced disease (Norberg et al. 2001). The same problem exists in evaluating p53 as a predictive marker for hormone treatment, chemotherapy and radiation. In addition, there are many other concurrent pathways activated by therapy and so tumor profiling correlated to choice of therapy is required to describe better predictive markers (or fingerprints) for cancer treatment. In one such correlative attempt in breast cancer, wild-type p53 was suggested to play a role in sensitivity to low-dose doxorubicin therapy (Borresen-Dale 2003). In contrast, other investigators found that mutant p53 was a better predictor of complete response to high-dose epirubicin and cyclophosphamide in advanced breast cancer (Bertheau et al. 2002). Thus, even within the same tumor group p53 may have alternative predictive value depending on the drug regime chosen. Importantly, the type of mutation appears to determine clinical outcome and mutations in the L2/L3 domain of the DNA-binding region of p53 seem to be especially ominous in several different cancer types (Soussi and Beroud 2001). A novel dimension is added by the finding that the p53 paralog p73 contributes to chemosensitivity, and interactions with mutant or polymorphic p53 can neutralize this effect (Bergamaschi et al. 2003; Irwin et al. 2003). In addition, antiapoptotic forms of p73 can mediate attenuation to apoptosis (Soussi and Beroud 2003).

The antimetabolite 5-FU is a uracil analog that is widely used in treatment of solid tumors such as breast and colorectal carcinoma. FU acts mainly by inhibiting thymidylate synthase (TS), a key enzyme in pyrimidine synthesis (Diasio and Johnson 2000; Relling and Dervieux 2001) but also by misincorporation resulting in aberrant DNA and RNA (Bunz et al. 1999; Pritchard et al. 1997). In vitro studies have proposed that the 5-FU response is p53-dependent (for a review see Longley et al. 2003). A low level of TS has been positively correlated to clinical responsiveness and overall survival in breast and colorectal cancer in several studies (Etienne et al. 2002). TS has also been suggested to be a direct target of repression by wild-type p53 (Lenz et al. 1998). Upregulation of TS and the UTPase gene, another enzyme involved in 5-FU metabolism, has been proposed as gain-of-functions to mutant p53 that may contribute to 5-FU resistance (Peters et al. 2002; Pugacheva et al. 2002).

In vitro cell systems and in vivo mouse model studies comparing genetically targeted isogenic colorectal cell lines ($p53^{+/+}$, $p53^{-/+}$, $p53^{-/-}$) showed a marked dependence on functional p53 for the apoptotic response after 5-FU exposure (Bunz et al. 1999). Both activation of extrinsic and intrinsic apoptotic pathways by 5-FU have been proposed to be mediated by p53 (Igney and Krammer 2002; Muller et al. 1998). In attempts to identify downstream targets essential for the p53-induced apoptotic response to 5-FU, ferredoxin reductase (FDXR) was identified as a potential candidate. FDXR is a mitochondrial cytochrome P-450 inner membrane protein participating in electron transportation and part of a ROS-generating system. Antioxidants which prevent ROS formation can promote clonogenic survival after 5-FU exposure in colorectal cells, suggesting that ROS formation is a possible mechanism for FU-induced toxicity (Hwang et al. 2001). In a phase III clinical trial the overall survival was significantly improved with 5-FU regimens in patients with colorectal cancer carrying c-*myc* amplification and wild-type p53 compared to patients with c-*myc* amplified/mutant p53 or only wild-type p53 status (Arango et al. 2001). This observation is in line with the hypothesis that Myc selects for p53-dependent apoptosis rather than growth arrest by repressing the induction of p21 in response to DNA damage in colon cancer cells (Seoane et al. 2002). However, whether p53 plays a pivotal role in response to 5-FU treatment in the clinical setting still needs to be adequately addressed (Longley et al. 2003).

In summary, the role of p53 in the therapeutic response of human cancers in vivo is still poorly understood and the available information is contradictory (for reviews see Brown and Wouters 1999; Schmitt and Lowe 2001). Future prospective clinical trials which include complete genomic sequencing of p53 correlated to treatment regime and clinical outcome are warranted to clarify this issue (Soussi and Beroud 2001).

7
p53 and Novel Therapeutic Strategies

Despite the current confusion as to the exact impact of p53 in conventional cancer therapy the prospect of exploiting the p53 pathway in novel therapeutic strategies in cancer is appealing. Such strategies may rely on reconstitution of p53-dependent apoptosis to eliminate tumor cells, or selective targeting of tumor cells that lack an intact p53 pathway.

7.1
Gene Therapy and Smart Viruses

Gene therapy is conceivable as a corrective therapy when p53 is disabled or lost in tumor cells. Replication defective adenoviruses which deliver a wild-type p53 allele have been used and improved systems are underway (Merritt et al. 2001). In most cases, proof of concept has been obtained in vitro and in vivo, but only a few approaches made it to the clinic. Phase I–II trials with intratumorally delivered Ad-p53 in lung cancer patients, either alone or in combination with chemotherapy or radiotherapy, have demonstrated acceptable side-effects and evidence of tumor regression (Swisher and Roth 2002). Unfortunately, these approaches seem only to offer local tumor control and such limitations may decrease the utility of this therapeutic approach.

An alternative strategy with trans-splicing ribozymes that can simultaneously reduce mutant p53 expression and restore wild-type p53 activity in various human cancer cells has been described (Watanabe and Sullenger 2000). The correction of aberrant p53 RNA by utilization of virally delivered trans-splicing ribozymes may be a better option in the gene therapy approach (Watanabe and Sullenger 2000). Alternatively, RNA interference technology may allow the targeting of specific mutant forms of p53 by virally delivered siRNA (Martinez et al. 2002; Shen et al. 2003).

Two novel alternative gene therapeutic routes to exploit the absence of functional p53 have been developed. First, vectors carrying a therapeutic gene designed to be repressed by wild-type p53 in the cell have been constructed. Such vectors, carrying a prodrug can in theory be expressed only in cells that are devoid of functional p53 and would be repressed in the wild-type p53 context (Zhu and Chen 2000). Another approach is the development of ONYX015, an adenovirus vector which can replicate only in p53-deficient cells because of a deletion of the E1B gene whose protein product binds and inactivates p53 (Bischoff et al. 1996). This virus should only be able to replicate in a p53-null environment. Later studies have shown, however, that ONYX015 does not show absolute fidelity to p53 status (Hay et al. 1999). For instance, a defective p53 pathway by loss of p14ARF in a wild-type p53 context can also support replication (Ries and Korn 2002). To this end, early clinical trials show that ONYX015 may have clinical effect in several tumor types, especially in combination with chemotherapeutics (Kirn 2001). Importantly, ONYX015 and similar viruses hold promise for systemic therapy. However, the efficiency of viral replication and delivery of viruses in solid tumors will still be issues for future development. In addition, the heterogeneity of the p53 status among cells within the same tumor may pose a problem for all tailored gene therapy techniques.

7.2
Inhibition of p53

Normal mouse tissues express variable levels of p53; radiosensitive tissues such as liver, spleen, thymus and small intestine were shown to have a higher p53-dependent activity upon exposure to DNA damage than other tissues (for a review see Gudkov and Komarova 2003). In humans, this poses a clinical problem during chemotherapy or radiotherapy as the rapidly growing cells in, for example, intestine and bone marrow are highly sensitive. In this setting, high wild-type p53 activity can be harmful and may limit the tolerated dose. By screening of a chemical library, Komarov et al. isolated a substance which suppressed p53-dependent transcription of a p53 promoter linked to lacZ in doxorubicin treated cells. The compound, named pifithrin-α (PFT-α; for p53-inhibitor alpha), was shown to rescue mice exposed to lethal doses of gamma radiation (Komarov et al. 1999). PFT-α does not affect p53 phosphorylation or specific DNA binding, but lowers the levels of nuclear but not cytoplasmic p53. One working hypothesis is that PFT-α acts downstreams of p53 by modulating stability or nuclear/cytoplasmic shuttling of the protein. Concerns have been raised about the potential mutagenicity of the drug. The gamma-irradiated mice treated with PFT-α showed no signs of abnormality 7 months postirradiation. However, a recent study suggested that PFT-α induces chromosomal abnormalities when applied with etoposide in human lymphoblastoid cells carrying wild-type p53 (Bassi et al. 2002). Future work is required to confirm the utility of PFT-α in humans.

7.3
Mutant p53 Reactivation

The tumor suppressor function of p53 is linked to apoptosis and require p53's activity as a transcription factor. p53-mediated transactivation of apoptosis-promoting genes like Bax, KILLER/DR5 and PUMA will trigger apoptosis while p53-dependent transrepression of survival genes such as survivin and Bcl-2 may lower the threshold for death inducing signals (Vousden and Lu 2002). Since most p53 mutations affect the conformation of the protein by disrupting the thermodynamic stability of its DNA-binding domain, several attempts have been made to isolate compounds which restore the wild-type conformation. This approach is attractive as mutant p53 molecules are usually very abundant in tumor cells due to the lack of transcriptional activation of the p53 antagonist MDM2. Reactivation of high levels of mutant p53 can be viewed as firing a 'loaded gun' which selectively kills cancer cells with mutant p53 and spares normal tissues (Selivanova 2001). The first demonstration that mutant p53 could be converted into a wild-type form came from several studies showing that short synthetic peptides derived from the p53 C-terminal region and monoclonal antibodies that recognize a C-terminal epitope can stimulate DNA-binding of wild-type p53. One peptide corresponding to amino acids 361–382 in human p53 was shown to induced apoptosis by binding to the core domain and restored the proper conformation and transactivation activity to mutant p53 (Selivanova et al. 1997). Recently, a nine-residue peptide, CDB3, derived from the crystal structure of the complex between ASPP7/p53BP2 was found to bind the core domain without interfering with p53's DNA-binding activity, and restore wild-type conformation and ability to transactivate p53 target genes to mutant p53 (Gorina and Pavletich 1996); Issaeva et al. 2003).

Several small molecular weight compounds have been shown to reactivate mutant p53. (Bykov et al. 2002b; Foster et al. 1999). CP31398, a styrylquinazoline, was derived from an in vitro screen for drugs that can modify de novo synthesized mutant p53 to wild-type. CP31398 could promote stabilization of the DNA-binding domain of mutant p53 and restore the transactivating function both in vitro and in vivo (Foster et al. 1999). Further studies have revealed that depending on genetic background CP31398 can induce either growth arrest or apoptosis. The compound also appears to be active in wild-type p53-carrying but not in p53-null cells, and to sensitize cells to killing by either chemotherapeutic agents or TRAIL. CP31398-induced apoptosis involves the Bax/mitochondrial/caspase-9 pathway in the colon carcinoma cell line HCT 116 (wild-type p53). However, microarray analysis of cells treated with CP31398 suggested that not only p53-regulated genes are altered by exposure to the compound (Luu et al. 2002; Takimoto et al. 2002). An independent report suggested that CP31398 may work rather by DNA intercalation and disruption of the DNA–p53 core domain interaction independently of p53 status, although the molecule was able to rescue the wild-type conformation of p53 probably during biosynthesis of the protein (Rippin et al. 2002).

In contrast to CP31398, the low molecular compound PRIMA-1 (for p53 reactivation and induction of massive apoptosis) was isolated using a cellular assay (Bykov et al. 2002b). p53-null Saos-2 cells transfected with an inducible His273 p53 mutant were screened with a chemical library consisting of low molecular compounds. PRIMA-1 restores DNA binding, wild-type conformation and transcriptional transactivation to mutant p53 in the absence of de novo protein synthesis. PRIMA-1 specifically inhibited growth of human tumor xenografts carrying mutant p53 in mice (Fig. 3). Data derived from the NCI screen of human tumor cell lines available at NCI confirm that mutant p53-carrying tumor cells are more sensitive to PRIMA-1 than tumor cells carrying wild-type p53 (Bykov et al. 2002a). This distinguishes PRIMA-1 from commonly used chemotherapeutic drugs.

Additionally, ellipticines and aminothiols have been found to confer wild-type properties on mutant p53 in cell culture systems (reviewed by (Bykov et al. 2003). In conclusion, the identification of small molecules that can restore biochemical and biological function to mutant p53 in vivo opens a novel paradigm for future cancer therapy. New drugs based on such molecules could be used alone or in combination with conventional chemotherapeutics to improve clinical outcome. However, much remains to be learned about the exact mechanism of action of mutant p53-reactivating molecules such as PRIMA-1, and drug optimization prior to clinical trials is a long

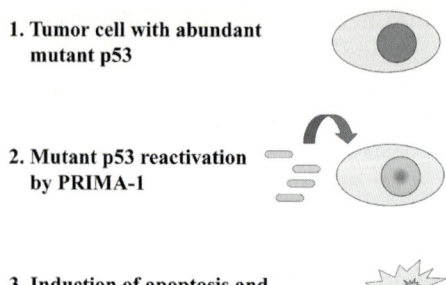

1. Tumor cell with abundant mutant p53

2. Mutant p53 reactivation by PRIMA-1

3. Induction of apoptosis and elimination of the tumor cell

Fig. 3 Mutant p53 reactivation by PRIMA-1: a novel strategy for cancer therapy. Tumor cells carrying mutant p53 often express high levels of mutant p53 due to failure to transactivate the p53 antagonist MDM2. Reactivation of mutant p53 by PRIMA-1 will result in p53 target gene activation and induction of massive apoptosis, leading to rapid elimination of the tumor. This strategy should selectively affect tumor cells, as normal cells do not normally express detectable levels of p53 and because tumor cells are already sensitized to apoptosis by oncogene activation and inactivation of the pRb pathway

process. The ultimate answer as to the clinical antitumor efficacy of these novel drug candidates is several years in the future.

7.4
Inhibition of MDM2

MDM2 regulates p53 levels in normal cells by ubiquitinating p53 and targeting p53 for proteasome-mediated degradation (see above). Therefore, illegitimate overexpression of MDM2 in a subset of tumors will mimic loss of p53. A therapeutic strategy in this situation could be to interfere with the MDM2 binding to p53 or to inhibit MDM2 in another way. Indeed, artificial fusion proteins or peptides that compete with p53 for MDM2 binding, or antisense inhibition of MDM2, can activate p53 in the absence of DNA damage (Bottger et al. 1997; Chen et al. 1998; Chene et al. 2000; Wasylyk et al. 1999). Leptomycin B was recently identified as a small molecule that stabilizes wild-type p53 by blocking MDM2 or HPV E6-mediated cytoplasmic export and degradation (Hietanen et al. 2000). Interestingly, the purine analog roscovitine and other CDK inhibitors have also been shown to stabilize p53 by downregulation of MDM2 mRNA and protein expression (Lu et al. 2001; Sausville 2002). The most promising approach to inhibit MDM2-mediated p53 degradation and thus activate p53 is probably to develop small molecules that inhibit MDM2 binding to p53. MDM2-p53 complexing seems particularly well suited for inhibition by a small molecule, due to the relatively small interacting interface in each protein (Chene 2003).

The ARF protein inhibits MDM2 but is not a competitive inhibitor of MDM2-p53 binding. Imitating the function of ARF by specific peptides is thus another conceivable method of stabilizing p53 (Midgley et al. 2000). A more facile approach would be the administration of low molecular weight compounds with the same function.

In summary, the identification of small cell permeable molecules that can activate wild-type p53 by mechanisms distinct from DNA damage may allow the development of more efficient anticancer drugs for the treatment of the large fraction of human tumors that carry wild-type p53. Such drugs may also act synergistically with conventional chemotherapy. However, the potential side effects of this type of drugs in normal cells carrying wild-type p53 need to be considered.

8
Future Perspectives

The possibility to develop more specific cancer therapeutics has become reality in the postgenomic era. The p53 pathway is critical for regulation of cell growth and survival. Exploitation of genetic defects in this pathway for the development of novel anticancer drugs is both rational and appealing, given its frequent inactivation in human tumors. The successful application of such drugs in clinical practice will require rapid and accurate methods for assessment of multiple components in the p53 pathway in clinical samples. Also, it is very likely that combination of novel therapy with conventional chemotherapy and/or radiotherapy will be necessary to impede evolution of drug resistance. Another prospect is the use of p53-based therapies in other diseases than cancer. Inflammation, neurodegenerative disorders and atherosclerosis share pathways where p53 may be a key player (Morrison and Kinoshita 2000; Ross et al. 2001; Vollmar et al. 2002).

Acknowledgements The authors are supported by the Swedish Cancer Society, the European Union (EU), the Cancer Society of Stockholm, the Gustaf V Jubilee Fund, the Ingabritt & Arne Lundberg Foundation, and the Karolinska Institute.

References

Anonymous (1994) DNA adducts: identification and biological significance. Proceedings of a meeting. Huddinge, Sweden, 18–21 November 1992. IARC Scientific Publication 125:1–478

Agarwal ML, Agarwal A, Taylor WR, Stark GR (1995) p53 controls both the G2/M and the G1 cell cycle checkpoints and mediates reversible growth arrest in human fibroblasts. Proc Natl Acad Sci USA 92:8493–8497

Almasan A, Yin Y, Kelly RE, Lee EY, Bradley A, Li W, Bertino JR, Wahl GM (1995) Deficiency of retinoblastoma protein leads to inappropriate S-phase entry, activation of E2F-responsive genes, and apoptosis. Proc Natl Acad Sci USA 92:5436–5440

An WG, Kanekal M, Simon MC, Maltepe E, Blagosklonny MV, Neckers LM (1998) Stabilization of wild-type p53 by hypoxia-inducible factor 1alpha. Nature 392:405–408

Appella E, Anderson CW (2001) Post-translational modifications and activation of p53 by genotoxic stresses. Eur J Biochem 268:2764–2772

Apple SK, Hecht JR, Lewin DN, Jahromi SA, Grody WW, Nieberg RK (1999) Immunohistochemical evaluation of K-ras, p53, and HER-2/neu expression in hyperplastic, dysplastic, and carcinomatous lesions of the pancreas: evidence for multistep carcinogenesis. Hum Pathol 30:123–129

Arango D, Corner GA, Wadler S, Catalano PJ, Augenlicht LH (2001) c-myc/p53 interaction determines sensitivity of human colon carcinoma cells to 5-fluorouracil in vitro and in vivo. Cancer Res 61:4910–4915

Ashcroft M, Ludwig RL, Woods DB, Copeland TD, Weber HO, MacRae EJ, Vousden KH (2002) Phosphorylation of HDM 2 by Akt. Oncogene 21:1955–1962

Asker C, Wiman KG, Selivanova G (1999) p53-induced apoptosis as a safeguard against cancer. Biochem Biophys Res Commun 265:1–6

Atadja P, Wong H, Garkavtsev I, Veillette C, Riabowol K (1995) Increased activity of p53 in senescing fibroblasts. Proc Natl Acad Sci USA 92:8348–8352

Attardi LD, Jacks T (1999) The role of p53 in tumour suppression: lessons from mouse models. Cell Mol Life Sci 55:48–63

Attardi LD, Reczek EE, Cosmas C, Demicco EG, McCurrach ME, Lowe SW, Jacks T (2000) PERP, an apoptosis-associated target of p53, is a novel member of the PMP-22/gas3 family. Genes Dev 14: 704–718

Ayhan A, Tuncer ZS, Ruacan S, Yasui W, Tahara E (1998) Abnormal expression of cripto and p53 protein in endometrial carcinoma and its precursor lesions. Eur J Gynaecol Oncol 19:316–318

Bakalkin G, Selivanova G, Yakovleva T, Kiseleva E, Kashuba E, Magnusson KP, Szekely L, Klein G, Terenius L, Wiman KG (1995) p53 binds single-stranded DNA ends through the C-terminal domain and internal DNA segments via the middle domain. Nucl Acids Res 23:362–369

Baker SJ, Fearon ER, Nigro JM, Hamilton SR, Preisinger AC, Jessup JM, vanTuinen P, Ledbetter DH, Barker DF, Nakamura Y, et al. (1989) Chromosome 17 deletions and p53 gene mutations in colorectal carcinomas. Science 244:217–221

Baptiste N, Friedlander P, Chen X, Prives C (2002) The proline-rich domain of p53 is required for cooperation with anti-neoplastic agents to promote apoptosis of tumor cells. Oncogene 21:9–21

Bargonetti J, Manfredi JJ (2002) Multiple roles of the tumor suppressor p53. Curr Opin Oncol 14:86–91

Bargonetti J, Manfredi JJ, Chen X, Marshak DR, Prives C (1993) A proteolytic fragment from the central region of p53 has marked sequence-specific DNA-binding activity when generated from wild-type but not from oncogenic mutant p53 protein. Genes Dev 7:2565–2574

Barlev NA, Liu L, Chehab NH, Mansfield K, Harris KG, Halazonetis TD, Berger SL (2001) Acetylation of p53 activates transcription through recruitment of coactivators/histone acetyltransferases. Mol Cell 8:1243–1254

Bassi L, Carloni M, Fonti E, Palma de la Pena N, Meschini R, Palitti F (2002) Pifithrin-alpha, an inhibitor of p53, enhances the genetic instability induced by etoposide (VP16) in human lymphoblastoid cells treated in vitro. Mutat Res 499:163–176

Bates S, Phillips AC, Clark PA, Stott F, Peters G, Ludwig RL, Vousden KH (1998) p14ARF links the tumour suppressors RB and p53. Nature 395:124–125

Ben-David Y, Bernstein A (1991) Friend virus-induced erythroleukemia and the multistage nature of cancer. Cell 66:831–834

Bennett M, Macdonald K, Chan SW, Luzio JP, Simari R, Weissberg P (1998) Cell surface trafficking of Fas: a rapid mechanism of p53-mediated apoptosis. Science 282:290–293

Bergamaschi D, Gasco M, Hiller L, Sullivan A, Syed N, Trigiante G, Yulug I, Merlano M, Numico G, Comino A, Attard M, Reelfs O, Gusterson B, Bell AK, Heath V, Tavassoli M, Farrell PJ, Smith P, Lu X, Crook T (2003) p53 polymorphism influences response in cancer chemotherapy via modulation of p73-dependent apoptosis. Cancer Cell 3:387–402

Bergh J (1999) Clinical studies of p53 in treatment and benefit of breast cancer patients. Endocr Relat Cancer 6:51–59

Bertheau P, Plassa F, Espie M, Turpin E, de Roquancourt A, Marty M, Lerebours F, Beuzard Y, Janin A, de The H (2002) Effect of mutated TP53 on response of advanced breast cancers to high-dose chemotherapy. Lancet 360:852–364

Birch JM, Blair V, Kelsey AM, Evans DG, Harris M, Tricker KJ, Varley JM (1998) Cancer phenotype correlates with constitutional TP53 genotype in families with the Li-Fraumeni syndrome. Oncogene 17:1061–1068

Bischoff JR, Kirn DH, Williams A, Heise C, Horn S, Muna M, Ng L, Nye JA, Sampson-Johannes A, Fattaey A, McCormick F (1996) An adenovirus mutant that replicates selectively in p53-deficient human tumor cells. Science 274:373–376

Blagosklonny MV (2000) p53 from complexity to simplicity: mutant p53 stabilization, gain-of-function, and dominant-negative effect. FASEB J 14:1901–1907

Blagosklonny MV (2002) P53: an ubiquitous target of anticancer drugs. Int J Cancer 98:161–166

Blaydes JP, Craig AL, Wallace M, Ball HM, Traynor NJ, Gibbs NK, Hupp TR (2000) Synergistic activation of p53-dependent transcription by two cooperating damage recognition pathways. Oncogene 19:3829–3839

Bond J, Haughton M, Blaydes J, Gire V, Wynford-Thomas D, Wyllie F (1996) Evidence that transcriptional activation by p53 plays a direct role in the induction of cellular senescence. Oncogene 13:2097–2104

Bond JA, Wyllie FS, Wynford-Thomas D (1994) Escape from senescence in human diploid fibroblasts induced directly by mutant p53. Oncogene 9:1885–1889

Borresen-Dale AL (2003) TP53 and breast cancer. Hum Mutat 21:292–300

Bottger A, Bottger V, Garcia-Echeverria C, Chene P, Hochkeppel HK, Sampson W, Ang K, Howard SF, Picksley SM, Lane DP (1997) Molecular characterization of the hdm2-p53 interaction. J Mol Biol 269:744–756

Brash DE, Rudolph JA, Simon JA, Lin A, McKenna GJ, Baden HP, Halperin AJ, Ponten J (1991) A role for sunlight in skin cancer: UV-induced p53 mutations in squamous cell carcinoma. Proc Natl Acad Sci USA 88:10124–10128

Brash DE, Ziegler A, Jonason AS, Simon JA, Kunala S, Leffell DJ (1996) Sunlight and sunburn in human skin cancer: p53, apoptosis, and tumor promotion. J Investig Dermatol Symp Proc 1:136–142

Bressac B, Kew M, Wands J, Ozturk M (1991) Selective G to T mutations of p53 gene in hepatocellular carcinoma from southern Africa. Nature 350:429–431

Brown JM, Wouters BG (1999) Apoptosis, p53, and tumor cell sensitivity to anticancer agents. Cancer Res 59:1391–1399

Buckbinder L, Talbott R, Velasco-Miguel S, Takenaka I, Faha B, Seizinger BR, Kley N (1995) Induction of the growth inhibitor IGF-binding protein 3 by p53. Nature 377:646–649

Bueso-Ramos CE, Yang Y, deLeon E, McCown P, Stass SA, Albitar M (1993) The human MDM-2 oncogene is overexpressed in leukemias. Blood 82:2617–2623

Bulavin DV, Demidov ON, Saito S, Kauraniemi P, Phillips C, Amundson SA, Ambrosino C, Sauter G, Nebreda AR, Anderson CW, Kallioniemi A, Fornace AJ, Jr., Appella E (2002) Amplification of PPM1D in human tumors abrogates p53 tumor-suppressor activity. Nat Genet 31:210–215

Bullock AN, Fersht AR (2001) Rescuing the function of mutant p53. Nat Rev Cancer 1:68–76

Bunz F, Dutriaux A, Lengauer C, Waldman T, Zhou S, Brown JP, Sedivy JM, Kinzler KW, Vogelstein B (1998) Requirement for p53 and p21 to sustain G2 arrest after DNA damage. Science 282:1497–1501

Bunz F, Hwang PM, Torrance C, Waldman T, Zhang Y, Dillehay L, Williams J, Lengauer C, Kinzler KW, Vogelstein B (1999) Disruption of p53 in human cancer cells alters the responses to therapeutic agents. J Clin Invest 104:263–269

Buschmann T, Lerner D, Lee CG, Ronai Z (2001) The Mdm-2 amino terminus is required for Mdm2 binding and SUMO-1 conjugation by the E2 SUMO-1 conjugating enzyme Ubc9. J Biol Chem 276:40389–40395

Bykov VJ, Issaeva N, Selivanova G, Wiman KG (2002a) Mutant p53-dependent growth suppression distinguishes PRIMA-1 from known anticancer drugs: a statistical analysis of information in the National Cancer Institute database. Carcinogenesis 23:2011–20118

Bykov VJ, Issaeva N, Shilov A, Hultcrantz M, Pugacheva E, Chumakov P, Bergman J, Wiman KG, Selivanova G (2002b) Restoration of the tumor suppressor function to mutant p53 by a low-molecular-weight compound. Nat Med 8:282–288

Bykov VJ, Jansen CT, Hemminki K (1998) High levels of dipyrimidine dimers are induced in human skin by solar-simulating UV radiation. Cancer Epidemiol Biomarkers Prev 7:199–202

Bykov VJN, Selivanova G, Wiman KG (2003) Small molecules that reactivate mutant p53. European Journal of Cancer 39:1828–1834

Cadwell C, Zambetti GP (2001) The effects of wild-type p53 tumor suppressor activity and mutant p53 gain-of-function on cell growth. Gene 277:15–30

Chan TA, Hermeking H, Lengauer C, Kinzler KW, Vogelstein B (1999) 14-3-3Sigma is required to prevent mitotic catastrophe after DNA damage. Nature 401:616–620

Chan TA, Hwang PM, Hermeking H, Kinzler KW, Vogelstein B (2000) Cooperative effects of genes controlling the G(2)/M checkpoint. Genes Dev 14:1584–1588

Chang C, Simmons DT, Martin MA, Mora PT (1979) Identification and partial characterization of new antigens from simian virus 40-transformed mouse cells. J Virol 31:463–471

Chao C, Saito S, Kang J, Anderson CW, Appella E, Xu Y (2000) p53 transcriptional activity is essential for p53-dependent apoptosis following DNA damage. EMBO J 19:4967–4975

Chehab NH, Malikzay A, Appel M, Halazonetis TD (2000) Chk2/hCds1 functions as a DNA damage checkpoint in G(1) by stabilizing p53. Genes Dev 14:278–288

Chen JM, Rosal R, Smith S, Pincus MR, Brandt-Rauf PW (2001) Common conformational effects of p53 mutations. J Protein Chem 20:101–105

Chen L, Agrawal S, Zhou W, Zhang R, Chen J (1998) Synergistic activation of p53 by inhibition of MDM 2 expression and DNA damage. Proc Natl Acad Sci USA 95:195–200

Chen X, Ko LJ, Jayaraman L, Prives C (1996) p53 levels, functional domains, and DNA damage determine the extent of the apoptotic response of tumor cells. Genes Dev 10:2438–2451

Chene P (2003) Inhibiting the p53-MDM 2 interaction: an important target for cancer therapy. Nat Rev Cancer 3:102–109

Chene P, Fuchs J, Bohn J, Garcia-Echeverria C, Furet P, Fabbro D (2000) A small synthetic peptide, which inhibits the p53-hdm2 interaction, stimulates the p53 pathway in tumour cell lines. J Mol Biol 299:245–253

Chin KV, Ueda K, Pastan I, Gottesman MM (1992) Modulation of activity of the promoter of the human MDR1 gene by Ras and p53. Science 255:459–462

Cho Y, Gorina S, Jeffrey PD, Pavletich NP (1994) Crystal structure of a p53 tumor suppressor-DNA complex: understanding tumorigenic mutations. Science 265:346–355

Clapp NK, Craig AW, Toya RE, Sr. (1968) Oncogenicity by methyl methanesulfonate in male RF mice. Science 161:913–914

Clarke AR, Purdie CA, Harrison DJ, Morris RG, Bird CC, Hooper ML, Wyllie AH (1993) Thymocyte apoptosis induced by p53-dependent and independent pathways. Nature 362:849–852

Clore GM, Ernst J, Clubb R, Omichinski JG, Kennedy WM, Sakaguchi K, Appella E, Gronenborn AM (1995) Refined solution structure of the oligomerization domain of the tumour suppressor p53. Nat Struct Biol 2:321–333

Cordon-Cardo C, Latres E, Drobnjak M, Oliva MR, Pollack D, Woodruff JM, Marechal V, Chen J, Brennan MF, Levine AJ (1994) Molecular abnormalities of mdm2 and p53 genes in adult soft tissue sarcomas. Cancer Res 54:794–799

Crook T, Brooks LA, Crossland S, Osin P, Barker KT, Waller J, Philp E, Smith PD, Yulug I, Peto J, Parker G, Allday MJ, Crompton MR, Gusterson BA (1998) p53 mutation with frequent novel condons but not a mutator phenotype in BRCA1- and BRCA2-associated breast tumours. Oncogene 17:1681–1689

Damalas A, Ben-Ze'ev A, Simcha I, Shtutman M, Leal JF, Zhurinsky J, Geiger B, Oren M (1999) Excess beta-catenin promotes accumulation of transcriptionally active p53. EMBO J 18:3054–3063

Damalas A, Kahan S, Shtutman M, Ben-Ze'ev A, Oren M (2001) Deregulated beta-catenin induces a p53- and ARF-dependent growth arrest and cooperates with Ras in transformation. EMBO J 20:4912–4922

Dameron KM, Volpert OV, Tainsky MA, Bouck N (1994) Control of angiogenesis in fibroblasts by p53 regulation of thrombospondin-1. Science 265:1582–1584

Daya-Grosjean L, Dumaz N, Sarasin A (1995) The specificity of p53 mutation spectra in sunlight induced human cancers. J Photochem Photobiol B 28:115–124

Deng Y, Wu X (2000) Peg3/Pw1 promotes p53-mediated apoptosis by inducing Bax translocation from cytosol to mitochondria. Proc Natl Acad Sci USA 97:12050–12055

Denissenko MF, Pao A, Tang M, Pfeifer GP (1996) Preferential formation of benzo[a]pyrene adducts at lung cancer mutational hotspots in P53. Science 274:430–432

Deppert W, Gohler T, Koga H, Kim E (2000) Mutant p53: 'gain of function' through perturbation of nuclear structure and function? J Cell Biochem Suppl Suppl 35:115–122

Diasio RB, Johnson MR (2000) The role of pharmacogenetics and pharmacogenomics in cancer chemotherapy with 5-fluorouracil. Pharmacology 61:199–203

Ding HF, Lin YL, McGill G, Juo P, Zhu H, Blenis J, Yuan J, Fisher DE (2000) Essential role for caspase-8 in transcription-independent apoptosis triggered by p53. J Biol Chem 275:38905–38911

Dittmer D, Pati S, Zambetti G, Chu S, Teresky AK, Moore M, Finlay C, Levine AJ (1993) Gain of function mutations in p53. Nat Genet 4:42–46

Donehower LA, Harvey M, Slagle BL, McArthur MJ, Montgomery CA, Jr., Butel JS, Bradley A (1992) Mice deficient for p53 are developmentally normal but susceptible to spontaneous tumours. Nature 356:215–221

Donehower LA, Harvey M, Vogel H, McArthur MJ, Montgomery CA, Jr., Park SH, Thompson T, Ford RJ, Bradley A (1995) Effects of genetic background on tumorigenesis in p53-deficient mice. Mol Carcinog 14:16–22

D'Orazi G, Cecchinelli B, Bruno T, Manni I, Higashimoto Y, Saito S, Gostissa M, Coen S, Marchetti A, Del Sal G, Piaggio G, Fanciulli M, Appella E, Soddu S (2002) Homeodomain-interacting protein kinase-2 phosphorylates p53 at Ser 46 and mediates apoptosis. Nat Cell Biol 4:11–19

Dumaz N, Meek DW (1999) Serine15 phosphorylation stimulates p53 transactivation but does not directly influence interaction with HDM2. EMBO J 18:7002–10

Eischen CM, Weber JD, Roussel MF, Sherr CJ, Cleveland JL (1999) Disruption of the ARF-Mdm2-p53 tumor suppressor pathway in Myc-induced lymphomagenesis. Genes Dev 13:2658–2669

Eisenstadt E, Warren AJ, Porter J, Atkins D, Miller JH (1982) Carcinogenic epoxides of benzo[a]pyrene and cyclopenta[cd]pyrene induce base substitutions via specific transversions. Proc Natl Acad Sci USA 79:1945–1949

Ekstrom K, Hjalgrim H, Brandt L, Baecklund E, Klareskog L, Ekbom A, Askling J (2003) Risk of malignant lymphomas in patients with rheumatoid arthritis and in their first-degree relatives. Arthritis Rheum 48:963–970

el-Deiry WS (1998) Regulation of p53 downstream genes. Semin Cancer Biol 8:345–357

el-Deiry WS, Kern SE, Pietenpol JA, Kinzler KW, Vogelstein B (1992) Definition of a consensus binding site for p53. Nat Genet 1:45–49

el-Deiry WS, Tokino T, Velculescu VE, Levy DB, Parsons R, Trent JM, Lin D, Mercer WE, Kinzler KW, Vogelstein B (1993) WAF1, a potential mediator of p53 tumor suppression. Cell 75:817–825

El-Hizawi S, Lagowski JP, Kulesz-Martin M, Albor A (2002) Induction of gene amplification as a gain-of-function phenotype of mutant p53 proteins. Cancer Res 62:3264–3270

Elmore LW, Rehder CW, Di X, McChesney PA, Jackson-Cook CK, Gewirtz DA, Holt SE (2002) Adriamycin-induced senescence in breast tumor cells involves functional p53 and telomere dysfunction. J Biol Chem 277:35509–35515

Elson A, Deng C, Campos-Torres J, Donehower LA, Leder P (1995) The MMTV/c-myc transgene and p53-null alleles collaborate to induce T-cell lymphomas, but not mammary carcinomas in transgenic mice. Oncogene 11:181–190

Espinosa JM, Emerson BM (2001) Transcriptional regulation by p53 through intrinsic DNA/chromatin binding and site-directed cofactor recruitment. Mol Cell 8:57–69

Etienne MC, Chazal M, Laurent-Puig P, Magne N, Rosty C, Formento JL, Francoual M, Formento P, Renee N, Chamorey E, Bourgeon A, Seitz JF, Delpero JR, Letoublon C, Pezet D, Milano G (2002) Prognostic value of tumoral thymidylate synthase and p53 in metastatic colorectal cancer patients receiving fluorouracil-based chemotherapy: phenotypic and genotypic analyses. J Clin Oncol 20:2832–2843

Faderl S, Kantarjian HM, Estey E, Manshouri T, Chan CY, Rahman Elsaied A, Kornblau SM, Cortes J, Thomas DA, Pierce S, Keating MJ, Estrov Z, Albitar M (2000) The prognostic significance of p16(INK4a)/p14(ARF) locus deletion and MDM-2 protein expression in adult acute myelogenous leukemia. Cancer 89:1976–1982

Fearon ER, Vogelstein B (1990) A genetic model for colorectal tumorigenesis. Cell 61:759–767

Ferreira CG, Tolis C, Giaccone G (1999) p53 and chemosensitivity. Ann Oncol 10:1011–1021

Fields S, Jang SK (1990) Presence of a potent transcription activating sequence in the p53 protein. Science 249:1046–1049

Finlay CA, Hinds PW, Levine AJ (1989) The p53 proto-oncogene can act as a suppressor of transformation. Cell 57:1083–1093

Firestein GS, Echeverri F, Yeo M, Zvaifler NJ, Green DR (1997) Somatic mutations in the p53 tumor suppressor gene in rheumatoid arthritis synovium. Proc Natl Acad Sci USA 94:10895–10900

Fiscella M, Zhang H, Fan S, Sakaguchi K, Shen S, Mercer WE, Vande Woude GF, O'Connor PM, Appella E (1997) Wip1, a novel human protein phosphatase that is induced in response to ionizing radiation in a p53-dependent manner. Proc Natl Acad Sci USA 94:6048–6053

Flores ER, Tsai KY, Crowley D, Sengupta S, Yang A, McKeon F, Jacks T (2002) p63 and p73 are required for p53-dependent apoptosis in response to DNA damage. Nature 416:560–564

Fogal V, Gostissa M, Sandy P, Zacchi P, Sternsdorf T, Jensen K, Pandolfi PP, Will H, Schneider C, Del Sal G (2000) Regulation of p53 activity in nuclear bodies by a specific PML isoform. EMBO J 19:6185–6195

Foster BA, Coffey HA, Morin MJ, Rastinejad F (1999) Pharmacological rescue of mutant p53 conformation and function. Science 286:2507–2510

Friedlander P, Haupt Y, Prives C, Oren M (1996) A mutant p53 that discriminates between p53-responsive genes cannot induce apoptosis. Mol Cell Biol 16:4961–4971

Futreal PA, Liu Q, Shattuck-Eidens D, Cochran C, Harshman K, Tavtigian S, Bennett LM, Haugen-Strano A, Swensen J, Miki Y, et al. (1994) BRCA1 mutations in primary breast and ovarian carcinomas. Science 266:120–122

Gao Y, Ferguson DO, Xie W, Manis JP, Sekiguchi J, Frank KM, Chaudhuri J, Horner J, DePinho RA, Alt FW (2000) Interplay of p53 and DNA-repair protein XRCC4 in tumorigenesis, genomic stability and development. Nature 404:897–900

Garcia-Cao I, Garcia-Cao M, Martin-Caballero J, Criado LM, Klatt P, Flores JM, Weill JC, Blasco MA, Serrano M (2002) 'Super p53' mice exhibit enhanced DNA damage response, are tumor resistant and age normally. EMBO J 21:6225–6235

Garkavtsev I, Grigorian IA, Ossovskaya VS, Chernov MV, Chumakov PM, Gudkov AV (1998) The candidate tumour suppressor p33ING1 cooperates with p53 in cell growth control. Nature 391:295–298

Goh HS, Yao J, Smith DR (1995) p53 point mutation and survival in colorectal cancer patients. Cancer Res 55:5217–5221

Gorina S, Pavletich NP (1996) Structure of the p53 tumor suppressor bound to the ankyrin and SH3 domains of 53BP2. Science 274:1001–1005

Gostissa M, Hengstermann A, Fogal V, Sandy P, Schwarz SE, Scheffner M, Del Sal G (1999) Activation of p53 by conjugation to the ubiquitin-like protein SUMO-1. EMBO J 18:6462–6471

Gottlieb TM, Leal JF, Seger R, Taya Y, Oren M (2002) Cross-talk between Akt, p53 and Mdm2: possible implications for the regulation of apoptosis. Oncogene 21:1299–1303

Gudkov AV, Komarova EA (2003) The role of p53 in determining sensitivity to radiotherapy. Nat Rev Cancer 3:117–129

Guimaraes DP, Hainaut P (2002) TP53: a key gene in human cancer. Biochimie 84:83–93

Hainaut P, Pfeifer GP (2001) Patterns of p53 G→T transversions in lung cancers reflect the primary mutagenic signature of DNA-damage by tobacco smoke. Carcinogenesis 22:367–374

Han Z, Boyle DL, Shi Y, Green DR, Firestein GS (1999) Dominant-negative p53 mutations in rheumatoid arthritis. Arthritis Rheum 42:1088–1092

Hanahan D, Weinberg RA (2000) The hallmarks of cancer. Cell 100:57–70

Harkin DP, Bean JM, Miklos D, Song YH, Truong VB, Englert C, Christians FC, Ellisen LW, Maheswaran S, Oliner JD, Haber DA (1999) Induction of GADD45 and JNK/SAPK-dependent apoptosis following inducible expression of BRCA1. Cell 97:575–586

Harvey M, McArthur MJ, Montgomery CA, Jr., Butel JS, Bradley A, Donehower LA (1993) Spontaneous and carcinogen-induced tumorigenesis in p53-deficient mice. Nat Genet 5:225–229

Harvey M, Vogel H, Morris D, Bradley A, Bernstein A, Donehower LA (1995) A mutant p53 transgene accelerates tumour development in heterozygous but not nullizygous p53-deficient mice. Nat Genet 9:305–311

Haupt Y, Maya R, Kazaz A, Oren M (1997) Mdm2 promotes the rapid degradation of p53. Nature 387:296–299

Hay JG, Shapiro N, Sauthoff H, Heitner S, Phupakdi W, Rom WN (1999) Targeting the replication of adenoviral gene therapy vectors to lung cancer cells: the importance of the adenoviral E1b-55kD gene. Hum Gene Ther 10:579–590

Hellborg F, Wang Q, Méndez-Vidal C, Asker C, Kost-Alimova M, Wilhelm M, Imreh S, Wiman KG (2001) Human wig-1, a p53-target gene that encodes a growth inhibitory zinc finger protein. Oncogene 20:5466–5474

Hermeking H, Lengauer C, Polyak K, He TC, Zhang L, Thiagalingam S, Kinzler KW, Vogelstein B (1997) 14-3-3 sigma is a p53-regulated inhibitor of G2/M progression. Mol Cell 1:3–11

Hietanen S, Lain S, Krausz E, Blattner C, Lane DP (2000) Activation of p53 in cervical carcinoma cells by small molecules. Proc Natl Acad Sci USA 97:8501–8506

Hinds PW, Finlay CA, Quartin RS, Baker SJ, Fearon ER, Vogelstein B, Levine AJ (1990) Mutant p53 DNA clones from human colon carcinomas cooperate with ras in transforming primary rat cells: a comparison of the 'hot spot' mutant phenotypes. Cell Growth Differ 1:571–580

Hofmann TG, Moller A, Sirma H, Zentgraf H, Taya Y, Droge W, Will H, Schmitz ML (2002) Regulation of p53 activity by its interaction with homeodomain-interacting protein kinase-2. Nat Cell Biol 4:1–10

Hollstein M, Shomer B, Greenblatt M, Soussi T, Hovig E, Montesano R, Harris CC (1996) Somatic point mutations in the p53 gene of human tumors and cell lines: updated compilation. Nucleic Acids Res 24:141–146

Hollstein M, Sidransky D, Vogelstein B, Harris CC (1991) p53 mutations in human cancers. Science 253:49–53

Honda R, Yasuda H (1999) Association of p19(ARF) with Mdm2 inhibits ubiquitin ligase activity of Mdm2 for tumor suppressor p53. EMBO J 18:22–27

Hsieh JK, Chan FS, O'Connor DJ, Mittnacht S, Zhong S, Lu X (1999) RB regulates the stability and the apoptotic function of p53 via MDM2. Mol Cell 3:181–193

Hsu B, Marin MC, el-Naggar AK, Stephens LC, Brisbay S, McDonnell TJ (1995) Evidence that c-myc mediated apoptosis does not require wild-type p53 during lymphomagenesis. Oncogene 11:175–179

Hsu IC, Metcalf RA, Sun T, Welsh JA, Wang NJ, Harris CC (1991) Mutational hotspot in the p53 gene in human hepatocellular carcinomas. Nature 350:427–428

Hupp TR, Meek DW, Midgley CA, Lane DP (1992) Regulation of the specific DNA binding function of p53. Cell 71:875–886

Hussain SP, Amstad P, Raja K, Sawyer M, Hofseth L, Shields PG, Hewer A, Phillips DH, Ryberg D, Haugen A, Harris CC (2001) Mutability of p53 hotspot codons to benzo(a)pyrene diol epoxide (BPDE) and the frequency of p53 mutations in nontumorous human lung. Cancer Res 61:6350–6355

Hussain SP, Harris CC (1998) Molecular epidemiology of human cancer: contribution of mutation spectra studies of tumor suppressor genes. Cancer Res 58:4023-4037

Hussain SP, Harris CC (1999) p53 mutation spectrum and load: the generation of hypotheses linking the exposure of endogenous or exogenous carcinogens to human cancer. Mutat Res 428:23-32

Hwang PM, Bunz F, Yu J, Rago C, Chan TA, Murphy MP, Kelso GF, Smith RA, Kinzler KW, Vogelstein B (2001) Ferredoxin reductase affects p53-dependent, 5-fluorouracil-induced apoptosis in colorectal cancer cells. Nat Med 7:1111-1117

Igney FH, Krammer PH (2002) Death and anti-death: tumour resistance to apoptosis. Nat Rev Cancer 2:277-288

Irwin M, Marin MC, Phillips AC, Seelan RS, Smith DI, Liu W, Flores ER, Tsai KY, Jacks T, Vousden KH, Kaelin WG, Jr. (2000) Role for the p53 homologue p73 in E2F-1-induced apoptosis. Nature 407:645-648

Irwin MS, Kaelin WG, Jr. (2001) Role of the newer p53 family proteins in malignancy. Apoptosis 6:17-29

Irwin MS, Kondo K, Marin MC, Cheng LS, Hahn WC, Kaelin WG, Jr. (2003) Chemosensitivity linked to p73 function. Cancer Cell 3:403-410

Issaeva N, Friedler A, Bozko P, Wiman KG, Fersht AR, Selivanova G (2003) Rescue of mutants of the tumor suppressor p53 in cancer cells by a designed peptide. Proc Natl Acad Sci USA:100:13303-13307

Javelaud D, Besancon F (2002) Inactivation of p21WAF1 sensitizes cells to apoptosis via an increase of both p14ARF and p53 levels and an alteration of the Bax/Bcl-2 ratio. J Biol Chem 277:37949-37954

Jayaraman J, Prives C (1995) Activation of p53 sequence-specific DNA binding by short single strands of DNA requires the p53 C-terminus. Cell 81:1021-1029

Jeffrey PD, Gorina S, Pavletich NP (1995) Crystal structure of the tetramerization domain of the p53 tumor suppressor at 1.7 angstroms. Science 267:1498-1502

Jimenez GS, Nister M, Stommel JM, Beeche M, Barcarse EA, Zhang XQ, O'Gorman S, Wahl GM (2000) A transactivation-deficient mouse model provides insights into Trp53 regulation and function. Nat Genet 26:37-43

Jirawatnotai S, Moons DS, Stocco CO, Franks R, Hales DB, Gibori G, Kiyokawa H (2003) The cyclin-dependent kinase inhibitors p27Kip1 and p21Cip1cooperate to restrict proliferative life span in differentiating ovarian cells. J Biol Chem 278:17021-17027

Johnsen JI, Aurelio ON, Kwaja Z, Jorgensen GE, Pellegata NS, Plattner R, Stanbridge EJ, Cajot JF (2000) p53-mediated negative regulation of stathmin/Op18 expression is associated with G(2)/M cell-cycle arrest. Int J Cancer 88:685-691

Jonason AS, Kunala S, Price GJ, Restifo RJ, Spinelli HM, Persing JA, Leffell DJ, Tarone RE, Brash DE (1996) Frequent clones of p53-mutated keratinocytes in normal human skin. Proc Natl Acad Sci U S A 93:14025-14029

Jonkers J, Meuwissen R, van der Gulden H, Peterse H, van der Valk M, Berns A (2001) Synergistic tumor suppressor activity of BRCA2 and p53 in a conditional mouse model for breast cancer. Nat Genet 29:418-425

Kahlenberg MS, Stoler DL, Basik M, Petrelli NJ, Rodriguez-Bigas M, Anderson GR (1996) p53 tumor suppressor gene status and the degree of genomic instability in sporadic colorectal cancers. J Natl Cancer Inst 88:1665-1670

Kannan K, Amariglio N, Rechavi G, Jakob-Hirsch J, Kela I, Kaminski N, Getz G, Domany E, Givol D (2001) DNA microarrays identification of primary and secondary target genes regulated by p53. Oncogene 20:2225-2234

Kastan MB, Zhan Q, el-Deiry WS, Carrier F, Jacks T, Walsh WV, Plunkett BS, Vogelstein B, Fornace AJ, Jr. (1992) A mammalian cell cycle checkpoint pathway utilizing p53 and GADD45 is defective in ataxia-telangiectasia. Cell 71:587–597

Kemp CJ, Donehower LA, Bradley A, Balmain A (1993) Reduction of p53 gene dosage does not increase initiation or promotion but enhances malignant progression of chemically induced skin tumors. Cell 74:813–822

Khosravi R, Maya R, Gottlieb T, Oren M, Shiloh Y, Shkedy D (1999) Rapid ATM-dependent phosphorylation of MDM 2 precedes p53 accumulation in response to DNA damage. Proc Natl Acad Sci USA 96:14973–14977

Kim MS, Kwon HJ, Lee YM, Baek JH, Jang JE, Lee SW, Moon EJ, Kim HS, Lee SK, Chung HY, Kim CW, Kim KW (2001) Histone deacetylases induce angiogenesis by negative regulation of tumor suppressor genes. Nat Med 7:437–443

Kirn D (2001) Clinical research results with dl1520 (Onyx-015), a replication-selective adenovirus for the treatment of cancer: what have we learned? Gene Ther 8:89–98

Knudson CM, Korsmeyer SJ (1997) Bcl-2 and Bax function independently to regulate cell death. Nat Genet 16:358–363

Knudson CM, Tung KS, Tourtellotte WG, Brown GA, Korsmeyer SJ (1995) Bax-deficient mice with lymphoid hyperplasia and male germ cell death. Science 270:96–99

Komarov PG, Komarova EA, Kondratov RV, Christov-Tselkov K, Coon JS, Chernov MV, Gudkov AV (1999) A chemical inhibitor of p53 that protects mice from the side effects of cancer therapy. Science 285:1733–1737

Koumenis C, Alarcon R, Hammond E, Sutphin P, Hoffman W, Murphy M, Derr J, Taya Y, Lowe SW, Kastan M, Giaccia A (2001) Regulation of p53 by hypoxia: dissociation of transcriptional repression and apoptosis from p53-dependent transactivation. Mol Cell Biol 21:1297–1310

Kress M, May E, Cassingena R, May P (1979) Simian virus 40-transformed cells express new species of proteins precipitable by anti-simian virus 40 tumor serum. J Virol 31:472–483

Kubbutat MH, Jones SN, Vousden KH (1997) Regulation of p53 stability by Mdm2. Nature 387:299–303

Kunz C, Pebler S, Otte J, von der Ahe D (1995) Differential regulation of plasminogen activator and inhibitor gene transcription by the tumor suppressor p53. Nucleic Acids Res 23:3710–17

Kwek SS, Derry J, Tyner AL, Shen Z, Gudkov AV (2001) Functional analysis and intracellular localization of p53 modified by SUMO-1. Oncogene 20:2587–2599

Lane DP, Crawford LV (1979) T antigen is bound to a host protein in SV40-transformed cells. Nature 278:261–263

Lavigueur A, Maltby V, Mock D, Rossant J, Pawson T, Bernstein A (1989) High incidence of lung, bone, and lymphoid tumors in transgenic mice overexpressing mutant alleles of the p53 oncogene. Mol Cell Biol 9:3982–3991

Lee H, Trainer AH, Friedman LS, Thistlethwaite FC, Evans MJ, Ponder BA, Venkitaraman AR (1999) Mitotic checkpoint inactivation fosters transformation in cells lacking the breast cancer susceptibility gene, Brca2. Mol Cell 4:1–10

Lee S, Elenbaas B, Levine A, Griffith J (1995) p53 and its 14 kDa C-terminal domain recognize primary DNA damage in the form of insertion/deletion mismatches. Cell 81:1013–1020

Legros Y, Meyer A, Ory K, Soussi T (1994) Mutations in p53 produce a common conformational effect that can be detected with a panel of monoclonal antibodies directed toward the central part of the p53 protein. Oncogene 9:3689–3694

Lenz HJ, Hayashi K, Salonga D, Danenberg KD, Danenberg PV, Metzger R, Banerjee D, Bertino JR, Groshen S, Leichman LP, Leichman CG (1998) p53 point mutations and thymidylate synthase messenger RNA levels in disseminated colorectal cancer: an analysis of response and survival. Clin Cancer Res 4:1243–1250

Li M, Luo J, Brooks CL, Gu W (2002a) Acetylation of p53 inhibits its ubiquitination by Mdm2. J Biol Chem 277:50607–50611

Li R, Waga S, Hannon GJ, Beach D, Stillman B (1994) Differential effects by the p21 CDK inhibitor on PCNA-dependent DNA replication and repair. Nature 371:534–537

Li Y, Dowbenko D, Lasky LA (2002b) AKT/PKB phosphorylation of p21Cip/WAF1 enhances protein stability of p21Cip/WAF1 and promotes cell survival. J Biol Chem 277:11352–11361

Lim DS, Vogel H, Willerford DM, Sands AT, Platt KA, Hasty P (2000) Analysis of ku80-mutant mice and cells with deficient levels of p53. Mol Cell Biol 20:3772–3780

Lin J, Chen J, Elenbaas B, Levine AJ (1994) Several hydrophobic amino acids in the p53 amino-terminal domain are required for transcriptional activation, binding to mdm-2 and the adenovirus 5 E1B 55-kD protein. Genes Dev 8:1235–1246

Lin Y, Ma W, Benchimol S (2000) Pidd, a new death-domain-containing protein, is induced by p53 and promotes apoptosis. Nat Genet 26:122–127

Lindsten T, Ross AJ, King A, Zong WX, Rathmell JC, Shiels HA, Ulrich E, Waymire KG, Mahar P, Frauwirth K, Chen Y, Wei M, Eng VM, Adelman DM, Simon MC, Ma A, Golden JA, Evan G, Korsmeyer SJ, MacGregor GR, Thompson CB (2000) The combined functions of proapoptotic Bcl-2 family members bak and bax are essential for normal development of multiple tissues. Mol Cell 6:1389–1399

Lindstrom MS, Klangby U, Wiman KG (2001) p14ARF homozygous deletion or MDM 2 overexpression in Burkitt lymphoma lines carrying wild type p53. Oncogene 20:2171–2177

Lindstrom MS, Wiman KG (2003) Myc and E2F1 induce p53 through p14ARF-independent mechanisms in human fibroblasts. Oncogene 22:4993–5005

Linzer DI, Levine AJ (1979) Characterization of a 54 K dalton cellular SV40 tumor antigen present in SV40-transformed cells and uninfected embryonal carcinoma cells. Cell 17:43–52

Liu G, McDonnell TJ, Montes de Oca Luna R, Kapoor M, Mims B, El-Naggar AK, Lozano G (2000) High metastatic potential in mice inheriting a targeted p53 missense mutation. Proc Natl Acad Sci USA 97:4174–4179

Lohmann DR (1999) RB1 gene mutations in retinoblastoma. Hum Mutat 14:283–288

Longley DB, Latif T, Boyer J, Allen WL, Maxwell PJ, Johnston PG (2003) The interaction of thymidylate synthase expression with p53-regulated signaling pathways in tumor cells. Semin Oncol 30:3–9

Lu W, Chen L, Peng Y, Chen J (2001) Activation of p53 by roscovitine-mediated suppression of MDM 2 expression. Oncogene 20:3206–3216

Ludwig RL, Bates S, Vousden KH (1996) Differential activation of target cellular promoters by p53 mutants with impaired apoptotic function. Mol Cell Biol 16:4952–4960

Ludwig T, Chapman DL, Papaioannou VE, Efstratiadis A (1997) Targeted mutations of breast cancer susceptibility gene homologs in mice: lethal phenotypes of Brca1, Brca2, Brca1/Brca2, Brca1/p53, and Brca2/p53 nullizygous embryos. Genes Dev 11:1226–1241

Luo J, Nikolaev AY, Imai S, Chen D, Su F, Shiloh A, Guarente L, Gu W (2001) Negative control of p53 by Sir2alpha promotes cell survival under stress. Cell 107:137–148

Luo Y, Hurwitz J, Massague J (1995) Cell-cycle inhibition by independent CDK and PCNA binding domains in p21Cip1. Nature 375:159–161

Luu Y, Bush J, Cheung KJ, Jr., Li G (2002) The p53 stabilizing compound CP-31398 induces apoptosis by activating the intrinsic Bax/mitochondrial/caspase-9 pathway. Exp Cell Res 276:214–222

MacLachlan TK, Somasundaram K, Sgagias M, Shifman Y, Muschel RJ, Cowan KH, El-Deiry WS (2000) BRCA1 effects on the cell cycle and the DNA damage response are linked to altered gene expression. J Biol Chem 275:2777–2785

MacLachlan TK, Takimoto R, El-Deiry WS (2002) BRCA1 directs a selective p53-dependent transcriptional response towards growth arrest and DNA repair targets. Mol Cell Biol 22:4280–4292

Mahyar-Roemer M, Roemer K (2001) p21 Waf1/Cip1 can protect human colon carcinoma cells against p53-dependent and p53-independent apoptosis induced by natural chemopreventive and therapeutic agents. Oncogene 20:3387–3398

Malkin D, Li FP, Strong LC, Fraumeni JF, Jr., Nelson CE, Kim DH, Kassel J, Gryka MA, Bischoff FZ, Tainsky MA, et al. (1990) Germ line p53 mutations in a familial syndrome of breast cancer, sarcomas, and other neoplasms. Science 250:1233–1238

Marchenko ND, Zaika A, Moll UM (2000) Death signal-induced localization of p53 protein to mitochondria. A potential role in apoptotic signaling. J Biol Chem 275:16202–16212

Martinez LA, Naguibneva I, Lehrmann H, Vervisch A, Tchenio T, Lozano G, Harel-Bellan A (2002) Synthetic small inhibiting RNAs: efficient tools to inactivate oncogenic mutations and restore p53 pathways. Proc Natl Acad Sci USA 99:14849–14854

Mashimo T, Watabe M, Hirota S, Hosobe S, Miura K, Tegtmeyer PJ, Rinker-Shaeffer CW, Watabe K (1998) The expression of the KAI1 gene, a tumor metastasis suppressor, is directly activated by p53. Proc Natl Acad Sci USA 95:11307–11311

Matas D, Sigal A, Stambolsky P, Milyavsky M, Weisz L, Schwartz D, Goldfinger N, Rotter V (2001) Integrity of the N-terminal transcription domain of p53 is required for mutant p53 interference with drug-induced apoptosis. EMBO J 20:4163–4172

Mayo LD, Donner DB (2001) A phosphatidylinositol 3-kinase/Akt pathway promotes translocation of Mdm2 from the cytoplasm to the nucleus. Proc Natl Acad Sci USA 98:11598–11603

McCormack SJ, Weaver Z, Deming S, Natarajan G, Torri J, Johnson MD, Liyanage M, Ried T, Dickson RB (1998) Myc/p53 interactions in transgenic mouse mammary development, tumorigenesis and chromosomal instability. Oncogene 16:2755–2766

Megha T, Ferrari F, Benvenuto A, Bellan C, Lalinga AV, Lazzi S, Bartolommei S, Cevenini G, Leoncini L, Tosi P (2002) p53 mutation in breast cancer. Correlation with cell kinetics and cell of origin. J Clin Pathol 55:461–466

Melero JA, Stitt DT, Mangel WF, Carroll RB (1979) Identification of new polypeptide species (48–55 K) immunoprecipitable by antiserum to purified large T antigen and present in SV40-infected and -transformed cells. Virology 93:466–480

Mendez-Vidal C, Wilhelm MT, Hellborg F, Qian W, Wiman KG (2002) The p53-induced mouse zinc finger protein wig-1 binds double-stranded RNA with high affinity. Nucleic Acids Res 30:1991–1996

Merajver SD, Pham TM, Caduff RF, Chen M, Poy EL, Cooney KA, Weber BL, Collins FS, Johnston C, Frank TS (1995) Somatic mutations in the BRCA1 gene in sporadic ovarian tumours. Nat Genet 9:439–443

Merritt JA, Roth JA, Logothetis CJ (2001) Clinical evaluation of adenoviral-mediated p53 gene transfer: review of INGN 201 studies. Semin Oncol 28:105–114

Midgley CA, Desterro JM, Saville MK, Howard S, Sparks A, Hay RT, Lane DP (2000) An N-terminal p14ARF peptide blocks Mdm2-dependent ubiquitination in vitro and can activate p53 in vivo. Oncogene 19:2312–2323

Mihara M, Erster S, Zaika A, Petrenko O, Chittenden T, Pancoska P, Moll UM (2003) p53 has a direct apoptogenic role at the mitochondria. Mol Cell 11:577–590

Miyaki M, Konishi M, Kikuchi-Yanoshita R, Enomoto M, Igari T, Tanaka K, Muraoka M, Takahashi H, Amada Y, Fukayama M, et al. (1994) Characteristics of somatic mutation of the adenomatous polyposis coli gene in colorectal tumors. Cancer Res 54:3011–3020

Miyashita T, Harigai M, Hanada M, Reed JC (1994a) Identification of a p53-dependent negative response element in the bcl-2 gene. Cancer Res 54:3131–3135

Miyashita T, Krajewski S, Krajewska M, Wang HG, Lin HK, Liebermann DA, Hoffman B, Reed JC (1994b) Tumor suppressor p53 is a regulator of bcl-2 and bax gene expression in vitro and in vivo. Oncogene 9:1799–1805

Miyashita T, Reed JC (1995) Tumor suppressor p53 is a direct transcriptional activator of the human bax gene. Cell 80: 293–299

Miyoshi Y, Nagase H, Ando H, Horii A, Ichii S, Nakatsuru S, Aoki T, Miki Y, Mori T, Nakamura Y (1992) Somatic mutations of the APC gene in colorectal tumors: mutation cluster region in the APC gene. Hum Mol Genet 1:229–233

Montes de Oca Luna R, Wagner DS, Lozano G (1995) Rescue of early embryonic lethality in mdm2-deficient mice by deletion of p53. Nature 378:203–206

Montesano R, Hainaut P, Hall J (1997) The use of biomarkers to study pathogenesis and mechanisms of cancer: oesophagus and skin cancer as models. IARC Sci Publ: 291–301

Morrison RS, Kinoshita Y (2000) The role of p53 in neuronal cell death. Cell Death Differ 7:868–879

Mowat M, Cheng A, Kimura N, Bernstein A, Benchimol S (1985) Rearrangements of the cellular p53 gene in erythroleukaemic cells transformed by Friend virus. Nature 314:633–636

Muller M, Wilder S, Bannasch D, Israeli D, Lehlbach K, Li-Weber M, Friedman SL, Galle PR, Stremmel W, Oren M, Krammer PH (1998) p53 activates the CD95 (APO-1/Fas) gene in response to DNA damage by anticancer drugs. J Exp Med 188:2033–2045

Muller S, Berger M, Lehembre F, Seeler JS, Haupt Y, Dejean A (2000) c-Jun and p53 activity is modulated by SUMO-1 modification. J Biol Chem 275:13321–13329

Murphy KL, Dennis AP, Rosen JM (2000) A gain of function p53 mutant promotes both genomic instability and cell survival in a novel p53-null mammary epithelial cell model. FASEB J 14:2291–2302

Murphy M, Ahn J, Walker KK, Hoffman WH, Evans RM, Levine AJ, George DL (1999) Transcriptional repression by wild-type p53 utilizes histone deacetylases, mediated by interaction with mSin3a. Genes Dev 13:2490–2501

Murphy M, Hinman A, Levine AJ (1996) Wild-type p53 negatively regulates the expression of a microtubule-associated protein. Genes Dev 10:2971–2980

Nacht M, Jacks T (1998) V(D)J recombination is not required for the development of lymphoma in p53-deficient mice. Cell Growth Differ 9:131–138

Nagashima M, Shiseki M, Miura K, Hagiwara K, Linke SP, Pedeux R, Wang XW, Yokota J, Riabowol K, Harris CC (2001) DNA damage-inducible gene p33ING2 negatively regulates cell proliferation through acetylation of p53. Proc Natl Acad Sci USA 98:9671–9676

Nakano K, Vousden KH (2001) PUMA, a novel proapoptotic gene, is induced by p53. Mol Cell 7:683–694

Nicholls CD, McLure KG, Shields MA, Lee PW (2002) Biogenesis of p53 involves cotranslational dimerization of monomers and posttranslational dimerization of dimers. Implications on the dominant negative effect. J Biol Chem 277:12937–12945

Nigro JM, Baker SJ, Preisinger AC, Jessup JM, Hostetter R, Cleary K, Bigner SH, Davidson N, Baylin S, Devilee P, et al. (1989) Mutations in the p53 gene occur in diverse human tumour types. Nature 342:705–708

Nishimori H, Shiratsuchi T, Urano T, Kimura Y, Kiyono K, Tatsumi K, Yoshida S, Ono M, Kuwano M, Nakamura Y, Tokino T (1997) A novel brain-specific p53-target gene, BAI1, containing thrombospondin type 1 repeats inhibits experimental angiogenesis. Oncogene 15:2145–2150

Noda A, Ning Y, Venable SF, Pereira-Smith OM, Smith JR (1994) Cloning of senescent cell-derived inhibitors of DNA synthesis using an expression screen. Exp Cell Res 211:90–98

Norberg T, Klaar S, Karf G, Nordgren H, Holmberg L, Bergh J (2001) Increased p53 mutation frequency during tumor progression–results from a breast cancer cohort. Cancer Res 61:8317–8321

Oda E, Ohki R, Murasawa H, Nemoto J, Shibue T, Yamashita T, Tokino T, Taniguchi T, Tanaka N (2000a) Noxa, a BH3-only member of the Bcl-2 family and candidate mediator of p53-induced apoptosis. Science 288:1053–1058

Oda K, Arakawa H, Tanaka T, Matsuda K, Tanikawa C, Mori T, Nishimori H, Tamai K, Tokino T, Nakamura Y, Taya Y (2000b) p53AIP1, a potential mediator of p53-dependent apoptosis, and its regulation by Ser-46-phosphorylated p53. Cell 102:849–862

Ohki R, Nemoto J, Murasawa H, Oda E, Inazawa J, Tanaka N, Taniguchi T (2000) Reprimo, a new candidate mediator of the p53-mediated cell cycle arrest at the G2 phase. J Biol Chem 275:22627–22630

Okamoto K, Li H, Jensen MR, Zhang T, Taya Y, Thorgeirsson SS, Prives C (2002) Cyclin G recruits PP2A to dephosphorylate Mdm2. Mol Cell 9:761–771

Okamura S, Arakawa H, Tanaka T, Nakanishi H, Ng CC, Taya Y, Monden M, Nakamura Y (2001) p53DINP1, a p53-inducible gene, regulates p53-dependent apoptosis. Mol Cell 8:85–94

Olivier M, Eeles R, Hollstein M, Khan MA, Harris CC, Hainaut P (2002) The IARC TP53 database: new online mutation analysis and recommendations to users. Hum Mutat 19:607–614

O'Rourke RW, Miller CW, Kato GJ, Simon KJ, Chen DL, Dang CV, Koeffler HP (1990) A potential transcriptional activation element in the p53 protein. Oncogene 5:1829–1832

Ouchi T, Monteiro AN, August A, Aaronson SA, Hanafusa H (1998) BRCA1 regulates p53-dependent gene expression. Proc Natl Acad Sci USA 95:2302–2306

Owen-Schaub LB, Angelo LS, Radinsky R, Ware CF, Gesner TG, Bartos DP (1995) Soluble Fas/APO-1 in tumor cells: a potential regulator of apoptosis? Cancer Lett 94:1–8

Palmero I, Pantoja C, Serrano M (1998) p19ARF links the tumour suppressor p53 to Ras. Nature 395:125–126

Pantoja C, Serrano M (1999) Murine fibroblasts lacking p21 undergo senescence and are resistant to transformation by oncogenic Ras. Oncogene 18:4974–4982

Parenti AR, Rugge M, Frizzera E, Ruol A, Noventa F, Ancona E, Ninfo V (1995) p53 overexpression in the multistep process of esophageal carcinogenesis. Am J Surg Pathol 19:1418–1422

Pavletich NP, Chambers KA, Pabo CO (1993) The DNA-binding domain of p53 contains the four conserved regions and the major mutation hot spots. Genes Dev 7:2556–2564

Pearson M, Carbone R, Sebastiani C, Cioce M, Fagioli M, Saito S, Higashimoto Y, Appella E, Minucci S, Pandolfi PP, Pelicci PG (2000) PML regulates p53 acetylation and premature senescence induced by oncogenic Ras. Nature 406:207–210

Peters GJ, Backus HH, Freemantle S, van Triest B, Codacci-Pisanelli G, van der Wilt CL, Smid K, Lunec J, Calvert AH, Marsh S, McLeod HL, Bloemena E, Meijer S, Jansen G, van Groeningen CJ, Pinedo HM (2002) Induction of thymidylate synthase as a 5-fluorouracil resistance mechanism. Biochim Biophys Acta 1587:194–205

Pettitt AR, Sherrington PD, Stewart G, Cawley JC, Taylor AM, Stankovic T (2001) p53 dysfunction in B-cell chronic lymphocytic leukemia: inactivation of ATM as an alternative to TP53 mutation. Blood 98:814–822

Pfeifer GP, Denissenko MF (1998) Formation and repair of DNA lesions in the p53 gene: relation to cancer mutations? Environ Mol Mutagen 31:197–205

Pinyol M, Hernandez L, Martinez A, Cobo F, Hernandez S, Bea S, Lopez-Guillermo A, Nayach I, Palacin A, Nadal A, Fernandez PL, Montserrat E, Cardesa A, Campo E (2000) INK4a/ARF locus alterations in human non-Hodgkin's lymphomas mainly occur in tumors with wild-type p53 gene. Am J Pathol 156:1987–1996

Pomerantz J, Schreiber-Agus N, Liegeois NJ, Silverman A, Alland L, Chin L, Potes J, Chen K, Orlow I, Lee HW, Cordon-Cardo C, DePinho RA (1998) The Ink4a tumor suppressor gene product, p19Arf, interacts with MDM 2 and neutralizes MDM 2's inhibition of p53. Cell 92:713–723

Ponten F, Williams C, Ling G, Ahmadian A, Nister M, Lundeberg J, Ponten J, Uhlen M (1997) Genomic analysis of single cells from human basal cell cancer using laser-assisted capture microscopy. Mutat Res 382:45–55

Pritchard DM, Watson AJ, Potten CS, Jackman AL, Hickman JA (1997) Inhibition by uridine but not thymidine of p53-dependent intestinal apoptosis initiated by 5-fluorouracil: evidence for the involvement of RNA perturbation. Proc Natl Acad Sci USA 94:1795–1799

Prives C (1998) Signaling to p53: breaking the MDM 2-p53 circuit. Cell 95:5–8

Prives C, Hall PA (1999) The p53 pathway. J Pathol 187:112–126

Pugacheva EN, Ivanov AV, Kravchenko JE, Kopnin BP, Levine AJ, Chumakov PM (2002) Novel gain of function activity of p53 mutants: activation of the dUTPase gene expression leading to resistance to 5-fluorouracil. Oncogene 21:4595–4600

Ramus SJ, Bobrow LG, Pharoah PD, Finnigan DS, Fishman A, Altaras M, Harrington PA, Gayther SA, Ponder BA, Friedman LS (1999) Increased frequency of TP53 mutations in BRCA1 and BRCA2 ovarian tumours. Genes Chromosomes Cancer 25:91–96

Relling MV, Dervieux T (2001) Pharmacogenetics and cancer therapy. Nat Rev Cancer 1:99–108

Ren ZP, Ponten F, Nister M, Ponten J (1996) Two distinct p53 immunohistochemical patterns in human squamous-cell skin cancer, precursors and normal epidermis. Int J Cancer 69:174–179

Ries S, Korn WM (2002) ONYX-015: mechanisms of action and clinical potential of a replication-selective adenovirus. Br J Cancer 86:5–11

Rippin TM, Bykov VJ, Freund SM, Selivanova G, Wiman KG, Fersht AR (2002) Characterization of the p53-rescue drug CP-31398 in vitro and in living cells. Oncogene 21:2119–2129

Robertson KD, Jones PA (1998) The human ARF cell cycle regulatory gene promoter is a CpG island which can be silenced by DNA methylation and down-regulated by wild-type p53. Mol Cell Biol 18:6457–6473

Robles AI, Bemmels NA, Foraker AB, Harris CC (2001) APAF-1 is a transcriptional target of p53 in DNA damage-induced apoptosis. Cancer Res 61:6660–6664

Rodriguez MS, Desterro JM, Lain S, Midgley CA, Lane DP, Hay RT (1999) SUMO-1 modification activates the transcriptional response of p53. EMBO J 18:6455–6461

Roemer K (1999) Mutant p53: gain-of-function oncoproteins and wild-type p53 inactivators. Biol Chem 380:879-887

Ross JS, Stagliano NE, Donovan MJ, Breitbart RE, Ginsburg GS (2001) Atherosclerosis and cancer: common molecular pathways of disease development and progression. Ann N Y Acad Sci 947:271-92; discussion 292-293

Rousseau D, Cannella D, Boulaire J, Fitzgerald P, Fotedar A, Fotedar R (1999) Growth inhibition by CDK-cyclin and PCNA binding domains of p21 occurs by distinct mechanisms and is regulated by ubiquitin-proteasome pathway. Oncogene 18:4313-4325

Rudolph KL, Chang S, Lee HW, Blasco M, Gottlieb GJ, Greider C, DePinho RA (1999) Longevity, stress response, and cancer in aging telomerase-deficient mice. Cell 96:701-712

Ruggeri B, Caamano J, Goodrow T, DiRado M, Bianchi A, Trono D, Conti CJ, Klein-Szanto AJ (1991) Alterations of the p53 tumor suppressor gene during mouse skin tumor progression. Cancer Res 51:6615-6621

Ruggeri B, DiRado M, Zhang SY, Bauer B, Goodrow T, Klein-Szanto AJ (1993) Benzo[a]pyrene-induced murine skin tumors exhibit frequent and characteristic G to T mutations in the p53 gene. Proc Natl Acad Sci USA 90:1013-1017

Russell JL, Powers JT, Rounbehler RJ, Rogers PM, Conti CJ, Johnson DG (2002) ARF differentially modulates apoptosis induced by E2F1 and Myc. Mol Cell Biol 22:1360-1368

Sakaguchi K, Herrera JE, Saito S, Miki T, Bustin M, Vassilev A, Anderson CW, Appella E (1998) DNA damage activates p53 through a phosphorylation-acetylation cascade. Genes Dev 12:2831-2841

Sakaguchi K, Sakamoto H, Lewis MS, Anderson CW, Erickson JW, Appella E, Xie D (1997) Phosphorylation of serine 392 stabilizes the tetramer formation of tumor suppressor protein p53. Biochemistry 36:10117-10124

Sampath J, Sun D, Kidd VJ, Grenet J, Gandhi A, Shapiro LH, Wang Q, Zambetti GP, Schuetz JD (2001) Mutant p53 cooperates with ETS and selectively up-regulates human MDR1 not MRP1. J Biol Chem 276:39359-39367

Samuels-Lev Y, O'Connor DJ, Bergamaschi D, Trigiante G, Hsieh JK, Zhong S, Campargue I, Naumovski L, Crook T, Lu X (2001) ASPP proteins specifically stimulate the apoptotic function of p53. Mol Cell 8:781-794

Sausville EA (2002) Complexities in the development of cyclin-dependent kinase inhibitor drugs. Trends Mol Med 8: S32-S37

Schmidt D, Muller S (2002) Members of the PIAS family act as SUMO ligases for c-Jun and p53 and repress p53 activity. Proc Natl Acad Sci USA 99:2872-2877

Schmitt CA, Fridman JS, Yang M, Baranov E, Hoffman RM, Lowe SW (2002a) Dissecting p53 tumor suppressor functions in vivo. Cancer Cell 1:289-298

Schmitt CA, Fridman JS, Yang M, Lee S, Baranov E, Hoffman RM, Lowe SW (2002b) A senescence program controlled by p53 and p16INK4a contributes to the outcome of cancer therapy. Cell 109:335-346

Schmitt CA, Lowe SW (2001) Apoptosis is critical for drug response in vivo. Drug Resist Updat 4: 132-134

Schmitt CA, Lowe SW (2002) Apoptosis and chemoresistance in transgenic cancer models. J Mol Med 80:137-146

Schmitt CA, McCurrach ME, de Stanchina E, Wallace-Brodeur RR, Lowe SW (1999) INK4a/ARF mutations accelerate lymphomagenesis and promote chemoresistance by disabling p53. Genes Dev 13:2670-2677

Selivanova G (2001) Mutant p53: the loaded gun. Curr Opin Investig Drugs 2:1136-1141

Selivanova G, Iotsova V, Okan I, Fritsche M, Strom M, Groner B, Grafstrom RC, Wiman KG (1997) Restoration of the growth suppression function of mutant p53 by a synthetic peptide derived from the p53 C-terminal domain. Nat Med 3:632–638

Seo YR, Fishel ML, Amundson S, Kelley MR, Smith ML (2002) Implication of p53 in base excision DNA repair: in vivo evidence. Oncogene 21:731–737

Seoane J, Le HV, Massague J (2002) Myc suppression of the p21(Cip1) Cdk inhibitor influences the outcome of the p53 response to DNA damage. Nature 419:729–734

Sharpless NE, DePinho RA (1999) The INK4A/ARF locus and its two gene products. Curr Opin Genet Dev 9:22–30

Shay JW, Pereira-Smith OM, Wright WE (1991) A role for both RB and p53 in the regulation of human cellular senescence. Exp Cell Res 196:33–39

Shen C, Buck AK, Liu X, Winkler M, Reske SN (2003) Gene silencing by adenovirus-delivered siRNA. FEBS Lett 539:111–114

Shieh SY, Ahn J, Tamai K, Taya Y, Prives C (2000) The human homologs of checkpoint kinases Chk1 and Cds1 (Chk2) phosphorylate p53 at multiple DNA damage-inducible sites. Genes Dev 14:289–300

Shikama N, Lee CW, France S, Delavaine L, Lyon J, Krstic-Demonacos M, La Thangue NB (1999) A novel cofactor for p300 that regulates the p53 response. Mol Cell 4:365–376

Somasundaram K, MacLachlan TK, Burns TF, Sgagias M, Cowan KH, Weber BL, el-Deiry WS (1999) BRCA1 signals ARF-dependent stabilization and coactivation of p53. Oncogene 18:6605–6614

Somasundaram K, Zhang H, Zeng YX, Houvras Y, Peng Y, Wu GS, Licht JD, Weber BL, El-Deiry WS (1997) Arrest of the cell cycle by the tumour-suppressor BRCA1 requires the CDK-inhibitor p21WAF1/CiP1. Nature 389:187–90

Soussi T, Beroud C (2001) Assessing TP53 status in human tumours to evaluate clinical outcome. Nat Rev Cancer 1:233–240

Soussi T, Beroud C (2003) Significance of TP53 mutations in human cancer: a critical analysis of mutations at CpG dinucleotides. Hum Mutat 21:192–200

Stambolic V, MacPherson D, Sas D, Lin Y, Snow B, Jang Y, Benchimol S, Mak TW (2001) Regulation of PTEN transcription by p53. Mol Cell 8:317–325

Stewart N, Hicks GG, Paraskevas F, Mowat M (1995) Evidence for a second cell cycle block at G2/M by p53. Oncogene 10:109–115

Stiewe T, Putzer BM (2000) Role of the p53-homologue p73 in E2F1-induced apoptosis. Nat Genet 26:464–469

Stott FJ, Bates S, James MC, McConnell BB, Starborg M, Brookes S, Palmero I, Ryan K, Hara E, Vousden KH, Peters G (1998) The alternative product from the human CDKN2A locus, p14(ARF), participates in a regulatory feedback loop with p53 and MDM 2. Embo J 17:5001–5014

Strano S, Munarriz E, Rossi M, Cristofanelli B, Shaul Y, Castagnoli L, Levine AJ, Sacchi A, Cesareni G, Oren M, Blandino G (2000) Physical and functional interaction between p53 mutants and different isoforms of p73. J Biol Chem 275:29503–29512

Sun Y, Cheung JM, Martel-Pelletier J, Pelletier JP, Wenger L, Altman RD, Howell DS, Cheung HS (2000) Wild type and mutant p53 differentially regulate the gene expression of human collagenase-3 (hMMP-13). J Biol Chem 275:11327–11332

Sun Y, Wenger L, Rutter JL, Brinckerhoff CE, Cheung HS (1999) p53 down-regulates human matrix metalloproteinase-1 (Collagenase-1) gene expression. J Biol Chem 274:11535–11540

Suzuki A, de la Pompa JL, Hakem R, Elia A, Yoshida R, Mo R, Nishina H, Chuang T, Wakeham A, Itie A, Koo W, Billia P, Ho A, Fukumoto M, Hui CC, Mak TW (1997)

Brca2 is required for embryonic cellular proliferation in the mouse. Genes Dev 11:1242–1252

Suzuki H, Tomida A, Tsuruo T (2001) Dephosphorylated hypoxia-inducible factor 1alpha as a mediator of p53-dependent apoptosis during hypoxia. Oncogene 20:5779–5788

Swisher SG, Roth JA (2002) p53 Gene therapy for lung cancer. Curr Oncol Rep 4:334–340

Takimoto R, Wang W, Dicker DT, Rastinejad F, Lyssikatos J, el-Deiry WS (2002) The mutant p53-conformation modifying drug, CP-31398, can induce apoptosis of human cancer cells and can stabilize wild-type p53 protein. Cancer Biol Ther 1:47–55

Tanaka H, Arakawa H, Yamaguchi T, Shiraishi K, Fukuda S, Matsui K, Takei Y, Nakamura Y (2000) A ribonucleotide reductase gene involved in a p53-dependent cell-cycle checkpoint for DNA damage. Nature 404:42–49

Tanaka H, Matsumura I, Ezoe S, Satoh Y, Sakamaki T, Albanese C, Machii T, Pestell RG, Kanakura Y (2002) E2F1 and c-Myc potentiate apoptosis through inhibition of NF-kappaB activity that facilitates MnSOD-mediated ROS elimination. Mol Cell 9:1017–1029

Tanikawa C, Matsuda K, Fukuda S, Nakamura Y, Arakawa H (2003) p53RDL1 regulates p53-dependent apoptosis. Nat Cell Biol 5:216–223

Taylor WR, Stark GR (2001) Regulation of the G2/M transition by p53. Oncogene 20:1803–1815

Tokino T, Nakamura Y (2000) The role of p53-target genes in human cancer. Crit Rev Oncol Hematol 33: 1–6

Tolbert D, Lu X, Yin C, Tantama M, Van Dyke T (2002) p19(ARF) is dispensable for oncogenic stress-induced p53-mediated apoptosis and tumor suppression in vivo. Mol Cell Biol 22:370–377

Tornaletti S, Pfeifer GP (1994) Slow repair of pyrimidine dimers at p53 mutation hotspots in skin cancer. Science 263:1436–1438

Tort F, Hernandez S, Bea S, Martinez A, Esteller M, Herman JG, Puig X, Camacho E, Sanchez M, Nayach I, Lopez-Guillermo A, Fernandez PL, Colomer D, Hernandez L, Campo E (2002) CHK2-decreased protein expression and infrequent genetic alterations mainly occur in aggressive types of non-Hodgkin lymphomas. Blood 100:4602–4608

Tretyakova N, Matter B, Jones R, Shallop A (2002) Formation of benzo[a]pyrene diol epoxide-DNA adducts at specific guanines within K-ras and p53 gene sequences: stable isotope-labeling mass spectrometry approach. Biochemistry 41:9535–9544

Tyner SD, Venkatachalam S, Choi J, Jones S, Ghebranious N, Igelmann H, Lu X, Soron G, Cooper B, Brayton C, Hee Park S, Thompson T, Karsenty G, Bradley A, Donehower LA (2002) p53 mutant mice that display early ageing-associated phenotypes. Nature 415:45–53

Vafa O, Wade M, Kern S, Beeche M, Pandita TK, Hampton GM, Wahl GM (2002) c-Myc can induce DNA damage, increase reactive oxygen species, and mitigate p53 function: a mechanism for oncogene-induced genetic instability. Mol Cell 9:1031–1044

Valenzuela MT, Guerrero R, Nunez MI, Ruiz De Almodovar JM, Sarker M, de Murcia G, Oliver FJ (2002) PARP-1 modifies the effectiveness of p53-mediated DNA damage response. Oncogene 21:1108–1116

Walker KK, Levine AJ (1996) Identification of a novel p53 functional domain that is necessary for efficient growth suppression. Proc Natl Acad Sci USA 93:15335–15340

Wang L, Wu Q, Qiu P, Mirza A, McGuirk M, Kirschmeier P, Greene JR, Wang Y, Pickett CB, Liu S (2001) Analyses of p53 target genes in the human genome by bioinformatic and microarray approaches. J Biol Chem 276:43604–43610

Wang Y, Prives C (1995) Increased and altered DNA binding of human p53 by S and G2/M but not G1 cyclin-dependent kinases. Nature 376:88–91

Varmeh-Ziaie S, Okan I, Wang Y, Magnusson KP, Warthoe P, Strauss M, Wiman KG (1997) Wig-1, a new p53-induced gene encoding a zinc finger protein. Oncogene 15:2699–26704.

Wasylyk C, Salvi R, Argentini M, Dureuil C, Delumeau I, Abecassis J, Debussche L, Wasylyk B (1999) p53 mediated death of cells overexpressing MDM 2 by an inhibitor of MDM 2 interaction with p53. Oncogene 18:1921–1934

Watanabe T, Sullenger BA (2000) Induction of wild-type p53 activity in human cancer cells by ribozymes that repair mutant p53 transcripts. Proc Natl Acad Sci USA 97:8490–8494

Vaziri H, Dessain SK, Ng Eaton E, Imai SI, Frye RA, Pandita TK, Guarente L, Weinberg RA (2001) hSIR2(SIRT1) functions as an NAD-dependent p53 deacetylase. Cell 107:149–159

Weinstein JN, Myers TG, O'Connor PM, Friend SH, Fornace AJ, Jr., Kohn KW, Fojo T, Bates SE, Rubinstein LV, Anderson NL, Buolamwini JK, van Osdol WW, Monks AP, Scudiero DA, Sausville EA, Zaharevitz DW, Bunow B, Viswanadhan VN, Johnson GS, Wittes RE, Paull KD (1997) An information-intensive approach to the molecular pharmacology of cancer. Science 275:343–349

Weller M (1998) Predicting response to cancer chemotherapy: the role of p53. Cell Tissue Res 292:435–445

Werner H, Karnieli E, Rauscher FJ, LeRoith D (1996) Wild-type and mutant p53 differentially regulate transcription of the insulin-like growth factor I receptor gene. Proc Natl Acad Sci U S A 93:8318–8323

Wilhelm MT, Mendez-Vidal C, Wiman KG (2002) Identification of functional p53-binding motifs in the mouse wig-1 promoter. FEBS Lett 524:69–72

Vogelstein B, Lane D, Levine AJ (2000) Surfing the p53 network. Nature 408:307–310

Vollmar B, El-Gibaly AM, Scheuer C, Strik MW, Bruch HP, Menger MD (2002) Acceleration of cutaneous wound healing by transient p53 inhibition. Lab Invest 82:1063–1071

Wong KB, DeDecker BS, Freund SM, Proctor MR, Bycroft M, Fersht AR (1999) Hot-spot mutants of p53 core domain evince characteristic local structural changes. Proc Natl Acad Sci USA 96:8438–8442

Woods DB, Vousden KH (2001) Regulation of p53 function. Exp Cell Res 264:56–66

Vousden KH, Farrell PJ (1994) Viruses and human cancer. Br Med Bull 50:560–581

Vousden KH, Lu X (2002) Live or let die: the cell's response to p53. Nat Rev Cancer 2:594–604

Wu GS, Burns TF, McDonald ER, 3rd, Jiang W, Meng R, Krantz ID, Kao G, Gan DD, Zhou JY, Muschel R, Hamilton SR, Spinner NB, Markowitz S, Wu G, el-Deiry WS (1997) KILLER/DR5 is a DNA damage-inducible p53-regulated death receptor gene. Nat Genet 17:141–143

Wu X, Bayle JH, Olson D, Levine AJ (1993) The p53-mdm-2 autoregulatory feedback loop. Genes Dev 7:1126–1132

Xu D, Wang Q, Gruber A, Bjorkholm M, Chen Z, Zaid A, Selivanova G, Peterson C, Wiman KG, Pisa P (2000) Downregulation of telomerase reverse transcriptase mRNA expression by wild type p53 in human tumor cells. Oncogene 19:5123–5133

Xu X, Qiao W, Linke SP, Cao L, Li WM, Furth PA, Harris CC, Deng CX (2001) Genetic interactions between tumor suppressors Brca1 and p53 in apoptosis, cell cycle and tumorigenesis. Nat Genet 28:266–271

Yang A, Kaghad M, Caput D, McKeon F (2002) On the shoulders of giants: p63, p73 and the rise of p53. Trends Genet 18:90–95

Yoshikawa R, Kusunoki M, Yanagi H, Noda M, Furuyama JI, Yamamura T, Hashimoto-Tamaoki T (2001) Dual antitumor effects of 5-fluorouracil on the cell cycle in colorectal carcinoma cells: a novel target mechanism concept for pharmacokinetic modulating chemotherapy. Cancer Res 61:1029–1037

Yu J, Zhang L, Hwang PM, Kinzler KW, Vogelstein B (2001) PUMA induces the rapid apoptosis of colorectal cancer cells. Mol Cell 7:673–682

Zhan Q, Chen IT, Antinore MJ, Fornace AJ, Jr. (1998) Tumor suppressor p53 can participate in transcriptional induction of the GADD45 promoter in the absence of direct DNA binding. Mol Cell Biol 18:2768–2778

Zhang H, Somasundaram K, Peng Y, Tian H, Bi D, Weber BL, El-Deiry WS (1998) BRCA1 physically associates with p53 and stimulates its transcriptional activity. Oncogene 16:1713–1721

Zhao R, Gish K, Murphy M, Yin Y, Notterman D, Hoffman WH, Tom E, Mack DH, Levine AJ (2000) Analysis of p53-regulated gene expression patterns using oligonucleotide arrays. Genes Dev 14:981–993

Zhou Y, Gwadry FG, Reinhold WC, Miller LD, Smith LH, Scherf U, Liu ET, Kohn KW, Pommier Y, Weinstein JN (2002) Transcriptional regulation of mitotic genes by camptothecin-induced DNA damage: microarray analysis of dose- and time-dependent effects. Cancer Res 62:1688–1695

Zhu J, Chen X (2000) MCG10, a novel p53 target gene that encodes a KH domain RNA-binding protein, is capable of inducing apoptosis and cell cycle arrest in G(2)-M. Mol Cell Biol 20:5602–5618

Ziegler A, Leffell DJ, Kunala S, Sharma HW, Gailani M, Simon JA, Halperin AJ, Baden HP, Shapiro PE, Bale AE, et al. (1993) Mutation hotspots due to sunlight in the p53 gene of nonmelanoma skin cancers. Proc Natl Acad Sci USA 90:4216–4220

Zindy F, Eischen CM, Randle DH, Kamijo T, Cleveland JL, Sherr CJ, Roussel MF (1998) Myc signaling via the ARF tumor suppressor regulates p53-dependent apoptosis and immortalization. Genes Dev 12:2424–2433

Zou Z, Gao C, Nagaich AK, Connell T, Saito S, Moul JW, Seth P, Appella E, Srivastava S (2000) p53 regulates the expression of the tumor suppressor gene maspin. J Biol Chem 275:6051–6054

The Role of Ets Transcription Factors in Mediating Cellular Transformation

G. Foos · C. A. Hauser (✉)

La Jolla Cancer Research Center, The Burnham Institute, 10901 North Torrey Pines Road, La Jolla, CA 92037, USA
CHauser@burnham.org

1	Introduction to the Ets Family of Transcription Factors.	260
1.1	The Ets Gene Families of Mice and Humans.	260
1.2	Ets Family Nomenclature	262
1.3	Ets Family Functions.	262
1.4	Many Different Ets Factors Can Be Present in a Specific Tissue or Cell Type .	263
1.5	Ets Target Genes	264
2	Evidence Implicating Ets Factors in Cellular Transformation and Cancer. .	264
2.1	Ets Factor Overexpression Resulting from Proviral Insertion	264
2.2	Chromosomal Translocations of Ets Genes Associated with Human Cancers.	265
2.3	Mutations in Ets Genes Associated with Human Cancers.	266
2.4	Signaling to Ets Factors from Oncogenes	266
2.5	Expression Correlation of Ets Factors with Tumors.	267
2.6	Reversal of Cellular Transformation by Altered Ets Factor Function	267
2.7	Genetic Loss-of-Function Studies of Ets Factors in Cancer.	269
3	Future Perspectives for Understanding the Role of Ets Factors in Transformation.	269
	References	270

Abstract The Ets family of transcription factors in mouse or humans is comprised of around 27 unique family members that contain an evolutionarily conserved DNA-binding domain called the Ets domain. The Ets family includes both transcriptional activators and repressors. The normal cellular Ets transcription factors have been implicated as mediators of a wide range of cellular processes, including oncogenic transformation. This chapter provides an overview of the Ets family, and describes each of the multiple lines of evidence that Ets transcription factors are mediators of cellular transformation. This evidence includes: (a) cancers resulting from Ets factor overexpression or chromosomal translocations that generate fusion proteins containing Ets factor domains; (b) signaling from oncogenes to Ets factors; (c) expression correlation of Ets factors with tumor formation; (d) reversal of cellular transformation by dominant inhibitory Ets constructs; (e) delayed tumor development after genetic disruption of an Ets factor; and (f) the potential role of many Ets target genes in transformation. A better understanding of the role of Ets factors and their target genes in cancer should provide the basis for more specific novel therapeutic approaches for the treatment of cancers.

Keywords Ets gene family · Transcription factors · Cellular transformation

1
Introduction to the Ets Family of Transcription Factors

The first Ets family member v-ets, was identified as part of a fusion oncogene in the E26 avian transforming retrovirus. The name 'ets' came simultaneously from E26 transformation-specific (Nunn et al. 1983) or E-twenty-six (Leprince et al. 1983). Since the initial identification of a v-ets cellular homolog in chickens (Leprince et al. 1983), and the recognition that other proteins have a related domain (Karim et al. 1990), Ets transcription factor families have been identified in a variety of organisms. The Ets family size ranges from 10 putative Ets factors in *Caenorhabditis elegans* (Hart et al. 2000) to 27 characterized Ets family members in humans (Oettgen et al. 2000). The Ets transcription factor family is defined by the presence of an evolutionarily conserved domain of about 85 amino acids—the Ets domain. The Ets domain mediates binding of Ets family members to DNA sequences containing a GGAA/T core sequence. While there is some specificity conferred by the nucleotides flanking the core sequence, there is considerable overlap of Ets factor DNA binding specificity. The functional specificity of Ets factors is thought to derive from a combination of their tissue-specific expression patterns, post-transcriptional modifications, and interactions with a variety of partner proteins (reviewed in Ghysdael and Boureux 1997; Graves and Petersen 1998; Sharrocks 2001; Verger and Duterque-Coquillaud 2002; Oikawa and Yamada 2003).

1.1
The Ets Gene Families of Mice and Humans

Mammalian Ets factors have been organized into subfamilies by several criteria, the most common based on the similarity of their Ets domains. Table 1 lists the 26 currently characterized human/mouse Ets family orthologs, and one human Ets factor (TEL2) where a mouse ortholog has not yet been reported. This subfamily grouping is based on the Ets domain molecular phylogeny analysis of (Laudet et al. 1999), with the addition of several more recently characterized Ets factors, as tabulated in (Oettgen et al. 2000). A bioinformatic study of the mouse genome sequence suggests that few additional Ets domain-containing genes remain to be discovered (Xuan et al. 2002). A second conserved domain found in 11 of the Ets family members is the pointed domain, named for the *Drosophila pointed* gene where this domain was first identified (Klambt 1993). The presence or absence of a pointed domain is indicated for each of the Ets factors in Table 1. Pointed domains are associated with highly divergent Ets domains (e.g., the Ets1/2 and TEL subfamilies), and thus arranging the Ets family by pointed domain homology would lead to an organization quite different from that shown in Table 1. The Ets1/2, Erg, and Elf/Ese subfamilies (based on Ets domain homol-

The Role of Ets Transcription Factors in Mediating Cellular Transformation

Table 1 The mouse and human Ets families

Ets Subfamily	Ets Gene[a]	Alternative Names[b]	Pointed Domain	Mouse UGRepAcc[c]	Mouse UGCluster[d]	Mouse Chrom.[e]	Human UGRepAcc[c]	Human UGCluster[d]	Human Chrom.[e]
Ets1/2	<u>Ets2</u>	c-ets-2	yes	NM_011809	Mm.22365	16	NM_005239	Hs.292477	21
	<u>Ets1</u>	c-ets-1	yes	NM_011808	Mm.14115	9	NM_005238	Hs.18063	11
	ER71	(m)<u>ETSRP71</u>, (h)ETV2	no	NM_007959	Mm.4829	7	XM_290831	Hs.194061	19
GABPa	<u>Gabpa</u>	E4TF1A	yes	NM_008065	Mm.18974	16	NM_002040	Hs.78	21
PEA3	PEA3	Etv4, (h)E1AF	no	NM_008815	Mm.5025	11	NM_001986	Hs.434059	17
	ER81	<u>Etv1</u>, (m)EtsRP81	no	NM_007960	Mm.4866	12	NM_004956	Hs.150011	7
	ERM	<u>Etv5</u>		NM_023794	Mm.155708	16	NM_004454	Hs.43697	3
Erg	<u>Fli1</u>	(h)ERGB, EWSR2	yes	NM_008026	Mm.258908	9	NM_002017	Hs.257049	11
	<u>Erg</u>		yes	NM_133659	Mm.164531	16	NM_182918	Hs.45514	21
	<u>Fev</u>	(h)HSRNAFEV, Pet1	no	NM_153111	Mm.150496	1	NM_017521	Hs.234759	2
ERF	<u>Erf</u>	(h)PE-2	no	NM_010155	Mm.8068	7	NM_006494	Hs.440332	19
	PE1	Etv3, METS	no	NM_012051	Mm.34510	3	NM_005240	Hs.352672	1
Elk/TCF	<u>Elk1</u>		no	NM_007922	Mm.3064	X	NM_005229	Hs.181128	X
	<u>Elk3</u>	Net, Sap-2, ERP	no	NM_013508	Mm.4454	10	NM_005230	Hs.288555	12
	<u>Elk4</u>	Sap1	no	NM_007923	Mm.195050	1	NM_021795	Hs.129969	1
Elf/Ese	<u>Elf1</u>	Elf-1	no	NM_007920	Mm.24876	14	NM_172373	Hs.124030	13
	<u>Elf2</u>	(h)NERF	no	NM_023502	Mm.46503	3	NM_006874	Hs.82143	4
	<u>Elf4</u>	MEF, ELFR	no	NM_019680	Mm.154274	X	NM_001421	Hs.151139	X
	Ese1	Elf3, ESX, jen, Ert	yes	NM_007921	Mm.3963	1	NM_004433	Hs.67928	1
	Ese2	<u>Elf5</u>	yes	NM_010125	Mm.20888	2	NM_001422	Hs.11713	11
	Ese3	<u>Ehf</u>	yes	NM_007914	Mm.10724	2	NM_012153	Hs.200228	11
	Pse	(m)<u>Spdef</u> (h)PDEF	yes	NM_013891	Mm.26768	17	NM_012391	Hs.79414	6
TEL	TEL	<u>Etv6</u>, TEL1	yes	NM_007961	Mm.269995	6	NM_001987	Hs.171262	12
	TEL2	Etv7, TELB	yes	None	None	–	NM_016135	Hs.272398	6
Spi	PU.1	(m)<u>Sfpi1</u>, (h)SPI1	no	NM_011355	Mm.1302	2	NM_003120	Hs.157441	11
	<u>SpiB</u>	Spi-B	no	U87620	Mm.8012	7	NM_003121	Hs.437905	19
	<u>SpiC</u>	Spi-C, Prf	no	NM_011461	Mm.21642	10	NM_152323	Hs.511791	12

[a] Names of Ets family members used in this work, based on wide usage or to emphasize subfamily relationships. Underline indicates current mouse UniGene symbols. Table data are compiled from Stanford SOURCE site (http://source.stanford.edu) and from UniGene (http://www.ncbi.nlm.nih.gov/UniGene).

[b] Other common names used for each Ets gene, with current UniGene symbols underlined. Names used primarily for mouse or human orthologs are designated (m) or (h), respectively. Where mouse/human UniGene symbols differ beyond capitalization, (human symbols are all capitals) both are shown.

[c] UniGene representative mRNA accession numbers

[d] UniGene Cluster

[e] Chromosomal location

ogy) are examples of Ets subfamilies of which only some subfamily members contain a pointed domain. Four of the seven Elf/Ese subfamily members contain a pointed domain, and this observation along with their epithelial pattern of expression, has led to the grouping of the Ese/Pse family as a distinct subfamily (Feldman et al. 2003b). The roles of Ets factor pointed domains in oncogenesis are discussed below.

1.2
Ets Family Nomenclature

One of the confounding problems of understanding the extensive literature on Ets transcription factors (currently more than 2000 publications) is the multiple names in use for each Ets factor. Table 1 includes alternative names used for the human and/or mouse Ets family members including their Uni-Gene symbols, representative transcript accession numbers, and cluster number. Additionally, both in common usage and even in UniGene symbols, there are sometimes different names for mouse and human orthologs. Finally, several UniGene symbols, particularly those based on involvement of seven sometimes unrelated Ets factors identified in chromosomal translocations (ETV 1–7), are not used by most researchers in the field. An example of the challenges in nomenclature is PEA3/E1AF/ETV4. PEA3 started out as a generic term for factors that bound to what later would be called an Ets-binding site in the polyoma enhancer (Gutman and Wasylyk 1990; Leprince et al. 1992). Subsequently, the name PEA3 was given to a specific Ets family member (Xin et al. 1992). Later, the human ortholog of PEA3 (with 94% total sequence identity) was discovered, but was named E1AF (Higashino et al. 1993). Finally, PEA3/E1AF was found to be the fourth Ets factor involved in chromosomal fusions with EWS (Kaneko et al. 1996), and was designated ETV4 in UniGene. A literature search revealed that for PEA3, E1AF, and ETV4, there were 136, 26, and 3 citations respectively, and this ratio has not substantially changed in the last 2 years.

1.3
Ets Family Functions

Ets transcription factors have been implicated in the regulation of virtually all cellular functions, including growth, development, differentiation, survival, and oncogenic transformation (reviewed in Dittmer and Nordheim 1998; Maroulakou and Bowe 2000; Oikawa and Yamada 2003). Gene products associated with all of these cellular functions are among the hundreds of putative Ets factor target genes already identified by a variety of criteria (reviewed in Sementchenko and Watson 2000). The involvement of some of these target genes in cellular transformation is discussed below. Despite the potential functional redundancy of Ets factors, gene disruption of most Ets factors studied thus far results in embryonic or perinatal lethality (Bartel et al. 2000; Oikawa and Yamada 2003). Such early lethality in knockout mice reveals essential early roles for Ets factors, but complicates the study of the role of individual Ets factors in oncogenesis.

The majority of Ets factors are transcriptional activators, which serve as downstream effectors for a variety of signal transduction pathways, as discussed below. However, at least five mammalian Ets factors have been reported to have repressor activity, including Erf, PE1/METS, Elk3/Net, TEL, and TEL2. (Mavrothalassitis and Ghysdael 2000; Gu et al. 2001; Klappacher et al. 2002). In addition, depending on the signaling inputs, several additional Ets factors possess both transcriptional activation and repression activities (reviewed in Sharrocks 2001). The mixed transcriptional role of Ets factors has been evolutionarily conserved from *Drosophila*, where several of the Ets factors are transcriptional activators (Hsu and Schulz 2000), but Yan is a negative regulator (O'Neill et al. 1994) which opposes the action activators such as pointed (Brunner et al. 1994; Gabay et al. 1996). The *C. elegans* Lin-1 Ets factor may also possess negative regulatory activity (Tan et al. 1998). Overall, in normal mammalian cells, there is a balance between positive and negative regulation of Ets-dependent gene expression, and there are multiple lines of evidence that changes in this balance can have a significant impact on oncogenic transformation.

1.4
Many Different Ets Factors Can Be Present in a Specific Tissue or Cell Type

Because of the similar DNA binding specificity of Ets factors, to understand how Ets target genes are regulated in a particular cell type, it is important to know which Ets factors are present. The normal course of gene discovery is that a new Ets factor is found, and its expression is analyzed in several tissues. Subsequently, other investigators may examine the expression of this Ets factor in tissues of their interest. The resulting expression data for each Ets factor is therefore rather anecdotal. When our studies led to the question of which Ets factors act as crucial mediators of cancer, we were surprised to find that the expression status of less than half of the Ets family members was known in any single cell type or tissue (Maroulakou and Bowe 2000; Barrett et al. 2002). Thus, we undertook a comprehensive study to determine which of the Ets factor mRNAs are expressed in normal mammary, mammary tumors, and mammary related cell lines. The unexpected result of this analysis was that 24 of the 25 mouse Ets factors analyzed were expressed in normal mammary tissue, and even in clonal cell lines, between 14 and 20 of the Ets factors were significantly expressed (Galang et al. 2004). These data show that identifying which Ets factors are regulating specific target genes is more complex than previously appreciated.

1.5
Ets Target Genes

There is substantial interest in Ets transcription factor target genes, in part, because of the potential role of these genes in the transformed phenotype. Over 200 genes with Ets factor-binding sites in their promoters have been established as Ets target genes by various criteria. The products of these target genes are associated with every aspect of cellular regulation, including growth, adhesion, motility, invasion, angiogenesis, and apoptosis (Sementchenko and Watson 2000). In addition, correlative evidence connects expression of various Ets factors to these cellular functions, and Ets factor-binding sites are found in the promoter of nearly every matrix metalloproteinase, molecules important in invasive behavior (Sato 2001; Singh et al. 2002; Oikawa and Yamada 2003). Clearly, gene products involved in controlling these diverse cellular functions are likely to be important downstream targets of oncogenic signaling. Because most of the Ets target genes have been characterized by reporter gene analysis upon overexpression of a few Ets factors, the physiological role of individual Ets factors in regulating these target genes remains unclear, as does the contribution of this observed regulation to the transformed phenotype.

2
Evidence Implicating Ets Factors in Cellular Transformation and Cancer

A variety of lines of evidence support the role of Ets factors as mediators of cellular transformation and tumor progression. These include: (a) erythroleukemias from viral-induced overexpression of mouse Ets factors; (b) chromosomal translocations involving at least six different Ets genes generate fusion proteins associated with a variety of tumors; (c) mutations in some Ets factors are associated with tumor development; (d) many Ets factors are downstream signaling targets for oncogenes; (e) correlation of Ets factor expression with tumor progression; (f) reversal of cellular transformation by dominant negative and positive Ets constructs or other reagents that interfere with Ets factor function; (g) impaired tumor development in mice with genetically altered function of a specific Ets factor. These seven lines of evidence are described below.

2.1
Ets Factor Overexpression Resulting from Proviral Insertion

There are two examples in where overexpression of mouse Ets factors due to nearby viral integration contributes to erythroleukemias. The Spi-1/PU.1 Ets factor was first identified in erythroleukemias as an oncogene frequently activated by Friend spleen focus forming virus insertion (Moreau-Gachelin et al. 1988). Similarly, elevated Fli1 expression resulting from Friend murine leukemia virus insertion was also found in erythroleukemias (Ben-David et al. 1991). The viral insertions did not alter the coding sequence of PU.1 or Fli1, but proximity of the strong viral enhancer elevated the transcription of these Ets factors. Transgenic mouse models were subsequently used to show that overexpression of PU.1, but not Fli1, was sufficient to induce erythroleukemia (Zhang et al. 1995; Moreau-Gachelin et al. 1996). In addition to these naturally occurring examples, experimental overexpression of several Ets factors has been reported to transform rodent cells (reviewed in Dittmer and Nordheim 1998).

2.2
Chromosomal Translocations of Ets Genes Associated with Human Cancers

Fusions of the N-terminal portion of EWS with the Ets domain (DNA-binding domain) of at least five different Ets factors (Fli, Erg, ER81, PEA3, FEV) are associated with Ewing's family tumors (reviewed in Arvand and Denny 2001). The ability of so many different Ets DNA-binding domains (Ets DBDs) to participate in these fusions with similar outcomes, suggests that the EtsDBD have similar DNA-binding specificities, and that critical Ets target gene expression is being altered by fusion to EWS. This is likely due in part to the enhanced transactivation activity of the Ets fusion proteins (Ohno et al. 1993; Bailly et al. 1994). Indeed, experimentally interfering with Ets-dependent gene expression by expression of the Fli1 EtsDBD fused to a repressor domain reverses the transformed phenotype of Ewing Sarcoma cells (Athanasiou et al. 2000). However, there is emerging evidence that other activities of EWS also mediate transformation, as the Ews–Fli1 fusion proteins can also negatively regulate Ets-dependent gene expression (Im et al. 2001) and EWS–Ets fusions exhibit both DNA-binding-dependent and -independent transformation mechanisms (Jaishankar et al. 1999; Knoop and Baker 2001; Welford et al. 2001).

The TEL gene is involved in several kinds of cancer associated gene fusions, which reveal distinct contributions of three different domains of this Ets family member. One type of TEL fusion associated with leukemias is the Ets domain of TEL fused to a transactivation domain of transcription factor MN1 (Buijs et al. 2000). This fusion protein presumably leads to inappropriate activation of Ets-dependent gene expression. A unique feature of TEL (and the recently discovered TEL2) among the Ets family members is its ability to homodimerize through its pointed domain. Fusions of the TEL dimerization domain to the kinase domain of variety of tyrosine kinase genes leads to dimerized and constitutively activated tyrosine kinases associated with leukemias (Golub et al. 1996). In addition to leukemias, such fusions can also lead to lymphomas (Yagasaki et al. 2001) and fibrosarcomas

(Knezevich et al. 1998). Finally, TEL also contains a repressor domain (Chakrabarti and Nucifora 1999), and gene fusion of this domain with AML1 generates a protein that may repress critical AML1 target genes leading to leukemias (Hiebert et al. 2001). Overall, the Ets fusion genes associated with cancers highlight the function of several Ets factor domains. These data, along with induction of erythroleukemias from elevated expression of PU.1 or Fli1, strongly suggest that altered regulation of Ets target genes contributes to a variety of malignancies.

2.3
Mutations in Ets Genes Associated with Human Cancers

There is not strong evidence that mutation of Ets family members is a widespread event in human cancers. Nonetheless, there are a few suggestive examples. In addition to participation of TEL in gene fusions, TEL also maps to a chromosomal region (12p12-p13) found deleted in about 5% of children with acute lymphoblastic leukemia (ALL), suggesting a possible role as a tumor suppressor (Stegmaier et al. 1995). Further analysis of TEL loss of heterozygosity (LOH) in ALL patients has generated mixed results, but loss of the unfused TEL allele in TEL-AML1-induced ALL is quite common, suggesting there is selective pressure for this LOH (Raynaud et al. 1996). Heterozygous mutations in PU.1 were recently identified in 9 of 126 acute myeloid leukemia (AML) patients, with most of these mutations disrupting PU.1 DNA-binding function. It was postulated that such mutations could inhibit PU.1 function and block early myeloid differentiation (analogous to the differentiation block observed in PU.1$^{-/-}$ mice), contributing to development of AML (Mueller et al. 2002). The Ese2/Elf5 and Ese3/EHF genes are closely linked and map to human chromosome 11p13–15. This chromosomal region has been found to exhibit LOH in breast and prostate carcinomas, suggestive of a possible negative role for these Ets factors in tumors (Zhou et al. 1998; Tugores et al. 2001).

2.4
Signaling to Ets Factors from Oncogenes

Ets transcription factors are downstream targets of multiple signaling pathways, and their activity can be modulated by a variety of post-transcriptional modifications. The Ras signaling pathway alters the activity of many Ets factors, and other oncogenic signaling also converges on Ets transcription factors (for review see Dittmer and Nordheim 1998; Wasylyk et al. 1998; Yordy and Muise-Helmericks 2000; Oikawa and Yamada 2003). As an example, Ets2 is transcriptionally activated by oncogenic Ras or Neu/ErbB-2 signaling, and this activation requires mitogen activated protein kinase-mediated phosphorylation of an evolutionarily conserved Ets2 threonine residue

(Galang et al. 1996; Yang et al. 1996; McCarthy et al. 1997). Another evolutionarily conserved function of oncogenic signaling is relief of negative regulation by Ets family repressors. This is seen from Ras signaling in flies (Gabay et al. 1996) to oncogenic signaling in mammals (Le Gallic et al. 1999; Lopez et al. 2003). In addition to phosphorylation, other reported regulatory modification of Ets family members include acetylation of Ets1 (Czuwara-Ladykowska et al. 2002), glycosylation of Elf1 (Tsokos et al. 2003), and SUMO modification of TEL (Wood et al. 2003). Overall, modifications of Ets factors resulting from oncogenic signaling may strongly influence their activity, through mechanisms including altered DNA binding, interactions with partner proteins, protein stability, or subcellular localization.

2.5
Expression Correlation of Ets Factors with Tumors

There have been many correlative studies demonstrating differences in the expression of many of the Ets factors in normal and tumor tissue. A recent comprehensive review on Ets1 cited 35 correlative studies of the expression of just this one Ets factor in tumors (Dittmer 2003). Our recent analysis of expression of the entire Ets family in mouse mammary tumor development found that expression of the mRNAs of nine different Ets factors was significantly elevated in mammary tumors as compared with normal mammary tissue (Galang et al. 2004). Some of this altered Ets factor expression was found to reflect changes the in cellular composition from normal mammary tissue to tumors (e.g., an increased epithelial cell content), whereas other differences were found to represent actual tumor-specific events. Another complicating factor in interpreting expression Ets correlation studies is that one cannot distinguish whether changes in Ets factor expression contribute to the tumor phenotype, or simply result from altered signaling in the tumors. Nonetheless, there is a wealth of suggestive evidence that alterations in expression of specific Ets factors correlates with the development or progression of specific types of tumors (Oikawa and Yamada 2003).

2.6
Reversal of Cellular Transformation by Altered Ets Factor Function

One of the most compelling lines of evidence that Ets factors play a causal role in specifically mediating cellular transformation comes from experimental alteration of Ets family function in transformed cells. In mouse cells, broadly inhibiting Ets factor activity by expression of a dominant negative Ets construct consisting of just the DNA-binding domain (DBD) of Ets1, Ets2, or PU.1 inhibits or reverses the Ras or Neu/ErbB-2 transformation of murine fibroblasts (Langer et al. 1992; Giovane et al. 1994; Galang et al. 1996; Foos et al. 1998). Transgenic expression of a PEA3 DBD also inhibits tumor formation in a mouse model (Shepherd et al. 2001), and cationic lipid delivery of a PEA3 DBD expression construct to tumors resulted in prolonged survival of the treated animals (Wang and Hung 2000). Additional evidence of the importance of Ets signaling in transformation came from reversal of Ras transformation by overexpression of an inhibitory mutant form of ERF (Le Gallic et al. 1999) or overexpression of TEL, a transcriptional repressor in the Ets family (Athanasiou et al. 2000).

Similar to rodent cells, reversal of aspects of the transformed phenotype was observed in human tumor cells upon interfering with Ets function in prostate, thyroid, breast, and Ewing sarcoma tumor cells (Kovar et al. 1996; Delannoy-Courdent et al. 1998; Sapi et al. 1998; Sementchenko et al. 1998; Athanasiou et al. 2000; Foos and Hauser 2000; de Nigris et al. 2001; G. Foos and C.A. Hauser, unpublished results). Interestingly, in either rodent or human tumor cells, while Ets DBD inhibition of cellular Ets function has strong effects on the transformed phenotype (e.g., loss of anchorage-independent growth) it does not usually impair normal cell growth. This indicates that cellular Ets factors mediate transformation-specific signaling not required for normal cell growth. Thus, intervening with this signaling could have the specificity desired for cancer therapy.

One must interpret the Ets factor DBD studies carefully with respect to which specific Ets factors are important. It has long been suspected that Ets DBD constructs (which bind to similar promoter sites) could broadly inhibit Ets family activity. We recently demonstrated such broad activity, showing that Ets2DBD expression strongly inhibits Ets-dependent gene expression even in an Ets2 knockout cell line (Hever et al. 2003). This study further showed that despite the ability the Ets2DBD to reverse Ras transformation in a variety of cells, that Ets2 knockout cells exhibited no defects in Ras transformation. Thus, due to the promiscuity of Ets domain DNA binding, Ets dominant negative experiments clearly do not identify which specific Ets factors mediate transformation, but they do reveal the critical role of the Ets family in mediating transformation-specific signaling.

Surprisingly, experimental overexpression of a variety of Ets family transcriptional activators can also reverse aspects of the transformed phenotype in mouse and human cells. Overexpression of Ets1, Ets2, PEA3, Ese1, or PDEF reverses aspects of the transformed phenotype in both Ras transformed NIH3T3 cells (Foos et al. 1998) and in human colon, prostate and breast tumor cell lines (Suzuki et al. 1995; Chang et al. 2000; Foos and Hauser 2000; Xing et al. 2000; Feldman et al. 2003a; G. Foos and C.A. Hauser, unpublished results). Such studies must also be carefully interpreted, as high-level expression of an Ets factor likely impacts on the physiological targets of other Ets family members. In summary, a balance of Ets function (mediated by one or more unidentified Ets factors) appears to be needed to provide signaling specifically required to maintain cellular transformation.

2.7
Genetic Loss-of-Function Studies of Ets Factors in Cancer

One of the most compelling ways that a gene product can be implicated in tumor formation or progression, is by genetic loss-of-function analysis. This approach has been difficult with Ets factors, because their homozygous disruption often leads to embryonic or perinatal lethality (Bartel et al. 2000; Oikawa and Yamada 2003). In light of the extensive literature connecting Ets transcription factors and cancer, it is surprising that only one Ets factor, Ets2, has been demonstrated to be specifically involved in tumor development in vivo. This analysis of Ets2 function was also complicated by embryonic lethality, but it was shown that heterozygote ets2 (+/−) mice exhibited delayed tumor onset in a transgenic mouse mammary tumor model (Neznanov et al. 1999). It was subsequently shown that mice homozygous for a hypomorphic ets2 allele (which could not be activated by Ras pathway signaling) also exhibited delayed mammary tumor formation (Man et al. 2003). Definitive genetic analysis of the requirement of Ets2 or the other 25 Ets family members may require the use of conditional gene disruption.

3
Future Perspectives for Understanding the Role of Ets Factors in Transformation

While there is fairly overwhelming evidence that Ets transcription factors are important mediators of cellular transformation, important questions still need to be addressed. One of these questions is which specific Ets family members mediate transformation? Given the size of the Ets family, identification of individual Ets factors mediating transformation in specific cellular contexts will likely require loss-of-function analysis. While several loss-of-function approaches are possible, the use of emerging RNA interference technologies holds great promise. If individual Ets members whose function is critical in transformation can be identified, then therapeutic approaches based on specifically interfering with their expression or interactions can be developed, or alternatively, approaches developed based on interfering with the signaling which modulates the Ets factor activity.

A second major question is what are the important target genes for the Ets-mediated transformation-specific signaling. One current problem is trying to determine which of the hundreds of identified putative Ets target genes are actually effectors of transformation. In addition, there may also be novel transformation-specific targets of Ets factors yet to be identified. Most of the broad functional analysis of Ets target genes by microarray analysis thus far, has focused on the role of Ets factors in differentiation. Such differentiation analysis includes targets of PU.1, TEL, and MEF in hematapoetic cells and Ets1 and ERG in HUVEC (McLaughlin et al. 2001; Teruyama et al. 2001; Yamada et al. 2001; Sakurai et al. 2003; Hedvat et al. 2004). As a start to the identification of Ets targets important in cancer, we have applied microarray analysis to the human breast tumor cell line system, comparing gene expression in tumor cells to subclones reverted by dominant-acting Ets constructs. This approach has identified at least one functionally important Ets target gene (interleukin-8) in these tumor cells, with several other intriguing candidates (G. Foos and C.A. Hauser, unpublished results). Overall, it is anticipated that important insights into the molecular events in oncogenic transformation and tumor progression will be made from future studies of the role of Ets transcription factors in cancers.

References

Arvand A, Denny CT (2001) Biology of EWS/ETS fusions in Ewing's family tumors. Oncogene 20:5747–5754

Athanasiou M, LeGallic L, Watson DK, Blair DG, Mavrothalassitis G (2000) Suppression of the Ewing's sarcoma phenotype by FLI1/ERF repressor hybrids. Cancer Gene Ther 7:1188–1195

Bailly RA, Bosselut R, Zucman J, Cormier F, Delattre O, Roussel M, Thomas G, Ghysdael J (1994) DNA-binding and transcriptional activation properties of the EWS-FLI-1 fusion protein resulting from the t(11;22) translocation in Ewing sarcoma. Mol Cell Biol 14:3230–3241

Barrett JM, Puglia MA, Singh G, Tozer RG (2002) Expression of Ets-related transcription factors and matrix metalloproteinase genes in human breast cancer cells. Breast Cancer Res Treat 72:227–232

Bartel FO, Higuchi T, Spyropoulos DD (2000) Mouse models in the study of the Ets family of transcription factors. Oncogene 19:6443–6454

Ben-David Y, Giddens EB, Letwin K, Bernstein A (1991) Erythroleukemia induction by Friend murine leukemia virus: insertional activation of a new member of the ets gene family, Fli-1, closely linked to c-ets-1. Genes Dev 5:908–918

Brunner D, Ducker K, Oellers N, Hafen E, Scholz H, Klambt C (1994) The ETS domain protein pointed-P2 is a target of MAP kinase in the sevenless signal transduction pathway. Nature 370:386–389

Buijs A, van Rompaey L, Molijn AC, Davis JN, Vertegaal AC, Potter MD, Adams C, van Baal S, Zwarthoff EC, Roussel MF, Grosveld GC (2000) The MN1-TEL fusion protein, encoded by the translocation (12;22)(p13;q11) in myeloid leukemia, is a transcription factor with transforming activity. Mol Cell Biol 20:9281–9293

Chakrabarti SR, Nucifora G (1999) The leukemia-associated gene TEL encodes a transcription repressor which associates with SMRT and mSin3A. Biochem Biophys Res Commun 264:871–877

Chang J, Lee C, Hahm KB, Yi Y, Choi SG, Kim SJ (2000) Over-expression of ERT(ESX/ESE-1/ELF3), an ets-related transcription factor, induces endogenous TGF-beta type II receptor expression and restores the TGF-beta signaling pathway in Hs578 t human breast cancer cells. Oncogene 19:151–154

Czuwara-Ladykowska J, Sementchenko VI, Watson DK, Trojanowska M (2002) Ets1 is an effector of the transforming growth factor beta (TGF-beta) signaling pathway and an antagonist of the profibrotic effects of TGF-beta. J Biol Chem 277:20399–20408

de Nigris F, Mega T, Berger N, Barone MV, Santoro M, Viglietto G, Verde P, Fusco A (2001) Induction of ETS-1 and ETS-2 transcription factors is required for thyroid cell transformation. Cancer Res 61:2267–2275

Delannoy-Courdent A, Mattot V, Fafeur V, Fauquette W, Pollet I, Calmels T, Vercamer C, Boilly B, Vandenbunder B, Desbiens X (1998) The expression of an Ets1 transcription factor lacking its activation domain decreases uPA proteolytic activity and cell motility, and impairs normal tubulogenesis and cancerous scattering in mammary epithelial cells. J Cell Sci 111:1521–1534

Dittmer J (2003) The biology of the Ets1 Proto-oncogene. Mol Cancer 2:29 (http://www.molecular-cancer.com/content/2/1/29)

Dittmer J, Nordheim A (1998) Ets transcription factors and human disease. Biochim Biophys Acta 1377: F1–F11

Feldman RJ, Sementchenko VI, Gayed M, Fraig MM, Watson DK (2003a) Pdef expression in human breast cancer is correlated with invasive potential and altered gene expression. Cancer Res 63:4626–4631

Feldman RJ, Sementchenko VI, Watson DK (2003b) The epithelial-specific Ets factors occupy a unique position in defining epithelial proliferation, differentiation and carcinogenesis. Anticancer Res 23:2125–2131

Foos G, García-Ramírez JJ, Galang CK, Hauser CA (1998) Elevated expression of Ets2 or distinct portions of Ets2 can reverse Ras-mediated cellular transformation. J Biol Chem 273:18871–18880

Foos G, Hauser CA (2000) Altered Ets transcription factor activity in prostate tumor cells inhibits anchorage-independent growth, survival, and invasiveness. Oncogene 19:5507–5516

Gabay L, Scholz H, Golembo M, Klaes A, Shilo BZ, Klambt C (1996) EGF receptor signaling induces pointed P1 transcription and inactivates Yan protein in the Drosophila embryonic ventral ectoderm. Development 122:3355–3362

Galang CK, Garcia-Ramirez J, Solski PA, Westwick JK, Der CJ, Neznanov NN, Oshima RG, Hauser CA (1996) Oncogenic Neu/ErbB-2 increases ets, AP-1, and NF-kappaB-dependent gene expression, and inhibiting ets activation blocks Neu-mediated cellular transformation. J Biol Chem 271:7992–7998

Galang CK, Muller WJ, Foos G, Oshima RG, Hauser CA (2004) Changes in the expression of many Ets family transcription factors and of potential target genes in normal mammary tissue and tumors. J Biol Chem 279:11281J. Biol. Chem. 11292

Ghysdael J, Boureux A (1997) The Ets family of transcriptional regulators. In: Yaniv M, Ghysdael J (eds) Oncogenes as transcriptional regulators, vo–6408

Hedvat CV, Yao J, Sokolic RA, Nimer SD (2004) Myeloid ELF1-like Factor Is a Potent Activator of Interleukin-8 Expression in Hematopoietic Cells. J Biol Chem 279:6395–6400

Hever A, Oshima RG, Hauser CA (2003) Ets2 is not required for Ras or Neu/ErbB-2 mediated cellular transformation in vitro. Exp Cell Res 290:132–143

Hiebert SW, Lutterbach B, Amann J (2001) Role of co-repressors in transcriptional repression mediated by the t(8;21), t(16;21), t(12;21), and inv(16) fusion proteins. Curr Opin Hematol 8:197–200

Higashino F, Yoshida K, Fujinaga Y, Kamio K, Fujinaga K (1993) Isolation of a cDNA encoding the adenovirus E1A enhancer binding protein: a new human member of the ets oncogene family. Nucl Acids Res 21:547–553

Hsu T, Schulz RA (2000) Sequence and functional properties of Ets genes in the model organism Drosophila. Oncogene 19:6409–6416

Im YH, Kim HT, Lee C, Poulin D, Welford S, Sorensen PH, Denny CT, Kim SJ (2001) EWS-FLI1, EWS-ERG, and EWS-ETV1 oncoproteins of Ewing tumor family all suppress transcription of transforming growth factor beta type II receptor gene. Cancer Res 60:1536–1540

Jaishankar S, Zhang J, Roussel MF, Baker SJ (1999) Transforming activity of EWS/FLI is not strictly dependent upon DNA-binding activity. Oncogene 18:5592–5597

Kaneko Y, Yoshida K, Handa M, Toyoda Y, Nishihira H, Tanaka Y, Sasaki Y, Ishida S, Higashino F, Fujinaga K (1996) Fusion of an ETS-family gene, EIAF, to EWS by t(17;22)(q12;q12) chromosome translocation in an undifferentiated sarcoma of infancy. Genes Chromosomes Cancer 15:115–121

Karim FD, Urness LD, Thummel CS, Klemsz MJ, McKercher SR, Celada A, Van Beveren C, Maki RA, Gunther CV, Nye JA, al. e (1990) The ETS-domain: a new DNA-binding motif that recognizes a purine-rich core DNA sequence [letter]. Genes Dev 4:1451–1453

Klambt C (1993) The Drosophila gene pointed encodes two ETS-like proteins which are involved in the development of the midline glial cells. Development 117:163–176

Klappacher GW, Lunyak VV, Sykes DB, Sawka-Verhelle D, Sage J, Brard G, Ngo SD, Gangadharan D, Jacks T, Kamps MP, Rose DW, Rosenfeld MG, Glass CK (2002) An induced Ets repressor complex regulates growth arrest during terminal macrophage differentiation. Cell 109:169–180

Knezevich SR, McFadden DE, Tao W, Lim JF, Sorensen PH (1998) A novel ETV6-NTRK3 gene fusion in congenital fibrosarcoma. Nat Genet 18:184–187

Knoop LL, Baker SJ (2001) EWS/FLI alters 5'-splice site selection. J Biol Chem 276:22317–22322

Kovar H, Aryee DN, Jug G, Henockl C, Schemper M, Delattre O, Thomas G, Gadner H (1996) EWS/FLI-1 antagonists induce growth inhibition of Ewing tumor cells in vitro. Cell Growth Differ 7:429–437

Langer SJ, Bortner DM, Roussel MF, Sherr CJ, Ostrowski MC (1992) Mitogenic signaling by colony-stimulating factor 1 and ras is suppressed by the ets-2 DNA-binding domain and restored by myc overexpression. Mol Cell Biol 12:5355–5362

Laudet V, Hanni C, Stehelin D, Duterque-Coquillaud M (1999) Molecular phylogeny of the ETS gene family. Oncogene 18:1351–1359

Le Gallic L, Sgouras D, Beal G, Jr., Mavrothalassitis G (1999) Transcriptional repressor ERF is a Ras/mitogen-activated protein kinase target that regulates cellular proliferation. Mol Cell Biol 19:4121–4133

Leprince D, Crepieux P, Stehelin D (1992) c-ets-1 DNA binding to the PEA3 motif is differentially inhibited by all the mutations found in v-ets. Oncogene 7:9–17

Leprince D, Gegonne A, Coll J, de Taisne C, Schneeberger A, Lagrou C, Stehelin D (1983) A putative second cell-derived oncogene of the avian leukaemia retrovirus E26. Nature 306:395–397

Lopez RG, Carron C, Ghysdael J (2003) v-SRC Specifically Regulates the Nucleo-cytoplasmic Delocalization of the Major Isoform of TEL (ETV6). J Biol Chem 278:41316–41325

Man AK, Young LJT, Tynan J, Lesperance J, Egeblad M, Werb Z, Hauser CA, Muller WJ, Cardiff RD, Oshima RG (2003) Ets2-dependent stromal regulation of mouse mammary tumors. Mol Cell Biol 23:8614–8625

Maroulakou IG, Bowe DB (2000) Expression and function of Ets transcription factors in mammalian development: a regulatory network. Oncogene 19:6432–6442

Mavrothalassitis G, Ghysdael J (2000) Proteins of the ETS family with transcriptional repressor activity. Oncogene 19:6524–6532

McCarthy SA, Chen D, Yang B-S, García-Ramírez JJ, Cherwinski H, Chen X-R, Klagsbrun ML, Hauser CA, Ostrowski MC, McMahon M (1997) Rapid phosphorylation of Ets-2 accompanies MAP kinase activation and the induction of HB-EGF gene expression by oncogenic Raf-1. Mol Cell Biol 17:2401–2412

McLaughlin F, Ludbrook VJ, Cox J, von Carlowitz I, Brown S, Randi AM (2001) Combined genomic and antisense analysis reveals that the transcription factor Erg is implicated in endothelial cell differentiation. Blood 98:3332–3339

Moreau-Gachelin F, Tavitian A, Tambourin P (1988) Spi-1 is a putative oncogene in virally induced murine erythroleukaemias. Nature 331:277–280

Moreau-Gachelin F, Wendling F, Molina T, Denis N, Titeux M, Grimber G, Briand P, Vainchenker W, Tavitian A (1996) Spi-1/PU.1 transgenic mice develop multistep erythroleukemias. Mol Cell Biol 16:2453–2463

Mueller BU, Pabst T, Osato M, Asou N, Johansen LM, Minden MD, Behre G, Hiddemann W, Ito Y, Tenen DG (2002) Heterozygous PU.1 mutations are associated with acute myeloid leukemia. Blood 100:998–1007

Neznanov N, Man AK, Yamamoto H, Hauser CA, Cardiff RD, Oshima RG (1999) A single targeted Ets2 allele restricts development of mammary tumors in transgenic mice. Cancer Res 59:4242–4246

Nunn MF, Seeburg PH, Moscovici C, Duesberg PH (1983) Tripartite structure of the avian erythroblastosis virus E26 transforming gene. Nature 306:391–395

O'Neill EM, Rebay I, Tjian R, Rubin GM (1994) The activities of two Ets-related transcription factors required for Drosophila eye development are modulated by the Ras/MAPK pathway. Cell 78:137–147

Oettgen P, Finger E, Sun ZJ, Akbarali Y, Thamrongsak U, Boltax J, GrallF., Dube A, Weiss A, Brown L, Quinn G, Kas K, Endress G, Kunsch C (2000) PDEF, a novel prostate epithelium-specific Ets transcription factor, interacts with the androgen receptor and activates prostate-specific antigen gene expression. J Biol Chem 275:1216–1225

Ohno T, Rao VN, Reddy ES (1993) EWS/Fli-1 chimeric protein is a transcriptional activator. Cancer Res 53:5859–5863

Oikawa T, Yamada T (2003) Molecular biology of the Ets family of transcription factors. Gene 303:11–34

Raynaud S, Cave H, Baens M, Bastard C, Cacheux V, Grosgeorge J, Guidal-Giroux C, Guo C, Vilmer E, Marynen P, Grandchamp B (1996) The 12;21 translocation involving TEL and deletion of the other TEL allele: two frequently associated alterations found in childhood acute lymphoblastic leukemia. Blood 87:2891–2899

Sakurai T, Yamada T, Kihara-Negishi F, Teramoto S, Sato Y, Izawa T, Oikawa T (2003) Effects of overexpression of the Ets family transcription factor TEL on cell growth and differentiation of K562 cells. Int J Oncol 22:1327–1333

Sapi E, Flick MB, Rodov S, Kacinski BM (1998) Ets-2 transdominant mutant abolishes anchorage-independent growth and macrophage colony-stimulating factor-stimulated invasion by BT20 breast carcinoma cells. Cancer Res 58:1027–1033

Sato Y (2001) Role of ETS family transcription factors in vascular development and angiogenesis. [Review] [52 refs]. Cell Struct Funct 26:19–24

Sementchenko VI, Schweinfest CW, Papas TS, Watson DK (1998) ETS2 function is required to maintain the transformed state of human prostate cancer cells. Oncogene 17:2883–2888

Sementchenko VI, Watson DK (2000) Ets target genes: past, present and future. Oncogene 19:6533–6548

Sharrocks AD (2001) The ETS-domain transcription factor family. [Review] [112 refs]. Nat Rev Mol Cell Biol 2:827–837

Shepherd TG, Kockeritz L, Szrajber MR, Muller WJ, Hassell JA (2001) The pea3 subfamily ets genes are required for HER2/Neu-mediated mammary oncogenesis. Curr Biol 11:1739–1748

Singh S, Barrett J, Sakata K, Tozer RG, Singh G (2002) ETS proteins and MMPs: partners in invasion and metastasis. Curr. Drug Targets 3:359–367

Stegmaier K, Pendse S, Barker GF, Bray-Ward P, Ward DC, Montgomery KT, Krauter KS, Reynolds C, Sklar J, Donnelly M (1995) Frequent loss of heterozygosity at the TEL gene locus in acute lymphoblastic leukemia of childhood. Blood 86:38–44

Suzuki H, Romano-Spica V, Papas TS, Bhat NK (1995) ETS1 suppresses tumorigenicity of human colon cancer cells. Proc Natl Acad Sci USA 92:4442–4446

Tan PB, Lackner MR, Kim SK (1998) MAP kinase signaling specificity mediated by the LIN-1 Ets/LIN-31 WH transcription factor complex during C. elegans vulval induction. Cell 93:569–580

Teruyama K, Abe M, Nakano T, Takahashi S, Yamada S, Sato Y (2001) Neurophilin-1 is a downstream target of transcription factor Ets-1 in human umbilical vein endothelial cells. FEBS Lett 504:1–4

Tsokos GC, Nambiar MP, Juang YT (2003) Activation of the Ets transcription factor Elf-1 requires phosphorylation and glycosylation: defective expression of activated Elf-1 is involved in the decreased TCR zeta chain gene expression in patients with systemic lupus erythematosus. Ann NY Acad Sci 987:240–245

Tugores A, Le J, Sorokina I, Snijders AJ, Duyao M, Reddy PS, Carlee L, Ronshaugen M, Mushegian A, Watanaskul T, Chu S, Buckler A, Emtage S, McCormick MK (2001) The epithelium-specific ETS protein EHF/ESE-3 is a context-dependent transcriptional repressor downstream of MAPK signaling cascades. J Biol Chem 276:20397–20406

Verger A, Duterque-Coquillaud M (2002) When Ets transcription factors meet their partners. Bioessays 24:362–370

Wang SC, Hung MC (2000) Transcriptional targeting of the HER-2/neu oncogene. Drugs Today 36:835–843

Wasylyk B, Hagman J, Gutierrez-Hartmann A (1998) Ets transcription factors - nuclear effectors of the Ras-MAP-kinase signaling pathway [review]. Trends Biochem Sci 23:213–216

Welford SM, Hebert SP, Deneen B, Arvand A, Denny CT (2001) DNA binding domain-independent pathways are involved in EWS/FLI1-mediated oncogenesis. J Biol Chem 276:41977–41984

Wood LD, Irvin BJ, Nucifora G, Luce KS, Hiebert SW (2003) Small ubiquitin-like modifier conjugation regulates nuclear export of TEL, a putative tumor suppressor. Proc Natl Acad Sci USA 100:3257–3262

Xin JH, Cowie A, Lachance P, Hassell JA (1992) Molecular cloning and characterization of PEA3, a new member of the Ets oncogene family that is differentially expressed in mouse embryonic cells. Genes Dev 6:481–496

Xing X, Wang SC, Xia W, Zou Y, Shao R, Kwong KY, Yu Z, Zhang S, Miller S, Huang L, Hung MC (2000) The ets protein PEA3 suppresses HER-2/neu overexpression and inhibits tumorigenesis. Nat Med 6:189–195

Xuan Z, McCombie WR, Zhang MQ (2002) GFScan: a gene family search tool at genomic DNA level. Genome Res 12:1142–1149

Yagasaki F, Wakao D, Yokoyama Y, Uchida Y, Murohashi I, Kayano H, Taniwaki M, Matsuda A, Bessho M (2001) Fusion of ETV6 to fibroblast growth factor receptor 3 in peripheral T-cell lymphoma with a t(4;12)(p16;p13) chromosomal translocation. Cancer Res 61: 8371–8374

Yamada T, Abe M, Higashi T, Yamamoto H, Kihara-Negishi F, Sakurai T, Shirai T, Oikawa T (2001) Lineage switch induced by overexpression of Ets family transcription factor PU.1 in murine erythroleukemia cells. Blood 97:2300–2307

Yang B-S, Hauser CA, Henkel G, Colman MS, Van Beveren C, Stacey KJ, Hume DA, Maki RA, Ostrowski MC (1996) Ras-mediated phosphorylation of a conserved threonine residue enhances the transactivation activities of c-Ets1 and c-Ets2. Mol Cell Biol 16:538–547

Yordy JS, Muise-Helmericks RC (2000) Signal transduction and the Ets family of transcription factors. Oncogene 19:6503–6513

Zhang L, Eddy A, Teng YT, Fritzler M, Kluppel M, Melet F, Bernstein A (1995) An immunological renal disease in transgenic mice that overexpress Fli-1, a member of the ets family of transcription factor genes. Mol Cell Biol 15:6961–6970

Zhou J, Ng AY, Tymms MJ, Jermiin LS, Seth AK, Thomas RS, Kola I (1998) A novel transcription factor, ELF5, belongs to the ELF subfamily of ETS genes and maps to human chromosome 11p13-15, a region subject to LOH and rearrangement in human carcinoma cell lines. Oncogene 17:2719–2732

Function of the E2F Transcription Factor Family During Normal and Pathological Growth

L. Hauck · R. von Harsdorf (✉)

Medizinische Klinik mit Schwerpunkt Kardiologie, Universitätsklinikum Charité, Humboldt-Universität zu Berlin, Augustenburger Platz 1, 13353 Berlin, Germany
ruediger.harsdorf@charite.de

1	Introduction	278
1.1	The Mammalian Cell Cycle	278
1.2	Cell Cycle Regulation by the E2F/pRb Pathway	279
2	The E2F Transcription Factor Family	282
2.1	The Activating E2Fs: E2F1–E2F3a	282
2.1.1	E2F1 and Cellular Proliferation	282
2.1.2	E2F1 Gene Knockouts	284
2.1.3	E2F1 and Apoptosis	288
2.1.4	E2F2 and E2F3	289
2.1.4.1	E2F2	289
2.1.4.2	E2F3	290
2.2	The Repressing E2Fs: E2F3b, E2F4, and E2F5	292
2.2.1	E2F3b	292
2.2.2	E2F4 and E2F5	293
2.2.2.1	Cell Cycle Regulation of E2F4 and E2F5	293
2.2.2.2	Cell Cycle-Dependent Regulation of Subcellular Localization of E2F4 and E2F5	295
2.2.2.3	E2F4 Gene Knockout Mice	297
2.2.2.4	E2F5 Gene Knockout Mice	298
2.2.2.5	Miscellaneous E2F4 Knockout Models	299
2.3	E2F6 and E2F7	300
2.3.1	E2F6	300
2.3.2	E2F6 Knockout Mice	301
2.3.3	E2F7	301
3	Concluding Remarks	301
	References	302

Abstract The E2F family of transcription factors elicits opposing roles in transcriptional activation and repression during the regulation of proliferation, apoptosis, and DNA repair and thus contributes to tumor suppression, oncogenesis, and differentiation. A vast array of studies have confirmed that the intracellular pathway controlling the activity of E2F and its main regulator, the retinoblastoma tumor suppressor protein (pRb) is disrupted in virtually all human cancers. Based on their ability to promote either proliferation or cell cycle withdrawal and differentiation, the individual E2Fs can be subdivided into three different groups. In this review, we focus on the relative contribution of two distinct E2F subclasses to gene activation and repression. However, critical questions remain to be answered as to the specific role played by individual E2Fs in growth control and whether the current concepts of E2F/pRb action explain all of the functions of these proteins in vivo.

Keywords E2F · pRb · Cell cycle · Apoptosis · Gene regulation

1
Introduction

The E2F family of transcription factors exerts fascinating and contrasting functions in transcriptional repression and activation of genes regulating proliferation, apoptosis and differentiation. This review will provide a compact survey of how the functional interplay of E2Fs and pocket proteins fulfill apparently opposite effects and how the deregulation of these activities contributes to tumor suppression and oncogenesis. To understand the physiological functions of E2F activities, the focus will be on genetic studies in current cell culture models and most intensively on the use of E2F mutant mouse models.

1.1
The Mammalian Cell Cycle

The mammalian cell cycle is a highly coordinated complex process that is under tight regulation of a diverse panel of factors. Cyclin-dependent kinases (CDKs) elicit important functions during the coordinated progression through the cell cycle (Sherr and Roberts 1999). The activities of CDKs are regulated by an intricate system of protein–protein interaction and phosphorylation. The binding of positive CDK regulators, the cyclins, activates the phosphotransferase of CDKs whereas negative regulators, the CDK inhibitors (CDKIs), inhibit CDK-dependent kinase activity (Sherr and Roberts 1999). Examples of these CDKIs are members of the INK4 family (e.g., $p16^{INK4a}$) and KIP family (e.g., $p21^{CIP1}$, $p27^{KIP1}$; thereafter referred to as p16, p21, and p27). Whereas p16 specifically inhibits CDK4/6 complexes, p21 and p27 generally target all of these G_1 CDKs. The transition from G_0 to mid-G_1 is regulated exclusively by CDK4/6 and D-type cyclins. CDK2 constitutes holoenzyme complexes with cyclin E1 and controls–in combination with CDK3–the progression from mid to late G_1 into S phase. Later on during cell cycle progression, CDK2/cyclin A2 regulates S phase transition. Through association with cyclins and CDKIs, CDKs cooperatively govern the mitogen-dependent transition through the first gap phase (G_0/G_1) and initiation of DNA synthesis (S phase) during the mammalian cell division cycle. When DNA replication begins during the G_1/S phase transition the cell must ensure that (a) its entire genome is replicated; (b) that replication of every DNA section occurs only once; and (c) that replication of the genome occurs only once during one cell cycle.

1.2
Cell Cycle Regulation by the E2F/pRb Pathway

The E2F transcription factor family is involved in the regulation of growth-related processes like cell cycle re-entry, G_1/S phase transition, and apoptosis (Wang et al. 1998). E2Fs bind to dimerization partner (DP) proteins to form a heterodimer constituting transcriptionally active E2F. The transcriptional activity of E2F is blocked through binding to the product of the retinoblastoma tumor suppressor gene (pRb) and its related family members p107 and p130, collectively referred to as pocket proteins. The pRb family carries a conserved motif, the pocket, necessary for the interaction with oncoproteins from several different DNA tumor viruses (Helt and Galloway 2003). The pocket consists of two domains (A pocket, B pocket) which are conserved among pRb family members. The A and B pocket domains are separated by a nonconserved spacer and are more similar to each other in p107 and p130 than to the corresponding domains of pRb. The repressor motif is formed by the association of domains A and B which are sufficient for the growth-inhibitory activity of pRb (Fig. 1).

Historically, the E2F transcription factor was originally identified through its ability to activate the Adenovirus E2 gene promoter. The induction of E2 gene transcription requires two additional Adenovirus gene products, E1A and E4. The physical interaction of E2F with pRb was revealed by the capacity of E1A (and other proteins from small DNA viruses) to release E2F from pRb and its related proteins p107 and p130 (Gosh and Harter 2003; reviewed by Helt and Galloway 2003).

A vast array of experimental data support the view that the growth-restraining activity of pRb, as well as that of p107 and p130, is based on the control of E2F-dependent transcriptional activity (Classon et al. 2000). The growth-suppressive ability of pRb is inactivated by CDK4/6- and CDK2-mediated phosphorylation converting pRb from an active hypophosphorylated isoform into an inactivated hyperphosphorylated protein variant (Knudsen et al. 1998). This post-translational modification of pRb is brought about by sequential action of D-type cyclin/CDK4,6 during mid-G_1 phase, and cyclin E/CDK2 (G_1/S phase specific), and cyclin A/CDK2 (S phase specific). The central importance of cyclin D1/CDK4 in pocket protein phosphorylation is emphasized by the finding that this kinase activity is dispensable for growth control in G_1 in retinoblastoma gene-deficient cells (Lukas et al. 1995). In a simplified model, in mid-G_1 phosphorylated pRb releases bound E2F which is now transcriptionally active. The hyperphosphorylation of pRb from late G_1 until mitosis appears to be essential for the continuous transcriptional activation of E2F-regulated genes. At the end of mitosis, pRb is dephospho-

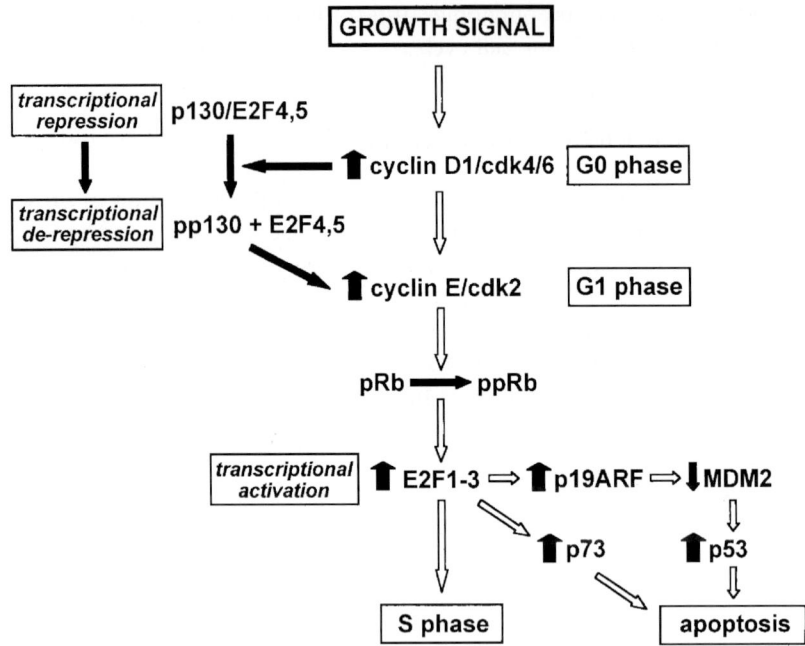

Fig. 1 Differential regulation of E2F family members during cell cycle entry and induction of S phase. In the absence of growth factors, hypophosphorylated, physiologically active retinoblastoma protein (*pRb*) and its relative p130 complex E2Fs thereby preventing E2F-dependent transcriptional activation. Stimulation of quiescent cells in G_0 results in the activation of CDK4/6 and CDK2, which phosphorylate and inactivate pRb and p130. Loss of E2F4,5/p130-mediated transrepression induces accumulation of E2F1–3, which in turn drives the expression of genes allowing either the initiation of S phase or apoptosis

rylated. Although it has often been suggested that pRb phosphorylation releases E2F from pRb during G_1 transition, no direct experimental proof has been delivered so far for this scenario. Additionally, phosphorylation of E2F1 may also contribute to the disruption of pRb/E2F complexes. Reportedly, phosphorylation of E2F1 at Ser332 and Ser337 prevents E2F1 from binding to pRb in vitro, but the physiological relevance of this phenomenon still remains to be determined (Fagan et al. 1994).

The time-separated sequential inactivation of pRb ensures that the growth inhibitory action of pRb is neutralized once a cell is induced by growth factors to re-enter the cell cycle (Knudsen et al. 1998). Interestingly, pRb does not require direct binding to (activator) E2F1 to suppress tumor cell growth, to activate transcription, and to promote differentiation in pRb-negative Saos-2 cells (Sellers et al. 1998). Most of the experimental evidence showing that E2F elicits key functions in G_1 progression is based on overexpression experiments. All of the activating E2Fs, E2F1–3, are capable of transforming cells. E2F1–3 share the capacity to efficiently induce S phase in resting G_0 cells (Lukas et al. 1996). The induction of wild-type (wt) E2F1 or mutant (mt) E2F1 (unable to bind pRb because of a point mutation located in the pocket-binding domain) led to S phase entry in quiescent fibroblasts which was significantly more pronounced in the cells expressing mt.E2F1 (Shan et al. 1996). In cultures infected by wt.E2F1 cells were rescued from E2F1-induced apoptosis by coexpression of wild-type pRb or a transdominant negative mutant species of p53. Collectively, the biological significance of the E2F/pRb interaction is based on direct genetic proof and strongly emphasizes that the functional interplay between E2Fs and pocket proteins is of critical importance for orderly regulation of cell cycle progression in G_1.

The E2F family of transcription factors comprises seven structurally related proteins: E2F1–7. E2F1–6 together with members of the DP family (DP1, DP2) bind to E2F DNA consensus binding sites. In general, this heterodimer is the functional transcription factor. Free E2F or DP bind only weakly to DNA in vitro using synthetic oligo deoxynucleotides containing an E2F recognition sequence whereas dimerization of E2F/DP leads to strong DNA binding and transcriptional activation (Helin et al. 1993). The two DP proteins bind to all E2Fs in vivo except E2F7 which lacks a DP interaction surface and does not require a DP subunit to interact with DNA. All of the different E2F/DP complexes recognize the same nucleotide sequence -TTTCCCGC- which, however, is not well conserved (Helin et al. 1993). Consequently, cellular functions that are specifically executed by a subgroup of E2Fs are probably not the result of the differential ability among the E2Fs to recognize particular DNA sequences within specific E2F-responsive promoters. Except for E2F6 and E2F7, both of which lack a trancritional transactivation domain, all of the various E2F/DP complexes are principally capable of activating gene transcription. The importance of the DP subunit of E2F complexes is underlined by the fact that expression of dominant-negative DP-1 lacking the E2F1-interaction surface blocks cell cycle progression in G_1 (Wu et al. 1996). Moreover, E2F/DP heterodimerization is required for the association of E2F with pRb and pRb-related proteins. The specificity of pocket protein binding is defined by the E2F subunit of E2F/DP heterodimers. The E2F family members display strong differences in their ability to bind to pRb, p107, and p130. The pRb protein interacts preferentially with E2F1–E2F3 whereas E2F4 and E2F5 specifically bind to p130 and E2F5 to p107 (Sardet et al. 1995). However, pRb can also associate with E2F4 and E2F5 (Moberg et al. 1996) and E2F4/pRb appears to be the major pRb-associated E2F in cells. Complexes consisting of p130 and E2F4 or E2F5 appear to function as transcriptional repressors in quiescent cells (Smith et al. 1996). E2F1–3, but not E2F4 and 5, are subjected to acetylation by P/CAF and p300/CBP resulting in increased activation of transcription. The acetylation of E2F can be reversed by pRb-associated deacetylases (Brehm et al. 1998; Rayman et al. 2002; Ross et al. 1999, 2001; Zhang et al. 2000).

Transcriptional studies in vitro have identified the existence of three types of E2F/pRb complexes. In 'activator' E2F complexes, the E2F transactivation domain drives transcription in the absence of pRb (Phillips et al. 1997). In 'inhibited' E2F/pRb complexes, the transactivation domain of E2F1 is blocked by bound pRb rendering E2F transcriptionally inactive (Ross et al. 1999, 2001). In 'repressor' E2F complexes, pRb is recruited to E2F-binding sites by E2F and inhibits promoter activation by E2F with or without further utilization of histone-modifying enzymes (Brehm et al. 1998; Magnaghi-Jaulin et al. 1998; Zhang et al. 2000).

Interestingly, pRb-mediated heterochromatin formation and silencing of E2F target genes is also found during cellular senescence (Narita et al. 2003). It is important to outline one crucial difference between quiescent cells and senescent cells. Whereas quiescent cells respond to serum stimulation with the activation of E2F-regulated promoters, senescent cells show a permanent insensitivity to mitogenic signaling. In this respect, senescent cells appear to resemble more closely the differentiated cell stage than do quiescent cells.

2
The E2F Transcription Factor Family

2.1
The Activating E2Fs: E2F1–E2F3a

2.1.1
E2F1 and Cellular Proliferation

Ectopic E2F1 (and E2F2 and 3) can block cells from entering G_0 and forces already quiescent cells into S phase (Lukas et al. 1996). This unusual feature of E2F1 to drive a G_0 cell all the way down through G_1 and into S phase is shared only by a very limited number of mammalian genes including the c-*myc* oncoprotein, and cyclin E (Leone et al. 2001; Lukas et al. 1997). Moreover, E2F1 protein accumulation bypasses a G_1 phase arrest resulting from the inhibition of G_1 CDKs by p16, p21, or p27 and by DNA damage or transforming growth factor (TGF)-β treatment (DeGregori et al. 1995b; Lukas et al. 1996; Zhang et al. 1999). Hence, E2F1 is located in a high position in the hierarchy that orchestrates the complex variety of events governing the G_1 phase (Wang et al. 1998). Deregulated E2F1 activity also leads to transformation of an established rat embryonic cell line and cooperates with activated *ras* in oncogenic transformation of primary rat embryo cells.

E2F1, E2F2, and E2F3a are not expressed in quiescent (G_0) and differentiated cells and only accumulate as cells are stimulated to enter G_1 phase. The expression of their heterodimerization partners DP1/2 do not fluctuate through the different stages of the cell cycle (Johnson et al. 1994; Leone et al. 1998; Sears et al. 1997). The cell cycle-dependent regulation of E2F1 gene expression is the consequence of E2F-mediated negative control of the E2F1 promoter in G_0 cells (Johnson et al. 1994). E2F4/p130 predominates in quiescent cells and most likely mediates specific repression of E2F1 gene transcription by occupying the two E2F sites in the E2F1 gene promoter. After growth factor stimulation, transderepression of the E2F1 promoter takes place through phosphorylation of p130 by CDK4/6 leading to dissociation of p130 from E2F4. This relief of repression by phosphorylation of p130 by G_1 CDKs allows the E2F1 gene promoter to become active. Positive transcriptional activity may be conferred through Sp1 and CCAAT binding sites upstream at the E2F sites in the E2F1 promoter, constituting an autoregulatory feedback loop model of E2F1 gene expression amplifying the expression of the E2F1 gene itself.

E2F-binding sites are found in the promoters of various genes. By microarray analyses of cDNAs derived from cell lines overexpressing E2F and complementary studies using chromatin immunoprecipitation (ChIP) assays with antibodies specific to E2F1/E2F4, the number of E2F-regulated genes was found to be relatively large and surprisingly diverse (Hurford et al. 1997; Ishida et al. 2001; Müller et al. 2001; Ren at al. 2002; Takahashi et al. 2000; Weinmann et al. 2002; Wells et al. 2000). E2F-dependent genes encode proteins necessary for DNA synthesis such as dihydrofolate reductase (DHFR), thymidine kinase (TK), proliferating cell nuclear antigen, DNA polymerase (pol) α, proteins involved in cell cycle control such as cyclin D1, cyclin E, cyclin A, CDK1, E2F1, E2F3, pRb, p107, HP1, epidermal growth factor receptor (erb-B) or proteins involved in proapoptotic signaling such as APAF1, caspase 3, caspase 7, bcl2, the latter of which is paradoxically an known inhibitor of apoptosis (DeGregori 1995a; Shan et al. 1994; Zhu et al. 1995) or are proto-oncogenes such as B-*myb*, c-*myb*, c-*myc*, N-myc. One question arising from these findings is whether the differences in the expression levels of these presumably E2F-dependent genes are directly or indirectly caused by E2Fs. Another caveat is that even when a promoter is E2F-responsive, E2F must not necessarily be rate-limiting for gene function. Moreover, artificially high E2F protein abundance due to overexpression does definitely not meet the fine-tuned stoichiometrical ratios of endogenous E2F/pRb-proteins. Therefore, it remains possible that a promoter is normally repressed by E2F/pRb complexes, but is activated when exposed to high levels of exogenous E2F.

The kinetics of transcriptional activation of the E2F1 gene during G_0/G_1 closely coincide temporally with the induction of E2F-regulated genes. Moreover, experimental data clearly established a link between the ability of E2F1 to induce S phase and the activation of many of the aforementioned genes required for cell cycle progression and DNA synthesis (Takahashi et al. 2000). That the E2F sites are indeed important for the correct temporal expression of many of the E2F-dependent genes has been demonstrated by mutation. However, for the majority of these E2F-regulated promoters, it is unknown which E2F/DP complex regulates expression or which pRb family member is actually responsible for silencing of E2F target genes. Clearly, the dependency on E2F for promoter activation needs to be determined for a greater number of E2F-responsive genes.

Accumulating experimental evidence supports the view, that cyclin D1 and CDK4/6-dependent kinase activity phosphorylate pRb and/or p107/p130 facilitating G_1/S phase transition (Lukas et al. 1995). In this model, the principal role of cyclin D-associated CDKs depends on functional pRb and entails the release of transcriptionally active E2F and other pRb-bound proteins from phosphorylated pRb (Ikeda et al. 1996). By use of the CDKI p16 and a phosphorylation-deficient constitutively active mt.pRb, Lukas et al. (1997) demonstrated, that ectopic expression of cyclin E induces S phase and completion of the cell division cycle in the absence of E2F-mediated transactivation, whereas cyclin D1 did not exhibit such an effect. The S phase-promoting activity of cyclin E occurs independently of pRb/E2F and can bypass the requirement for E2F transactivation capacity during G_1/S phase progression. Cyclin E also overrides a G_1 block induced by dominant-negative mt.DP1, the heterodimeric partner of E2Fs, indicating that cyclin E acts either downstream of or parallel to the pRb/E2F pathway during G_1/S transition.

2.1.2
E2F1 Gene Knockouts

At the time of writing functional gene disruption of all E2Fs except E2F7 have been reported. One important issue when considering the phenotypes caused by the disruption of gene function in mice is that the entire phenotype reflects only the outcome of the loss of nonredundant functions of the inactivated gene under consideration. Moreover, the observed phenotype reports only the relative contribution of the disrupted gene in a developmental context which might differ critically from its function in adult mice physiology. In addition, the understanding of how the gene of interest is physiologically integrated in the observed phenotype is often very difficult given the complexity of mouse embryonic development and differentiation.

Two groups have independently undertaken the inactivation of E2F1 in the mouse germline (Field et al. 1996, Yamasaki et al. 1996). To date, E2F1–5 have been identified and both groups expected some functional redundancy—thus, the expected outcomes from these knockout studies were clearly formulated. As E2F1 drives cell cycle progression during G_1, the absence of E2F1 should yield absent or underdeveloped tissues. Surprisingly, mice generated to lack E2F1 exhibit few if any of these outcomes and the observed consequences of E2F1 loss are directly opposite to those that were expected—homozygous loss of E2F1 evokes tumorigenesis.

Interbreeding of heterozygous mutant E2F1$^{+/-}$ mice yields homozygous mutant E2F1$^{-/-}$ pups with a normal Mendelian frequency, indicating that E2F1 has no crucial role in embryonic viability (Field et al. 1996, Yamasaki et al. 1996). E2F1$^{-/-}$ mice develop and reproduce normally. No obvious abnormalities are found in the liver, brain, gut, cardiac and skeletal muscles, bone, spleen, kidneys or pancreas at 4–6 weeks of age. According to these findings, E2F1 is not required for overall development, survival, or reproduction of mice. Examination of E2F1$^{-/-}$ primary embryonic fibroblasts does not reveal any differences in doubling time, cell cycle distribution or timing of re-entry into the cell cycle from G_0 when compared with their wild-type counterparts. Likewise, lymph node cells from 2.5-month-old E2F1$^{-/-}$ mice respond equally well to mitogenic signals. Therefore, under various circumstances, E2F1 is obviously not required for cellular proliferation or cell cycle progression. Instead, the disruption of E2F1 in the mouse results in the genesis of a diverse range of tumors in older adults.

Three abnormalities become obvious with increasing age in the E2F1$^{-/-}$ mice: testicular atrophy, exocrine gland dysplasia, and the development of tumors. Tumors observed include reproductive tract sarcomas, lung tumors, lymphomas, and a variety of other tumor types that are encountered at a lower frequency. Usually, the expected frequency of tumors in wild-type animals of various genetic backgrounds is 0%–2% by 12 months and rises to 10% by 18 months of age. In comparison, the total tumor incidence of autopsied E2F1$^{-/-}$ mice is 34% within 15 months. One-third of the tumors detected in E2F1$^{-/-}$ mice are reproductive tract sarcomas which is an extremely rare tumor in the 129xC57BL/6 hybrid genetic background used in that study. E2F1$^{-/-}$ cells of certain exocrine glands are also dysplastic. Their larger cell size and occasional binucleate appearance are definitely not indicative of a proliferative failure caused by the lack of E2F1.

E2F1 mutant mice also exhibit decreased T-cell apoptosis. The interaction of major histocompatibility complexes/peptide antigen-presenting cells with T-cell receptors induces proliferation and differentiation of T cells. The protein levels of E2F1 and E2F2 are very low in naive T cells but are induced in late G_1 following T-cell activation. In E2F1$^{-/-}$ mice, the basal proliferation of peripheral T cells and thymocytes from 2.5-month-old animals is not affected. In contrast, 6–12-month-old E2F1$^{-/-}$ mice display increased proliferation of thymocytes, but not mature lymphocytes. These data imply that cultured cortical thymocytes from the E2F1$^{-/-}$ mice are less susceptible to apoptosis in vitro than those from their wild-type counterparts and suggest an unknown role of E2F1 in negative control of proliferation.

Through its proapoptotic activity (see below), E2F1 suppresses cell proliferation, thereby functioning as a tumor suppressor in some cellular settings—yet E2F1 can act as oncogene in other settings. In this respect, E2F1 has commonly been thought of as an oncogene with growth-promoting properties, rather than acting as a tumor-suppressor gene. The tumor spectrum as it is seen in E2F1$^{-/-}$ mice at 12–18 months of age is not observed in wild-type animals. The late age of tumor onset in E2F1-null mice suggests that additional mutations might have occurred during tumorigenesis. Mice lacking E2F1 are viable, but certain tissues are sensitive to the absence of E2F1, predisposing homozygous E2F1$^{-/-}$ mutants to develop tumors. Yamasaki et al. (1996) concluded that in some tissues E2F1 promotes proliferation whereas E2F1 must act in other tissues to suppress proliferation. The work on E2F1-null mice has reclassified E2F1 as a tumor suppressor as well as an oncogene meaning that E2F1 owns the ability to induce tumors either by overexpression of the wild-type protein or by a loss of function mutation. Excess of E2F1 or loss of E2F1 could be tumorigenic by leading to inappropriate amplification of one or more target genes. On the other hand, if either E2F1-activated or E2F1-repressed genes are growth promoting genes, then gain or loss of E2F1 function is tumorigenic. Furthermore, E2F1 may act as a tumor-suppressor gene by forming part of an pRb/E2F growth-inhibitory complex. As a consequence, pRb cannot repress the expression of proliferation-promoting genes in the absence of E2F1. This view appears to agree with the recent finding that suppression of pRb-dependent G_1 checkpoint control is necessary for E2F-induced S phase entry (Lomazzi et al. 2002).

A more detailed understanding of E2F1/pRb-dependent regulation of cellular genes is helpful regarding the interpretation of physiological effects due to changes in the biological activities of E2F and pRb family members. Interestingly, in contrast to pRb, naturally occurring mutations in E2F genes have not been detected in human tumors. The product of the retinoblastoma susceptibility gene was the first tumor suppressor identified (Jacks et al. 1992). Loss of function of such a gene is, as the name implies, tumorigenic. It should be noted that the growth-suppression function (like p53) is consistent with pRb's role as a tumor suppressor. The death protection function, however, is seemingly at odds with pRb being a tumor suppressor which should induce cell death (like p53). Generally, protection against cell death is the function of oncogenes, as exemplified by bcl2. A central component of the Knudson model of haploinsufficient tumor suppressor genes is that loss of the gene is recessive, i.e., a single active allele of pRb is sufficient to suppress the development of tumors (Cook and McCaw 2000). As opposed to E2F$^{+/-}$ mice, those heterozygous for a pRb mutation are cancer prone (but not predisposed to retinoblastoma). Semiquantitative PCR analysis has demonstrated that both alleles of pRb at extremely early stages of those tumors are already inactivated, implying that loss of function of the second allele is rate-limiting for the onset of tumorigenesis (Cook and McCaw 2000). pRb$^{-/-}$ mice suffer from midgestational embryonic lethality, accompanied by impaired erythroid and neuronal differentiation and widespread cell cycle defects (Jacks et al. 1992). Remarkably, the cellular consequences of pRb mutation in the mouse embryo resemble the effects of ectopic overexpression of E2F1. In the central nervous system (CNS), increased E2F levels in its free forms as well as enhanced expression levels of E2F-regulated genes and ectopic S phase entry are present. Significantly, pRb$^{-/-}$ embryos show an increased rate of apoptosis in the CNS. Loss of pRb leads to excessive apoptosis in the developing mouse fetus, whereas apoptosis is far less pronounced in mice lacking one or both of p107 and p130. These data confirm the anticipation of important physiological consequences of pRb mutations through dysregulated E2F1 activity.

As stated above, the inactivation of pRb in mice results in unscheduled cellular proliferation, and apoptosis with widespread developmental defects resulting in early embryonic death at day 14.5 of gestation. However, as shown in a recent study, the loss of pRb function leads to deleterious defects in placental development (Wu et al. 2003). This is an interesting example of how cell autonomous functions contribute to phenotypic development in genetic model systems. One of the earliest differentiation events in mammalian development is the commitment of embryonic cells to either an extra-embryonic cell lineage forming the placenta or to the inner-cell mass that eventually makes up the embryo. In pRb-deficient placentas, excessive proliferation of trophoblast cells and severe disruption of the normal labyrinth architecture was accompanied by a decrease in vascularization and a marked reduction in placental transport function (Wu et al. 2003). Remarkably, the labyrinth is the site of oxygen and nutrient exchange between the mother and the fetus. pRb-null embryos when supplied with a wt.pRb placenta through tetraploid aggregation and conditional knockout strategies can be carried to term, but die soon after birth. Importantly, most of the neurological and erythroid abnormalities thought to be responsible for the embryonic lethality of pRb$^{-/-}$ mice. were virtually absent in these rescued pRb-null pups. These cell-autonomous roles of pRb during development must be considered when evaluating the numerous previous studies that were based on the use of pRb knockout animals and cells.

The analysis of pRb$^{+/-}$,E2F1$^{+/-}$ and pRb$^{+/-}$,E2F1$^{-/-}$ animals showed that loss of one E2F1 allele is sufficient to extend the lifespan of pRb$^{+/-}$ mice by roughly 3 months and to reduce the incidence of hyperplastic lesions (Tsai et al. 1998). In contrast, pRb$^{-/-}$,E2F1$^{-/-}$ mutant mice die at around embryonic day 17 with anemia and defective skeletal muscle and lung development, suggesting that E2F1 regulation is not the sole function of pRb in murine development. Reportedly, nullizygosity for Id2 also allows pRb$^{-/-}$ embryos to survive to birth, although the pups are stillborn presumably due to the absent of muscle fibers (Lasorella et al. 2000). In summary, mice embryos mutant for E2F1 and pRb show significant suppression of apoptosis and ectopic S phase entry in certain tissues as compared to Rb$^{-/-}$ mutants, confirming E2F1 as a critical mediator of these cellular effects.

2.1.3
E2F1 and Apoptosis

Considerable experimental evidence exists that E2F1 participates in the induction of apoptosis and that E2F1 directly couples the cell cycle and cell death machinery (Nahle et al. 2003). Deregulated E2F1 activity achieved either by overexpression or mutant mouse models can provoke apoptosis. E2F1-triggered cell death is mediated through both p53-dependent and p53-independent mechanisms (Pan et al. 1998). p53 can augment E2F-induced apoptosis as shown in several experimental systems. However, the growth-inhibitory effects of E2F1 do not have an absolute requirement for p53. For example, pRb protects p53-null cells from apoptosis in an E2F1-dependent manner, and deregulation of E2F1 by deletion of pRb leads to apoptosis in both p53-positive and p53-negative mouse embryos (Tsai et al. 1998).

One mechanism of p53-independent induction of apoptosis by E2F1 involves the p73 homolog of p53. The p73 protein shares a high degree of sequence homology and functional similarity with p53. E2F1 stimulates the expression of p73 which induces apoptosis and the proapoptotic capacity of E2F1 is significantly less pronounced in cells deficient for p73 (Stiewe and Pützer 2000). In addition, E2F1 specifically induces the expression of Apaf1 which activates the downstream executioner caspase 3, the final effector of cell death (Moroni et al. 2001). Moreover, E2F1 is able to inhibit anti-apoptotic signaling by nontranscriptional mechanisms. The best understood of these signaling events is the tumor necrosis factor receptor (TNFR)-associated survival response (Phillips et al. 1999). TNFR signals through nuclear factor κB (NF-κB) and JNK/SAPK which then inhibit the intracellular caspase cascade. E2F1 appears to prevent anti-apoptotic signaling from TNFR probably by inducing the degradation of the signaling complex at the receptor.

The proapoptotic activity of E2F1 can be dissociated from its cell cycle-promoting capacity. The studies of Hsieh et al. (1997) and Phillips et al. (1997) revealed, that induction of DNA synthesis and apoptosis are separable functions of E2F1. Coexpression of pRb with wt.E2F1 rescued cells from apoptosis whereas coexpression of pRb and mt.E2F1(1–374) lacking the pRb-interaction surface did not exhibit such an effect. mt.E2F1 deficient for DNA-binding failed to induce apoptosis in pRb/p53-functionally negative Saos cells. Importantly, the transactivation deficient mt.E2F1(1–374) retained the ability to trigger apoptosis. Therefore, the transcriptional transactivation capacity of E2F1 is not necessary for the induction of apoptosis by E2F1. Based on these results, E2F1-dependent cell death may be mediated through transderepression of E2F-regulated promoters of proapoptotic genes.

E2F1 triggers p53-dependent apoptosis through transcriptional activation of p19^{ARF} the murine homolog of human p14^{ARF} (Rowland et al. 2002). E2F1 induces the expression of ARF and ARF blocks MDM2 allowing p53 to accumulate finally leading to the induction of apoptosis (Bates et al. 1998). E2F1 can also induce p53-dependent apoptosis in the absence of ARF implying that death-inducing pathways operate in parallel with ARF. The p53 protein elicits key functions in the cellular decision of whether to undergo growth arrest or apoptosis. MDM2 cuts off the apoptotic function of p53 by targeting p53 for unbiquitin-mediated proteasome degradation. In turn, MDM2, which also physically interacts with E2F1, is negatively regulated by ARF (Martin et al. 1995). By this mechanism, ARF directly connects the pRb/E2F pathway to p53 providing a link between cell cycle progression and cell fate determination through activation of apoptosis.

A variety of studies delivered contradictory results about the identity of the E2Fs that induce apoptosis. Overexpression studies suggest that the induction of apoptosis is a specific property of E2F1 and not E2F2 or E2F3 (Moroni et al. 2001; Pan et al. 1998; Wang et al. 2000). Other investigations point out that all three of the activating E2Fs can activate proapoptotic pathways (Müller et al. 2001; Ziebold et al. 2001). Because the fundamental regulation of proapoptotic mechanisms have been specifically elaborated for E2F1, this chapter focuses on E2F1.

2.1.4
E2F2 and E2F3

2.1.4.1
E2F2

Similar to E2F1, the E2F2 gene is tightly regulated in a growth stage-dependent fashion during cell cycle entry from G_0 to S phase (Sears et al. 1997). The expression of E2F2 also involves an E2F-dependent negative control of gene transcription in quiescent cells. This transrepression is probably mediated by E2F4 and/or E2F5/p130 complex. In contrast to E2F1, the E2F2 promoter lacks Sp1 and CCAAT-binding sites. Instead, the transcriptional activation of the E2F2-gene promoter presumably depends on a series of E-box elements absent in the E2F1-promoter. Although other E-box-binding activities cannot be ruled out, E2F2-promoter studies suggest a direct role for c-Myc in the activation of the E2F2-gene whereas the E2F1-gene promoter is not c-Myc responsive (Leone et al. 2001).

E2F1, E2F2, and E2F3a can be activated by overexpressed c-myc in G_0-cells through CDK-independent mechanisms (Leone et al. 2001). In contrast to E2F1$^{-/-}$ mouse embryonic fibroblasts (MEF) where the S phase inducing capacity of c-Myc is not unimpaired, the ability of c-Myc is essentially abolished in E2F2$^{-/-}$ MEF (Leone 2001). Introduction of wt.E2F2 or E2F3 into E2F2$^{-/-}$ MEF restores c-Myc responsiveness leading to the induction

DNA synthesis in these cells. Thus, E2F2 and E2F3 are necessary for c-Myc to stimulate S phase entry in quiescent fibroblasts.

Mutant E2F2$^{-/-}$ mice frequently develop autoimmunity and tumors (Zhu et al. 2001). Double knockout E2F1$^{-/-}$,E2F2$^{-/-}$ mice show modest hyperproliferation of hematopoietic cells (Li et al. 2003). Especially, loss of E2F1 and E2F2 impedes B-cell differentiation and bone marrow-derived hematopoietic progenitor cells undergo DNA synthesis more rapidly. Significantly, gene disruption of E2F1 and E2F2 results in markedly decreased expression of E2F-regulated genes, including the cyclin A2 necessary for S phase progression. Based on these data, E2F1 and E2F2 elicit essential (and redundant) functions in the coordination of cell cycle progression and differentiation of hematopoietic cells.

2.1.4.2
E2F3

Viable mutant E2F3$^{-/-}$ mice arise at approximately 25% of the expected Mendelian frequency implying that E2F3 is essential for normal embryonic development (Humbert et al. 2000b). This high incidence of perinatal lethality is caused by an intrinsic defect in the E2F3$^{-/-}$ mice. E2F3 affects proper development of the adult animal in a dose-dependent manner since growth of E2F3$^{+/-}$ animals was found to be intermediate between that of wild-type and E2F3$^{-/-}$ littermates. In contrast to E2F1-deficient mice, E2F3$^{-/-}$ mice did not show a higher incidence of tumors suggesting that E2F3 is not functioning as a tumor suppressor in a manner similar to E2F1. However, the life span of E2F3 knockout animals was significantly shortened. E2F3$^{-/-}$ mice show increased mortality starting at 7 months of age vs. 23 months for the wt.E2F3 control. Importantly, more than 85% of the E2F3$^{-/-}$ deficient mice showed indications of congestive heart failure. Massive accumulation of thrombi were observed in the left ventricle and left atria of E2F3$^{-/-}$ hearts whereas fibric thrombi were not detected in any other organs indicating that clot formation was not due to a general hypercoagulable condition. The defects found in E2F3 knockout mice are completely distinct from those arising in E2F1 mutant animals (and E2F2, E2F4, E2F5 mice). Thus, genetic inactivation of E2F3 deteriorates heart function, which is astonishing when compared to the phenoptypes of other E2F knockout mice.

MEF prepared from mutant E2F3$^{-/-}$ mice proliferate with a greatly reduced rate as compared to wt.MEF (Humbert et al. 2000b). The respective E2F3 mutant mouse strain is deficient for both E2F3a and E2F3b. The average doubling time was approximately twice that of wild-type littermate controls. No difference in the occurrence of apoptosis was observed between wild-type and E2F3$^{-/-}$ MEF. Therefore, the proliferative disorder of E2F3$^{-/-}$ MEF is due to a defect in cell division and is not based on the induction of apoptosis. Thus, E2F3 elicits functions important in governing the proliferation rate of MEF. As evaluated by BrdU- and [^{3}H]thymidine incorporation, E2F3$^{-/-}$ MEF are significantly impaired in their ability to enter S phase compared with E2F1$^{-/-}$ which behave like wt.MEF (Humbert et al. 2000b). These findings suggest that E2F1 is fully dispensable for cell cycle control in G_1/S phase, whereas E2F3 is a rate-limiting factor for cell cycle progression in response to mitogenic stimulation of quiescent cells. On a molecular level, the loss of E2F3 has profound impact on the expression of E2F-responsive genes, e.g., *cyclin E1, cyclin A2*, B-myb, *cdc6, DHFR*, and *E2F1* (Leone et al. 1998). The cellular consequences of E2F3 loss was found to affect both the maximal transcript levels and the time point of peak expression of these genes (Humbert et al. 2000b). In MEF lacking E2F3, the cell cycle-dependent expression of E2F-regulated genes is severely impaired and this defect correlates with cell cycle and proliferation perturbations. The proliferative disorder of E2F3-null cells is rescued by ectopic overexpression of E2F3 (or E2F1) confirming that the decreased rate of proliferation in E2F3$^{-/-}$ MEF is actually caused by the absence of E2F3.

The disruption of E2F3 gene function (but not E2F1, E2F2, E2F4, and E2F5) in MEF also results in defects in nucleophosmin B association with centrosomes, premature centriole separation and duplication, finally leading to centrosome amplification, mitotic spindle defects, and aneuploidy (Saavedra et al. 2003). In sharp contrast to the other E2Fs, E2F3 appears to participate in the regulation of the faithful transmission of genetic material to daughter cells.

E2F3 is not essential for the immortalization of primary mouse cells or anchorage-independent growth of these cells which serves as marker for their tumorigenic potential in vivo (Humbert et al. 2000b). However, E2F3 deficiency compromises the capacity of transformed murine cells to proliferate. Furthermore, E2F3 might be responsible to a large extent for inappropriate proliferation resulting from the loss of pRb as concluded from the analysis of pRb$^{+/-}$,E2F3$^{-/-}$ double-deficient mice (Ziebold et al. 2001). In conclusion, E2F3 is critical for the transcriptional transactivation of growth-associated genes essential for the control of cell cycle progression of both primary and tumor cells. Through control of the expression of genes rate limiting for initiation of DNA synthesis, E2F3 imparts a stringent control of entry into S phase.

E2F1$^{-/-}$,E2F2$^{-/-}$ mutant mice are viable and develop to reach adulthood (Wu et al. 2001). In contrast, E2F1$^{-/-}$,E2F3$^{-/-}$ and E2F2$^{-/-}$,E2F3$^{-/-}$ knockout mice die early at or just before embryonic day 9.5 (Wu et al. 2001). The combined loss of all three activating E2Fs, E2F1, E2F2, and E2F3 severely weakened the expression of E2F target genes in response to serum stimulation including DHFR, cdc6, MCM3, TK, DNA-polα, and cyclin A but not cyclin E (Wu et al. 2001). Moreover, loss of E2F function in these MEF resulted in increased expression of p21 concomitant of decreased CDK-activity and impaired inactivating pRb phosphorylation.

Combined loss of these three E2Fs completely abolishes the ability of MEF to undergo DNA synthesis, progress through G_2/M phase and proliferate (Wu et al. 2001). These fundamental outcome provides a direct genetic proof that the activating E2Fs are required for cell cycle progression, proliferation and development.

The histological inspection of double mutant $E2F1^{-/-}$, $E2F3^{-/-}$ mice revealed, that loss of E2F1 shortened the life span of $E2F3^{-/-}$ mice dying between embryonic days 9.5 and 10.5 (Cloud et al. 2002). The developmental and age-related defects arising in the individual E2F1- and E2F3-deficient mice were made more severe by the mutation of both of the other E2Fs defining redundant functions in tissue development and maintenance. Given the combined incidence of heart failure in the double knockouts, normal heart function appears to depend on a critical level of activating E2Fs. There is one important difference in the properties of E2F1 and E2F3 in that E2F3 did not enhance tumor incidence either alone or in combination with E2F1 deficiency. This finding again points out that tumor suppression is a specific feature of E2F1 and not E2F3. However, recent findings argue against that view. Ziebold et al. (2003) analyzed $pRb^{-/-}$, $E2F3^{-/-}$ deficient mice. In these animals, loss of E2F3 completely suppresses the defect in proliferation and the p53-dependent apoptosis in the pRb mutant embryos. These data imply that E2F3 is a major contributor of apoptosis arising from pRb deficiency, and this raises an important issue: either E2F3 is equally capable of inducing ARF in vivo or E2F3 activates p53-dependent apoptosis through mechanisms unrelated to that utilized by E2F1. It also remains formally possible that E2F1 and E2F3 use the same pathway for the execution of apoptosis operating independently from ARF. These crucial questions have yet to be answered.

2.2
The Repressing E2Fs: E2F3b, E2F4, and E2F5

2.2.1
E2F3b

The *Drosophila* genome encodes two E2F genes, *dE2F1* and *dE2F2* (Frolov et al. 2001). Both dE2F heterodimerize with dDp and bind to promoters of E2F-responsive genes. dE2F1 is a strong activator of E2F-regulated promoters whereas dE2F2 represses the transcription of E2F-dependent target genes. Therefore, in *Drosophila*, one repressor E2F and one activator E2F is used to pattern gene transcription. We will now turn to the repressive E2Fs present in mammalian cells and discuss how the antagonistic E2F3b, E2F4, and E2F6 operate the transcriptional program.

Recently, a second from of E2F3, called E2F3b, has been identified (Adams et al. 2000). In E2F3b, the amino-terminal 121 amino acids of E2F3a are substituted by six novel amino acids (He et al. 2000). E2F3b retains the nuclear localization signal but lacks the cyclin A/cdk2-binding site. Therefore, E2F3b protein abundance is not–unlike E2F3a–controlled by post-translational modifications. Additionally, E2F3a and E2F3b are transcribed from distinct promoters at distinct stages of the cell cycle. The E2F3a promoter is activated in mitogen-stimulated quiescent cells with kinetics similar to those of E2F1 and E2F3. In contrast, E2F3b transcript levels remain constant throughout growth stimulation comparable to the expression pattern of E2F4 and E2F5. Taken together, the E2F3a promoter is silenced in quiescent cells, whereas the E2F3b promoter is active in G_0 cells and remains constant throughout the cell cycle. As a result, the E2F3b gene product is synthesized in resting cells whereas E2F3a is detectable only in growing cells. Coimmunoprecipitation analysis revealed that E2F3b predominantly associates with pRb in quiescent cells, whereas E2F4 and E2F5 are found to be complexed with p130 (Leone et al. 2000). This finding is important, as the promoter specific-selection in pocket protein-mediated repression most likely resides in the DNA-binding component of such complexes—E2F. Thus, the specificity of pocket protein binding by E2F3b may provide a mechanism by which specificity of pRb-mediated repression is accomplished on an individual promoter. However, these biological properties of E2F3b have yet to be defined more precisely.

2.2.2
E2F4 and E2F5

2.2.2.1
Cell Cycle Regulation of E2F4 and E2F5

The cDNAs encoding E2F4 and E2F5 were identified by their ability to interact with p130 in a yeast-based two-hybrid screen of a serum starved WI138 human fibroblast cell line (Sardet et al. 1995). In this assay, E2F4 and E2F5 failed to bind to pRb. The homology between E2F4 and E2F5 (80% similarity) is closer than to the other E2F family members (52% and 60% similarity, respectively). The major differences of E2F4 and E2F5 as compared to the activating E2Fs is that both lack the amino-terminal amino acids comprising the cyclin A/cdk2-interaction site and the nuclear localization signal sequence. In sharp contrast to the activating E2Fs, E2F4/5 are already expressed in G_0-cells (Sardet et al. 1995). Protein expression of E2F4/5 reaches maximal levels in mid-G_1 phase prior to accumulation of E2F1 to E2F3. E2F4 protein expression returns to G_0 phase levels toward the end of the cell cycle. Therefore, E2F4/5 may be important regulators of the G_0/G_1 transition.

In quiescent cells, E2F4/5 form complexes with p130 which is unique to this particular cell cycle stage and essentially absent in asynchronously growing cells (Moberg et al. 1996; Smith et al. 1996). Because E2F4 is the major E2F expressed in G_0 we will focus on this E2F family member. As discussed above, the predominant E2F4/130 complex in quiescent cells plays an important role in the active repression of E2F-regulated promoters leading to transcriptional silencing. This complex appears to be responsible for the low levels of E2F-dependent genes in G_0 cells. Hurford et al. (1997) showed that pRb and p107/p130 complexes are required for the proper expression of distinct subgroups of E2F-regulated promoters. The E2F4/p130 complex is suddenly lost during mid-to-late G_1 phase during mitogen stimulation coinciding with the relief of E2F-mediated transrepression (Moberg et al. 1996; Smith et al. 1996, 1998). During cell cycle re-entry the p130 protein becomes phosphorylated and is subsequently degraded in a proteosome-dependent manner as cells enter S phase (Smith et al. 1998). Thus, p130 protein expression is hardly detected in asynchronously dividing cells. Overexpression of the CDKI p16 prevents cell cycle progression from G_0/G_1 and degradation of p130 suggesting that loss of the E2F4/p130 is dependent on CDK4/6 kinase activity in G_0/G_1 (Bruce et al. 2000). Conversely, the p130 protein level increases dramatically as cells begin to enter quiescence (Smith et al. 1996). In contrast, the p107 protein is regulated in a fashion opposite to that of p130. In serum-starved cells, p107 is undetectable but accumulates as cells re-enter the cell cycle. As demonstrated for various types of cell line, the p130 protein reflects the G_0 phase and the accumulation of an E2F4/p130 transcriptional repressor complex clearly distinguishes a G_0 cell state from a G_1 cell state. This view is important, as all mammalian tissues consist basically of post-mitotic, nondividing, and differentiated cells confined to the G_0 phase. E2F1, E2F2, and E2F3 cannot execute the repression of transcription of E2F-regulated promoters in G_0 cells simply because of their absence at this particular stage of the cell cycle. The same is applicable to p107 which is found only in proliferating cells. It is important to note that in contrast to cells or tissues of human and rat origin in which p107 protein expression is not detectable, p107 is most likely expressed in G_0 cells in mice (Hauck and Harsdorf, unpublished results).

Smith et al. (1996) showed that the regulation of E2F sites in the E2F1 gene promoter are subjected to repression by E2F4/p130 in G_0 cells but play no apparent role in cycling cells. They concluded that E2F4/p130 mediates the negative control of certain (but not all) E2F-regulated gene promoters. In a simplified model, dedicated promoter repression in G_0 and early G_1 is brought about by E2F4/p130, but by late G_1 these proteins are replaced by E2F1 and E2F3. Remarkably and unexpectedly, no pRb binding (or E2F5 binding) to promoters (p107, E2F1, cdc25A, cdc6, B-myb, cyclin A, cdc2) was observed in G_0 cells as studied by ChIP assays (Hurford et al. 1997). At the G_1/S phase boundary, the activating E2Fs play a pivotal role in inducing gene transcription. During cell cycle progression in G_1, a rapid diminution of E2F4 binding to promoters occurs which cannot be fully explained by the initiation of degradation of p130 at this time. Obviously, the sudden disappearance of E2F4 DNA-binding activity is not caused by protein abundance as the E2F4 protein level is constant in G_1 and S phase even though no binding of E2F4 to any promoter analyzed was detected during this interval (Moberg et al. 1996). Whether E2F4 participates in the stimulation of other promoters remains to be investigated.

E2F4 accounts for the vast majority of endogenous E2F activity in quiescent cells. As analyzed by gel retardation assays, pRb-containing E2F complexes are not found until cells have traversed the G_1/S phase boundary (Moberg et al. 1996; Smith et al. 1996). Immunoprecipitation assays revealed that pRb-associated E2Fs are detected specifically in G_1/S cells. The formation of pRb/E2F complexes is highly consistent with the increase in pRb protein level that is observed as cells enter S phase. These data suggest that pRb, albeit present in G_0/G_1 cells, is not associated with E2F-DNA binding activity. In G_1/S phase cells, E2F2 contributes very little to the endogenous pool of E2F activity, whereas E2F1 and E2F3 are distinctly detectable (Moberg et al. 1996). Most of E2F3 is complexed with pRb throughout S and G_2 phase, whereas E2F1 is found predominantly in within a free E2F complex. Therefore, E2F1 and E2F3 make unequal contributions to the pRb-associated and free E2F activity in post-G_1/S phase cells. While cells are entering S phase E2F4, consisting of p130-released and newly synthesized E2F4, associates with pRb and p107 implying that E2F4 switches from p130 to pRb and p107 in response to cell cycle re-entry. However, the free mid-G_1 E2F4 probably does not activate E2F-responsive genes. Despite that, a considerable amount of free E2F4 exists in S phase cells, and total S phase levels of E2F4 greatly exceed those of E2F1 (E2F4 comprises more than 80% of total E2F activity). Remarkably, despite their huge differences in expression levels E2F1 and E2F4 contribute equally to the amount of free E2F DNA-binding activity that is present at this stage of the cell cycle (Moberg et al. 1996). In summary, these findings are in favor of an important function of E2F4 and p130 in maintenance of the quiescent and terminally differentiated state of cells. A thorough understanding of how promoter activity of the E2F4/5 and p130 genes is regulated will contribute important insights into the early events of cell cycle exit that are necessary for differentiation-associated processes.

2.2.2.2
Cell Cycle-Dependent Regulation of Subcellular Localization of E2F4 and E2F5

E2F1, E2F2, and E2F3 induce S phase in serum starved fibroblasts whereas this ability is grossly impaired in E2F4 and E2F5 (Müller et al. 1997; Wang et al. 2000). This result is in good agreement with the inability of ectopic E2F4 and E2F5 to transactivate a synthetic E2F-responsive promoter efficiently. Endogenously expressed E2F4 and E2F5 proteins are nuclear until mid-to-late G_1 phase and are translocated to the cytoplasm after p130 has been phosphorylated (Lindeman et al. 1997). In contrast, E2F1, E2F2, and E2F3 are constitutively in the nucleus (Verona et al. 1997). Therefore, the transcriptional activities of the two E2F subgroups depend on their nuclear localization (Müller et al. 1997). A chimeric E2F4 carrying the amino-terminal nuclear localizations signal (NLS) sequence of E2F1, and an E2F4 fused to a short NLS derived from simian virus 40 large T antigen (both designated as NLS.E2F4) accumulate in the nucleus of transiently transfected cells. NLS.E2F4 activates an E2F-regulated promoter and induced S phase in quiescent cells as effectively as E2F1. NLS.E2F4 also overcomes a G_1 block in these cells exposed by ectopic p16. These data imply that the major physiological differences between E2F1 and E2F4 can be explained simply by their different intracellular localization signals. Exciting reports of Magae et al. (1996) and de la Luna et al. (1996) provided a more detailed insight into the regulation of the intracellular localization of E2F4 and E2F5. DP2, p107, and p130 are able to shuttle cotransfected E2F4 and E2F5 into the nucleus. In contrast, DP1 and pRb do not shuttle E2F4/5 from the cytosol to the nuclear compartment. When overexpressed, ectopic E2F4/5 expression is restricted exclusively to the cytoplasm, whereas exogenous E2F1, E2F2, and E2F3 are predominantly nuclear. The heterodimeric binding partner DP2, like p107 and p130, contains an amino-terminal NLS that mediates the nuclear uptake of E2F4/5. DP1 lacks an NLS sequence, explaining its inability to confer nuclear uptake of E2F4/5. Instead, the NLS of E2F1 to E2F3 is responsible for the nuclear import of DP1. These differences help to explain why coexpression of E2F1, E2F2, and E2F3 with DP1 but not DP2 dramatically increased the nuclear accumulation of these proteins. These studies raise the possibility that nuclear entry of E2F4/5-containing complexes is an active process regulated specifically by association with DP2 and p130 during the cell cycle.

On a molecular level, E2F4 and E2F5 harbor two nuclear export signals and are actively transported from the nucleus by the nuclear export receptor CRM1 (Gaubatz et al. 2001). In contrast, E2F1, E2F2, E2F3, E2F6, and pRb and p107 are not excluded from the nucleus by CRM1. Thus, the nuclear export of E2F4/5–as opposed to their import mechanism–is not mediated by association with pocket proteins but is an intrinsic function of E2F4/5. Unfortunately, no experimental data are available regarding the effect of Leptomycin B on the intracellular localization of p130. Leptomycin B, a specific inhibitor of CRM1-mediated nuclear export, prevents the cytoplasmic translocation of E2F4/5 (Gaubatz et al. 2001). E2F4/5 shuttles from the nucleus to the cytoplasm during mid-to-late G_1. The cytoplasmic concentration of E2F4/5 is not caused by proteasome-dependent nuclear degradation because proteasome inhibitors such as lactacystin have no effect on the cytoplasmic sequestration of E2F4/5. Interestingly, coexpression of CRM1 and p16 impaired the ability of p16 to impose a G_1 arrest by roughly 50%. Notably, ectopic CRM1 was without any effect on a p21-induced growth arrest as this is being a pocket protein-independent process in contrast with p16-induced cell cycle block which is pocket protein dependent. These findings emphasize, that either E2F4 or E2F5 is required for p16-mediated cell cycle arrest in G_1 (Bruce at al. 2000). The p130-mediated G_1 arrest depends on its ability to physically interact with E2F4 thereby forming transcriptional repressor complexes. As E2F4 is cytoplasmically sequestered before the onset of S phase, and p130 is proteasome-dependent degraded at this time, E2F4/p130 may well mediate transcriptional repression in G_0/G_1. Conceivably, E2F1 and E2F3 constantly remaining in the nucleus may activate E2F-responsive genes during mid-to-late G_1.

E2F4 is held in the nucleus of G_0 cells through physical association with p130 and/or DP2. The stringent control of subcellular localization of E2F4 suggests that E2F4 may act as a scaffolding protein, inducing gene silencing either through recruitment of p130 blocking the transactivation capacity of E2F4 or by directing chromatin remodeling factors associated with p130 to E2F-regulated promoters. Activator E2Fs recruit histone acetyltransferase to E2F-regulated promoters enabling gene expression through changes in chromatin structure. In contrast, histone deacetylase (HDAC) activity is held responsible for the relatively underacetylated state of histones H3 and H4 of E2F-regulated promoters and transcriptional repression during G_0 (Rayman et al. 2002). p107 and p130, but not pRb, are strictly required for HDAC binding to promoters as observed by ChIP analysis using pRb family protein deficient cells. p130 together with E2F4 constitutes nuclear corepressor complexes containing mSin3B and HDAC1 capable of directly inhibiting E2F-dependent transcriptional activation as cells withdraw from the cell cycle. Therefore it is possible, that recruitment of activating E2Fs in G_1 phase displaces E2F4/p130-associated HDAC activity recruited to E2F-dependent promoters in G_0. According to this model, p107 and p130 deficiency might cause a sustained loss of nuclear E2F4 leading to transactivation of E2F-responsive promoters and cell cycle re-entry.

2.2.2.3
E2F4 Gene Knockout Mice

Mice deficient for E2F4 are born at the expected Mendelian frequency (Humbert et al. 2000a; Rempel et al. 2000). However, most of the E2F4$^{-/-}$ animals die within the first postnatal days with more than 85% failing to reach the weaning age. The E2F4$^{-/-}$ mice die of an increased susceptibility to opportunistic infections as the result of craniofacial defects, suggesting that E2F4 is necessary for normal development in mice. Mutant E2F4$^{-/-}$ mice display a variety of erythroid abnormalities suggesting that proper E2F4 function contributes to the control of erythrocyte development in a pRb-dependent manner. MEF derived from mutant E2F4$^{-/-}$ mice arrest in G_0 as efficiently as wild-type or E2F4$^{+/-}$ MEF under low serum conditions (Humbert et al. 2000a; Rempel et al. 2000). Upon mitogen stimulation, quiescent E2F4$^{-/-}$ MEF re-enter the cell cycle with kinetics similar to those of wild-type or E2F4$^{+/-}$ MEF. Thus, E2F4 appears to be fully dispensable for the regulation of growth arrest and proliferation in these cells. Presumably, other E2F activities, particularly E2F5, may compensate for the loss of E2F4. Interestingly, loss of E2F4 results in the loss of the majority of total cellular E2F activity implying that other E2F family members do not compensate this decline at either the level of expression or DNA-binding activity. As judged by the expression of E2F-responsive genes in E2F4$^{-/-}$ MEF, E2F4 is fully dispensable for the normal cell cycle-dependent regulation of these factors.

Overexpression studies led to the assumption that the repressive E2F pocket protein complexes exert essential functions in controlling the expression of E2F-responsive genes. pRb$^{+/-}$ Heterozygous mice die between 8.5 and 13.9 months of age. In contrast, the lifespan of an significant fraction of pRb$^{+/-}$,E2F4$^{-/-}$ animals is extended to 20–27 months (Lee et al. 2002). Loss of E2F4 also suppresses the incidence of pituitary and thyroid tumors, and the aberrant proliferation caused by inappropriate gene expression in pRb$^{+/-}$ mice. In mutant pRb$^{+/-}$ MEF, p107 and p130 appear to substitute for pRb in the regulation of the activating E2Fs with which they fail to associate in wt.MEF. The additional loss of E2F4 in these cells further enhances the formation of these novel E2F pocket protein complexes probably through increased levels of p107 and p130. These data provide a direct evidence for a crucial role of E2F4 in regulating pRb activity participating in tumorigenic processes in vivo.

2.2.2.4
E2F5 Gene Knockout Mice

Homozygous null E2F5 mice arise with a normal Mendelian frequency suggesting that E2F5 has no role in embryonic viability (Lindemann et al. 1998). Knockout E2F5 MEF display no cell cycle pertubations as compared with their wild-type and E2F5$^{+/-}$ counterparts and enter S phase with similar kinetics. Therefore, loss of E2F5 has no influence on the growth capacity of MEF. However, by 3–4 weeks of age, E2F5$^{-/-}$ mice begin to develop enlarged, domed craniae reminiscent of animals with congenital hydrocephalus. Later on, most of the E2F5$^{-/-}$ deficient animals die with a median survival rate of 6 weeks. Histological examination confirmed that these mice indeed develop nonobstructive hydrocephalus evoked by excessive cerebrospinal fluid production. Mice deficient for E2F5 are healthy except for the hydrocephalus. This finding implies that E2F5 may lack a function in tissues other than the choroid plexus or that the absence of E2F5 is compensated for by other E2F species.

2.2.2.5
Miscellaneous E2F4 Knockout Models

Mice homozygous mutant for p107 or p130 are viable, fertile, and display no obvious abnormalities (Lee et al. 1996; LeCouter et al. 1998). Triple knockout mouse (TKO) of pRb, p107, and p130 are completely resistant to G_1 arrest induced by serum starvation or contact inhibition and exhibit a shorter cell cycle as compared to wild-type, single, or double knockout control MEF (Dannenberg et al. 2000; Classon et al. 2000; Sage et al. 2000). TKO MEF fail to undergo senescence and indeed show some characteristics of transformed cells. These findings emphasize the particular importance of the pRb family in the control of the G_0/S phase transition.

Constitutively expressed E2F4 and E2F5 exhibit a different expression profile during mouse development. Whereas E2F5 expression is found to be restricted to terminally differentiated tissues, E2F4 is widely expressed. In light of their extensive structural similarities, E2F4 and E2F5 may well perform common physiological properties. The analysis of double knockout E2F4$^{-/-}$<E2F5$^{-/-}$ mice shows that these animals are nonviable postnatally and die during late mouse embryonic development (Gaubatz et al. 2000). However, viable siblings carrying at least one intact allele of E2F4 or E2F5 can be obtained. Therefore, at least one functional allele of E2F4 or E2F5 is necessary for postnatal viability corroborating that E2F4 and E2F5 perform overlapping functions during mouse development. Given that E2F4- and E2F5/p130 complexes play a pivotal role in the regulation of cell cycle transition from G_0/G_1, it is very surprising that E2F4$^{-/-}$,E2F5$^{-/-}$ MEF in comparison to their wild-type counterparts arrest normally in G_0 in response to mitogen starvation and re-enter the G_1/S phase with normal kinetics after serum restimulation. In E2F4$^{-/-}$,E2F5$^{-/-}$ mice the expression of the E2F-responsive genes E2F1 and E2F3 is not altered when compared to their wild-type counterparts. Therefore, lack of E2F4 and E2F5 is not compensated for by the deregulation of other E2F family members. MEF with E2F4$^{-/-}$,E2F5$^{-/-}$ genotype do not differ in their doubling times from E2F4$^{-/-}$ and E2F5$^{-/-}$ deficient MEF indicating that E2F4 and E2F5, at least in mouse fibroblasts, are not required for normal proliferation. Obviously, E2F4 and E2F5 are dispensable for entry into the quiescent state and for normal cell cycle progression induced by serum stimulation. However, E2F proteins are required for cell cycle arrest, as demonstrated by the finding that dominant-negative alleles of E2F block cell cycle inhibition exerted by p16 or TGF-β (Chen et al. 2002; Zhang et al. 1999). Growth suppression by p16 in G_1 phase requires functional retinoblastoma protein (Medema et al. 1995). Ectopic p16 fails to prevent DNA synthesis in E2F4$^{-/-}$,E2F5$^{-/-}$ MEF. As reintroduction of either E2F4 or E2F5 into these MEF is sufficient to restore sensitivity to p16, E2F4 and E2F5 display functional redundancy in this regard. Despite the fact that E2F4 and E2F5 are dispensable for proliferation, they are required for (pock-

et protein-dependent) p16-mediated G_1 arrest of cycling cells. This view was further corroborated by a recent study analyzing the requirement of pRb family proteins in p16-dependent cell cycle arrest (Bruce et al. 2000). Overexpression of p16 in MEF containing wt.pRb, p107, and p130 results in dramatic reduction of cells in S phase in the presence of mitogens. In contrast, p16 could not block cell cycle progression in G_1 in pRb-deficient MEF. Therefore, pRb$^{-/-}$ MEF are refractory to p16. In contrast, MEF devoid of p107 or p130 arrest in G_1 phase when p16 is overexpressed. Unexpectedly, in MEF containing wt.pRb but lacking p107 and p130 (pRb$^{+/+}$,p107$^{-/-}$,p130$^{-/-}$) p16 does not lead to inhibition of DNA synthesis. Obviously, pRb-deficient MEF and p107$^{-/-}$,p130$^{-/-}$ MEF possess the same insensitivity to p16-induced cell cycle arrest. These exciting data imply that p16-mediated growth arrest does not depend entirely on pRb, but requires the nonredundant functions of p107 and p130.

2.3
E2F6 and E2F7

2.3.1
E2F6

High levels of E2F6 transcripts are found in the testis and small intestine, reduced levels were detected in placenta, heart, and spleen, whereas E2F6 mRNA is absent in thymus, ovary, and brain (Gaubatz et al. 1998). E2F6 lacks the carboxy-terminal amino acids that mediate pocket protein binding and transactivation capacity but retains the residues that are crucial for DNA binding and heterodimerzation with DPs (Morkel et al. 1997). Overexpression of E2F6 inhibit S phase entry in an established mouse fibroblast cell line (Cartwright et al. 1998). Moreover, E2F6 can repress E2F-responsive genes. Therefore, E2F6 contributes to gene silencing in a manner independently of pRb protein family members. E2F6 promotes gene silencing during the G_0 phase of the cell cycle by recruiting histone methyltransferases to promoters (Ogawa et al. 2002). In addition, polycomb group-related proteins are known to interact with E2F6 which has been implicated in the maintenance of inactive chromatin. Similar to E2F, the transcription factor Myc forms heterodimers with Max. Myc/Max dimers activate transcription whereas Mad/Max dimers block transcription. Another Myc family member, Mga binds to Max and blocks Myc activity. E2F6 complexes contain Max/Mga heterodimers rendering E2F6 capable of binding (in addition to E2F sites) to c-*myc* sites in the DNA of target gene promoter. The T-box binding site of Mga also dictates the binding of the E2F6 complex to unrelated T-box binding sites. These different heterologous protein–protein interactions direct E2F6 to cell cycle regulated genes during G_0. Interestingly, E2F6 appears to be replaced by E2F1 and E2F4 as cells move from G_0 into G_1. These findings are indicative, that E2F6 plays an important role in the regulation of gene activity in quiescent cells.

2.3.2
E2F6 Knockout Mice

E2F6 mutant mice are healthy implying that E2F6 is dispensable for mouse embryonic development and postnatal viability (Storre et al. 2002). E2F6$^{-/-}$ MEF proliferate normally and display an unperturbed entry and exit from the cell cycle. Therefore, E2F6 is not required for growth arrest following serum starvation and timely G_0/S phase transition in mouse fibroblasts. However, E2F6 gene deficient mice display overt homeotic transformations of the axial skeleton similar to the skeletal transformations previously observed in polycomb mutant animals. Based on the analysis of E2F6 knockout mice, E2F7 may participate in the specific recruitment of polycomb proteins to distinct target genes during mouse embryonic development.

2.3.3
E2F7

Most recently, a novel E2F family member was identified and termed E2F7 (De Bruin et al. 2003). E2F7 contains two distinct DNA-binding domains. However, E2F6 lacks the dimerization domain as well as a transcriptional activation and a pocket protein-binding domain. The expression of E2F7 is growth regulated and E2F7 inhibits E2F-dependent activation of a subset of E2F target genes. Therefore, E2F7 exhibits properties reminiscent of transcriptional repressors that impair cellular proliferation. Future studies are warranted to elucidate how E2F7 is integrated in the cellular network governing cell cycle progression. In particular, mutant mouse models will provide an insight into how E2F7 contributes to the regulation of development and differentiation.

3
Concluding Remarks

In contrast to pRb, mutations in the E2F1 gene and in genes of the other members of the E2F family have never been detected. Germline mutations in the pRb gene are associated with childhood retinoblastomas and a predisposition to osteosarcomas. Somatic mutations in the pRb gene contribute to the development of retinoblastomas, osteosarcomas, lung carcinomas, renal carcinomas, and bladder carcinomas. The functional role of p107 and p130 in proliferative disorders remains unclear as no human tumors are known to carry mutations in these genes. These observations raise the following interesting question: why is pRb and not E2F(1) targeted for mutations in human diseases caused by proliferative disorders?

Pocket proteins associate with E2Fs in a defined manner strictly depending on the cell cycle stage. This observation gives rise to the following interesting conclusion. In contrast to p130 and p107, pRb is a tumor suppressor mutated in 30% of all human tumors. This is somewhat unexpected because the three pRb family members share many characteristics. All three pocket proteins are involved in the regulation of E2Fs and all are targeted by the transforming proteins of small DNA tumor viruses. The unique property of pRb in comparison to p130 is that pRb/E2F complexes target those E2F-responsive genes that confer a growth advantage (Cobrinik et al. 1996). Thus, pRb may be the tumor suppressor (and not p130) through its ability to control E2F4 in addition to E2F1/3 and this feature is not shared by p107 and p130. According to this model, loss of pRb is sufficient to release the vast majority of endogenous E2F activity necessary for unrestrained cell cycling.

Another plausible answer may be found in the analysis of the pRb growth control pathway. The CDKI *p16* gene is regularly found to be targeted by inactivating mutations in human melanomas. The *cyclin D1* gene is either amplified and/or accumulates due to deregulated expression in a wide spectrum of human proliferative disorders. In addition, CDK4/6 mutations have been detected in human cancer cells rendering these CDKs unable to bind p16. All of these alterations finally lead to steady inactivating phosphorylation of pRb (and p130) and loss of control of E2F1 activity. Therefore, alterations in the E2F1 gene appear to be simply unnecessary for the development of tumorigenesis. Once the p16/pRb pathway acquires a mutation resulting in deactivation of pRb, deregulated E2F might lead to pathological situations. Unrestrained E2F1 activity is suspected to contribute to defects e.g., in T-cell development, Down syndrome, or postischemic brain infarct. Therefore, molecular strategies targeting E2F function may have therapeutic implications in the near future.

References

Adams M, Sears R, Nuckolls F, Leone G, Nevins JR (2000) Complex transcriptional regulatory mechanism control expression of the E2F3 locus. Mol Cell Biol 20:3633–3639

Bates S, Phillips AC, Clark PA, Stott F, Peters G, Ludwig RL, Vousden KH (1998) p14ARF links the tumour suppressors RB and p53. Nature 395:124–125

Brehm A, Miska EA, McCance DJ, Reid JL, Bannister AJ, Kouzarides T (1998) Retinoblastoma protein recruits histone deacetylase to repress transcription. Nature 391:597–601

Bruce JL, Hurford RK, Classon M, Koh J, Dyson N (2000) Requirements for cell cycle arrest by p16INK4 Mol Cell 6:737–742

Cartwright P, Müller H, Wagener C, Holm K, Helin K (1998) E2F-6: a novel member of the E2F family is an inhibitor of E2F-dependent transcription. Oncogene 17:611–623

Chen CR, Kang Y, Siegel PM, Massague J (2002) E2F4/5 and p107 as Smad cofactors linking the TGF-receptor to c-myc repression. Cell 110:19–32

Classon M, Salama S, Gorka C, Mulloy R, Braun P, Harlow E (2000) Combinatorial roles for pRB, p107, and p130 in E2F-mediated cell cycle control. Proc Natl Acad Sci USA 97:10820–10825

Cloud JE, Rogers C, Reza TL, Ziebold U, Stone JR, Picard MH, Caron AM, Bronson RT, Lees JA (2002) Mutant mouse models reveal the relative roles of E2F1 and E2F3 in vivo. Mol Cell Biol 22:2663–2672

Cobrinik D, Lee MH, Hannon G, Mulligan G, Bronson RT, Dyson N, Harlow E, Beach D, Weinberg RA, Jacks T (1996) Shared role of the pRb-related p130 and p107 proteins in limb development. Genes Dev 10:1633–1644

Cook WD, McCaw BJ (2000) Accommodating haploinsufficient tumour suppressor genes in Knudsons model. Oncogene 19:3434–3438

Dannenberg JH, van Rossum A, Schuijff L, te Riele H (2000) Ablation of the retinoblastoma gene family deregulates G1 control causing immortilization and increased cell turnover under growth-restricting conditions. Genes Dev 14:3051–3064

De Bruin A, Maiti B, Jakoi L, Timmers C, Leone G (2003) Identification and characterization of E2F7, a novel mammalian E2F family member capable of blocking cellular proliferation. J Biol Chem 278:42041–42049

DeGregori J, Kowalik T, Nevins JR (1995a) Cellular targets for activation by the E2F1 transcripion factor include DNA synthesis- and G1/S-regulatory genes. Mol Cell Biol 15:4215–4224

DeGregori J, Leone G, Ohtani K, Miron A, Nevins JR (1995b) E2F-1 accumulation bypasses a G1 arrest resulting from the inhibition of G1 cyclin-dependent kinase activity. Genes Dev 9:2873–2887

De la Luna S, Burden MJ, Lee CW, Thangue NB (1996) Nuclear accumulation of the E2F heterodimer regulated by subunit composition and alternative splicing of a nuclear localization signal. J Cell Sci 109:2443–2452

Fagan R, Flint KJ, Jones N (1994) Phosphorylation of E2F-1 modulates its interaction with the retinoblastoma gene product and the adenoviral E4 19 kDa protein. Cell 78:799–811

Field SJ, Tsai FY, Kuo F, Zubiaga AM, Kaelin WG, Livingston DM, Orkin SH, Greenberg ME (1996) E2F-1 functions in mice to promote apoptosis and suppress proliferation. Cell 85:549–561

Frolov MV, Huen DS, Stevaux O, Dessislava D, Balsczarek-Strang K, Elsdon M, Dyson NJ (2001) Functional antagonism between E2F family members. Genes Dev 15:2146–2160

Gaubatz S, Lees JA, Lindeman GJ, Livingston DM (2001) E2F4 is exported from the nucleus in a CRM1-dependent manner. Mol Cell Biol 21:1384–1392

Gaubatz S, Lindemann GJ, Ishida S, Jakoi L, Nevins JR, Livingston DM, Rempel RE (2000) E2F4 and E2F5 play an essential role in pocket protein-mediated G1 control. Mol Cell 6:729–735

Gaubatz S, Wood JG, Livingston DM (1998) Unusual proliferation arrest and transcriptional control properties of a newly discovered E2F family member, E2F-6. PROC NATL ACAD SCI USA 95:9190–9195

Gosh MK, Harter ML (2003) A viral mechanism for remodeling chromatin structure in G0 cells. Mol Cell 12:255–260

He Y, Armanious MK, Thomas MJ, Cress WD (2000) Identificaton of E2F3B, an alternative from of E2F-3 lacking a conserved N-terminal region. Oncogene 19:3422–3433

Helin K, Wu CL, Fattaey AR, Lees JA, Dynlacht BD, Ngwu C, Harlow E (1993) Heterodimerization of the transcription factors E2F-1 and DP-1 leads to cooperative trans-activation. Genes Dev 7:1850–1861

Helt AM, Galloway DA (2003) Mechanisms by which DNA tumor virus oncoproteins target the Rb family of pocket proteins. Carcinogenesis 24:159–169

Hsieh JK, Fredersdorf S, Kouzarides T, Martin K, Lu X (1997) E2F1-induced apoptosis requires DNA binding but not transactivation and is inhibited by the retinoblastoma protein through direct interaction. Genes Dev 11:1840–1852

Humbert PO, Rogers C, Ganiatsas S, Landsberg RL, Trimarchi JM, Dandapani S, Brugnara C, Erdman S, Schrenzel M, Bronson R, Lees JA (2000a) E2F4 is essential for normal erythrocyte maturation and neonatal viability. Mol Cell 6:281–291

Humbert PO, Verona R, Trimarchi JM, Rogers C, Dandapani S, Lees JA (2000b) E2f3 is critical for normal cellular proliferation. Genes Dev 14:690–703

Hurford RK, Cobrinik D, Lee MH, Dyson N (1997) pRb and p107/p130 are required for the regulated expression of different sets of E2F responsive genes. Genes Dev 11:1447–1463

Ikeda MA, Lakoi L, Nevins JR (1996) A unique role for the Rb protein in controlling E2F accumulation during cell growth and differentiation. Proc Natl Acad Sci USA 93:3215–3220

Ishida S, Huang E, Zuzan H, Spang R, Leone G, West M, Nevins JR (2001) Role for E2F in control of both DNA replication and mitotic functions as revealed from DNA microarray analysis. Mol Cell Biol 21:4684–4699

Jacks T, Fazeli A, Schmitt EM, Bronson RT, Goodell MA, Weinberg RA (1992) Effects of an Rb mutation in the mouse. Nature 359:295–300

Johnson DG, Ohtani K, Nevins JR (1994) Autoregulatory control of E2F1 expression in response to positive and negative regulators of cell cycle progression. Genes Dev 8:1514–1525

Knudsen ES, Buckmaster C, Chen TT, Feramisco JR, Wang JYJ (1998) Inhibition of DNA synthesis by RB: effects on G1/S transition and S-phase progression. Genes Dev 12:2278–2292

Lasorella A, Noseda M, Beyna M, Iavarone A (2000) Id2 is a retinoblastoma protein target that mediates signalling by myc. Nature 407:592–598

LeCouter JE, Kablar B, Whyte PFM, Ying C, Rudnicki M (1998) Strain-dependent embryonic lethality in mice lacking the retinoblastoma-related p130 gene. Development 125:4669–4679

Lee EY, Cam H, Ziebold U, Rayman JB, Lees JA (2002) E2F4 loss suppresses tumorigenesis in Rb mutant mice. Cancer Cell 2:463–472

Lee MH, Williams BO, Mulligan G, Mukai S, Bronson RT, Dyson N, Harlow E, Jacks T (1996) Targeted disruption of p107: functional overlap between p107 and pRb. Genes Dev 10:1621–1632

Leone G, Sears R, Huang E, Rempel R, Nuckolls F, Park CH, Giangrande P, Wu L, Saavedra HI, Field SJ, Thompson MA, Yang H, Fujiwara Y, Greenberg ME, Orkin S, Smith C, Nevins JR (2001) Myc requires distinct E2F activities to induce S phase and apoptosis. Mol Cell 8:105–113

Leone G, Nuckolls F, Ishida S, Adams M, Sears R, Jakoi L, Miron A, Nevins JR (2000) Identification of a novel E2F3 product suggests a mechanism for determining specificity of repression by Rb proteins. Mol Cell Biol 20:3626–3632

Leone G, DeGregori J, Yan Z, Jakoi L, Ishida S, Williams RS, Nevins JR (1998) E2F3 activity is regulated during the cell cycle and is required for the induction of S phase. Genes Dev 12:2120–2130

Li FX, Zhu JW, Hogan CJ, DeGregori J (2003) Defective gene expression; S phase progression, and Maturation during hematopoiesis in E2F1/E2F2 mutant mice. Mol Cell Biol 23:3607–3622.

Lindemann GJ, Dagnino L, Gaubatz S, Xu Y, Bronson RT, Warren HB, Livingston DM (1998) A specific, nonproliferative role for E2F-5 in choroid plexus function revealed by gene targeting. Genes Dev 12:1092–1098

Lindeman GJ, Gaubatz S, Livingston DM, Ginsberg D (1997) The subcellular localization of E2F-4 is cell-cycle dependent. Proc Natl Acad Sci USA 94:5095–5100

Lomazzi M, Moronie MC, Jensen MR, Frittoli E, Helin K (2002) Suppression of the p53- or pRb-mediated G1 checkpoint is required for E2F-induced S-phase entry. Nat Genet 31:190–194

Lukas J, Herzinger T, Hansen K, Moroni MC, Resnitzky D, Helin K, Reed SI, Bartek J (1997) Cyclin E-induced S phase without activation of th pRb/E2F pathway. Genes Dev 11:1479–1492

Lukas J, Peterson BO, Holm K, Bartek J, Helin K (1996) Deregulated expression of E2F family members induces S-phase entry and overcomes p16INK4-mediated growth suppression. Mol Cell Biol 16:1047–1057

Lukas J, Bartkova J, Rohde M, Strauss M, Bartek J (1995) Cyclin D1 is dispensable for G1 control in retinoblastoma gene-deficient cells independently of cdk4 activity. Mol Cell Biol 15:2600–2611

Magae J, Wu CL, Illenye S, Harlow E, Heintz NH (1996) Nuclear localization of DP and E2F transcription factors by heterodimeric partners and retinoblastoma protein family members. J Cell Sci 109:1717–1726

Magnaghi-Jaulin L, Groisman R, Naguibneva I, Robin P, Lorain S, Le Villain JP, Troalen F, Trouche D, Harel-Bellan A (1998) Retinoblastoma protein represses transcription by recruiting a histone deacetylase. Nature 391:601–604

Martin K, Trouche D, Hagemeier C, Sorenson TS, La Thangue NB, Kouzarides T (1995) Stimulation of E2F1/DP1 transcriptional activity by MDM 2 oncoprotein. Nature 375:691–694

Medema RH, Herrera RE, Lam F, Weinberg RA (1995) Growth suppression by p16INK4 requires functional retinoblastoma protein. Proc Natl Acad Sci USA 92:6289–6293

Moberg K, Starz MA, Lees JA (1996) E2F-4 switches from p130 to p107 and pRb in response to cell cycle reentry. Mol Cell Biol 16:1436–1449

Morkel M, Wenkel J, Bannister AJ, Kouzarides T, Hagemeier C (1997) An E2F-like repressor of transcription. Nature 390:567–568

Moroni MC, Hickman ES, Denchi EL, Caprara G, Colli E, Cecconi F, Müller H, Helin K (2001) Apaf-1 is a transcriptional target for E2F and p53. Nat Cell Biol 3:552–558

Müller H, Bracken AP, Vernell R, Moroni MC, Christians F, Grassilli E, Prosperini E, Vigo E, Oliner JD, Helin K (2001) E2Fs regulate the expression of genes involved in differentiation, development, proliferation, and apoptosis. Genes Dev 15:267–285

Müller H, Moroni MC, Vigo E, Peterson BO, Bartek J, Helin K (1997) Induction of S-phase entry by E2F transcription factors depends on their nuclear localization. Mol Cell Biol 17:5508–5520

Nahle Z, Polakoff J, Davuluri RV, McCurrach ME, Jacobson MD, Narita M, Zhang MQ, Lazebnik Y, Bar Sagi D, Lowe SW (2002) Direct coupling of the cell cycle and cell death machinery by E2F. Nat Cell Biol 4:859–864

Narita M, Nun S, Heard E, Narita M, Lin AW, Hearn SA, Spector DL, Hannon GJ, Lowe SW (2003) Rb-mediated heterochromatin formation and silencing of E2F target genes during cellular senescence. Cell 113:703–716

Ogawa H, Ishiguro KI, Gaubatz S, Livingston DM, Nakatani Y (2002) A complex with chromatin modifiers that occupies E2F- and myc-responsive genes in G0 cell. Science 296:1132–1136

Pan H, Yin C, Dyson NJ, Harlow E, Yamasaki L, Van Dyke T (1998) Key roles for E2F1 in signaling p53-dependent apoptosis and in cell division within developing tumors. Mol Cell 2:283–292

Phillips CP, Ernst MK, Bates S, Rice NR, Vousden KH (1999) E2F-1 potentiates cell death by blocking antiapoptotic signalling pathways. Mol Cell 4:771–781

Phillips AC, Bates S, Ryan KM, Helin K, Vousden KH (1997) Induction of DNA synthesis and apoptosis are separable functions of E2F-1. Genes Dev 11:1853–1863

Rayman JB, Takahashi Y, Indjeian VB, Dannenberg JH, Catchpole S, Watson RJ, te Riele H, Dynlacht, BD (2002) E2F mediates cell cycle-dependent transcriptional repression in vivo by recruitment of an HDAC1/mSin3B corepressor complex. Genes Dev 16:933–947

Rempel RE, Saenz-Robles MT, Storms R, Morham S, Ishida S, Engel A, Jakoi L, Melhem MF, Pipas JM, Smith C, Nevins JR (2000) Loss of E2F4 activity leads to abnormal development of multiple cellular lineages. Mol Cell 6:293–306

Ren B, Cam H, Takahashi Y, Volkert T, Terragni J, Young RA, Dynlacht BD (2002) E2F integrates cell cycle progression with DNA repair, replication, and G2/M checkpoints. Genes Dev 16:245–256

Ross JF, Näär A, Cam H, Gregory R, Dynlacht BD (2001) Active repression and E2F inhibition by pRB are biochemically distinguishable. Genes Dev 15:392–397

Ross JF, Liu X, Dynlacht BD (1999) Mechanism of transcriptional repression of E2F by the retinoblastoma tumor suppressor protein. Mol Cell 3:195–205

Rowland BD, Denissov SG, Douma S, Stunnenberg HG, Bernards R, Peeper DS (2002) E2F transcriptional repressor complexes are critical downstream targets of p19ARF/p53-induced proliferative arrest. Cancer Cell 2:55–65

Saavedra HI, Maiti B, Timmers C, Altura R, Tokuyama Y, Fukasawa K, Leone G (2003) Inactivation of E2F3 results in centrosome amplification. Cancer Cell 3:333–346

Sage J, Mulligan GJ, Attardi LD, Miller A, Chen SQ, Williams B, Theodorou E, Jacks T (2000) Targeted disruption of the three Rb-related genes leads to loss of G1 control and immortalization. Genes Dev 14:3037–3050

Sardet C, Vidal M, Cobrinik D, Geng Y, Onufryk C, Chien A, Weinberg RA (1995) E2F-4 and E2F-5, two members of the E2F family, are expressed in the early phases of the cell cycle. Proc Natl Acad Sci USA 92:2403–2407

Sears, R, Ohtani K, Nevins JR (1997) Identification of positively and negatively acting elements regulating expression of the E2F2 gene in response to cell growth signals. Mol Cell Biol 17:5227–5235

Sellers WR, Novitch BG, Miyake S, Heith A, Otterson GA, Kaye FJ, Lassar AB, Kaelin WG (1998) Stable binding to E2F is not required for the retinoblastoma protein to activate transcription, promote differentiation, and suppress tumor cell growth. Genes Dev 12:95–106

Shan B, Durfee T, Lee WH (1996) Disruption of RB/E2F-1 interaction by single point mutations in E2F-1 enhances S-phase entry and apoptosis. Proc Natl Acad Sci USA 93:679–684

Shan B, Chang CY, Jones D, Lee WH (1994) The transcription factors E2F-1 mediates the autoregulation of RB gene expression. Mol Cell Biol 14:299–309

Sherr CJ, Roberts JM (1999) CDK inhibitors: positive and negative regulators of G1-phase progression. Genes Dev 13:1501–1512

Smith EJ, Leone G, Nevins JR (1998) Distinct mechanisms control the accumulation of the Rb-related p107 and p130 proteins during cell growth. Cell Growth Diff 9:297–303

Smith EJ, Leone G, DeGregori J, Jakoi L, Nevins JR (1996) The accumulation of an E2F-p130 transcriptional repressor distinguishes a G0 cell state from a G1 cell state. Mol Cell Biol 16:6965–6976

Stiewe T, Pützer BM (2000) Role of the p53-homologue p73 in E2F1-induced apoptosis. Nat Genet 26:464–469

Storre J, Elsässer HP, Fuchs M, Ullmann D, Livingston DM, Gaubatz S (2002) Homeotic transformations of the axial skeleton that accompany a targeted deletion of E2f6. EMBO J 3:695–700

Takahashi Y, Rayman JB, Dynlacht BD (2000) Analysis of promoter binding by the E2F and pRb families in vivo: distinct E2F proteins mediate activation and repression. Trimarchi JM, Lees JA (2001) Sibling rivalry in the E2F family. Nat Rev Mol Cell Biol 3:11–20

Tsai KY, Hu Y, Macleod KF, Crowley D, Yamasaki L, Jacks T (1998) Mutation of E2F-1 suppresses apoptosis and inappropriate S phase entry and extends survival of Rb-deficient mouse embryos. Mol Cell 2:293–304

Verona R, Moberg K, Estes S, Starz M, Vernon JP, Lees JA (1997) E2F activity is regulated by cell cycle-dependent changs in subcellular localization. Mol Cell Biol 17:7268–7282

Wang D, Russel JL, Johnson (2000) E2F4 and E2F1 have similar proliferative properties but different apoptotic and oncogenic properties in vivo. Mol Cell Biol 20:3417–3424

Wang ZM, Yang H, Livingston D (1998) Endogenous E2F-1 promotes timely G0 exit of resting mouse embryo fibroblasts. Proc Natl Acad Sci USA 95:15583–15586

Weinmann AS, Yan PS, Oberley MJ, Huang THM, Farnham PJ (2002) Isolating human transcription factor targets by coupling chromatin immunoprecipitation and CpG island microarray analysis. Genes Dev 16:235–244

Wells J, Boyd KE, Fry C, Bartley SM, Farnham PJ (2000) Target gene specificity of E2F and pocket protein family members in living cells. Mol Cell Biol 20:5797–5807

Wu CH, Classon M, Dyson N, Harlow E (1996) Exprssion of dominant-negative mutant DP-1 blocks cell cycle progression in G1. Mol Cell Biol 16:3698–3706

Wu L, de Bruin A, Saavedra HI, Starovic M, Trimboli A, Yang Y, Ostrowski MC, Rosol TJ, Woollett LA, Weinstein M, Cross JC, Robinson ML, Leone G (2003) Extra-embryonic function of Rb is essential for embryonic development and viability. Nature 421:942–947

Wu L, Timmers C, Maiti B, Saavedra HI, Sang L, Chong GT, Nuckolls F, Giangrande P, Wright FA, Field SJ, Greenberg ME, Orkin S, Nevins JR, Robinson ML, Leone G (2001) The E2F1–3 transcription factors are essential for cellular proliferation. Nature 414:457–462

Yamasaki L, Jacks T, Bronson R, Goillot E, Harlow E, Dyson NJ (1996) Tumor induction and tissue atrophy in mice laking E2F-1. Cell 85:537–548

Zhang HS, Gavin M, Dahiya A, Postigo AA, Ma D, Luo RX, Harbour JW, Dean DC (2000) Exit from G1 and S phase of the cell cycle is regulated by repressor complexes containing HDAC-RB-hSWI/SNF and Rb-hSWI/SNF. Cell 101:79–89

Zhang HS, Postigo AA, Dean DC (1999) Active transcriptional repression by the Rb-E2F complex mediates G1 arrest triggered by p16INK4a, TGFβ, and contact inhibition. Cell 97:53–61

Zhu JW, Field SJ, Gore L, Thompson M, Yang H, Fujiwara Y, Cardiff RD, Greenberg M, Orkin SH, DeGregori J (2001) E2F1 and E2F2 determine thresholds for antigen-induced T-cell proliferation and suppress tumorigenesis. Mol Cell Biol 21:8547–8564

Zhu L, Zhu L, Xie E, Chang LS (1995) Differential roles of two tandem E2F sites in repression of the p107 promoter by retinoblastoma and p107 proteins. Mol Cell Biol 15:3552–3562

Ziebold U, Lee EY, Bronson RT, Lees JA (2003) E2F3 loss has opposing effects on different pRb-deficient tumors, resulting in suppression of pituitary tumors but metastasis of medullary thyroid carcinomas. Mol Cell Biol 23:6542–6552

Ziebold U, Reza T, Caron A, Lees JA (2001) E2F3 contributes both to the inappropriate proliferation and to the apoptosis arising in Rb mutant embryos. Genes Dev 15:386–391

c-Myc in Cellular Transformation and Cancer

J.-H. Sheen · R. B. Dickson (✉)

Department of Oncology and Lombardi Comprehensive Cancer Center,
Georgetown University, Washington, DC 20007, USA
dicksonr@georgetown.edu

1	Human Cancer and the *c-myc* Proto-oncogene	309
2	Structure and Functions of the c-Myc Oncoprotein	310
3	c-Myc, Genomic Instability, and Tumorigenesis	315
4	c-Myc Both Induces DNA Damage and Attenuates the DNA Damage-Induced Cell Cycle Checkpoints	317
5	c-Myc and Anchorage-Independent Growth	320
6	Summary and Future Directions	320
References		321

Abstract The c-myc gene plays a central role in a wide diversity of cellular processes, including cell cycle, cell growth (including protein synthesis), DNA dynamics, and apoptosis. As one of the most common oncogenes in human cancer, *c-myc* is commonly gene amplified in solid tumors and deregulated by chromosomal translocation in hematologic malignancies. A significant body of work has already defined pathologic effects of deregulated *c-myc* on cell proliferation, apoptosis, adhesion, and immortality in cancer. However, recent studies have begun to shed light on two, key, additional processes in cancer: cell transformation (anchorage independent growth), and genomic instability (including loss of cell cycle DNA damage-dependent checkpoint control). This review first covers structure–function relationships of *c-myc*; it then focuses on the emerging aspects of transformation and genomic instability in the pathophysiologic function of this very important oncogene.

Keywords c-Myc · Oncogene · Transformation · Genetic instability · Cell cycle checkpoint · Cancer

1
Human Cancer and the *c-myc* Proto-oncogene

Since its discovery as the cellular homolog to a transforming gene (v-*myc*) of avian *my*elo*c*ytomatosis virus in 1982 (Vennstrom et al. 1982), the c-*myc* proto-oncogene has emerged as one of the prototypical oncogenic switches in many human cancers. Deregulated expression of *c-myc* has been shown in a wide variety of human tumors, including carcinomas of the breast, colon, and cervix, glioblastomas, myeloid leukemias, osteosarcomas, and small cell lung carcinomas (Spencer and Groudine 1991). Overexpression of c-*myc* is probably the primary mode of activation for the proto-oncogene; genetic alterations include gene amplification, gene rearrangement, translocation, or proviral insertion. For breast cancer, c-*myc* has been particularly important as a major oncogene, because its overexpression has been demonstrated in 50% to 100% of cases; the major genetic mechanism at work appears to be gene amplification (Deming et al. 2000; Liao and Dickson 2000).

2
Structure and Functions of the c-Myc Oncoprotein

The c-*myc* gene, located at chromosome 8q24 in human, consists of three exons: a long first exon (exon I) and two exons containing the protein coding sequences (exons II and III). Transcription of the proto-oncogene is initiated primarily from two major promoters, P1 and P2, in exon I, producing transcripts of 2.4 and 2.2 kb, respectively. The P2-initiated transcripts comprise the majority of c-*myc* mRNA. The protein product of human c-*myc* is a highly conserved nuclear phosphoprotein, and the P1- and P2-initiated transcripts both have the capacity to encode the two major species of the oncoprotein: a 67-kDa c-Myc1 (453 amino acids) and a 64-kDa c-Myc-2 (439 amino acids) (reviewed in Dang 1999; Dang et al. 1999; Facchini and Penn 1998; Grandori et al. 2000) (Fig. 1). The former c-Myc species contains an additional 14 amino acids at its amino terminus. Although the relative abundance of c-Myc1 vs. c-Myc2 varies among tissues and cell lines, c-Myc2 is known to be a major isoform in most cases. A third isoform of c-Myc, named c-MycS, has been recently identified. This short version of c-Myc originates from an internal translation initiation site, 300 base pairs downstream of the amino terminus, and appears to have a molecular weight of 46 kDa. Unless otherwise indicated, c-Myc refers to the abundant, 64-kDa c-Myc2, and c-MycS refers to the 46-kDa protein.

The protein structure of c-Myc is highly conserved among different species. Homologs to human c-Myc have been isolated from mammals, birds, frogs, and the fruit fly. However, no homologous genes or proteins have yet been identified in yeast or nematodes. The oncoprotein consists largely of an amino-terminal transregulatory domain (TRD) and a carboxyl-terminal heterodimerization domain (HD) (Fig. 1). The TRD (the first 143 amino acids) was initially defined through a domain study, using yeast GAL4 reporter constructs (Kato et al. 1990). When the region is linked to a GAL4 DNA-binding domain, the fusion protein acts as a transcription factor. In its substructure, the TRD has two amino acid sequences that are highly conserved among the different Myc family proteins throughout evolution: an upstream

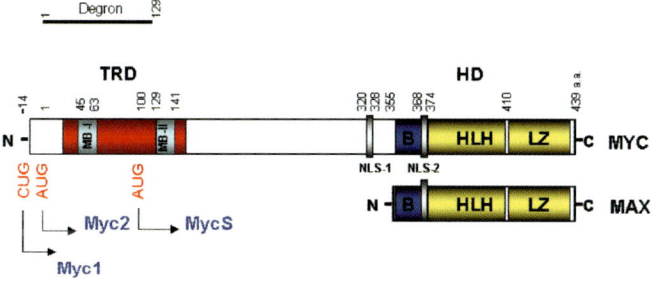

Fig. 1 Protein structure of human c-Myc. The c-Myc oncoprotein consists of 439 amino acid residues. The overall structure of the protein is divided into two parts: the amino-terminal transcriptional regulation domain (*TRD*) and the carboxy-terminal heterodimerization (*HD*) domain. A flexible middle region connects these two functional domains. Three isoforms with different lengths (*Myc1*, *Myc2*, and *MycS*) have been identified which have different translation initiation sites. TRD contains two highly conserved motifs, MB-I and MB-II, that are critical for transcriptional regulation. The TRD also contains a degron sequence that provides a recognition site for the proteosome-mediated degradation of the oncoprotein. The carboxy-terminal HD contains a basic (*B*), a helix-loop-helix (*HLH*), and a leucine-zipper (*LZ*) region. HLH and LZ are required to form a stable heterodimer transcription factor complex with a partner protein, Max. The B region recognizes and binds to specific sequences on the DNA. Two nuclear localization signal (*NLS*) sequences facilitate nuclear trafficking of this nuclear protein

MB-I (Myc-Box I; amino acids 45–63) and a downstream MB-II (Myc-Box II; amino acids 129–141) (Atchley and Fitch 1995) (Fig. 1). Deletion of either MB-I or MB-II dramatically diminished transactivation activity by 10- or 50-fold, respectively. Both MB-I and MB-II domains are, therefore, required for transactivation of c-Myc targets. Interestingly, the internal translation-initiated c-MycS isoform does not have MB-I, but retains MB-II, because its AUG translation initiation codon is placed after MB-I and before MB-II (Spotts et al. 1997). c-MycS did not demonstrate transactivation activity but retained transrepression activity in a study using an artificial reporter construct (Xiao et al. 1998). This might be consistent with previous domain studies, suggesting that MB-I is dispensable for the c-Myc-mediated transrepression. As a nuclear oncoprotein, the protein has two nuclear localization signal (NLS) sequence motifs: a primary motif (amino acids 320–328) and a secondary motif (amino acids 364–374). The second NLS overlaps with a basic region (B) (amino acids 355–368) that is implicated in recognition of specific DNA sequences. Immediately downstream of the basic region are contiguous, helix-loop-helix (HLH) (amino acids 368–410) and leucine zipper (LZ) (amino acids 411–439) motifs, responsible for a specific protein–protein interaction. This B–HLH–LZ region is essential for c-Myc to form a heterodimer with its partner protein Max (*Myc-associated factor-X*). Het-

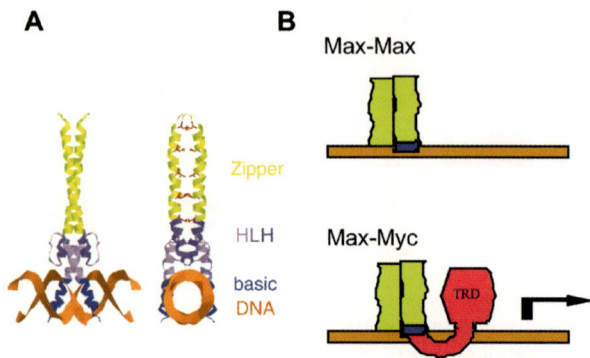

Fig. 2A, B Structural model for DNA binding of the c-Myc/Max transcription factor. **A** Molecular modeling for a Max/Max homodimer bound to DNA. The basic region recognizes a specific sequence on the DNA double helix. The helix-loop-helix and the leucine zipper mediate the formation of stable homodimer structure. **B** Hypothetical structure for a Myc/Max heterodimer transcription factor. Formation of heterodimer between Max and the carboxy terminus of Myc allows its stable binding to DNA double helix and a globular structure of TRD regulates transcription of target genes

erodimerization is essential for a functional Myc/Max transcription factor complex that is responsible for most of the c-Myc-mediated phenotypes.

The crystal structure of c-Myc has not yet been reported. Structural modeling however, predicts overall, a three-dimensional protein structure of c-Myc that is organized into three domains: a globular amino-terminal half, from amino acid 1 to 203, a short, unstructured middle region, from amino acid 204 to 237, and an α-helical, carboxyl-terminal half, from amino acid 238 to 439. Moreover, the crystal structure of the Max homodimer provides insights as to how B–HLH–LZ-mediated heterodimerization contributes to recognition of the E-box target sequence (CACGTG) in promoter regions of c-Myc targets (Fig. 2). In these Max studies, two B–HLH–LZ motifs from each partner associate to generate a 'four-helix bundle fold,' stabilizing the heterodimer as it binds DNA. Binding to a specific DNA sequence is mediated primarily by the basic region. Each basic region makes four DNA base contacts, within the E-box sequence, as well as numerous phosphate backbone contacts.

As 'deregulated expression' appears to be the primary mode of oncogenic activation for the c-*myc* proto-oncogene in human carcinoma, and because its expression correlates closely with the proliferative potential of the cells, it is important to understand how the level of the c-Myc oncoprotein is regulated. In quiescent cells, c-Myc is virtually undetectable. Upon mitogen or serum stimulation, its mRNA and protein levels are rapidly increased, as cells enter the G_1 phase. The proto-oncogene is therefore well established as one of the 'immediate early genes,' controlled by mitogenic signaling cas-

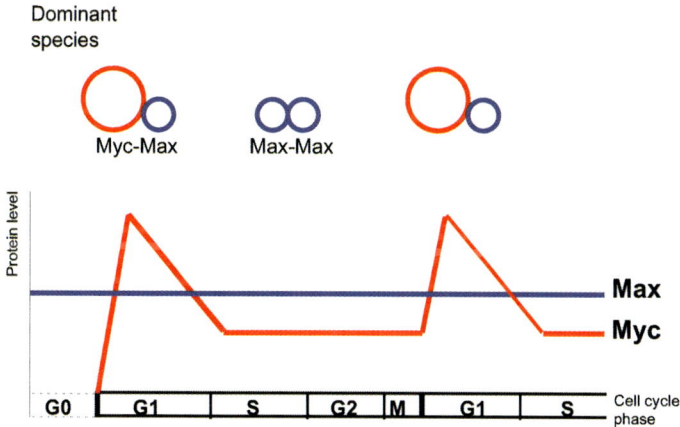

Fig. 3 Cell cycle phase-specific regulation of c-Myc expression. The level of c-Myc is tightly coupled with cell cycle progression. The protein is almost absent in noncycling cells. However, growth factor stimulation for the re-initiation of cell cycle dramatically increases the level of c-Myc. As the cell cycle progresses, c-Myc gradually decreases and maintains a low level for the rest of the cell cycle in proliferating cells. As the protein levels of its partner Max is constant throughout the cell cycle, dominance of the Myc/Max heterodimer is regulated by the level of c-Myc

cades; it is 'tightly' regulated throughout the cell cycle, with a scheduled peak at the G_1/S boundary (Blackwood et al. 1991) (Fig. 3). Following a gradual decline after this peak, low but steady levels of the protein are maintained in proliferating cells. Various overlapping mechanisms cooperate to regulate c-Myc expression, presumably to maintain the responsiveness of cells to various intra- and extracellular signals. The homeostatic control of mRNA transcription of c-*myc* is intrinsically maintained by a negative feedback, autorepression mechanism. c-Myc autorepresses its mRNA transcription when protein levels over a certain threshold occur. c-Myc expression is also regulated post-transcriptionally at the levels of protein translation and mRNA stability. Finally, the mRNA transcripts and protein products are both subject to rapid degradation. Specially, c-Myc family proteins have extremely short half-lives, in the order of 20–30 min (Hann and Eisenman 1984). Recent studies report that the stability of c-Myc is actively regulated by the ubiquitin-mediated proteosome pathway, similar to the majority of short-lived transcription factors (Ciechanover et al. 1991; Gross-Mesilaty et al. 1998; Salghetti et al. 1999). Proteosome-mediated protein degradation depends on ubiquitination of the target protein, which is a highly specific, multi-step process. Ubiquitination starts when an ubiquitin-conjugating enzyme, either alone, or in conjunction with an ubiquitin-protein ligase, recognizes 'degron'—a destruction element within the substrate protein. After the degron has been recognized and bound, ubiquitin is then transferred to

Table 1 Representative c-Myc targets

Myc phenotype	Regulation	Transregulation of target genes
Cell cycle	↑ Up	Cdc2
	↑ Up	Cdk4
	↑ Up	Cdc25A
	↑ Up	Cyclin A
	↑ Up	Cyclin E
	↑ Up	E2F-2
	↑ Up/↓ Down	Cyclin D1
	↓ Down	c-Myc
	↓ Down	p21/Cip1
Apoptosis and immortality	↑ Up	p53
	↑ Up	p14/ARF
	↑ Up	Telomerase/hTERT
Cell growth	↑ Up	Carbamoyl phosphate synthase
	↑ Up	Dihydrofolate reductase
	↑ Up	Ornithine decarboxylase
	↑ Up	eIF4E
	↑ Up	α-Prothymosin
	↑ Up	RCC1
	↑ Up	Lactate dehydrogenase-A
Cell adhesion	↓ Down	Collagen
	↓ Down	Thrombospondin
Genomic instability		?

a lysine residue within the target protein substrate. Next, repeated rounds of ubiquitination produce a highly ubiquitinated target protein that is destined to be destroyed rapidly by the proteosome. Interestingly, a functional c-Myc degron overlaps closely with the transcription regulation domain of c-Myc, and was mapped within the first 128 amino acids of the protein (Salghetti et al. 1999) (Fig. 1). This suggests that the prolonged protein stability of c-MycS, compared to full-length c-Myc, is due to the absence of degron in the c-MycS. Furthermore, some tumor-related mutations in c-*myc* result in significant stabilization of the protein product of the gene (Gavine et al. 1999; Salghetti et al. 1999). Although the molecular consequences of abnormal stabilization of c-Myc should be further investigated, the inappropriate increase of protein stability would be expected to contribute to an overall deregulation of the c-Myc level, and is likely to activate uncontrolled c-Myc activity. The deregulated c-Myc activity, in turn, will induce c-Myc-mediated phenotypes.

As a powerful oncogenic transcription factor, c-Myc/Max has been thought to regulate transcription of various target genes to exert its well-established functions: cell cycle regulation, induction of apoptosis, cell growth and metabolism, anti-differentiation, and cell adhesion (Table 1). Among these phenotypes, alteration of the cell cycle is one of the more dramatic results of deregulated c-Myc expression. This phenotype has been examined primarily with a focus on early G_1 to the G_1/S boundary, because c-*myc* has been established as an 'immediate early gene' and because its expression is rapidly stimulated by mitogenic signals. Overexpression of exogenous c-Myc in fibroblasts shortens G_1 and facilitates S phase entry. Also, expression of c-Myc was sufficient to drive quiescent cells into the cell cycle (Eilers et al. 1989; Karn et al. 1989). These observations were explained, at least in part, by enhanced cyclin-dependent kinase (CDK) activity and inappropriate phosphorylation of the tumor suppressor retinoblastoma protein (Rb) by c-Myc. However, target(s) of c-Myc responsible for these phenotypes are still controversial. Although a direct involvement of components of the cell cycle machine (e.g., cyclin D1, cyclin D2, CDK4, cyclin E, cyclin A, CDK2, Cdc25A, p21/Cip1, and p27/Kip1) has been suggested, none of them solely can explain the dramatic effects of c-Myc on the cell cycle. A recent study using c-*myc* (–/–) rat fibroblasts, obtained through somatic knockout techniques, has provided new insights into the role of c-Myc in the cell cycle progression (Mateyak et al. 1997, 1999). Loss of c-Myc led to the significant lengthening of 'both' the G_1 and G_2 phases of the cell cycle, and this could be explained by reduced activities of 'all' CDK complexes in exponentially cycling, c-*myc* null cells. This study also confirmed that c-Myc could downregulate p27/Kip1, a CKI (*C*yclin-dependent *K*inase *I*nhibitor), and implicated the involvement of Cdk7, a component of the CAK (*C*DK *A*ctivating *K*inase). Inappropriately increased CDK activity results in deregulation of downstream cell cycle events, such as Rb phosphorylation and an inappropriate release of the E2F transcription factor, a major inducer for S phase entry. In addition to those well-represented effects of c-Myc, recent studies have suggested that c-Myc can induce genetic instability. Various genomic alterations have been observed in c-Myc-induced mouse mammary tumor models. In vitro expression of c-Myc also induced genomic instability, such as gene amplification, chromosomal aberration, and aneuploidy. Because genomic instability has been proposed as a driving force for tumorigenesis, the c-Myc associated genomic instability may provide the genetic basis for the well-known oncogenicity of c-Myc.

3
c-Myc, Genomic Instability, and Tumorigenesis

Cancer is a genetic disease, arising from the accumulation of somatic mutations (Vogelstein and Kinzler 1993). Although several cancer syndromes with strongly inherited predispositions to cancer exist, the sporadic nature of most cancers could be explained by the above notion. This model also explains why solid tumors are more frequently detected in the older people, who have experienced more time, allowing for the step-wise accumulation of somatic mutations. Therefore, it is generally accepted that tumors initiate as a result of mutations in a single gene in a single progenitor cell, and then additional somatic mutations would subsequently accumulate to allow the malignant tumor phenotype (Nowell 1976). Selection of somatic mutations might be based on the degree to which the mutations are useful for the tumorigenic processes, such as unregulated cellular proliferation, nonresponsiveness to extracellular, antiproliferative factors, anchorage independent growth, metastases, and drug resistance. However, although this clonal origin of cancer is well accepted, the mechanism(s) driving the natural history of cancer have been the subject of continuing debate. Therefore, study of the underlying mechanism(s) through which tumorigenic mutations are created and selected should be a central and immediate goal for understanding and prevention of cancer.

Unavoidable, normal rates of mutations, coupled with waves of clonal expansion, may allow the tumorigenic process in humans (Tomlinson and Bodmer 1999; Tomlinson et al. 1996). Alternatively, 'genomic instability' or the 'mutator phenotype' has been also proposed to explain a driving force for the accumulation of tumorigenic mutations (Loeb 1991; Tlsty et al. 1995). One of hallmarks of cancer is the unstable genome, and it is well established that all tumor cells contain genetic alterations, ranging from DNA sequence changes to gross chromosomal changes (chromosome loss, gain, and translocation). Since the mutation rate in the normal cell is extremely low (estimated 1 in 10^9 nucleotides per cell division), genomic instability may explain observed tumor frequencies through its capacity to produce diverse genetic alterations. With this genetic heterogeneity, tumor-related mutations would be more available to cancer cells to evade various challenges, such as differentiation, senescence, extracellular antiproliferative signals, and hypoxia, imposed in the in vivo microenvironment. Therefore, although there is no definitive evidence yet as to whether genomic instability drives tumor development, the hypothesis is a fascinating and fundamentally important one for cancer research.

Recent studies on c-Myc-induced genomic instability have suggested c-*myc* as a new candidate genomic instability gene. Notably, studies using advanced cytogenetic methodologies (spectral karyotyping and comparative genomic hybridization) have demonstrated various chromosomal aberrations, formerly undetected by classical cytogenetic techniques, in c-Myc overexpressing mouse mammary tumor cells (Liyanage et al. 1996; McCormack et al. 1998). These alterations are likely to indicate the dynamic instability of genomes in c-Myc overexpressing mammary tumor cells, as previously indicated in the cervical intraepithelial neoplasia studies. Independently, several in vitro studies have also demonstrated, in rat fibroblasts, a rapid increase in chromosomal aberrations and an induction of gene amplification by c-Myc (Felsher and Bishop 1999; Mai 1994, 1996). In particular, in a study using a regulatable c-Myc, c-MycER^{TM}, Rat1A cells (an immortal pseudodiploid rat fibroblast line) and normal human fibroblasts (NHF) demonstrated a selective increase of genomic alterations following activation of c-MycER^{TM}. Estrogen ligand-mediated activation of c-MycER^{TM} for as little as 2 days (roughly corresponding to one to two cell cycles) led to a significant percentage of aneuploidy (about 22% in Rat1A cells and 21% in NHF) (Felsher and Bishop 1999). Furthermore, the same treatment induced a significant amount of two distinct types of karyotypic abnormality (polycentric chromosomes 21% and double minutes 7%) in Rat1A -MycER^{TM} cells. Interestingly, these abnormal structures are known as a central intermediates for gene amplification. c-Myc-induced genomic instability is important, because c-*myc* has been established as a powerful oncogene. These data could, therefore, provide direct evidence for the oncogenicity of a genomic instability phenotype if (1) c-Myc directly increases the rate for genomic changes, and (2) the oncogenicity of c-Myc can be blocked by selectively suppressing its genomic instability function. These studies will provide an important link between an established oncogene and its genomic instability phenotype that provides 'genomic heterogeneity' and 'growth advantage for clonal expansion,' which are two components required to evade ever-changing, in vivo antiproliferative challenges.

4
c-Myc Both Induces DNA Damage and Attenuates the DNA Damage-Induced Cell Cycle Checkpoints

Regarding the underlying mechanisms for the c-Myc-induced genomic instability, questions were raised as to whether c-Myc directly induces DNA damage or indirectly induces the accumulation of DNA damage. According to recent studies, deregulated c-Myc both induces DNA damage and creates permissive conditions for cells containing damaged DNA to undergo cell cycle (Sheen and Dickson 2002; Vafa et al. 2002). When c-Myc is transiently deregulated in NHF by activating MycER it induces DNA damage, particularly in resting cells (Vafa et al. 2002). The induction of DNA damage correlated with production of reactive oxygen species (ROS) and an antioxidant reduced the ROS production and decreased DNA damage. Furthermore, the c-Myc-induced DNA damage appears to precede the c-Myc-induced apoptosis in the absence of growth factor stimuli, potentially suggesting that DNA damage induce apoptosis in c-Myc overexpressing cells following serum starvation. The production of ROS in c-Myc overexpressing cells was also observed in an independent study performed in NIH3T3 murine embryonic fibroblasts and Saos-2 cells (Tanaka et al. 2002). In this study, overexpression of c-Myc, as well as E2F-1, induced accumulation of ROS during serum starvation. The mRNA of superoxide dismutase was induced by a transcription factor nuclear factor κB (NF-κB) in NIH3T3 cells following serum star-

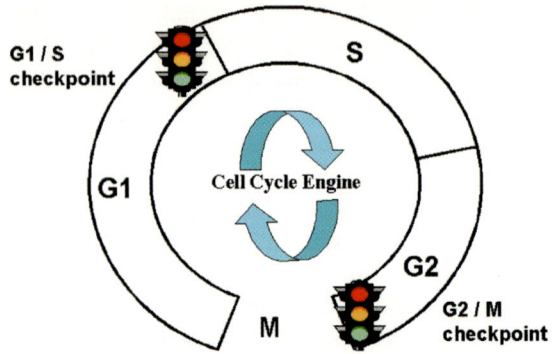

Fig. 4 Cell cycle checkpoints at G_1/S and G_2/M boundaries

vation, while superoxide dismutase induction was inhibited in c-Myc or E2F-1-overexpressing cells. Taken together, ROS may promote apoptosis, possibly through induction of DNA damage, under serum-starved conditions.

Although DNA damage-induced apoptosis usually depends on the function of the p53 tumor suppressor, c-Myc-induced apoptosis is unlikely to occur via the p53-dependent pathway because NIH3T3 cells lack ARF and Saos cells lack p53 (Tanaka et al. 2002). Furthermore, in the study using the NHF-MycER system, the transient activation of c-Myc attenuates the DNA damage-induced G_1/S cell cycle checkpoint, creating conditions permissive for replicating damaged DNA in S phase (Vafa et al. 2002). Consistently, a transient excess of c-Myc activity is sufficient to alter the G_1/S checkpoint in nontransformed immortal human mammary epithelial cell (HMEC) lines and normal mortal HMECs (Sheen and Dickson 2002). The cell cycle checkpoints ensure the proper execution of sequential events of the eukaryotic cell cycle; unprepared entry into either S phase or M phase can be fatal. The DNA damage-induced checkpoint is a prototypic class of cell cycle checkpoints that acts as a cellular surveillance system for DNA damage, and which ensures that cells containing damaged DNA do not proceed into S phase or into mitosis (Fig. 4). Many biochemical components of the G_1/S checkpoint have been identified, such as the tumor suppressor p53 and the cyclin-dependent kinase inhibitor (CDKI) p21/Cip1 (cyclin-dependent kinase inhibitor protein 1). When stabilized by a DNA damage signal, p53 is homotetramerized and acts as a transcription factor. Importantly, CDKI p21/Cip1 links p53 and the DNA damage-induced inhibition of cell cycle progression. These roles of p53 and p21/Cip1 in the response to DNA damage have been firmly established through recent studies using gene knockout approaches. Inhibition of cyclin/CDK then leads to hypophosphorylation of Rb. The hypophosphorylated form of Rb (pRb) actively sequesters the E2F-1 transcription factor, preventing transactivation of E2F-1 targets. Since E2F-1-mediated transactivation is essential for progression into S phase, Rb-mediated inhibition of E2F-1 activity causes cell cycle arrest at G_1/S. Therefore, there are multiple points where c-Myc may interfere the p53-mediated cell cycle arrest upon DNA damage. One interesting clue came from a study of Miz-1. Upon ultraviolet (UV) irradiation (IR), Miz-1, a POZ domain transcription factor, binds to the p21/Cip1 core promoter to induce the CDKI protein. However, in c-Myc overexpressing cells, c-Myc negatively regulates p21/Cip1 induction by binding to Miz-1 and attenuates UV-induced cell cycle arrest (Herold et al. 2002).

In addition to the G_1/S boundary, deregulated c-Myc appears to attenuate the cellular safeguard mechanisms against DNA damage at multiple points, such as G_2/M and post-M checkpoints (Li and Dang 1999; Sheen et al. 2003; Yin et al. 1999). In particular, effects of deregulated c-Myc on the G_2/M checkpoint are interesting because this, together with the G_1/S checkpoint, constitutes a critical safeguard against amplification of DNA damage. An unexpected, intriguing finding of our previous G_1/S study (Sheen and Dickson 2002) was an induction of micronuclei and the development of the sub-G_1 and the >4 n populations in irradiated c-Myc overexpressing HMECs. According to a previous study using a computerized video time-lapse method c-Myc overexpressing rat embryonic cells (REC:Myc) were susceptible to IR-induced apoptosis after 4–9.5 Gy. Cell death occurred primarily by postmitotic apoptosis, following several divisions of the irradiated cells. These results suggest the possibility of an altered G_2/M entry in these cells, because the irradiated c-Myc overexpressing RECs were allowed to undergo several mitotic divisions. In addition, a study by McKenna et al. reported a minimal delay in G_2/M of Myc-expressing fibroblasts, following IR. These studies provide interesting clues to stimulate further, rigorous investigations of the effects of deregulated c-Myc on DNA damage-induced arrest at the G_2/M boundary. To determine whether cells undergo mitosis following IR-induced DNA damage, we performed a mitotic trap assay with γ-irradiated HMECs (Sheen et al. 2003): cells with an intact G_2/M checkpoint arrested at G_2/M while cells without a G_2/M checkpoint progressed through the G_2/M boundary and into mitosis in spite of the presence of damaged DNA. In our results, γ-irradiated c-Myc overexpressing HMECs demonstrated an accumulation in mitosis as indicated by M-phase specific marker, while the control HMECs did not show mitotic figures following IR-induced DNA damage. Furthermore, c-Myc overexpressing HMECs displayed the upregulated expression of cyclin B1 and the expression level was maintained even after IR-induced DNA damage (Sheen et al. 2003). Taken together, deregulated c-Myc appears to attenuate DNA damage-induced checkpoints both at G_1/S and at G_2/M boundaries, and thus promotes genomic instability that may facilitate the accumulation of oncogenic mutations.

5
c-Myc and Anchorage-Independent Growth

In addition to genomic instability, another hallmark of cellular transformation is anchorage-independent growth. Normal epithelial cells require distinct types of extracellular stimuli, such as growth factors, and adhesion to the extracellular matrix, in order to proliferate. When deprived of attachment to a solid substrate, even in the presence of growth factors, normal epithelial cells are unable to progress in the cell cycle and arrest at the G_1 phase. However, c-Myc overexpression significantly reduces the requirement of cell adhesion to the extracellular matrix, allowing anchorage-independent growth of cells overexpressing c-Myc. Previous studies suggest that the cell cycle progression and cell adhesion are directly linked through the cyclins and their associated CDKs. Therefore, deregulation of c-Myc, a critical regulator of cell cycle progression, may also provide mechanistic insights into anchorage-independent growth of epithelial cells. In particular, adhesion-regulated G_1 arrest is correlated with downregulation of c-Myc: nontransformed mammary epithelial cells rapidly decrease their level of c-Myc mRNA and upregulate the cell cycle inhibitor p27/Kip1, while enforced expression of c-Myc in nonadherent epithelial cells reverses the G_1 arrest (Benaud and Dickson 2001). The effects of c-Myc on anchorage-independent growth may be through the downregulation of p27/Kip1 following c-Myc overexpression. Importantly, an independent study demonstrated that c-Myc promotes ubiquitin-dependent proteolysis by directly activating expression of Cul1 gene, encoding a critical component of the ubiquitin ligase SCF^{SKP2}. Furthermore, p27/Kip1 is a known target of the SCF^{SKP2} complex, and the c-Myc-mediated Cul1 expression matched well with the kinetics of declining p27Kip1 protein (O'Hagan et al. 2000).

6
Summary and Future Directions

c-*myc* is one of the prototypic proto-oncogenes, found about 20 years ago. The significance of the c-*myc* oncogene in human cancers is evident because there is a wide-spread occurrence of deregulated expression of the oncogene in many types of human cancers, including breast cancers. The protein product of the proto-oncogene constitutes a hetero-dimeric transcription factor with the partner protein, Max, suggesting the importance of transcriptional targets of c-Myc/Max in the biological effects of c-Myc in cellular transformation and in vivo tumor development and growth. Despite extensive efforts to elucidate the downstream targets of c-Myc, it appears that no single target explains the diverse effects of c-Myc. Rather, c-Myc appears to promote tumorigenesis in vivo through a complex interaction among downstream targets. The complexity of Myc targets is further confirmed by recent genomic screening of the downstream targets, demonstrating a high regulatory and biological diversity among the downstream Myc target genes (Fernandez et al. 2003). Therefore, to determine the primary effectors of Myc functions in human cancers it is necessary to dissect the complex web of c-Myc functions into specific subgroups (e.g., phenotypic units), such as genomic instability and anchorage-independence. This could lead to more specific molecular targets for prevention and treatment of the human cancers caused by c-Myc.

References

Atchley WR, Fitch WM (1995) Myc and Max: molecular evolution of a family of proto-oncogene products and their dimerization partner. Proc Natl Acad Sci USA 92:10217–10221

Benaud CM, Dickson RB (2001) Adhesion-regulated G1 cell cycle arrest in epithelial cells requires the downregulation of c-Myc. Oncogene 20:4554–4567

Blackwood EM, Luscher B, Kretzner L, Eisenman RN (1991) The Myc:Max protein complex and cell growth regulation. Cold Spring Harb Symp Quant Biol 56:109–117

Ciechanover A, DiGiuseppe JA, Bercovich B, Orian A, Richter JD, Schwartz AL, and Brodeur GM (1991) Degradation of nuclear oncoproteins by the ubiquitin system in vitro. Proc Natl Acad Sci USA 88:139–143

Dang CV (1999) c-Myc target genes involved in cell growth, apoptosis, and metabolism. Mol Cell Biol 19:1–11

Dang CV, Resar LM, Emison E, Kim S, Li Q, Prescott JE, Wonsey D, Zeller K (1999) Function of the c-Myc oncogenic transcription factor. Exp Cell Res 253:63–77

Deming SL, Nass SJ, Dickson RB, Trock BJ (2000) C-myc amplification in breast cancer: a meta-analysis of its occurrence and prognostic relevance. Br J Cancer 83:1688–1695

Eilers M, Picard D, Yamamoto KR, Bishop JM (1989) Chimaeras of myc oncoprotein and steroid receptors cause hormone-dependent transformation of cells. Nature 340:66–68

Facchini LM, Penn LZ (1998) The molecular role of Myc in growth and transformation: recent discoveries lead to new insights FASEB J 12:633–651

Felsher DW, Bishop JM (1999) Transient excess of MYC activity can elicit genomic instability and tumorigenesis. Proc Natl Acad Sci USA 96:3940–3944

Fernandez PC, Frank SR, Wang L, Schroeder M, Liu S, Greene J, Cocito A, Amati B (2003) Genomic targets of the human c-Myc protein. Genes Dev 17:1115–1129

Gavine PR, Neil JC, Crouch DH (1999) Protein stabilization: a common consequence of mutations in independently derived v-Myc alleles. Oncogene 18:7552–7558

Grandori C, Cowley SM, James LP, Eisenman RN (2000) The Myc/Max/Mad network and the transcriptional control of cell behavior. Annu Rev Cell Dev Biol 16:653–699

Gross-Mesilaty S, Reinstein E, Bercovich B, Tobias KE, Schwartz AL, Kahana C, Ciechanover A (1998). Basal and human papillomavirus E6 oncoprotein-induced degradation of Myc proteins by the ubiquitin pathway. Proc Natl Acad Sci USA 95:8058–8063

Hann SR, Eisenman RN (1984) Proteins encoded by the human c-myc oncogene: differential expression in neoplastic cells. Mol Cell Biol 4:2486–2497

Herold S, Wanzel M, Beuger V, Frohme C, Beul D, Hillukkala T, Syvaoja J, Saluz HP, Haenel F, Eilers M (2002) Negative regulation of the mammalian UV response by Myc through association with Miz-1. Mol Cell 10:509–521

Karn J, Watson JV, Lowe AD, Green SM, Vedeckis W (1989) Regulation of cell cycle duration by c-myc levels. Oncogene 4:773–787

Kato GJ, Barrett J, Villa-Garcia M, Dang CV (1990) An amino-terminal c-myc domain required for neoplastic transformation activates transcription. Mol Cell Biol 10:5914–5920

Li Q, Dang CV (1999) c-Myc overexpression uncouples DNA replication from mitosis. Mol Cell Biol 19:5339–5351

Liao DJ, Dickson RB (2000) c-Myc in breast cancer. Endocr Relat Cancer 7:143–164

Liyanage M, Coleman A, du Manoir S, Veldman T, McCormack S, Dickson RB, Barlow C, Wynshaw-Boris A, Janz S, Wienberg J, Ferguson-Smith MA, Schrock E, Ried T (1996) Multicolour spectral karyotyping of mouse chromosomes. Nat Genet 14:312–315

Loeb LA (1991) Mutator phenotype may be required for multistage carcinogenesis. Cancer Res 51:3075–3079

Mai S (1994) Overexpression of c-myc precedes amplification of the gene encoding dihydrofolate reductase. Gene 148:253–260

Mai S, Fluri M, Siwarski D, Huppi K (1996) Genomic instability in MycER-activated Rat1A-MycER cells. Chromosome Res 4:365–371

Mateyak MK, Obaya AJ, Adachi S, Sedivy JM (1997) Phenotypes of c-Myc-deficient rat fibroblasts isolated by targeted homologous recombination. Cell Growth Differ 8:1039–1048

Mateyak MK, Obaya AJ, Sedivy JM (1999) c-Myc regulates cyclin D-Cdk4 and -Cdk6 activity but affects cell cycle progression at multiple independent points. Mol Cell Biol 19:4672–4683

McCormack SJ, Weaver Z, Deming S, Natarajan G, Torri J, Johnson MD, Liyanage M, Ried T, Dickson RB (1998). Myc/p53 interactions in transgenic mouse mammary development, tumorigenesis and chromosomal instability. Oncogene 16:2755–2766

Nowell P (1976). The clonal evolution of tumor cell populations. Science 194:23–28

O'Hagan RC, Ohh M, David G, de Alboran M, Alt FW, Kaelin WG, DePinho AR (2000) Myc-enhanced expression of Cul1 promotes ubiquitin-dependent proteolysis and cell cycle progression. Genes Dev 14:2185–2191

Salghetti SE, Kim SY, and Tansey WP (1999) Destruction of Myc by ubiquitin-mediated proteolysis: cancer-associated and transforming mutations stabilize Myc. EMBO J 18:717–726

Sheen JH, Dickson RB (2002) Overexpression of c-Myc alters G(1)/S arrest following ionizing radiation. Mol Cell Biol 22:1819–1833

Sheen JH, Woo JK, Dickson RB (2003) c-Myc alters the DNA damage-induced G2/M arrest in human mammary epithelial cells. Br J Cancer 89:1479–1485

Spencer CA, Groudine M (1991) Control of c-myc regulation in normal and neoplastic cells. Adv Cancer Res 56:1–48

Spotts GD, Patel SV, Xiao Q, Hann SR (1997) Identification of downstream-initiated c-Myc proteins which are dominant-negative inhibitors of transactivation by full-length c-Myc proteins. Mol Cell Biol 17:1459–1468

Tanaka H, Matsumura I, Ezoe S, Satoh Y, Sakamaki T, Albanese C, Machii T, Pestell RG, Kanakura Y (2002) E2F1 and c-Myc potentiate apoptosis through inhibition of NF-kappaB activity that facilitates MnSOD-mediated ROS elimination. Mol Cell 9:1017–1029

Tlsty TD, Briot A, Gualberto A, Hall I, Hess S, Hixon M, Kuppuswamy D, Romanov S, Sage M, White A (1995) Genomic instability and cancer. Mutat Res 337:1–7

Tomlinson I, Bodmer W (1999) Selection, the mutation rate and cancer: ensuring that the tail does not wag the dog. Nat Med 5:11–12

Tomlinson IP, Novelli MR, Bodmer WF (1996). The mutation rate and cancer. Proc Natl Acad Sci USA 93:14800–14803

Vafa O, Wade M, Kern S, Beeche M, Pandita TK, Hampton GM, Wahl GM (2002) c-Myc can induce DNA damage, increase reactive oxygen species, and mitigate p53 function: a mechanism for oncogene-induced genetic instability. Mol Cell 9:1031–1044

Vennstrom B, Sheiness D, Zabielski J, Bishop JM (1982) Isolation and characterization of c-myc, a cellular homolog of the oncogene (v-myc) of avian myelocytomatosis virus strain 29. J Virol 42:773–779

Vogelstein B, Kinzler KW (1993) The multistep nature of cancer. Trends Genet 9:138–141

Xiao Q, Claassen G, Shi J, Adachi S, Sedivy J, Hann SR (1998) Transactivation-defective c-MycS retains the ability to regulate proliferation and apoptosis. Genes Dev 12:3803–3808

Yin XY, Grove L, Datta NS, Long MW, Prochownik EV (1999) C-myc overexpression and p53 loss cooperate to promote genomic instability. Oncogene 18:1177–1184

NF-κB: Critical Regulator of Inflammation and the Immune Response

A. Lasar · R. Marienfeld · T. Wirth (✉) · B. Baumann

Department of Physiological Chemistry, Ulm University, Albert-Einstein-Allee 11, 89081 Ulm, Germany
thomas.wirth@medizin.uni-ulm.de

1	Key Components of the NF-κB System: NF-κB Subunits, IκB Proteins and the IKK Signalosome	326
1.1	NF-κB Subunits	326
1.2	IκB Proteins	327
1.3	IKK Signalosome	329
2	Role of NF-κB in Innate and Adaptive Immunity	331
2.1	NF-κB Activation by Toll-Like Receptor Signaling	331
2.2	TCR/BCR-Mediated NF-κB Activation	334
2.3	NF-κB Activation by Proinflammatory Cytokines: TNF and IL-1 Receptor Signaling	335
3	The Alternative Pathway of NF-κB Activation: Contribution to Development of Secondary Lymphoid Organs	337
4	Role of NF-κB in Inflammation—The Integral Part of Innate Immunity	338
4.1	Phases of Inflammation	339
4.2	Involvement of NF-κB in Inflammatory Processes	341
4.3	Inflammatory Phenotypes of NF-κB Mutant Mice	342
4.4	Crosstalk Between NF-κB and Other Signaling Pathways During the Development of Inflammation	344
5	Chronic Inflammatory Diseases	344
5.1	Rheumatoid Arthritis	345
5.2	Atherosclerosis	347
6	Pharmacological Strategies to Inhibit NF-κB Activity in Chronic Inflammation	351
7	Conclusions	356
	References	356

Abstract The nuclear factor-κB (NF-κB) was first discovered by Sen and Baltimore in 1986 as a transcription factor that binds to the intronic enhancer of the immunglobulin κ-light chain gene of mature B cells and plasma cells. Since then, the NF-κB/Rel family of dimeric transcription factors has been the subject of intense research documented by thousands of publications. It became clear that NF-κB is present in virtually all cell types where it is held in the cytoplasm in an inactive state by association with specific inhibitor proteins (IκBs). In response to a great variety of extracellular stimuli, e.g., inflammatory cytokines, NF-κB is activated and translocates to the nucleus where it controls the expression of a large number of target genes. Target genes are involved in the regulation of multiple biological processes ranging from inflammation, development, cell survival/apoptosis, cell growth, cellular transformation, to neuronal differentiation. Here we focus on the central role of NF-κB in promoting inflammation and mediating innate and adaptive immune responses. In particular, inflammatory diseases associated with dysregulated NF-κB activity are highlighted and the NF-κB system as target for anti-inflammatory drugs is discussed.

Keywords NF-κB · Inflammation · Cytokines · Pharmacological intervention

1
Key Components of the NF-κB System:
NF-κB Subunits, IκB Proteins and the IKK Signalosome

1.1
NF-κB Subunits

In mammals five structurally related and evolutionarily conserved members of the nuclear factor-κB (NF-κ) B transcription factor family are known (reviewed by Karin and Ben-Neriah 2000; Ghosh and Karin 2002): RelA (p65), RelB, c-Rel, NF-κB1 (p105/p50) and NF-κB2 (p100/p52). NF-κB1 and NF-κB2 are synthesized as precursor proteins which are proteolytically processed to obtain the mature proteins, p50 and p52, respectively. The mechanism of p105 processing and p50 generation is a matter of debate. Both co-translational (Lin et al. 1998) and post-translational mechanisms involving the ubiquitin-proteasome pathway (Orian et al. 2000; Ciechanover et al. 2001; Heissmeyer et al. 2001) have been suggested. P100 processing and p52 generation however, appears to be an inducible process which is tightly connected to RelB and B-cell function. This issue will be discussed below in more detail.

The other three family members (RelA, RelB, and c-Rel) do not need proteolytic processing for their synthesis. All family members are characterized by a conserved domain of about 300 amino acids, the Rel-homology-domain (RHD). The RHD contains a nuclear localization sequence (NLS) and is essential for dimerization, DNA-binding to κB motifs and interaction with the inhibitory IκB proteins. The active DNA-binding form of NF-κB is dimeric and members of the NF-κB family can form homo- or heterodimers with a high degree of promiscuity except for RelB which has a strong preference for p50/NF-κB1 and p52/NF-κB2. The Rel subunits bear potent transcriptional activation domains in their C-terminal domains (RelA and c-Rel) or in both N- and C-terminal domains (RelB). Therefore they are largely responsible for the transactivation capacity of dimers containing these subunits. In contrast, p50 and p52 lack transactivation domains, and as a consequence, p50 and p52 homodimers are usually considered to function as transcriptional repressors (May and Ghosh 1997). Furthermore the various NF-κB dimers exhibit some preferences for distinct κB motifs which can result in differential regulation of target genes (Menetski 2000). An additional level of regulation comes into play by the tissue specific expression of the NF-κB proteins. RelA and p50 are more or less ubiquitously expressed and in many cell types, p50/RelA heterodimers represent the prototypical NF-κB dimer activated by proinflammatory cytokines. In contrast, RelB, c-Rel and NF-κB2 are predominantly expressed in hematopoietic cells and tissues and they are largely responsible for the constitutive nuclear NF-κB activity originally found in mature B and plasma cells.

1.2
IκB Proteins

The main mechanism regulating NF-κB activity is based on the control of the subcellular localization of NF-κB. In most cells, NF-κB is localized in the cytoplasm in an inactive form due to the interaction with the IκB (inhibitor of κB) proteins. These IκB proteins inhibit nuclear translocation of NF-κB dimers by masking the NLS. In addition, they also directly block NF-κB DNA-binding due to the interaction with the RHD. The IκB proteins again represent a protein family of at least eight members in mammals: IκBα, IκBβ, IκBγ, IκBε, IκBζ, IκBNS, and the precursor proteins p105 and p100.

All IκBs are characterized by N- and C-terminal domains of variable length flanking either six or seven copies of a conserved protein domain, the ankyrin repeat, an amino acid motif that mediates protein–protein interaction (see Fig. 1). IκBα, IκBβ and IκBε, the most important regulators of NF-κB, contain an N-terminal signal response domain (SRD) with specific serine residues (Ser32 and 36 in IκBα, Ser19 and 23 in IκBβ, Ser18 and 22 in IκBε) that get phosphorylated de novo upon stimulation (e.g., proinflammatory cytokines). This essential step in NF-κB activation targets the IκB proteins for polyubiquitination on critical lysine residues also located in the SRD. This polyubiquination step is performed by a protein–ubiquitin ligase complex called Skp1/cullin-1/F-box protein ($SCF^{\beta-TrCP}$). $SCF^{\beta-TrCP}$ is composed of the adapter proteins Skp1, cullin-1, the RING finger containing protein Roc1 and the F-box protein β-TrCP which is responsible for the initial terminally recognition of the N-terminal phosphorylated IκB proteins. Roc1 recruits the E2 ubiquitin-conjugating enzyme UbcH5 which then covalently links ubiquitin to specific lysine residues on IκBs (Karin and Ben-Neriah 2000; Ben-Neriah 2002). Upon polyubiquitination, the IκB proteins get rapidly degraded via the 26S proteasome pathway. This finally results in the release of NF-κB dimers which can migrate into the nucleus and activate the transcription of target genes.

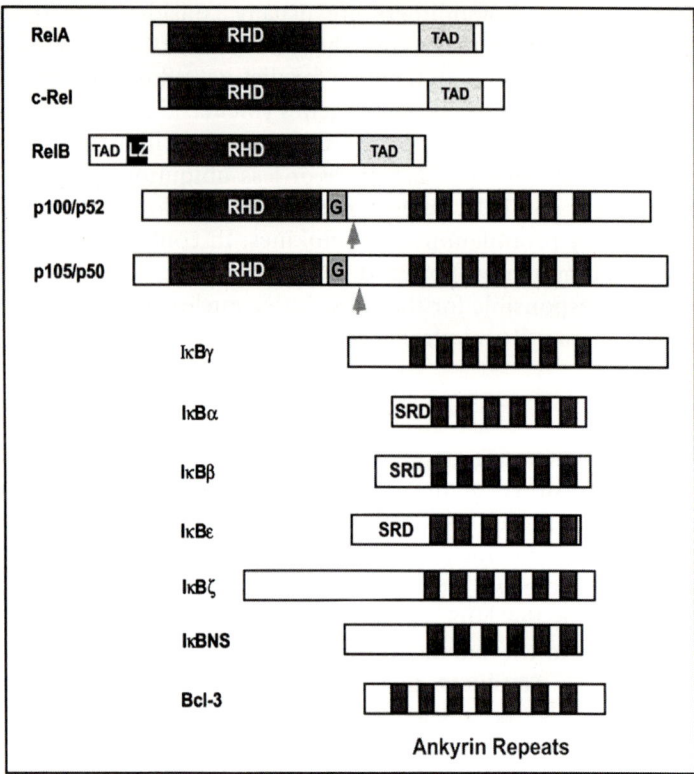

Fig. 1 The family of NF-κB transcription factors and their inhibitors, the IκB proteins. *RHD*, Rel homology domain; *TAD*, transactivation domain; *LZ*, leucine zipper; *G*, glycine-rich region; *SRD*, signal response domain. The *arrow* indicates the C-terminal ends of the p50 and p52 proteins, respectively

Importantly, IκBα is a prominent NF-κB target gene that is newly transcribed upon NF-κB activation (Sun et al. 1993; Chiao et al. 1994) and thereby initiates a negative feedback loop that ensures a fast turn-off of the NF-κB response. Upon sustained NF-κB stimulation continued cycles of IκBα degradation and resynthesis can create cycling oscillations in nuclear NF-κB activity (Hoffmann et al. 2002). However, IκBβ and IκBε expression levels are not regulated by NF-κB and their steady-state expression leads to the reduction of the oscillatory potential of the system and stabilizes NF-κB responses during prolonged stimulation (Hoffmann et al. 2002). Independently of the classical view that IκB proteins mask the NLS of NF-κB dimers and prevent their nuclear translocation, recent findings indicate a continuous shuttling of NF-κB/IκBα complexes between cytoplasm and nucleus even in the absence of any stimulus. Both incomplete masking of one of the two NLSs in a NF-κB dimer as well as the nuclear export signal (NES) found in the N terminus of IκBα are involved in this process. However, the predominant localization of the complex is cytoplasmic due to the potent NES of IκBα (Arenzana-Seisdedos et al. 1997; Johnson et al. 1999; Huang et al. 2000; Malek et al. 2001).

The proto-oncoprotein Bcl-3 is a unique member of the IκB family because it interacts specifically with p50 and p52 homodimers. It is primarily found in the nucleus and in contrast to the inhibitory function of other IκB proteins, Bcl-3 can enhance NF-κB dependent transcription (Bours et al. 1993; Fujita et al. 1993; May and Ghosh 1997).

IκBα, IκBβ, IκBε, Bcl-3, IκBζ and IκBNS represent independent genes, whereas IκBγ is encoded by a mRNA which is identical to the C-terminal part of the NF-κB1 gene. IκBζ and IκBNS represent newly identified members of the IκB family that were shown to control NF-κB activity during proinflammatory stimulation and negative selection of thymocytes (Yamazaki et al. 2001; Fiorini et al. 2002; Muta et al. 2003).

1.3
IKK Signalosome

In response to numerous and diverse stimuli including tumor necrosis factor (TNF)-α, interleukin (IL)-1, bacterial products and viruses (Pahl 1999), cells initiate signal transduction cascades which converge in the activation of a multisubunit protein kinase complex, the IκB kinase (IKK) complex. The 700–900-kDa IKK complex consists of at least three distinct polypeptides: two related protein kinases called IKK1 and IKK2 (also known as IKKα and IKKβ, respectively) which are the catalytic subunits of the complex and a structural and regulatory component named NF-κB essential modulator (NEMO) or IKKγ (Zandi and Karin 1999; Israel 2000). The absence of NEMO leads to a complete inhibition of NF-κB activation indicating that this noncatalytic subunit is essential for the kinase activity of the canonical pathway (Yamaoka et al. 1998; Makris et al. 2000; Rudolph et al. 2000; Schmidt-Supprian et al. 2000). IKK1 and IKK2 are 85-kDa and 87-kDa proteins, respectively. They show a high similarity in their structural organization. Both contain an N-terminal kinase domain as well as a leucine zipper and a helix–loop–helix (HLH) motif in the C-terminal portion of the protein. The C-terminal region mediates the interaction with NEMO. IKK1 and IKK2 dimerize via their leucine zipper to form homo- and heterodimers and this dimerization is a prerequisite for their kinase activity. Thus, leucine zipper mutations that prevent dimer formation also abolish kinase activity (Mercurio et al. 1997; Woronicz et al. 1997; Zandi et al. 1997). In response to extracellular stimuli, IKK1/2 are phosphorylated on two conserved serine residues in the activation loop, a small motif located in the kinase domain. This post-translational modification leads to the activation of the kinases and is mediated either by upstream kinases or through *trans*-autophosphorylation between the IKK proteins. Consistent with the phosphorylation dependent activation mechanism, replacement of these serine residues with alanines prevented kinase activation. Conversely, mutation to glutamic acid, which mimics the effect of phosphorylated serine, resulted in constitutive kinase activity (Mercurio et al. 1997; Ling et al. 1998; Delhase et al. 1999). In addition, mutagenesis studies revealed that the C-terminal region of the kinases including the HLH motif is involved in both positive and negative regulation of the kinase activity (Zandi et al. 1997, 1998; Delhase et al. 1999).

The structure of NEMO is primarily helical and contains two coiled-coil domains, a C-terminal leucine zipper and putative zinc finger at the extreme C terminus. The N-terminal domain is responsible for the interaction with IKK1/2, whereas the C-terminal region including the coiled-coil, leucine zipper and zinc finger motif is critical for oligomerization and interaction with upstream activators of the IKK complex (Agou et al. 2002; Makris et al. 2002). The stoichiometry of the 700–900-kDa large IKK complex is not resolved in detail but it is likely that there is a core unit including two NEMO molecules per IKK dimer (Miller and Zandi 2001). However, the exact number of these core units in the functional IKK complex is not clear. In addition, the heat shock protein Hsp90 and Hsp90-associated Cdc37 protein have been suggested to be components of the IKK complex and thus seem to play a critical role at least in TNF-α-stimulated IKK activation (Chen et al. 2002).

The activated IKK complex is able to phoshorylate the conserved serine residues in the SRD of the IκB proteins and thereby target them for degradation. From in vitro experiments it became evident that both kinases are able to phosphorylate these serine residues implying redundant functions. However, gene deletion studies clearly demonstrated that IKK1 and IKK2 are nonredundant in their functions and have distinct physiological roles. IKK2 deficient mice die at embryonic day (E)12.5–E14 due to extensive apoptosis of fetal hepatocytes (Li et al. 1999b; Tanaka et al. 1999). Disruption of the IKK1 locus results in perinatal lethality. Remarkably, IKK1 deficient animals were found to have strong defects in keratinocyte differentiation and proliferation which leads to abnormal morphogenesis (Hu et al. 1999; Li et al. 1999a; Takeda et al. 1999). Most interestingly, the development of this phenotype is independent of the kinase function, the interaction with NEMO and NF-κB activity, but depends on the integrity of the leucine zipper and HLH domain of IKK1. It was suggested that IKK1, but not IKK2, controls the production of a soluble polypeptide factor that induces keratinocyte differentiation (Hu et al. 2001).

Analysis of NEMO deficient cells revealed that NEMO is absolutely necessary for cytokine- and lipopolysaccharide (LPS)-induced NF-κB activation. Mouse embryonic fibroblasts (MEF) derived from IKK2$^{-/-}$ animals also show a strongly impaired IκBα phosphorylation and NF-κB activation upon TNF-α stimulation. Furthermore, IKK2$^{-/-}$ mice crossed to TNF receptor 1 deficient animals display a suppression of the phenotype indicating that apoptosis is most likely mediated by TNF-α action (Senftleben et al. 2001b). In contrast, TNF-α-induced IκBα phosphorylation, NF-κB nuclear translocation and DNA binding of NF-κB are only marginally affected in IKK1-deficient MEFs (Hu et al. 1999; Takeda et al. 1999). These results indicate, that NEMO and IKK2, but not IKK1 are essential for NF-κB activation by inflammatory stimuli. However, there is also evidence that loss of IKK1 function results in deficiencies in NF-κB transcriptional activity and target gene expression in response to TNF-α and IL-1, although induction of IκBα degradation and NF-κB DNA binding is not affected (Li et al. 1999a, 2002; Sizemore et al. 2002). In support of these findings, two recent publications report on a nuclear function of IKK1 involved in TNF-α-induced gene expression (Anest et al. 2003; Yamamoto et al. 2003c). The new findings suggest that IKK1 migrates to the nucleus, associates with NF-κB regulated promoters and mediates the phosphorylation of histone H3. Thereby, histone function is modified in a way facilitating gene expression in response to TNF-α. Taken together, these studies provide new mechanistic insights into the regulation of cytokine-induced NF-κB dependent gene expression. NEMO is the central integration point of signaling cascades initiated by TNF-α, IL-1 and other upstream signaling pathways. In addition, IKK2 function is essential for IκBα phosphorylation and efficient DNA-binding of NF-κB whereas IKK1 primarily acts as chromatin modifier through histone H3 phosphorylation. Thus, IKK1 and IKK2 have nonredundant functions in the regulation of NF-κB dependent gene expression. Consistent with this model, TNF-α-induced NF-κB activation and target gene expression is completely abrogated in MEFs lacking both IKK1 as well as IKK2 (Li et al. 2000b).

In summary, NEMO, IKK1 and IKK2 were found to be the basic components of the IKK signalosome which guarantee optimal cytokine-induced NF-κB dependent gene expression by distinct, but independent functions.

2
Role of NF-κB in Innate and Adaptive Immunity

2.1
NF-κB Activation by Toll-Like Receptor Signaling

The mammalian immune system can generally be subdivided into innate and adaptive immunity. Innate immune responses represent the first line of defense in preventing infections by invading pathogens. The adaptive immunity is composed of humoral and cell-mediated responses. Adaptive immune responses depend on the differentiation and clonal expansion of B and T cells and they are characterized by high specificity and selective memory (Medzhitov and Janeway 1998).

Innate immunity is characterized by rapid defense mechanisms with lower specificity including inflammation and phagocytosis of microorganisms by neutrophils and macrophages. In an initial step a few highly conserved structures present on bacteria, viruses and other microorganisms are recognized by a set of receptors called pattern recognition receptors. In mammals this initial recognition of pathogen-associated molecular patterns (PAMPs) is mediated by the family of Toll-like receptors (TLRs). TLRs are widely expressed on host defense cells like macrophages, dendritic cells and neutrophilic granulocytes. The stimulation of the various TLRs triggers intracellular signaling cascades that mostly culminate in the activation of NF-κB, but also activate the mitogen-activated protein kinases (MAPK) p38 and Jun N-terminal kinase (JNK). Thereby a genetic program that is responsible for the production of proinflammatory cytokines such as TNF-α, IL-1, IL-12 and interferons, is initiated. Moreover, TLR signaling activates antigen-presenting cells, a process that is essential for adaptive immune responses.

TLRs are type 1 transmembrane proteins. They are defined by the presence of a conserved intracellular signaling motif called the Toll/IL-1 receptor (TIR) domain and an extracellular portion which contains several leucine-rich repeats. Currently, ten TLR family members (TLR1–TLR10) are known which recognize various ligands ranging from double-stranded RNA (virus), CpG DNA (bacteria), peptidoglycans, flagellin, zymosan and porins (Barton and Medzhitov 2002). Upon ligand binding, cytosolic adapter molecules and protein kinases can associate with the receptors and initiate signal transduction cascades funneling into the activation of the IKK signalosome. This principle of intracellular signal transduction resembles that of IL-1 receptor signaling.

TLR4, the first member of the TLR family identified in mammals, induces the expression of several inflammatory cytokines (Medzhitov et al. 1997). The most important ligand recognized by TLR4 is the bacterial compound LPS, the causative agent of endotoxic shock. Recognition of LPS by TLR4 also requires a serum protein named LPS-binding protein (LBP). LPS bound to LBP is able to interact with TLR4 which is associated with its coreceptor CD14 and the accessory protein MD2. MD2 was found to enhance LPS responsiveness (Nagai et al. 2002). Upon LPS binding, the TIR domain of TLR4 is able to recruit the TIR domain containing adaptor proteins myeloid differentiation factor 88 (MyD88) and MyD88-adapter-like (MAL) via homophilic interaction. Recruitment of MyD88 promotes the association with the IL-1R associated kinases (IRAK1, IRAK4) via an interaction between the so-called death domains of each molecule. As a consequence autophosphorylation of the IRAKs can take place which allows the engagement and activation of TRAF6, a member of the TNF receptor-associated factor (TRAF) family. In an intermediate step IRAK-1/TRAF6 dissociate from the receptor complex and interact at the membrane with a ternary complex composed of TAK-1 [transforming growth factor-β-activated kinase 1] and the TAK-1-

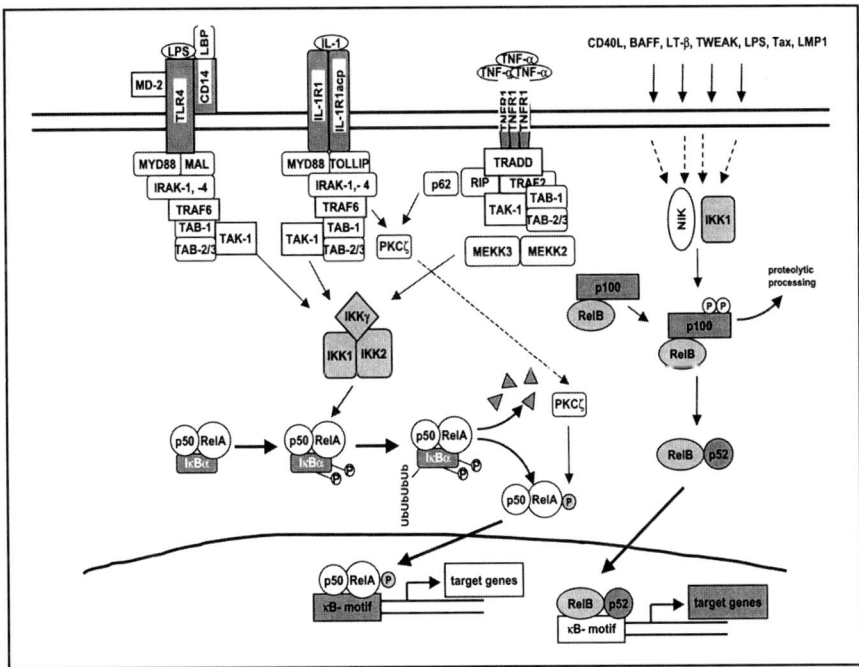

Fig. 2 Classical and alternative signaling pathways leading to NF-κB activation. Only selected components of the LPS, IL-1 and TNF-α signaling cascades are depicted. All three pathways funnel in the activation of the classical IKK-complex and the degradation of IκBα. In contrast, stimuli (CD40L, BAFF, LT-β, TWEAK, LPS, Tax and LMP1) initiating the alternative pathway mediate the IKK1-dependent processing of NF-κB/p100. Phosphorylation of a given protein is indicated by a *circled P*. *UbUbUb*, Polyubiquitin

binding proteins TAB-1 and TAB-2. In a poorly understood manner including phosphorylation and ubiquitination events, a multimeric cytosolic complex is generated, which contains TRAF6, TAK-1, TAB-1 and TAB-2. The activation of TAK-1 then mediates the stimulation the IKK-signalosome (see Fig. 2). In addition, activated TAK1 is able to phosphorylate mitogen-activated protein kinase kinase (MKK)3 and MKK6, the kinases upstream of p38 and JNK (Janssens and Beyaert 2003).

This scenario of NF-κB activation is most probably common to all TLRs since MyD88 is an universal adapter for TIR domain containing receptors. However, in studies with MyD88-deficient mice, it became clear that a second, MyD88-independent pathway exists which is induced upon engagement of TLR3 and TLR4. The stimulation of TLR3 and TLR4, which recognize dsRNA and LPS, respectively, leads to the MyD88-independent expression of interferon β (IFNβ) and activation of NF-κB (Kawai et al. 1999). The secreted IFNβ induces the expression of a host of anti-viral and anti-bacterial cytokines. Initially it was thought that the TIR-domain containing adaptor protein Mal/TIRAP might be the key component of this MyD88-inde-pen-dent pathway, but it turned out that Mal/TIRAP acts in conjunction with MyD88, thus transmitting signals from TLR2 and TLR4 (Shinobu et al. 2002; Yamamoto et al. 2002). Recently, additional TIR domain containing adaptor proteins have been identified: TRIF/TICAM1 and TRAM. Mice with a loss of function mutation of TRIF or TRAM, either by conventional gene targeting or by chemical mutagenesis, showed that TRIF is necessary for IFNβ expression and NF-κB activation after dsRNA (TLR3) as well as LPS (TLR4) treatment, whereas TRAM is required only for the TLR4-mediated signaling pathways including NF-κB (Fitzgerald et al. 2003b; Hoebe et al. 2003; Yamamoto et al. 2003a, 2003b). TRIF and TRAM mediate signals which lead to the induction of two IKK-related kinases, IKKϵ and TRAF-as-sociated NF-κB activator (TANK)-binding kinase 1 (TBK1). Upon stimulation via TLR3 or TLR4 IKKϵ and TBK1, which are in a complex with the adaptor protein TANK, are induced and mediate the phosphorylation of the transcription factors IFN response factor 3 and 7 (IRF3 and IRF7). The phosphorylation of IRF3 and IRF7 is necessary for the activation of these transcription factors, and IRF3 and IRF7 are crucial for the expression of IFNβ (Fitzgerald et al. 2003a; Sharma et al. 2003). Interestingly, a link between IKKϵ/TBK1 and the conventional IKK-complex is mediated by TANK, which is an interaction partner for NEMO (Chariot et al. 2002). Whether the TANK:NEMO interaction is the molecular basis for the TLR3 and TLR4-mediated NF-κB activation is still unclear. However, the LPS-induced NF-κB activation remains unaltered in IKKϵ-deficient MEFs, arguing for an IKKϵ-independent activation of NF-κB (Kravchenko et al. 2003).

2.2
TCR/BCR-Mediated NF-κB Activation

It has been known for some time that NF-κB activation in response to stimulation of the T cell receptor and the B cell receptor is critical for development, function, and survival of T and B lymphocytes (Esslinger et al. 1997, 1998; Hettmann and Leiden 2000; Voll et al. 2000; Kaisho et al. 2001; Senftleben et al. 2001b; Pasparakis et al. 2002b; Ren et al. 2002; Li et al. 2003; Schmidt-Supprian et al. 2003). Although a lot of progress has been made in recent years in elucidating this pathway, there are still several open questions. Stimulation of the antigen receptor results in the activation of protein kinase C isoforms, PKCθ in T cells and PKCβ in B cells (Sun et al. 2000; Saijo et al. 2002; Pfeifhofer et al. 2003). In a way not understood in detail, the CARD-family member (CARD = caspase recruitment domain) CARMA1 (Bimp3/ CARD11) then associates with lipid rafts in the membrane and can activate, by homophilic interactions, the CARD-family member BCL10 (Gaide et al. 2001; Wang et al. 2002a; Thome and Tschopp 2003). This interaction results in the recruitment of the paracaspase MALT1, which then can activate the IKK complex (Ruefli-Brasse et al. 2003; Ruland et al. 2003). The molecular mechanism underlying the activation process is based on the Bcl10-dependent ubiquitination of NEMO, which is mediated by MALT1 and the ubiquitin conjugating factor UBC-13 (Zhou et al. 2003). Importantly, NEMO/IKKγ is recruited to the T cell receptor in the immunological synapse and it was demonstrated that the membrane targeting of NEMO/IKKγ itself is sufficient to activate the IKK complex (Weil et al. 2003). Given that PKCθ is also present in this immunological synapse (Bi et al. 2001) it is conceivable that the ordered association of this supramolecular activation cluster is the key step in activating the IKK-signalosome.

2.3
NF-κB Activation by Proinflammatory Cytokines: TNF and IL-1 Receptor Signaling

TNF-α and IL1 are the key cytokines for the initiation of inflammatory responses which is mainly due to their ability to induce NF-κB. However, the signal transduction pathways leading to the activation of NF-κB by TNF-α or IL-1 differ significantly (see Fig. 2). TNF-α induces NF-κB by the engagement of the two different TNF receptors, TNFR1 (p55) and TNFR2 (p75), although TNFR1 has been shown to be the key player in most cell types with the exception of lymphoid cells. Active TNF-α is a homotrimer of 157 amino acid subunits, expressed either as a soluble or a membrane-bound form. Upon binding of TNF-α the TNFR1 trimerizes creating a signaling platform necessary for the recruitment of a cascade of adaptor proteins and signaling intermediates. As a first step the adaptor protein TNF receptor-associated cell death domain (TRADD) binds via its death domain to the TNFR1 causing the recruitment of the Fas-associated death domain(FADD), TNFR-associated factor 2 (TRAF2) and the receptor-interacting protein (RIP), a serine/threonine protein kinase (Hsu et al. 1996a, 1996b; Wiegmann et al. 1999). RIP is essential for the TNF-α-mediated NF-κB activation, however, its kinase activity is dispensable for this process. Finally, the IKK complex itself binds to the TNF receptor and gets activated. It still remains unknown if the IKK complex interacts via the IKKs or via NEMO. However, TRAF2 seems to be crucial for the binding of the IKK complex to the TNFR (Devin et al. 2000, 2001; Zhang et al. 2000b). The precise mechanism leading to the IKK activation is currently unknown. A recent study reveals a crucial role of the MAP3Ks MEKK2 and MEKK3 for the TNF-α-mediated degradation of IκBβ and IκBα, respectively (Schmidt et al. 2003). In addition, recent findings revealed a critical role for ubiquitination steps in the regulation of NF-κB following TNF-α stimulation, probably including the ubiquitination of NEMO (Brummelkamp et al. 2003; Kovalenko et al. 2003; Trompouki et al. 2003).

In contrast to TNFR1, the TNFR2 does not recruit TRADD. However, NF-κB is activated by TRAF2, which is able to interact directly with TNFR2

(Rothe et al. 1995). And this TRAF2:TNFR2-interaction is sufficient to induce NF-κB. In addition, the NF-κB activity is affected due to an intense cross-talk between TNFR1 and TNFR2.

A TRADD independent NF-κB activation seems to be caused by the adaptor protein FAN, which binds to the TNFR1 outside its death domain. FAN activates the neutral sphingomyelinase (N-Smase), which in turn leads to the production of ceramide. The role of ceramide for the TNF-α-induced NF-κB activity is still a matter of discussion. Conflicting data demonstrated a positive or a negative effect of ceramide on NF-κB. This discrepancy might reflect cell type specific differences or the consequences of different ceramide concentrations in the experimental setup (Adam-Klages et al. 1996; Boland et al. 1996; Kitajima et al. 1996; Boland and O'Neill 1998).

IL-1β is another important proinflammatory cytokine known to induce NF-κB. However, as mentioned above, the IL-1-induced signaling pathways leading to NF-κB activity differ significantly from those used by TNF-α. Stimulation with IL-1β induces the binding of a subset of adaptor proteins and kinases to the IL-1β receptor which resemble those used by the TLRs. As an initial step the IL-1β receptor accessory protein is engaged to the IL-1β receptor followed by the binding of the adaptor protein MyD88 (Huang et al. 1997). MyD88 subsequently recruits the protein kinases IRAK1 and IRAK4. IRAK4 phosphorylates and activates IRAK1 upon recruitment which in turn leads to the binding of the adaptor protein TRAF6 (Boland et al. 1996; Muzio et al. 1997; Burns et al. 1998; Qian et al. 2001). At this stage the interaction of IRAK with Tollip (Toll-interacting protein) is resolved and Tollip, which is thought to be a silencer of the IL-1/TLR-signaling pathways, is degraded by the proteasome (Burns et al. 2000). The activated IRAK1/IRAK4/TRAF6-complex dissociates from the IL-1β receptor and subsequently induces MAPKs and NF-κB. The molecular mechanism underlying the activation of NF-κB has not been resolved yet. One scenario involves the activation of the MAP3 K MEKK1 by an endoproteolytical step which involves the factor ECSIT (Kopp et al. 1999). Activated MEKK1 has been implicated to activate the IKK complex (Lee et al. 1998; Baumann et al. 2000). However, the TAK1/TAB1/TAB2-complex, which is recruited by TRAF6, similar to TLR signaling seems to play a crucial role in the IL-1-mediated activation of NF-κB (Ninomiya-Tsuji et al. 1999; Ishitani et al. 2003; Takaesu et al. 2003).

Although the pathways leading to the IKK-activation and subsequently the degradation of the IκB proteins differ between IL-1β and TNF-α, they also overlap in part. One example is the TAK1/TAB complex, which is involved in the activation of IKK by both cytokines (Ninomiya-Tsuji et al. 1999; Holtmann et al. 2001; Sakurai et al. 2003; Takaesu et al. 2003). The importance of the TAB/TAK1 complex for IL-1 as well as TNF-α-mediated signaling is supported by studies with specific small inhibitory RNAs. In addition, TAB3, another component of the TAB/TAK1 complex, has been identified. TAB3 and TAB2 seem to have redundant functions. A further important step in the activation of NF-κB is the phosphorylation of p65/RelA itself. A phosphorylation of p65 at serine 311 in the RHD by the atypical protein kinase C (PKC) isoform PKCζ affects the binding of p65 to DNA. Both IL-1β (via TRAF6) and TNF-α (via RIP and p62) induce PKCζ which in turn phosphorylates p65 (Sanz et al. 2000; Leitges et al. 2001; Martin et al. 2002; Duran et al. 2003).

3
The Alternative Pathway of NF-κB Activation: Contribution to Development of Secondary Lymphoid Organs

The most striking result obtained from the studies with IKK-deficient mice is the finding that, despite their shared ability to phosphorylate IκBα in vitro, these kinases exert different, nonredundant functions. IKK2 mediates the cytokine-induced phosphorylation and degradation of IκBα whereas IKK1 is dispensable for this process (Hu et al. 1999; Li et al. 1999a, 1999b, 1999c). However, both IKKs, IKK1 as well as IKK2, are necessary for the full activation of NF-κB due to the phosphorylation of the p65/RelA transactivation domain (Li et al. 2002). Besides this function within the classical IKK complex, studies with a knockin mouse line that expresses exclusively a dominant negative isoform of IKK1 (with serine to alanine substitutions in the activation loop) revealed an essential role of IKK1 in a second, alternative NF-κB-pathway. Different members of the TNF family (CD40L, BAFF, LT-β and TWEAK) as well as LPS or viral proteins (Tax, LMP1) induce the processing of the precursor NF-κB2/p100 to the mature p52 protein (Kaisho et al. 2001; Senftleben et al. 2001a; Xiao et al. 2001; Claudio et al. 2002; Dejardin et al. 2002; Eliopoulos et al. 2003; Mordmuller et al. 2003; Muller and Siebenlist 2003). NF-κB2, either as p100 or as p52, forms a complex with RelB and p52:RelB heterodimers are specifically induced by the alternative pathway (see Fig. 2). This process has a slow kinetic when compared to the classical NF-κB pathway and requires hours rather then minutes. In addition, it has been shown to be dependent on de novo protein synthesis, which is in contrast to the classical pathway. Nearly all alternative pathways identified so far are NEMO independent, with the exception of the LPS-mediated p100 processing (Mordmuller et al. 2003), but require the protein kinase NF-κB-inducing kinase (NIK). The molecular mechanism–how NIK and IKK1 act together–is not fully understood, but as a consequence of their action p100 is phosphorylated at its C terminus and is subsequently processed in the proteasome (Senftleben et al. 2001a; Xiao et al. 2001). Interestingly, the alternative pathways play a crucial role in developmental processes, especially in the development of secondary lymphoid organs. This is also highlighted by the fact that several mouse lines with targeted disruption of components of the alternative pathway, like RelB-, NF-κB2- or lymphotoxin β (LTβ) deficient mice, as well as the *aly/aly*-mouse line which expresses exclusively a dominant negative mutant of NIK, lack secondary lymphoid organs. These mice show a defective micro-architecture of the thymus and spleen and lack lymph nodes as well as Peyer's patches (Koni et al. 1997; Caamano et al. 1998; Xiao et al. 2001; Weih and Caamano 2003). Despite these differences between the two distinct NF-κB activation pathways it is clear that the classical and alternative pathways are connected. For example, the NF-κB/p100 and RelB are regulated by NF-κB and it has been demonstrated that the expression of RelB is dramatically increased after induction of the classical pathway by various agonists (Murakami et al. 1995; de Wit et al. 1998; Bren et al. 2001; Solan et al. 2002). Furthermore, the complex formation with RelB is necessary to stabilize the p100 protein (Maier et al. 2003). Taken together, it seems that the activation of the classical pathway is required to achieve the full activation of the alternative pathway.

4
Role of NF-κB in Inflammation—The Integral Part of Innate Immunity

Inflammation is not necessarily a result of an invasion of microbial organisms, it is rather a general response to an insult of the body integrity, which might also be caused by a wound, stress or necrosis. However, microbial organisms are the main cause of acute inflammatory responses. An inflammatory response is generally caused by a shift in cytokine expression. This initial cytokine expression and secretion is amplified by a subsequent expression of more proinflammatory mediators (cytokines and low molecular weight mediators like nitric oxide) and other factors (for a detailed discussion see below). As discussed above, NF-κB is not only induced by various proinflammatory cytokines, but also by the engagement of all known TLRs, the key receptors for the regulation of the innate immunity (see above). Furthermore, NF-κB is the master transcription factor involved in the regulation of the vast majority of cytokines and acute response factors. Therefore, the activity of NF-κB integrates innate immunity into the inflammatory response and adaptive immunity. Inflammation is a fine-tuned response both in intensity and duration. A failure to terminate the inflammatory response has dramatic consequences for the morbidity and mortality of the organism. The importance of NF-κB for septic shock and chronic inflammatory diseases are highlighted in the section below. It is noteworthy, however, that NF-κB plays a role in all aspects of inflammation. It is also necessary for the resolution of an inflammatory response (Lawrence et al. 2001). This is caused either by a negative effect mediated by p50-homodimers or RelB:p50-heterodimers, or by the expression of anti-inflammatory gene products mediated by p65:p50-heterodimers (Xia et al. 1997; Saccani et al. 2003).

4.1
Phases of Inflammation

Inflammation is usually a beneficial host response to tissue damage caused by a wound or the invasion of a pathogenic microorganism. It involves the complex interplay of different cell types, which in the end, leads to the restoration of the tissue integrity and function. The sites of inflammation show four characteristic features: redness (rubor), heat (calor), swelling (tumor) and pain (dolor).

In the development of an inflammatory response to microbes, distinct phases can be distinguished: in the first phase, bacteria have to exit from the blood stream and enter the tissue by crossing the endothelial cell layer. Normally, the bacteria infect the endothelium thereby disturbing the endothelial barrier function. In the tissue, they get phagocytosed by resident macrophages (Fig. 3A). This results in activation of macrophages and consequently synthesis of inflammatory mediators like cytokines, enzymes and leukotrienes that are secreted. They act on endothelial cells to increase the blood flow to the inflammatory site by vasodilation. Also the adhesiveness and the permeability of the vasculature are increased leading to the uptake of leukocytes and proteins from the blood. Subsequently, leukocytes get to the inflammatory site where they exert their effects. On the other hand, the blood vessels leading away from the site are constricted to prevent the spreading of the tissue damage or infection which would ultimately lead to sepsis (for a review see Nathan 2002).

Attraction and migration of leukocytes through the endothelial cell layer itself also can be divided into distinct phases (Fig. 3B): the first step is the reversible adhesion to the endothelium which is mediated through members of the selectin family on the endothelial cells and their carbohydrate ligands on the leukocytes. This interaction is weak and allows leukocytes to 'roll' along the vascular endothelial surface by making and breaking contacts. The strong and irreversible contact is achieved by interactions of induced intracellular and vascular adhesion molecules (ICAM-1 and VCAM-1) on the endothelium and their integrin receptors LFA-1, Mac-1 and VLA-4 on the rolling leukocyte. The expression of the integrins and their clustering at the leukocyte cell surface is induced by chemokines such as IL-8 and monocyte chemoattractant protein (MCP)-1 released by the endothelium. Once a strong interaction between leukocytes and endothelial cells is achieved, the leukocytes can transmigrate through the endothelial cell layer, most likely through endothelial junctions (reviewed by Ebnet and Vestweber 1999). The first cell types that emigrate from the blood are the phagocytic short-lived neutrophils and macrophages. In later stages of inflammation, also cells of the adaptive immune response, B and T cells enter the site. The transmigrated cells follow a chemokine gradient through the tissue to the inflammatory site where they remove the causative agents.

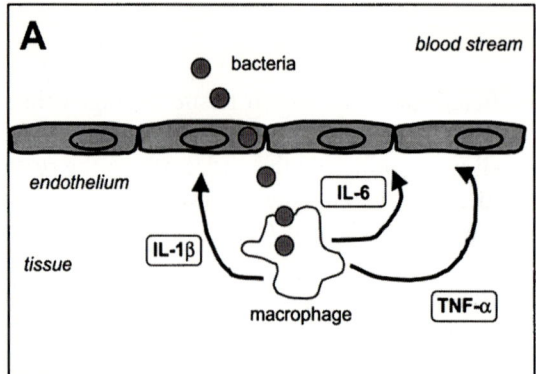

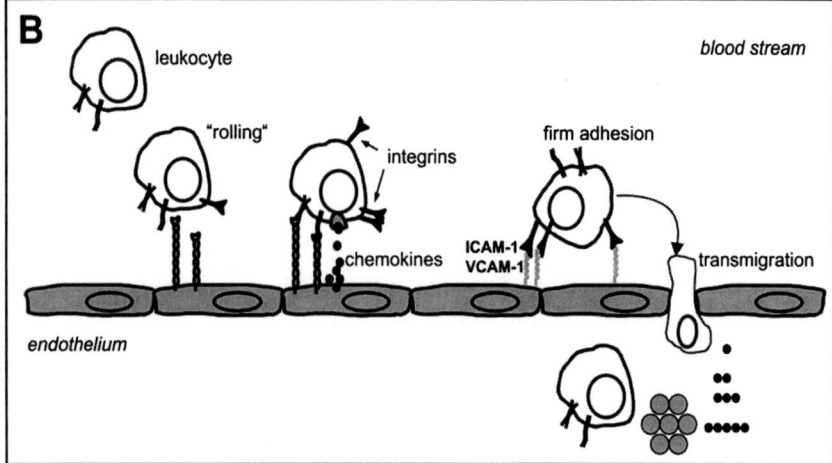

Fig. 3A, B Phases of inflammation. **A** Initial events following invasion of bacteria. **B** Later stages of inflammation: leukocyte attraction and activation. See text for details

To repair inflammatory lesions after the causative agent has been eliminated, the normal tissue architecture has to be restored. Blood clots are removed by fibrinolysis and the tissue is remodeled and repaired by the action of matrix metalloproteinases (MMPs). If it is not possible to return the tissue to its original form, scarring occurs. A granuloma might form if the inflammation-causing agent could not be removed completely.

In addition to the above mentioned, many other cellular and chemical mediators have been shown to be involved in inflammatory responses, such as mast cells, eosinophils and the complement system. But due to the limited space we are focusing on just some of the key players.

4.2
Involvement of NF-κB in Inflammatory Processes

The transcription factor NF-κB is the most critical transcriptional regulator of inflammation. It is involved in the control of virtually every step in this process. The activation of NF-κB correlates with the recruitment of leukocytes and the development of inflammation (Manning et al. 1995).

First of all, NF-κB is activated by many of the viral and bacterial products that are taken up by macrophages (Pahl 1999). Necrotic cells resulting from tissue damage are able to activate NF-κB and proinflammatory gene expression also (Li et al. 2001).

The pathway initiated by the bacterial compound LPS that leads to NF-κB activation has been elucidated in detail (see subsection 2.1). The activated NF-κB dimer is translocated to the nucleus and induces the transcription of proinflammatory genes produced by the macrophage (O'Neill 2002). Among those are the proinflammatory cytokines TNF-α, IL-1β and IL-6, which can diffuse through the tissue and activate endothelial cells. In addition to the proinflammatory cytokines, macrophages also secrete lipid-derived chemical mediators like the prostaglandins and the leukotrienes. The prostaglandins are synthesized from phospholipids by the action of the enzymes phospholipase A2 and cyclooxygenase, whose expression is controlled by NF-κB. They increase vasodilation and vascular permeability and act as a chemoattractant for neutrophils. The leukotriens are also synthesized by NF-κB-controlled enzymes, phospholipase A2 and 5-lipoxygenase, and serve as chemoattractants for neutrophils.

The proinflammatory cytokines TNF-α and IL-1β released by the macrophages activate endothelial cells by an NF-κB dependent mechanism meaning that NF-κB is both upstream and downstream of these proteins. This could lead to a positive feedback loop potentiating and perpetuating the inflammatory response and could be one of the reasons for chronic inflammatory diseases. However, there are also negative feedback loops which prevent excessive NF-κB activation. First of all, the expression of IκBα is NF-κB dependent and newly synthesized IκBα is able to enter the nucleus and to remove NF-κB from the DNA (Sun et al. 1993; Chiao et al. 1994; Arenzana-Seisdedos et al. 1995). Recently it was shown that twist proteins are induced by proinflammatory cytokines via an NF-κB dependent pathway and interact with RelA to repress NF-κB-dependent transcription and consequently exaggerated inflammatory responses (Sosic et al. 2003).

The activation of the endothelial cells leads to the synthesis of adhesion molecules such as ICAM-1, VCAM-1 and E-selectin and of chemokines such as MCP-1 and IL-8, and NF-κB-binding sites have been identified in the promoters of these molecules (Whelan et al. 1991; Iademarco et al. 1992; Kunsch and Rosen 1993; Ueda et al. 1994; Ledebur and Parks 1995). Moreover, approaches with modulated NF-κB activity have shown that NF-κB is

Table 1 Inflammatory diseases with NF-κB involvement

Disease	Reference
Rheumatoid arthritis	Handel et al. 1995
Atherosclerosis	Brand et al. 1996
Multiple sclerosis	Gveric et al. 1998
Chronic inflammatory demyelinating polyradi-culoneuritis	Andorfer et al. 2001
Asthma	Hart et al. 1998
Cystic fibrosis	Venkatakrishnan et al. 2000
Acute respiratory distress syndrome	Schwartz et al. 1996
Inflammatory bowel disease	Rogler et al. 1998
Helicobacter pylori-associated gastritis	van Den Brink et al. 2000
Atopic dermatitis	Nakamura et al. 2002
Systemic inflammatory response syndrome	Christman et al. 1998
Glomerulonephritis	Sakurai et al. 1996
Acute pancreatitis	Dunn et al. 1997

required for their expression. For example, the expression of a kinase-inactive mutant of IKK2 leads to a decrease in the expression of several activation markers and consequently also in the adhesion of monocytes (Denk et al. 2001; Oitzinger et al. 2001; Meiler et al. 2002). The treatment of endothelial cells with the NF-κB inhibitor salicylate results in an inhibition of adhesion molecule expression and neutrophil transmigration and a dominant negative mutant of RelA also prevents proinflammatory gene expression (Pierce et al. 1996; Soares et al. 1998). However, NF-κB is not only essential, but also sufficient for the expression of ICAM-1, VCAM-1, E-selectin, MCP-1 and IL-8 (Denk et al. 2001). This finding additionally stresses the importance of NF-κB in the development of an inflammatory response. Consequently, NF-κB is involved in a variety of inflammatory diseases affecting different tissues (see Table 1).

NF-κB is not only essential for the activation of different cell types during inflammation, but it also regulates the differentiation of cells of the adaptive immune system which are recruited at later stages of the inflammatory response. This issue will be discussed in the second part of this chapter.

It should be noted that recent reports suggested that NF-κB might be involved not only in the progression of inflammation, but also in its resolution by promoting leukocyte apoptosis. However, to date this process has been investigated relatively little (Lawrence et al. 2001, 2002).

4.3
Inflammatory Phenotypes of NF-κB Mutant Mice

In agreement with its prominent role in the regulation of inflammation, mice lacking certain components of the NF-κB pathway show inflammatory phenotypes or an effect on experimentally induced inflammation.

If the gene for IκBα, which usually downregulates NF-κB activity, is deleted, the mice develop a severe dermatitis due to infiltration of macrophages and neutrophils into the skin which leads to premature death. Moreover, the spleen is infiltrated and enhanced granulopoiesis takes place (Beg et al. 1995; Klement et al. 1996; Huber et al. 2002). When RAG2$^{-/-}$, IκB$\alpha^{-/-}$ chimeras were generated, the mice developed a psoriasisform skin disease, thereby directly linking deregulated NF-κB activity to disease (Chen et al. 2000).

The loss of NEMO, in humans or in mice, results in a skin phenotype resembling the disease incontinentia pigmenti (Makris et al. 2000; Schmidt-Supprian et al. 2000; Smahi et al. 2000). It is characterized by inflammatory infiltration of the skin mediated mainly by eosinophils. Furthermore, keratinocytes show increased proliferation and apoptosis.

The role of NF-κB especially in inflammation of the skin is further strengthened by the phenotype of mice with a skin-specific deletion of the IKK2 gene. They show a severe inflammatory disease of the skin that is mediated by TNF-α and involves macrophages, granulocytes and CD4$^+$ T cells (Pasparakis et al. 2002a).

The fact that mutations both activating and inactivating NF-κB trigger an inflammatory skin disease reflects the complex interplay of different cell types and cytokines in this organ which has to be correctly balanced by the action of NF-κB.

The lack of the ankyrin repeats of NF-κB1 results in a chronic perivascular inflammation of the lung and the liver due to B and T cell recruitment (Ishikawa et al. 1998).

The strongest inflammatory phenotype is observed in mice that are not able to express RelB: they show multifocal, mixed inflammatory cell infiltration in various organs which is T cell dependent (Burkly et al. 1995; Weih et al. 1995, 1996). This might be due to the failure of RelB-deficient fibroblasts to terminate the expression of certain chemokines that is a prerequisite for the resolution of an acute inflammation (Xia et al. 1997). Again, the skin is particularly affected showing symptoms of the disease atopic dermatitis such as infiltration with CD4$^+$ T cells and eosinophils, hyperkeratosis and epidermal hyperplasia (Barton et al. 2000).

In addition to the NF-κB deficient animals developing spontaneous inflammation, other NF-κB subunits have been shown to be involved in inflammation by animal models of inflammatory diseases.

In this respect, it could be shown that both p50 and c-Rel are involved in the development of allergen-induced asthma as mice lacking any one of them show decreased airway hyperresponsiveness and reduced eosinophilic infiltration (Yang et al. 1998; Donovan et al. 1999). Also, mice that do not express either p50 or c-Rel are resistant to autoimmune encephalomyelitis, an animal model of multiple sclerosis, due to a Th2 or Th1 deficiency, respectively (Hilliard et al. 1999, 2002). Distinct roles of p50 and c-Rel have been shown for the development of inflammatory arthritis: C-Rel seems to be required only for the chronic, but not for the acute joint disease whereas p50 is essential for both (Campbell et al. 2000).

4.4
Crosstalk Between NF-κB and Other Signaling Pathways During the Development of Inflammation

The promoters of many genes involved in inflammation do not only contain binding sites for NF-κB but are often coregulated by the AP-1 transcription factor family (Okamoto et al. 1994; Roebuck et al. 1995; Martin et al. 1997; Read et al. 1997; Ahmad et al. 1998). Members of this family are activated by the same stimuli as NF-κB, but by a signaling pathway involving the stress activated protein kinase JNK. It has been shown that for most of the inflammatory genes, NF-κB and AP-1 have to cooperate to induce maximal gene induction (Holtmann et al. 1999). However, expression of a kinase-inactive mutant of SEK, the AP-1 activating kinase, did not result in a severe block of gene induction by TNF-α in human umbilical vein endothelial cells (HUVEC) (Denk et al. 2001). This suggests that AP-1 might play a minor role in regulating gene expression in comparison to NF-κB or that there is a stimulus-specific regulation of the genes.

Another MAPK pathway involved in proinflammatory gene induction is the p38 pathway. It could be shown that p38 stabilizes IL-8 RNA whose expression was stimulated by NF-κB and AP-1 (Holtmann et al. 1999). Moreover, p38 stimulates NF-κB transcriptional activity by the recruitment of coactivators and it is able to activate histone H3 kinase that then phosphorylates histone H3 in the promoters of genes induced by proinflammatory stimuli (Goebeler et al. 2001; Saccani et al. 2002).

These data indicate that the NF-κB pathway is not the only pathway induced by proinflammatory stimuli, although it is probably one of the most important.

The glucocorticoid receptor negatively interferes with proinflammatory NF-κB activity, but this will be discussed in the pharmacological inhibitors section.

5
Chronic Inflammatory Diseases

As mentioned above, inflammation is usually a beneficial host response that leads to the clearance of microbial invasion or damaged tissue from the body. However, if the antigen cannot be cleared efficiently, inflammation persists and becomes chronic. Chronic inflammation can be distinguished from acute inflammation by the composition of the inflammatory infiltrate:

in contrast to acute inflammation, low numbers of neutrophils and high numbers of T cells and macrophages can be detected in the affected tissue. Because of the resulting tissue damage and the attempts to repair it, it can lead to the development of scarred tissue, fibrosis, and granuloma formation, which may influence the quality of life of the affected individuals. As NF-κB is one of the main transcriptional regulators of inflammation, it is not surprising that it also plays a role in the initiation and progression of various chronic inflammatory diseases. In the following, we will discuss the role of NF-κB in the development of two of the most extensively studied examples.

5.1
Rheumatoid Arthritis

Rheumatoid arthritis (RA) is an autoimmune disease which leads to chronic inflammation of the joints (reviewed by Feldmann et al. 1996); it affects about 1% of the US population. Important cells in the development of the disease are the synoviocytes that line the joint capsule and produce synovial fluid. In RA, leukocytes–mainly T cells and activated macrophages–invade the synovium and the macrophages produce high amounts of IL-1β and TNF-α. The cells of the synovium are stimulated to grow and to divide abnormally. This leads to the formation of a tumor-like structure called pannus. During the progression of the disease, the abnormal synovial cells begin to invade and to irreversibly destroy the cartilage and bone within the joint. This is accompanied by a reduced proteoglycan synthesis of chondrocytes, enhanced proteoglycan degradation and the induction of the cartilage-degrading MMP-1 and MMP-13 in synoviocytes. Due to this, the surrounding muscles, ligaments and tendons become weak and do not work properly. The bone destruction that takes place is due to the action of osteoclasts which develop from monocytes and which are responsible for bone resorption. These destructive processes lead to progressive disability and crippling of RA patients.

NF-κB is a key regulator in the development of RA since it is involved in all of the pathogenic steps (reviewed in Makarov 2001). First of all, activated NF-κB can be detected in the synovial tissue in both early and later stages of inflammation by immunohistochemistry (Handel et al. 1995; Marok et al. 1996; Gilston et al. 1997). Also, electrophoretic mobility shift analyses demonstrated an increased DNA-binding activity of NF-κB in synovial explants and the infiltrating T cells (Asahara et al. 1995; Collantes et al. 1998). The use of a NF-κB dependent reporter gene in single clones of primary synovial cells revealed that a constitutive NF-κB activity correlated with a high production of the proinflammatory cytokine IL-6 (Miyazawa et al. 1998). In a mouse model of RA, increased synovial NF-κB binding precedes the development of clinical joint involvement and further increases during the evolution of the disease (Han et al. 1998). These data provide evidence that RA is accompanied by an increased NF-κB activity which might be caused by TNF-α in the synovial fluid (Lehmann et al. 2000).

Besides regulating the expression of inflammatory mediators, NF-κB is also involved in the enhanced growth of synoviocytes. It could be shown that inhibition of NF-κB by a transdominant IκBα mutant leads to enhanced apoptosis of RA synoviocytes in response to FasL and TNF-α treatment (Miagkov et al. 1998; Zhang et al. 2000a; Yamasaki et al. 2001). This correlation of proliferation and apoptosis was also shown in platelet-derived growth factor (PDGF) treated cells: NF-κB induces the expression of c-Myc and simultaneously protects the cells from the cytotoxic cells of c-Myc (Romashkova and Makarov 1999). This indicates that NF-κB regulates the balance between proliferation and apoptosis: if NF-κB is active, it protects from apoptosis and proliferation takes place; if NF-κB is inhibited, apoptosis occurs and proliferation is blocked.

Several lines of evidence indicate that NF-κB is also involved in the destruction of cartilage and bone, the latter of which is the distinctive feature of RA. The destruction of the joint takes place via the action of MMPs that are synthesized by the synoviocytes and degrade the extracellular matrix. The promoters of MMP-1 and MMP-9 have a NF-κB-binding site that is required for their transcriptional induction (Bond et al. 1998; Vincenti et al. 1998). By the use of transdominant IκBα it could be shown that NF-κB activity is necessary for the expression of MMP-1, MMP-3, MMP-9 and MMP-13 (Bond et al. 1999, 2001; Bondeson et al. 1999, 2000; Mengshol et al. 2000). This indicates that constitutive NF-κB activity in RA contributes to cartilage destruction. The cell type that is suggested to be responsible for the destruction of the bones is the osteoclast. These are derived from hematopoietic stem cells and develop together with the myeloid lineage. Their main function is bone resorption by decalcification and degradation of bone matrix proteins. The osteoclast precursors are recruited to the synovium by the NF-κB dependent proinflammatory cytokines. When they have reached their target site, they are activated by interaction of RANK (receptor activator of NF-κB) with its ligand which is expressed on activated T cells upon stimulation with proinflammatory cytokines (for a review see Rehman and Lane 2001). The importance of NF-κB in the maturation of osteoclasts is demonstrated by the phenotype of p50/p52-deficient mice that develop osteopetrosis due to a maturation defect of the osteoclasts (Iotsova et al. 1997). The importance of osteoclasts for the bone destruction in RA has been nicely studied in a genetic model. When mice transgenic for TNF-α that develop RA were crossed to c-Fos deficient mice which do not possess mature osteoclasts the bone destruction was inhibited while inflammation still took place (Redlich et al. 2002).

The role of NF-κB in the pathogenesis of RA is further stressed by numerous experiments using modulators of NF-κB activity. When an inhibitor of the proteasome was used in a rat model of arthritis, the degradation of IκBα and the progression of the disease was blocked (Palombella et al. 1998). Introduction of the specific NF-κB decoy oligonucleotides into the synovial cells of RA patients leads to a reduction in the expression of several inflammatory genes and to an inhibition of synovial cell proliferation (Tomita et al. 2000). Consequently, the delivery of NF-κB decoy oligodeoxynucleotides into the joints of rats with experimentally induced arthritis results in a decrease in joint destruction (Tomita et al. 1999). Infection of rheumatoid joints with an adenovirus carrying a nondegradable IκBα mutant leads to a reduction of the expression of proinflammatory cytokines and matrix degrading enzymes (Foxwell et al. 1998; Bondeson et al. 1999). In a similar approach, variants of the IKK proteins were overexpressed in synoviocytes. In this case, wild-type IKK2 increased IL-6 and IL-8 production, ICAM-1 and collagenase synthesis, whereas a dominant-negative mutant of IKK2 inhibited the expression of all of these genes (Aupperle et al. 2001). In this study it was also shown that IKK2 is the main IκB kinase in RA as a dominant-negative mutant of IKK1 had no effect on proinflammatory gene expression. When wild-type IKK2 was introduced into normal rat joints, the animals developed synovial inflammation and clinical signs of arthritis, whereas a dominant negative IKK2 mutant attenuated the severity of experimentally induced arthritis (Tak et al. 2001). Taken together, these results convincingly show that NF-κB activation by IKK2-induced IκBα degradation is a key process in the development of RA.

5.2
Atherosclerosis

Atherosclerosis is a multistep, chronic inflammatory disease that involves several cell types and chemical mediators. In the western world, about 50% of people die from atherosclerosis-related disease which means that atherosclerosis is the most common cause of death (Biedermann 2001). Multiple risk factors, such as hypertension, diabetes mellitus, smoking and lipoprotein disorders have been identified that are involved in the development of atherosclerosis.

The development of early atherosclerotic lesions can be divided into three different stages: initiation, expansion and progression to plaques.

Initiation involves the increased uptake of low density lipoproteins (LDL) at lesion prone sites. These are mainly located in areas where the blood flow is disturbed, like branch points and bifurcations. In the arterial wall the LDLs are modified to oxidized LDL (oxLDL) by any of the cell types present there. Also monocytes are recruited from the blood stream to the intima of the vascular wall by the enhanced expression of adhesion molecules and chemokines on endothelial cells. The activated monocytes take up oxLDL via the scavenger receptor leading to their transition into foam cells and the formation of fatty streaks underlying the endothelium of large arteries. Foam cells produce cytokines, growth factors and a variety of stress proteins. Activated T helper cells also immigrate into the intima where they interact with each other and the macrophages via cytokines and induce an immune response.

In the expansion phase, the fatty streaks grow vertically and laterally and coalesce. The progression to atherosclerotic plaques is associated with the immigration of smooth muscle cells into the intima of the artery wall where they proliferate and synthesize extracellular matrix proteins which form a fibrous cap. This leads to the narrowing of the arteries and the impairment of blood flow at the site of the plaque. A late event in atherosclerosis is the formation of a lipid-rich necrotic core due to the death of macrophages and smooth muscle cells at the base and the margin of the plaque.

The clinical consequences of atherosclerosis like stroke or myocardial infarction are caused by plaque rupture and the formation of occlusive thrombi. Plaque rupture usually occurs in areas with a thin fibrous cap and may be due to the actions of MMPs secreted by the macrophages. After plaque rupture, plaque cells expressing tissue factor are exposed to the blood stream; this leads to coagulation, the recruitment of platelets and the formation of a thrombus (for review on atherosclerosis development see Rosenfeld 2000; Glass and Witztum 2001).

A first hint that NF-κB could be critical for the development of atherosclerosis came from the detection of activated NF-κB in atherosclerotic lesions whereas there is almost no NF-κB activity in healthy vessels (Brand et al. 1996). This NF-κB activity was present in smooth muscle cells, macrophages and endothelial cells, the key cell types in atherosclerosis development. The prerequisite for the NF-κB activity is the high expression of RelA in regions with a high probability for the development of atherosclerotic plaques (Hajra et al. 2000). Moreover, active NF-κB can also be detected in the coronary arteries of pigs with a hypercholesterolemic diet and in the arteries of patients with unstable angina pectoris (Wilson et al. 2000, 2002).

NF-κB can be involved in different cell types and in several steps of atherosclerosis (for review see Collins and Cybulsky 2001). First, it regulates the expression of the majority of adhesion molecules and chemokines by endothelial cells, but also the expression of growth regulatory proteins that may influence proliferation and cell death of macrophages and smooth muscle cells. Furthermore, the expression of some coagulant proteins such as tissue factor is under the control of NF-κB (for review on target genes see Pahl 1999). Together these target genes may have a substantial impact on atherosclerosis development.

The importance of some NF-κB target genes for the development of atherosclerosis has been demonstrated in genetic model systems. In these models mainly apoE$^{-/-}$ or LDLR$^{-/-}$ mice are used which show hypercholesterolemia and are therefore susceptible to atherosclerosis. These mice were crossed to mice deficient for adhesion molecules or chemokines and the development of lesions was investigated. Thereby it was shown that a reduction in VCAM-1 expression protects mice from the development of atherosclerosis (Cybulsky et al. 2001). This corresponds to the expression pattern of VCAM-1 in areas predisposed to lesion formation (Iiyama et al. 1999). In line with this, the combined absence of E-selectin and P-selectin leads to a 40%–60% decrease in atherosclerosis (Dong et al. 1998). In addition, the NF-κB dependent chemokine MCP-1 has been shown to be critically involved in the progression of the disease. Consequently, mice lacking MCP-1 or its receptor CCR2 are protected from the formation of atherosclerotic lesions (Boring et al. 1998; Gu et al. 1998). Recently it was shown that aspirin, an inhibitor of NF-κB activity that is used to prevent vascular complications of atherosclerosis, leads to a significant reduction of NF-κB activity in the aorta of animals susceptible to atherosclerosis. This was accompanied by a reduction of proinflammatory gene expression and stabilization of the atherosclerotic plaque (Cyrus et al. 2002).

Importantly, mediators or risk factors in atherosclerosis simultaneously activate NF-κB. As mentioned above, one hallmark of atherosclerosis is the generation of intracellular oxidative stress which functions as an activator of NF-κB activity (reviewed in Li and Karin 1999). Therefore, treatment of cells with anti-oxidant substances or enzymes leads to the inhibition of NF-κB activity. When hypercholesteremic pigs were treated with the antioxidative vitamins E and C, NF-κB activation was reduced and this corresponded to a protection of normal endothelial functions such as vasoreactivity (Rodriguez-Porcel et al. 2002). However, the mechanism of how reactive oxygen species activate NF-κB remains so far unresolved. Elevated levels of LDL as well as the product of LDL oxidation, oxLDL, are able to induce of NF-κB activity in monocytes and endothelial cells leading to the expression of chemokines and adhesion molecules (Brand et al. 1997; Calara et al. 1998; Palmetshofer et al. 1999). Other risk factors for atherosclerosis, like hypertension and hyperglycaemia, lead to the upregulation of mediators such as angiotensin II and advanced glycation end products which induce oxidative stress and thereby activate NF-κB (Morigi et al. 1998; Han et al. 1999; Golovchenko et al. 2000; Pueyo et al. 2000; Tham et al. 2002). As atherosclerosis is considered to be a chronic inflammatory disease of the arterial wall, the inflammatory response might be triggered by bacteria and/or viruses. In particular, cytomegalovirus and *Chlamydia pneumonia* infection have been associated with atherosclerosis (reviewed in Leowattana 2001); both of these microbial pathogens have been shown to activate NF-κB (Yurochko et al. 1995; Donath et al. 2002).

Also disturbed laminar flow in regions predisposed for the development of atherosclerotic lesions leads to the activation of NF-κB through an IKK-dependent mechanotransduction mechanism (Mohan et al. 1997; Bhullar et al. 1998; Gimbrone 1999).

The role of NF-κB in the activation of vascular smooth muscle cells that proliferate abnormally in atherosclerosis and are involved in the formation of the fibrous cap has been mostly studied in vitro. In the normal vessel wall, smooth muscle cells do not contain active NF-κB, but in atherogenic vessels it is present in the nuclei of these cells (Bourcier et al. 1997). It has been shown that NF-κB is activated in response to mitogens and cytokines in smooth muscle cells and regulates their proliferation (Lawrence et al. 1994; Obata et al. 1996; Hishikawa et al. 1997). Consequently, the inhibition of NF-κB by mutants of the pathway or chemicals leads to a decrease in cytokine- or growth factor-induced proliferation (Nakajima et al. 1994; Autieri et al. 1995; Bellas et al. 1995; Bretschneider et al. 1997; Selzman et al. 1999a, 1999b; Sasu and Beasley 2000; Wang et al. 2002b). Another function of NF-κB linked to intimal hyperplasia is the protection of smooth muscle cells from apoptosis as inhibition of NF-κB by a transdominant IκBα dramatically increases spontaneous and TNF-α-induced apoptosis (Erl et al. 1999; Obara et al. 2000; Wang et al. 2002b). Migration of smooth muscle cells to the intima of the artery is also regulated by NF-κB as TNF-α- and vascular endothelial growth factor-induced migration could be inhibited by a transdominant IκBα (Wang et al. 2001a, 2001b). Furthermore, expression of the MMPs that degrade the extracellular matrix at the shoulder regions of atherosclerotic plaques seems to be NF-κB-dependent (Bond et al. 2001).

Taken together, these data indicate that NF-κB is involved in the pathogenesis of atherosclerosis by regulating different reactions in different cell types.

Although a lot of data exist about the role of NF-κB in atherosclerosis, all of them are only correlative. They all state that NF-κB activity is higher in atherosclerotic lesions and that NF-κB target genes are upregulated by stimuli that are able to activate NF-κB. But direct evidence for the critical involvement of NF-κB in the initiation and progression of this disease has not yet been obtained. A first hint that influencing NF-κB function in vivo could have an effect on vascular function came from the use of NF-κB decoy oligonucleotides in a rat model of balloon injury. Vascular injuries like this often lead to the formation of lesions in the arteries and the subsequent development of atherosclerosis. When the NF-κB decoy oligonucleotides were transfected into rat carotid arteries that were injured before, the formation of intimal hyperplasia was inhibited by induction of apoptosis. Moreover, the expression of adhesion molecules was markedly decreased (Yoshimura et al. 2001).

However, the establishment of mouse systems which have repressed or activated NF-κB and their crossing to atherosclerosis-susceptible animals would be necessary to investigate the involvement of NF-κB in the future.

6
Pharmacological Strategies to Inhibit NF-κB Activity in Chronic Inflammation

So far, about 200 natural and chemical agents have been identified which are able to regulate NF-κB activity by interfering with different steps in the signal transduction pathway. We will focus on some of those that either are commonly used to treat inflammation or that have given results promising for new treatments in the future (Table 2).

Table 2 Pharmacological inhibitors of NF-κB in inflammatory diseases and their use in treating chronic inflammation

Drug	Modes of action	Treated disease	Reference
Gold compounds	Inhibition of IKK activation by thiol reactivity	Rheumatoid arthritis	Jeon et al. 2000
Thalidomide	Inhibition of IKK activity	Rheumatoid arthritis	Keifer et al. 2001
Ibuprofen	Inhibition of IκBα phosphorylation	Rheumatoid arthritis	Scheuren et al. 1998
Salicylates (aspirin)	Inhibition of IKKβ kinase activity by inhibition of ATP binding	Rheumatoid arthritis	Yin et al. 1998
Sulfasalazines	Inhibition of IKKα and IKKβ kinase activity by inhibition of ATP binding	Rheumatoid arthritis, inflammatory bowel disease	Weber et al. 2000
α-TNF-α, α-IL-1β	Inhibition of IKK activation	Rheumatoid arthritis, inflammatory bowel disease	Reviewed in Reimold 2003
Mesalamine	Inhibition of RelA phosphorylation	Inflammatory bowel disease	Egan et al. 1999
PPAR activators (e.g., thiazolidinedions)	Induction of IκBα synthesis	Atherosclerosis	Delerive et al. 2000; Delerive et al. 2002
Glucocorticoids	Induction of IκBα synthesis; direct binding to NF-κB	Various inflammatory diseases	Reviewed in Almawi and Melemedjian 2002
Atorvastatin	Induction of IκBα synthesis	Atherosclerosis	Ortego et al. 1999
Proteasome inhibitors	Inhibition of IκB degradation	Psoriasis, multiple sclerosis, asthma	Elliott et al. 1999; Vanderlugt et al. 2000; Zollner et al. 2002
Anti-oxidants (PDTC, vitamin E)	?	Arthritis	Cuzzocrea et al. 2002
Dehydroxymethylepoxyquinomicin	Inhibition of nuclear translocation	Rheumatoid arthritis	Ariga et al. 2002
NF-κB decoy oligonucleotides	Inhibition of DNA binding	Rheumatoid arthritis, atopic dermatitis	Miagkov et al. 1998; Nakamura et al. 2002

The glucocorticoids are drugs that have proved to be reliable and effective in the treatment of various inflammatory diseases. However, as one of their main functions is the regulation of glucose metabolism, systemic treatment results in many side effects such as redistribution of fat, muscle degeneration and growth retardation—collectively known as the Cushing syndrome. Therefore, only short-term treatment is reasonable, when there is an inflammatory crisis. The molecular mechanism of glucocorticoid action was elucidated only after they had been used for the treatment of inflammation for quite a long time. It could be shown that they influence the transcriptional activity of NF-κB in different ways—each of which might be cell type specific. As the glucocorticoid receptor (GR) belongs to the transcription factor family of nuclear hormone receptors it was reasonable to ask whether the activated GR is able to induce inhibitory genes. Indeed, activated GR was able to upregulate the expression of IκBα that then in turn might associate with NF-κB and downregulate its activity (Auphan et al. 1995; Scheinman et al. 1995). However, mutants of GR unable to transactivate transcription are still able to repress NF-κB activity suggesting that there are additional mechanisms regulating NF-κB function (Heck et al. 1997). As the activated GR is able to bind NF-κB (Adcock et al. 1995; Kleinert et al. 1996; De Bosscher et al. 1997), two different models of action exist. The simplest model would be that interaction of the proteins prevents NF-κB DNA binding, and several hints in this direction exist (Mukaida et al. 1994; Adcock et al. 1995; Steer et al. 2000). Another model suggests that GR-bound NF-κB is able to bind DNA, but it is not able to transactivate transcription as the interaction with the basal transcription apparatus is disturbed (De Bosscher et al. 2000; Nissen and Yamamoto 2000). As GR is a transcription factor like NF-κB it could also be that the two of them compete with each other for a limited amount of transcriptional coactivators such as CBP or p300 (Sheppard et al. 1998; Kino et al. 1999; Nissen and Yamamoto 2000). Whatever the exact mechanism of GR action in a particular cell type is, it is clear that the transactivating and the anti-inflammatory functions of the GR can be separated. In line with this, it could be shown that mice expressing a GR that is not able to bind to DNA are able to suppress inflammatory responses without the side effects on glucose metabolism (Reichardt et al. 2001). Therefore, future studies should concentrate on finding ligands for the GR that do not induce transcription, but are still able to activate the receptor and its transrepressing function.

Another class of drugs is the non-steroidal anti-inflammatory drugs (NSAIDs) whose main use is in arthritis. Aspirin and sodium salicylate are the most prominent members of this class. These agents have been shown to inhibit the enzyme activity of COX2 that is responsible for the synthesis of proinflammatory prostaglandins. But they also seem to have effects on NF-κB activity as aspirin and sodium salicylate can reduce ATP binding to IKK2 which inhibits its kinase activity (Kopp and Ghosh 1994; Yin et al.

1998). Although these drugs are widely used to treat inflammatory diseases they have severe side effects on the gastrointestinal tract and blood clotting.

Sulfasalazine is composed of an anti-inflammatory and an anti-bacterial part which are produced by cleavage carried out by colonic bacteria. It inhibits NF-κB activation at the level of the IκB kinases similarly to aspirin (Wahl et al. 1998; Yin et al. 1998; Weber et al. 2000). In contrast to the NSAIDs it acts as a disease-modifying drug in RA meaning that it not only fights inflammation, but also slows down bone and cartilage destruction. However, treated patients may suffer from severe side effects like infertility in men and sensitivity to sunlight.

One class of drugs which has attracted much interest with respect to vascular biology and atherosclerosis are the PPAR activators which bind and activate the nuclear transcription factors of the PPAR (peroxisome proliferator-activated receptors) family, PPAR α, γ and δ. The receptors activated by fatty acids or eicosanoids play an important role in the regulation of lipid and glucose metabolism. Thiazoledinediones (TZDs) are ligands for PPARγ that are already used as insulin-sensitizers for the treatment of diabetes (Olefsky and Saltiel 2000). In addition, these PPARγ ligands are able to retard atherosclerotic processes by the action on the main cell types in the vasculature (reviewed by Hsueh and Law 2001).

This influence on atherosclerosis is, at least in part, due to the interaction with NF-κB. PPARγ ligands induce apoptosis of cultured macrophages by inhibiting the transcriptional activity of RelA (Chinetti et al. 1998) and PPARα ligands inhibit the activation of human aortic smooth muscle cells by repression of NF-κB signaling (Staels et al. 1998). Furthermore, the activation of PPARs leads to a reduction in the expression of inflammatory cytokines and adhesion molecules (Marx et al. 1998, 1999; Delerive et al. 1999; Su et al. 1999; Kintscher et al. 2000; Pasceri et al. 2000; Rival et al. 2002). One postulated mechanism by which activated PPARs downregulate NF-κB activity is the induction of IκBα synthesis (Delerive et al. 2000, 2002). Another possibility is by influencing the cellular redox status which itself has an impact on NF-κB activity (Poynter and Daynes 1998). Consistent with the suggested role in atheroprotection, the activation of PPARγ in mice susceptible for atherosclerosis leads to an inhibition of lesion formation (Poynter and Daynes 1998; Li et al. 2000a). However, some PPAR ligands are able to modulate NF-κB activity without binding to their receptor as shown for 15-deoxy-$\Delta^{12,14}$-prostaglandin J$_2$. This cyclopentone prostaglandin is able to covalently modify critical cysteine residues in IKK, p65 and p50 thereby inhibiting NF-κB-mediated transactivation (Castrillo et al. 2000; Rossi et al. 2000; Straus et al. 2000; Cernuda-Morollon et al. 2001). This may be part of a negative feedback loop leading to a repression of COX2 expression that is required for prostaglandin synthesis.

Another strategy to inhibit NF-κB specifically in atherosclerotic situations is the use of HMG-CoA reductase inhibitors that are established drugs for the treatment of hypercholesterolemia, but which can also induce the regression of atherosclerosis. These drugs are able to inhibit NF-κB transcriptional activity by upregulation of IκBα expression and consequently downregulate chemokine expression (Ortego et al. 1999; Dichtl et al. 2003). Furthermore, it could be shown that HMG-CoA reductase inhibitors reduce NF-κB activity in atherosclerotic plaques and inhibit the expression of proinflammatory molecules (Hernandez-Presa et al. 2003). This provides an additional explanation for the efficacy of these drugs in the treatment of atherosclerosis.

After phosphorylation by the IκB kinase complex there is one last step necessary to liberate NF-κB dimers from IκB inhibition, namely the ubiquitination and the degradation of IκBs via the 26S proteasome. Therefore, inhibitors of the proteasome should be able to inhibit NF-κB activation. There are three different molecular classes of proteasome inhibitors: peptide aldehydes, acylating agents and boronic acid peptides. Of these, mainly the third class was used to treat inflammatory diseases in animal models. The proteasome inhibitor PS-519 has been used successfully in animal models of inflammatory diseases such as asthma, psoriasis and multiple sclerosis (Elliott et al. 1999; Vanderlugt et al. 2000; Zollner et al. 2002). Thus, targeting the proteasome might be a promising approach to the treatment of inflammatory diseases.

Many inflammatory diseases have been shown to correlate with oxidative stress that is able to activate the redox-sensitive NF-κB (Li and Karin 1999; Rahman 2003). However, the exact pathway to NF-κB activation by oxidative stress has not yet been elucidated and may differ between the different inducing agents. To influence oxidative stress, either synthetic (e.g., pyrrolidinedithiocarbamate; PDTC) or naturally occurring (e.g., vitamin E) anti-oxidants can be used. Indeed, nutritional components which have been shown to be protective against atherosclerosis (e.g., vitamins E and C, garlic compounds and red wine) are acting as anti-oxidants and are able to inhibit NF-κB activity (Staal et al. 1993; Suzuki and Packer 1993; Geng et al. 1997; Blanco-Colio et al. 2000). In animal models of inflammation, PDTC was able to reduce several effects of inflammation (Cuzzocrea et al. 2002). Additionally, vitamin E inhibits NF-κB activation and monocyte adhesion and could be used in the systemic inflammatory response syndrome and in endotoxin-induced airway inflammation (Erl et al. 1997; Bulger and Maier 2003; Rocksen et al. 2003). Therefore it might be possible to modulate inflammation by a special diet.

One big problem that arises when someone attempts to modulate NF-κB to prevent or to cure inflammation is that NF-κB has many beneficial effects besides the effects on inflammation, for example on the development of immune cells or the regulation of apoptosis. The consequences of a defective NF-κB system have been studied in NF-κB deficient mice. The most dramatic effects were seen in mice lacking RelA, IKK2 or NEMO which all die of liver apoptosis during embryonic development. Even if blocking NF-κB activity might not have such a strong effect, it could still render the treated individual more susceptible to bacterial infections as seen in mice expressing a transdominant IκBα mutant in the liver (Lavon et al. 2000). Therefore, very specific inhibitors would be the best way to interfere with NF-κB activation. As the constitutive and the inducible NF-κB activity seem to be controlled by different signaling pathways (see 'Introduction'), it might also be possible to interfere only with the inducible pathway.

One approach to specifically inhibit NF-κB is the use of NF-κB decoy oligonucleotides that provide binding sites for activated NF-κB dimers. They have been used quite often in animal models of inflammation like RA, chronic intestinal inflammation and atopic dermatitis and have given some promising results (Miagkov et al. 1998; Tomita et al. 1999; Nakamura et al. 2002). However, there are substantial problems with the delivery and the controlled release of these molecules. Only recently the use of biodegradable polymer microparticles was suggested to solve these problems (Gill et al. 2002; Zhu et al. 2002).

As NEMO is an essential component of the IκB kinase complex, blocking its interaction with IKK2 by a cell-permeable peptide of the NEMO-binding domain abrogates NF-κB activation and the inflammatory responses in animal models of acute inflammation (May et al. 2000). Additionally, an allosteric inhibitor of IKK1 and IKK2 has been developed which efficiently and selectively blocks NF-κB activation and TNF-α production in mice (Burke et al. 2003).

Another class of drugs for treatment of inflammatory diseases is already on the market and attacks the proinflammatory cytokines that activate NF-κB. Ethanercept is a fusion protein of the soluble TNF-α receptor with an antibody backbone, whereas infliximab is a chimeric anti-TNF-α monoclonal antibody containing a murine TNF-α-binding region and a human IgG1 backbone. Both of them bind TNF-α selectively and efficiently thereby preventing TNF-α binding to its receptor on the target cells. In contrast, anakinra is a recombinant human IL-1 receptor antagonist (IL-1Ra) that binds to IL-1 receptors but does not stimulate any intracellular responses (reviewed by Arend 2002). The advantage of these drugs is that they inhibit NF-κB activation and also the expression of the endogenous cytokines resulting from enhanced NF-κB activity.

However, these agents are not specific for the NF-κB pathway, but can also inhibit other signal transduction pathways initiated by TNF-α or IL-1β treatment. Additionally, NF-κB could be activated by another pathway that might explain the variation of the results in anti-TNF-α therapy (Moreland et al. 1997).

7
Conclusions

NF-κB is a key player in inflammatory diseases as it regulates the expression of different classes of genes involved in all stages of inflammation in a huge variety of cell types. Consequently, dysregulated NF-κB activity is a key feature of various chronic inflammatory diseases. Many anti-inflammatory drugs have been shown to inhibit NF-κB activity by distinct mechanisms. The main focus in the future should be to develop highly selective inhibitors to rule out undesired side effects.

References

Adam-Klages S, Adam D, Wiegmann K, Struve S, Kolanus W, Schneider-Mergener J, Kronke M (1996) FAN, a novel WD-repeat protein, couples the p55 TNF-receptor to neutral sphingomyelinase. Cell 86:937–947

Adcock IM, Brown CR, Gelder CM, Shirasaki H, Peters MJ, Barnes PJ (1995) Effects of glucocorticoids on transcription factor activation in human peripheral blood mononuclear cells. Am J Physiol 268: C331–C338

Agou F, Ye F, Goffinont S, Courtois G, Yamaoka S, Israel A, Veron M (2002) NEMO trimerizes through its coiled-coil C-terminal domain. J Biol Chem 277:17464–17475

Ahmad M, Theofanidis P, Medford RM (1998) Role of activating protein-1 in the regulation of the vascular cell adhesion molecule-1 gene expression by tumor necrosis factor-alpha. J Biol Chem 273:4616–4621

Almawi WY, Melemedjian OK (2002) Negative regulation of nuclear factor-kappaB activation and function by glucocorticoids. J Mol Endocrinol 28:69–78

Andorfer B, Kieseier BC, Mathey E, Armati P, Pollard J, Oka N, Hartung HP (2001) Expression and distribution of transcription factor NF-kappaB and inhibitor IkappaB in the inflamed peripheral nervous system. J Neuroimmunol 116:226–232

Anest V, Hanson JL, Cogswell PC, Steinbrecher KA, Strahl BD, Baldwin AS (2003) A nucleosomal function for IkappaB kinase-alpha in NF-kappaB-dependent gene expression. Nature 423:659–663

Arend WP (2002) The mode of action of cytokine inhibitors. J Rheumatol Suppl 65:16–21

Arenzana-Seisdedos F, Thompson J, Rodriguez MS, Bachelerie F, Thomas D, Hay RT (1995) Inducible nuclear expression of newly synthesized I kappa B alpha negatively regulates DNA-binding and transcriptional activities of NF-kappa B. Mol Cell Biol 15:2689–2696

Arenzana-Seisdedos F, Turpin P, Rodriguez M, Thomas D, Hay RT, Virelizier JL, Dargemont C (1997) Nuclear localization of I kappa B alpha promotes active transport of NF-kappa B from the nucleus to the cytoplasm. J Cell Sci 110: 369–378

Ariga A, Namekawa J, Matsumoto N, Inoue J, Umezawa K (2002) Inhibition of tumor necrosis factor-alpha -induced nuclear translocation and activation of NF-kappa B by dehydroxymethylepoxyquinomicin. J Biol Chem 277:24625–24630

Asahara H, Asanuma M, Ogawa N, Nishibayashi S, Inoue H (1995) High DNA-binding activity of transcription factor NF-kappa B in synovial membranes of patients with rheumatoid arthritis. Biochem Mol Biol Int 37:827–832

Auphan N, DiDonato JA, Rosette C, Helmberg A, Karin M (1995) Immunosuppression by glucocorticoids: inhibition of NF-kappa B activity through induction of I kappa B synthesis. Science 270:286–290

Aupperle K, Bennett B, Han Z, Boyle D, Manning A, Firestein G (2001) NF-kappa B regulation by I kappa B kinase-2 in rheumatoid arthritis synoviocytes. J Immunol 166:2705–2711

Autieri MV, Yue TL, Ferstein GZ, Ohlstein E (1995) Antisense oligonucleotides to the p65 subunit of NF-kB inhibit human vascular smooth muscle cell adherence and proliferation and prevent neointima formation in rat carotid arteries. Biochem Biophys Res Commun 213:827–836

Barton D, HogenEsch H, Weih F (2000) Mice lacking the transcription factor RelB develop T cell-dependent skin lesions similar to human atopic dermatitis. Eur J Immunol 30:2323–2332

Barton GM, Medzhitov R (2002) Toll-like receptors and their ligands. Curr Top Microbiol Immunol 270:81–92

Baumann B, Weber CK, Troppmair J, Whiteside S, Israel A, Rapp UR, Wirth T (2000) Raf induces NF-kappaB by membrane shuttle kinase MEKK1, a signaling pathway critical for transformation. Proc Natl Acad Sci USA 97:4615–4620

Beg AA, Sha WC, Bronson RT, Baltimore D (1995) Constitutive NF-kappa B activation, enhanced granulopoiesis, and neonatal lethality in I kappa B alpha-deficient mice. Genes Dev 9:2736–2746

Bellas RE, Lee JS, Sonenshein GE (1995) Expression of a constitutive NF-kappa B-like activity is essential for proliferation of cultured bovine vascular smooth muscle cells. J Clin Invest 96:2521–2527

Ben-Neriah Y (2002) Regulatory functions of ubiquitination in the immune system. Nat Immunol 3:20–26

Bhullar IS, Li YS, Miao H, Zandi E, Kim M, Shyy JY, Chien S (1998) Fluid shear stress activation of IkappaB kinase is integrin-dependent. J Biol Chem 273:30544–30549

Bi K, Tanaka Y, Coudronniere N, Sugie K, Hong S, van Stipdonk MJ, Altman A (2001) Antigen-induced translocation of PKC-theta to membrane rafts is required for T cell activation. Nat Immunol 2:556–563

Biedermann BC (2001) Vascular endothelium: checkpoint for inflammation and immunity. News Physiol Sci 16:84–88

Blanco-Colio LM, Valderrama M, Alvarez-Sala LA, Bustos C, Ortego M, Hernandez-Presa MA, Cancelas P, Gomez-Gerique J, Millan J, Egido J (2000) Red wine intake prevents nuclear factor-kappaB activation in peripheral blood mononuclear cells of healthy volunteers during postprandial lipemia. Circulation 102:1020–1026

Boland MP, Foster SJ, O'Neill LA (1996) Activation of NF kappa B and potentiation of TNF-induced NF kappa B activation by ceramide analogues in leukemic cell lines despite the absence of an observed sphingomyelinase signalling event. Biochem Soc Trans 24:1S

Boland MP, O'Neill LA (1998) Ceramide activates NFkappaB by inducing the processing of p105. J Biol Chem 273:15494–15500

Bond M, Baker AH, Newby AC (1999) Nuclear factor kappaB activity is essential for matrix metalloproteinase-1 and -3 upregulation in rabbit dermal fibroblasts. Biochem Biophys Res Commun 264:561–567

Bond M, Chase AJ, Baker AH, Newby AC (2001) Inhibition of transcription factor NF-kappaB reduces matrix metalloproteinase-1, -3 and -9 production by vascular smooth muscle cells. Cardiovasc Res 50:556–565

Bond M, Fabunmi RP, Baker AH, Newby AC (1998) Synergistic upregulation of metalloproteinase-9 by growth factors and inflammatory cytokines: an absolute requirement for transcription factor NF-kappa B. FEBS Lett 435:29–34

Bondeson J, Brennan F, Foxwell B, Feldmann M (2000) Effective adenoviral transfer of IkappaBalpha into human fibroblasts and chondrosarcoma cells reveals that the induction of matrix metalloproteinases and proinflammatory cytokines is nuclear factor-kappaB dependent. J Rheumatol 27:2078–2089

Bondeson J, Foxwell B, Brennan F, Feldmann M (1999) Defining therapeutic targets by using adenovirus: blocking NF-kappaB inhibits both inflammatory and destructive mechanisms in rheumatoid synovium but spares anti-inflammatory mediators. Proc Natl Acad Sci U S A 96:5668–5673

Boring L, Gosling J, Cleary M, Charo IF (1998) Decreased lesion formation in CCR2-/- mice reveals a role for chemokines in the initiation of atherosclerosis. Nature 394:894–897

Bourcier T, Sukhova G, Libby P (1997) The nuclear factor kappa-B signaling pathway participates in dysregulation of vascular smooth muscle cells in vitro and in human atherosclerosis. J Biol Chem 272:15817–15824

Bours V, Franzoso G, Azarenko V, Park S, Kanno T, Brown K, Siebenlist U (1993) The oncoprotein Bcl-3 directly transactivates through kappa B motifs via association with DNA-binding p50B homodimers. Cell 72:729–739

Brand K, Eisele T, Kreusel U, Page M, Page S, Haas M, Gerling A, Kaltschmidt C, Neumann FJ, Mackman N, Baeurele PA, Walli AK, Neumeier D (1997) Dysregulation of monocytic nuclear factor-kappa B by oxidized low-density lipoprotein. Arterioscler Thromb Vasc Biol 17:1901–1909

Brand K, Page S, Rogler G, Bartsch A, Brandl R, Knuechel R, Page M, Kaltschmidt C, Baeuerle PA, Neumeier D (1996) Activated transcription factor nuclear factor-kappa B is present in the atherosclerotic lesion. J Clin Invest 97:1715–1722

Bren GD, Solan NJ, Miyoshi H, Pennington KN, Pobst LJ, Paya CV (2001) Transcription of the RelB gene is regulated by NF-kappaB. Oncogene 20:7722–7733

Bretschneider E, Wittpoth M, Weber AA, Glusa E, Schror K (1997) Activation of NFkappaB is essential but not sufficient to stimulate mitogenesis of vascular smooth muscle cells. Biochem Biophys Res Commun 235:365–368

Brummelkamp TR, Nijman SM, Dirac AM, Bernards R (2003) Loss of the cylindromatosis tumour suppressor inhibits apoptosis by activating NF-kappaB. Nature 424:797–801

Bulger EM, Maier RV (2003) An argument for Vitamin E supplementation in the management of systemic inflammatory response syndrome. Shock 19:99–103

Burke JR, Pattoli MA, Gregor KR, Brassil PJ, MacMaster JF, McIntyre KW, Yang X, Iotzova VS, Clarke W, Strnad J, Qiu Y, Zusi FC (2003) BMS-345541 is a highly selective inhibitor of I kappa B kinase that binds at an allosteric site of the enzyme and blocks NF-kappa B-dependent transcription in mice. J Biol Chem 278:1450–1456

Burkly L, Hession C, Ogata L, Reilly C, Marconi LA, Olson D, Tizard R, Cate R, Lo D (1995) Expression of relB is required for the development of thymic medulla and dendritic cells. Nature 373:531–536

Burns K, Clatworthy J, Martin L, Martinon F, Plumpton C, Maschera B, Lewis A, Ray K, Tschopp J, Volpe F (2000) Tollip, a new component of the IL-1RI pathway, links IRAK to the IL-1 receptor. Nat Cell Biol 2:346–351

Burns K, Martinon F, Esslinger C, Pahl H, Schneider P, Bodmer JL, Di Marco F, French L, Tschopp J (1998) MyD88, an adapter protein involved in interleukin-1 signaling. J Biol Chem 273:12203–12209

Caamano JH, Rizzo CA, Durham SK, Barton DS, Raventos-Suarez C, Snapper CM, Bravo R (1998) Nuclear factor (NF)-kappa B2 (p100/p52) is required for normal splenic microarchitecture and B cell-mediated immune responses. J Exp Med 187:185–196

Calara F, Dimayuga P, Niemann A, Thyberg J, Diczfalusy U, Witztum JL, Palinski W, Shah PK, Cercek B, Nilsson J, Regnstrom J (1998) An animal model to study local oxidation of LDL and its biological effects in the arterial wall. Arterioscler Thromb Vasc Biol 18:884–893

Campbell IK, Gerondakis S, O'Donnell K, Wicks IP (2000) Distinct roles for the NF-kappaB1 (p50) and c-Rel transcription factors in inflammatory arthritis. J Clin Invest 105:1799–1806

Castrillo A, Diaz-Guerra MJ, Hortelano S, Martin-Sanz P, Bosca L (2000) Inhibition of IkappaB kinase and IkappaB phosphorylation by 15-deoxy-Delta(12,14)-prostaglandin J(2) in activated murine macrophages. Mol Cell Biol 20:1692–1698

Cernuda-Morollon E, Pineda-Molina E, Canada FJ, Perez-Sala D (2001) 15-Deoxy-Delta 12,14-prostaglandin J2 inhibition of NF-kappaB-DNA binding through covalent modification of the p50 subunit. J Biol Chem 276:35530–35536

Chariot A, Leonardi A, Muller J, Bonif M, Brown K, Siebenlist U (2002) Association of the adaptor TANK with the I kappa B kinase (IKK) regulator NEMO connects IKK complexes with IKK epsilon and TBK1 kinases. J Biol Chem 277:37029–37036

Chen CL, Yull FE, Cardwell N, Singh N, Strayhorn WD, Nanney LB, Kerr LD (2000) RAG2-/-, I kappa B-alpha-/- chimeras display a psoriasiform skin disease. J Invest Dermatol 115:1124–1133

Chen G, Cao P, Goeddel DV (2002) TNF-induced recruitment and activation of the IKK complex require Cdc37 and Hsp90. Mol Cell 9:401–410

Chiao PJ, Miyamoto S, Verma IM (1994) Autoregulation of IκBα activity. Proc Natl Acad Sci USA 91:28–32

Chinetti G, Griglio S, Antonucci M, Torra IP, Delerive P, Majd Z, Fruchart JC, Chapman J, Najib J, Staels B (1998) Activation of proliferator-activated receptors alpha and gamma induces apoptosis of human monocyte-derived macrophages. J Biol Chem 273:25573–25580

Christman JW, Lancaster LH, Blackwell TS (1998) Nuclear factor kappa B: a pivotal role in the systemic inflammatory response syndrome and new target for therapy. Intensive Care Med 24:1131–1138

Ciechanover A, Gonen H, Bercovich B, Cohen S, Fajerman I, Israel A, Mercurio F, Kahana C, Schwartz AL, Iwai K, Orian A (2001) Mechanisms of ubiquitin-mediated, limited processing of the NF-kappaB1 precursor protein p105. Biochimie 83:341–349

Claudio E, Brown K, Park S, Wang H, Siebenlist U (2002) BAFF-induced NEMO-independent processing of NF-kappa B2 in maturing B cells. Nat Immunol 3:958–965

Collantes E, Valle Blazquez M, Mazorra V, Macho A, Aranda E, Munoz E (1998) Nuclear factor-kappa B activity in T cells from patients with rheumatic diseases: a preliminary report. Ann Rheum Dis 57:738–741

Collins T, Cybulsky MI (2001) NF-kappaB: pivotal mediator or innocent bystander in atherogenesis? J Clin Invest 107:255–264

Cuzzocrea S, Chatterjee PK, Mazzon E, Dugo L, Serraino I, Britti D, Mazzullo G, Caputi AP, Thiemermann C (2002) Pyrrolidine dithiocarbamate attenuates the development of acute and chronic inflammation. Br J Pharmacol 135:496–510

Cybulsky MI, Iiyama K, Li H, Zhu S, Chen M, Iiyama M, Davis V, Gutierrez-Ramos JC, Connelly PW, Milstone DS (2001) A major role for VCAM-1, but not ICAM-1, in early atherosclerosis. J Clin Invest 107:1255–1262

Cyrus T, Sung S, Zhao L, Funk CD, Tang S, Pratico D (2002) Effect of low-dose aspirin on vascular inflammation, plaque stability, and atherogenesis in low-density lipoprotein receptor-deficient mice. Circulation 106:1282–1287

De Bosscher K, Schmitz ML, Vanden Berghe W, Plaisance S, Fiers W, Haegeman G (1997) Glucocorticoid-mediated repression of nuclear factor-kappaB-dependent transcription involves direct interference with transactivation. Proc Natl Acad Sci USA 94:13504–13509

De Bosscher K, Vanden Berghe W, Vermeulen L, Plaisance S, Boone E, Haegeman G (2000) Glucocorticoids repress NF-kappaB-driven genes by disturbing the interaction of p65 with the basal transcription machinery, irrespective of coactivator levels in the cell. Proc Natl Acad Sci USA 97:3919–3924

de Wit H, Dokter WH, Koopmans SB, Lummen C, van der Leij M, Smit JW, Vellenga E (1998) Regulation of p100 (NFKB2) expression in human monocytes in response to inflammatory mediators and lymphokines. Leukemia 12:363–370

Dejardin E, Droin NM, Delhase M, Haas E, Cao Y, Makris C, Li ZW, Karin M, Ware CF, Green DR (2002) The lymphotoxin-beta receptor induces different patterns of gene expression via two NF-kappaB pathways. Immunity 17:525–535

Delerive P, De Bosscher K, Besnard S, Vanden Berghe W, Peters JM, Gonzalez FJ, Fruchart JC, Tedgui A, Haegeman G, Staels B (1999) Peroxisome proliferator-activated receptor alpha negatively regulates the vascular inflammatory gene response by negative cross-talk with transcription factors NF-kappaB and AP-1. J Biol Chem 274:32048–32054

Delerive P, De Bosscher K, Vanden Berghe W, Fruchart JC, Haegeman G, Staels B (2002) DNA binding-independent induction of IkappaBalpha gene transcription by PPARalpha. Mol Endocrinol 16:1029–1039

Delerive P, Gervois P, Fruchart JC, Staels B (2000) Induction of IkappaBalpha expression as a mechanism contributing to the anti-inflammatory activities of peroxisome proliferator-activated receptor-alpha activators. J Biol Chem 275:36703–36707

Delhase M, Hayakawa M, Chen Y, Karin M (1999) Positive and negative regulation of IkappaB kinase activity through IKKbeta subunit phosphorylation. Science 284:309–313

Denk A, Goebeler M, Schmid S, Berberich I, Ritz O, Lindemann D, Ludwig S, Wirth T (2001) Activation of NF-kappa B via the Ikappa B kinase complex is both essential and sufficient for proinflammatory gene expression in primary endothelial cells. J Biol Chem 276:28451–28458

Devin A, Cook A, Lin Y, Rodriguez Y, Kelliher M, Liu Z (2000) The distinct roles of TRAF2 and RIP in IKK activation by TNF-R1: TRAF2 recruits IKK to TNF-R1 while RIP mediates IKK activation. Immunity 12:419–429

Devin A, Lin Y, Yamaoka S, Li Z, Karin M, Liu Z (2001) The alpha and beta subunits of IkappaB kinase (IKK) mediate TRAF2-dependent IKK recruitment to tumor necrosis factor (TNF) receptor 1 in response to TNF. Mol Cell Biol 21:3986–3994

Dichtl W, Dulak J, Frick M, Alber HF, Schwarzacher SP, Ares MP, Nilsson J, Pachinger O, Weidinger F (2003) HMG-CoA reductase inhibitors regulate inflammatory transcription factors in human endothelial and vascular smooth muscle cells. Arterioscler Thromb Vasc Biol 23:58–63

Donath B, Fischer C, Page S, Prebeck S, Jilg N, Weber M, da Costa C, Neumeier D, Miethke T, Brand K (2002) Chlamydia pneumoniae activates IKK/I kappa B-mediated signaling, which is inhibited by 4-HNE and following primary exposure. Atherosclerosis 165:79–88

Dong ZM, Chapman SM, Brown AA, Frenette PS, Hynes RO, Wagner DD (1998) The combined role of P- and E-selectins in atherosclerosis. J Clin Invest 102:145–152

Donovan CE, Mark DA, He HZ, Liou HC, Kobzik L, Wang Y, De Sanctis GT, Perkins DL, Finn PW (1999) NF-kappa B/Rel transcription factors: c-Rel promotes airway hyperresponsiveness and allergic pulmonary inflammation. J Immunol 163:6827–6833

Dunn JA, Li C, Ha T, Kao RL, Browder W (1997) Therapeutic modification of nuclear factor kappa B binding activity and tumor necrosis factor-alpha gene expression during acute biliary pancreatitis. Am Surg 63:1036–1043; discussion 1043–1034

Duran A, Diaz-Meco MT, Moscat J (2003) Essential role of RelA Ser311 phosphorylation by zetaPKC in NF-kappaB transcriptional activation. EMBO J 22:3910–3918

Ebnet K, Vestweber D (1999) Molecular mechanisms that control leukocyte extravasation: the selectins and the chemokines. Histochem Cell Biol 112:1–23

Egan LJ, Mays DC, Huntoon CJ, Bell MP, Pike MG, Sandborn WJ, Lipsky JJ, McKean DJ (1999) Inhibition of interleukin-1-stimulated NF-kappaB RelA/p65 phosphorylation by mesalamine is accompanied by decreased transcriptional activity. J Biol Chem 274:26448–26453

Eliopoulos AG, Caamano JH, Flavell J, Reynolds GM, Murray PG, Poyet JL, Young LS (2003) Epstein-Barr virus-encoded latent infection membrane protein 1 regulates the processing of p100 NF-kappaB2 to p52 via an IKKgamma/NEMO-independent signalling pathway. Oncogene 22:7557–7569

Elliott PJ, Pien CS, McCormack TA, Chapman ID, Adams J (1999) Proteasome inhibition: A novel mechanism to combat asthma. J Allergy Clin Immunol 104:294–300

Erl W, Hansson GK, de Martin R, Draude G, Weber KS, Weber C (1999) Nuclear factor-kappa B regulates induction of apoptosis and inhibitor of apoptosis protein-1 expression in vascular smooth muscle cells. Circ Res 84:668–677

Erl W, Weber C, Wardemann C, Weber PC (1997) alpha-Tocopheryl succinate inhibits monocytic cell adhesion to endothelial cells by suppressing NF-kappa B mobilization. Am J Physiol 273: H634–H640

Esslinger CW, Jongeneel CV, MacDonald HR (1998) Survival-independent function of NF-kappaB/Rel during late stages of thymocyte differentiation. Mol Immunol 35:847–852

Esslinger CW, Wilson A, Sordat B, Beermann F, Jongeneel CV (1997) Abnormal T lymphocyte development induced by targeted overexpression of IkappaB alpha. J Immunol 158:5075–5078

Feldmann M, Brennan FM, Maini RN (1996) Rheumatoid arthritis. Cell 85:307–310

Fiorini E, Schmitz I, Marissen WE, Osborn SL, Touma M, Sasada T, Reche PA, Tibaldi EV, Hussey RE, Kruisbeek AM, Reinherz EL, Clayton LK (2002) Peptide-induced negative selection of thymocytes activates transcription of an NF-kappa B inhibitor. Mol Cell 9:637–648

Fitzgerald KA, McWhirter SM, Faia KL, Rowe DC, Latz E, Golenbock DT, Coyle AJ, Liao SM, Maniatis T (2003a) IKKepsilon and TBK1 are essential components of the IRF3 signaling pathway. Nat Immunol 4:491–496

Fitzgerald KA, Rowe DC, Barnes BJ, Caffrey DR, Visintin A, Latz E, Monks B, Pitha PM, Golenbock DT (2003b) LPS-TLR4 signaling to IRF-3/7 and NF-kappaB involves the toll adapters TRAM and TRIF. J Exp Med 198:1043–1055

Foxwell B, Browne K, Bondeson J, Clarke C, de Martin R, Brennan F, Feldmann M (1998) Efficient adenoviral infection with IkappaB alpha reveals that macrophage tumor necrosis factor alpha production in rheumatoid arthritis is NF-kappaB dependent. Proc Natl Acad Sci USA 95:8211–8215

Fujita T, Nolan GP, Liou HC, Scott ML, Baltimore D (1993) The candidate proto-oncogene bcl-3 encodes a transcriptional coactivator that activates through NF-kappa B p50 homodimers. Genes Dev 7:1354–1363

Gaide O, Martinon F, Micheau O, Bonnet D, Thome M, Tschopp J (2001) Carma1, a CARD-containing binding partner of Bcl10, induces Bcl10 phosphorylation and NF-kappaB activation. FEBS Lett 496:121–127

Geng Z, Rong Y, Lau BH (1997) S-allyl cysteine inhibits activation of nuclear factor kappa B in human T cells. Free Radic Biol Med 23:345–350

Ghosh S, Karin M (2002) Missing pieces in the NF-kappaB puzzle. Cell 109 Suppl: S81–S96

Gill JS, Zhu X, Moore MJ, Lu L, Yaszemski MJ, Windebank AJ (2002) Effects of NFkappaB decoy oligonucleotides released from biodegradable polymer microparticles on a glioblastoma cell line. Biomaterials 23:2773–2781

Gilston V, Jones HW, Soo CC, Coumbe A, Blades S, Kaltschmidt C, Baeuerle PA, Morris CJ, Blake DR, Winyard PG (1997) NF-kappa B activation in human knee-joint synovial tissue during the early stage of joint inflammation. Biochem Soc Trans 25:518S

Gimbrone MA, Jr. (1999) Vascular endothelium, hemodynamic forces, and atherogenesis. Am J Pathol 155:1–5

Glass CK, Witztum JL (2001) Atherosclerosis. the road ahead. Cell 104:503–516

Goebeler M, Gillitzer R, Kilian K, Utzel K, Brocker EB, Rapp UR, Ludwig S (2001) Multiple signaling pathways regulate NF-kappaB-dependent transcription of the monocyte chemoattractant protein-1 gene in primary endothelial cells. Blood 97:46–55

Golovchenko I, Goalstone ML, Watson P, Brownlee M, Draznin B (2000) Hyperinsulinemia enhances transcriptional activity of nuclear factor-kappaB induced by angiotensin II, hyperglycemia, and advanced glycosylation end products in vascular smooth muscle cells. Circ Res 87:746–752

Gu L, Okada Y, Clinton SK, Gerard C, Sukhova GK, Libby P, Rollins BJ (1998) Absence of monocyte chemoattractant protein-1 reduces atherosclerosis in low density lipoprotein receptor-deficient mice. Mol Cell 2:275–281

Gveric D, Kaltschmidt C, Cuzner ML, Newcombe J (1998) Transcription factor NF-kappaB and inhibitor I kappaBalpha are localized in macrophages in active multiple sclerosis lesions. J Neuropathol Exp Neurol 57:168–178

Hajra L, Evans AI, Chen M, Hyduk SJ, Collins T, Cybulsky MI (2000) The NF-kappa B signal transduction pathway in aortic endothelial cells is primed for activation in regions predisposed to atherosclerotic lesion formation. Proc Natl Acad Sci USA 97:9052–9057

Han Y, Runge MS, Brasier AR (1999) Angiotensin II induces interleukin-6 transcription in vascular smooth muscle cells through pleiotropic activation of nuclear factor-kappa B transcription factors. Circ Res 84:695–703

Han Z, Boyle DL, Manning AM, Firestein GS (1998) AP-1 and NF-kappaB regulation in rheumatoid arthritis and murine collagen-induced arthritis. Autoimmunity 28:197–208

Handel ML, McMorrow LB, Gravallese EM (1995) Nuclear factor-kappa B in rheumatoid synovium. Localization of p50 and p65. Arthritis Rheum 38:1762–1770

Hart LA, Krishnan VL, Adcock IM, Barnes PJ, Chung KF (1998) Activation and localization of transcription factor, nuclear factor-kappaB, in asthma. Am J Respir Crit Care Med 158:1585–1592

Heck S, Bender K, Kullmann M, Gottlicher M, Herrlich P, Cato AC (1997) I kappaB alpha-independent downregulation of NF-kappaB activity by glucocorticoid receptor. EMBO J 16:4698–4707

Heissmeyer V, Krappmann D, Hatada EN, Scheidereit C (2001) Shared pathways of IkappaB kinase-induced SCF(betaTrCP)-mediated ubiquitination and degradation for the NF-kappaB precursor p105 and IkappaBalpha. Mol Cell Biol 21:1024–1035

Hernandez-Presa MA, Ortego M, Tunon J, Martin-Ventura JL, Mas S, Blanco-Colio LM, Aparicio C, Ortega L, Gomez-Gerique J, Vivanco F, Egido J (2003) Simvastatin reduces NF-kappaB activity in peripheral mononuclear and in plaque cells of rabbit atheroma more markedly than lipid lowering diet. Cardiovasc Res 57:168–177

Hettmann T, Leiden JM (2000) NF-kappa B is required for the positive selection of CD8+ thymocytes. J Immunol 165:5004–5010

Hilliard B, Samoilova EB, Liu TS, Rostami A, Chen Y (1999) Experimental autoimmune encephalomyelitis in NF-kappa B-deficient mice:roles of NF-kappa B in the activation and differentiation of autoreactive T cells. J Immunol 163:2937–2943

Hilliard BA, Mason N, Xu L, Sun J, Lamhamedi-Cherradi SE, Liou HC, Hunter C, Chen YH (2002) Critical roles of c-Rel in autoimmune inflammation and helper T cell differentiation. J Clin Invest 110:843–850

Hishikawa K, Oemar BS, Yang Z, Luscher TF (1997) Pulsatile stretch stimulates superoxide production and activates nuclear factor-kappa B in human coronary smooth muscle. Circ Res 81:797–803

Hoebe K, Du X, Georgel P, Janssen E, Tabeta K, Kim SO, Goode J, Lin P, Mann N, Mudd S, Crozat K, Sovath S, Han J, Beutler B (2003) Identification of Lps2 as a key transducer of MyD88-independent TIR signalling. Nature 424:743–748

Hoffmann A, Levchenko A, Scott ML, Baltimore D (2002) The IkappaB-NF-kappaB signaling module: temporal control and selective gene activation. Science 298:1241–1245

Holtmann H, Enninga J, Kalble S, Thiefes A, Dorrie A, Broemer M, Winzen R, Wilhelm A, Ninomiya-Tsuji J, Matsumoto K, Resch K, Kracht M (2001) The MAPK kinase kinase TAK1 plays a central role in coupling the interleukin-1 receptor to both transcriptional and RNA-targeted mechanisms of gene regulation. J Biol Chem 276:3508–3516

Holtmann H, Winzen R, Holland P, Eickemeier S, Hoffmann E, Wallach D, Malinin NL, Cooper JA, Resch K, Kracht M (1999) Induction of interleukin-8 synthesis integrates effects on transcription and mRNA degradation from at least three different cytokine- or stress-activated signal transduction pathways. Mol Cell Biol 19:6742–6753

Hsu H, Huang J, Shu HB, Baichwal V, Goeddel DV (1996a) TNF-dependent recruitment of the protein kinase RIP to the TNF receptor-1 signaling complex. Immunity 4:387–396

Hsu H, Shu HB, Pan MG, Goeddel DV (1996b) TRADD-TRAF2 and TRADD-FADD interactions define two distinct TNF receptor 1 signal transduction pathways. Cell 84:299–308

Hsueh WA, Law RE (2001) PPARgamma and atherosclerosis: effects on cell growth and movement. Arterioscler Thromb Vasc Biol 21:1891–1895

Hu Y, Baud V, Delhase M, Zhang P, Deerinck T, Ellisman M, Johnson R, Karin M (1999) Abnormal morphogenesis but intact IKK activation in mice lacking the IKKalpha subunit of IkappaB kinase. Science 284:316–320

Hu Y, Baud V, Oga T, Kim KI, Yoshida K, Karin M (2001) IKKalpha controls formation of the epidermis independently of NF-kappaB. Nature 410:710–714

Huang J, Gao X, Li S, Cao Z (1997) Recruitment of IRAK to the interleukin 1 receptor complex requires interleukin 1 receptor accessory protein. Proc Natl Acad Sci USA 94:12829–12832

Huang TT, Kudo N, Yoshida M, Miyamoto S (2000) A nuclear export signal in the N-terminal regulatory domain of IkappaBalpha controls cytoplasmic localization of inactive NF-kappaB/IkappaBalpha complexes. Proc Natl Acad Sci USA 97:1014–1019

Huber MA, Denk A, Peter RU, Weber L, Kraut N, Wirth T (2002) The IKK-2/Ikappa Balpha/NF-kappa B pathway plays a key role in the regulation of CCR3 and eotaxin-1 in fibroblasts. A critical link to dermatitis in Ikappa Balpha -deficient mice. J Biol Chem 277:1268–1275

Iademarco MF, McQuillan JJ, Rosen GD, Dean DC (1992) Characterization of the promoter for vascular cell adhesion molecule-1 (VCAM-1). J Biol Chem 267:16323–16329

Iiyama K, Hajra L, Iiyama M, Li H, DiChiara M, Medoff BD, Cybulsky MI (1999) Patterns of vascular cell adhesion molecule-1 and intercellular adhesion molecule-1 expression in rabbit and mouse atherosclerotic lesions and at sites predisposed to lesion formation. Circ Res 85:199–207

Iotsova V, Caamano J, Loy J, Yang Y, Lewin A, Bravo R (1997) Osteopetrosis in mice lacking NF-κB1 and NF-κB2. Nature Med 3:1285–1289

Ishikawa H, Claudio E, Dambach D, Raventos-Suarez C, Ryan C, Bravo R (1998) Chronic inflammation and susceptibility to bacterial infections in mice lacking the polypeptide (p)105 precursor (NF-kappaB1) but expressing p50. J Exp Med 187:985–996

Ishitani T, Takaesu G, Ninomiya-Tsuji J, Shibuya H, Gaynor RB, Matsumoto K (2003) Role of the TAB2-related protein TAB3 in IL-1 and TNF signaling. EMBO J 22:6277–6288

Israel A (2000) The IKK complex: an integrator of all signals that activate NF-kappaB? Trends Cell Biol 10:129–133

Janssens S, Beyaert R (2003) Functional diversity and regulation of different interleukin-1 receptor-associated kinase (IRAK) family members. Mol Cell 11:293–302

Jeon KI, Jeong JY, Jue DM (2000) Thiol-reactive metal compounds inhibit NF-kappa B activation by blocking I kappa B kinase. J Immunol 164:5981–5989

Johnson C, Van Antwerp D, Hope TJ (1999) An N-terminal nuclear export signal is required for the nucleocytoplasmic shuttling of IkappaBalpha. EMBO J 18:6682–6693

Kaisho T, Takeda K, Tsujimura T, Kawai T, Nomura F, Terada N, Akira S (2001) IkappaB kinase alpha is essential for mature B cell development and function. J Exp Med 193:417–426

Karin M, Ben-Neriah Y (2000) Phosphorylation meets ubiquitination: the control of NF-[kappa]B activity. Annu Rev Immunol 18:621–663

Kawai T, Adachi O, Ogawa T, Takeda K, Akira S (1999) Unresponsiveness of MyD88-deficient mice to endotoxin. Immunity 11:115–122

Keifer JA, Guttridge DC, Ashburner BP, Baldwin AS, Jr. (2001) Inhibition of NF-kappa B activity by thalidomide through suppression of IkappaB kinase activity. J Biol Chem 276:22382–22387

Kino T, Nordeen SK, Chrousos GP (1999) Conditional modulation of glucocorticoid receptor activities by CREB-binding protein (CBP) and p300. J Steroid Biochem Mol Biol 70:15–25

Kintscher U, Goetze S, Wakino S, Kim S, Nagpal S, Chandraratna RA, Graf K, Fleck E, Hsueh WA, Law RE (2000) Peroxisome proliferator-activated receptor and retinoid X receptor ligands inhibit monocyte chemotactic protein-1-directed migration of monocytes. Eur J Pharmacol 401:259–270

Kitajima I, Soejima Y, Takasaki I, Beppu H, Tokioka T, Maruyama I (1996) Ceramide-induced nuclear translocation of NF-kappa B is a potential mediator of the apoptotic response to TNF-alpha in murine clonal osteoblasts. Bone 19:263–270

Kleinert H, Euchenhofer C, Ihrig-Biedert I, Forstermann U (1996) Glucocorticoids inhibit the induction of nitric oxide synthase II by down-regulating cytokine-induced activity of transcription factor nuclear factor-kappa B. Mol Pharmacol 49:15–21

Klement JF, Rice NR, Car BD, Abbondanzo SJ, Powers GD, Bhatt PH, Chen CH, Rosen CA, Stewart CL (1996) IκB-α deficiency results in a sustained NF-κB response and severe widespread dermatitits in mice. Mol Cell Biol 16:2341–2349

Koni PA, Sacca R, Lawton P, Browning JL, Ruddle NH, Flavell RA (1997) Distinct roles in lymphoid organogenesis for lymphotoxins alpha and beta revealed in lymphotoxin beta-deficient mice. Immunity 6:491–500

Kopp E, Ghosh S (1994) Inhibition of NF-kappa B by sodium salicylate and aspirin. Science 265:956–959

Kopp E, Medzhitov R, Carothers J, Xiao C, Douglas I, Janeway CA, Ghosh S (1999) ECSIT is an evolutionarily conserved intermediate in the Toll/IL-1 signal transduction pathway. Genes Dev 13:2059–2071

Kovalenko A, Chable-Bessia C, Cantarella G, Israel A, Wallach D, Courtois G (2003) The tumour suppressor CYLD negatively regulates NF-kappaB signalling by deubiquitination. Nature 424:801–805

Kravchenko VV, Mathison JC, Schwamborn K, Mercurio F, Ulevitch RJ (2003) IKKi/IKKepsilon plays a key role in integrating signals induced by pro-inflammatory stimuli. J Biol Chem 278:26612–26619

Kunsch C, Rosen CA (1993) NF-kappa B subunit-specific regulation of the interleukin-8 promoter. Mol Cell Biol 13:6137–6146

Lavon I, Goldberg I, Amit S, Landsman L, Jung S, Tsuberi BZ, Barshack I, Kopolovic J, Galun E, Bujard H, Ben-Neriah Y (2000) High susceptibility to bacterial infection, but no liver dysfunction, in mice compromised for hepatocyte NF-kappaB activation. Nat Med 6:573–577

Lawrence R, Chang LJ, Siebenlist U, Bressler P, Sonenshein GE (1994) Vascular smooth muscle cells express a constitutive NF-kappa B-like activity. J Biol Chem 269:28913–28918

Lawrence T, Gilroy DW, Colville-Nash PR, Willoughby DA (2001) Possible new role for NF-kappaB in the resolution of inflammation. Nat Med 7:1291–1297

Lawrence T, Willoughby DA, Gilroy DW (2002) Anti-inflammatory lipid mediators and insights into the resolution of inflammation. Nat Rev Immunol 2:787–795

Ledebur HC, Parks TP (1995) Transcriptional regulation of the intercellular adhesion molecule-1 gene by inflammatory cytokines in human endothelial cells. Essential roles of a variant NF-kappa B site and p65 homodimers. J Biol Chem 270:933–943

Lee FS, Peters RT, Dang LC, Maniatis T (1998) MEKK1 activates both IkappaB kinase alpha and IkappaB kinase beta. Proc Natl Acad Sci USA 95:9319–9324

Lehmann T, Nguyen LQ, Handel ML (2000) Synovial fluid induced nuclear factor-kappaB DNA binding in a monocytic cell line. J Rheumatol 27:2769–2776

Leitges M, Sanz L, Martin P, Duran A, Braun U, Garcia JF, Camacho F, Diaz-Meco MT, Rennert PD, Moscat J (2001) Targeted disruption of the zetaPKC gene results in the impairment of the NF-kappaB pathway. Mol Cell 8:771–780

Leowattana W (2001) Chronic infections and atherosclerosis. J Med Assoc Thai 84 Suppl 3: S650–S657

Li AC, Brown KK, Silvestre MJ, Willson TM, Palinski W, Glass CK (2000a) Peroxisome proliferator-activated receptor gamma ligands inhibit development of atherosclerosis in LDL receptor-deficient mice. J Clin Invest 106:523–531

Li M, Carpio DF, Zheng Y, Bruzzo P, Singh V, Ouaaz F, Medzhitov RM, Beg AA (2001) An essential role of the NF-kappa B/Toll-like receptor pathway in induction of inflammatory and tissue-repair gene expression by necrotic cells. J Immunol 166:7128–7135

Li N, Karin M (1999) Is NF-kappaB the sensor of oxidative stress? FASEB J 13:1137–1143

Li Q, Estepa G, Memet S, Israel A, Verma IM (2000b) Complete lack of NF-kappaB activity in IKK1 and IKK2 double-deficient mice: additional defect in neurulation. Genes Dev 14:1729–1733

Li Q, Lu Q, Hwang JY, Buscher D, Lee KF, Izpisua-Belmonte JC, Verma IM (1999a) IKK1-deficient mice exhibit abnormal development of skin and skeleton. Genes Dev 13:1322–1328

Li Q, Van Antwerp D, Mercurio F, Lee KF, Verma IM (1999b) Severe liver degeneration in mice lacking the IkappaB kinase 2 gene. Science 284:321–325

Li X, Massa PE, Hanidu A, Peet GW, Aro P, Savitt A, Mische S, Li J, Marcu KB (2002) IKKalpha, IKKbeta, and NEMO/IKKgamma are each required for the NF-kappa B-mediated inflammatory response program. J Biol Chem 277:45129–45140

Li ZW, Chu W, Hu Y, Delhase M, Deerinck T, Ellisman M, Johnson R, Karin M (1999c) The IKKbeta subunit of IkappaB kinase (IKK) is essential for nuclear factor kappaB activation and prevention of apoptosis. J Exp Med 189:1839–1845

Li ZW, Omori SA, Labuda T, Karin M, Rickert RC (2003) IKK beta is required for peripheral B cell survival and proliferation. J Immunol 170:4630–4637

Lin L, DeMartino GN, Greene WC (1998) Cotranslational biogenesis of NF-kappaB p50 by the 26S proteasome. Cell 92:819–828

Ling L, Cao Z, Goeddel DV (1998) NF-kappaB-inducing kinase activates IKK-alpha by phosphorylation of Ser-176. Proc Natl Acad Sci USA 95:3792–3797

Maier HJ, Marienfeld R, Wirth T, Baumann B (2003) Critical role of RelB serine 368 for dimerization and p100 stabilization. J Biol Chem 278:39242–39250

Makarov SS (2001) NF-kappa B in rheumatoid arthritis: a pivotal regulator of inflammation, hyperplasia, and tissue destruction. Arthritis Res 3:200–206

Makris C, Godfrey VL, Krahn-Senftleben G, Takahashi T, Roberts JL, Schwarz T, Feng L, Johnson RS, Karin M (2000) Female mice heterozygous for IKK gamma/NEMO deficiencies develop a dermatopathy similar to the human X-linked disorder incontinentia pigmenti. Mol Cell 5: 969–979

Makris C, Roberts JL, Karin M (2002) The carboxyl-terminal region of IkappaB kinase gamma (IKKgamma) is required for full IKK activation. Mol Cell Biol 22:6573–6581

Malek S, Chen Y, Huxford T, Ghosh G (2001) IkappaBbeta, but not IkappaBalpha, functions as a classical cytoplasmic inhibitor of NF-kappaB dimers by masking both NF-kappaB nuclear localization sequences in resting cells. J Biol Chem 276:45225–45235

Manning AM, Bell FP, Rosenbloom CL, Chosay JG, Simmons CA, Northrup JL, Shebuski RJ, Dunn CJ, Anderson DC (1995) NF-kappa B is activated during acute inflammation in vivo in association with elevated endothelial cell adhesion molecule gene expression and leukocyte recruitment. J Inflamm 45:283–296

Marok R, Winyard PG, Coumbe A, Kus ML, Gaffney K, Blades S, Mapp PI, Morris CJ, Blake DR, Kaltschmidt C, Baeuerle PA (1996) Activation of the transcription factor nuclear factor-kappaB in human inflamed synovial tissue. Arthritis Rheum 39:583–591

Martin P, Duran A, Minguet S, Gaspar ML, Diaz-Meco MT, Rennert P, Leitges M, Moscat J (2002) Role of zeta PKC in B-cell signaling and function. EMBO J 21:4049–4057

Martin T, Cardarelli PM, Parry GC, Felts KA, Cobb RR (1997) Cytokine induction of monocyte chemoattractant protein-1 gene expression in human endothelial cells depends on the cooperative action of NF-kappa B and AP-1. Eur J Immunol 27:1091–1097

Marx N, Schonbeck U, Lazar MA, Libby P, Plutzky J (1998) Peroxisome proliferator-activated receptor gamma activators inhibit gene expression and migration in human vascular smooth muscle cells. Circ Res 83:1097–1103

Marx N, Sukhova GK, Collins T, Libby P, Plutzky J (1999) PPARalpha activators inhibit cytokine-induced vascular cell adhesion molecule-1 expression in human endothelial cells. Circulation 99:3125–3131

May MJ, D'Acquisto F, Madge LA, Glockner J, Pober JS, Ghosh S (2000) Selective inhibition of NF-kappaB activation by a peptide that blocks the interaction of NEMO with the IkappaB kinase complex. Science 289:1550–1554

May MJ, Ghosh S (1997) Rel/NF-kappa B and I kappa B proteins: an overview. Semin Cancer Biol 8:63–73

Medzhitov R, Janeway CA, Jr. (1998) Innate immune recognition and control of adaptive immune responses. Semin Immunol 10:351–353

Medzhitov R, Preston-Hurlburt P, Janeway CA, Jr. (1997) A human homologue of the Drosophila Toll protein signals activation of adaptive immunity. Nature 388:394–397

Meiler SE, Hung RR, Gerszten RE, Gianetti J, Li L, Matsui T, Gimbrone MA, Jr., Rosenzweig A (2002) Endothelial IKK beta signaling is required for monocyte adhesion under laminar flow conditions. J Mol Cell Cardiol 34:349–359

Menetski JP (2000) The structure of the nuclear factor-kappaB protein-DNA complex varies with DNA-binding site sequence. J Biol Chem 275:7619–7625

Mengshol JA, Vincenti MP, Coon CI, Barchowsky A, Brinckerhoff CE (2000) Interleukin-1 induction of collagenase 3 (matrix metalloproteinase 13) gene expression in chondrocytes requires p38, c-Jun N-terminal kinase, and nuclear factor kappaB: differential regulation of collagenase 1 and collagenase 3. Arthritis Rheum 43:801–811

Mercurio F, Zhu H, Murray BW, Shevchenko A, Bennett BL, Li J, Young DB, Barbosa M, Mann M, Manning A, Rao A (1997) IKK-1 and IKK-2: cytokine-activated IkappaB kinases essential for NF-kappaB activation. Science 278:860–866

Miagkov AV, Kovalenko DV, Brown CE, Didsbury JR, Cogswell JP, Stimpson SA, Baldwin AS, Makarov SS (1998) NF-kappaB activation provides the potential link between inflammation and hyperplasia in the arthritic joint. Proc Natl Acad Sci USA 95:13859–13864

Miller BS, Zandi E (2001) Complete reconstitution of human IkappaB kinase (IKK) complex in yeast. Assessment of its stoichiometry and the role of IKKgamma on the complex activity in the absence of stimulation. J Biol Chem 276:36320–36326

Miyazawa K, Mori A, Yamamoto K, Okudaira H (1998) Constitutive transcription of the human interleukin-6 gene by rheumatoid synoviocytes: spontaneous activation of NF-kappaB and CBF1. Am J Pathol 152:793–803

Mohan S, Mohan N, Sprague EA (1997) Differential activation of NF-kappa B in human aortic endothelial cells conditioned to specific flow environments. Am J Physiol 273: C572–C578

Mordmuller B, Krappmann D, Esen M, Wegener E, Scheidereit C (2003) Lymphotoxin and lipopolysaccharide induce NF-kappaB-p52 generation by a co-translational mechanism. EMBO Rep 4:82–87

Moreland LW, Heck LW, Jr., Koopman WJ (1997) Biologic agents for treating rheumatoid arthritis. Concepts and progress. Arthritis Rheum 40:397–409

Morigi M, Angioletti S, Imberti B, Donadelli R, Micheletti G, Figliuzzi M, Remuzzi A, Zoja C, Remuzzi G (1998) Leukocyte-endothelial interaction is augmented by high glucose concentrations and hyperglycemia in a NF-kB-dependent fashion. J Clin Invest 101:1905–1915

Mukaida N, Morita M, Ishikawa Y, Rice N, Okamoto S, Kasahara T, Matsushima K (1994) Novel mechanism of glucocorticoid-mediated gene repression. Nuclear factor-kappa B is target for glucocorticoid-mediated interleukin 8 gene repression. J Biol Chem 269:13289–13295

Muller JR, Siebenlist U (2003) Lymphotoxin beta receptor induces sequential activation of distinct NF-kappa B factors via separate signaling pathways. J Biol Chem 278: 12006–12012

Murakami T, Hirai H, Suzuki T, Fujisawa J, Yoshida M (1995) HTLV-1 Tax enhances NF-kappa B2 expression and binds to the products p52 and p100, but does not suppress the inhibitory function of p100. Virology 206:1066–1074

Muta T, Yamazaki S, Eto A, Motoyama M, Takeshige K (2003) IkappaB-zeta, a new anti-inflammatory nuclear protein induced by lipopolysaccharide, is a negative regulator for nuclear factor-kappaB. J Endotoxin Res 9:187–191

Muzio M, Ni J, Feng P, Dixit VM (1997) IRAK (Pelle) family member IRAK-2 and MyD88 as proximal mediators of IL-1 signaling. Science 278:1612–1615

Nagai Y, Akashi S, Nagafuku M, Ogata M, Iwakura Y, Akira S, Kitamura T, Kosugi A, Kimoto M, Miyake K (2002) Essential role of MD-2 in LPS responsiveness and TLR4 distribution. Nat Immunol 3:667–672

Nakajima T, Kitajima I, Shin H, Takasaki I, Shigeta K, Abeyama K, Yamashita Y, Tokioka T, Soejima Y, Maruyama I (1994) Involvement of NF-kappa B activation in thrombin-induced human vascular smooth muscle cell proliferation. Biochem Biophys Res Commun 204:950–958

Nakamura H, Aoki M, Tamai K, Oishi M, Ogihara T, Kaneda Y, Morishita R (2002) Prevention and regression of atopic dermatitis by ointment containing NF-kB decoy oligodeoxynucleotides in NC/Nga atopic mouse model. Gene Ther 9:1221–1229

Nathan C (2002) Points of control in inflammation. Nature 420:846–852

Ninomiya-Tsuji J, Kishimoto K, Hiyama A, Inoue J, Cao Z, Matsumoto K (1999) The kinase TAK1 can activate the NIK-I kappaB as well as the MAP kinase cascade in the IL-1 signalling pathway. Nature 398:252–256

Nissen RM, Yamamoto KR (2000) The glucocorticoid receptor inhibits NFkappaB by interfering with serine-2 phosphorylation of the RNA polymerase II carboxy-terminal domain. Genes Dev 14:2314–2329

O'Neill LA (2002) Signal transduction pathways activated by the IL-1 receptor/toll-like receptor superfamily. Curr Top Microbiol Immunol 270:47–61

Obara H, Takayanagi A, Hirahashi J, Tanaka K, Wakabayashi G, Matsumoto K, Shimazu M, Shimizu N, Kitajima M (2000) Overexpression of truncated IkappaBalpha induces TNF-alpha-dependent apoptosis in human vascular smooth muscle cells. Arterioscler Thromb Vasc Biol 20:2198–2204

Obata H, Biro S, Arima N, Kaieda H, Kihara T, Eto H, Miyata M, Tanaka H (1996) NF-kappa B is induced in the nuclei of cultured rat aortic smooth muscle cells by stimulation of various growth factors. Biochem Biophys Res Commun 224:27–32

Oitzinger W, Hofer-Warbinek R, Schmid JA, Koshelnick Y, Binder BR, de Martin R (2001) Adenovirus-mediated expression of a mutant IkappaB kinase 2 inhibits the response of endothelial cells to inflammatory stimuli. Blood 97:1611–1617

Okamoto S, Mukaida N, Yasumoto K, Rice N, Ishikawa Y, Horiguchi H, Murakami S, Matsushima K (1994) The interleukin-8 AP-1 and kappa B-like sites are genetic end targets of FK506-sensitive pathway accompanied by calcium mobilization. J Biol Chem 269:8582–8589

Olefsky JM, Saltiel AR (2000) PPAR gamma and the treatment of insulin resistance. Trends Endocrinol Metab 11:362–368

Orian A, Gonen H, Bercovich B, Fajerman I, Eytan E, Israel A, Mercurio F, Iwai K, Schwartz AL, Ciechanover A (2000) SCF(beta)(-TrCP) ubiquitin ligase-mediated processing of NF-kappaB p105 requires phosphorylation of its C-terminus by IkappaB kinase. EMBO J 19:2580–2591

Ortego M, Bustos C, Hernandez-Presa MA, Tunon J, Diaz C, Hernandez G, Egido J (1999) Atorvastatin reduces NF-kappaB activation and chemokine expression in vascular smooth muscle cells and mononuclear cells. Atherosclerosis 147:253–261

Pahl HL (1999) Activators and target genes of Rel/NF-kappaB transcription factors. Oncogene 18:6853–6866

Palmetshofer A, Robson SC, Nehls V (1999) Lysophosphatidic acid activates nuclear factor kappa B and induces proinflammatory gene expression in endothelial cells. Thromb Haemost 82:1532–1537

Palombella VJ, Conner EM, Fuseler JW, Destree A, Davis JM, Laroux FS, Wolf RE, Huang J, Brand S, Elliott PJ, Lazarus D, McCormack T, Parent L, Stein R, Adams J, Grisham MB (1998) Role of the proteasome and NF-kappaB in streptococcal cell wall-induced polyarthritis. Proc Natl Acad Sci USA 95:15671–15676

Pasceri V, Willerson JT, Yeh ET (2000) Direct proinflammatory effect of C-reactive protein on human endothelial cells. Circulation 102:2165–2168

Pasparakis M, Courtois G, Hafner M, Schmidt-Supprian M, Nenci A, Toksoy A, Krampert M, Goebeler M, Gillitzer R, Israel A, Krieg T, Rajewsky K, Haase I (2002a) TNF-mediated inflammatory skin disease in mice with epidermis-specific deletion of IKK2. Nature 417:861–866

Pasparakis M, Schmidt-Supprian M, Rajewsky K (2002b) IkappaB kinase signaling is essential for maintenance of mature B cells. J Exp Med 196:743–752

Pfeifhofer C, Kofler K, Gruber T, Tabrizi NG, Lutz C, Maly K, Leitges M, Baier G (2003) Protein kinase C theta affects Ca2+ mobilization and NFAT cell activation in primary mouse T cells. J Exp Med 197:1525–1535

Pierce JW, Read MA, Ding H, Luscinskas FW, Collins T (1996) Salicylates inhibit I kappa B-alpha phosphorylation, endothelial-leukocyte adhesion molecule expression, and neutrophil transmigration. J Immunol 156:3961–3969

Poynter ME, Daynes RA (1998) Peroxisome proliferator-activated receptor alpha activation modulates cellular redox status, represses nuclear factor-kappaB signaling, and reduces inflammatory cytokine production in aging. J Biol Chem 273:32833–32841

Pueyo ME, Gonzalez W, Nicoletti A, Savoie F, Arnal JF, Michel JB (2000) Angiotensin II stimulates endothelial vascular cell adhesion molecule-1 via nuclear factor-kappaB activation induced by intracellular oxidative stress. Arterioscler Thromb Vasc Biol 20:645–651

Qian Y, Commane M, Ninomiya-Tsuji J, Matsumoto K, Li X (2001) IRAK-mediated translocation of TRAF6 and TAB2 in the interleukin-1-induced activation of NFkappa B. J Biol Chem 276:41661–41667

Rahman I (2003) Oxidative stress, chromatin remodeling and gene transcription in inflammation and chronic lung diseases. J Biochem Mol Biol 36:95–109

Read MA, Whitley MZ, Gupta S, Pierce JW, Best J, Davis RJ, Collins T (1997) Tumor necrosis factor alpha-induced E-selectin expression is activated by the nuclear factor-kappaB and c-JUN N-terminal kinase/p38 mitogen-activated protein kinase pathways. J Biol Chem 272:2753–2761

Redlich K, Hayer S, Ricci R, David JP, Tohidast-Akrad M, Kollias G, Steiner G, Smolen JS, Wagner EF, Schett G (2002) Osteoclasts are essential for TNF-alpha-mediated joint destruction. J Clin Invest 110:1419–1427

Rehman Q, Lane NE (2001) Bone loss. Therapeutic approaches for preventing bone loss in inflammatory arthritis. Arthritis Res 3:221–227

Reichardt HM, Tuckermann JP, Gottlicher M, Vujic M, Weih F, Angel P, Herrlich P, Schutz G (2001) Repression of inflammatory responses in the absence of DNA binding by the glucocorticoid receptor. EMBO J 20:7168–7173

Reimold AM (2003) New Indications for Treatment of Chronic Inflammation by TNF-alpha Blockade. Am J Med Sci 325:75–92

Ren H, Schmalstieg A, van Oers NS, Gaynor RB (2002) I-kappa B kinases alpha and beta have distinct roles in regulating murine T cell function. J Immunol 168:3721–3731

Rival Y, Beneteau N, Taillandier T, Pezet M, Dupont-Passelaigue E, Patoiseau JF, Junquero D, Colpaert FC, Delhon A (2002) PPARalpha and PPARdelta activators inhibit cytokine-induced nuclear translocation of NF-kappaB and expression of VCAM-1 in EAhy926 endothelial cells. Eur J Pharmacol 435:143–151

Rocksen D, Ekstrand-Hammarstrom B, Johansson L, Bucht A (2003) Vitamin E reduces transendothelial migration of neutrophils and prevents lung injury in endotoxin-induced airway inflammation. Am J Respir Cell Mol Biol 28:199–207

Rodriguez-Porcel M, Lerman LO, Holmes DR, Jr., Richardson D, Napoli C, Lerman A (2002) Chronic antioxidant supplementation attenuates nuclear factor-kappa B activation and preserves endothelial function in hypercholesterolemic pigs. Cardiovasc Res 53:1010–1018

Roebuck KA, Rahman A, Lakshminarayanan V, Janakidevi K, Malik AB (1995) H2O2 and tumor necrosis factor-alpha activate intercellular adhesion molecule 1 (ICAM-1) gene transcription through distinct cis-regulatory elements within the ICAM-1 promoter. J Biol Chem 270:18966–18974

Rogler G, Brand K, Vogl D, Page S, Hofmeister R, Andus T, Knuechel R, Baeuerle PA, Scholmerich J, Gross V (1998) Nuclear factor kappaB is activated in macrophages and epithelial cells of inflamed intestinal mucosa. Gastroenterology 115:357–369

Romashkova JA, Makarov SS (1999) NF-kappaB is a target of AKT in anti-apoptotic PDGF signalling. Nature 401:86–90

Rosenfeld ME (2000) An overview of the evolution of the atherosclerotic plaque: from fatty streak to plaque rupture and thrombosis. Z Kardiol 89 (Suppl 7):2–6

Rossi A, Kapahi P, Natoli G, Takahashi T, Chen Y, Karin M, Santoro MG (2000) Anti-inflammatory cyclopentenone prostaglandins are direct inhibitors of IkappaB kinase. Nature 403:103–108

Rothe M, Sarma V, Dixit VM, Goeddel DV (1995) TRAF2-mediated activation of NF-kappa B by TNF receptor 2 and CD40. Science 269:1424–1427

Rudolph D, Yeh WC, Wakeham A, Rudolph B, Nallainathan D, Potter J, Elia AJ, Mak TW (2000) Severe liver degeneration and lack of NF-kappaB activation in NEMO/IKKgamma-deficient mice. Genes Dev 14:854–862

Ruefli-Brasse AA, French DM, Dixit VM (2003) Regulation of NF-kappaB-dependent lymphocyte activation and development by paracaspase. Science 302:1581–1584

Ruland J, Duncan GS, Wakeham A, Mak TW (2003) Differential requirement for Malt1 in T and B cell antigen receptor signaling. Immunity 19:749–758

Saccani S, Pantano S, Natoli G (2002) p38-Dependent marking of inflammatory genes for increased NF-kappa B recruitment. Nat Immunol 3:69–75

Saccani S, Pantano S, Natoli G (2003) Modulation of NF-kappaB activity by exchange of dimers. Mol Cell 11:1563–1574

Saijo K, Mecklenbrauker I, Santana A, Leitger M, Schmedt C, Tarakhovsky A (2002) Protein kinase C beta controls nuclear factor kappaB activation in B cells through selective regulation of the IkappaB kinase alpha. J Exp Med 195:1647–1652

Sakurai H, Hisada Y, Ueno M, Sugiura M, Kawashima K, Sugita T (1996) Activation of transcription factor NF-kappa B in experimental glomerulonephritis in rats. Biochim Biophys Acta 1316:132–138

Sakurai H, Suzuki S, Kawasaki N, Nakano H, Okazaki T, Chino A, Doi T, Saiki I (2003) Tumor necrosis factor-alpha-induced IKK phosphorylation of NF-kappaB p65 on serine 536 is mediated through the TRAF2, TRAF5, and TAK1 signaling pathway. J Biol Chem 278:36916–36923

Sanz L, Diaz-Meco MT, Nakano H, Moscat J (2000) The atypical PKC-interacting protein p62 channels NF-kappaB activation by the IL-1-TRAF6 pathway. EMBO J 19:1576–1586

Sasu S, Beasley D (2000) Essential roles of IkappaB kinases alpha and beta in serum- and IL-1-induced human VSMC proliferation. Am J Physiol Heart Circ Physiol 278: H1823–H1831

Scheinman RI, Cogswell PC, Lofquist AK, Baldwin AS, Jr. (1995) Role of transcriptional activation of I kappa B alpha in mediation of immunosuppression by glucocorticoids. Science 270:283–286

Scheuren N, Bang H, Munster T, Brune K, Pahl A (1998) Modulation of transcription factor NF-kappaB by enantiomers of the nonsteroidal drug ibuprofen. Br J Pharmacol 123:645–652

Schmidt C, Peng B, Li Z, Sclabas GM, Fujioka S, Niu J, Schmidt-Supprian M, Evans DB, Abbruzzese JL, Chiao PJ (2003) Mechanisms of proinflammatory cytokine-induced biphasic NF-kappaB activation. Mol Cell 12:1287–1300

Schmidt-Supprian M, Bloch W, Courtois G, Addicks K, Israel A, Rajewsky K, Pasparakis M (2000) NEMO/IKK gamma-deficient mice model incontinentia pigmenti. Mol Cell 5:981–992

Schmidt-Supprian M, Courtois G, Tian J, Coyle AJ, Israel A, Rajewsky K, Pasparakis M (2003) Mature T cells depend on signaling through the IKK complex. Immunity 19:377–389

Schwartz MD, Moore EE, Moore FA, Shenkar R, Moine P, Haenel JB, Abraham E (1996) Nuclear factor-kappa B is activated in alveolar macrophages from patients with acute respiratory distress syndrome. Crit Care Med 24:1285–1292

Selzman CH, Shames BD, McIntyre RC, Jr., Banerjee A, Harken AH (1999a) The NFkappaB inhibitory peptide, IkappaBalpha, prevents human vascular smooth muscle proliferation. Ann Thorac Surg 67:1227–1231; discussion 1231–1222

Selzman CH, Shames BD, Reznikov LL, Miller SA, Meng X, Barton HA, Werman A, Harken AH, Dinarello CA, Banerjee A (1999b) Liposomal delivery of purified inhibitory-kappaBalpha inhibits tumor necrosis factor-alpha-induced human vascular smooth muscle proliferation. Circ Res 84:867–875

Senftleben U, Cao Y, Xiao G, Greten FR, Krahn G, Bonizzi G, Chen Y, Hu Y, Fong A, Sun SC, Karin M (2001a) Activation by IKKalpha of a second, evolutionary conserved, NF-kappa B signaling pathway. Science 293:1495–1499

Senftleben U, Li ZW, Baud V, Karin M (2001b) IKKbeta is essential for protecting T cells from TNFalpha-induced apoptosis. Immunity 14:217–230

Sharma S, tenOever BR, Grandvaux N, Zhou GP, Lin R, Hiscott J (2003) Triggering the interferon antiviral response through an IKK-related pathway. Science 300:1148–1151

Sheppard KA, Phelps KM, Williams AJ, Thanos D, Glass CK, Rosenfeld MG, Gerritsen ME, Collins T (1998) Nuclear integration of glucocorticoid receptor and nuclear factor-kappaB signaling by CREB-binding protein and steroid receptor coactivator-1. J Biol Chem 273:29291–29294

Shinobu N, Iwamura T, Yoneyama M, Yamaguchi K, Suhara W, Fukuhara Y, Amano F, Fujita T (2002) Involvement of TIRAP/MAL in signaling for the activation of interferon regulatory factor 3 by lipopolysaccharide. FEBS Lett 517:251–256

Sizemore N, Lerner N, Dombrowski N, Sakurai H, Stark GR (2002) Distinct roles of the Ikappa B kinase alpha and beta subunits in liberating nuclear factor kappa B (NF-kappa B) from Ikappa B and in phosphorylating the p65 subunit of NF-kappa B. J Biol Chem 277:3863–3869

Smahi A, Courtois G, Vabres P, Yamaoka S, Heuertz S, Munnich A, Israel A, Heiss NS, Klauck SM, Kioschis P, Wiemann S, Poustka A, Esposito T, Bardaro T, Gianfrancesco F, Ciccodicola A, D'Urso M, Woffendin H, Jakins T, Donnai D, Stewart H, Kenwrick SJ, Aradhya S, Yamagata T, Levy M, Lewis RA, Nelson DL (2000) Genomic rearrangement in NEMO impairs NF-kappaB activation and is a cause of incontinentia pigmenti. The International Incontinentia Pigmenti (IP) Consortium. Nature 405:466–472

Soares MP, Muniappan A, Kaczmarek E, Koziak K, Wrighton CJ, Steinhauslin F, Ferran C, Winkler H, Bach FH, Anrather J (1998) Adenovirus-mediated expression of a dominant negative mutant of p65/RelA inhibits proinflammatory gene expression in endothelial cells without sensitizing to apoptosis. J Immunol 161:4572–4582

Solan NJ, Miyoshi H, Carmona EM, Bren GD, Paya CV (2002) RelB cellular regulation and transcriptional activity are regulated by p100. J Biol Chem 277:1405–1418

Sosic D, Richardson JA, Yu K, Ornitz DM, Olson EN (2003) Twist regulates cytokine gene expression through a negative feedback loop that represses NF-kappaB activity. Cell 112:169–180

Staal FJ, Roederer M, Raju PA, Anderson MT, Ela SW, Herzenberg LA (1993) Antioxidants inhibit stimulation of HIV transcription. AIDS Res Hum Retroviruses 9:299–306

Staels B, Koenig W, Habib A, Merval R, Lebret M, Torra IP, Delerive P, Fadel A, Chinetti G, Fruchart JC, Najib J, Maclouf J, Tedgui A (1998) Activation of human aortic smooth-muscle cells is inhibited by PPARalpha but not by PPARgamma activators. Nature 393:790–793

Steer JH, Kroeger KM, Abraham LJ, Joyce DA (2000) Glucocorticoids suppress tumor necrosis factor-alpha expression by human monocytic THP-1 cells by suppressing transactivation through adjacent NF-kappa B and c-Jun-activating transcription factor-2 binding sites in the promoter. J Biol Chem 275:18432–18440

Straus DS, Pascual G, Li M, Welch JS, Ricote M, Hsiang CH, Sengchanthalangsy LL, Ghosh G, Glass CK (2000) 15-deoxy-delta 12,14-prostaglandin J2 inhibits multiple steps in the NF-kappa B signaling pathway. Proc Natl Acad Sci USA 97:4844–4849

Su CG, Wen X, Bailey ST, Jiang W, Rangwala SM, Keilbaugh SA, Flanigan A, Murthy S, Lazar MA, Wu GD (1999) A novel therapy for colitis utilizing PPAR-gamma ligands to inhibit the epithelial inflammatory response. J Clin Invest 104:383–389

Sun S-C, Ganchi PA, Ballard DW, Greene WC (1993) NF-κB controls expression of inhibitor IκB: evidence for an inducible autoregulatory pathway. Science 259:1912–1915

Sun Z, Arendt CW, Ellmeier W, Schaeffer EM, Sunshine MJ, Gandhi L, Annes J, Petrzilka D, Kupfer A, Schwartzberg PL, Littman DR (2000) PKC-theta is required for TCR-induced NF-kappaB activation in mature but not immature T lymphocytes. Nature 404:402–407

Suzuki YJ, Packer L (1993) Inhibition of NF-kappa B activation by vitamin E derivatives. Biochem Biophys Res Commun 193:277–283

Tak PP, Gerlag DM, Aupperle KR, van de Geest DA, Overbeek M, Bennett BL, Boyle DL, Manning AM, Firestein GS (2001) Inhibitor of nuclear factor kappaB kinase beta is a key regulator of synovial inflammation. Arthritis Rheum 44:1897–1907

Takaesu G, Surabhi RM, Park KJ, Ninomiya-Tsuji J, Matsumoto K, Gaynor RB (2003) TAK1 is critical for IkappaB kinase-mediated activation of the NF-kappaB pathway. J Mol Biol 326:105–115

Takeda K, Takeuchi O, Tsujimura T, Itami S, Adachi O, Kawai T, Sanjo H, Yoshikawa K, Terada N, Akira S (1999) Limb and skin abnormalities in mice lacking IKKalpha. Science 284:313–316

Tanaka M, Fuentes ME, Yamaguchi K, Durnin MH, Dalrymple SA, Hardy KL, Goeddel DV (1999) Embryonic lethality, liver degeneration, and impaired NF-kappa B activation in IKK-beta-deficient mice. Immunity 10:421–429

Tham DM, Martin-McNulty B, Wang YX, Wilson DW, Vergona R, Sullivan ME, Dole W, Rutledge JC (2002) Angiotensin II is associated with activation of NF-kappaB-mediated genes and downregulation of PPARs. Physiol Genomics 11:21–30

Thome M, Tschopp J (2003) TCR-induced NF-kappaB activation: a crucial role for Carma1, Bcl10 and MALT1. Trends Immunol 24:419–424

Tomita T, Takano H, Tomita N, Morishita R, Kaneko M, Shi K, Takahi K, Nakase T, Kaneda Y, Yoshikawa H, Ochi T (2000) Transcription factor decoy for NFkappaB inhibits cytokine and adhesion molecule expressions in synovial cells derived from rheumatoid arthritis. Rheumatology (Oxford) 39:749–757

Tomita T, Takeuchi E, Tomita N, Morishita R, Kaneko M, Yamamoto K, Nakase T, Seki H, Kato K, Kaneda Y, Ochi T (1999) Suppressed severity of collagen-induced arthritis by in vivo transfection of nuclear factor kappaB decoy oligodeoxynucleotides as a gene therapy. Arthritis Rheum 42:2532–2542

Trompouki E, Hatzivassiliou E, Tsichritzis T, Farmer H, Ashworth A, Mosialos G (2003) CYLD is a deubiquitinating enzyme that negatively regulates NF-kappaB activation by TNFR family members. Nature 424:793–796

Ueda A, Okuda K, Ohno S, Shirai A, Igarashi T, Matsunaga K, Fukushima J, Kawamoto S, Ishigatsubo Y, Okubo T (1994) NF-kappa B and Sp1 regulate transcription of the human monocyte chemoattractant protein-1 gene. J Immunol 153:2052–2063

van Den Brink GR, ten Kate FJ, Ponsioen CY, Rive MM, Tytgat GN, van Deventer SJ, Peppelenbosch MP (2000) Expression and activation of NF-kappa B in the antrum of the human stomach. J Immunol 164:3353–3359

Vanderlugt CL, Rahbe SM, Elliott PJ, Dal Canto MC, Miller SD (2000) Treatment of established relapsing experimental autoimmune encephalomyelitis with the proteasome inhibitor PS-519. J Autoimmun 14:205–211

Venkatakrishnan A, Stecenko AA, King G, Blackwell TR, Brigham KL, Christman JW, Blackwell TS (2000) Exaggerated activation of nuclear factor-kappaB and altered IkappaB-beta processing in cystic fibrosis bronchial epithelial cells. Am J Respir Cell Mol Biol 23:396–403

Vincenti MP, Coon CI, Brinckerhoff CE (1998) Nuclear factor kappaB/p50 activates an element in the distal matrix metalloproteinase 1 promoter in interleukin-1beta-stimulated synovial fibroblasts. Arthritis Rheum 41:1987–1994

Voll RE, Jimi E, Phillips RJ, Barber DF, Rincon M, Hayday AC, Flavell RA, Ghosh S (2000) NF-kappa B activation by the pre-T cell receptor serves as a selective survival signal in T lymphocyte development. Immunity 13:677–689

Wahl C, Liptay S, Adler G, Schmid RM (1998) Sulfasalazine: a potent and specific inhibitor of nuclear factor kappa B. J Clin Invest 101:1163–1174

Wang D, You Y, Case SM, McAllister-Lucas LM, Wang L, DiStefano PS, Nunez G, Bertin J, Lin X (2002a) A requirement for CARMA1 in TCR-induced NF-kappa B activation. Nat Immunol 3:830–835

Wang Z, Castresana MR, Detmer K, Newman WH (2002b) An IkappaB-alpha mutant inhibits cytokine gene expression and proliferation in human vascular smooth muscle cells. J Surg Res 102:198–206

Wang Z, Castresana MR, Newman WH (2001a) NF-kappaB is required for TNF-alpha-directed smooth muscle cell migration. FEBS Lett 508:360–364

Wang Z, Castresana MR, Newman WH (2001b) Reactive oxygen and NF-kappaB in VEGF-induced migration of human vascular smooth muscle cells. Biochem Biophys Res Commun 285:669–674

Weber CK, Liptay S, Wirth T, Adler G, Schmid RM (2000) Suppression of NF-kappaB activity by sulfasalazine is mediated by direct inhibition of IkappaB kinases alpha and beta. Gastroenterology 119:1209–1218

Weih F, Caamano J (2003) Regulation of secondary lymphoid organ development by the nuclear factor-kappaB signal transduction pathway. Immunol Rev 195:91–105

Weih F, Carrasco D, Durham SK, Barton DS, Rizzo CA, Ryseck R-P, Lira SA, Bravo R (1995) Multiorgan Inflammation and hematopoietic abnormalities in mice with a targeted disruption of RelB, a member of the NF-κB/Rel family. Cell 80:331–340

Weih F, Durham SK, Barton DS, Sha WC, Baltimore D, Bravo R (1996) Both multiorgan inflammation and myeloid hyperplasia in RelB-deficient mice are T cell dependent. J Immunol 157:3974–3979

Weil R, Schwamborn K, Alcover A, Bessia C, Di Bartolo V, Israel A (2003) Induction of the NF-kappaB cascade by recruitment of the scaffold molecule NEMO to the T cell receptor. Immunity 18:13–26

Whelan J, Ghersa P, Hooft van Huijsduijnen R, Gray J, Chandra G, Talabot F, DeLamarter JF (1991) An NF kappa B-like factor is essential but not sufficient for cytokine induction of endothelial leukocyte adhesion molecule 1 (ELAM-1) gene transcription. Nucl Acids Res 19:2645–2653

Wiegmann K, Schwandner R, Krut O, Yeh WC, Mak TW, Kronke M (1999) Requirement of FADD for tumor necrosis factor-induced activation of acid sphingomyelinase. J Biol Chem 274:5267–5270

Wilson SH, Best PJ, Edwards WD, Holmes DR, Jr., Carlson PJ, Celermajer DS, Lerman A (2002) Nuclear factor-kappaB immunoreactivity is present in human coronary plaque and enhanced in patients with unstable angina pectoris. Atherosclerosis 160:147–153

Wilson SH, Caplice NM, Simari RD, Holmes DR, Jr., Carlson PJ, Lerman A (2000) Activated nuclear factor-kappaB is present in the coronary vasculature in experimental hypercholesterolemia. Atherosclerosis 148:23–30

Woronicz JD, Gao X, Cao Z, Rothe M, Goeddel DV (1997) IkappaB kinase-beta: NF-kappaB activation and complex formation with IkappaB kinase-alpha and NIK. Science 278:866–869

Xia Y, Pauza ME, Feng L, Lo D (1997) RelB regulation of chemokine expression modulates local inflammation. Am J Pathol 151:375–387

Xiao G, Cvijic ME, Fong A, Harhaj EW, Uhlik MT, Waterfield M, Sun SC (2001) Retroviral oncoprotein Tax induces processing of NF-kappaB2/p100 in T cells: evidence for the involvement of IKKalpha. EMBO J 20:6805–6815

Yamamoto M, Sato S, Hemmi H, Hoshino K, Kaisho T, Sanjo H, Takeuchi O, Sugiyama M, Okabe M, Takeda K, Akira S (2003a) Role of adaptor TRIF in the MyD88-independent toll-like receptor signaling pathway. Science 301:640–643

Yamamoto M, Sato S, Hemmi H, Sanjo H, Uematsu S, Kaisho T, Hoshino K, Takeuchi O, Kobayashi M, Fujita T, Takeda K, Akira S (2002) Essential role for TIRAP in activation of the signalling cascade shared by TLR2 and TLR4. Nature 420:324–329

Yamamoto M, Sato S, Hemmi H, Uematsu S, Hoshino K, Kaisho T, Takeuchi O, Takeda K, Akira S (2003b) TRAM is specifically involved in the Toll-like receptor 4-mediated MyD88-independent signaling pathway. Nat Immunol 4:1144–1150

Yamamoto Y, Verma UN, Prajapati S, Kwak YT, Gaynor RB (2003c) Histone H3 phosphorylation by IKK-alpha is critical for cytokine-induced gene expression. Nature 423:655–659

Yamaoka S, Courtois G, Bessia C, Whiteside ST, Weil R, Agou F, Kirk HE, Kay RJ, Israel A (1998) Complementation cloning of NEMO, a component of the IkappaB kinase complex essential for NF-kappaB activation. Cell 93:1231–1240

Yamasaki S, Kawakami A, Nakashima T, Nakamura H, Kamachi M, Honda S, Hirai Y, Hida A, Ida H, Migita K, Kawabe Y, Koji T, Furuichi I, Aoyagi T, Eguchi K (2001) Importance of NF-kappaB in rheumatoid synovial tissues: in situ NF-kappaB expression and in vitro study using cultured synovial cells. Ann Rheum Dis 60:678–684

Yamazaki S, Muta T, Takeshige K (2001) A novel IkappaB protein, IkappaB-zeta, induced by proinflammatory stimuli, negatively regulates nuclear factor-kappaB in the nuclei. J Biol Chem 276:27657–27662

Yang L, Cohn L, Zhang DH, Homer R, Ray A, Ray P (1998) Essential role of nuclear factor kappaB in the induction of eosinophilia in allergic airway inflammation. J Exp Med 188:1739–1750

Yin MJ, Yamamoto Y, Gaynor RB (1998) The anti-inflammatory agents aspirin and salicylate inhibit the activity of I(kappa)B kinase-beta. Nature 396:77–80

Yoshimura S, Morishita R, Hayashi K, Yamamoto K, Nakagami H, Kaneda Y, Sakai N, Ogihara T (2001) Inhibition of intimal hyperplasia after balloon injury in rat carotid artery model using cis-element 'decoy' of nuclear factor-kappaB binding site as a novel molecular strategy. Gene Ther 8:1635–1642

Yurochko AD, Kowalik TF, Huong SM, Huang ES (1995) Human cytomegalovirus upregulates NF-kappa B activity by transactivating the NF-kappa B p105/p50 and p65 promoters. J Virol 69:5391–5400

Zandi E, Chen Y, Karin M (1998) Direct phosphorylation of IkappaB by IKKalpha and IKKbeta: discrimination between free and NF-kappaB-bound substrate. Science 281:1360–1363

Zandi E, Karin M (1999) Bridging the gap: composition, regulation, and physiological function of the IkappaB kinase complex. Mol Cell Biol 19:4547–4551

Zandi E, Rothwarf DM, Delhase M, Hayakawa M, Karin M (1997) The IkappaB kinase complex (IKK) contains two kinase subunits, IKKalpha and IKKbeta, necessary for IkappaB phosphorylation and NF-kappaB activation. Cell 91:243–252

Zhang HG, Huang N, Liu D, Bilbao L, Zhang X, Yang P, Zhou T, Curiel DT, Mountz JD (2000a) Gene therapy that inhibits nuclear translocation of nuclear factor kappaB results in tumor necrosis factor alpha-induced apoptosis of human synovial fibroblasts. Arthritis Rheum 43:1094–1105

Zhang SQ, Kovalenko A, Cantarella G, Wallach D (2000b) Recruitment of the IKK signalosome to the p55 TNF receptor: RIP and A20 bind to NEMO (IKKgamma) upon receptor stimulation. Immunity 12:301–311

Zhou H, Wertz I, O'Rourke K, Ultsch M, Seshagiri S, Eby M, Xiao W, Dixit VM (2004) Bcl10 activates the NF-kappaB pathway through ubiquitination of NEMO. Nature 427:167–171

Zhu X, Lu L, Currier BL, Windebank AJ, Yaszemski MJ (2002) Controlled release of NFkappaB decoy oligonucleotides from biodegradable polymer microparticles. Biomaterials 23:2683–2692

Zollner TM, Podda M, Pien C, Elliott PJ, Kaufmann R, Boehncke WH (2002) Proteasome inhibition reduces superantigen-mediated T cell activation and the severity of psoriasis in a SCID-hu model. J Clin Invest 109:671–679

Hepatitis B Virus X Protein: Structure–Function Relationships and Role in Viral Pathogenesis

V. Kumar[1] (✉) · D. P. Sarkar[2]

[1] Virology Group, International Centre for Genetic Engineering and Biotechnology, Aruna Asaf Ali Marg, P.O. Box 10504, New Delhi 110067, India
vijay@icgeb.res.in
[2] Department of Biochemistry, University of Delhi South Campus, Benito Juarez Marg, New Delhi 110021, India

1	History	378
2	Homology and Domain Structure	378
3	Properties of HBx	380
3.1	Physicochemical Properties	380
3.2	Domain Structure and Function	380
3.3	Phosphorylation and Kinase Activities	382
4	Role in Transactivation	383
4.1	Interaction with Transcriptional Apparatus and Transactivation	384
4.2	Signal Transduction and Transactivation	388
4.3	Inhibition of Protease Activity and Transactivation	390
5	Regulation of the Cell Cycle	390
6	Anti- and Proapoptotic Effects of HBx	391
7	Role in Cell Transformation	392
8	Role in the Virus Life Cycle	393
9	Role in Hepatocarcinogenesis	393
10	Therapeutic Applications	395
11	Concluding Remarks	396
	References	396

Abstract The hepatitis B virus (HBV) genome codes for a 16.5-KDa protein termed pX or HBx which is a prevalent marker in the liver of patients with hepatitis B-associated hepatocellular carcinoma (HCC). Although the specific function of HBx in natural infection remains elusive, it is considered to play an important role in the etiology of HBV-induced HCC. It is a multifunctional regulatory protein that is best characterized as promiscuous transactivator. It can induce different signaling pathways, deregulate cell cycle checkpoints or even adapt cell responses to genotoxic stress and protein degradation. The oncogenic nature of HBx is evident from its ability to transform permanent cell lines ex vivo and induce liver-specific tumors in transgenic animals. RNA interference against the expression of HBx could be an effective therapeutic tool in the management of hepatitis B-associated liver diseases.

Keywords Cell cycle deregulation · Hepadnavirus · Hepatocellular carcinoma · Signal transduction · Transactivation

1
History

Molecular cloning and DNA sequencing of the genomes of hepatitis B virus (HBV) (Galibert et al. 1979) and the related woodchuck (WHV) and ground squirrel (GSHV) hepatitis viruses (Galibert et al. 1982; Seeger at al. 1984) revealed the conservation of four open reading frames (ORFs). Three of these ORFs could be assigned to known viral proteins—polymerase (P), surface antigen (S) and core antigen (C). However, no known protein or function could be assigned to the smallest fourth ORF that could potentially code for a basic polypeptide of 154 amino acids. Originally called as region 5 (Galibert et al. 1979), and later as region '*X*' this small ORF was postulated to have a role in viral transcription (Tiollais et al. 1981). The existence of the '*X*' gene was finally established in 1985 by the cross-reaction of sera from HBV-infected patients with the recombinant hepatitis B virus X protein (HBx) expressed in a mammalian cell line or in *Escherichia coli* (Kay et al. 1985; Moriarty et al. 1985). Although *X* sequences are found in both genomic and subgenomic mRNA forms, it is encoded primarily by a 0.7-kb mRNA species (Suzuki et al. 1989; Guo et al. 1991). As HBV shows extreme host tropism and cannot be grown easily in heterologous systems, structure–function studies of HBx have been carried out in human hepatocyte cultures.

2
Homology and Domain Structure

The *X* ORF is present in all mammalian hepadnaviruses (hepatotropic DNA viruses) including HBV (Ganem and Schneider 2001). A homology search of X protein across species has made it possible to functionally dissect the X protein into various regions. Comparison of primary sequence of human HBx with X proteins of other hepadnaviruses including four nonhuman primates (gorilla, gibbon, chimpanzee and woolly monkey) and other small mammals like woodchuck, arctic ground squirrel and ground squirrel indicates a high level (72%) of conservation. The X protein sequences can be subdivided into six regions (A–F) where the more conserved regions A, C and E are interspaced by the less conserved B, D and F regions (Fig. 1). The amino-terminal A (residues 1–20) and carboxy-terminal E regions (residues 120–140) are the two most conserved segments of HBx with 90% homology followed by 86% homology in the middle region C (residues 58–84). Region D (residues 85–119) is the most divergent part (40% homology) while re-

Hepatitis B Virus X Protein: Structure-Function Relationships

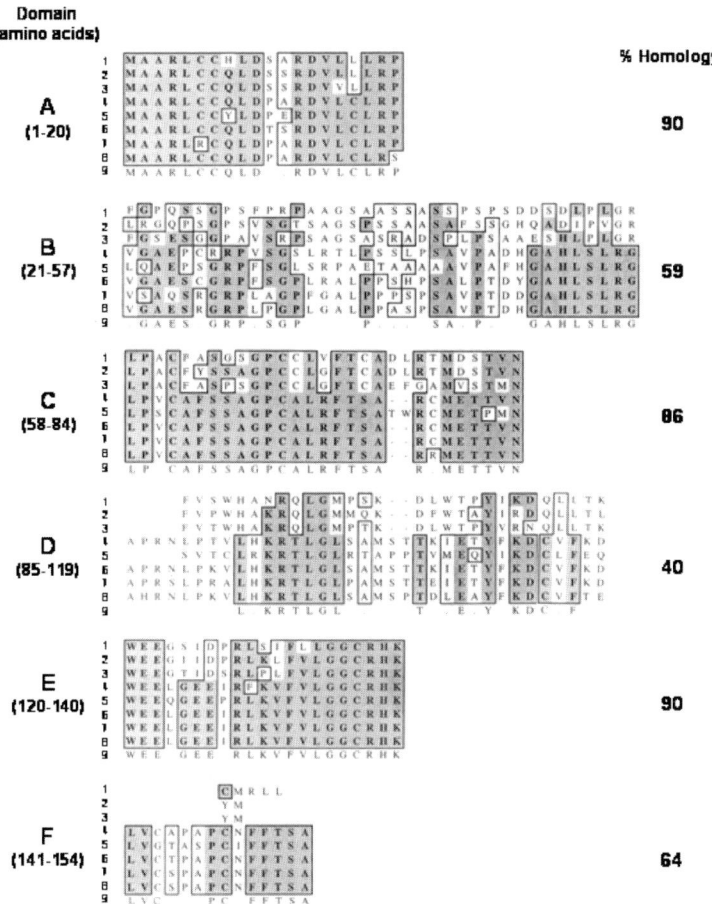

Fig. 1 Homology of HBx. Domain wise (**A–F**) clustal alignment of the X proteins of different mammalian hepadnaviruses (*1–8*). Different domains and their demarcations are shown on the *left*, while percent homology is shown on the *right*. The gene accession numbers for different viral sequence are: *1*, woodchuck hepatitis virus (M11082); *2*, ground squirrel hepatitis virus (K02715); *3*, Arctic ground squirrel hepatitis virus (U29144); *4*, gorilla hepatitis virus (AJ131567); *5*, woolly monkey hepatitis virus (AF046996); *6*, chimpanzee hepatitis virus (D00220); *7*, gibbon hepatitis virus (AJ1311568); *8*, human hepatitis B virus (X00715); *9*, consensus

gions B (residues 21–57) and F (residues 141–154) show a homology close to 60%. HBx has 14 conserved basic amino acids and nine cysteine residues. The conserved cysteines have a distribution of three in region A, two each in regions C and F, and one each in regions D and E. There is no evidence for a phylogenetic evolution of HBx from hepadnaviruses of nonhuman primates. The uprooted tree analysis of 29 *HBV X* genes and their predicted proteins has indicated a common ancestral sequence that has given rise to two main groups of *X* genes representing the two predominant HBV strains found in the eastern and western hemispheres (Kidd-Ljunggren et al. 1995).

3
Properties of HBx

3.1
Physicochemical Properties

To date, only a few studies on the biochemical properties of HBx have been undertaken. The low level expression of HBx in infected liver and in transfected cells has hindered detailed analysis of its biochemical properties. Moreover, the biophysical properties of recombinant HBx have been extremely difficult to determine and are suspect owing to artifacts caused by its poor solubility and nonspecific aggregation problems. In human hepatoma cells, the transiently expressed HBx decays with a bimodal half-life of 15 min and 3 h depending on whether it is associated with detergent-soluble or -insoluble fractions, respectively (Schek et al. 1991). As predicted, the purified X protein from *E. coli* shows a basic isoelectric point of 8.3 (Jameel et al. 1990). The disulphide linkage analysis of HBx produced in two heterologous systems has yielded different results. Tryptic fragment analysis of the air oxidized HBx produced in *E. coli* has shown that eight of the nine cysteines of HBx are disulphide linked—each cysteine is linked to the fourth cysteine in a sequential manner with the carboxy-terminal cysteine being free (Gupta et al. 1995). While the mass spectrometric analysis of a recombinant HBx (with ten cysteines) produced in insect cells showed five intra-molecular disulphide bridges (Urban et al. 1997), HBx synthesized in a cell-free system is reported to form a homodimer with the four amino-terminal cysteine residues playing a key role in the dimerization process (Lin and Lo 1989). The homo-dimerization of HBx has also been demonstrated by far-Western blotting and the region of interaction has been mapped to residues 21–50 (Murakami et al. 1994). The homo-dimerization of HBx has also been observed in the yeast cellular environment and mapped to its carboxy-terminal region (Reddi and Kumar 2004). The physiological significance of disulphide linkages and oligomerization is not clear.

3.2
Domain Structure and Function

HBx exerts a pleiotropic effect on common biochemical pathways such as gene transcription, signal transduction, and protein degradation that are ultimately associated with cell transformation and development of hepatocel-

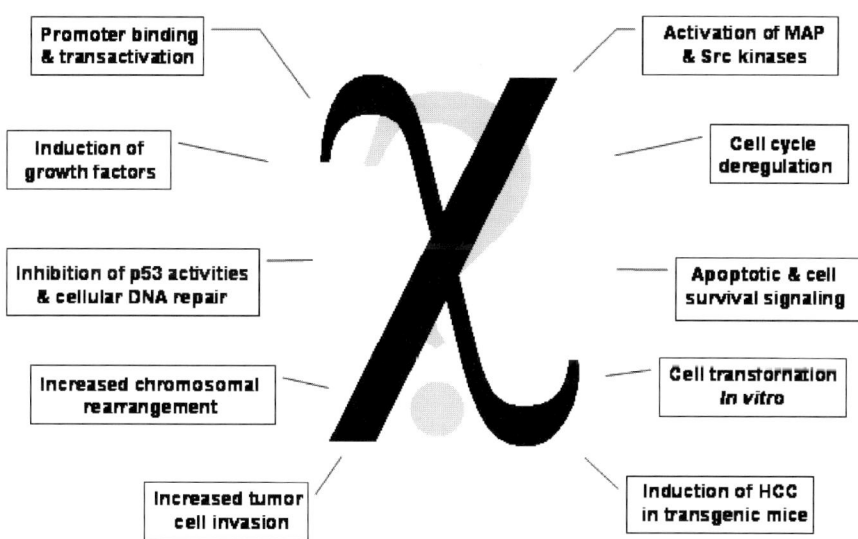

Fig. 2 Pleiotropic X protein (HBx) of hepatitis B virus

lular carcinoma (HCC) (Fig. 2). A serial analysis of gene expression in freshly isolated human primary hepatocytes infected with a replication-defective adenovirus containing the X gene of HBV showed HBx-specific induction of a total of 57 transcripts and repression of 46 transcripts by more than fivefold (Wu et al. 2002). The HBx-induced transcripts fall into three major categories, including genes that encode ribosomal proteins, transcription factors with zinc-finger motifs, and proteins associated with polypeptide degradation pathways (Wu et al. 2001, 2002). Detailed mutational analysis of the X gene has led to the identification of the regions that appear to perform some discrete functions such as nuclear transactivation, signal transduction and transrepression (Fig. 3). Approximately 13 amino acids from the carboxy terminus and at least 50 amino acids of the amino terminus of HBx can be deleted without affecting its transactivation function (Renner 1995; Ritter et al. 1991; Runkel et al 1993). Murakami et al. (1994) have mapped the transactivation domain to residues 52–148 and showed a negative regulatory domain in its amino-terminal region. In our laboratories, the HBx-related transactivation and transgenic studies have been based largely on the homology parameters discussed in Sect. 2. We have identified the transactivation domain of HBx in the region amino acids 58–140 in which residues 120–140 are critical for the nuclear transactivation (Kumar et al. 1996; Reddi et al. 2003) while amino acids 58–119 are required for signal transduction (Nijhara et al. 2001a). More recently, we have identified a strong regulatory function in the highly conserved amino-terminal A region (residues 1–20) of

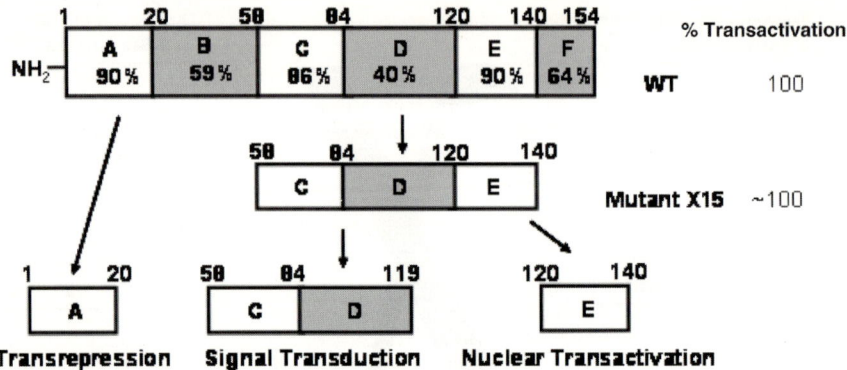

Fig. 3 Domain structure of HBx with some putative functions. *WT*, wild-type

HBx. Expression of this 20-amino acid region shows a general transrepression function (Reddi et al. 2003).

3.3
Phosphorylation and Kinase Activities

HBx appears to be phosphorylated in hepatoma cells as evident from its metabolic labeling with [^{32}P]-orthophosphate (Schek et al. 1991) and in insect cells as evident from mass spectrometric analysis (Urban et al. 1997). Further, purified recombinant HBx serves as a substrate for mitogen activated protein (MAP) kinase and protein kinase C in vitro suggesting a possible post-translational regulation of this viral protein (Lee et al. 2001a). The phosphorylation site(s) for both of these protein kinases have yet to be mapped. Human HBx can bind to cyclin-dependent kinase 2 (CDK2) and become phosphorylated at its amino-terminal region where there is a putative '^{41}SPSA^{44}' motif for CDK2 (Mukherji and Kumar, unpublished results).

Interestingly, HBx shares limited structural similarity with the nucleoside diphosphate kinases and shows a phosphotransferase activity in vitro by generating a phosphoryl-HBx intermediate in the presence of ATP and EDTA. The conserved histidyl residue (H^{52}) present in the B region of HBx may be involved in the phosphotransfer process (de-Medina and Shaul 1994). However, the relevance of transphosphorylation in the virus life cycle is not clear. Recombinant HBx also shows a potent AMP kinase activity that converts AMP to ADP using ATP as the exclusive phosphate donor (Dopheide and Azad 1996). The X protein also possesses ATPase and GTPase activities that may be active during the regulation of nuclear transcription and signal transduction in the host cell (de-Medina et al. 1994; Dopheide and Azad 1996). While ATP hydrolysis may be required to regulate the helicase and RNA polymerase II activities (Oelgeschlager 2002; Qadri et al. 1996; Wang et al. 1992), GTP may be involved in the modulation of signal transduction by HBx (Benn and Schneider 1994; Kim et al. 2001), modulation of Ca^{2+} signaling for viral replication (Bouchard et al. 2001, 2003) and nuclear import machinery (Forgues et al. 2001, 2003).

4
Role in Transactivation

The transactivation function of HBx was first postulated in 1981 soon after the HBV genome was cloned and sequenced (Tiollais et al. 1981). Based on the reverse transcriptase activity of the HBV polymerase and its ability to synthesize an RNA intermediate during viral replication, Miller and Robinson (1986) speculated that the *X* gene of HBV is analogous to the *pX* genes of some retroviruses like the human T lymphotropic viruses (HTLV) I and II and may have a role in transactivation, cell transformation and tumor development. Interestingly, altogether the *X* gene shows a cellular codon usage yet it does not have a cellular homolog like many retroviral oncogenes. Later studies have confirmed many functional similarities between *HBx* and *pX* gene products. Nevertheless, HBx has been best characterized as a broad-spectrum activator of transcription with the ability to regulate all three classes of promoters (Aufiero and Schneider 1990; Wang et al. 1995, 1997,

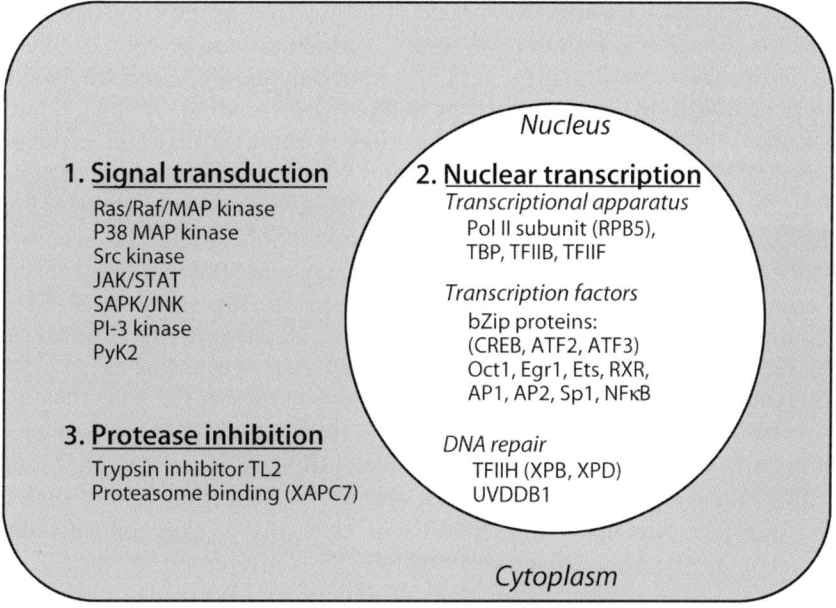

Fig. 4 Transactivation by HBx: possible mechanisms and interlinks

1998) through a direct interaction with nuclear transcription factors, via signal transduction or by way of inhibition of the cellular proteolytic machinery (Fig. 4).

4.1
Interaction with Transcriptional Apparatus and Transactivation

The transactivation function of HBx has been studied extensively in cell culture by transfecting the HBx expression plasmid along with a reporter gene construct (such as firefly luciferase or bacterial chloramphenicol acetyl transferase) under the control of a wide array of promoters (reviewed in Caselmann 1996; Murakami 2001; Rossner 1992). The ability of the X gene product to activate transcription was first demonstrated on the human beta-interferon promoter in Vero cells (Twu and Schloemer 1987). In the past few years, several autologous and heterologous viral and cellular regulatory sequences have been identified as targets of HBx-mediated transactivation (Table 1). Not only can HBx stimulate its own enhancer/promoter sequences but it can also activate transcription from a wide range of viral regulatory elements. The retroviral long terminal repeats (LTRs) of human immunodeficiency virus 1, HTLV-I, mouse mammary tumor virus, Rous sarcoma virus, murine sarcoma virus and Visna, are known to be positively regulated by the X protein. The enhancer/promoter elements of several DNA viruses such as simian virus 40 early promoter, adenovirus E3 and E4 and herpes simplex virus thymidine kinase are activated by HBx. Several different mammalian promoters are effective targets of transactivation by HBx. These include the promoters of c-*fos*, c-*jun*, c-*myc*, retinoblastoma protein, α- and β-interferon, TATA-binding protein (TBP), RNA polymerase II and III. The inability of HBx to directly bind to any defined DNA-binding sequences, however, has suggested that the transactivation mechanism may not involve a known DNA sequence-specific interaction (Avantaggiati et al. 1993; Maguire et al. 1991). These and other reports show the involvement of protein–protein interaction with components of the transcriptional apparatus (Haviv et al. 1996; Williams and Andrisani 1995) and increased DNA-binding capacity of some transcription factors in the presence of HBx (Choi et al. 1999; Maguire et al. 1991; Perini et al. 1999; Reddi et al. 2003). General transcription factors like RNA polymerase II (Cheong et al. 1995, Lin et al. 1997), TBP (Qadri et al. 1995; Wang et al. 1997, 1998), TFIIB (Haviv et al. 1998; Lin et al. 1997), TFIIF (Wei et al. 2001), and TFIIH (Jaitovich-Groisman et al. 2001; Qadri et al. 1996) are known to interact directly with HBx. This could facilitate the transactivation process. Common transcription factors such as AP-1 (*jun/fos*) (Kim et al. 2001; Natoli et al. 1994), AP-2 (Kim and Rho 2002; Seto et al. 1990), ATF2 (Maguire 1991), NF-AT (Carretero et al. 2002; Lara-Pezzi et al. 1998), nuclear factor (NF)-κB (Lucito and Schneider 1992; Mahe et al. 1991; Su and Schneider 1996), ATF/CREB (Barnabas and Andrisani

Table 1 Viral and cellular promoters transactivated by HBx

Class/promoter	Cell line	References
A. Viral regulatory elements		
DNA viral elements		
Adenovirus E3 promoter	PLC/PRF/5	Colgrove et al. 1989
Adenovirus E4 promoter	HepG2	Balsano et al. 1991
CMV immediate early promoter	HepG2, MRC5	Assogba et al. 2002
	CV1	Cross et al. 1993
	CHO-K1	Reddi et al. 2003
HBV X gene promoter/Enh1	HepG2	Siddiqui et al. 1989
	HepG2	Nakatake et al. 1993
	Chang	Wollersheim et al. 1988
HBV Core promoter/Enh2	HepG2	Siddiqui et al. 1989
	Huh7	Colgrove et al. 1989
HBV preS1 promoter	HepG2, Huh7	Rossner et al. 1990
HBV preS2 promoter	Huh7	Colgrove et al. 1989
	HepG2, Huh7	Rossner et al. 1990
HSV tk promoter	Vero	Twu and Schloemer 1987
	Chang	Zahm et al. 1988
SV40 early promoter/Enh	HepG2	Siddiqui et al. 1989
	HepG2	Zahm et al. 1988
	Chang	Siddiqui et al. 1989
	Chang	Twu and Robinson 1989
	Chang	Aufiero and Schneider 1990
	Chang	Kwee et al. 1992
Retroviral elements		
HIV1-LTR	HepG2	Siddiqui et al. 1989
	HepG2	Twu and Robinson 1989
	Jurkat	Seto et al. 1988
HIV2-LTR	HepG2	Levrero et al. 1990
HTLV1-LTR	HepG2	Woolersheim et al. 1988
	HepG2	Siddiqui et al. 1989
	HeLa	Marriott et al. 1996
HTLV2-LTR	HepG2	Siddiqui et al. 1989
MMTV-LTR	Chang	Wollersheim at al. 1988
	Chang	Zahm et al. 1988
RSV-LTR	Huh7	Colgrove et al. 1989
	HepG2	Balsano et al. 1991
Visna-LTR	CHO-K1	Reddi et al. 2003
B. Cellular genes (Pol I-dependent)		
rRNA	S2	Wang et al. 1998
C. Cellular genes (Pol II-dependent)		
Cell signaling/transcription factors		
c-Fos	HeLa	Avantaggiati et al. 1993
	HepG2	Natoli et al. 1994
c-Jun	HepG2	Natoli et al. 1994
	HepG2	Twu et al. 1993
	HepG2	Zhou et al. 1994
c-Myc	HepG2	Balsano et al. 1991
N-Myc2	HepG2	Flajolet et al. 1997
TATA binding protein	S2, Rat 1A	Wang et al. 1995, 1997
iNOS	HepG2	Amaro et al. 1999
	HepG2, Chang	Majano et al. 2001
Cell cycle regulators		
p21$^{waf1/cip1}$	CHO-K1	Leach et al. 2003
	Mouse hepatocyte	Qiao et al. 2001
	Hep3B	Park et al. 2000

Table 1 (continued)

Class/promoter	Cell line	References
p27^{kip1}	CHO-K1	Leach et al. 2003
	Mouse hepatocyte	Qiao et al, 2001
Retinoblastoma	HepG2	Farshid et al. 1997
Growth factors		
Epidermal growth factor receptor	HepG2	Menzo et al. 1993
IGF-I receptor	SNU368	Kim et al. 1996
Insulin like growth factor II	HepG2	Lee et al. 1998
	"	Kang-Park et al. 2001
Transforming growth factor α	HepG2	Kim and Rho 2002
Transforming growth factor β	AML12	Barnabas et al. 1997
	HepG2	Yoo et al. 1996
Immune network/apoptosis		
FasL	PLC/PRF/5	Shin et al. 1999
β-Interferon	Vero	Twu and Schloemer 1987
Interleukin 6	HepG2, Huh7	Lee et al. 1998
Interleukin 8	8387	Mahe et al. 1991
Interleukin 18	Chang, SNU354	Lee et al. 2002
MHC-I (H2 k)	HepG2	Zhou et al. 1990
MHC-II (HLA-DR)	HepG2, Huh7	Hu et al. 1992
Nur77	Chang X-34	Lee et al. 2001
Tumor necrosis factor α	HepG2, liver biopsy	Gonzalez-Amaro et al. 1994
Cytoskeleton/extracellular matrix		
Actin	CHO-K1	Reddi et al. 2003
ICAM-1	HepG2	Hu et al. 1992
Integrin-β1	Chang	Lara-Pezzi et al. 2001
Troponin	CHO-K1	Reddi et al. 2003
Others		
α1-Antitrypsin	HepG2	Lee et al. 1990
α-Fetoprotein	HepG2	Zhou et al. 1994
	HepG2, Huh7	Arima et al. 2002
Galectin-3	HepG2, Huh7	Hsu et al. 1999
MDR1	HepG2, Hep3B	Doong et al. 1998
Metallothionein	Chang	Zahm et al. 1988
mtATPase	HepG2	Oh et al. 1998
D. Cellular genes (Pol III-dependent)		
tRNA^{ala}	Chang	Aufiero and Schneider 1990
tRNA^{alrg}	S2, Rat 1A	Wang et al. 1995, 1997
U6	HeLa	Antunovic et al. 1993
	S2	Wang et al. 1995, 1997
VA1RNA	Chang	Kwee et al. 1992
5S RNA	S2	Wang et al. 1995

CMV, cytomegalovirus; Enh, enhancer; HBV, hepatitis B virus; HIF-1α, hypoxia-inducible factor-1 alpha; HIV, human immunodeficiency virus; HSV, herpes simplex virus; HTLV, human T cell leukemia virus; IGF, insulin-like growth factor; iNOS, inducible nitric oxide synthase; LTR, long terminal repeat; MDR, multidrug resistance; MHC, major histocompatibility complex; SV40, simian virus 40.

2000; Maguire et al. 1991; Reddi et al. 2003), C/EBPα (Choi et al. 1999), Egr-1 (Yoo et al. 1996), Sp1 (Lee et al. 1998; Li and Ou 2001), Oct-1 (Antunovic et al. 1993), and retinoic acid receptor (Kong et al. 2000) are reported to mediate the activities of HBx. '*Cis*' elements such as the serum response ele-

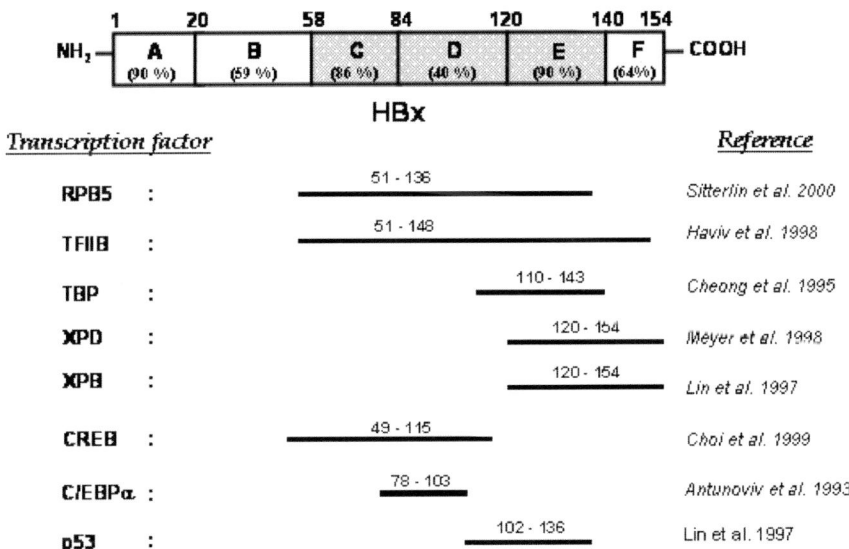

Fig. 5 HBx domains delineated for interaction with various transcription factors

ment (Goto et al. 2003; Kim et al. 2001), AP1-binding site (Henkler et al. 1998; Natoli et al. 1994), Ets-responsive element (Yoo et al. 1996), NF-κB-binding site (Assogba et al. 2002; Kim et al. 2001; Mahe et al. 1991; Majano et al. 2001), and CAAT box (Mahe et al. 1991) appear to mediate promoter activation by HBx. The interacting sites for some of these transcription factors on HBx have been mapped and a consensus study indicates that a two-third portion of its C-terminal region is the preferred site of interaction (Fig. 5). Mutational analysis of HBx in our laboratories using reporter gene activation in cell culture (Kumar et al. 1996; Reddi et al. 2003) and MAP kinase signaling in vivo (Nijhara et al. 2001a) have related very well with the importance of region C to E in nuclear transactivation and signal transduction.

During the normal course of infection, the X gene can possibly be translated into polypeptides of three different lengths from a single mRNA using alternate translation initiation from one of the three in-frame AUG codons (Met1, Met79, and Met103). It is interesting to note that while all the three isoforms, i.e., full-length (16.5 kDa), 8- and 6.6-kDa forms can individually stimulate polymerase III-dependent promoters, some cellular promoters such as NF-κB are activated exclusively by full-length HBx suggesting differential regulatory mechanisms (Kwee et al. 1992). It would be interesting to identify the cellular factors that differentially regulate the activities of HBx. Many questions on the modulation of general transcription at proximal promoters by HBx still remain unanswered because it does not contribute to basal transcription and yet activates transcription. As HBx does not show promoter selectivity, it is likely to have only a co-activator function. Once recruited to the transcriptional machinery, HBx may interact with several kinds of transactivators and stimulate a sustained transcription. As there are multiple interaction partners of HBx in transactivation, it may piggyback on different cellular factors and present an activation domain to the transcriptional apparatus being assembled at the promoter. Regardless of the actual mechanism(s) of transactivation by HBx, accumulating lines of evidence clearly indicate that HBx augments expression of many endogenous genes through different pathways that may confer advantages for viral replication and cell proliferation.

There is one report that suggests that suppression of the translation initiation factor hu-sui1 by HBx might also be associated with abrogation of negative growth regulation and consequent development of HCC. This view gains credibility from the fact that introduction of hu-sui1 into HepG2 cells inhibits cell growth in soft agar culture and interferes with tumor formation in nude mice (Lian et al. 1999).

4.2
Signal Transduction and Transactivation

The predominant cytoplasmic location of HBx is consistent with its activation of cytoplasmic signaling cascades that are well known to play a crucial role in regulating cellular growth, differentiation and apoptosis (Hoare et al. 2001; Sirma et al. 1998). The first report showing diacylglycerol-dependent activation of protein kinase C by HBx (Kekule et al. 1993) does not seem to be a major signaling mechanism. HBx is now known to stimulate a number of other cell signaling cascades including Ras-Raf-MAP kinase (Benn and Schneider 1994; Tarn et al. 2001), p38 MAP kinase (Tarn et al. 2002), Janus kinases/signal transducers and activators of transcription (JAK/STAT) pathways (Arbuthnot et al. 2000; Lee and Yun 1998), stress-activated protein kinase/c-*jun* N-terminal kinase (SAPK/JNK) (Tarn et al 2001; Benn et al. 1996), phosphatidylinositol 3-kinase (PI3 K) (Lee et al. 2001), Src tyrosine kinases (Klein and Schneider 1997) and calcium-dependent proline rich tyrosine kinase-2 (PyK2) (Bouchard et al. 2001a). Activation of MAP kinases is necessary for modulation of the activity of a wide range of transcriptional factors such as c-*jun*, c-*fos* and c-*myc* that in turn regulate cellular growth, differentiation and apoptosis (Hazzalin and Mahadevan 2002). Constitutive activation of MAP kinase pathway relates to enhanced kinase activity in many tumors and in primary cell culture leading to the transformed phenotypes (Shapiro 2002). HBx activation of Ras/Raf/MEK/ERK in cell culture has been reported by many laboratories (Benn and Schneider 1994; Cross et al. 1993). Stimulation of the MAP kinase pathway is critical for the activation of AP-1 proteins (Benn et al. 1996; Natoli et al 1994). We demonstrated

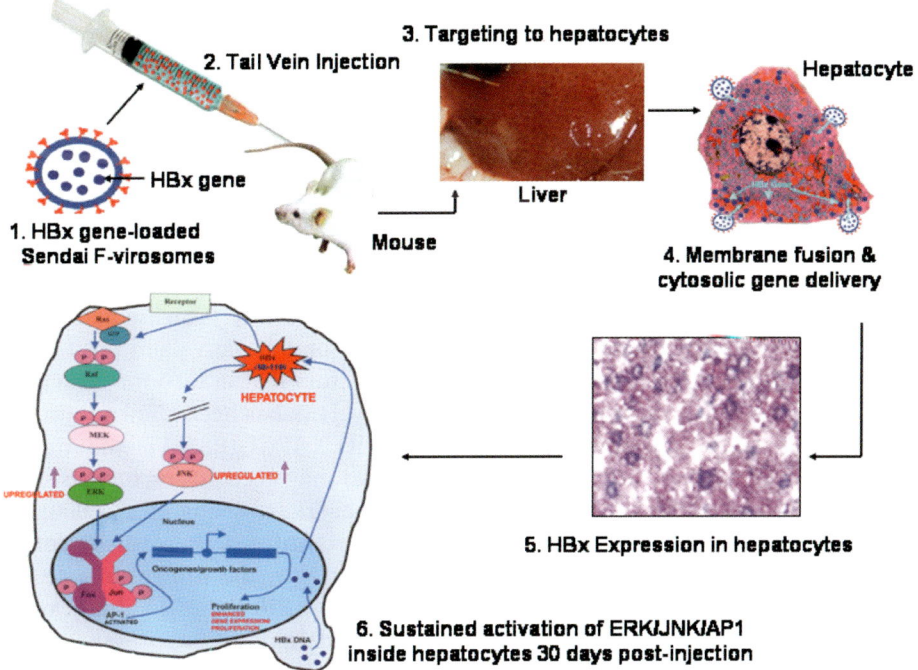

Fig. 6 In vivo modulation of MAP kinase activity by HBx leading to sustained activation of Erk 1/2, JNK and AP-1 factors in mouse liver

for the first time that expression of HBx in the mouse liver can cause long-term activation of MAP kinases and AP-1 proteins (Fig. 6) which may precede cell transformation and tumor development in vivo (Nijhara et al. 2001b). The activation of MAP kinase by HBx involves increased GTP uptake and enhanced association among Shc, Grb2 and Sos (Klein and Schneider 1997). The X-stimulated Ras is also required for the modulation of cellular TBP levels and activation of transcription from RNA pol I- and pol III-dependent promoters (Wang et al. 1995, 1997, 1998). The requirement of Ras signaling for the activation of NF-κB still remains a controversial issue. While one study suggests the requirement of Ras signaling for NF-κB activation (Su and Schneider 1996), the other seems to suggest that this may not be the case (Chirillo et al. 1996). The binding of HBx to the inhibitory protein IκB, may be one of the mechanisms of release and nuclear translocation of NF-κB for further transactivation (Weil et al. 1999).

HBx can also activate the nonreceptor tyrosine kinases of the Src family which are upstream activators of Ras GTPases (Klein and Schneider 1997). Src kinases are constitutively activated in hepatoma cells during the replication of WHV and HBV only in the presence of a functional HBx (Klein et al. 1999). HBx may involve the upstream calcium-dependent PyK2 to activate

Src kinases (Bouchard et al. 2001a). Src kinases can also activate the JAK/STAT cell signaling pathways that are associated with increased hepatocyte proliferation (Cressman et al. 1995).

4.3
Inhibition of Protease Activity and Transactivation

HBx bears a striking sequence homology in its C and E regions to the functionally essential domain (F-V/I-X-G-G-C-R/K) of Kunitz-type serine protease inhibitors and can block the action of a liver-specific protease tryptase TL_2 (Takada et al. 1994). A general inhibition in the serine protease activity can either suppress the proteolytic activation of some negative and positive factors or stabilize liver-specific transcription factors that in turn may cause a global increase in transcription. More recently, HBx has been shown to inhibit proteasomal activity by interaction with its XAPC7 (PSMA7) subunit. Inhibition of general proteases and proteasomes may account for increased transactivation in the presence of HBx (Hu et al. 1999; Zhang et al. 2000). Ubiquitination of transcription factors is now regarded a prerequisite for augmenting their transcriptional ability. Interestingly, the proteasome-dependent proteolytic processing of the p105 precursor of the p50 subunit of the NF-κB transcription factor slows down considerably in the presence of HBx (Sirma et al. 1998).

5
Regulation of the Cell Cycle

Cell cycle control by oncogenic viruses usually involves disruption of the normal restraints on cellular proliferation via abnormal proteolytic degradation and malignant transformation of cells. The cell cycle regulatory molecules (i.e.,. cyclins, CDKs and inhibitors of CDKs, as well as the transcriptional targets of signaling pathways) induce cells to move through the cell cycle checkpoints. These checkpoints are often found to be deregulated in tumor cells and in cells infected by DNA tumor viruses predisposing them towards transformation (Op De Beeck and Caillet-Fauquet 1997). HBx has been shown to override the replicative senescence of cells in G_0 phase by binding to $p55^{sen}$ (Sun et al. 1998). It stimulates G_0 cells to transit through G_1 phase by activating Src kinases and the cyclin A–CDK2 complexes, that in turn induces the cyclin A promoter (Bouchard et al. 2001b). AML12-immortalized hepatocytes, when transformed by HBx, also show G_1 to S and G_2 to M progression as shown by the activation of cyclins and cdc2 kinase (Lee et al. 2002). The expression of HBx in mouse fibroblasts also causes progression from the G_0/G_1 phase into the S phase (Koike et al. 1994a). This effect of HBx on the G_1 to S checkpoint control appears to be related to the cellular levels of p53 and downstream activation of p21$^{wafl/cipl}$ (Ahn et al. 2002). Driving the HBV-infected cells to grow continuously may be essential for active viral replication that could facilitate the full manifestation of the oncogenic potential of HBx.

6
Anti- and Proapoptotic Effects of HBx

Controlled cell growth and apoptosis are essential for maintaining normal tissue homeostasis in an organism. HBx regulates this cellular homeostasis in two diametrically opposite ways: it appears to inhibit apoptosis during early hepatocyte infection and viral replication, and to activate apoptosis in later stages to facilitate viral spread and immune evasion. HBx can efficiently block the p53-mediated apoptosis either by reducing p21WAF1 expression or by interfering with binding of p53 to the TFIIH transcription-nucleotide excision repair (NER) complex (Wang et al. 1995). Further, HBx contributes to the process by downregulating the transcription of the *p53* gene (Lee and Rho 2000) as well as by directly interacting with p53 (Truant et al. 1995). Because the *X* gene is frequently integrated into the genome of HCC cells, inhibition of p53-mediated apoptosis by HBx may give a clonal selective advantage for these cells during the early stages of hepatocarcinogenesis (Huo et al. 2001). Thus, functional inactivation (rather than structural mutation) of p53 by HBx may be contributing towards the development of HCC (Ueda et al. 1995). HBx may also interfere with the NER pathway through both p53-dependent and -independent mechanisms (Groisman et al. 1999). Because HBx binds to TFIIH-associated proteins, it may interfere with the NER pathway by binding to and altering the activities of helicases necessary for NER activity. This can increase the mutation rate induced by chemical carcinogens, such as aflatoxin B1, during human liver carcinogenesis (Bressac et al. 1991; Hsu et al. 1991; Jia et al. 1999; Sohn et al. 2000). Further, HBx can significantly inhibit the ability of cells to repair damaged DNA by interacting with XAP-1 (Becker et al. 1998; Jia et al. 1999). Co-expression of transforming growth factor beta (TGFβ) and HBx has been reported in the dysplastic nodules induced in the liver of HBx transgenic mice, suggesting at least functional interaction between the viral protein and growth factors during tumor development (Yoo et al. 1996). It protects Hep3B hepatoma cells from TGFβ-induced apoptosis by inducing a survival signal that links Src to the PI3 K/Akt signaling pathway (Lee et al. 2001b; Shih et al. 2003). HBx has been shown to inhibit the caspase-3 activity (Gottlob et al. 1998a). Thus, abrogation of the antiproliferative and apoptotic effects of HBx or its naturally occurring mutants might tune the hepatocytes to uncontrolled growth and contribute to multi-step hepatocarcinogenesis associated with HBV infection (Sirma et al. 1999).

The X protein also shows proapoptotic effect on liver cells that could be both p53-mediated as well as p53-independent (Chirillo et al. 1997; Elmore et al. 1997; Kim et al. 1998; Terradillos et al. 1998). The proapoptotic activity of HBx helps to overcome the inhibitory effect of Bcl-2 against Fas cytotoxicity (Terradilos et al. 2002). The HBx-induced loss of mitochondrial membrane potential has also been suggested as a cause of cell death (Shirakata and Koike 2003). The interaction of HBx with c-FLIP and enhancement of death-inducing signals are considered important for its proapoptotic function (Kim and Seong 2003). Myc protein also appears to sensitize cells to killing by HBx in the presence of tumor necrosis factor α(Su et al. 2001). The proapoptotic action of X protein may have a role in viral spread during the acute phase of HBV infection.

7
Role in Cell Transformation

The X protein seems to predispose cells to genetic damage. The oncogenic potential of HBx is evident from its ability to transform immortalized cell lines. In sharp contrast, the duck hepatitis B virus that lacks an *X* ORF fails to transform primary hepatocytes in culture and is nontumorigenic (Tuttleman et al. 1986). Overexpression of HBx can accelerate the growth characteristics of NIH 3T3 cells and accentuate their ability to develop tumors in nude mice (Shirakata et al. 1989). Similarly, the immortalized mouse hepatocyte line FMH202 transformed by the *X* gene displays malignant growth characteristics in soft agar and is tumorigenic upon inoculation in nude mice (Hohne et al. 1990; Schaefer and Gerlich 1995). The amino-terminal 50 amino acid region of HBx has been considered to be essential for cell transformation because HBx mutants that lack the amino-terminal segment, but retain most of the transactivating function of HBx, are unable to alter the growth characteristics of REV-2 cells (Gottlob et al. 1998b). The altered genetic background in immortalized cell lines due to insertion of certain transgenes such as simian virus 40 T antigen in FMH-202 and E1A in REV-2 may genetically predispose them to easy transformation by HBx.

Recently, the X protein has been shown to induce the synthesis of matrix metalloproteinase and cyclooxygenase-2 that may be required for tumor cell invasion (Lara-Pezzi et al. 2002). HBx can also work cooperatively with other viruses such as hepatitis C virus to transform mouse fibroblast NIH3T3 cells. The cells expressing both the core protein of hepatitis C virus and X protein of HBV exhibit anchorage-independent cell growth, secrete more matrix metalloproteases, and show higher tumorigenicity in athymic nude mice (Jung et al. 2003). HBx also collaborates with activated H-*ras* to transform immortalized rat embryonic fibroblast REF52 cells. Injection of nude mice with REF52 cells transfected with both HBx and activated H-*ras* leads to tumor development, whereas cells transfected with either gene alone appear nontumorigenic (Kim et al. 2001). Interestingly, the naturally occurring carboxy-terminally truncated HBx mutants enhance the transforming ability of *ras* and *myc* and thus control cell proliferation, viability, and transformation (Tu et al. 2001). Cell lines stably expressing HBx also show augmented levels of endogenous c-*myc* mRNA transcripts (Balsano et al. 1991).

8
Role in the Virus Life Cycle

In spite of the discovery of various biochemical properties and pleiotropic behaviors inherent to HBx, its precise role in the virus life cycle and in pathogenesis has remained enigmatic. In cell culture, HBx appears to be dispensable as HBx-null mutants synthesize wild-type levels of viral proteins, and show wild-type levels of replicative intermediates and virion export (Blum et al. 1992). The transgenic mouse models, carrying either the wild-type HBV genome or a mutated HBV genome incapable of expressing HBx, indicate that while the X protein is not essential for viral replication or its maturation, it can enhance viral replication, apparently by activating viral gene expression (Reifenberg et al. 2002; Xu et al. 2002). However, none of these experiments investigated productive infection. On the other hand, for WHV, the X protein is essential for replication and productive infection in vivo (Chen et al. 1993; Zoulim et al. 1994). It has been shown that HBx stimulation of calcium signaling and Src tyrosine kinases promotes a high level of viral replication in cell culture (Bouchard et al. 2003; Klein et al. 1999) which can be abolished by the MEK1/2 inhibitor U0126. (Zheng et al. 2003).

9
Role in Hepatocarcinogenesis

It is evident from previous discussions that HBx can directly or indirectly support a number of mitogenic and mutagenic cellular events that may eventually give rise to primary HCC (Table 2). A piece of key supporting evidence is the fact that HBx is a more prominent marker of HBV infection in HCC than are the HBV surface or core antigens (Cromlish 1996). The transactivation functions of HBV—mediated either by HBx or truncated envelope proteins—could be relevant to hepatocarcinogenesis. Incidentally, HBV transactivator sequences are present at a higher frequency as genomic integrations in HCC tissues or hepatoma derived cell lines than structural (surface and core) gene sequences (Caselmann 1995; Paterlini et al. 1995; Poussin et al. 1999; Zhou et al. 1987). Further, analysis of the integrated HBV sequences in HCC patients indicate that complete or 3' truncated functional *X*

Table 2 Mitogenic and mutagenic signals of HBx

Biochemical/cellular events	References
1. Activation of Ras/Raf/MEK/ERK cascade	Benn and Schneider 1994
	Tarn et al. 2001
	Nijhara et al. 2001b
2. Activation of Src/PI 3 kinases	Klein and Schneider 1997
	Lee et al. 2001
3. Transactivation of growth factor genes/proto-oncogenes	Balsano et al. 1991
	Barnabas et al. 1997
	Kim et al. 1996
	Kim and Rho 2002
	Lee et al. 1998
	Menzo et al. 1993
	Yoo et al. 1996
4. Increased tumor cell invasion and metastasis	Lara-Pezzi et al. 2001, 2002
5. Inhibition of p53 activities and cellular DNA repair	Groisman et al. 1999
	Truant et al. 1995
	Wang et al. 1995
6. Increased sensitivity to genotoxic agents	Jia et al. 1999
	Madden et al. 2001
7. Cell cycle deregulation	Bouchard et al. 2001b
	Koike et al. 1994a
	Lee et al. 2002
8. Cell survival signaling (antiapoptosis)	Gottlob et al. 1998a
	Lee et al. 2001b
	Shih et al. 2003
9. Increased chromosomal rearrangement and micronuclei formation	Livezey et al. 2002
10. Induction of supernumerary centrosome	Forgues et al. 2003
11. Cell transformation in vitro	Gottlob et al. 1998b
	Hohne et al. 1990
	Shirakata et al. 1989
12. Induction of hepatocellular carcinoma in transgenic animal models	Kim et al. 1991
	Terradillos et al. 1997
	Lakhtakia et al. 2003

ORFs are often retained in those cases whereas viral core, envelope and polymerase encoding genes are often rearranged (Balsano et al. 1991; Schluter et al. 1994; Takada and Koike 1990; Tsuei et al. 1994; Wollersheim et al. 1988, Yamamoto et al. 1993). Selective expression of HBx in the tumor cells could be a reason for chromosomal rearrangement (Livezey et al 2002; Su et al. 1998).

The first direct evidence of the role of HBx in cell transformation/HCC came from a transgenic mouse model: constitutive expression of the *HBx* gene under the control of its own regulatory elements led to the development multi-focal areas of altered hepatocytes, progressing to benign adenomas and then to malignant carcinomas (Kim et al. 1991; Koike et al. 1994b). In contrast, another transgenic mouse model expressing HBx under the control of the fetal alpha-1-antitrypsin promoter did not develop tumors (Lee et al. 1990) presumably because of the lack of sustained expression of the viral protein after birth. Similarly, HBx transgenic mice in which HBx is under the control of the antithrombin III promoter also do not develop tumors despite expression of functional viral protein. Nevertheless, these mice are sensitive to genotoxic reagents that are tumorigenic (Madden et al. 2001; Slagle et al. 1996). Further, a genetic cross between two separate transgenic lines, expressing either HBx or c-*myc* that otherwise do not show a tumor phenotype, show liver-specific tumor formation (Terradillos et al. 1997). We have developed a transgenic mouse model of HCC using a HBx–myc bicistronic construct (Kumar et al. 2001). Co-expression of HBx and c-*myc* facilitates early development of HCC in these mice (Lakhtakia et al. 2003; Singh and Kumar 2003). Thus, just like its in vitro cell transformation ability, HBx also displays oncogenic cooperativity for malignant transformation in vivo although its own oncogenic potential may be weak.

10
Therapeutic Applications

An important goal in treatment for hepatitis B is to prevent the development of HCC for which there is presently no effective therapy. As the integration/expression of the *X* gene has been observed in majority of the HBV-related HCC (Caselmann 1995; Paterlini et al. 1995; Poussin et al. 1999; Zhou et al. 1987) an effective way to control this disease would be to interfere with HBx expression. Attempts have been made in the recent past to regulate the expression of HBx in cell culture and experimental animals using anti-HBx antibody, antisense oligonucleotide, ribozyme or even small inhibitory RNAs. According to a report from China, treatment of nude mice (having human HCC grafts) with anti-HBx antibody leads to significant tumor regression and prolonged survival (Li et al. 1996). Further, HCC patients treated with anti-HBx antibody also show reduction in tumor size and a decrease in the serum level of alpha fetoprotein (Li et al. 1996). The antisense phosphorothioate oligonucleotides directed against the initiation codon of the *X* gene can prevent the development of preneoplastic lesions in a transgenic mouse model of HCC (Moriya et al. 1996). Some hammerhead ribozymes have also been designed that can inhibit the expression of functional HBx and show antireplicative effect in cell culture (Kim et al. 1999; Weinberg et al. 2000). This illustrates the potential of ribozymes in the treatment of chronic liver disease and HCC. Nowadays, RNA interference (RNAi) is being used as a powerful antiviral strategy to silence virus-specific genes. Expression of short hairpin RNAs directed against the *X* gene appears to induce an RNAi response (Hung and Kumar 2004) that can dramatically reduce the HBV DNA replication (Shlomai and Shaul 2003) and secretion levels of surface antigen in cell culture (McCaffrey et al. 2003). Thus, RNAi could be used as an effective therapeutic tool in the management of hepatitis B-associated liver diseases.

11
Concluding Remarks

Despite numerous biochemical studies and development of cell transformation and transgenic animal models, the precise role of HBx in the development of HCC is not fully understood. The main point of whether a regulated and stage-specific expression of HBx in the normal physiological micro-environment of liver cells is enough to induce neoplastic growth, is the most relevant question to be addressed. In the case that HBx is shown to have this ability in adult liver cells any interference in its expression is likely to be of therapeutic importance. New transgenic animal models with the inducible HBx expression system, or the liver-specific F-virosome gene delivery system (Nijhara et al. 2001a, 2001b) should be used to address these crucial questions. Moreover, F-virosome delivery of various therapeutic agents could be tested for an effective intervention of HBx-induced HCC.

References

Ahn JY, Jung EY, Kwun HJ, Lee CW, Sung YC, Jang KL (2002) Dual effects of hepatitis B virus X protein on the regulation of cell-cycle control depending on the status of cellular p53. J Gen Virol 83:2765–2772

Amaro MJ, Bartolome J, Carreno V (1999) Hepatitis B virus X protein transactivates the inducible nitric oxide synthase promoter. Hepatology 29:915–923

Antunovic J, Lemieux N, Cromlish JA (1993) The 17 kDa HBx protein encoded by hepatitis B virus interacts with the activation domains of Oct-1, and functions as a coactivator in the activation and repression of a human U6 promoter. Cell Mol Biol Res 39:463–482

Arbuthnot P, Capovilla A, Kew M (2000) Putative role of hepatitis B virus X protein in hepatocarcinogenesis: effects on apoptosis, DNA repair, mitogen-activated protein kinase and JAK/STAT pathways. J Gastroenterol Hepatol 15:357–368

Arima T, Nakao K, Nakata K, Ishikawa H, Ichikawa T, Hamasaki K, Ishii N, Eguchi K (2002) Transactivation of human alpha-fetoprotein gene by X-gene product of hepatitis B virus in human hepatoma cells. Int J Mol Med 9:397–400

Assogba BD, Choi BH, Rho HM (2002) Transcriptional activation of the promoter of human cytomegalovirus immediate early gene (CMV-IE) by the hepatitis B viral X protein (HBx) through the NF-kappaB site. Virus Res 84:171–179

Aufiero B, Schneider RJ (1990) The hepatitis B virus X-gene product trans-activates both RNA polymerase II and III promoters. EMBO J 9:497–504

Avantaggiati ML, Natoli G, Balsano C, Chirillo P, Artini M, De Marzio E, Collepardo D, Levrero M (1993) The hepatitis B virus (HBV) pX transactivates the c-fos promoter through multiple cis-acting elements. Oncogene 8:1567–1574

Balsano C, Avantaggiati ML, Natoli G, De Marzio E, Will H, Perricaudet M, Levrero M (1991) Full-length and truncated versions of the hepatitis B virus (HBV) X protein (pX) transactivate the c-myc protooncogene at the transcriptional level. Biochem Biophys Res Commun 176:985–992

Barnabas S, Hai T, Andrisani OM (1997) The hepatitis B virus X protein enhances the DNA binding potential and transcription efficacy of bZip transcription factors. J Biol Chem 272:20684–20690

Barnabas S, Andrisani OM (2000) Different regions of hepatitis B virus X protein are required for enhancement of bZip-mediated transactivation versus transrepression. J Virol 74:83–90

Becker SA, Lee TH, Butel JS, Slagle BL (1998) Hepatitis B virus X protein interferes with cellular DNA repair. J Virol 72:266–272

Benn J, Schneider RJ (1994) Hepatitis B virus HBx protein activates Ras-GTP complex formation and establishes a Ras, Raf, MAP kinase signaling cascade. Proc Natl Acad Sci USA 91:10350–10354

Benn J, Su F, Doria M, Schneider RJ (1996) Hepatitis B virus HBx protein induces transcription factor AP-1 by activation of extracellular signal-regulated and c-Jun N-terminal mitogen-activated protein kinases. J Virol 70:4978–4985

Blum HE, Zhang ZS, Galun E, von Weizsacker F, Garner B, Liang TJ, Wands JR (1992) Hepatitis B virus X protein is not central to the viral life cycle in vitro. J Virol 66: 1223–1227

Bouchard MJ, Wang LH, Schneider RJ (2001a) Calcium signaling by HBx protein in hepatitis B virus DNA replication. Science 294:2376–2378

Bouchard M, Giannakopoulos S, Wang EH, Tanese N, Schneider RJ (2001b) Hepatitis B virus HBx protein activation of cyclin A-cyclin-dependent kinase 2 complexes and G1 transit via a Src kinase pathway. J Virol 75:4247–4257

Bouchard MJ, Puro RJ, Wang L, Schneider RJ (2003) Activation and inhibition of cellular calcium and tyrosine kinase signaling pathways identify targets of the HBx protein involved in hepatitis B virus replication. J Virol 77:7713–7719

Bressac B, Kew M, Wands J, Ozturk M (1991) Selective G to T mutations of p53 gene in hepatocellular carcinoma from southern Africa. Nature 350:429–431

Carretero M, Gomez-Gonzalo M, Lara-Pezzi E, Benedicto I, Aramburu J, Martinez-Martinez S, Redondo J, Lopez-Cabrera M (2002) The hepatitis B virus X protein binds to and activates the NH(2)-terminal trans-activation domain of nuclear factor of activated T cells-1. Virology 299:288–300

Caselmann WH (1995) Transactivation of cellular gene expression by hepatitis B viral proteins: a possible molecular mechanism of hepatocarcinogenesis. J Hepatol 22 (1 Suppl):34–37

Caselmann WH (1996) Trans-activation of cellular genes by hepatitis B virus proteins: a possible mechanism of hepatocarcinogenesis. Adv Virus Res 47:253–302

Chen HS, Kaneko S, Girones R, Anderson RW, Hornbuckle WE, Tennant BC, Cote PJ, Gerin JL, Purcell RH, Miller RH (1993) The woodchuck hepatitis virus X gene is important for establishment of virus infection in woodchucks. J Virol 67:1218–1226

Cheong JH, Yi M, Lin Y, Murakami S (1995) Human RPB5, a subunit shared by eukaryotic nuclear RNA polymerases, binds human hepatitis B virus X protein and may play a role in X transactivation. EMBO J 14:143–150

Chirillo P, Falco M, Puri PL, Artini M, Balsano C, Levrero M, Natoli G. (1996) Hepatitis B virus pX activates NF-kappa B-dependent transcription through a Raf-independent pathway. J Virol 70:641–646

Chirillo P, Pagano S, Natoli G, Puri PL, Burgio VL, Balsano C, Levrero M (1997) The hepatitis B virus X gene induces p53-mediated programmed cell death. Proc Natl Acad Sci USA 94:8162–8167

Choi BH, Park GT, Rho HM (1999) Interaction of hepatitis B viral X protein and CCAAT/enhancer-binding protein alpha synergistically activates the hepatitis B viral enhancer II/pregenomic promoter. J Biol Chem 274:2858–2865

Colgrove R, Simon G, Ganem D (1989) Transcriptional activation of homologous and heterologous genes by the hepatitis B virus X gene product in cells permissive for viral replication. J Virol 63:4019–4026

Cressman DE, Diamond RH, Taub R (1995) Rapid activation of the Stat3 transcription complex in liver regeneration. Hepatology 21:1443–1449

Cromlish JA (1996) Hepatitis B virus-induced hepatocellular carcinoma: possible roles for HBx. Trends Microbiol 4:270–274

Cross JC, Wen P, Rutter WJ (1993) Transactivation by hepatitis B virus X protein is promiscuous and dependent on mitogen-activated cellular serine/threonine kinases. Proc Natl Acad Sci USA 90:8078–8082

de-Medina T, Shaul Y (1994) Functional and structural similarity between the X protein of hepatitis B virus and nucleoside diphosphate kinases. FEBS Lett 351:423–426

de-Medina T, Haviv I, Noiman S, Shaul Y (1994) The X protein of hepatitis B virus has a ribo/deoxy ATPase activity. Virology 202:401–407

Doong SL, Lin MH, Tsai MM, Li TR, Chuang SE, Cheng AL (1998) Transactivation of the human MDR1 gene by hepatitis B virus X gene product. J Hepatol 29:872–878

Dopheide TA, Azad AA (1996) The hepatitis B virus X protein is a potent AMP kinase. J Gen Virol 77:173–176

Elmore LW, Hancock AR, Chang SF, Wang XW, Chang S, Callahan CP, Geller DA, Will H, Harris CC (1997) Hepatitis B virus X protein and p53 tumor suppressor interactions in the modulation of apoptosis. Proc Natl Acad Sci USA 94:14707–4712

Farshid M, Nedjar S, Mitchell F, Biswas R (1997) Effect of hepatitis B virus X protein on the expression of retinoblastoma gene product. Acta Virol 41:125–129

Flajolet M, Gegonne A, Ghysdael J, Tiollais P, Buendia MA, Fourel G (1997) Cellular and viral trans-acting factors modulate N-myc2 promoter activity in woodchuck liver tumors. Oncogene 15:1103–1110

Forgues M, Difilippantonio MJ, Linke SP, Ried T, Nagashima K, Feden J, Valerie K, Fukasawa K, Wang XW (2003) Involvement of Crm1 in hepatitis B virus X protein-induced aberrant centriole replication and abnormal mitotic spindles. Mol Cell Biol 23:5282–5292

Galibert F, Mandart E, Fitoussi F, Tiollais P, Charnay P (1979) Nucleotide sequence of the hepatitis B virus genome (subtype ayw) cloned in *E. coli*. Nature 281:646–650

Galibert F, Chen TN, Mandart E (1982) Nucleotide sequence of a cloned woodchuck hepatitis virus genome: comparison with the hepatitis B virus sequence. J Virol 41:51–65

Ganem D, Schneider RJ (2001) Hepadnaviridae: The viruses and their replication. In: Knipe DM, Howley PM (eds) Fields Virology (4th edn). Lippincott Williams and Wilkins, Philadephia, pp 2923–2969

Gonzalez-Amaro R, Garcia-Monzon C, Garcia-Buey L, Moreno-Otero R, Alonso JL, Yague E, Pivel JP, Lopez-Cabrera M, Fernandez-Ruiz E, Sanchez-Madrid F. (1994) Induction of tumor necrosis factor alpha production by human hepatocytes in chronic viral hepatitis. J Exp Med 179:841–848.

Goto T, Kato N, Yoshida H, Otsuka M, Moriyama M, Shiratori Y, Koike K, Matsumura M, Omata M (2003) Synergistic activation of the serum response element-dependent pathway by hepatitis B virus x protein and large-isoform hepatitis delta antigen. J Infect Dis 187:820–828

Gottlob K, Fulco M, Levrero M, Graessmann A (1998a) The hepatitis B virus HBx protein inhibits caspase-3 activity. J Biol Chem 273:33347–33353

Gottlob K, Pagano S, Levrero M, Graessmann A (1998b) Hepatitis B virus X protein transcription activation domains are neither required nor sufficient for cell transformation. Cancer Res 58:3566–3570

Groisman IJ, Koshy R, Henkler F, Groopman JD, Alaoui-Jamali MA (1999) Downregulation of DNA excision repair by the hepatitis B virus-x protein occurs in p53-proficient and p53-deficient cells. Carcinogenesis 20:479–483

Guo WT, Wang J, Tam G, Yen TS, Ou JS (1991) Leaky transcription termination produces larger and smaller than genome size hepatitis B virus X gene transcripts. Virology 181:630–636

Gupta A, Mal TK, Jayasuryan N, Chauhan VS (1995) Assignment of disulphide bonds in the X protein (HBx) of hepatitis B virus. Biochem Biophys Res Commun 212:919–924

Haviv I, Vaizel D, Shaul Y (1996) pX, the HBV-encoded coactivator, interacts with components of the transcription machinery and stimulates transcription in a TAF-independent manner. EMBO J 15:3413–3420

Haviv I, Shamay M, Doitch G, Shaul Y (1998) Hepatitis B virus pX targets TFIIB in transcription coactivation. Mol Cell Biol 18:1562–1569

Hazzalin CA, Mahadevan LC (2002) MAPK-regulated transcription: a continuously variable gene switch? Nat Rev Mol Cell Biol 3:30–40

Henkler F, Lopes AR, Jones M, Koshy R (1998) Erk-independent partial activation of AP-1 sites by the hepatitis B virus HBx protein. J Gen Virol 79:2737–2742

Hoare J, Henkler F, Dowling JJ, Errington W, Goldin RD, Fish D, McGarvey MJ (2001) Subcellular localisation of the X protein in HBV infected hepatocytes. J Med Virol 64:419–426

Hohne M, Schaefer S, Seifer M, Feitelson MA, Paul D, Gerlich WH (1990) Malignant transformation of immortalized transgenic hepatocytes after transfection with hepatitis B virus DNA. EMBO J 9:1137–1145

Hsu DK, Dowling CA, Jeng KC, Chen JT, Yang RY, Liu FT (1999) Galectin-3 expression is induced in cirrhotic liver and hepatocellular carcinoma. Int J Cancer 81:519–526

Hsu IC, Metcalf RA, Sun T, Welsh JA, Wang NJ, Harris CC (1991) Mutational hotspot in the p53 gene in human hepatocellular carcinomas. Nature 350:427–428

Hu KQ, Yu CH, Vierling JM (1992) Up-regulation of intercellular adhesion molecule 1 transcription by hepatitis B virus X protein. Proc Natl Acad Sci USA 89:11441–11445

Hu Z, Zhang Z, Doo E, Coux O, Goldberg AL, Liang TJ (1999) Hepatitis B virus X protein is both a substrate and a potential inhibitor of the proteasome complex. J Virol 73:7231–7240

Hung L, Kumar V (2004) Specific inhibition of gene expression and transactivation functions of hepatitis B virus X protein and c-myc by small interfering RNAs. FEBS Lett 560:210–214

Huo TI, Wang XW, Forgues M, Wu CG, Spillare EA, Giannini C, Brechot C, Harris CC (2001) Hepatitis B virus X mutants derived from human hepatocellular carcinoma retain the ability to abrogate p53-induced apoptosis. Oncogene 20:3620–3628

Jaitovich-Groisman I, Benlimame n, Slagle BL, Perez MH, Alpert L, Song DJ, Fotouhi-Ardakani N, Galipeau J, Alaoui-Jamali MA (2001) Transcriptional regulation of the

TFIIH transcription repair components XPB and XPD by the hepatitis B virus X protein in liver cells and transgenic liver tissue. J Biol Chem 276:14124–14132

Jameel S, Siddiqui A, Maguire HF, Rao KV (1990) Hepatitis B virus X protein produced in Escherichia coli is biologically functional. J Virol 64:3963–3966

Jia L, Wang XW, Harris CC (1999) Hepatitis B virus X protein inhibits nucleotide excision repair. Int J Cancer 80:875–879

Jung EY, Kang HK, Chang J, Yu DY, Jang KL (2003) Cooperative transformation of murine fibroblast NIH3T3 cells by hepatitis C virus core protein and hepatitis B virus X protein. Virus Res 94:79–84

Kang-Park S, Lee JH, Shin JH, Lee YI (2001) Activation of the IGF-II gene by HBV-X protein requires PKC and p44/p42 map kinase signalings. Biochem Biophys Res Commun 283:303–307

Kay A, Mandart E, Trepo C, Galibert F (1985) The HBV HBX gene expressed in *E. coli* is recognised by sera from hepatitis patients. EMBO J 4:1287–1292

Kekule AS, Lauer U, Weiss L, Luber B, Hofschneider PH (1993) Hepatitis B virus transactivator HBx uses a tumour promoter signalling pathway. Nature 361:742–745

Kidd-Ljunggren K, Oberg M, Kidd AH (1995) The hepatitis B virus X gene: analysis of functional domain variation and gene phylogeny using multiple sequences. J Gen Virol 76:2119–2130

Kim CM, Koike K, Saito I, Miyamura T, Jay G (1991) HBx gene of hepatitis B virus induces liver cancer in transgenic mice. Nature 351:317–320

Kim H, Lee H, Yun Y (1998) X-gene product of hepatitis B virus induces apoptosis in liver cells. J Biol Chem 273:381–385

Kim H, Lee YH, Won J, Yun Y (2001) Through induction of juxtaposition and tyrosine kinase activity of Jak1, X-gene product of hepatitis B virus stimulates Ras and the transcriptional activation through AP-1, NF-kappaB, and SRE enhancers. Biochem Biophys Res Commun 286:886–894

Kim JH, Rho HM (2002). Activation of the human transforming growth factor alpha (TGF-alpha) gene by hepatitis B viral X protein (HBx) through AP-2 sites. Mol Cell Biochem 231:155–161

Kim KH, Seong BL (2003) Pro-apoptotic function of HBV X protein is mediated by interaction with c-FLIP and enhancement of death-inducing signal. EMBO J 22:2104–2116

Kim SO, Park JG, Lee YI (1996) Increased expression of the insulin-like growth factor I (IGF-I) receptor gene in hepatocellular carcinoma cell lines: implications of IGF-I receptor gene activation by hepatitis B virus X gene product. Cancer Res 56:3831–3836

Kim YC, Song KS, Yoon G, Nam MJ, Ryu WS (2001) Activated ras oncogene collaborates with HBx gene of hepatitis B virus to transform cells by suppressing HBx-mediated apoptosis. Oncogene 20:16–23

Kim YK, Junn E, Park I, Lee Y, Kang C, Ahn JK (1999) Repression of hepatitis B virus X gene expression by hammerhead ribozymes. Biochem Biophys Res Commun 257:759–765

Klein NP, Schneider RJ (1997) Activation of Src family kinases by hepatitis B virus HBx protein and coupled signaling to Ras. Mol Cell Biol. 17:6427–6436

Klein NP, Bouchard MJ, Wang LH, Kobarg C, Schneider RJ (1999) Src kinases involved in hepatitis B virus replication. EMBO J 18:5019–5027

Koike K, Moriya K, Yotsuyanagi H, Iino S, Kurokawa K (1994a) Induction of cell cycle progression by hepatitis B virus HBx gene expression in quiescent mouse fibroblasts. J Clin Invest 94:44–49

Koike K, Moriya K, Iino S, Yotsuyanagi H, Endo Y, Miyamura T, Kurokawa K (1994b) High-level expression of hepatitis B virus HBx gene and hepatocarcinogenesis in transgenic mice. Hepatology 19:810–819

Kong HJ, Hong SH, Lee MY, Kim HD, Lee JW, Cheong J (2000) Direct binding of hepatitis B virus X protein and retinoid X receptor contributes to phosphoenolpyruvate carboxykinase gene transactivation. FEBS Lett 483:114–118

Kumar V, Jayasuryan N, Kumar R (1996) A truncated mutant (residues 58–140) of the hepatitis B virus X protein retains transactivation function. Proc Natl Acad Sci USA 93:5647–5652

Kumar V, Singh M, Totey SM Anand RK (2001) Bicistronic DNA construct comprising X-myc transgene for use in production of transgenic animal model systems for human hepatocellular carcinoma and transgenic animal model systems so produced. US Patent No. 6,274,788 B1

Kwee L, Lucito R, Aufiero B, Schneider RJ (1992) Alternate translation initiation on hepatitis B virus X mRNA produces multiple polypeptides that differentially transactivate class II and III promoters. J Virol 66:4382–4389

Lakhtakia R, Kumar V, Reddi H, Mathur M, Dattagupta S, Panda SK (2003) Hepatocellular carcinoma in a hepatitis B 'x' transgenic mouse model: A sequential pathological evaluation. J Gastroenterol Hepatol 18:80–91

Lara-Pezzi E, Armesilla AL, Majano PL, Redondo JM, Lopez-Cabrera M (1998) The hepatitis B virus X protein activates nuclear factor of activated T cells (NF-AT) by a cyclosporin A-sensitive pathway. EMBO J 17:7066–7077

Lara-Pezzi E, Majano PL, Yanez-Mo M, Gomez-Gonzalo M, Carretero M, Moreno-Otero R, Sanchez-Madrid F, Lopez-Cabrera M (2001) Effect of the hepatitis B virus HBx protein on integrin-mediated adhesion to and migration on extracellular matrix. J Hepatol 34:409–415

Lara-Pezzi E, Gomez-Gaviro MV, Galvez BG, Mira E, Iniguez MA, Fresno M, Martinez-A C, Arroyo AG, Lopez-Cabrera M (2002) The hepatitis B virus X protein promotes tumor cell invasion by inducing membrane-type matrix metalloproteinase-1 and cyclooxygenase-2 expression. J Clin Invest 110:1831–1838

Leach JK, Qiao L, Fang Y, Han SL, Gilfor D, Fisher PB, Grant S, Hylemon PB, Peterson D, Dent P (2003) Regulation of p21 and p27 expression by the hepatitis B virus X protein and the alternate initiation site X proteins, AUG2 and AUG3. J Gastroenterol Hepatol 18:376–385

Lee MO, Kang HJ, Cho H, Shin EC, Park JH, Kim SJ (2001) Hepatitis B virus X protein induced expression of the Nur77 gene. Biochem Biophys Res Commun 288:1162–1168

Lee S, Tarn C, Wang WH, Chen S, Hullinger RL, Andrisani OM (2002) Hepatitis B virus X protein differentially regulates cell cycle progression in X-transforming versus nontransforming hepatocyte (AML12) cell lines. J Biol Chem 277:8730–8740

Lee SG, Rho HM (2000) Transcriptional repression of the human p53 gene by hepatitis B viral X protein. Oncogene 19:468–471

Lee TH, Finegold MJ, Shen RF, DeMayo JL, Woo SL, Butel JS (1990) Hepatitis B virus transactivator X protein is not tumorigenic in transgenic mice. J Virol 64:5939–5947

Lee Y, Park US, Choi I, Yoon SK, Park YM, Lee YI (1998) Human interleukin 6 gene is activated by hepatitis B virus-X protein in human hepatoma cells. Clin Cancer Res 4:1711–1717

Lee YH, Yun Y (1998) HBx protein of hepatitis B virus activates Jak1-STAT signaling. J Biol Chem 273:25510–25515

Lee YI, Lee S, Lee Y, Bong YS, Hyun SW, Yoo YD, Kim SJ, Kim YW, Poo HR (1998) The human hepatitis B virus transactivator X gene product regulates Sp1 mediated transcription of an insulin-like growth factor II promoter 4. Oncogene 16:2367–2380

Lee YI, Kim SO, Kwon HJ, Park JG, Sohn MJ, Jeong SS (2001a) Phosphorylation of purified recombinant hepatitis B virus-X protein by mitogen-activated protein kinase and protein kinase C in vitro. J Virol Methods 95:1–10

Lee YI, Kang-Park S, Do SI, Lee YI (2001b) The hepatitis B virus-X protein activates a phosphatidylinositol 3-kinase-dependent survival signaling cascade. J Biol Chem 276:16969–16977

Levrero M, Balsano C, Natoli G, Acantaggiati ML, Elfassi E (1990) Hepatitis B virus X protein transactivates the long terminal repeats of human immunodeficiency virus type 1 and 2. J Virol 64:3082–3086

Li J, Tang Z, Liu K (1996) Application of radiolabeled anti-HBx monoclonal antibody for HCC targeting therapy. Zhonghua Yi Xue Za Zhi 76:271–274

Li J, Ou JH (2001) Differential regulation of hepatitis B virus gene expression by the Sp1 transcription factor. J Virol 75:8400–8406

Lian Z, Pan J, Liu J, Zhang S, Zhu M, Arbuthnot P, Kew M, Feitelson MA (1999) The translation initiation factor, hu-Sui1 may be a target of hepatitis B X antigen in hepatocarcinogenesis. Oncogene 18:1677–1687

Lin MH, Lo SC (1989) Dimerization of hepatitis B viral X protein synthesized in a cell-free system. Biochem Biophys Res Commun 164:14–21

Lin Y, Nomura T, Cheong J, Dorjsuren D, Iida K, Murakami S (1997) Hepatitis B virus X protein is a transcriptional modulator that communicates with transcription factor IIB and the RNA polymerase II subunit 5. J Biol Chem 272:7132–7139

Livezey KW, Negorev D, Simon D (2002) Increased chromosomal alterations and micronuclei formation in human hepatoma HepG2 cells transfected with the hepatitis B virus HBX gene. Mutat Res 505:63–74

Lucito R, Schneider RJ (1992) Hepatitis B virus X protein activates transcription factor NF-kappa B without a requirement for protein kinase C. J Virol 66:983–991

Madden CR, Finegold MJ, Slagle BL (2001) Hepatitis B virus X protein acts as a tumor promoter in development of diethylnitrosamine-induced preneoplastic lesions. J Virol 75:3851–3858

Maguire HF, Hoeffler JP, Siddiqui A (1991) HBV X protein alters the DNA binding specificity of CREB and ATF-2 by protein-protein interactions. Science 252:842–844

Majano P, Lara-Pezzi P, Lopez-Cabrera M, Apolinario A, Moreno-Otero R, Garcia-Monzon C (2001) Hepatitis B virus X protein transactivates inducible nitric oxide synthase gene promoter through the proximal nuclear factor kappaB-binding site: Evidence that cytoplasmic location of X protein is essential for gene transactvation. Hepatology 34:1218–1224

Mahe Y, Mukaida N, Kuno K, Akiyama M, Ikeda N, Matsushima K, Murakami S (1991) Hepatitis B virus X protein transactivates human interleukin-8 gene through acting nuclear factor kB and CCAAT/enhancer-binding protein-like cis elements. J Biol Chem 266:13759–13763

Marriott SJ, Lee TH, Slagle BL, Butel JS (1996) Activation of the HTLV-I long terminal repeat by the hepatitis B virus X protein. Virology 224:206–213

McCaffrey AP, Nakai H, Pandey K, Huang Z, Salazar FH, Xu H, Wieland SF, Marion PL, Kay MA (2003) Inhibition of hepatitis B virus in mice by RNA interference. Nat Biotechnol 21:639–644

Menzo S, Clementi M, Alfani E, Bagnarelli P, Iacovacci S, Manzin A, Dandri M, Natoli G, Levrero M, Carloni G (1993) Trans-activation of epidermal growth factor receptor gene by the hepatitis B virus X-gene product. Virology 196:878–882

Miller RH, Robinson WS (1986) Common evolutionary origin of hepatitis B virus and retroviruses. Proc Natl Acad Sci USA 83:2531–2535

Moriarty AM, Alexander H, Lerner RA, Thornton GB (1985) Antibodies to peptides detect new hepatitis B antigen: serological correlation with hepatocellular carcinoma. Science 227:429–433

Moriya K, Matsukura M, Kurokawa K, Koike K (1996) In vivo inhibition of hepatitis B virus gene expression by antisense phosphorothioate oligonucleotides. Biochem Biophys Res Commun 218:217–223

Murakami S, Cheong JH, Kaneko S (1994) Human hepatitis virus X gene encodes a regulatory domain that represses transactivation of X protein. J Biol Chem 269:15118–15123

Murakami S (2001) Hepatitis B virus X protein: a multifunctional viral regulator. J Gastroenterol 36:651–660

Nakatake H, Chisaka O, Yamamoto S, Matsubara K, Koshy R (1993) Effect of X protein on transactivation of hepatitis B virus promoters and on viral replication. Virology 195:305–314

Natoli G, Avantaggiati ML, Chirillo P, Costanzo A, Artini M, Balsano C, Levrero M (1994). Induction of the DNA-binding activity of c-jun/c-fos heterodimers by the hepatitis B virus transactivator pX. Mol Cell Biol 14:989–998

Nijhara R, Jana SS, Goswami SK, Kumar V, Sarkar DP (2001a) An internal segment (residues 58–119) of the hepatitis B virus X protein is sufficient to activate MAP kinase pathways in mouse liver. FEBS Lett. 504:59–64

Nijhara R, Jana SS, Goswami SK, Rana A, Majumdar SS, Kumar V, Sarkar DP (2001b) Sustained activation of mitogen-activated protein kinases and activator protein 1 by the hepatitis B virus X protein in mouse hepatocytes in vivo. J Virol 75:10348–10358

Oelgeschlager T (2002) Regulation of RNA polymerase II activity by CTD phosphorylation and cell cycle control. J Cell Physiol 190:160–169

Oh S, Kim N, Lee Y (1998) Identification of differentially expressed genes in human hepatoblastoma cell line (HepG2) and HBV-X transfected hepatoblastoma cell line (HepG2-4x). Mol Cell 8:212–218

Op De Beeck A, Caillet-Fauquet P (1997) Viruses and the cell cycle. Prog Cell Cycle Res 3:1–19

Park US, Park SK, Lee YI, Park JG, Lee YI (2000) Hepatitis B virus-X protein upregulates the expression of p21waf1/cip1 and prolongs G1→S transition via a p53-independent pathway in human hepatoma cells. Oncogene 19:3384–3394

Paterlini P, Poussin K, Kew M, Franco D, Brechot C (1995) Selective accumulation of the X transcript of hepatitis B virus in patients negative for hepatitis B surface antigen with hepatocellular carcinoma. Hepatology. 21:313–321

Perini G, Oetjen E, Green MR (1999) The hepatitis B pX protein promotes dimerization and DNA binding of cellular basic region/leucine zipper proteins by targeting the conserved basic region. J Biol Chem 274:13970–13977

Poussin K, Dienes H, Sirma H, Urban S, Beaugrand M, Franco D, Schirmacher P, Brechot C, Paterlini Brechot P (1999) Expression of mutated hepatitis B virus X genes in human hepatocellular carcinomas. Int J Cancer 80:497–505

Qadri I, Maguire HF, Siddiqui A (1995) Hepatitis B virus transactivator protein X interacts with the TATA-binding protein. Proc Natl Acad Sci USA 92:1003–1007

Qadri I, Conaway JW, Conaway RC, Schaack J, Siddiqui A (1996) Hepatitis B virus transactivator protein, HBx, associates with the components of TFIIH and stimulates the DNA helicase activity of TFIIH. Proc Natl Acad Sci USA 93:10578–10583

Qiao L, Leach K, McKinstry R, Gilfor D, Yacoub A, Park JS, Grant S, Hylemon PB, Fisher PB, Dent P (2001) Hepatitis B virus X protein increases expression of p21(Cip-1/WAF1/MDA6) and p27(Kip-1) in primary mouse hepatocytes, leading to reduced cell cycle progression. Hepatology 34:906–917

Reddi H, Kumar R, Jain SK, Kumar V (2003) A carboxy-terminal region of the hepatitis B virus X protein promotes DNA interaction of CREB and mimics the native protein for transactivation function. Virus Genes 26:227–238

Reddi H, Kumar V (2004) Self-association of the hepatitis B virus X protein in the yeast two-hybrid system. Biochem Biophys Res Commun 317:1017–1022

Reifenberg K, Nusser P, Lohler J, Spindler G, Kuhn C, von Weizsacker F, Kock J (2002) Virus replication and virion export in X-deficient hepatitis B virus transgenic mice. J Gen Virol 83:991–996

Renner M, Haniel A, Burgelt E, Hofschneider PH, Koch W (1995) Transactivating function and expression of the x gene of hepatitis B virus. J Hepatol 23:53–65

Ritter SE, Whitten TM, Quets AT, Schloemer RH (1991) An internal domain of the hepatitis B virus X antigen is necessary for transactivating activity. Virology 182:841–845

Rossner MT, Jackson RJ, Murray K (1990) Modulation of expression of the hepatitis B virus surface antigen gene by the viral X-gene product. Proc R Soc Lond B Biol Sci 241:51–58

Rossner MT (1992) Hepatitis B virus X-gene product: a promiscuous transcriptional activator. J Med Virol 36:101–117

Runkel L, Fischer M, Schaller H (1993) Two-codon insertion mutations of the HBx define two separate regions necessary for its trans-activation function. Virology 197:529–536

Schaefer S, Gerlich WH (1995) In vitro transformation by hepatitis B virus DNA. Intervirology 38:143–154

Schek N, Bartenschlager R, Kuhn C, Schaller H (1991) Phosphorylation and rapid turnover of hepatitis B virus X-protein expressed in HepG2 cells from a recombinant vaccinia virus. Oncogene 6:1735–1744

Schluter V, Meyer M, Hofschneider PH, Koshy R, Caselmann WH (1994) Integrated hepatitis B virus X and 3' truncated preS/S sequences derived from human hepatomas encode functionally active transactivators. Oncogene 9:3335–3344

Seeger C, Ganem D, Varmus HE (1984) The cloned genome of ground squirrel hepatitis virus is infectious in the animal. Proc Natl Acad Sci USA 81:5849–5852

Seto E, Yen TS, Peterlin BM, Ou JH (1988) Trans-activation of the human immunodeficiency virus long terminal repeat by the hepatitis B virus X protein. Proc Natl Acad Sci USA 85:8286–8290

Seto E, Mitchell PJ, Yesn TS (1990) Tranactivation by the hepatitis B virus X protein depends on AP-2 and other transcription factors. Nature 344:72–74

Shapiro P (2002) Ras-MAP kinase signaling pathways and control of cell proliferation: relevance to cancer therapy. Crit Rev Clin Lab Sci 39:285–330

Shih WL, Kuo ML, Chuang SE, Cheng AL, Doong SL (2003) Hepatitis B virus X protein activates a survival signaling by linking SRC to phosphatidylinositol 3-kinase. J Biol Chem 278:31807–31813

Shin EC, Shin JS, Park JH, Kim H, Kim SJ (1999) Expression of fas ligand in human hepatoma cell lines: role of hepatitis-B virus X (HBX) in induction of Fas ligand. Int J Cancer 82:587–591

Shirakata Y, Kawada M, Fujiki Y, Sano H, Oda M, Yaginuma K, Kobayashi M, Koike K (1989) The X gene of hepatitis B virus induced growth stimulation and tumorigenic transformation of mouse NIH3T3 cells. Jpn J Cancer Res 80:617–621

Shirakata Y, Koike K (2003) Hepatitis B virus X protein induces cell death by causing loss of mitochondrial membrane potential. J Biol Chem 278:22071–22078

Shlomai A, Shaul Y (2003) Inhibition of hepatitis B virus expression and replication by RNA interference. Hepatology 37:764–770

Singh M, Kumar V (2003) Transgenic mouse models of hepatitis B virus-associated hepatocellular carcinoma. Rev Med Virol 13:243–253

Siddiqui A, Gaynor R, Srinivasan A, Mapoles J, Farr RW (1989) Trans-activation of viral enhancers including long terminal repeat of the human immunodeficiency virus by the hepatitis B virus X protein. Virology 169:479–484

Sirma H, Weil R, Rosmorduc O, Urban S, Israel A, Kremsdorf D, Brechot C (1998) Cytosol is the prime compartment of hepatitis B virus X protein where it colocalizes with the proteasome. Oncogene 16:2051–2063

Sirma H, Giannini C, Poussin K, Paterlini P, Kremsdorf D, Brechot C (1999) Hepatitis B virus X mutants, present in hepatocellular carcinoma tissue abrogate both the antiproliferative and transactivation effects of HBx. Oncogene 18:4848–4859

Slagle BL, Lee TH, Medina D, Finegold MJ, Butel JS (1996) Increased sensitivity to the hepatocarcinogen diethylnitrosamine in transgenic mice carrying the hepatitis B virus X gene. Mol Carcinog 15:261–269

Sohn S, Jaitovitch-Groisman I, Benlimame N, Galipeau J, Batist G, Alaoui-Jamali MA (2000) Retroviral expression of the hepatitis B virus x gene promotes liver cell susceptibility to carcinogen-induced site specific mutagenesis. Mutat Res 460:17–28

Spandau DF, Lee CH (1988) Trans-activation of viral enhancers by the hepatitis B virus X protein. J Virol 62:427–434

Su F, Schneider RJ (1996). Hepatitis B virus HBx protein activates transcription factor NF-kappaB by acting on multiple cytoplasmic inhibitors of rel-related proteins. J. Virol 70:4558–4566

Su F, Theodosis CN, Schneider RJ (2001) Role of NF-kappaB and myc proteins in apoptosis induced by hepatitis B virus HBx protein. J Virol 75:215–225

Su Q, Schroder CH, Hofmann WJ, Otto G, Pichlmayr R, Bannasch P (1998) Expression of hepatitis B virus X protein in HBV-infected human livers and hepatocellular carcinomas. Hepatology 27:1109–1120

Sun BS, Zhu X, Clayton MM, Pan J, Feitelson MA (1998) Identification of a protein isolated from senescent human cells that binds to hepatitis B virus X antigen. Hepatology 27:228–239

Suzuki T, Masui N, Kajino K, Saito I, Miyamura T (1989) Detection and mapping of spliced RNA from a human hepatoma cell line transfected with the hepatitis B virus genome. Proc Natl Acad Sci USA 86:8422–8426

Takada S, Koike K (1990) Trans-activation function of a 3' truncated X gene-cell fusion product from integrated hepatitis B virus DNA in chronic hepatitis tissues. Proc Natl Acad Sci USA 87:5628–5632

Takada S, Kido H, Fukutomi A, Mori T, Koike K (1994) Interaction of hepatitis B virus X protein with a serine protease, tryptase TL2 as an inhibitor. Oncogene 9:341–348

Takada S, Tsuchida N, Kobayashi M, Koike K (1995) Disruption of the function of tumor-suppressor gene p53 by the hepatitis B virus X protein and hepatocarcinogenesis. J Cancer Res Clin Oncol 121:593–601

Tarn C, Lee S, Hu Y, Ashendel C, Andrisani OM (2001) Hepatitis B virus X protein differentially activates RAS-RAF-MAPK and JNK pathways in X-transforming versus non-transforming AML12 hepatocytes. J Biol Chem 276:34671–34680

Tarn C, Zou L, Hullinger RL, Andrisani OM (2002) Hepatitis B virus X protein activates the p38 mitogen-activated protein kinase pathway in dedifferentiated hepatocytes. J Virol 76:9763–9772

Terradillos O, Billet O, Renard CA, Levy R, Molina T, Briand P, Buendia MA (1997) The hepatitis B virus X gene potentiates c-myc-induced liver oncogenesis in transgenic mice. Oncogene 14:395–404

Terradillos O, Pollicino T, Lecoeur H, Tripodi M, Gougeon ML, Tiollais P, Buendia MA (1998) p53-independent apoptotic effects of the hepatitis B virus HBx protein in vivo and in vitro. Oncogene 17:2115–2123

Terradillos O, de La Coste A, Pollicino T, Neuveut C, Sitterlin D, Lecoeur H, Gougeon ML, Kahn A, Buendia MA (2002) The hepatitis B virus X protein abrogates Bcl-2-mediated protection against Fas apoptosis in the liver. Oncogene 21:377–386

Tiollais P, Charnay P, Vyas GN (1981) Biology of hepatitis B virus. Science 213:406–411

Truant R, Antunovic J, Greenblatt J, Prives C, Cromlish JA (1995) Direct interaction of the hepatitis B virus HBx protein with p53 leads to inhibition by HBx of p53 response element-directed transactivation. J Virol 69:1851–1859

Tsuei DJ, Hsu TY, Chen JY, Chang MH, Hsu HC, Yang CS (1994) Analysis of integrated hepatitis B virus DNA and flanking cellular sequences in a childhood hepatocellular carcinoma. J Med Virol 42:287–293

Tu H, Bonura C, Giannini C, Mouly H, Soussan P, Kew M, Paterlini-Brechot P, Brechot C, Kremsdorf D (2001) Biological impact of natural COOH-terminal deletions of hepatitis B virus X protein in hepatocellular carcinoma tissues. Cancer Res 61:7803–7810

Tuttleman JS, Pugh JC, Summers JW (1986) In vitro experimental infection of primary duck hepatocyte cultures with duck hepatitis B virus. J Virol 58:17–25

Twu JS, Schloemer RH (1987) Transcriptional trans-activating function of hepatitis B virus. J Virol 61:3448–3453

Twu JS, Robinson WS (1989) Hepatitis B virus X gene can transactivate heterologous viral sequences. Proc Natl Acad Sci USA 86:2046–2050

Twu JS, Lai MY, Chen DS, Robinson WS (1993) Activation of protooncogene c-jun by the X protein of hepatitis B virus. Virology 192:346–350

Ueda H, Ullrich SJ, Gangemi JD, Kappel CA, Ngo L, Feitelson MA, Jay G (1995) Functional inactivation but not structural mutation of p53 causes liver cancer. Nat Genet 9:41–47

Urban S, Hildt E, Eckerskorn C, Sirma H, Kekule A, Hofschneider PH (1997) Isolation and molecular characterization of hepatitis B virus X-protein from a baculovirus expression system. Hepatology 26:1045–1053

Wang HD, Yuh CH, Dang CV, Johnson DL (1995) The hepatitis B virus X protein increases the cellular level of TATA-binding protein, which mediates transactivation of RNA polymerase genes. Mol Cell Biol 15:6720–6728

Wang HD, Trivedi A, Johnson DL (1997) Hepatitis B virus X protein induces RNA polymerase III-dependent gene transcription and increases cellular TATA-binding protein by activating the Ras signaling pathway. Mol Cell Biol 17:6838–6846

Wang HD, Trivedi A, Johnson DL (1998) Regulation of RNA polymerase I-dependent promoters by the hepatitis B virus X protein via activated Ras and TATA-binding protein. Mol Cell Biol 18:7086–7094

Wang W, Carey M, Gralla JD (1992) Polymerase II promoter activation: closed complex formation and ATP-driven start site opening. Science 255:450–453

Wang XW, Gibson MK, Vermeulen W, Yeh H, Forrester K, Sturzbecher HW, Hoeijmakers JH, Harris CC (1995) Abrogation of p53-induced apoptosis by the hepatitis B virus X gene. Cancer Res 55:6012–6016

Wei w, Dorjsuren D, Lin Y, Qin W, Nomura T, Hayashi N, Murakami S (2001) Direct interaction between the subunit RAP30 of transcription factor IIF (TFIIF) and RNA polymerase subunit 5, which contributes to the association between TFIIF and RNA polymerase II. J Biol Chem 276:12266–12273

Weil R, Sirma H, Giannini C, Kremsdorf D, Bessia C, Dargemont C, Brechot C, Israel A (1999) Direct association and nuclear import of the hepatitis B virus X protein with the NF-kappaB inhibitor IkappaBalpha. Mol Cell Biol 19:6345–6354

Weinberg M, Passman M, Kew M, Arbuthnot P (2000) Hammerhead ribozyme-mediated inhibition of hepatitis B virus X gene expression in cultured cells. J Hepatol 33:142–151

Williams JS, Andrisani OM (1995) The hepatitis B virus X protein targets the basic region-leucine zipper domain of CREB. Proc Natl Acad Sci USA 92:3819–3823

Wollersheim M, Debelka U, Hofschneider PH (1988) A transactivating function encoded in the hepatitis B virus X gene is conserved in the integrated state. Oncogene 3:545–552

Wu CG, Salvay DM, Forgues M, Valerie K, Farnsworth J, Markin RS, Wang XW (2001) Distinctive gene expression profiles associated with Hepatitis B virus x protein. Oncogene 20:3674–3682

Wu CG, Forgues M, Siddique S, Farnsworth J, Valerie K, Wang XW (2002) SAGE transcript profiles of normal primary human hepatocytes expressing oncogenic hepatitis B virus X protein. FASEB J 16:1665–1667

Xu Z, Yen TS, Wu L, Madden CR, Tan W, Slagle BL, Ou JH (2002) Enhancement of hepatitis B virus replication by its X protein in transgenic mice. J Virol 76:2579–2584

Yamamoto S, Mita E, Nakatake H, Takimoto M, Koshy R, Matsubara K (1993) Transactivating function of integrated hepatitis B virus. Biochem Biophys Res Commun 197:1209–1215

Yoo YD, Ueda H, Park K, Flanders KC, Lee YI, Jay G, Kim SJ (1996) Regulation of transforming growth factor-beta 1 expression by the hepatitis B virus (HBV) X transactivator. Role in HBV pathogenesis. J Clin Invest 97:388–395

Zahm P, Hofschneider PH, Koshy R (1988) The HBV X-ORF encodes a transactivator: a potential factor in viral hepatocarcinogenesis. Oncogene 3:169–177

Zhang Z, Torii N, Furusaka A, Malayaman N, Hu Z, Liang TJ (2000) Structural and functional characterization of interaction between hepatitis B virus X protein and the proteasome complex. J Biol Chem 2000 275:15157–15165

Zheng Y, Li J, Johnson DL, Ou JH (2003) Regulation of hepatitis B virus replication by the Ras-mitogen-activated protein kinase signaling pathway. J Virol 77:7707–7712

Zhou DX, Taraboulos A, Ou JH, Yen TS (1990) Activation of class I major histocompatibility complex gene expression by hepatitis B virus. J Virol 64:4025–4028

Zhou MX, Watabe M, Watabe K (1994) The X-gene of human hepatitis B virus transactivates the c-jun and alpha-fetoprotein genes. Arch Virol 134:369–378

Zhou YZ, Butel JS, Li PJ, Finegold MJ, Melnick JL (1987) Integrated state of subgenomic fragments of hepatitis B virus DNA in hepatocellular carcinoma from mainland China. J Natl Cancer Inst 79:223–231

Zoulim F, Saputelli J, Seeger C (1994) Woodchuck hepatitis virus X protein is required for viral infection in vivo. J Virol 68:2026–2030

Regulation of Xenobiotic Detoxification by PXR, CAR, GR, VDR and SHP Receptors: Consequences in Physiology

J. M. Pascussi[1] · Z. Dvorák[1] · S. Gerbal-Chaloin[2] · E. Assenat[1] · L. Drocourt[1] · P. Maurel[1] · M. J. Vilarem[1] (✉)

[1] UMR 632, 1919 Route de Mende, 34293 Montpellier Cedex 5, France
vilarem@montp.inserm.fr
[2] CRBM-CNRS, FRE2593, Montpellier, France

1	Introduction	410
2	Nuclear Receptors and Xenobiotic Detoxification	412
2.1	Pregnane X Receptor	412
2.2	Constitutive Androstane Receptor	414
2.3	Vitamin D Receptor	415
2.4	Short Heterodimer Partner	416
2.5	Glucocorticoid Receptor	418
3	PXR, CAR and VDR Signal by Means of the Same Response Elements	420
4	Consequences of Crosstalk Between PXR/CAR and Other Nuclear Receptors in Detoxification and Physiology	423
4.1	Crosstalk Between PXR/CAR and VDR	423
4.2	Crosstalk Between PXR/CAR and GR	423
4.2.1	Inhibition of Detoxification by Inflammatory Response	424
4.2.2	Inhibition of Detoxification by Microtubule Disrupting Agent	426
5	Conclusion	427
	References	428

Abstract Cytochromes P450 (CYP) from families CYP1/2/3 are involved in xenobiotic detoxification and are regulated by numerous agents including xenobiotics, plant and fungal toxins, bile salts, etc. Nuclear receptors PXR (pregnane X receptor, NR1I3) and CAR (constitutive androstane receptor, NR1I2) control the expression of distinct but overlapping arrays of genes including the *CYP2/3* families. These receptors are involved in a tangle of regulatory networks including those pathways controlling vitamin D and cholesterol/bile salt homeostasis. They are expressed under the control of the glucocorticoid receptor and are inhibited by the short heterodimer partner (SHP). The crosstalk between both PXR and CAR and other nuclear receptors allows for the establishing and prediction of functional interferences between the detoxification and other signaling and metabolic pathways. These interferences provide new clues for the interpretation of xenobiotic adverse effects, such as bone demineralization, and for understanding how physiopathological factors, such as biliary salts or inflammation and infections, affect an organism's ability to detoxify xenobiotics.

Keywords CYP induction · Nuclear receptors · Detoxification · Inflammatory response · Pathophysiology

1
Introduction

The body defends itself against potentially harmful compounds, such as drugs and toxic endogenous compounds and their metabolites, by inducing the expression of cytochromes P450 and the transporters involved in their metabolism and elimination. Cytochrome P450 (CYP) is the generic name of a superfamily of genes encoding monooxygenases that catalyze the oxidative metabolism of xenobiotics (drugs, environmental pollutants, derivatives from industrial and domestic combustion, plant and fungal toxins, etc.) and numerous endogenous compounds (steroids, fatty acids, retinoids, bile acids) (Roy-Chowdhury et al. 2003). Human beings have 17 known CYP gene families, but only the first three, CYP1, CYP2 and CYP3, which are expressed mainly in the liver, intestine and kidney (Gonzalez and Gelboin 1992) (Shimada et al. 1994) are involved in the metabolism of drugs and xenobiotics. CYP-mediated detoxification converts xenobiotics to more hydrophilic compounds, which are more easily conjugated and excreted. In addition, these CYPs also play an important role in cholesterol biosynthesis, bile acid metabolism and the biosynthesis and catabolism of steroids (Nelson 1999).

In the CYP2B subfamily, CYP2B6 has long been thought to play a minor role in human drug metabolism and has therefore received little attention. However, several recent findings have generated an increased interest in this isoenzyme: identification of ethnic differences in its expression (Ariyoshi et al. 2001), the identification of new substrates for CYP2B6 and perhaps a shared specificity with CYP3A4 (Ekins et al. 1999), and the suggestion that its transcriptional activation is regulated by mechanisms similar to those affecting CYP3A4 (Goodwin et al. 2001, 2002; Smirlis et al. 2001; Xie et al. 2000).

CYP2C9 is a member of the CYP2C subfamily, which includes in humans at least three other members: CYP2C8, CYP2C18 and CYP2C19. CYP2C9 is involved in the metabolism of numerous substrates including phenytoin, tolbutamide, torsemide, S-warfarin and numerous nonsteroidal anti-inflammatory drugs (Goldstein 2001).

The CYP3A subfamily includes the most abundant cytochromes P450 present in the adult human liver, comprising approximately 30% of the total content. The human CYP3A family consists of four enzymes which show variable levels of expression in the population, CYP3A4, CYP3A43, CYP3A5 and CYP3A7. Among them, the CYP3A4 isoform is the most prevalent in adults. It has been estimated that about 50% of currently marketed drugs are metabolized by CYP3A4 (Bertz and Granneman 1997). The substrates for this enzyme include drugs such as quinidine, nifedipine, diltiazem, lidocaine, lovastatin, erythromycin, cyclosporin, triazolam, and midazolam and endogenous substances, including testosterone, progesterone, androstenediol and bile acids (Guengerich 1999). CYP3A4 also activates procarcinogens, including aflatoxinB1, PAHs, NNK and 6-aminochrysene (Guengerich 1999). CYP3A4 is induced in human hepatocytes by rifampicin (Pichard et al. 1995; Schuetz et al. 1993a), dexamethasone, (Pascussi et al. 2001; Pichard et al. 1992); calcium channel modulators such as nifedipine and derivatives (Drocourt et al. 2001), phenobarbital (PB; Schuetz et al. 1993a; Corcos and Lagadic-Gossmann 2001; Kocarek et al. 1995; Pascussi et al. 2000c) and vitamin D (Schmiedlin-Ren et al. 2001; Thummel et al. 2001; Drocourt et al. 2002). Induction of CYPs and other xenobiotic metabolizing and transporting systems (XMTS) leads to clinically important drug–drug interactions (Lin and Lee 1998). Understanding the molecular events leading to CYP2 and CYP3A induction, and determining whether these genes share identical regulatory mechanisms, would ultimately lead to better models for the screening and prediction of drug interactions.

Induction of CYPs is mediated primarily at the transcriptional level by the interaction of ligand–nuclear receptor complexes with enhancer sequences located upstream of CYP gene promoters. Two xenoreceptors, the pregnane X receptor (PXR) (Bertilsson et al. 1998; Blumberg et al. 1998; Lehmann et al. 1998), and the constitutive androstane receptor (CAR), (Honkakoski and Negishi 1998) members of the nuclear receptor superfamily, have been shown to control the inducibility of these CYP genes and other XMTS in response to xenobiotics (Kliewer et al. 1998; Sueyoshi et al. 1999). Recently, several other members of the nuclear hormone receptor superfamily such as the vitamin D receptor (VDR) (Drocourt et al. 2002; Thummel et al. 2001) and the retinoid X receptor alpha (RXR) (Lehmann et al. 1998; Wan et al. 2000) as well as the glucocorticoid receptor (GR) (Pascussi et al. 2001; Gerbal-Chaloin et al. 2002) and the short heterodimer partner (SHP) (Ourlin et al. 2003), an atypical orphan member of the nuclear receptor superfamily that lacks a conventional DNA-binding domain (DBD), interacts with and inhibits the transcriptional activity of several members of the nuclear receptor superfamily (Seol et al. 1997, 1998) have been shown to be responsible for and/or involved in the endobiotic- and xenobiotic-mediated expression of *CYP2/3A* genes.

Domains that regulate nuclear import and DNA binding have been found to overlap in many transcription factors and may be the outcome of functional coevolution (Cokol et al. 2000). CAR is closely related to VDR with 40% of amino acid identity in their ligand-binding domain (LDB). PXR is also structurally and functionally related to VDR and CAR. PXR cDNA encodes a protein of 434 amino acids that is 68% and 66% identical to hVDR and hCAR, respectively, in the DBD (Blumberg et al. 1998). Recently, a receptor related to mammalian CAR and PXR was identified in the chicken and named CXR (Handschin et al. 2000). Sequence comparison showed that this receptor was roughly equally distant from both the mammalian PXR and CAR receptors. These data suggest that CAR and PXR are derived from the same ancestral gene CXR. Also, two receptors from *Xenopus laevis* have been identified which resemble both the mammalian CAR and PXR (Blumberg et al. 1998).

In spite of the clarification of the mechanism of CYP gene induction following the discovery of PXR and CAR receptors, including the elucidation of the species-specificity of this induction, the mechanism by which dexamethasone regulates CYP genes is still unclear, and the possibility that GR might be involved is an important question. Induction of *CYP3A* by dexamethasone and other glucocorticoids has been investigated for many years. The GR has been suspected to play a major part in this process. (Burger et al. 1992; Ogg et al. 1999; Pascussi et al. 2001). However, on the basis of unexpected observations made with antiglucocorticoids (stimulation instead of inhibition of *CYP3A* induction) and of the lack of correlation between expression of *CYP3A* and other genes typically regulated by GR (tyrosine aminotransferase, *TAT*, for example), the contribution of this receptor has been the object of numerous controversies (Burger et al. 1992; Lehmann et al. 1998).

In this chapter, we review studies made in order to understand: (a) the mechanism of regulation of the expression of CYPs by nuclear receptors; (b) the mechanism of nuclear receptor crosstalk and the clinical/toxicological consequences thereof.

2
Nuclear Receptors and Xenobiotic Detoxification

2.1
Pregnane X Receptor

The PXR (NR1I3, also designated SXR or PAR) is a class III nuclear receptor which forms functional heterodimers with RXRα. It is expressed mainly in the liver, the small intestine, and the colon as well as in the kidney (Kliewer et al. 1998; Blumberg et al. 1998). This receptor binds generally with low affinity [50% effective concentration (EC_{50}) in the micromolar range) a wide variety of structurally diverse exogenous and endogenous compounds including drugs such as rifampicin, phenobarbital, nifedipine and other calcium channel blockers, clotrimazole, mifepristone, metyrapone (Drocourt et al. 2001; Bertilsson et al. 1998; Blumberg et al. 1998; Lehmann et al. 1998; Ogg et al. 1999; Moore et al. 2000a; El-Sankary et al. 2000; Harvey et al. 2000), steroid hormones and metabolites such as progesterone, estrogens, corticosterone, and androstenol, and dietary compounds such as coumestrol and hyperforin (Moore et al. 2000a). The latter compound is a constituent of St. John's wort, a herbal remedy for depression, which appears to be the most potent PXR activator with an EC_{50} of 23 nM. After some controversial studies of PXR activation by dexamethasone (Bertilsson et al. 1998; Kliewer et al. 1998; Lehmann et al. 1998), this compound has been shown to be a real ligand of the human PXR but only at supramicromolar concentrations (Pascussi et al. 2001). This is consistent with *CYP3A4* induction.

A well-known particularity of *CYP3A* inducibility is its species specificity (Maurel 1996). For example, rifampicin is a good inducer in man but not in the rat, while pregnenolone 16α-carbonitrile is a good inducer in the rat but not in man. Similarly, dexamethasone is a potent inducer in the rat but a moderate inducer of human *CYP3A* (Pascussi et al. 2001). Interestingly, the same species specificity is observed when comparing the activation of human and rat PXR by these compounds (LeCluyse 2001). Crystal structure analyses reveal that the ligand-binding domain of human PXR is highly hydrophobic and flexible, allowing lipophilic molecules of different sizes to bind in multiple orientations (Watkins et al. 2001). PXR appears to be responsible for the xenobiotic-mediated induction of a battery of genes including *CYP3A4*, *CYP3A7*, *CYP2B6*, *CYP2C9*, *MDR1*, *BSEP* and *MRP2* (Lehmann et al. 1998; Sueyoshi et al. 1999; Pascussi et al. 2001; Kast et al. 2002). Two distinct PXR-responsive elements have been identified in the 5'-flanking region of *CYP3A4*. A proximal response element, located at –160 consisting of an everted repeat of the nuclear receptor half-site AGGTCA separated by 6 nucleotides (ER6) (Lehmann et al. 1998), and a distal response element located between –7800 and –7200 (Goodwin et al. 1999). A cooperative interaction between both response elements appears to be necessary for full PXR-mediated induction of *CYP3A4*. Ligand binding induces interaction between PXR and the coactivator protein SRC-1 (Kliewer et al. 1998). Studies in PXR-null mice suggest that PXR is not only a xenobiotic sensor, but that it also plays a role in the protective mechanism against liver injury produced by secondary bile acids, in part by upregulating *CYP3A* (Staudinger et al. 2001; Xie et al. 2000). Indeed, bile acids such as lithocholic acid and 6-ketolithocholic acid are PXR ligands (Moore et al. 2002). Wild mice pretreated with pregnenolone 16α-carbonitrile (a potent PXR activator in rodents) are protected against lithocholic acid-induced liver injury, while PXR-null mice are not. In addition, experiments on PXR-null mice demonstrated that PXR is a negative regulator of *CYP7A1* (Staudinger et al. 2001), the enzyme controlling the rate-limiting step in bile acid biosynthesis.

2.2
Constitutive Androstane Receptor

CAR (NR1I2) was originally characterized as a transactivator of retinoid acid response elements, in the absence of ligand (Baes et al. 1994). Like PXR, CAR is predominantly expressed in the liver and to a lesser extent in the intestine and the stomach (Wei et al. 2000; Baes et al. 1994). In contrast to most nuclear receptors which contain five domains, the human CAR protein contains three: a highly conserved DNA-binding domain (C domain); a hinge region (D domain); and a divergent ligand-binding/dimerization/transcriptional activation domain (E domain). CAR possesses neither an A/B domain that typically confers a ligand-independent transactivation response, nor the less well characterized hypervariable F domain. In CAR, the LBD-AF2 domain of the protein interacts with the coactivator SRC-1 in a ligand-independent manner. However, some ligands, which are inverse agonists (Forman et al. 1998) affect the protein upon binding in such a way that an inactive conformation is induced. CAR is located in the cytosol under normal conditions, i.e., in mice hepatocytes in vivo and in primary human hepatocytes (Kawamoto et al. 1999; Pascussi et al. 2000c). In response to activators such as phenobarbital, CAR translocates to the nucleus where it forms functional heterodimers with RXRα. The molecular mechanism responsible for this translocation is still unclear, but seems to involve specific phosphorylation-sensitive steps, as suggested by the interfering effect of okadeic acid, an inhibitor of protein the phosphatases 1 and 2A (Kawamoto et al. 1999). In support of this observation, the importance of phosphorylation in *CYP2B* induction has been reported previously (Sidhu et al. 1993; Nirodi et al. 1996). In addition, the PB-inducible translocation activity of the receptor has been mapped to a xenochemical response signal corresponding to a leucin-rich peptide near the C terminus of the CAR protein (Zelko et al. 2001; Maglich et al. 2003). Nuclear translocation appears to be a general process by which CAR regulates gene induction, since various PB-type inducers (e.g., chlorpromazine, chlorinated biphenyls, and methoxychlor) are also capable of inducing the translocation of CAR into the nucleus in the liver. In addition, ligand binding to CAR alone is not sufficient to induce CAR translocation into the nucleus and induction of CAR target genes.

Because CAR exhibits an intrinsically high transcriptional activity, its nuclear localization provokes the activation of target gene expression in the absence of ligand binding (Honkakoski and Negishi 1998). Very little is known about the mechanism of CAR activation by chemicals or its transcriptional regulation. The mouse CAR gene has two identified mRNA isoforms (mCAR1 and mCAR2). mCAR1 is closely related to human CAR. In contrast, mCAR2 is truncated, lacking a C-terminal region of the ligand-binding domain, resulting from alternative exon splicing, leading to the loss of exon 8 (Choi et al. 1997). In man, CAR is expressed predominantly in hepatocytes, and the most prominent mRNA band migrates as a rather broad band spanning approximately 1.3–1.7 kb (Baes et al. 1994). It is curious that certain CAR activators have not been identified as ligands of the reference CAR isoform (Tzameli et al. 2000; Moore et al. 2000a). Thus, it is possible that certain inducing compounds, such as PB or PB-like compounds, may interact with the ligand-binding domain of a other CAR isoforms, as recently hypothesized (Auerbach et al. 2003). These CAR isoforms could be part of a range to enlarge the number of xenobiotics recognized by the body. The recent production of a CAR knockout mouse confirms that this orphan nuclear receptor mediates the induction of *CYP2B*, as well as the increase of both liver weight and DNA synthesis in response to PB (Wei et al. 2000). Like PXR, CAR exhibits a pronounced species-specificity of ligands, activators and inverse agonists. Some of the CAR activators like TCPOBOP and estrogens are able to reverse the inhibition induced by the presence of inverse agonists. Interestingly, some CAR activators are not ligands in vitro. This is notably the case for PB, which does not influence CAR–SRC-1 binding. Only a few molecules among CYP inducers have been shown to bind directly to human CAR. These include clotrimazole and 5-β-pregnane-3,20-dione (Moore et al. 2000a).

In man, recent data show that CAR mediates the PB induction of UDP-glucuronosyltransferase 1A1, *CYP2B6*, *CYP3A4* and *CYP2C9* (Sueyoshi et al. 1999; Gerbal-Chaloin et al. 2002; Sugatani et al. 2001; Goodwin et al. 2002). Several groups have identified a CAR-response element which consists of two nuclear receptor-binding sites (termed NR1 and NR2) and one NF1-binding site (Sueyoshi et al. 1999; Kawamoto et al. 1999; Trottier et al. 1995). NR1 and NR2 are both imperfect DR4 motifs essential for the PB induction of *CYP2B* genes. In human *CYP2B6*, this element is located between −1,684 and −1,733 (Goodwin et al. 2001). The heterodimer can bind to an array of prototypic nuclear receptor-binding sites, including DR-5 (direct repeat-5), DR-4, DR-3, ER-6 (everted repeat-6) and IR-8 (inverted repeat-8) motifs (Kast et al. 2002; Tzameli et al. 2000). Such elements have been found in several major hepatic CYPs involved in drug metabolism in human beings. These include *CYP2B6* (Sueyoshi et al. 1999), *CYP2C9* (Gerbal-Chaloin et al. 2002) and *CYP3A4* (Sueyoshi et al. 1999). CAR also regulates the expression of cytochrome P450 reductase (Ueda et al. 2002), an essential component of CYP-dependent metabolic activity.

2.3
Vitamin D Receptor

The biologically active form of vitamin D3, 1,25-dihydroxy vitamin D3, is an important regulator of cell growth, differentiation, and death. The cellular action of 1,25-dihydroxy vitamin D3 is known to be mediated via an intracellular receptor, the VDR, a member of the superfamily of steroid receptors

(VDR, NR1I1). Ligand-activated VDR provokes partial arrest in G_0/G_1 of the cell cycle, the induction of differentiation and the control of calcium homeostasis (Carlberg and Polly 1998). Although the liver is the site of the 25-hydroxylation of vitamin D, it has been shown to have a very low proportion of VDRs and, consequently, has not been considered as a target site of vitamin D action. However, further studies have demonstrated that calcium and/or vitamin D deficiency has a significant effect on liver cell physiology (Takahashi et al. 1997).

After ligand binding, VDR forms a heterodimer with RXRα which transactivates vitamin D response elements (VDRE) present in the regulatory region of the target genes (Katoh et al. 2000; Danielsson et al. 1998). Although the consensus VDRE is an imperfect direct repeat of (G/A)GGT(G/C)A with a three-nucleotide spacer (DR3), previous investigations identified other VDRE motifs including a DR4 and an inverted palindrome IP9 (Danielsson et al. 1998). As we shall see in Sect. 2, VDR has been shown to bind to and transactivate response elements previously characterized as PXR and CAR response elements in *CYP2* and *CYP3A* genes. This is likely to generate functional crosstalk between vitamin D homeostasis and xenobiotic detoxification pathways.

2.4
Short Heterodimer Partner

SHP is an atypical orphan member of the nuclear receptor superfamily that lacks a conventional DBD. In spite of the lack of a DBD or a conventional nuclear localization signal, it appears to be located in the nucleus of mammalian transfected cells (Seol et al. 1998). A series of reports have shown that SHP interacts with and inhibits the transcriptional activity of several members of the nuclear receptor superfamily including the CAR, estrogen receptor, thyroid receptor, RXR, retinoic acid receptor, androgen receptor (AR), liver X receptor (LXR α/β), liver receptor homolog-1, and hepatocyte nuclear factor-4 (Seol et al. 1996, 1998; Goodwin et al. 2000; Gobinet et al. 2001; Johansson et al. 1999; Klinge et al. 2001; Lee et al. 2000; Lu et al. 2000; Brendel et al. 2002). It is mostly expressed in the liver under the control of FXR (farnesoid X receptor) (Seol et al. 1996, 1997). FXR is activated by primary bile acids such as cholic acid and chenodeoxycholic acid (Parks et al. 1999). Moreover, as SHP contains a strong transcriptional repression domain in its C terminus (Seol et al. 1996) it has been suspected to act as a direct transcriptional repressor by recruiting conventional corepressors such as N-coR, as reported with DAX-1 (Crawford et al. 1998). SHP has recently been proposed to represent a new category of nuclear receptor coregulator, interfering directly with AF-1 and AF-2 coactivator factors such RIP140, TIF2, SRC-1 and FHL2 (Seol et al. 1998; Gobinet et al. 2001; Johansson et al. 2000).

Recently, SHP has been shown to be involved in the control of bile acid biosynthesis (Chiang et al. 2000). CYP7α, the rate-limiting enzyme in this pathway, has been shown to be subject to primary bile acid feedback regulation via SHP (Goodwin et al. 1999, 2000). On the other hand, secondary bile acid derivatives are ligands of PXR and regulate the expression of genes involved in bile acid homeostasis including CYP7α and Oatp2 (Staudinger et al. 2001; Moore et al. 2002).

A series of recent reports suggest that CYP3A expression is negatively regulated by FXR ligands: (a) taurochenodeoxycholic acid, a FXR ligand (Parks et al. 1999), reduced CYP3A-associated monooxygenase activities in vivo in rats (Paolini et al. 2000); (b) cyp3a11 upregulation observed in cyp27$^{-/-}$ mice is reversed if mice are fed a diet containing cholic acid or chenodeoxycholic acid(Goodwin et al. 2003). However, the molecular mechanism involved remains largely unknown. As FXR has been shown not to bind to PXR or CAR responsive elements (Handschin et al. 2002), we suspected that these negative effects on *CYP3A* expression could be mediated by FXR-induced SHP upregulation. Indeed, we demonstrated that SHP is a negative regulator of PXR (Ourlin et al. 2003). Using primary hepatocytes we observed that chenodeoxycholic acid and cholic acid, two FXR ligands, induce SHP mRNA expression and concomitantly decrease the expression of CYP7α and reduce rifampicin-mediated induction of *CYP3A4*, *CYP2B6* and *CYP2C9*. Using pulldown assays, we showed that SHP interacts with PXR in a ligand-dependent manner. From transient transfection assays, SHP is shown to be a potent repressor of PXR transactivation. This inhibition is reversed by SRC-1, a transcriptional coactivator of PXR (Kliewer et al. 1998). These results suggest either that SHP interacts directly with PXR and weakens its binding to DNA as proposed previously for the retinoic acid receptor, or that SHP blocks the AF-2 activation domain as proposed previously for the estrogen receptor (Johansson et al. 1999). Previous observations on the modulation of CYP3A expression by bile acids and FXR invalidation support these conclusions (Schuetz et al. 2001; Handschin et al. 2002; Paolini et al. 2000).

The conversion of cholesterol to bile acids occurs exclusively in the liver. Many nuclear receptors (LXR, FXR, PXR) control the conversion of cholesterol to the primary bile acids, CDCA and CA. According to our results, PXR function is inhibited in the presence of CDCA or CA through the upregulation of SHP (Fig. 1). The reason why primary bile acids repress PXR activity and, consequently, the expression of CYP3A involved in the clearance of bile acids is still unclear. As CA and CDCA are nontoxic bile acid derivatives, in contrast to some bile acid precursors or secondary bile acid products, one explanation is that this process could prevent the overconversion of cholesterol into such toxic compound through a regulatory feedback loop similar to that observed with CYP7α.

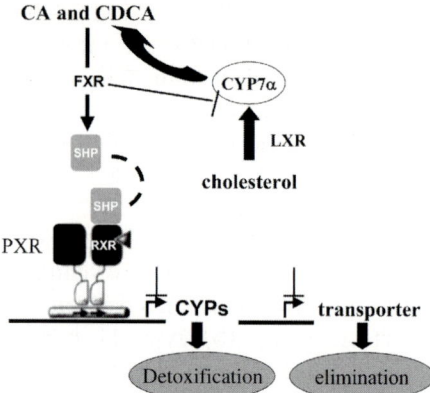

Fig. 1 SHP is a negative regulator of PXR transcriptional activity. Nuclear receptors LXR, FXR and PXR, control the conversion of cholesterol to primary bile acids CDCA and CA. PXR function is inhibited in the presence of CDCA or CA through the up-regulation of SHP

Thus, SHP interacts with and inhibits the transcriptional activity of both PXR and CAR. As PXR and CAR control genes are involved not only in the metabolism and elimination of xenobiotics but also in the biosynthesis and catabolism of bile acids, this is likely to generate functional interferences between bile acid homeostasis and xenobiotic detoxification pathways.

2.5
Glucocorticoid Receptor

Previous work from the group of Guzelian on the rat suggested that induction of *CYP3A* by glucocorticoids, and paradoxically also by antiglucocorticoids, is dependent on a nonclassical glucocorticoid-mediated induction process (Kocarek et al. 1995). These studies revealed an atypical profile of *CYP3A23* induction, compared to the classical glucocorticoid-mediated induction of known GR-dependent genes (Schuetz et al. 1993a; Burger et al. 1992). For example, induction of *CYP3A23* by dexamethasone requires a concentration 100 times higher than that necessary for maximum induction of the *TAT* gene. Further, the potency of various glucocorticoids for inducing *CYP3A23* does not correlate with their potency for inducing *TAT*. On the other hand, it had previously been reported that GR binds to a glucocorticoid response element (GRE) present in the rat *CYP3A1* gene, and it has been suggested that cooperation of the upstream GRE and downstream elements (e.g., PXR element) may be required for the maximal response of *CYP3A* to glucocorticoids (Wrighton et al. 1985). Schuetz et al. (1993a) proposed that the GR is not required for the induction of *CYP3A* by glucocorticoids in mouse while its expression is essential for the induction of *CYP2B*.

However, extrapolating data from rodents to man is hazardous because of the species-specificity in the response of CYPs and nuclear receptors to xenobiotics (LeCluyse 2001; Wrighton et al. 1985).

Although computer analysis of approximately 1 kb of the human *CYP3A4* proximal promoter revealed the presence of putative binding sites for the estrogen receptor, COUP-TF, HNF4, HNF5 and Oct-1 (Hashimoto et al. 1993), no consensus binding site for the GR was identified within this region. However, the human *CYP3A5* gene promoter contains two GRE, separated by 160 bp, which confer glucocorticoid response to reporter genes in HepG2 cells (Schuetz et al. 1993b). Interestingly, the *CYP3A5* promoter has no functional PXR-binding site in its proximal region (Blumberg et al. 1998) but is still inducible by glucocorticoids (Hukkanen et al. 2000). In contrast, a functional GRE in the *CYP2C9* gene promoter has recently been reported to trigger the GR-mediated *CYP2C9* gene induction by glucocorticoids, while PXR failed to transactivate this element (Gerbal-Chaloin et al. 2002).

Several lines of evidence suggest that CAR and PXR may not totally account for the steroid induction of *CYP2s* and *CYP3A*, and the possibility exists that the GR is involved indirectly in this process: (a) induction of endogenous human *CYP2B6*, *CYP2C8/9* and *CYP3A4* in cultured hepatocytes (Honkakoski and Negishi 1998; Pascussi et al. 2000c) is potentiated by pretreatment of cells with dexamethasone; (b) the response to glucocorticoids of a *CYP3A4* promoter-dependent gene reporter is increased in the presence of cotransfected hGR (El-Sankary et al. 2000); (c) transcriptional activation by dexamethasone of rat *CYP3A1* and human *CYP3A4* promoters (Pascussi et al. 2001; Ogg et al. 1999; Burger et al. 1992) is blocked by the antiglucocorticoid mifepristone (RU486), a mouse and human PXR activator (Lehmann et al. 1998); (d) induction of *CYP3A4* and *CYP2C9* mRNA expression by dexamethasone in cultured human hepatocytes is inhibited by submicromolar concentrations of RU486 while this compound induces *CYP3A4* mRNA only at micromolar concentrations, suggesting the involvement of at least two distinct pathways in the response of this gene to agonist and antagonist glucocorticoids (Pascussi et al. 2001; Gerbal-Chaloin et al. 2002).

In previous work, we have shown that expression of CAR and PXR in human hepatocytes is regulated by glucocorticoid hormones (Pascussi et al. 2000a) and made the hypothesis that GR controls the expression of PXR and CAR on the basis of the following arguments: (a) dexamethasone does not affect the degradation of CAR and PXR mRNA; (b) CAR and PXR mRNA induction is blocked by the glucocorticoid antagonist RU 486; (c) the induction is not suppressed by cycloheximide treatment, indicating that it is mediated by preexisting GR; and (e) the RNA synthesis inhibitor actinomycin D abolishes the stimulatory effect of dexamethasone. This hypothesis was confirmed recently by a systematic analysis of 4.7 kb of the human CAR regulatory region (Pascussi et al. 2003). The results revealed the existence of a functional GRE between $-4,447$ and $-4,432$. This element has a classical

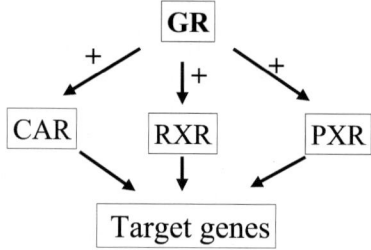

Fig. 2 Activation cascade of signal transmission GR–CAR/PXR–target genes. This cascade involves: activation of the GR, increasing PXR, CAR and RXR expression, induction of PXR/RXR and CAR/RXR heterodimers by its cognate activators and induction of the expression of PXR and CAR target genes

GRE structure, i.e., two half-sites separated by three nucleotides, and mutations of this GRE in either half-site drastically decreased both binding and transactivation by hGR. These in vitro experiments were confirmed by a chromatin immunoprecipitation assay revealing that dexamethasone treatment of cultured human hepatocytes causes binding of GR to this DNA region of the CAR promoter in intact cells. Of particular significance is that the homologous region of the murine CAR promoter gene precisely matches to the human GRE (starting at chromosome 1 position 172,485,157), suggesting that similar glucocorticoid-dependent regulation should be observed in this species. Whether PXR is a primary glucocorticoid responsive gene as well remains to be confirmed.

Interaction of this GR regulatory pathway with the PXR and CAR-mediated xenobiotic signaling pathway suggests the existence of an activation cascade of signal transmission: GR-CAR/PXR-CYP2/3 (Fig. 2). According to this model, both glucocorticoids and PXR and CAR activators have to be present for the cascade to work. Hence, it may explain the cooperative effect observed between glucocorticoids and PB or rifampicin on the expression of several PXR or CAR target genes (Honkakoski et al. 1998; Kliewer et al. 1998; Pascussi et al. 2000a, 2000c), as well as the inhibitory effect of RU486 on rifampicin or phenobarbital-mediated CYP induction (Bertilsson et al. 1998; Lehmann et al. 1998; Tsai and O'Malley 1994).

3
PXR, CAR and VDR Signal by Means of the Same Response Elements

The fact that *CYP2B* and *CYP3A*, members of two distinct CYP subfamilies, can be induced by structurally unrelated compounds such as PB, pregnenolone 16α-carbonitrile (PCN) or rifampicin (RIF) is intriguing. Induction of *CYP2B* in response to PB has been shown to be mediated by the interaction of CAR with a PB response element (PBRE; Trottier et al. 1995; Honkakoski et al. 1998) while that of *CYP3A* in response to PCN or RIF is mediated by the interaction of PXR with a pregnane X response element (PXRE) (Bertilsson et al. 1998; Kliewer et al. 1998; Blumberg et al. 1998). However, no typical PBRE has been identified in the flanking regions of the *CYP3A* gene, and no typical PXRE has been identified in the *CYP2B* promoter. The question is thus, how can the same class of chemicals induce two distinct CYP subfamily genes and how can distinct classes of chemicals induce the same CYP gene?

Recently, different authors have shown that CAR and PXR can transactivate CYP genes via the same response element in a xenobiotic-specific manner: PXR can regulate *CYP2B* both in cultured cells and in transgenic mice via adaptative recognition of the PB response element (Xie et al. 2000), and reciprocally, CAR binds to and transactivates the PXR response elements of *CYP3A4* (Goodwin et al. 1999). Similarly, we characterized in the promoter of *CYP2C9* a response element that can be transactivated by both PXR and CAR (Gerbal-Chaloin et al. 2002). Consistent with this, we observed a detectable accumulation of *CYP3A4* mRNA in HepG2 cells transfected with a CAR expression plasmid, in the absence of exogenous ligand. These data confirm that besides PXR, CAR plays a significant role in both basal and inducible expression of *CYP3A4* gene (Pascussi et al. 2001).

In addition, many compounds are able to interact with both receptors as ligands, activators and/or inverse agonists (Moore et al. 2002; Ekins et al. 2002). For example, PB and 5β-pregnane-3,20-dione are activators of both CAR and PXR. On the other hand, PXR activators such as clotrimazole, androstanol and progesterone appear to be inverse agonists of CAR. Therefore, differential response of CYPs to inducers may be due, in part, to the relative abundance of these receptors, as they compete for cofactors such as SCR-1 and RXRα, and ligands. Consequently, the net effect of a xenobiotic on gene transcription will often be complex and depend on both CAR and PXR (Moore et al. 2000b). Thus, one frequently encountered problem is to be able to discriminate between the two receptors. In this respect, several strategies have been used including the use of transgenic mice in which one of the receptors has been knocked out. One attractive alternative should be the use of specific agonists or antagonists. Recently 6-(4-chlorophenyl)imidazo[2,1-b][1,3]thiazole-5-carbaldehyde-*O*-(3,4-dichlorobenzyl)oxime has been demonstrated to act as a novel human CAR agonist displaying very low activity toward PXR (Maglich et al. 2003) . This compound has proven that CAR and PXR regulate differently the expression of several genes. Notably, the *CYP2A6* gene was shown to be more responsive to CAR than to PXR activators. As the CYP2A6 enzyme hydroxylates a variety of steroid hormones, including androgens and estrogens (Tyndale et al. 1999), and because CAR is activated by estrogens but inactivated by androgens (Kawamoto et al. 2000), it is possible that CAR plays an important role in metabolizing endogenous steroids.

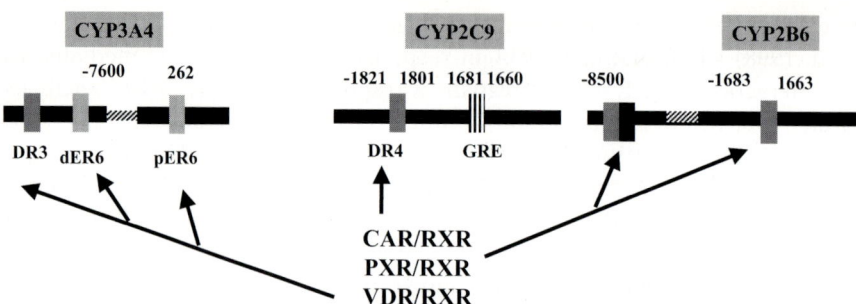

Fig. 3 PXR, CAR and VDR signal through the same response elements. The crosstalk between CAR, PXR and VDR signaling pathways is an important factor in appropriate response to a xenobiotics and also in hepatic physiology

Recent reports have revealed that 1α, 25-dihydroxy vitamin D3, behaves as a transcriptional inducer of CYP3A4 in the colic carcinoma Caco-2 cell line and the human intestinal LS180 cell line (Schuetz et al. 2001; Thummel et al. 2001). In addition, we have shown that this derivative induces the expression of *CYP3A4* and to a lesser extent that of *CYP2B6* and *CYP2C9* genes in normal differentiated primary human hepatocytes (Drocourt et al. 2001). Electrophoretic mobility shift assays and cotransfection in HepG2 cells using wild-type and mutated oligonucleotides revealed that the VDR binds and transactivates those xenobiotic-responsive elements (ER6, DR3 and DR4) previously identified in *CYP3A4*, *CYP2B6* and *CYP2C9* promoters and shown to be targeted by PXR and/or CAR. Cotransfection of a *CYP3A4* homologous promoter–reporter construct (including distal and proximal PXR-binding motifs) and of PXR or CAR expression vectors in HepG2 cells revealed the ability of these receptors to compete with VDR for transcriptional regulation of CYP3A4 (Drocourt et al. 2002). These results have been confirmed by Makishima et al. (2002). In addition these authors showed that VDR also functions as a receptor for secondary bile acid lithocholic acid, which is hepatotoxic and a potential carcinogen.

The present and previous results showing that vitamin D affects CYP gene expression increase the list of those physiological compounds able to interfere with the metabolism of xenobiotics. These observations suggest that CAR, PXR and VDR receptors are capable of regulating a same series of genes through the same *cis*-acting elements (Fig. 3). Crosstalk between these signaling pathways provides new clues on the toxic effects of xenobiotics.

4
Consequences of Crosstalk Between PXR/CAR and Other Nuclear Receptors in Detoxification and Physiology

The previous observations on the control of PXR and CAR expression by GR on one hand, and the crosstalk between PXR/CAR and VDR on the other hand, led us to envisage new ways of considering the relationship between detoxification and hepatic physiology because of crosstalk connecting various signaling pathways in which PXR and CAR play a central role.

4.1
Crosstalk Between PXR/CAR and VDR

Prolonged therapy with RIF causes vitamin D deficiency (Brodie et al. 1982; Chan 1996; D'Erasmo et al. 1998). In eight healthy subjects, RIF treatment reduced circulating levels of 25-hydroxy vitamin D and 1,25-dihydroxy vitamin D by 34% and 23%, respectively. In addition, RIF and PB are two of the drugs most commonly associated with osteomalacia, a metabolic bone disease characterized by a defect in bone mineralization frequently due to an alteration of vitamin D metabolism.

Because we have shown that VDR is able to activate PXR- and CAR-responsive genes including *CYP3A4*, *CYP2C9* and *CYP2B6* (Drocourt et al. 2002), the clinical observations above suggest the hypothesis that the reciprocal proposition is true, that is, PXR and CAR are able to activate VDR-responsive genes. In particular, this suggests that CAR and/or PXR might be involved in the control of the genes responsible for the synthesis or catabolism of vitamin D.

The major route of degradation of vitamin D is oxidation of the side chain of the molecule, catalyzed by vitamin D-24 hydroxylase (CYP24), an enzyme which is induced by 1,25 dihydroxy vitamin D and its other derivatives (Zehnder and Hewison 1999). It is therefore possible that xenobiotics stimulate the expression of VDR target genes, notably CYP24, (J.M. Pascussi, A. Robert, M. Nguyen, M. Garabedian, J. Saric, F. Navarro, P. Maurel and M.J. Vilarem, unpublished results) through PXR/CAR thus producing the observed deficiency in Vitamin D leading to osteomalacia. This is important as several such drugs are currently in use, including notably barbiturates, antibiotics, carbamazepine, isophosphamide and HIV protease inhibitors. This possibility is under current evaluation in our laboratory.

4.2
Crosstalk Between PXR/CAR and GR

As previously demonstrated, GR controls detoxification via the signal transmission cascade GR-PXR/CAR-P450. These results lead to the anticipation that any pathological or physiological stimulus which affects negatively the expression or activity of GR, produces a downregulation of CYPs and other XMTS and inhibits the detoxification function.

4.2.1
Inhibition of Detoxification by Inflammatory Response

The decreased capacity of the liver to metabolize drugs and other xenobiotics during episodes of inflammation or infection has previously been documented in many reports on animals and on man in vivo and in vitro (Morgan et al. 1998). Numerous experimental studies have shown that this is related to a decreased level of CYP enzymes in response to cytokines (Morgan 1993; Muntane et al. 1995). For example, we and others have previously reported that interleukin (IL)-6 strongly represses the inducibility of CYP3A4 mRNA and protein in primary human hepatocytes (Muntane et al. 1995). Indeed, exposure to proinflammatory cytokines decreased the expression of *CYP3A4* mRNA in the Caco-2 intestinal cells (Bertilsson et al. 2001). Treatment with lipopolysacharide (LPS) decreases CYP3A11 in hepatocytes (Panesar et al. 1999; Sever 1997) and treatment with IL-2 downregulates the expression of CYP2C11 and CYP3A in the rat. In addition, Khatsenko et al. reported that LPS inhibits the PB-induced expression of CYP2B1 and CYP2B2 in the rat at both the mRNA and protein levels. Moreover, they found that these effects were attenuated in rats treated with inhibitors of nitric-oxide synthase, which are known to interfere with nuclear factor-κB (NF-κB) activation (Khatsenko et al. 1997).

Cytokines are known to activate a cascade of genes and factors which lead to the transcriptional activation of different transcriptional factors (Schule et al. 1990). Tumor necrosis factor-α is believed to repress CYP1A1 via redox regulation of nuclear factor 1 (Morel and Barouki 1998), IL-2 downregulates rat CYP2C11 and CYP3A2 expression, probably via induction of the protooncogene c-*myc* (Tinel et al. 1999; Thummel et al. 2001), IL-1 inhibits *CYP2C11* transcription via binding of NF-κB to a low-affinity site in the promoter of the gene (Iber et al. 2000). IL-6 inhibits CYP3A4 through the induction of C/EBPβ-LIP which competes with other constitutive C/EBP-activating factors such as C/EBPα. These genes have been shown to interact with GR and to inhibit its transcriptional activity (Jover et al. 2002). On the other hand, nuclear receptors have been proposed to be relevant factors in the mechanism of CYP repression, as reduced expression of *PXR* and *CAR* has been observed after stimulation with IL-6 (Pascussi et al. 2000b) or LPS (Beigneux et al. 2002). More recently, Beigneux et al. (2000) reported that LPS represses *RXR*α, *CAR* and *PXR* gene expression in mouse liver, while we observed that IL-1β decreases CAR expression and abolishes PB- or bilirubin-mediated induction of drug-metabolizing enzymes and bilirubin clearance in human hepatocytes (unpublished results).

Numerous studies indicate that NF-κB-p65 and GR interact physically, resulting in a mutual transcription antagonism (Ray and Prefontaine 1994). On the other hand, it is known that transcriptional coregulators, such as p300/CBP or SRC-1, function as co-activators for both the GR and the NF-κB pathways (Sheppard et al. 1998). These co-activators have been shown to possess histone acetyl transferase (HAT) activity which is involved in chromatin remodeling, a key step in transcriptional activation. Our data suggest that activation of NF-κB is a critical event in LPS- or IL-1β-mediated inhibition of CAR expression through the repression of ligand-activated GR action. We observed that NF-κB p65 interferes with the enhancer function of the distal GRE of the CAR promoter gene which we recently have identified. Recent data show that: (a) overexpression of LPS, IL-1β or p65RelA down-regulate GR-induced CAR expression; (b) these suppressive effects can be blocked both by pyrrolidine dithiocarbamate, a well known inhibitor of NF-κB activation, or by the overexpression of the NF-κB repressor, SRIκBα; (c) the glucocorticoid-induced histone H4 acetylation of the CAR promoter is markedly inhibited by LPS or IL-1β in human hepatocytes (E. Assenat, S. Gerbal-Chaloin, D. Larrey, P. Maurel, M.J. Vilarem and J.M. Pascussi, unpublished results). It is possible that this is due to inhibition of GR-associated HAT activity and/or recruitment of histone deacetylases to the GR activation complex by NF-κB. A similar mechanism has been described recently for the inhibition of IL-1β-stimulated histone acetylation of granulocyte-macrophage colony stimulating factor expression by dexamethasone (Ito 2000). This suggests that anti-inflammatory effects of glucocorticoids and anti-glucocorticoid effects of pro-inflammatory cytokines are based on similar mechanisms. In summary, we propose that NFκB inhibits the transcriptional activity of GR and thus repress the expression of the genes down-stream of GR in the GR-PXR/CAR-CYP cascade.

NFκB activation by proinflammatory cytokines inhibits the GR transactivation function and leads to a decrease in CAR expression; this could be the mechanism by which inflammation induces intrahepatic cholestasis and hyperbilirubinemia in man, as previously observed (Trauner et al. 1999): (a) CAR has been reported to regulate directly the organic anion transporting polypeptide 2 (Oatp2) (Guo et al. 2002), as well as the multidrug resistance-associated protein 2 (MRP2, ABCC2) (Kast et al. 2002) which regulate the bilirubin clearance; (b) elevated bilirubin levels can be decreased by treatment with PB in man (Yaffe et al. 1966); and (c) CAR has been recently shown to participate in bilirubin clearance by the use of CAR null mice (Huang et al. 2003). We hypothesized that a NF-κB-induced CAR deficit may contribute to inflammation-induced hyperbilirubinemia (Fig. 4). Indeed, we observed that basal expression and bilirubin-induced upregulation of these genes were drastically decreased by IL-1β pretreatment of human hepatocytes (J.M. Pascussi, E. Assenat, S. Gerbal-Chaloin, D. Larrey, P. Maurel and M.J. Vilarem, unpublished results).

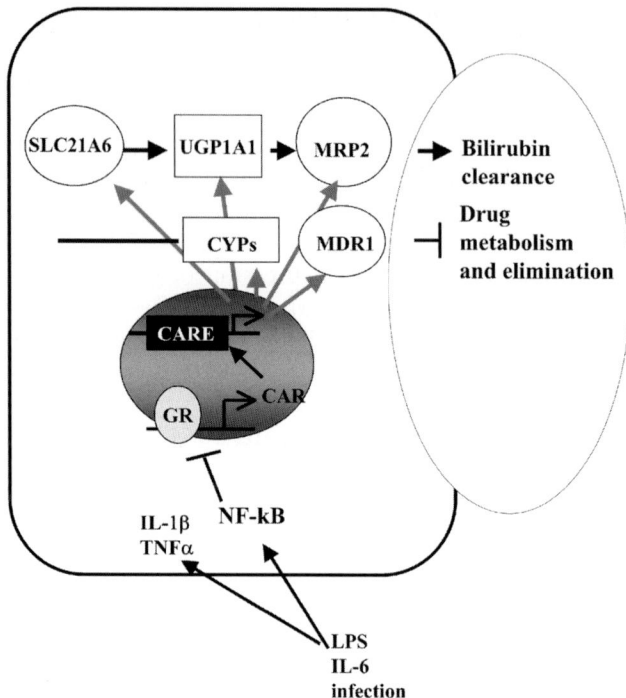

Fig. 4 NF-κB inhibits drug metabolism/elimination and induces intrahepatic cholestasis. NF-κB activation inhibits the GR transactivation function leading to a decrease in CAR expression. CAR regulates CYP induction and the organic anion transporting polypeptide 2 and the multidrug resistance-associated protein 2 that regulate the bilirubin clearance. The clinical consequences are induction of hyperbilirubinemia and inhibition of drug metabolism and elimination

4.2.2
Inhibition of Detoxification by Microtubule Disrupting Agent

Microtubules serve as an intracellular scaffold, and their unique polymerization dynamics are critical for many intracellular functions (Pratt et al. 1999). Microtubule disrupting agents (MIAs) interfere with the dynamic process of microtubule assembly. For example, colchicine strongly binds to tubulin, which results in cytoskeleton disruption (Ben-Chetrit and Levy 1998). GR is a nucleocytoplasmic shuttling protein and, as such, is one of the prime subjects to be influenced by cytoskeleton status. The inactive form of GR is located in the cytoplasm where it is bound to hsp90, a known chaperone protein. Upon hormone binding, GR dissociates from hsp90, homodimerizes and is rapidly imported into the nucleus (Madan and DeFranco 1993). It has been proposed that this process requires functional microtubules (Pratt 1993). MIAs are therefore expect to perturb the GR-PXR/CAR-CYP cascade.

We recently reported that MIAs such as colchicine or nocodazole decrease *CYP2B6*, *CYP2C9*, and *CYP3A4* basal and phenobarbital-induced expression in human hepatocytes (Dvorak et al. 2003). In sharp contrast, colchicein, a colchicine derivative which cannot bind to β-tubulin, has no effect on CYP expression or induction by xenobiotics, or on *TAT* gene expression (Dvorak et al. 2003). Several lines of evidence suggest that these effects are the direct consequence of MIA-induced GR inhibition: (a) the decrease in CYP expression is accompanied by a concomitant and massive decrease in TAT and CAR mRNAs and a significant, but more modest, decrease in PXR mRNA expression; (b) neither GR mRNA expression nor dexamethasone-GR binding are affected by colchicine; (c) dexamethasone-inducible transcriptional activity of GR in HeLa cells stably transfected with a GRE-driven reporter gene is inhibited by colchicine; (d) the basal expression of *CYP2C9*, a GR primary-responsive gene (Gerbal-Chaloin et al. 2002), is strongly repressed by colchicine at the mRNA level. The potent inhibition of the transcriptional activity of GR by MIAs, most likely due to the alteration of GR trafficking between cytoplasm and nucleus (Fig 4), could explain—at least partially—the observed downregulation of *CYPs*, *CAR*, *PXR* and *TAT*, by cytoskeleton disruption (Dvorak et al. 2003).

5
Conclusion

The multiple possibilities of crosstalk between PXR and CAR and other nuclear receptors such as GR, VDR and SHR lead us to visualize the adaptive system of protection against xenobiotics and drugs within a tangle of regulatory networks where receptors share partners, ligands, DNA response elements and target genes. Adverse effects of xenobiotics are ascribed primarily to the chemical reactivity of compounds or metabolites which bind covalently to proteins, DNA, or membrane constituents (lipids). This review illustrates that the crosstalk between PXR and CAR and other nuclear receptors allows functional interferences between detoxification and other signaling and metabolic pathways to be established and predicted. These interferences provide new clues for the interpretation of xenobiotic adverse effects (e.g., bone demineralization) and for understanding how physiopathological factors, such as biliary salts or inflammation and infections, affect an organism's ability to detoxify xenobiotics.

References

Ariyoshi N, Miyazaki M, Toide K, Sawamura Y, Kamataki T (2001) A single nucleotide polymorphism of CYP2b6 found in Japanese enhances catalytic activity by autoactivation. Biochem Biophys Res Commun 281:1256–1260

Auerbach SS, Ramsden R, Stoner MA, Verlinde C, Hassett C, Omiecinski CJ (2003). Alternatively spliced isoforms of the human constitutive androstane receptor. Nucl Acids Res 31:3194–3207

Baes M, Gulick T, Choi HS, Martinoli MG, Simha D, Moore DD (1994) A new orphan member of the nuclear hormone receptor superfamily that interacts with a subset of retinoic acid response elements. Mol Cell Biol 14:1544–1551

Beigneux AP, Moser AH, Shigenaga JK, Grunfeld C, Feingold KR (2000) The acute phase response is associated with retinoid X receptor repression in rodent liver. J Biol Chem 275:16390–16499

Beigneux AP, Moser AH, Shigenaga JK, Grunfeld C, Feingold KR (2002) Reduction in cytochrome P-450 enzyme expression is associated with repression of CAR (constitutive androstane receptor) and PXR (pregnane X receptor) in mouse liver during the acute phase response. Biochem Biophys Res Commun 293:145–149

Ben-Chetrit E, Levy M (1998) Colchicine: 1998 update. Semin Arthritis Rheum 28:48–59

Bertilsson G, Berkenstam A, Blomquist P (2001) Functionally conserved xenobiotic responsive enhancer in cytochrome P450 3A7. Biochem Biophys Res Commun 280:139–144

Bertilsson G, Heidrich J, Svensson K, Asman M, Jendeberg L, Sydow-Backman M, Ohlsson R, Postlind H, Blomquist H, Berkenstam A (1998) Identification of a human nuclear receptor defines a new signaling pathway for CYP3A induction. Mol Pharmacol 54:1113–1117

Bertz RJ, Granneman GR (1997) Use of in vitro and in vivo data to estimate the likelihood of metabolic pharmacokinetic interactions. Clin Pharmacokinet 32:210–258

Blumberg B, Sabbagh WJ, Juguilon H, Bolado JJ, van MC, Ong ES, Evans RM (1998) SXR, a novel steroid and xenobiotic-sensing nuclear receptor. Genes Dev 12:3195–3205

Brendel C, Schoonjans K, Botrugno OA, Treuter E, Auwerx J (2002) The small heterodimer partner interacts with the liver X receptor alpha and represses its transcriptional activity. Mol Endocrinol 16:2065–2076

Brodie MJ, Boobis AR, Hillyard CJ, Abeyasekera G, Stevenson JC, MacIntyre I, Park BK (1982) Effect of rifampicin and isoniazid on vitamin D metabolism. Clin Pharmacol Ther 32:525–530

Burger HJ, Schuetz JD, Schuetz EG, Guzelian PS (1992) Paradoxical transcriptional activation of rat liver cytochrome P-450 3A1 by dexamethasone and the antiglucocorticoid pregnenolone 16 alpha-carbonitrile: analysis by transient transfection into primary monolayer cultures of adult rat hepatocytes. Proc Natl Acad Sci USA 89:2145–2149

Carlberg C, Polly P (1998) Gene regulation by vitamin D3. Crit Rev Eukaryot Gene Expr 8:19–42

Chan TY (1996) Osteomalacia during rifampicin and isoniazid therapy is rare in Hong Kong. Int J Clin Pharmacol Ther 34:533–534

Chiang JY, Kimmel R, Weinberger C, Stroup D (2000) Farnesoid X receptor responds to bile acids and represses cholesterol 7alpha-hydroxylase gene (CYP7A1) transcription. J Biol Chem 275:10918–10924

Choi HS, Chung M, Tzameli I, Simha D, Lee YK, Seol W, Moore DD (1997) Differential transactivation by two isoforms of the orphan nuclear hormone receptor CAR. J Biol Chem 272:23565–23571

Cokol M, Nair R, Rost B (2000) Finding nuclear localization signals. EMBO Rep 1:411–415

Corcos L, Lagadic-Gossmann D (2001) Gene induction by phenobarbital: an update on an old question that receives key novel answers. Pharmacol Toxicol 89:113–122

Crawford PA, Dorn C, Sadovsky Y, Milbrandt J (1998) Nuclear receptor DAX-1 recruits nuclear receptor corepressor N-CoR to steroidogenic factor 1. Cell 93:445–454

D'Erasmo E, Ragno A, Raejntroph N, Pisani D (1998) Drug-induced osteomalacia. Recenti Prog Med 89: 529–533

Danielsson C, Fehsel K, Polly P, Carlberg C (1998) Differential apoptotic response of human melanoma cells to 1 alpha,25- dihydroxyvitamin D3 and its analogues. Cell Death Differ 5: 946–952

Drocourt L, Ourlin JC, Pascussi JM, Maurel P, Vilarem MJ (2002) Expression of CYP3A4, CYP2B6 and CYP2C9 is regulated by the vitamin D receptor pathway in primary human hepatocytes. J Biol Chem 277:25125–25132

Drocourt L, Pascussi JM, Assenat E, Fabre JM, Maurel P, Vilarem MJ (2001) Calcium channel modulators of the dihydropyridine family are human pregnane X receptor activators and inducers of CYP3A, CYP2B, and CYP2C in human hepatocytes. Drug Metab Dispos 29:1325–1331

Dvorak Z, Modriansky M, Pichard-Garcia L, Balaguer P, Vilarem MJ, Ulrichova J, Maurel P, Pascussi JM (2003) Colchicine down-regulates cytochrome P450 2B6, 2C8, 2C9, and 3A4 in human hepatocytes by affecting their glucocorticoid receptor-mediated regulation. Mol Pharmacol 64:160–169

Ekins S, Bravi G, Wikel JH, Wrighton SA (1999) Three-dimensional-quantitative structure activity relationship analysis of cytochrome P-450 3A4 substrates. J Pharmacol Exp Ther 291:424–433

Ekins S, Mirny L, Schuetz EG (2002) A ligand-based approach to understanding selectivity of nuclear hormone receptors PXR, CAR, FXR, LXRalpha, and LXRbeta. Pharm Res 19:1788–1800

El-Sankary W, Plant NJ, Gibson GG, Moore DJ (2000) Regulation of the CYP3A4 gene by hydrocortisone and xenobiotics: role of the glucocorticoid and pregnane X receptors. Drug Metab Dispos 28:493–496

Forman BM, Tzameli I, Choi HS, Chen J, Simha D, Seol W, Evans RM, Moore DD (1998) Androstane metabolites bind to and deactivate the nuclear receptor CAR-beta [see comments]. Nature 395:612–615

Gerbal-Chaloin S, Daujat M, Pascussi JM, Pichard-Garcia L, Vilarem MJ, Maurel P (2002) Transcriptional Regulation of CYP2C9 Gene. Role of glucocorticoid receptor and constitutive androstane receptor. J Biol Chem 277:209–217

Gobinet J, Auzou G, Nicolas JC, Sultan C, Jalaguier S (2001) Characterization of the interaction between androgen receptor and a new transcriptional inhibitor, SHP. Biochemistry 40:15369–15377

Goldstein JA (2001) Clinical relevance of genetic polymorphisms in the human CYP2C subfamily. Br J Clin Pharmacol 52: 349–355

Gonzalez FJ, Gelboin HV (1992) Human cytochromes P450: evolution and cDNA-directed expression. Environ Health Perspect 98:81–85

Goodwin B, Gauthier KC, Umetani M, Watson MA, Lochansky MI, Collins JL, Leitersdorf E, Mangelsdorf DJ, Kliewer SA, Repa JJ (2003) Identification of bile acid precursors as endogenous ligands for the nuclear xenobiotic pregnane X receptor. Proc Natl Acad Sci USA 100: 223–228

Goodwin B, Hodgson E, D'Costa DJ, Robertson GR, Liddle C (2002) Transcriptional regulation of the human CYP3A4 gene by the constitutive androstane receptor. Mol Pharmacol 62:359–365

Goodwin B, Hodgson E, Liddle C (1999) The orphan human pregnane X receptor mediates the transcriptional activation of CYP3A4 by rifampicin through a distal enhancer module. Mol Pharmacol 56:1329–1339

Goodwin B, Jones SA, Price RR, Watson MA, McKee DD, Moore LB, Galardi C, Wilson JG, Lewis MC, Roth ME, Maloney PR, Willson TM, Kliewer SA (2000) A regulatory cascade of the nuclear receptors FXR, SHP-1, and LRH-1 represses bile acid biosynthesis [in process citation]. Mol Cell 6:517–526

Goodwin B, Moore LB, Stoltz CM, McKee DD, Kliewer SA (2001) Regulation of the human CYP2B6 gene by the nuclear pregnane X receptor. Mol Pharmacol 60:427–431

Guengerich FP (1999) Cytochrome P-450 3A4: regulation and role in drug metabolism. Annu Rev Pharmacol Toxicol 39:1–17

Guo Z, Hou F, Zhang X, Liu Z, Wang L (2002) Advanced glycation end products inhibit production of nitric oxide by human endothelial cells through activation of the p38 signal pathway. Zhonghua Yi Xue Za Zhi 82:1328–1331

Handschin C, Podvinec M, Amherd R, Looser R, Ourlin JC, Meyer UA (2002) Cholesterol and bile acids regulate xenosensor signaling in drug- mediated induction of cytochromes P450. J Biol Chem 277:2951–2957

Handschin C, Podvinec M, Meyer UA (2000) CXR, a chicken xenobiotic-sensing orphan nuclear receptor, is related to both mammalian pregnane X receptor (PXR) and constitutive androstane receptor (CAR) Proc Natl Acad Sci USA 97:10769–10774

Harvey JL, Paine AJ, Maurel P, Wright MC (2000) Effect of the adrenal 11-beta-hydroxylase inhibitor metyrapone on human hepatic cytochrome P-450 expression: induction of cytochrome P-450 3A4. Drug Metab Dispos 28:96–101

Hashimoto H, Toide K, Kitamura R, Fujita M, Tagawa S, Itoh S, Kamataki T (1993) Gene structure of CYP3A4, an adult-specific form of cytochrome P450 in human livers, and its transcriptional control. Eur J Biochem 218:585–595

Honkakoski P, Moore R, Washburn KA, Negishi M (1998) Activation by diverse xenochemicals of the 51-base pair phenobarbital-responsive enhancer module in the CYP2B10 gene. Mol Pharmacol 53:597–601

Honkakoski P, Negishi M (1998) Regulatory DNA elements of phenobarbital-responsive cytochrome P450 CYP2B genes. J Biochem Mol Toxicol 12:3–9

Huang W, Zhang J, Chua SS, Qatanani M, Han Y, Granata R, Moore DD (2003) Induction of bilirubin clearance by the constitutive androstane receptor (CAR) Proc Natl Acad Sci USA 100:4156–4161

Hukkanen J, Lassila A, Paivarinta K, Valanne S, Sarpo S, Hakkola J, Pelkonen O, Raunio H (2000) Induction and regulation of xenobiotic-metabolizing cytochrome P450s in the human A549 lung adenocarcinoma cell line. Am J Respir Cell Mol Biol 22: 360–366

Iber H, Chen Q, Cheng PY, Morgan ET (2000) Suppression of CYP2C11 gene transcription by interleukin-1 mediated by NF-kappaB binding at the transcription start site. Arch Biochem Biophys 377:187–194

Johansson L, Bavner A, Thomsen JS, Farnegardh M, Gustafsson JA, Treuter E (2000) The orphan nuclear receptor SHP utilizes conserved LXXLL-related motifs for interactions with ligand-activated estrogen receptors. Mol Cell Biol 20:1124–1133

Johansson L, Thomsen JS, Damdimopoulos AE, Spyrou G, Gustafsson JA, Treuter E (1999) The orphan nuclear receptor SHP inhibits agonist-dependent transcriptional activity of estrogen receptors ERalpha and ERbeta. J Biol Chem 274:345-353

Jover R, Bort R, Gomez-Lechon MJ, Castell JV (2002) Down-regulation of human CYP3A4 by the inflammatory signal interleukin-6: molecular mechanism and transcription factors involved. FASEB J 16:1799-1801

Kast HR, Goodwin B, Tarr PT, Jones SA, Anisfeld AM, Stoltz CM, Tontonoz P, Kliewer S, Willson TM, Edwards PA (2002) Regulation of multidrug resistance-associated protein 2 (ABCC2) by the nuclear receptors pregnane X receptor, farnesoid X-activated receptor, and constitutive androstane receptor. J Biol Chem 277:2908-2915

Katoh M, Nakajima M, Yamazaki H, Yokoi T (2000) Inhibitory potencies of 1,4-dihydropyridine calcium antagonists to P- glycoprotein-mediated transport: comparison with the effects on CYP3A4.[In Process Citation]. Pharm Res 17:1189-1197

Kawamoto T, Kakizaki S, Yoshinari K, Negishi M (2000) Estrogen activation of the nuclear orphan receptor CAR (constitutive active receptor) in induction of the mouse Cyp2b10 gene [in process citation]. Mol Endocrinol 14:1897-1905

Kawamoto T, Sueyoshi T, Zelko I, Moore R, Washburn K, Negishi M (1999) Phenobarbital-responsive nuclear translocation of the receptor CAR in induction of the CYP2B gene. Mol Cell Biol 19:6318-6322

Khatsenko OG, Boobis AR, Gross SS (1997) Evidence for nitric oxide participation in down-regulation of CYP2B1/2 gene expression at the pretranslational level. Toxicol Lett 90:207-216

Kliewer SA, Moore JT, Wade L, Staudinger JL, Watson MA, Jones SA, McKee DD, Oliver BB, Willson TM, Zetterstrom RH, Perlmann T, Lehmann JM (1998) An orphan nuclear receptor activated by pregnanes defines a novel steroid signaling pathway. J Clin Invest 102:1016-1023

Klinge CM, Jernigan SC, Risinger KE, Lee JE, Tyulmenkov VV, Falkner KC, Prough RA (2001) Short heterodimer partner (SHP) orphan nuclear receptor inhibits the transcriptional activity of aryl hydrocarbon receptor (AHR)/AHR nuclear translocator (ARNT). Arch Biochem Biophys 390:64-70

Kocarek TA, Schuetz EG, Strom SC, Fisher RA, Guzelian PS (1995) Comparative analysis of cytochrome P4503A induction in primary cultures of rat, rabbit, and human hepatocytes. Drug Metab Dispos 23:415-421

LeCluyse EL (2001) Pregnane X receptor: molecular basis for species differences in CYP3A induction by xenobiotics. Chem Biol Interact 134:283-289

Lee YK, Dell H, Dowhan DH, Hadzopoulou-Cladaras M, Moore DD (2000) The orphan nuclear receptor SHP inhibits hepatocyte nuclear factor 4 and retinoid X receptor transactivation: two mechanisms for repression. Mol Cell Biol 20:187-195

Lehmann JM, McKee DD, Watson MA, Willson TM, Moore JT, Kliewer SA (1998) The human orphan nuclear receptor PXR is activated by compounds that regulate CYP3A4 gene expression and cause drug interactions. Proc Natl Acad Sci USA 95:12208-12213

Lin T, Lee H (1998) Parallel control mechanisms underlying locomotor activity and sexual receptivity of the female German cockroach, Blattella germanica (L.) J Insect Physiol 44:1039-1051

Lu TT, Makishima M, Repa JJ, Schoonjans K, Kerr TA, Auwerx J, Mangelsdorf DJ (2000) Molecular basis for feedback regulation of bile acid synthesis by nuclear receptors. Mol Cell 6:507-515

Madan AP, DeFranco DB (1993) Bidirectional transport of glucocorticoid receptors across the nuclear envelope. Proc Natl Acad Sci USA 90:3588-3592

Maglich JM, Parks DJ, Moore LB, Collins JL, Goodwin B, Billin AN, Stoltz CA, Kliewer SA, Lambert MH, Willson TM, Moore JT (2003) Identification of a novel human constitutive androstane receptor (CAR) agonist and its use in the identification of CAR target genes. J Biol Chem 278:17277–17283

Makishima M, Lu TT, Xie W, Whitfield GK, Domoto H, Evans RM, Haussler MR, Mangelsdorf DJ (2002) Vitamin D receptor as an intestinal bile acid sensor. Science 296:1313–1316

Maurel P (1996) The CYP3 family. In: C. Ioannides (ed.) Cytochromes P450 metabolic and toxicological aspects. CRC Press, Boca Raton, New York, London, Tokyo, 241–270

Moore LB, Goodwin B, Jones SA, Wisely GB, Serabjit SC, Willson TM, Collins JL, Kliewer SA (2000a) St. John's wort induces hepatic drug metabolism through activation of the pregnane X receptor. Proc Natl Acad Sci USA 97:7500–7502

Moore LB, Maglich JM, McKee DD, Wisely B, Willson TM, Kliewer SA, Lambert MH, Moore JT (2002) Pregnane X receptor (PXR), constitutive androstane receptor (CAR), and benzoate X receptor (BXR) define three pharmacologically distinct classes of nuclear receptors. Mol Endocrinol 16:977–986

Moore LB, Parks DJ, Jones SA, Bledsoe RK, Consler TG, Stimmel JB, Goodwin B, Liddle C, Blanchard SG, Willson TM, Collins JL, Kliewer SA (2000b) Orphan nuclear receptors constitutive androstane receptor and pregnane X receptor share xenobiotic and steroid ligands [in process citation]. J Biol Chem 275:15122–15127

Morel Y, Barouki R (1998) Down-regulation of cytochrome P450 1A1 gene promoter by oxidative stress. Critical contribution of nuclear factor 1. J Biol Chem 273:26969–26976

Morgan ET (1993) Down-regulation of multiple cytochrome P450 gene products by inflammatory mediators in vivo. Independence from the hypothalamo-pituitary axis. Biochem Pharmacol 45:415–419

Morgan ET, Sewer MB, Iber H, Gonzalez FJ, Lee YH, Tukey RH, Okino S, Vu T, Chen YH, Sidhu JS, Omiecinski CJ (1998) Physiological and pathophysiological regulation of cytochrome P450. Drug Metab Dispos 26:1232–1240

Muntane RJ, Ourlin JC, Domergue J, Maurel P (1995) Differential effects of cytokines on the inducible expression of CYP1A1, CYP1A2, and CYP3A4 in human hepatocytes in primary culture. Hepatology 273:1143–1153

Nelson DR (1999) Cytochrome P450 and the individuality of species. Arch Biochem Biophys 369:1–10

Nirodi CS, Sultana S, Ram N, Prabhu L, Padmanaban G (1996) Involvement of synthesis and phosphorylation of nuclear protein factors that bind to the positive cis-acting element in the transcriptional activation of the CYP2B1/B2 gene by phenobarbitone in vivo. Arch Biochem Biophys 331:79–86

Ogg MS, Williams JM, Tarbit M, Goldfarb PS, Gray TJ, Gibson GG (1999) A reporter gene assay to assess the molecular mechanisms of xenobiotic- dependent induction of the human CYP3A4 gene in vitro. Xenobiotica 29:269–279

Ourlin JC, Lasserre F, Pineau T, Fabre JM, Sa-Cunha A, Maurel P, Vilarem MJ, Pascussi JM (2003) The Small Heterodimer Partner Interacts with the Pregnane X Receptor and Represses Its Transcriptional Activity. Mol Endocrinol 17:1693–1703

Panesar N, Tolman K, Mazuski JE (1999) Endotoxin stimulates hepatocyte interleukin-6 production. J Surg Res 85:251–258

Paolini M, Pozzetti L, Piazza F, Guerra MC, Speroni E, Cantelli-Forti G, Roda A (2000) Mechanism for the prevention of cholestasis involving cytochrome P4503A overexpression. J Invest Med 48:49–59

Parks DJ, Blanchard SG, Bledsoe RK, Chandra G, Consler TG, Kliewer SA, Stimmel JB, Willson TM, Zavacki AM, Moore DD, Lehmann JM (1999) Bile acids: natural ligands for an orphan nuclear receptor. Science 284:1365–1368

Pascussi JM, Busson-Le Coniat M, Maurel P, Vilarem MJ (2003) Transcriptional analysis of the orphan nuclear receptor constitutive androstane receptor (NR1I3) gene promoter: identification of a distal glucocorticoid response element. Mol Endocrinol 17:42–55

Pascussi JM, Drocourt L, Fabre JM, Maurel P, Vilarem MJ (2000a) Dexamethasone induces pregnane X receptor and retinoid X receptor-alpha expression in human hepatocytes: synergistic increase of CYP3A4 induction by pregnane X receptor activators. Mol Pharmacol 58:361–372

Pascussi JM, Drocourt L, Gerbal-Chaloin S, Fabre JM, Maurel P, Vilarem MJ (2001) Dual effect of dexamethasone on CYP3A4 gene expression in human hepatocytes. Sequential role of glucocorticoid receptor and pregnane X receptor. Eur J Biochem 268: 6346–6358

Pascussi JM, Gerbal CS, Pichard GL, Daujat M, Fabre JM, Maurel P, Vilarem MJ (2000b) Interleukin-6 negatively regulates the expression of pregnane X receptor and constitutively activated receptor in primary human hepatocytes. Biochem Biophys Res Commun 274:707–713

Pascussi JM, Gerbal-Chaloin S, Fabre JM, Maurel P, Vilarem MJ (2000c) Dexamethasone enhances constitutive androstane receptor expression in human hepatocytes: consequences on cytochrome P450 gene regulation [in process citation]. Mol Pharmacol 58:1441–1450

Pichard L, Fabre I, Daujat M, Domergue J, Joyeux H, Maurel P (1992) Effect of corticosteroids on the expression of cytochromes P450 and on cyclosporin A oxidase activity in primary cultures of human hepatocytes. Mol Pharmacol 41:1047–1055

Pichard L, Fabre I, Daujat M, Domergue J, Joyeux H, Maurel P (1995) Effect of corticosteroids on the expression of cytochromes P450 and on cyclosporin A oxidase activity in primary cultures of human hepatocytes. Mol Pharmacol 47:410–418

Pratt WB (1993) The role of heat shock proteins in regulating the function, folding, and trafficking of the glucocorticoid receptor. J Biol Chem 268:21455–21458

Pratt WB, Silverstein AM, Galigniana MD (1999) A model for the cytoplasmic trafficking of signalling proteins involving the hsp90-binding immunophilins and p50cdc37. Cell Signal 11:839–851

Ray A, Prefontaine KE (1994) Physical association and functional antagonism between the p65 subunit of transcription factor NF-kappa B and the glucocorticoid receptor. Proc Natl Acad Sci USA 91:752–756

Roy-Chowdhury J, Locker J, Roy-Chowdhury N (2003) Nuclear receptors orchestrate detoxification pathways. Dev Cell 4:607–608

Schmiedlin-Ren P, Thummel KE, Fisher JM, Paine MF, Watkins PB (2001) Induction of CYP3A4 by 1 alpha,25-dihydroxyvitamin D3 is human cell line-specific and is unlikely to involve pregnane X receptor. Drug Metab Dispos 29:1446–1453

Schuetz EG, Schuetz JD, Strom SC, Thompson MT, Fisher RA, Molowa DT, Li D, Guzelian PS (1993a) Regulation of human liver cytochromes P-450 in family 3A in primary and continuous culture of human hepatocytes. Hepatology 18:1254–1262

Schuetz EG, Strom S, Yasuda K, Lecureur V, Assem M, Brimer C, Lamba J, Kim RB, Ramachandran V, Komoroski BJ, Venkataramanan R, Cai H, Sinal CJ, Gonzalez FJ, Schuetz JD (2001) Disrupted bile acid homeostasis reveals an unexpected interaction among nuclear hormone receptors, transporters, and cytochrome P450. J Biol Chem 276:39411–39418

Schuetz JD, Kauma S, Guzelian PS (1993b) Identification of the fetal liver cytochrome CYP3A7 in human endometrium and placenta. J Clin Invest 92:1018-1024

Schule R, Rangarajan P, Kliewer S, Ransone LJ, Bolado J, Yang N, Verma IM, Evans RM (1990) Functional antagonism between oncoprotein c-Jun and the glucocorticoid receptor. Cell 62:1217-1226

Seol W, Choi HS, Moore DD (1997) An orphan nuclear hormone receptor that lacks a DNA binding domain and heterodimerizes with other receptors. J Mol Biol 270:26-35

Seol W, Chung M, Moore DD (1998) Novel receptor interaction and repression domains in the orphan receptor SHP. J Biol Chem 273:14398-14402

Seol W, Mahon MJ, Lee YK, Moore DD (1996) Two receptor interacting domains in the nuclear hormone receptor corepressor RIP13/N-CoR. FEBS Lett 398:151-154

Sever JL (1997) Major technological advances affecting clinical and diagnostic immunology. Clin Diagn Lab Immunol 4:1-3

Sheppard KA, Phelps KM, Williams AJ, Thanos D, Glass CK, Rosenfeld MG, Gerritsen ME, Collins T (1998) Nuclear integration of glucocorticoid receptor and nuclear factor- kappaB signaling by CREB-binding protein and steroid receptor coactivator-1. J Biol Chem 273:29291-29294

Shimada K, Moriyama H, Ikeda M, Tomita H, Shigihara S, Gasser RF (1994) Peripheral communication of the facial nerve at the angle of the mouth. Eur Arch Otorhinolaryngol:S110-S112

Sidhu JS, Farin FM, Omiecinski CJ (1993) Influence of extracellular matrix overlay on phenobarbital-mediated induction of CYP2B1, 2B2, and 3A1 genes in primary adult rat hepatocyte culture. Arch Biochem Biophys 301:103-113

Smirlis D, Muangmoonchai R, Edwards M, Phillips IR, Shephard EA (2001) Orphan receptor promiscuity in the induction of cytochromes p450 by xenobiotics. J Biol Chem 276:12822-12826

Staudinger JL, Goodwin B, Jones SA, Hawkins-Brown D, MacKenzie KI, LaTour A, Liu Y, Klaassen CD, Brown KK, Reinhard J, Willson TM, Koller BH, Kliewer SA (2001) The nuclear receptor PXR is a lithocholic acid sensor that protects against liver toxicity. Proc Natl Acad Sci USA 98:3369-3374

Sueyoshi T, Kawamoto T, Zelko I, Honkakoski P, Negishi M (1999) The repressed nuclear receptor CAR responds to phenobarbital in activating the human CYP2B6 gene. J Biol Chem 274:6043-6046

Sugatani J, Kojima H, Ueda A, Kakizaki S, Yoshinari K, Gong QH, Owens IS, Negishi M, Sueyoshi T (2001) The phenobarbital response enhancer module in the human bilirubin UDP- glucuronosyltransferase UGT1A1 gene and regulation by the nuclear receptor CAR. Hepatology 33:1232-1238

Takahashi F, Finch JL, Denda M, Dusso AS, Brown AJ, Slatopolsky E (1997) A new analog of 1,25-(OH)2D3, 19-NOR-1,25-(OH)2D2, suppresses serum PTH and parathyroid gland growth in uremic rats without elevation of intestinal vitamin D receptor content. Am J Kidney Dis 30:105-112

Thummel KE, Brimer C, Yasuda K, Thottassery J, Senn T, Lin Y, Ishizuka H, Kharasch E, Schuetz J, Schuetz E (2001) Transcriptional control of intestinal cytochrome P-4503A by 1alpha,25- dihydroxy vitamin D3. Mol Pharmacol 60:1399-1406

Tinel M, Elkahwaji J, Robin MA, Fardel N, Descatoire V, Haouzi D, Berson A, Pessayre D (1999) Interleukin-2 overexpresses c-myc and down-regulates cytochrome P-450 in rat hepatocytes. J Pharmacol Exp Ther 289:649-655

Trauner M, Fickert P, Stauber RE (1999) Inflammation-induced cholestasis. J Gastroenterol Hepatol 14:946-959

Trottier E, Belzil A, Stoltz C, Anderson A (1995) Localization of a phenobarbital-responsive element (PBRE) in the 5'-flanking region of the rat CYP2B2 gene. Gene 158:263–268

Tsai MJ, O'Malley BW (1994) Molecular mechanisms of action of steroid/thyroid receptor superfamily members. Annu Rev Biochem 63:451–486

Tyndale RF, Pianezza ML, Sellers EM (1999) A common genetic defect in nicotine metabolism decreases risk for dependence and lowers cigarette consumption. Nicotine Tob Res 1 Suppl 2: S63–S67; discussion S69–S70

Tzameli I, Pissios P, Schuetz EG, Moore DD (2000) The xenobiotic compound 1,4-bis[2-(3,5-dichloropyridyloxy)]benzene is an agonist ligand for the nuclear receptor CAR. Mol cell Biol 20:2951–2958

Ueda A, Hamadeh HK, Webb HK, Yamamoto Y, Sueyoshi T, Afshari CA, Lehmann JM, Negishi\M (2002) Diverse roles of the nuclear orphan receptor CAR in regulating hepatic genes in response to phenobarbital. Mol Pharmacol 61:1–6

Wan YJ, Cai Y, Lungo W, Fu P, Locker J, French S, Sucov HM (2000) Peroxisome proliferator-activated receptor alpha-mediated pathways are altered in hepatocyte-specific retinoid X receptor alpha-deficient mice. J Biol Chem 275:28285–28290

Watkins RE, Wisely GB, Moore LB, Collins JL, Lambert MH, Williams SP, Willson TM, Kliewer SA, Redinbo MR (2001) The human nuclear xenobiotic receptor PXR: structural determinants of directed promiscuity. Science 292:2329–2333

Wei P, Zhang J, Egan-Hafley M, Liang S, Moore DD (2000) The nuclear receptor CAR mediates specific xenobiotic induction of drug metabolism. Nature 407:920–923

Wrighton SA, Schuetz EG, Watkins PB, Maurel P, Barwick J, Bailey BS, Hartle HT, Young B, Guzelian P (1985) Demonstration in multiple species of inducible hepatic cytochromes P- 450 and their mRNAs related to the glucocorticoid-inducible cytochrome P-450 of the rat. Mol Pharmacol 28:312–321

Xie W, Barwick JL, Simon CM, Pierce AM, Safe S, Blumberg B, Guzelian PS, Evans RM (2000) Reciprocal activation of xenobiotic response genes by nuclear receptors SXR/PXR and CAR [in process citation]. Genes Dev 14:3014–3023

Yaffe SJ, Levy G, Matsuzawa T, Baliah T (1966) Enhancement of glucuronide-conjugating capacity in a hyperbilirubinemic infant due to apparent enzyme induction by phenobarbital. N Engl J Med 275:1461–1466

Zehnder D, Hewison M (1999) The renal function of 25-hydroxyvitamin D3-1alpha-hydroxylase. Mol Cell Endocrinol 151:213–220

Zelko I, Sueyoshi T, Kawamoto T, Moore R, Negishi M (2001) The peptide near the C terminus regulates receptor CAR nuclear translocation induced by xenochemicals in mouse liver. Mol Cell Biol 21:2838–2846

Part III
Exogenous Control of Transcription Factor Activity

Part III

Chapter 9

Potential of Transcription Factor Decoy Oligonucleotides as Therapeutic Approach

R. Morishita[1] (✉) · N. Tomita[2] · Y. Kaneda[1] · T. Ogihara[2]

[1] Division of Gene Therapy, Osaka University Medical School, 2-2 Yamada-oka, Suita, 565 Osaka, Japan
morishit@gts.med.osaka-u.ac.jp
[2] Department of Geriatric Medicine, Osaka University Medical School, Osaka, Japan

1	Introduction	440
2	Principles	441
3	Application	443
3.1	Gene Therapy Using E2F Decoy ODN	443
3.2	Gene Therapy Using NFκB Decoy ODN	445
4	Unresolved Issues in ODN-Based Gene Therapy	448
References		450

Abstract Molecular therapy is emerging as a potential strategy for the treatment of cardiovascular diseases such as restenosis after angioplasty, vascular bypass graft occlusion, and transplant coronary vasculopathy, for which no effective therapy is known. One strategy for combating disease processes has been to target the transcriptional process. Two approaches have been used to accomplish this. One is the use of antisense RNA that is complimentary to the mRNA of interest. The second approach is the use of ribozymes, a unique class of RNA molecules that not only store information but also possess catalytic activity. Ribozymes are known to catalytically cleave specific target RNA, leading to their degradation, whereas antisense molecules inhibit translation by binding to mRNA sequences on a stoichiometric basis. In theory, ribozymes would be the more effective means of inhibiting expression of a target gene. The application of DNA technology such as the antisense strategy to regulate the transcription of disease-related genes in vivo has important therapeutic potential. Transfection of *cis*-element double-stranded (ds) oligodeoxynucleotides (ODN) (=decoy) has been reported as a powerful tool in a new class of anti-gene strategies for gene therapy. Transfection of decoy ODN will result in attenuation of authentic *cis–trans* interaction, leading to the removal of *trans*-factors from the endogenous *cis*-elements, with subsequent modulation of gene expression. This decoy ODN strategy is not only a novel strategy for gene therapy as an anti-gene strategy, but is also a powerful tool for the study of endogenous gene regulation in vivo as well as in vitro. In this review, we focus on the future potential of decoy ODN-based therapy.

Keywords Decoy · Atopic dermatitis · Arthritis · Gene therapy · Restenosis

1
Introduction

As a consequence of the revolutionary developments in the field of molecular biology and their impact on our understanding of the mechanisms of disease processes, treatment strategies that exploit our expanding knowledge of the structures and functions of molecules are being pursued. Recent progress in molecular biology has provided new techniques to inhibit target gene expression. In particular, the application of DNA technology such as the antisense strategy to regulate the transcription of disease-related genes in vivo has important therapeutic potential. Antisense oligodeoxynucleotides (ODN) are widely used as inhibitors of specific gene expression, because they offer the exciting possibility of blocking the expression of a particular gene without any changes in functions of other genes (see Fig. 1) (Stein and Cohen 1988). Therefore, antisense ODN are useful tools in the study of gene function and may be potential therapeutic agents. Indeed, the first approved drug for ODN-based therapy was an antisense drug to treat cytomegalovirus retinopathy in 1999. The second approach is the use of ribozymes, a unique class of RNA molecules that not only store information but also process catalytic activity (Kiehntopf et al. 1995). Ribozymes are

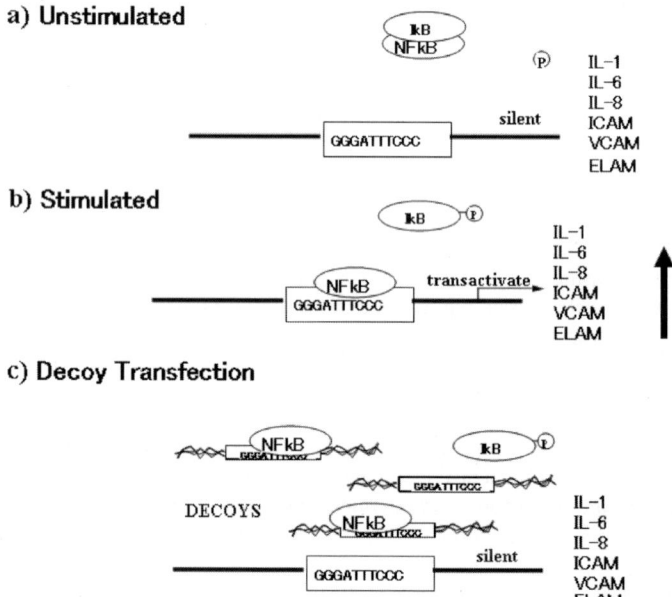

Fig. 1a–c Scheme of 'decoy' strategy. **a** In the basal state, TF is bound to *cis*-element, resulting in continuous activation of target gene expression. **b** TF 'decoy' *cis*-element dsODN bind to TF, resulting in prevention of TF interaction and transactivation of TF-promoting target gene expression. *TF*, transcription factor; *cis-E*, *cis*-element

known to catalytically cleave specific target RNA, leading to their degradation, whereas antisense molecules inhibit translation by binding to mRNA sequences on a stoichiometric basis. Theoretically, ribozymes are the more effective means of inhibiting target gene expression. On the other hand, more recently, we have developed a novel molecular strategy in which synthetic double-stranded (ds) DNA with high affinity for a target transcription factor may be introduced into target cells as a 'decoy' cis-element to bind the transcriptional factor and alter gene transcription (Morishita et al. 1998). Transfection of cis-element dsODN(=decoy) has been reported as a powerful tool in a new class of anti-gene strategies for gene therapy and the study of transcriptional regulation. Transfection of ds ODN corresponding to a cis-sequence will result in the attenuation of authentic cis–trans interaction, leading to the removal of trans-factors from the endogenous cis-element with subsequent modulation of gene expression (Fig. 1). Therefore, the decoy approach may enable us to treat diseases by modulation of endogenous transcriptional regulation. In this chapter, we review (a) the mechanisms and (b) the potential applications of the decoy strategy.

2
Principles

Correct regulation of gene expression is essential both to normal development and to the correct functioning of the adult organism. Such regulation is usually achieved at the level of DNA transcription, a process that controls which genes are transcribed into RNA by the enzyme RNA polymerase, although post-transcriptional regulation is also important. The transcription of specific genes is controlled by regulatory proteins known as transcription factors. Transcription factors have been grouped in families on the basis of shared DNA-binding motifs. Other regions of the factors interact with RNA polymerase and its associated proteins to increase or decrease the rate of transcription. The vital role of these factors, together with the fact that a single factor can affect the expression of many genes, suggests that the inactivation of a transcription factor as a result of an inherited mutation is incompatible with survival. Initially, overexpression of TAR-containing sequences (TAR decoys) in a double copy murine retroviral vector was used to render cells resistant to HIV replication (Sullenger et al. 1990). Currently, TAR decoys, short RNA oligonucleotides corresponding to the HIV TAR sequence, are used to inhibit HIV expression and replication by blocking the binding of the HIV regulatory protein Tat to the authentic TAR region (Bahner et al. 1996; Lee et al. 1995; Sullenger et al. 1990). However, such RNA decoys are very difficult to use in vivo. In addition, the regulation of decoy expression is also problematic. To overcome these issues, we hypothesized that synthetic ds DNA with high affinity for transcription factors may be introduced in vivo as a 'decoy' cis-element to bind transcription factors and block the acti-

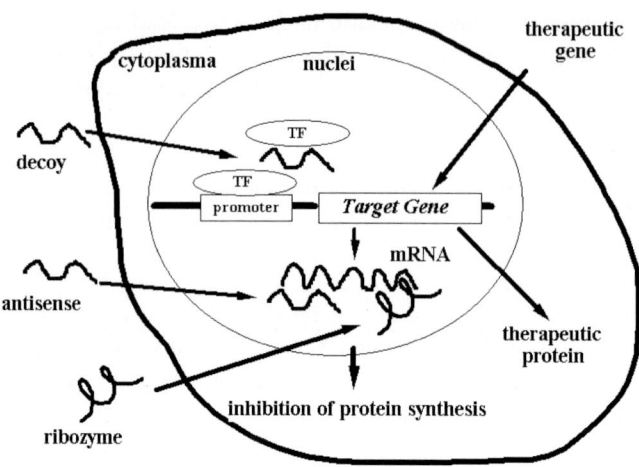

Fig. 2 Target sites for antisense, ribozyme and decoy strategies. *antisense*, Antisense ODN; *ribozyme*, ribozyme ON; *decoy*, decoy ODN; *TF*, transcription factor

vation of genes mediating certain diseases, thus providing effective therapy: for treating disease transfection of ds ODN corresponding to the *cis*-sequence will result in attenuation of the removal of the *trans*-factors from the endogenous *cis*-element with subsequent modulation of gene expression. This approach is particularly attractive for several reasons: (a) the potential drug targets (transcription factors) are plentiful and readily identifiable: (b) the synthesis of the sequence-specific decoy is relatively simple and can be targeted to specific tissues; (c) knowledge of the exact molecular structure of the target transcription factor is unnecessary; and (d) decoy ODN may be more effective than antisense ODN in blocking constitutively expressed factors as well as multiple transcription factors that bind to the same *cis*-element. Although the mechanisms of actions of antisense ODN are still unclear, the principle of the transcription factor decoy approach is simply the reduction of promoter activity due to the inhibition of binding of a transcription factor to a specific sequence in the promoter region (Fig. 2). Alternatively, this strategy also provides a powerful tool to study endogenous gene regulation in vivo as well as in vitro by modulation of endogenous transcriptional regulation.

3
Application

3.1
Gene Therapy Using E2F Decoy ODN

Potential targets for decoy strategy as gene therapy are summarized in Table 1. One important disease potentially amenable to gene therapy based on decoy strategy is restenosis after angioplasty, since the long-term effectiveness of this procedure is limited by the development of restenosis in more than 40% of patients (Gibbons and Dzau 1994). Intimal hyperplasia after angioplasty develops in large part as a result of vascular smooth muscle cell (VSMC) proliferation and migration induced by a complex interaction of multiple growth factors that are activated by vascular injury (Gibbons and Dzau 1994). The process of VSMC proliferation is dependent on the coordinated activation of a series of cell cycle regulatory genes which results in mitosis. A critical element in regulation of cell cycle progression is the complex formed by E2F, cyclin A and cyclin dependent kinase (CDK)2. The dissociation of the transcription factor E2F from retinoblastoma (Rb) gene product is proposed to play a pivotal role in the regulation of cell proliferation by inducing a coordinated transactivation of genes involved in cell cycle regulation including proto-oncogenes such as *c-myc*, *c-myb*, *cdc2*, proliferating cell nuclear antigen (*PCNA*) and thymidine kinase. Accordingly, inhibition of *c-myc* by antisense ODN was reported to inhibit neointimal formation in several animal models (Shi et al. 1994). Recently, the results from a phase II trial using antisense *c-myc* to treat restenosis have been reported (Kutryk et al. 2002). However, treatment with 10 mg of phosphorothioate-modified ODN directed against *c-myc* does not reduce neointimal volume obstruction or the angiographic restenosis rate (Kutryk et al. 2002). As this trial utilized intra-coronary infusion of antisense c-myc ODN without any vectors, sever-

Table 1 Potential targets for gene therapy in cardiovascular disease using decoy strategy (adapted from Ahn et al. 2002b)

Gene therapy	Target transcription factor
1. Restenosis after angioplasty	E2F
	NFκB
	AP-1
2. Hypertension	AGE (angiotensinogen)
3. Transplant vasculopathy	E2F
	NFκB
4. Myocardial reperfusion injury	NFκB
5. Inflammation	NFκB
6. Graft failure	E2F
	NFκB

al issues such as low transfection efficiency may limit the efficacy of this strategy. In addition, this result may also explain the importance of the inhibition of multiple points of the cell cycle. Indeed, our previous data revealed that a single administration of antisense ODN against PCNA and cdc2 kinase genes inhibited neointimal formation after angioplasty for at least 8 weeks after transfection (Morishita et al. 1993), while inhibition of a single gene partially inhibited neointimal formation (Morishita et al. 1994a, 1994b).

The antiproliferative effects of the *Rb* gene product appear to depend on its capacity to bind to E2F and thereby prevent this transcription factor from binding to the E2F *cis*-element within the promoters of these essential cell cycle regulatory genes. Accordingly, we hypothesized that transfection of VSMC with a sufficient quantity of the decoy ODN containing the E2F *cis*-element (consensus sequence TTTTCGGCGC) would effectively bind E2F, prevent it from transactivating the gene expression of essential cell cycle regulatory proteins and thereby inhibit VSMC proliferation and neointimal formation (Morishita et al. 1995). Synthesized 14-mer ds ODN containing the consensus sequence effectively competed with binding of E2F to its binding site as assessed by gel mobility shift assay (Morishita et al. 1995). Transfection of E2F decoy ODN into rat balloon-injured carotid arteries using the HVJ-liposome method resulted in almost complete inhibition of neointimal formation at 2 weeks after balloon injury, accompanied by a reduction in mRNA of PCNA and cdc2 kinase, but not β-actin, whereas mismatched ODN had no effect on neointimal hyperplasia (Morishita et al. 1995). Of importance, sustained inhibition of neointimal formation by a single administration of E2F decoy ODN was observed at least up to 8 weeks after treatment. This is the first successful in vivo transfer of a decoy *cis*-element to bind E2F, modulate gene expression, and consequently inhibit smooth muscle proliferation and vascular lesion formation as gene therapy for restenosis. Inhibition of neointimal formation by E2F decoy with hydrogel-coated catheter delivery was also demonstrated using a porcine coronary artery model (Nakamura et al. 2002). Modification of the delivery of ODN by a hydrogel catheter may overcome issues such as the low transfection efficiency observed with coronary infusion. Based on these results, we started a clinical trial using hydrogel catheter delivery of E2F decoy to treat restenosis after angioplasty in April 2000. As of October 2002, we have treated five patients with E2F decoy ODN. We did not observe any side effects up to 6 months post-treatment, although the clinical outcome has not yet been evaluated.

In addition, in 1996, the clinical trial of 'decoy' against E2F by Dr. Dzau at Harvard University was approved by the Food and Drugs Administration (FDA) to treat neointimal hyperplasia in vein bypass grafts, which results in failure in up to 50% of grafts within a period of 10 years. They demonstrated successful inhibition of graft occlusion, accompanied by selective inhibition of PCNA and c-*myc* expression (Mann et al. 1999). More recently, similar results were obtained in a double-blind placebo-controlled study to treat cardiac vein graft failure. At 12 months, fewer graft occlusions, revisions, or critical stenoses were seen in the E2F-decoy group than in the untreated group (hazard ratio 0.34; 95% confidence interval, 0.12–0.99) (Mann et al. 1999). They concluded that intraoperative transfection of human bypass vein grafts with E2F-decoy oligodeoxynucleotide is safe, feasible, and can achieve sequence-specific inhibition of cell cycle gene expression and DNA replication. Currently, phase III trials are underway in the USA and in European countries.

Similarly, the potential of the transcription factor decoy approach to treat renal diseases such as glomerulonephritis has been assessed (Maeshima et al. 1998). Although numerous growth factors including platelet-derived growth factor and angiotensin II regulate this process, the proliferation of mesangial cells is also regulated by cell cycle regulatory genes. As discussed above, the transcription factor E2F has been reported to play a pivotal role in the regulation of cell cycle regulatory genes. Indeed, intrarenal arterial perfusion of E2F decoy ODN inhibited the mesangial cell proliferation induced by anti-Thy 1 antibody which specifically injures glomerular mesangial cells, resulting in a proliferative glomerular lesion (Maeshima et al. 1998). As E2F has been postulated to play an important role in the pathogenesis of numerous diseases, for example cancer and arthritis, the development of the E2F decoy strategy may provide a useful therapeutic tool for treating these proliferative diseases. For the treatment of systemic diseases, tissue-specific inhibition of E2F activity might be important, because replication of cells would be necessary as 'wound healing' in certain physiological conditions. Thus, modifications of the oligonucleotide composition that can prolong decoy stability in vivo and tissue-specific delivery will be critical to enhancing potential therapeutic efficacy.

3.2
Gene Therapy Using NFκB Decoy ODN

The transcription factor nuclear factor κB (NFκB) plays a pivotal role in the coordinated transactivation of cytokine and adhesion molecule genes, the activation of which has been postulated to be involved in numerous diseases, such as myocardial infarction and glomerulonephritis. These diseases are, importantly, potentially amenable to ODN-based gene therapy because at present there are no effective pharmacological agents. The pathophysiology of myocardial infarction and glomerulonephritis is quite complicated. Numerous cytokines including interleukin (IL)-1, 2, 6, 8 and tumor necrosis factor (TNF)-α, to name a few, regulate this process. However, gene regulation of many cytokines is relatively simple, because the transcription factor NFκB has been reported to upregulate these cytokines. Interestingly, adhesion molecules such as vascular cellular adhesion molecule (VCAM) and intercellular adhesion molecule (ICAM) are also known to be upregulated by

NFκB. Accordingly, we hypothesize that myocardial infarction and glomerulonephritis could be prevented by the blockade of genes regulating cell inflammation—the final common pathway that is induced by NFκB binding. The necessity of blocking cytokine and adhesion molecule genes at more than one point to achieve maximum inhibitory effects may be due to the redundancy and complexity of the interactions of these genes.

Myocardial reperfusion injury develops to a large degree as a result of severe damage of myocytes and endothelial cells, probably induced by the complex interaction of multiple cytokines and adhesion molecules that are activated by reperfusion. The process of ischemic reperfusion may be dependent on the coordinated activation of a series of cytokine and adhesion molecule genes that results in the attachment of leukocytes and release of cytotoxic molecules. Importantly, increased NFκB binding activity was confirmed in hearts with myocardial infarction (Morishita et al. 1997). Our previous study provided the first evidence of the feasibility of decoy strategy against NFκB in treating myocardial reperfusion injury (Morishita et al. 1997; Sawa et al. 1997). Transfection of NFκB decoy ODN into rat coronary artery prior to left ascending artery occlusion markedly reduced the damaged area of myocytes at 24 h after reperfusion, whereas no difference was observed between scrambled decoy ODN-treated and untransfected rats. The therapeutic efficacy of this strategy via intra-coronary administration immediately after reperfusion, similar to the clinical situation, was also examined. NFκB decoy ODN reduced the damage of myocytes due to reperfusion in contrast to rats treated with scrambled control decoy or vehicle. The selectivity of the NFκB decoy ODN effect was confirmed further by the demonstration that reduction of the damaged myocardial area was not observed in rats treated with antisense ODN directed against the rat inducible nitric oxide synthase gene. The specificity of the NFκB decoy in the inhibition of cytokine and adhesion molecule expression was also confirmed by in vitro experiments using human and rat coronary artery endothelial cells. Transfection of NFκB decoy ODN markedly inhibited the protein expression of cytokines (IL-6 and 8) and adhesion molecules (VCAM, ICAM and endothelial adhesion molecule) in response to TNF-α stimulation in human aortic endothelial cells. In contrast, the control scrambled decoy ODN failed to inhibit the induction of these protein expressions. Cell numbers after transfection were not changed, indicating that the NFκB decoy induces a specific inhibitory effect rather than nonspecific cytotoxicity. Treatment of glomerulonephritis by means of NFκB decoy ODN is also reported (Tomita et al. 2000a, 2000b, 2001; Azuma et al. 2003).

Importantly, increased NFκB binding activity has been confirmed in balloon-injured blood vessels (Yoshimura et al. 2001). Our recent study provided the first evidence of the feasibility of a decoy strategy against NFκB in treating restenosis (Yoshimura et al. 2001; Yamasaki et al. 2003). Transfection of NFκB decoy ODN into balloon-injured carotid artery or porcine coronary artery markedly reduced neointimal formation, whereas no difference was observed between scrambled decoy ODN-treated and untransfected blood vessels (Yoshimura et al. 2001; Yamasaki et al. 2003). Based upon the therapeutic efficacy of this strategy, we obtained permission for a second clinical trial using the decoy strategy to treat restenosis from 2001. In addition, the inhibition of VSMC replication was confirmed by the observation that transfection of NFκB decoy ODN inhibited the progression of vasculopathy in cardiac transplantation models (Suzuki et al. 2000; Yokoseki et al. 2001).

As NFκB has been postulated to play an important role in the pathogenesis of numerous diseases, for example cancer and arthritis, the development of the NFκB decoy strategy may provide a useful therapeutic tool for treating these diseases (Tomita et al. 1999, 2000; Kawamura et al. 2001). In particular we have focused on the treatment of atopic dermatitis—a chronic inflammatory disease which results from complex interactions between genetic and environmental mechanisms. Atopic dermatitis afflicts 10%–15% of children and adolescents in the Western world. Keratinocytes of patients with atopic dermatitis exhibit a propensity to exaggerated production of cytokines and chemokines, a phenomenon that can have a major role in promoting and maintaining inflammation. Especially, NF-κB is the center of interest, as it promotes the transcription of Th2 cytokines such as IL-6 as well as adhesion molecules such as ICAM-1. In addition, the delayed allergic component of atopic dermatitis, dependent on the Th1 dominant pathway, is also believed to be regulated by NF-κB-induced cytokines such as IFN-γ. Therefore, in this study, we focused on the role of NF-κB in the pathogenesis of atopic dermatitis. Interestingly, topical administration of NF-κB decoy ODN twice a month resulted in a significant reduction in clinical skin condition score and marked improvement of histological findings: there was a significant decrease in migration of mast cells into the dermis and an increase in apoptotic cells (Nakamura et al. 2002). What is the clinical relevance of NFκB decoy ODN? Currently, the pharmaceutical drugs that can be used to treat atopic dermatitis are limited: despite the rapid and proven efficacy of topical corticosteroids, their side effects limit their clinical usefulness. Topically active macrolide immunosuppressants such as ascomycin and tacrolimus appear to provide comparable therapeutic potency. Although the data from ongoing studies will be crucial in determining the safety of these agents in the long term, and also their place within the current therapeutic armamentarium available for patients with atopic dermatitis, these agents still have potential side effects such as systemic absorption. In contrast, NF-κB decoy ODN exerts only local pharmacological effects for the following reasons: (a) ODN in serum were rapidly destroyed (within several hours), leading to a lack of systemic effects (Roque et al. 2001); (b) the toxicity of systemic administration of ODN was reported only at more than 100-fold higher concentration than that used in the study. From December

2001, we started human clinical trial using NFκB decoy ODN to treat atopic dermatitis at Hirosaki University as Phase I/IIa trial. Topical application of NFκB decoy ODN to the faces of atopic dermatitis patients resulted in marked therapeutic effects (data not shown). Because side effects can limit the clinical usefulness of corticosteroids, topically applied NF-κB decoy ODN appears to provide novel therapeutic potency without significant local or adverse effects.

4
Unresolved Issues in ODN-Based Gene Therapy

ODN-based gene therapy still has many unsolved problems such as the short half-life, low efficiency of uptake, and degradation by endocytosis and nucleases. Therefore, currently many groups are focusing on modifications of the gel approach using a catheter delivery system. Further modification of ODN pharmacokinetics will facilitate the potential clinical utility of the agents by: (a) allowing a shorter intraluminal incubation time to preserve organ perfusion' (b) prolonging the duration of biological action; and (c) enhancing efficacy such that the nonspecific effects of high doses of ODN can be avoided. Although direct transfer of 'naked' decoy ODN can be achieved via passive uptake, the transfection efficiency seems to be lower than that with single-stranded antisense ODN. To enhance the transfection efficiency of decoy ODN, the cationic liposome, HVJ (Sendai virus)-liposome method or other vector systems are generally used (Kaneda et al. 2002). The majority of ODN is sequestered and degraded in lysosomes and never reaches the nucleus. As the site of decoy effects is apparently in the nucleus, bypassing the endocytotic pathway and translocation of decoy ODN from the cytoplasm are extremely important in the practical application of therapeutics.

Regarding the ODN-based strategy as gene therapy, one of the major concerns is nonspecific effects, particularly those of phosphorothioate-substituted ODN. This concern is related not only to the antisense and decoy strategies, but to all ODN-mediated therapy. Nonsequence-specific inhibition may operate through blockade of cell surface receptor activity or interference with other proteins (Gibson 1996). At the same time, ODN containing GC dinucleotides may bring about immune activation (Khaled et al. 1996). In addition, sequence-specific binding of nontranscriptional factor proteins to ODN has been reported to result in nonspecific effects of ODN-based gene therapy (Henry et al. 1997a). Moreover, Burgess et al. reported that the antiproliferative activity of c-*myb* and c-*myc* antisense ODN in VSMC is caused by a nonantisense mechanism (Burgess et al. 1995). To overcome these issues, careful controlled experiments must be performed to eliminate the potential nonspecific effects of ODN-mediated therapy. For gene therapy

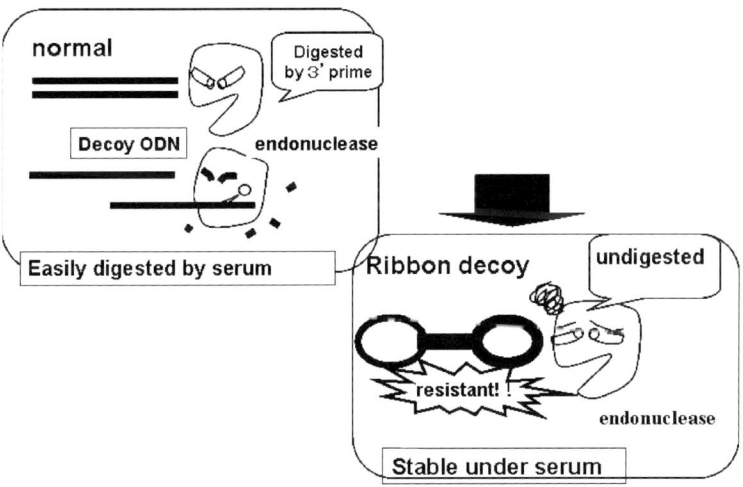

Fig. 3 Development of the ribbon type of decoy ODN that has potent resistance to endonuclease

using an ODN-based strategy, the toxicity of phosphorothioate ODN may also be important. Although low dosage administration does not seem to cause any toxicity, bolus infusions may be dangerous. Higher doses over prolonged periods of time may cause kidney damage, as evidenced by proteinuria and leukocytes in the urine in animals (Henry et al. 1997a). Liver enzymes may also be increased in animals treated with moderate to high doses. Several phosphorothioate ODN have been shown to cause acute hypotensive events in monkeys (Cornish et al. 1993; Srinivasan and Iversen 1995), probably due to complement activation (Henry et al. 1997b). These effects are transient, if managed appropriately, and relatively uncommon. This toxicity can be avoided by giving intravenous infusions rather than bolus injections. More recently, prolongation of prothrombin, partial thromboplastin, and bleeding times has been reported in monkeys (Crooke 1995). Alternatively, we recently developed a new modification of decoy ODNs in order to increase their stability against nucleases. Although the chemical modifications of ODN such as phosphorothioation and methylphosphonation were used, problems with these modified ODNs became apparent including insensitivity to RNaseH, lack of sequence specificity, and immune activation, as described above. To overcome these limitations, covalently modified ODN were developed by enzymatically ligating two identical molecules, thereby preventing their degradation by exonucleases (Fig. 3) (Ahn et al. 2002b). In fact, these modified decoy (ribbon) ODN possess increased nuclease resistance and are transported more efficiently into cells than chemically modified linear ODN (Abe et al. 1998; Chu BCF and Orgal 1992). More recently, we reported that the transfection of novel AP-1 decoy ODN with circular ribbon structure, prior to the balloon injury procedure, prevented neointimal formation in the rat balloon injured artery more effectively than nonmodified decoy ODN (Ahn et al. 2002a).

Molecular therapy based on decoy ODN would be useful for the treatment of many diseases, including the prevention of restenosis after angioplasty, myocardial infarction and rejection in heart transplantation. The first federally approved antisense drug was launched in 1999 for retinopathy patients. Although there are still many unresolved issues, decoy ODN drugs may become a reality.

References

Abe T, Takai K, Nakada S, Yokota T, Takaku H (1998) Specific inhibition of influenza virus RNA polymerase and nucleoprotein gene expression by circular dumbbell RNA/DNA chimeric oligonucleotides containing antisense phosphodiester oligonucleotides. FEBS Lett 425:91–96

Ahn JD, Morishita R, Kaneda Y, Kim HS, Chang YC, Lee KU, Park JY, Lee HW, Kim YH, Lee IK (2002a) 2F decoy oligodeoxynucleotides inhibit in vitro vascular smooth muscle cell proliferation and in vivo neointimal hyperplasia. Gene Ther 9:1682–1692

Ahn JD, Morishita R, Kaneda Y, Lee SJ, Kwon KY, Choi SY, Lee KU, Park JY, Moon IJ, Park JG, Yoshizumi M, Ouchi Y, Lee IK (2002b) Inhibitory effects of novel AP-1 decoy oligodeoxynucleotides on vascular smooth muscle cell proliferation in vitro and neointimal formation in vivo. Circ Res 90:1325–1332

Azuma H, Tomita N, Kaneda Y, Koike H, Ogihara T, Katsuoka Y, Morishita R (2003) Transfection of NFκB decoy oligodeoxynucleotides using efficient ultrasound-mediated gene transfer into donor kidneys prolonged survival of rat renal allografts. Gene Ther 10:415–425

Bahner I, Kearns K, Hao QL, Smogorzewska EM, Kohn DB (1996) Transduction of human CD34+ hematopoietic progenitor cells by a retroviral vector expressing an RRE decoy inhibits human immunodeficiency virus type 1 replication in myelomonocytic cells produced in long-term culture. J Virol 70:4352–4360

Burgess TL, Fisher EF, Ross SL, Bready JV, Qian YX, Bayewitch LA, Cohen AM, Herrera CJ, Hu SS, Kramer TB, Lott FD, Martin FH, Pierce GF, Simonet L, Farrell CL (1995) The antiproliferative activity of c-myb and c-myc antisense oligonucleotides in smooth muscle cells is caused by a nonantisense mechanism. Proc Natl Acad Sci USA 92:4051–4055

Chu BCF, Orgal L (1992) The stability of different forms of double-stranded decoy DNA in serum and nuclear extracts. Nucleic Acids Res 20:5857–5858

Cornish KG, Iversen P, Smith L, Arneson M, Bayever E (1993) Cardiovascular effects of a phosphorothioate oligonucleotide with sequence antisense to p53 in the conscious rhesus monkey. Pharmacol Commun 3:239

Crooke ST. Progress in antisense therapeutics (1995) Hematol Pathol 9:59–72

Gibbons GH, Dzau VJ (1994) The emerging concept of vascular remodeling. N Engl J Med 330:1431–1438

Gibson I (1996) Antisense approaches to the gene therapy of cancer – 'Recnac'. Cancer Metastasis Rev 15:287–299

Henry SP, Bolte H, Auletta C, Kornbrust DJ (1997a) Evaluation of the toxicity of ISIS 2302, a phosphorothioate oligonucleotide, in a four-week study in cynomolgus monkeys. Toxicology 120:145–155

Henry SP, Giclas PC, Leeds J, Pangburn M, Auletta C, Levin AA, Kornbrust DJ (1997b) Activation of the alternative pathway of complement by a phosphorothioate oligonucleotide: potential mechanism of action. J Pharmacol Exp Ther 281:810–816

Kaneda Y, Nakajima T, Nishikawa T, Yamamoto S, Ikegami H, Suzuki N, Nakamura H, Morishita R, Kotani T (2002) Hemagglutinating Virus of Japan (HVJ) envelope vector as a versatile gene delivery system. Mol Ther 6:219–226

Kawamura I, Morishita R, Tsujimoto S, Manda T, Tomoi M, Tomita N, Goto T, Ogihara T, Kaneda Y (2001) Intravenous injection of oligodeoxynucleotides to the NF-kappaB binding site inhibits hepatic metastasis of M5076 reticulosarcoma in mice. Gene Ther 8:905–912

Khaled Z, Benimetskaya L, Zeltser R, Khan T, Sharma HW, Narayanan R, Stein CA (1996) Multiple mechanisms may contribute to the cellular anti-adhesive effects of phosphorothioate oligodeoxynucleotides. Nucl Acids Res 24:737–745

Kiehntopf M, Esquivel EL, Brach MA, Herrmann F (1995) Clinical applications of ribozymes. Lancet 345:1027–1031

Kutryk MJ, Foley DP, van den Brand M, Hamburger JN, van der Giessen WJ, deFeyter PJ, Bruining N, Sabate M, Serruys PW (2002) Local intracoronary administration of antisense oligonucleotide against c-myc for the prevention of in-stent restenosis: results of the randomized investigation by the Thoraxcenter of antisense DNA using local delivery and IVUS after coronary stenting (ITALICS) trial. J Am Coll Cardiol 39:281–287

Lee SW, Gallardo HF, Gaspar O, Smith C, Gilboa E (1995) Inhibition of HIV-1 in CEM cells by a potent TAR decoy. Gene Ther 2:377–384

Maeshima Y, Kashihara N, Yasuda T, Sugiyama H, Sekikawa T, Okamoto K, Kanao K, Watanabe Y, Kanwar YS, Makino H (1998) Inhibition of mesangial cell proliferation by E2F decoy oligodeoxynucleotide in vitro and in vivo. J Clin Invest 101:2589–2597

Mann MJ, Whittemore AD, Donaldson MC, Belkin M, Conte MS, Polak JF, Orav EJ, Ehsan A, Dell'Acqua G, Dzau VJ (1999) Ex-vivo gene therapy of human vascular bypass grafts with E2F decoy: the PREVENT single-centre, randomised, controlled trial. Lancet 354:1493–1498

Morishita R, Gibbons GH, Ellison KE, Nakajima M, Zhang L, Kaneda Y, Ogihara T, Dzau VJ (1993) Single intraluminal delivery of antisense cdc 2 kinase and PCNA oligonucleotides results in chronic inhibition of neointimal hyperplasia. Proc Natl Acad Sci USA 90:8474–8479

Morishita R, Gibbons GH, Ellison KE, Nakajima M, Leyen HVL, Zhang L, Kaneda Y, Ogihara T, Dzau VJ (1994a) Intimal hyperplasia after vascular injury is inhibited by antisense cdk 2 kinase oligonucleotides. J Clin Invest 93:1458–1464

Morishita R, Gibbons GH, Kaneda Y, Ogihara T, Dzau VJ (1994b) Pharmacokinetics of antisense oligonucleotides (cyclin B1 and cdc 2 kinase) in the vessel wall: enhanced therapeutic utility for restenosis by HVJ-liposome method. Gene 149:13–19

Morishita R, Gibbons GH, Horiuchi M, Ellison KE, Nakajima M, Zhang L, Kaneda Y, Ogihara T, Dzau VJ (1995) A novel molecular strategy using cis element 'decoy' of E2F binding site inhibits smooth muscle proliferation in vivo. Proc Natl Acad Sci USA 92:5855–5859

Morishita R, Sugimoto T, Aoki M, Kida I, Tomita N, Moriguchi A, Maeda K, Sawa Y, Kaneda Y, Higaki J, Ogihara T (1997) In vivo transfection of cis element 'decoy' against NFκB binding site prevented myocardial infarction as gene therapy. Nat Med 3:894–899

Morishita R, Higaki J, Tomita N, Ogihara T (1998) Application of transcription factor 'decoy' strategy as means of gene therapy and study of gene expression in cardiovascular disease. Circ Res 82:1023–1028

Nakamura H, Aoki M, Tamai K, Oishi M, Ogihara T, Kaneda Y, Morishita R (2002) Prevention and regression of atopic dermatitis by ointment containing NF-kB decoy oligodeoxynucleotides in NC/Nga atopic mouse model. Gene Ther 9:1221–1229

Nakamura T, Morishita R, Asai T, Tsuboniwa N, Aoki M, Sakonjo H, Yamasaki K, Hashiya N, Kaneda Y, Ogihara T (2002) Molecular strategy using cis-element 'decoy' of E2F binding site inhibits neointimal formation in porcine balloon-injured coronary artery model. Gene Ther 9:488–494

Roque F, Mon G, Belardi J, Rodriguez A, Grinfeld L, Long R, Grossman S, Malcolm A, Zon G, Ormont ML, Fischman DL, Shi Y, Zalewski A (2001) Safety of intracoronary administration of c-myc antisense oligomers after percutaneous transluminal coronary angioplasty (PTCA). Antisense Nucleic Acid Drug Dev 2:99–106

Sawa Y, Morishita R, Suzuki K, Kagisaki K, Kaneda Y, Maeda K, Kadoba K, Matsuda H (1997) A novel strategy for myocardial protection using in vivo transfection of cis element 'decoy' against NFκB binding site: evidence for a role of NFκB in ischemic-reperfusion injury. Circulation 965:280–285

Shi Y, Fard A, Galeo A, Hutchinson HG, Vermami P, Dodge GR, Hall DJ, Shaheen F, Zalewski A (1994) Transcatheter delivery of c-myc antisense oligomers reduces neointimal formation in a porcine model of coronary artery balloon injury. Circulation 90:944–951

Srinivasan SK, Iversen P (1995) Review of in vivo pharmacokinetics and toxicology of phosphorothioate oligonucleotides. J Clin Lab Anal 9:129–137

Stein CA, Cohen JS (1988) Oligodeoxynucleotides as inhibitors of gene expression: a review. Cancer Res 48:2659–2668

Sullenger BA, Gallardo HF, Ungers GE, Giboa E (1990) Overexpression of TAR sequence renders cells resistant to human immunodeficiency virus replication. Cell 63:601–608

Suzuki J, Morishita R, Amano J, Kaneda Y, Isobe M (2000) Decoy against nuclear factor-kappa B attenuates myocardial cell infiltration and arterial neointimal formation in murine cardiac allografts. Gene Ther 7:1847–1852

Tomita N, Morishita R, Tomita S, Gibbons GH, Zhang L, Horiuchi M, Kaneda Y, Higaki J, Ogihara T, Dzau VJ (2000b) Transcription factor decoy for NFkB inhibits TNF-α-induced cytokine and adhesion molecule expression in vivo. Gene Ther 7:1326–1332

Tomita N, Morishita R, Lan HY, Yamamoto K, Hashizume M, Notakae M, Toyosawa K, Fujitani B, Mu W, Nikolic-Paterson DJ, Atkins RC, Kaneda Y, Higaki J, Ogihara T (2002a) In vivo administration of a nuclear transcription factor-κB decoy suppresses experimental crescentic glomerulonephritis. J Am Soc Nephrol 11:1244–1252

Tomita N, Morishita R, Tomita S, Kaneda Y, Higaki J, Ogihara T, Horiuchi M (2001) Inhibition of TNF-alpha, induced cytokine and adhesion molecule expression in glomerular cells in vitro and in vivo by transcription factor decoy for NFκB. Exp Nephrol 9:181–190

Tomita T, Takeuchi E, Tomita N, Morishita R, Kaneko M, Yamamoto K, Nakase T, Seki H, Kato K, Kaneda Y, Ochi T (1999) Suppressed severity of collagen-induced arthritis by in vivo transfection of nuclear factor kappaB decoy oligodeoxynucleotides as a gene therapy. Arthritis Rheum 42:2532–2542

Tomita T, Takano H, Tomita N, Morishita R, Kaneko M, Shi K, Takahi K, Nakase T, Kaneda Y, Yoshikawa H, Ochi T (2000) Transcription factor decoy for nuclear factor kappaB inhibits cytokine and adhesion molecule expressions in synovial cells derived from rheumatoid arthritis. Rheumatology 39:749–757

Yamasaki K, Asai T, Shimizu M, Aoki M, Hashiya N, Sakonjo H, Makino H, Kaneda Y, Ogihara T, Morishita R (2003) Inhibition of NFκB activation using cis-element 'decoy' of NFκB binding site reduces neointimal formation in porcine balloon-injured coronary artery model. Gene Ther 10:356–364

Yokoseki O, Suzuki J, Kitabayashi H, Watanabe N, Wada Y, Aoki M, Morishita R, Kaneda Y, Ogihara T, Futamatsu H, Kobayashi Y, Isobe M (2001) Cis element decoy against nuclear factor-kappaB attenuates development of experimental autoimmune myocarditis in rats. Circ Res 89:899–906

Yoshimura S, Morishita R, Hayashi K, Yamamoto K, Nakagami H, Kaneda Y, Sakai N, Ogihara T (2001) Inhibition of intimal hyperplasia after balloon injury in rat carotid artery model using cis-element 'decoy' of nuclear factor-kB binding site as a novel molecular strategy. Gene Ther 8:1635–1642

Selective Estrogen Receptor Modulators as Therapeutic Agents in Breast Cancer Treatment

C. M. Klinge

Department of Biochemistry and Molecular Biology, Center for Genetics and Molecular Medicine, James Graham Brown Cancer Center, University of Louisville School of Medicine, Louisville, KY 40292, USA
carolyn.klinge@louisville.edu

1	Introduction	456
2	Primary Effects of SERMs	458
2.1	ER Binding Affinities of Representative SERMs	459
3	Primary Effects of Steroidal Antiestrogens	460
4	The Effect of SERMS on the Structural and Functional Features of ERα and ERβ	461
4.1	A/B Domain and AF-1	464
4.2	ER DNA-Binding Domain	464
4.3	D Domain/Hinge Region	465
4.4	LBD and AF-2	465
4.5	F Domain	467
5	Heterodimerization of ERα and ERβ	467
6	Conformational Difference in SERM vs. E_2-Occupied ER	467
7	Proteasomal Degradation of ER	469
8	Mechanisms by Which Activated ER Increases Target Gene Transcription	469
8.1	Ligand-Induced Hsp90 Dissociation and ERα Activation	469
8.2	The Tethering Model of ER Transactivation	471
8.3	Inactive ER Bound to EREs	471
8.4	The 'Tripartite' Model of ER Action	472
8.5	ER-ERE Binding	473
8.6	ERE as an Allosteric Effector of ER Action	474
9	Effect of SERMs on ER Interaction with Coactivators	475
10	Effect of SERMs on ER Interaction with Corepressors	477
11	New Insights into the Mechanism for the Mixed Agonist/Antagonist Activity of Tamoxifen	478
12	ER-Independent Activities of SERMs	479
13	Summary	480
References		481

Abstract Breast cancer is the most commonly diagnosed cancer in women of the Western world. Breast tumors are initially dependent upon estrogens for their growth. Drugs with antiestrogenic activity, e.g., tamoxifen, are used clinically both to prevent and treat breast cancer. Estrogens stimulate breast cancer cell proliferation by binding to estrogen receptors alpha and beta (ERα and ERβ) that are ligand-activated transcription factors. Estrogen-activated ER either binds directly to DNA sequences called estrogen response elements (EREs), or interacts with other DNA-bound transcription factors in the promoters of target genes to regulate transcription. Selective estrogen receptor modulators (SERMs) compete with estrogens for binding ER and act as a mixed ER agonist/antagonists depending on the cell type, e.g., antiestrogenic in breast and estrogenic in bone. Examples of SERMs in current clinical use are tamoxifen and raloxifene. This chapter discusses the effect of SERMs on the structure and function of ERα and ERβ. Structural studies indicate that ER occupied by SERMs has a different conformation from estradiol-occupied ER. This structural difference prevents the interaction of SERM-occupied ER with coactivator proteins necessary for chromatin remodeling involved in the initiation of gene transcription. Instead, SERM-occupied ER interacts with the corepressors NCoR and SMRT in vitro and when SERM-occupied ER is bound to ERE-containing genes in breast cancer cells. Corepressors interact with histone deacetylase (HDAC) complexes that maintain chromatin in a repressed state that precludes initiation of target gene transcription. Recent data from chromatin immunoprecipitation assays indicate that SERM-occupied ERα interacts with coactivators when tethered to the promoters of AP-1-regulated estrogen target genes in endometrial cancer cells, thus providing a mechanism for the agonist activity of tamoxifen in endometrial cells. The applicability of these findings to other cell types in which tamoxifen has agonist activity remains to be determined.

Keywords Estrogen receptor · Antiestrogens · Breast cancer · SERM · Transcription

1
Introduction

Breast cancer is the most commonly diagnosed malignancy in women in the Western world. Lifetime exposure to endogenous estrogens, i.e., resulting from early menarche combined with late menopause, postmenopausal obesity, and the use of hormone replacement therapy (estrogen plus progesterone), are risk factors for breast cancer development (Russo et al. 2001). Studies suggest that endogenous estrogens and their metabolites, e.g., 4-hydroxyestradiol can promote breast cancer cell carcinogenesis (Samuni et al. 2003). Estrogens exert a wide variety of effects on growth, development, and differentiation, including important regulatory functions within the reproductive systems of both females and males, in mammary gland development and differentiation, as anti-atherosclerotic agents, in central nervous system functions, and in the regulation of hypothalamic-gonadal axis. In breast cancer cells, 17β-estradiol (E$_2$) activates the transcription of genes involved in cell proliferation and G$_1$ progression, e.g., progesterone receptor (PR), c-*fos*, c-*myc*, cathespsin D, cyclin D1, pS2, and SDF-1 (Hall and Korach 2003). Estrogens mediate these activities through binding to a specific intranuclear receptor protein, the estrogen receptor (ER), encoded by two genes: alpha and beta (*ERα* and *ERβ*) that function both as signal transducers and transcription factors to modulate expression of target genes (Nilsson et al. 2001). It has been suggested that ER is the ancestral steroid receptor because estrogen regulates reproductive maturation and function (Thornton 2001). Recent evidence that the 'nongenomic' or estrogen activation of a cell membrane-mediated intracellular signaling cascades, i.e., MAPK/ERK and PI3K/AKT, may also contribute to breast cancer is intriguing (Chen et al. 2000), but will not be the focus of this review.

For approximately 30 years, the measurement of ER, now called ERα, has been a useful prognostic indicator predicting the risk of metastatic recurrence of breast cancer and the response of disseminated disease to endocrine therapy (Jensen 1996). ERα is expressed at elevated levels in approximately 75% of clinical breast cancer samples and selection of patients with ERα-positive breast tumors increases the patient response rates from approximately 33% to 50% (McGuire et al. 1986). Further, as PR expression is dependent on ERα expression, patients whose breast tumors express both ERα and PR show an 80% response rate to antiestrogen therapy (McGuire et al. 1986). Patients whose tumors are ERα-positive and express Her2/neu/ErbB2 show reduced response to antiestrogen treatment (Dowsett et al. 2001).

Tamoxifen (TAM, Nolvadex) is currently the endocrine treatment of choice for all stages of breast cancer and is the gold standard for antiestrogen treatment because, as an adjuvant therapy, it prevents the recurrence of breast cancer by eradicating micro-metastases (Jordan 1997). Five years of adjuvant TAM (20 mg/day) treatment has been shown to extend a survival benefit to both pre- and postmenopausal women (Fisher et al. 1996). Additionally, preoperative TAM therapy has been shown to benefit elderly patients with ER-positive breast tumors (Toi et al. 2003). Tamoxifen is an example of a 'selective estrogen receptor modulator' (SERM) that acts as a mixed ER agonist/antagonist depending on the cell or tissue type. Patients whose breast tumors express ERα have the best prognosis for survival because TAM blocks most ERα activities in breast tumor cells (Jordan 2001). TAM treatment reduces the annual odds of recurrence by 30% and the annual odds of death by 19% (Shapiro and Henderson 1994). It is estimated that there are 400,000 women alive today because of long-term adjuvant TAM treatment (Jordan and Pappas 2003). In addition, TAM has chemopreventive activity, i.e., TAM reduced the incidence of breast cancer in high-risk women by 50% (Smigel 1998), making it the first Food and Drug Administration (FDA)-approved drug to prevent a type of human cancer. It is estimated that 10.2 million of the 65.8 million US women aged 35–79 years without breast cancer would be eligible for TAM chemoprevention (Freedman et al. 2003).

Unfortunately, despite its beneficial activities in breast cancer, TAM therapy has drawbacks. The agonist activity of TAM increases the risk for the development of endometrial cancer by two- to fivefold (Fisher et al. 1996). Although the exact mechanism for the carcinogenic effect of TAM is unknown, there is evidence that TAM metabolites form DNA adducts in the endometrium of women on TAM therapy (Shibutani et al. 2000). TAM also increases the risk of thromboembolic disease (Cuzick et al. 2003). TAM's central antiestrogenic effects result in hot flashes. Moreover, the use of TAM as a chemopreventive agent may allow TAM-resistant or TAM-stimulated subpopulations of cells to arise (Horwitz 1994). These untoward effects of TAM have led to the development of newer SERMs including raloxifene (RAL, Evista) that have estrogen antagonist activity in both breast and uterus, and estrogen agonist activity in bone making it suitable for use in osteoporosis prevention and treatment (Jordan 2001). The Study of Tamoxifen and Raloxifene (*STAR* trial) will compare the value of TAM vs. RAL as chemopreventative agents in postmenopausal women with results anticipated by 2005 (Jordan and Pappas 2003).

Blocking the synthesis of estrogens by anastrozole, an aromatase inhibitor, is an alternative pathway to block estrogen's stimulatory effects in breast tissue (Buzdar 2003). Indeed, an early clinical report showed that anastrozole was more effective than TAM in reducing breast cancer recurrence (Allred et al. 2003). However, the long-term testing of aromatase inhibitors is required to avoid concerns about osteoporosis, Alzheimer's disease and coronary heart disease (Jordan 2002).

In the past decade, numerous insights into the molecular mechanisms of estrogen/SERM action have been reported (reviewed by Jordan and Pappas 2003). This review will summarize briefly the current state of understanding of estrogen/SERM action as it pertains primarily to breast cancer. The body of literature in this area is enormous and therefore it is impossible to cite every report pertaining to SERM activities. I apologize to authors whose reports were omitted in this review, and I assure the reader that there has been no deliberate attempt to exclude any work from this review.

2
Primary Effects of SERMs

More than 100 years ago, removal of the ovaries was shown to benefit premenopausal patients with breast cancer (Beatson 1896). Thus, the concept of estrogen ablation as a treatment for breast cancer was born. The first nonsteroidal antiestrogen, MER-25, was developed in 1954 as part of a contraceptive program and was shown to block estrogen-induced uterine weight gain in mice (Lerner and Jordan 1990). TAM, a triphenylethylene compound derived from MER-25 by ICI Pharmaceuticals (now Zeneca; reviewed by Jordan and Pappas 2003), is currently the most widely clinically used antiestrogen (Lerner and Jordan 1990). TAM was the first clinically proven and

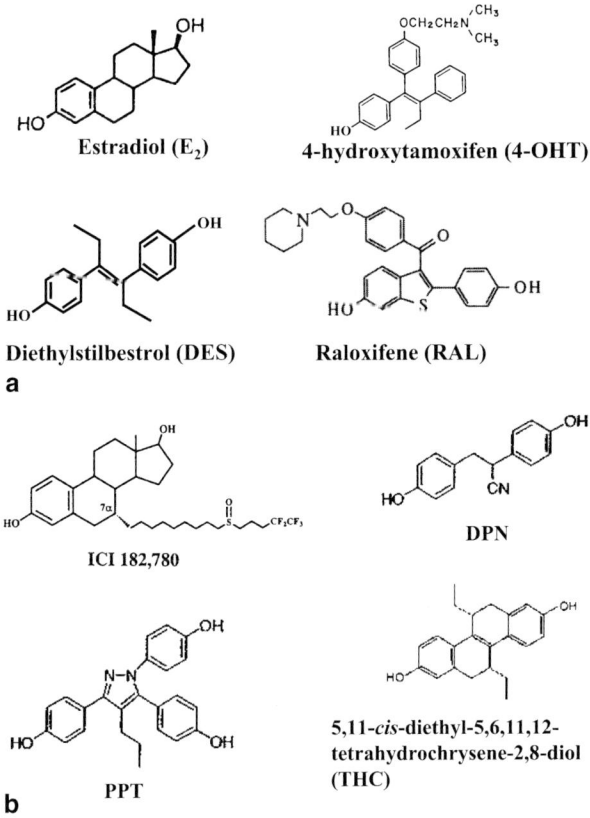

Fig. 1a, b Structures of ligands and SERMS. The structures of (**a**) 17β-estradiol (*E₂*), diethylstilbestrol (*DES*), 4-hydroxytamoxifen (*4-OHT*), raloxifene (*RAL*), ICI 182,780, propyl pyrazole triol (*PPT*, an ERα-selective agonist; Kraichely et al. 2000); (**b**) 2,3-bis(4-hydroxyphenyl)propionitrile (*DPN*, an ERβ-selective agonist; Meyers et al. 2001); and the R,R enantiomer of 5,11-*cis*-diethyl-5,6,11,12-tetrahydrochrysene-2,8-diol (*R,R-THC*, an ERβ-selective antagonist; Shiau et al. 2002) are shown

FDA-approved (in 1977) antiestrogen for the treatment of advanced breast cancer. TAM is hydroxylated by P450 CYP 2D6, 2C9 and 3A4 in the liver forming 4-hydroxytamoxifen (4-OHT; see Fig. 1a and b) (Crewe et al. 1997).

2.1
ER Binding Affinities of Representative SERMs

4-OHT is considered to be the most biologically active TAM metabolite because it binds ERα with a relative binding affinity that is 310% of that of E₂ (Katzenellenbogen et al. 1984). In contrast, TAM binds ERα with 3% of the affinity of that of E₂ (Lieberman et al. 1983) and 500-fold more slowly than

E_2 (Rich et al. 2002). For a patient taking 40 mg TAM/day, the mean serum concentration is ~3 μM and when the N-desmethyl and 4-OHT- metabolites are included, the total TAM concentration is ~10 μM (Clarke et al. 2001). For a patient taking 20 mg TAM/day, steady-state plasma levels of TAM are ~320 nM (Coward et al. 2001). TAM induces G_0/G_1 arrest (Carroll et al. 2000) and apoptosis in breast cancer cells (Ellis et al. 1997).

The definition of an ideal SERM is that it has mixed ER agonist/antagonist activities depending on the tissue, i.e., producing beneficial estrogen-like effects on bone and lipid metabolism, while antagonizing estrogen in breast cancer and in reproductive tissues (Jordan and Pappas 2003). In addition to TAM, other examples of SERMS are toremifene, droloxifene, idoxifene, GW5638 and its metabolite GW7604, and RAL.

RAL, formerly called keoxifene or LY156,758, is derived from the SERM LY117018 has benzothiophene structure (Fig. 1a). RAL binds ERα with an affinity similar to that of E_2, i.e., K_d=0.11–0.54 nM (Fuchs-Young et al. 1995) and binds ERβ with lower affinity than ERα, i.e., K_d=2.43 nM (Zou et al. 1999). RAL differs from TAM in that it has estrogen antagonist activity in uterine endometrial cells whereas TAM is an agonist in uterus (Fuchs-Young et al. 1995). RAL blocks E_2 agonist activity in breast cancer cells (Fuchs-Young et al. 1995). Like TAM, RAL lowers serum cholesterol levels (De Leo et al. 2001). RAL has been shown to decrease total cholesterol and increase high density lipoprotein (HDL)-cholesterol without increasing serum triglyceride levels whereas TAM increased triglycerides levels (Tsai et al. 2001). RAL also decreases serum homocysteine levels in post-menopausal women (De Leo et al. 2001). RAL is FDA-approved for the treatment of osteoporosis because of its estrogen agonist activity in bone.

EM-652 (acolbifene) is a fourth-generation SERM exerting complete anti-estrogenic effects on the breast and uterus (Simoncini et al. 2002). EM-652 inhibits bone resorption and induces positive lipid modifications in estrogen-deficient animals (reviewed by Simoncini et al. 2002). Similar to other SERMs, EM-652 blocks ERα and ERβ interaction with the receptor interaction domains (RIDs) of the coactivator SRC-1 in vitro (Bai and Giguere 2003), providing a mechanism for the complete antagonist activity of this new SERM.

3
Primary Effects of Steroidal Antiestrogens

A series of 7 alpha-alkylamide analogs of estradiol with pure antiestrogenic activity, exemplified by ICI164,384 and ICI182,780 were characterized by Wakeling and colleagues (Wakeling 1989). ICI164,384 was shown to have greater maximal inhibitory activity than 4-OHT in cultured breast cancer cells (Wakeling 1989) and in nude mice implanted with MCF-7 cells (Osborne et al. 1995). ICI182,780 (Fulvestrant, Faslodex, see Fig. 1b) has more potent antiestrogenic activity than ICI164,384 (Wakeling and Bowler 1992). Both ICI164,384 and ICI182,780 bind ERα and ERβ. The relative binding affinities of ICI182,780 and ICI164,384 for ERα are 0.89 and 0.19, respectively, compared with that of E_2 (1.0) (Wakeling et al. 1991). Importantly, cell culture studies show that TAM-resistant cells remain sensitive to the growth-inhibitory action of ICI164,384 and ICI182,780 (Wakeling 1993). Some reports indicate that the binding site of ICI182,780 within the ERα ligand-binding domain (LBD) is not identical to that occupied by E_2 since E_2 did not displace ICI182,780 in a fluorescence resonance energy transfer (FRET) assay (Llopis et al. 2000) and ICI182,780 acted as an allosteric effector that increased E_2 binding to the ERα LBD in yeast (Dudley et al. 2000). However, the LBD of ERβ occupied by ICI164,384 was crystallized (Pike et al. 2000) and showed that that the antiestrogenic side chain follows the same path in the ER complex as the side chain of RAL (Brzozowski et al. 1997). To do this, the steroid must flip over 180° in the LBD (Jordan 2001). In addition to ER binding, ICI182,780 inhibits aromatase activity in normal human breast and other carcinoma cell lines, albeit with a lower potency than letrozole and anastrozole (Long et al. 1998). ICI182,780 is on the market as Faslodex (Fulvestrant) and is the current pure steroidal antiestrogen of choice that was approved by the FDA in 2002 for the treatment of postmenopausal women with hormone-sensitive advanced or metastatic breast cancer who have progressed on prior antiestrogen therapy (reviewed by Morris et al. 2002).

4
The Effect of SERMS on the Structural and Functional Features of ERα and ERβ

Significant progress has been made in elucidating the mechanisms of estrogen regulation of growth and differentiation of target cells and tissues including breast, endometrium, ovaries, liver, pituitary, cervix, vagina, and bone and in the pathogenesis of breast cancer. In each of these tissues, the mechanism of estrogen action mediated by the nuclear ER is thought to be identical. The current model is that E_2 diffuses freely into the cytoplasm and enters the nucleus by passive diffusion. Inside the nucleus, E_2 binds ER with high affinity. Please note that the term ER refers to both ERα and ERβ and that referrals to ERα or ERβ indicate that particular ER subtype. Unless otherwise indicated, the data presented are from studies that used human ER.

ERα and ERβ have unique physiological roles (reviewed by Couse and Korach 1999). Targeted disruption of the mouse *ERα* gene (*ERα* knockout = ERKO) revealed severe defects in mammary gland structure in the female mice (Couse and Korach 1999). Therefore, ERα is essential for normal mammary gland development. In contrast, targeted disruption of the mouse *ERβ* gene (*ERβ* knockout = BERKO) revealed no effect of ERβ deletion on mammary gland development or lactation (Couse and Korach 1999), but as the mice aged, abnormal epithelial growth, overexpression of Ki67, and severe cystic breast disease were detected (Gustafsson and Warner 2000). This indicates that loss of ERβ expression may lead to abnormal mammary cell proliferation. Studies examining the expression of ERα and ERβ in normal and malignant human breast tissue by reverse transcription–PCR showed that ERβ expression predominates in normal breast tissue (Iwao et al. 2000). There is evidence from experiments in which the ER-negative MDA-MB-231 human breast cancer cell line was infected with adenoviruses expressing ERα or ERβ that ERβ inhibits breast cancer cell proliferation in a ligand-independent manner (Lazennec et al. 2001). These data and other experimental data support the 'yin/yang relationship' model of ERα/ERβ that ERα is proliferative and ERβ is antiproliferative (Weihua et al. 2003).

At the protein level, many breast cancer cells lines, even those that are 'ER-negative,' e.g., MDA-MB-435 and MDA-MB-231, express full-length and/or truncated ERβ (Fuqua et al. 1999). In contrast, most human breast tumors express ERα, either alone or in combination with ERβ (Omoto et al. 2001). Because conflicting findings regarding correlation between ERβ expression, tumor marker expression, and lymph node status have been reported, the utility of ERβ analysis as a prognostic factor in patients with breast cancer remains unclear. Further complicating our understanding of ER action, a 46-kDa ERα isoform has been shown to be expressed at the protein level in breast cancer cells (Flouriot et al. 2000); and at least five isoforms of ERβ, derived from alternate initiation or splice sites, exist at the mRNA level in human breast cancer cells (Leygue et al. 1999). The functional importance of these ERα and ERβ variants remains to be determined.

ERα (NR3A1) and ERβ (NR3A2), located on chromosomes 6q25 and 14q11.1, are members of the steroid/nuclear receptor (SR/NR) superfamily of enhancer proteins that includes 48 members sharing a highly conserved structure and common mechanisms affecting gene transcription (Maglich et al. 2001). SR/NR regulate the transcription of a diverse array of target genes during development and in response to specific physiological and pathological signals. The SR/NR superfamily includes the class I NR: the steroid receptors, e.g., glucocorticoid, mineralocorticoid, progesterone, and androgen receptors (GR, MR, PR, and AR) and the class II NR, e.g., retinoic acid receptor, retinoid X receptor, vitamin D receptor, thyroid receptor, and peroxisome proliferator activated receptor (RAR, RXR, VDR, TR, and PPAR, respectively). Additionally, the NR superfamily includes a number of 'orphan receptors,' denoted as such because their endogenous ligands, if necessary, are either unknown or are not required, e.g., chicken ovalbumin upstream promoter transcription factor (COUP-TF).

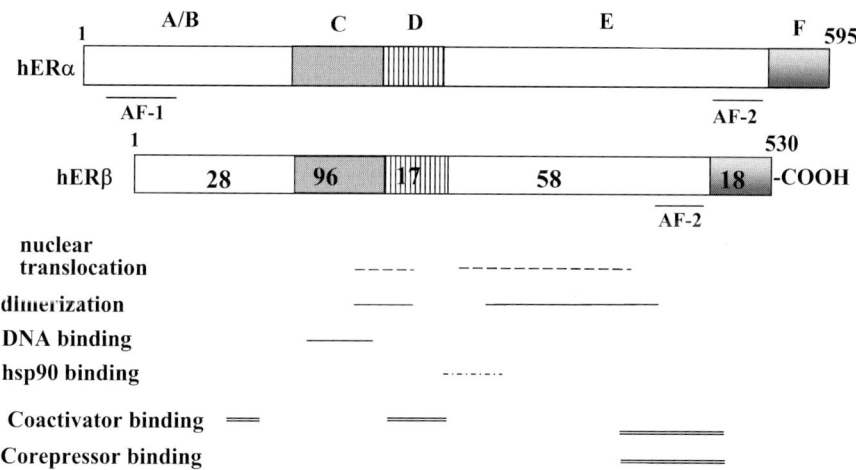

Fig. 2 Structure of ERα and ERβ. ERα and ERβ are shown schematically as linearized proteins with common structural and functional domains. The *numbers* in the domains of ERβ indicate the percent of amino acid homology between ERα and ERβ (Kuiper et al. 1996). Variability between members of the steroid hormone receptor family is due primarily to differences in the length and amino acid sequence of the N-terminal domain. (Adapted from Wahli and Martinez 1991)

The evolutionary relationship among the SR/NR superfamily has been deduced by the high conservation in their DNA-binding domains (DBDs) and the lesser-conservation in their LBDs and indicates that this large group of proteins arose from a common ancestral molecule (Laudet 1997). Interestingly, ER was deduced to be the first steroid receptor, having evolved from ancient receptors by two large-scale genome expansions, one before the advent of jawed vertebrates and one after (Thornton 2001).

ERα and ERβ have six domains named A–F from N to C termini, encoded by eight exons (Kuiper et al. 1998) (Fig. 2). The three major functional domains of the ER are: (1) a variable N-terminus (domains A and B) that modulates transcription in a gene and cell-specific manner through its N-terminal activation function-1 (AF-1); (2) a central DBD, consisting of the C domain, comprised of two functionally distinct zinc fingers through which the receptor interacts directly with the DNA helix or interacts with other DNA-bound transcription factors; and (3) the LBD, domain E, that contains ligand-dependent AF-2. E_2 and SERMs interact directly with the LBD by hydrogen bonding of hydroxyl groups on the ligand to critical amino acids in the LBD, i.e., Glu 353 with the 3-hydroxyl and His535 for the 17β-hydroxyl of E_2 (Tanenbaum et al. 1998).

4.1
A/B Domain and AF-1

The N termini of ERα and ERβ are not conserved. Experimental evidence indicates that ERα AF-1 activity is much stronger than that of ERβ (Cowley and Parker 1999; Hall and McDonnell 1999). Studies using chimeric ERα–ERβ proteins with switched N-terminal or C-terminal domains revealed that replacing the N terminus of ERβ with the N terminus of ERα had no effect on E_2 or TAM activity (Weatherman and Scanlan 2001). However, when the N terminus of ERβ was transferred to ERα, TAM agonist activity was lost, indicating that AF-1 of ERα is responsible for TAM agonist activity (Weatherman and Scanlan 2001).

In addition to AF-1, the N terminus of ERα plays a role in DNA and coactivator interaction. ERα AF-1 interacts with the coactivators SRC-1 and SRC-2 (TIF-2/GRIP1) and these coactivators increase 4-OHT agonist activity in transfected HeLa cells (Webb et al. 1998). In contrast, ERβ activity was only weakly stimulated by GRIP1/SRC-2 and ERβ-AF-1 activity was not responsive to GRIP1/SRC-2 (Webb et al. 1998). These data are in agreement with the finding that in contrast to the 'stand alone' transcriptional activity of ERα AF-1 in a yeast two-hybrid assay, the N terminus of ERβ has no intrinsic transcriptional activity (Delaunay et al. 2000).

AF-1 function may depend on phosphorylation and AF-1 interacting coactivators. The N terminus of ERα is the target of phosphorylation mediated by various growth factor cascades (reviewed by Lannigan 2003). For example, in response to E_2 binding, ERα is phosphorylated predominately on Ser118 and to a lesser extent on Ser104 and Ser106 (Lannigan 2003). 4-OHT and ICI164,384 also cause the phosphorylation of ERα at Ser118, although phosphorylation is less than that seen with E_2. In response to growth factor, e.g., epidermal growth factor activation of the mitogen-activated protein kinase (MAPK) pathway, ERα is phosphorylated on Ser118 and Ser167 (Lannigan 2003). Cyclin A2-CDK2 phosphorylates Ser104 and Ser106 and stimulates ERα transactivation whether cells are treated with E_2 or 4-OHT (Hodges et al. 2003). The transcriptional activity of ERα AF-1, but not AF-2 was potentiated by a coactivator called p68 that interacts with ERα when phosphorylated on Ser118 (Endoh et al. 1999). In contrast, p68 had no effect on either AF-1 or AF-2 of ERβ.

4.2
ER DNA-Binding Domain

The most conserved region between ERα and ERβ is the DBD featuring two Cys–Cys zinc fingers (CI and CII) with which the receptor interacts with the major groove and phosphate backbone of DNA, respectively (Kuiper et al. 1997). CII also plays a role in ER dimerization (Schwabe et al. 1993). The specificity of the DBD in targeting ER to specific DNA sequences was demonstrated by domain-swapping experiments in which the DBD of ERα was switched with that of the GR. The chimeric receptor, containing AF-1 and AF-2 of ERα and the DBD of GR, bound to glucocorticoid response elements, but upregulated transcription in response to E_2 (Green and Chambon 1987), thus demonstrating the specificity of the DBD in target gene regulation. ERα and ERβ bind with high affinity to specific sequences called estrogen response elements (EREs, the minimal sequence of which is 5'-GGT-CAnnnTGACC-3', where n=any nucleotide (reviewed by Klinge 2001). We have shown that addition of an extra nucleotide to each arm of the palindrome, i.e., 5'-AGGTCAnnnTGACCT-3', increases ERα and ERβ binding affinity and transcriptional activation by at least tenfold because of the added stability of ER binding (Kulakosky et al. 2002).

4.3
D Domain/Hinge Region

The hinge region or D domain is a 40–50-amino acid sequence separating the DBD and the LBD. The D domain contains sequences for receptor dimerization and nuclear localization sequences (Nilsson et al. 2001). In addition, the D domain is essential for stabilizing ER–ERE binding. The hinge region interacts with nuclear corepressor proteins and with a 27-kDa protein called L7/SPA that binds TAM–ERα, but not ICI164,384–ERα, and enhances the partial agonist activity of TAM by 3–10-fold without affecting E_2-stimulated transcription (Jackson et al. 1997). The hinge domain is also the site of interaction of the nuclear matrix protein/scaffold attachment factor HET/SAF-B with ERα (Oesterreich et al. 2000). Interestingly, interaction of ERα with HET/SAF-B is increased by TAM (Oesterreich et al. 2000).

4.4
LBD and AF-2

ER has been called 'promiscuous' because of its ability to bind a wide range of ligands (Elsby et al. 2000). E_2 binds ERα and ERβ with a K_d of 0.1 and 0.4 nM, respectively (Kuiper et al. 1997). ERα and ERβ show similar binding preference for synthetic estrogens and SERMs: diethylstilbestrol (DES)>hexestrol>dienestrol>4-OHT>17β-estradiol>coumestrol, ICI164,384>estrone, 17α-estradiol>nafoxidine, moxestrol>clomifene>estriol, 4-OH-estradiol> TAM, 2-OH-estradiol, 5-androstene-3 beta, 17 beta-diol, genistein for ERα; and dienestrol>4-OH-TAM>DES>hexestrol>coumestrol, ICI164,384>17β-estradiol>estrone, genistein>estriol>nafoxidine, 5-androstene-3 beta, 17 beta-diol>17α-estradiol, clomifene, 2-OH-estradiol >4-OH-estradiol, TAM, moxestrol for ERβ (Kuiper et al. 1997). Phytoestrogens such as coumestrol and genistein bind ERβ with higher affinity than ERα; likewise, the xenoestrogen bisphenol A, found in plastics, binds ERβ with relatively higher affinity than ERα (Kuiper et al. 1997).

The protein architecture around the A-ring ligand-binding pocket imposes an absolute requirement that effective ER ligands contain a planar ring group; however, the remainder of the cavity is quite accommodating (Pike et al. 2000). The X-ray crystal structure of the ERα LBD in the presence E_2 or RAL revealed that the LBD has a compact structure consisting of 12 α-helices with a 'pocket' into which the ligand fits (Brzozowski et al. 1997). Binding of E_2 within the pocket alters the conformation of the LBD with helix 12 forming a 'lid' over the pocket, trapping E_2 in a hydrophobic environment and forming a surface with a groove within the LBD with which co-activator proteins interact through a conserved six amino acid sequence, i.e., LXXLL, where X=any amino acid, called the RID (Chen et al. 1999). Correct positioning of helix 12 is indispensable for AF-2 function (Shiau et al. 1998). AF-2, localized to the most C-terminal end of the E domain, is highly conserved within the SR/NR superfamily. AF-2 is recognized by transcriptional coactivators (reviewed by Klinge 2000).

ERα and ERβ selective agonists have been developed in the laboratory of John Katzenellenbogen (Kraichely et al. 2000; Sun et al. 2003). These ligands have already provided new information regarding the roles of ERα vs. ERβ in mouse uterus in vivo, i.e., showing that ERα is the predominant isoform involved in the uterotropic effects of estrogen, but that there is complex crosstalk between ERα and ERβ that is gene dependent (Frasor et al. 2003). Propyl pyrazole triol (PPT) is a selective ERα agonist (Kraichely et al. 2000) and 2,3-bis(4-hydroxyphenyl)propionitrile (DPN) acts as an agonist on both ER subtypes, but has a 70-fold higher relative binding affinity and 170-fold higher relative potency in transcription assays with ERβ compared to those with ERα (Meyers et al. 2001). The structures of PPT and DPN are shown in Fig. 1b. The R,R enantiomer of 5,11-*cis*-diethyl-5,6,11,12-tetrahydrochrysene-2,8-diol (R,R-THC, structure in Fig. 1b), is an ERα agonist and an ERβ-selective antagonist (Shiau et al. 2002). These compounds provide useful tools for examining ERα- and ERβ-selective activities.

The ER LBD has functions in addition to ligand and coactivator binding. The N terminus of the ERα LBD (amino acids 374–385, AF-2a) contains charged amino acid residues that impact DNA binding and transcriptional activity by altering the affinity of E_2 binding (Pakdel et al. 1993b). The LBD is also the region of ER shown to associate with hsp90 and other chaperonin proteins in vitro, presumably by assisting ER to assume the appropriate conformation for ligand binding (Segnitz and Gehring 1995).

4.5
F Domain

The 42-amino acid F domain of ERα, which is not well conserved among different vertebrate ER species, plays a role in distinguishing estrogen agonists vs. antagonists, perhaps through interaction with cell-specific factors (Montano et al. 1995). There is experimental evidence showing that the N and C termini of ERα functionally interact and that the coactivator SRC-1 enhances this interaction (McInerney et al. 1996). Deletion of the ERα F domain was shown to increase ERα basal activity, implying a negative regulatory role for the F domain in unliganded ERα (Weatherman and Scanlan 2001). Similar results were observed for the ERβ F domain deletion. Studies with chimeric ERα and ERβ proteins showed that attachment of the ERβ F domain onto ERα increased E_2 response, but inhibited the agonist activity of TAM (Weatherman and Scanlan 2001). As AF-1 of ERα mediates the agonist activity of TAM, these data further support a role for N–C terminal interaction in the AF-1 activity.

5
Heterodimerization of ERα and ERβ

In addition to ERα and ERβ homodimers, ERα and ERβ form heterodimers in vitro (Cowley et al. 1997) and in vivo (Matsuda et al. 2002). E_2 treatment of cells causes a relocalization of ERα or ERβ to identical discrete nuclear foci and an overlap with Brg-1, a component of the ATP-dependent human SWI/SNF-hBRM chromatin remodeling complex (Aalfs and Kingston 2000). ERα or ERβ also overlap with acetylated histone H4, indicating that E_2-occupied ER clusters to regions of activate chromatin remodeling (Matsuda et al. 2002). The biological activities of ERα/ERβ heterodimers, if distinct from those of the ERα and ERβ homodimers, remain to be determined.

6
Conformational Difference in SERM vs. E_2-Occupied ER

A number of studies using a variety of experimental techniques have documented conformational differences in ERα occupied by TAM (or 4-OHT) vs. E_2. Crystal structure studies on the purified ER LBD occupied by DES, 4-OHT, ICI164,384, and genistein have provided important information about the molecular mechanisms by which agonists, SERMs, and the pure steroidal antagonist ICI164,384 impact ER conformation and interaction with coactivator RIDs (Brzozowski et al. 1997; Pike et al. 1999, 2000; Shiau et al. 1998, 2002). Not surprisingly, these studies revealed that ERα and ERβ LBDs have very similar dimer interfaces (Pike et al. 2000). The ligand-binding cavity is formed by residues from H3, H6, the loop region between H7 and H8, H8, H11, and H12. All the ligands tested bind across the cavity between H3 and H11 (Pike et al. 2000). E_2 and RAL form different amino acid contacts within the ERα ligand-binding pocket (Brzozowski et al. 1997). This results in different positioning of helix 12 that is thought to permit ERα interaction with coactivators, e.g., SRC-1, in the presence of E_2, but not RAL, or by inference other SERMs (Brzozowski et al. 1997). The structural information corresponds to experimental data showing that TAM inactivates ERα AF-2 but not AF-1 (Tzukerman et al. 1994).

The coactivator binding site in ERα is proportioned to bind the LXXLL core consensus motif of the RIDs of coactivators and comprises a shallow hydrophobic grove about 10 Å in length and 6 Å wide that is formed by residues between H3, H4, H5, and H12 (Pike et al. 2000). The RID modules bind in a helical conformation so that all three leucines of the LXXLL motif are in contact with the LBD. The peptide conformation is stabilized by a 'charge clamp' with N- and C-capping interactions provided by a glutamic acid residue form H12 and a Lys located at the C-terminal end of H3, respectively (Pike et al. 2000).

A study of the crystal structure of the rat ERβ LBD occupied by ICI164,384 revealed complete abolition of the association between the transactivation helix (H12) and the rest of the LBD (Pike et al. 2001). The 7α-alkylamide bulky side chain of ICI164,384 protrudes from the hormone-binding pocket and binds along the coactivator recruitment site, producing a confirmation distinct from that adopted when ERβ is bound by other ligands and physically prevents H12 from adopting either its characteristic agonist or AF2 antagonist orientation (Pike et al. 2001). Further, binding of ICI164,384 resulted in the complete destabilization of H12 (Pike et al. 2001).

Electrophoretic mobility shift assays (EMSA) visually revealed differences in the migration of ERα and ERβ in the absence of ligand, or when the ERE-bound receptors were occupied by E_2, TAM, 4-OHT, ICI182,780, and other ligands. SERM-occupied ERs form specific complexes with EREs that migrate more slowly than the corresponding E_2–ER–ERE complex (Klinge et al. 1992). Despite differences in migration, the extent of SERM–ER–ERE binding is comparable to that of E_2–ER and both E_2 and 4-OHT stabilize ERα–ERE binding in vitro (Pace et al. 1997).

More recently, phage display has demonstrated unique surface conformations of ERα and ERβ when bound by agonist and antagonist ligands (Norris et al. 1999). Importantly, these studies revealed differences in the conformation of unoccupied as well as E_2, 4-OHT, and ICI182,780-occupied ERα and ERβ (Hall et al. 2000). These differences offer the possibility of using selective peptides as therapeutic modalities to target breast cancer. The conclusion from all these experiments is that there are conformational differences in E_2- vs. SERM-occupied ER and these differences underlie the observed differences in ER function.

7
Proteasomal Degradation of ER

In the absence of ligand, ERα has a half-life of 24 h in cells and E_2 reduces this to 3 h (Pakdel et al. 1993a). Binding of E_2 results in ubiquitination of ERα and targeted degradation of ERα by the 26S proteasome (Wijayaratne and McDonnell 2001). Recent evidence suggests a role for ligand-dependent acetylation of ERα in directing ERα to the proteasome for degradation, a necessary step in promoter clearance allowing re-initiation of estrogen target gene transcription (reviewed by Reid et al. 2003). One mechanism of pure steroidal antiestrogen action is accelerated ER degradation, e.g., ICI182,780 reduced ERα half-life to less than 1 h (Pink and Jordan 1996). In contrast, 4-OHT stabilizes ERα (Pink and Jordan 1996).

8
Mechanisms by Which Activated ER Increases Target Gene Transcription

Initiation of transcription is a complex event occurring through the cooperative interaction of multiple factors at the target gene promoter. At present, the precise intra-nuclear location of ER in the absence of ligand is uncertain and may differ depending on the cell type. Data from a study in which HeLa cells were transfected with green fluorescent protein (GFP)–ERα indicate that 85%–90% of GFP–ERα was nuclear, ERα shuttles between the nucleus and cytoplasm, and E_2 and 4-OHT increase nuclear retention of GFP–ERα (Maruvada et al. 2003).

8.1
Ligand-Induced Hsp90 Dissociation and ERα Activation

The predominant model for ERα activation is that in the absence of ligand ERα is sequestered in an inactivate complex with hsp90, hsp70 and other proteins (Segnitz and Gehring 1995). According to this model, in response to ligand binding, ERα undergoes conformational changes, termed 'activation,' accompanied by dissociation of hsp90, hsp70, and other proteins (Fig. 3a), forming a ligand-occupied ER dimer that then binds to EREs (Devlin-Leclerc et al. 1998). In contrast to ERα, ERβ is bound constitutively to DNA in the absence of ligand (Hall et al. 2000).

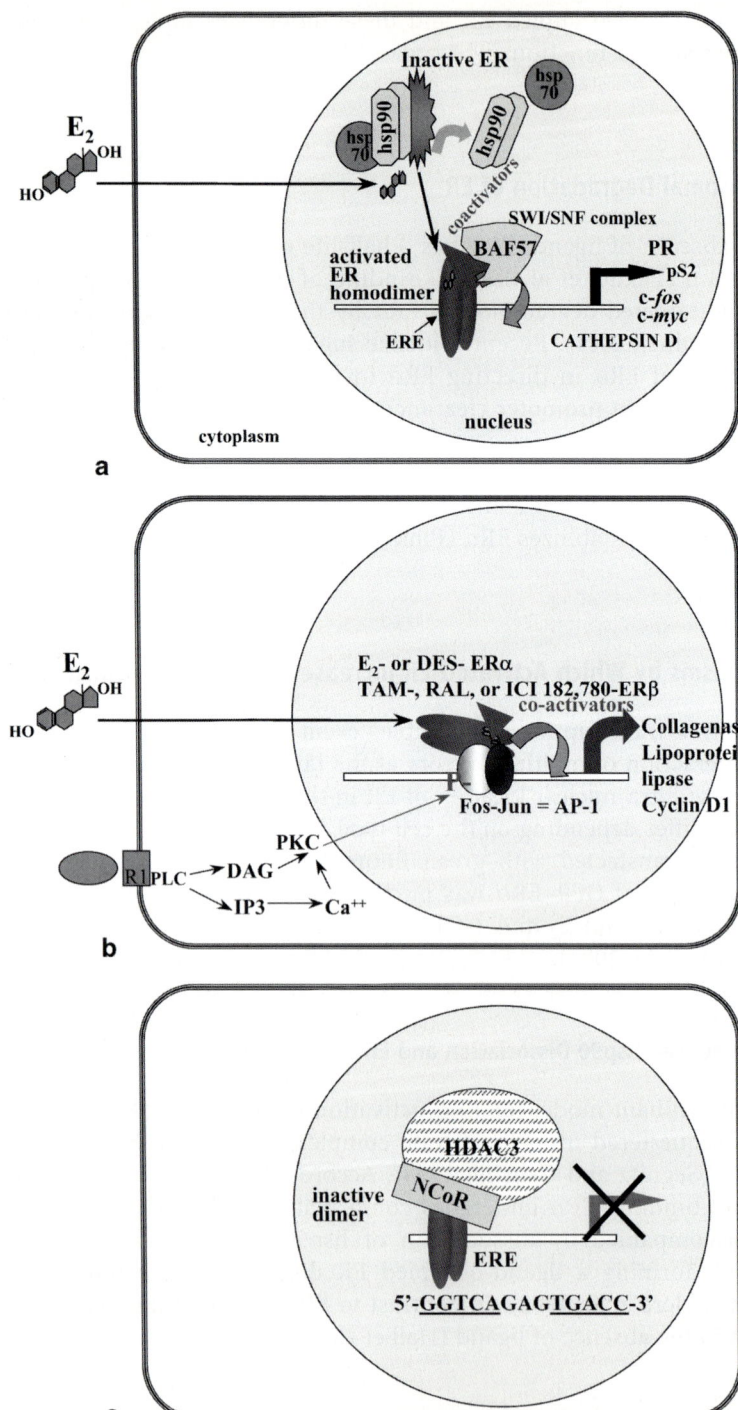

8.2
The Tethering Model of ER Transactivation

A variation of this model is the 'tethering mechanism' in which the activated E_2–ERα interacts with AP-1 (Kushner et al. 2000), Sp1 (Safe 2001), or other transcription factors to upregulate gene transcription (Fig. 3b).

8.3
Inactive ER Bound to EREs

Recent chromatin immunoprecipitation (ChIP) assays challenge the model that inactive ER is not bound to DNA. ChIP is a powerful technique that allows investigators to determine what proteins are in close proximity or bound to the promoter of a target gene. For example, ChIP assays revealed that ERα was bound to the EREs located in the human pS2 and cathepsin D1 gene promoters in association with BRCA1 in the absence of ligand (Zheng

◀───

Fig. 3a–c Models of ER activation of gene transcription. In an estrogen target cell most ER resides within the nucleus. In the absence of ligand there are controversies over the proteins associated with unactivated ERα. **a** Purification studies showed that unliganded ERα is associated with hsp90, hsp70, and other proteins in a chaperonin complex (Segnitz and Gehring 1995). Estrogens are bound to steroid hormone binding globulin (SHBG) with high affinity or to albumin with low affinity in the blood. Estrogens freely enter the cell and gain access to the nucleus where they bind to ER with high affinity, thus 'activating' the receptor by inducing alterations in the conformation of ERα that allow dissociation of the hsp90/hsp70 chaperonin complex. This allows the receptors to form homodimers and bind to specific DNA sequences called ERE (in which the minimal consensus ERE is 5'-GGTCAnnnTGACC-3' where n=any nucleotide in the regulatory region of the target gene). Agonist ligand-dependent alterations in ERα enhance direct interaction of ERα via AF-2 in the LBD and AF-1 with coactivator proteins, e.g., SRC-1, that have HAT activity resulting in histone acetylation (Klinge 2000), recruitment of the SWI/SNF ATP-dependent chromatin remodeling complex via interaction with Baf57 (Belandia et al. 2002), and alterations in nucleosomal structure. This results in a 'loosening' of chromatin structure that facilitates the recruitment of RNA polymerase II that initiates transcription of target genes including *pS2, cathepsin D, fos,* and *myc*. (Adapted from Naar et al. 1998 and Zwijsen et al. 1998). **b** The 'tethering mechanism' of estrogen-regulated gene transcription is that ER interacts with another DNA-bound transcription factor and does not directly bind to DNA itself. As shown in this panel, E_2- or DES-occupied ERα; or TAM-, RAL-, or ICI182,780-occupied ERβ; interact with AP-1 (the Fos–Jun heterodimer activated by PKC signaling as shown; Kushner et al. 2000). This results in increased transcription of the indicated genes. **c** New model of an estrogen responsive cell with inactive ERα bound to an ERE with a corepressor complex. Recent ChIP experiments indicate that ERα is bound to EREs in the absence of ligand (Shang and Brown 2002). This keeps chromatin in a repressed state and blocks gene transcription. E_2 binding causes dissociation of the NCoR and HDACs from the gene promoter

et al. 2001). Germline inactivation of the BRCA1 gene accounts for ~15%–45% of hereditary breast cancer (Martin and Weber 2000). The N terminus of BRCA1 interacts directly with the LBD of ERα and suppresses ERα transcriptional activity (Fan et al. 2001). Addition of E_2 resulted in a dissociation of BRCA1, but not ERα, from the pS2 and cathepsin D promoters, prompting the authors to speculate that BRCA1 forms a ligand-reversible barrier to block transcriptional activation by unliganded ERα (Zheng et al. 2001). Putting this new role for BRCA1 into the existing model of ER action suggests that upon E_2 binding, conformational changes in ERα allow BRCA1 dissociation (Zheng et al. 2001), and frees AF-1 and AF-2 to synergistically recruit coactivators, the SWI/SNF complex, and the RNA polymerase II holoenzyme for transcriptional activation.

Other ChIP data indicate that unliganded ERα is bound to EREs in association with NCoR on the pS2 and cyclin D1 promoters (Metivier et al. 2002). This model is shown in Fig. 3c. Corepressors were first identified as binding to class II NR including RAR, RXR, and TR that are bound to DNA in the absence of ligand and are constitutively bound by the corepressors NCoR and SMRT (reviewed by Klinge 2000). Unlike coactivators, which have intrinsic histone acetyl transferase (HAT) activity, corepressors do not have enzymatic activity, but rather recruit a complex of proteins having HDAC activity that is believed to repress gene expression by limiting access of critical transcription factors to the template, by maintaining chromatin in a more condensed state (reviewed by Klinge 2000).

8.4
The 'Tripartite' Model of ER Action

The 'tripartite' model of ER action is that ligands bind ER with different affinities, that distinct ligands alter ER conformation (depicted in Fig. 4), and ligands impact the association of ER with coregulators (Katzenellenbogen et al. 2000). Stimulation of target gene expression in response to E_2, or other agonists, is mediated by two mechanisms. (1) Direct binding: E_2–ER binds directly to a specific ERE and interacts directly with coactivator proteins and components of the RNA polymerase II transcription initiation complex resulting in enhanced transcription (Fig. 3a); (2) Tethering: ER interacts with another DNA-bound transcription factor in a way that stabilizes the DNA binding of that transcription factor and/or recruits coactivators to the complex (Fig. 3b). In mechanism (2), ER does not bind directly to DNA. Examples of genes regulated by ER tethering to the promoter through protein–protein interactions with another transcription factor include: ERα interaction with Sp1 in conferring estrogen responsiveness on a number of genes such as DNA polymerase *α*, *E2F1*, *bcl-2*, *c-fos*, adenosine deaminase, insulin-like growth factor binding protein 4 (*IGF-BP4*), and *RARα1* (reviewed by Safe 2001); ERα interaction with USF-1 and USF-2 in the cathep-

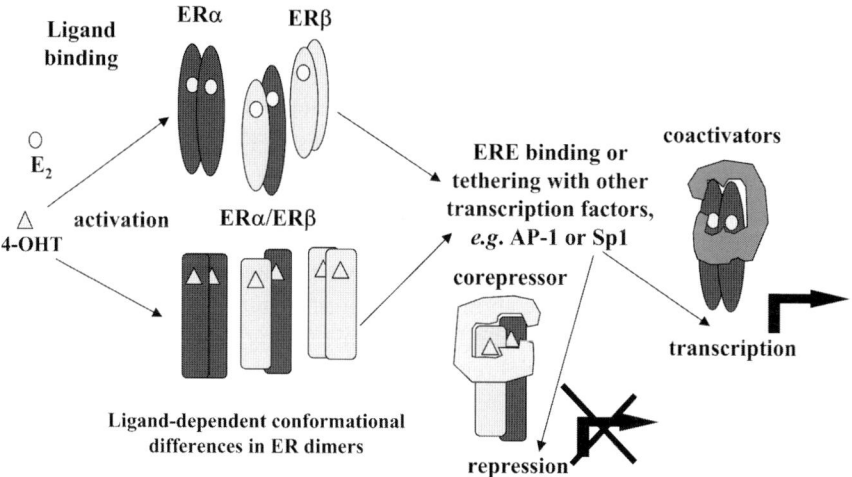

Fig. 4 Tripartite model of ER pharmacology (Katzenellenbogen et al. 2000). Binding of E_2 and 4-OHT to ERα or ERβ result in receptor dimerization forming homodimers or heterodimers as shown. E_2- and 4-OHT-occupied ERs have different conformations. Upon directly binding to EREs or interaction with another DNA-bound transcription factor such as AP-1 or Sp1, the conformation of the E_2- or 4-OHT-occupied ER is altered in turn determining the ability of the receptor to interactive with coregulators and either activate or repress target gene transcription

sin D promoter (Xing and Archer 1998); and ERα and ERβ interaction with AP-1 (Kushner et al. 2000). Both E_2 and TAM activate ERα transcription at AP-1 sites, but TAM, RAL, and ICI182,380 and not E_2 activate ERβ at AP-1 sites (Kushner et al. 2000).

8.5
ER-ERE Binding

ER binds directly to EREs as a dimer with each ER monomer bound to DNA in the major groove with the ER dimer located predominantly on one face of the DNA helix (Koszewski and Notides 1991). Sequence analysis of the 5'-regulatory regions of various estrogen responsive genes revealed a 13-bp palindromic consensus ERE sequence: 5'-GGTCAnnnTGACC-3', where n=any nucleotide, that conferred estrogen responsiveness to reporter genes analyzed by transfection assay (Klein-Hitpass et al. 1986). Most estrogen responsive genes identified to date contain one or more imperfect copies of this ERE or multiple copies of the ERE half-site (reviewed by Klinge 2001). ERα and ERβ bind with the highest affinity, in the subnanomolar range (K_d=0.11 or 0.68 nM for ERα and ERβ, respectively) to a 1-bp longer, i.e., 15-bp ERE palindrome 5'-AGGTCAnnnTGACCT-3' (reviewed by Klinge 2001). Upon ERE binding, ERα bends DNA (Nardulli and Shapiro 1992), perhaps resulting in enhanced recruitment of other factors, e.g., HMG-1, that aide in transcriptional initiation (Romine et al. 1998).

Most estrogen-regulated genes contain imperfect, nonpalindromic EREs. I recently reviewed studies on ER interaction with 38 estrogen-responsive genes whose promoters or 3′ untranslated regions (UTRs) contain functional EREs (Klinge 2001). One conclusion from that review is that the more nucleotide changes there are from the consensus within a half-site of the ERE palindrome, the lower the ERα binding affinity and the lower the transcriptional activity (Klinge 2001). Further, EREs in which nucleotides are altered in both arms of the palindrome show lower transcriptional activity than those containing alterations in only one ERE half-site (Klinge 2001). Differences in ERE sequence allow differential regulation of estrogen target gene expression with genes containing a palindromic ERE more highly estrogen-responsive than genes containing many nucleotide differences from the 15-bp ERE palindrome.

8.6
ERE as an Allosteric Effector of ER Action

Lefstin and Yamamoto proposed that response elements recognized by nuclear transcription factors, including members of the SR/NR superfamily, contain information that is interpreted by bound regulator factors (Lefstin and Yamamoto 1998). Based on Lefstin and Yamamoto's model, we postulated that DNA acts as an allosteric effector whose binding alters ER conformation, resulting in altered ERE binding affinity and transcriptional activity. In turn, ERE-induced changes in ER conformation were predicted to alter ER affinity for other 'ligands,' such as coactivators or corepressors (Klinge et al. 2001). Thus, ER–ERE binding should produce changes in ER conformation that alter ER interaction with other proteins. In concordance with this hypothesis, we observed differences in ERα conformation, assessed by α-chymotrypsin or trypsin digestion, upon interaction with a palindromic ERE or nonpalindromic EREs from the human pS2, PR, and c-*fos* genes and a direct repeat of the ERE half-site separated by 5 bp (DR5) (Klinge et al. 2001; Ramsey and Klinge 2001). The ERE-mediated increase in ERα sensitivity to protease digestion correlated with E_2-stimulated transcription from the same EREs in transiently transfected cells and with the affinity of ERα–ERE binding in vitro (Kulakosky et al. 2002).

Similarly, other investigators reported differential recognition of ERα by select antibodies when ERα was bound to the *Xenopus laevis* vitellogenin A1 ERE (vit A2 ERE) vs. the nonpalindromic ERE from the human pS2 gene in EMSAs (Wood et al. 1998). Limited protease digestion of ERα bound to vit A2 ERE or the nonpalindromic pS2, vitellogenin B1, and oxytocin EREs revealed different sized [^{32}P]DNA fragments in EMSA (Wood et al. 2001). These data indicated that ERα binding to different EREs changes ERα conformation (Wood et al. 2001). Likewise, experiments using phage display revealed conformational differences in ERα and ERβ when bound to vit A2, pS2, lactoferrin, and complement C3 EREs in vitro (Hall et al. 2002). We suggest that these ERE-specific differences in ER conformation allow gene-specific regulation in response to E_2 and other ligands by selective coactivator recruitment.

9
Effect of SERMs on ER Interaction with Coactivators

Ligand-dependent activation of gene transcription by ER requires ligand-dependent ER-coactivator interaction. Over the past 7 years, at least 39 different ERα coactivator proteins have been identified (Klinge 2000; McKenna and O'Malley 2002). Less information is available regarding ERβ–coactivator interaction, although a recent study comparing the interaction of purified ERα and ERβ LBDs with coactivator peptides by FRET revealed that the coactivator peptides from SRC-1, PGC-1, and CBP have different affinities but the same rank order preference for ERα and ERβ (Liu et al. 2003). The current model to account for cell- and ligand-specific regulation of estrogen target gene expression suggests that responsive cells express different levels of coactivators and corepressors. This, along with different concentrations of ERα, ERβ, and ligand, allows fine-tuning of target gene transcription in response to estrogens.

Coactivators have six known functions: (1) acetylation of the N-terminal tails of lysine residues in histones H3 and H4 leading to 'relaxed' chromatin structure (reviewed by Struhl 1998); (2) acetylation of other transcription factors, including ERα (Wang et al. 2001), and coactivators (Huang et al. 2003); (3) recruitment of secondary coactivators including coactivator associated arginine methyltransferase 1 (CARM1) and protein arginine methyltransferase 1 (PRMT1) that methylate histones (Koh et al. 2001); (4) recruitment of ATP-dependent chromatin remodeling complexes, e.g. SWI/SNF (Belandia et al. 2002); (5) direct interaction with and stabilization of basal transcription factor binding (reviewed by Hager et al. 1998); and (6) targeting ER to the proteasome for degradation (Lonard et al. 2000). Studies have demonstrated that coactivators rapidly associate with estrogen target gene promoters upon E_2 treatment in a cyclic manner (Shang et al. 2000). Once the transcription initiation complex is complete, RNA polymerase II is recruited to the transcription start site and begins transcription. This model is depicted in Fig. 5a. The ERE sequence to which ERα is bound alters receptor conformation and in turn impacts the interaction between coactivators and E_2–ERα such that certain coactivators exhibit greater stimulatory activity on one ERE sequence than another (Fig. 5b).

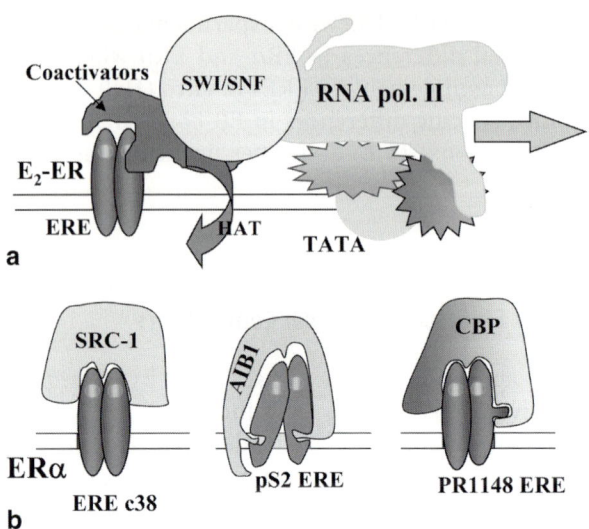

Fig. 5a, b Coactivators interact with E_2–ER and upregulate target gene transcription. a Coactivators such as SRC-1 are recruited to E_2-occupied ER bound to EREs in E_2-regulated genes. Coactivators have HAT activity that along with the ATP-dependent chromatin-remodeling activity of the SWI/SNF complex allows chromatin opening so that components of the transcriptional initiation complex can be recruited and RNA Polymerase II can transcribe the target gene. b E_2–ERα has different conformations in response to binding different ERE sequences such as those found in the *Xenopus* vitellogenin A2 (*EREc38*), *pS2*, and *PR* genes (Klinge et al. 2001; Kulakosky et al. 2002) and differentially recruits coactivators in an ERE-specific manner (Klinge, unpublished results)

The function of coactivators is to enhance SR/NR, and other transcription factor-mediated transcriptional activation by acetylating histones, bringing the ATP–chromatin remodeling complex to the target DNA, and remodeling chromatin structure (Urnov and Wolffe 2001). ERα interaction with members of the p160 family of coactivators, i.e., SRC-1, SRC-2 (GRIP1, TIF2), and SRC-3 (AIB1, ACTR, RAC3, p/CIP), is mediated by interaction between AF-2 in the LBD and the coactivator RID that takes place in agonist liganded-ER, but not antagonist-liganded ER (reviewed by Klinge 2000). Thus, a major mechanism of SERM antiestrogen action is to block ER–coactivator interaction. The inability of coactivators to bind SERM-occupied ER blocks the interaction of ER with chromatin remodeling enzyme complexes the activity of which is necessary for the alteration of nucleosomal structure that allows RNA polymerase II to efficiently transcribe genes (Urnov and Wolffe 2001).

10
Effect of SERMs on ER Interaction with Corepressors

SERMS have been shown to enhance ERα interaction with the corepressors NCoR and SMRT in vitro (reviewed by Klinge 2000) and in transfected cells (Metivier et al. 2002). This is modeled in Fig. 6a. NCoR and SMRT interact with helices H3 and H5 of the ERα LBD in a 4-OHT- or RAL-dependent manner (Yamamoto et al. 2001). In accordance with their antiestrogenic activity, ICI18,780 and RAL are more efficient than TAM in promoting ERα–NCoR interaction in vitro and in vivo (Webb et al. 2003). In turn, NCoR and SMRT associate with HDACs to maintain repressive chromatin structure (Shang and Brown 2002). Treatment of cells with the HDAC inhibitor trichostatin A enhanced ligand-independent as well as E_2-, TAM-, and RAL-stimulated ERα transcription of an ERE reporter gene, indicating that repression of ERα transcriptional activity is mediated by a corepressor–HDAC complex (Webb et al. 2003).

It is likely that additional corepressor proteins that interact with SERM-occupied ER and aide in repression of ER-mediated transcriptional activity remain to be identified. Two recent examples of newly identified repressor are RTA (repressor of tamoxifen transcriptional activity; Norris et al. 2002) and Smad4 (Wu et al. 2003). RTA was isolated by its interaction with the N

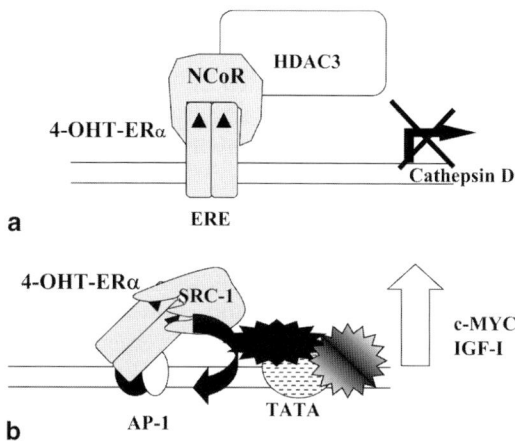

Fig. 6a, b DNA and cell-specific recruitment of corepressors or coactivators to 4-OHT-ERα. **a** ChIP experiments indicated that 4-OHT-ERα and the corepressor NCoR were bound to ERE-containing target genes, e.g., *cathepsin D* as shown, in MCF-7 breast cancer cells along with HDACs, thus repressing gene transcription (Shang and Brown 2002). **b** In Ishikawa human endometrial cancer cells, in which 4-OHT has agonist activity, 4-OHT-ERα was associated with the coactivator SRC-1 at the *c-myc* and *IGF-1* gene promoters, which contain AP-1 binding sites through which ER regulates the transcription of these genes, resulting in gene transcription (Shang and Brown 2002)

terminus of ERα (Norris et al. 2002). RTA decreased TAM, but not E_2, agonist activity on two different ERE–luciferase reporters in transiently transfected HeLa and HepG2 cells (Norris et al. 2002). Smad4, a signal transducer in the TGFβ pathway that acts as a tumor suppressor, was shown to act as a corepressor of ERα occupied by TAM, RAL, and droloxifene (Wu et al. 2003). Further experiments are necessary to understand the roles of these proteins in SERM activity.

11
New Insights into the Mechanism for the Mixed Agonist/Antagonist Activity of Tamoxifen

TAM has agonist activity in human endometrium and in cultured endometrial cells (Shang and Brown 2002). This is why TAM increases the risk of endometrial cancer development in women taking TAM to prevent or treat breast cancer (Jordan and Pappas 2003). The agonist activity of 4-OHT-ERα is determined by AF-1 which is cell type- and promoter context-dependent. Recently, the paradoxical ability of 4-OHT to increase ERα-mediated transcription despite the interaction of 4-OHT–ERα with the corepressors NCoR and SMRT (reviewed by Klinge 2000), has been, in part, elucidated (Shang and Brown 2002). In Ishikawa human endometrial carcinoma cells, both TAM and E_2 upregulated *myc* and *IGF-I* gene transcription, but only E_2 upregulated *cathepsin D* transcription, indicating gene-specific TAM agonist activity (Shang and Brown 2002). In MCF-7 breast cancer cells, only E_2 upregulated these genes (Shang and Brown 2002). These observations mimic the known mixed agonist/antagonist activity of TAM in endometrial and breast tissues. ChIP experiments showed that the corepressors NCoR and SMRT, and HDAC2 and HDAC4 interacted with *c-myc* promoter in MCF-7 cells treated with TAM or RAL, but that only RAL treatment resulted in the occupation of the *myc* promoter by these corepressors and HDACs in Ishikawa cells. With the *cathepsin D* gene promoter, the results with TAM and RAL were identical in both MCF-7 and Ishikawa cells: NCoR and SMRT, and HDAC2 and HDAC4 were recruited, in agreement with the transcription data. Surprisingly, ChIP revealed that TAM-occupied ERα recruited the coactivators SRC-1, AIB1, and CBP to the *c-myc* and *IGF-1* promoters of Ishikawa human endometrial carcinoma, but not MCF-7 human breast cancer cells (Shang and Brown 2002). The human *c-myc* and *IGF-1* genes are not regulated by direct interaction of ER with an ERE, but rather by the tethering of ER to AP-1 bound to its response elements in each promoter. Thus, these studies show that TAM-occupied ERα can interact functionally with the coactivators SRC-1, AIB1, and CBP and upregulate the expression of non-ERE-regulated target genes in Ishikawa and also in ECC-1 human endometrial carcinoma cells (Shang and Brown 2002). This is depicted in Fig. 6b.

Importantly, transfection of Ishikawa cells with small inhibitory RNA for SRC-1 eliminated TAM-stimulated expression of c-Myc and IGF-I. Shang and Brown speculated that binding of coactivators to TAM–ERα is blocked when ERα is directly bound to DNA through a classical ERE, but that when interacting with promoters indirectly, e.g., via an AP-1 site, TAM-bound ERα adopts a conformation that promotes SRC-1 binding (Shang and Brown 2002).

12
ER-Independent Activities of SERMs

Although most research focuses on ER-mediated SERM action, SERMs have non-ER-mediated activities as well. Early studies showed that both ERα-expressing and nonexpressing breast cancer cells were susceptible to TAM-induced cytotoxicity, albeit a higher concentration of TAM was required to inhibit the proliferation of the ER-negative cells (Reddel et al. 1985). TAM inhibited the replication of MDA-MB-231 cells, which are ERα-negative but ERβ-positive, perhaps by inhibiting telomerase activity (Aldous et al. 1999). However, there may be crosstalk between ER-independent and ER-dependent activities of SERMs because TAM blocks E_2 upregulation of telomerase (*hTERT*) gene expression in MCF-7 cells, but TAM increases *hTERT* expression in Ishikawa endometrial cells, reflecting the mixed antagonist/agonist activity of TAM in these cell types (Wang et al. 2002).

TAM activates protein kinases and other cell membrane proteins. TAM stimulated phospholipase D (PLD) in both ER-positive and -negative cells in vivo, whereas RAL inhibited PLD activity in these same cell types (Eisen and Brown 2002). The mechanisms that mediate the effects of TAM, 4-OHT, and RAL on PLD isoforms is unknown. TAM inhibits global protein kinase C (PKC) activity, primarily via the PKC epsilon isozyme, with an apparent 50% inhibitory concentration (IC_{50}) of 25 μM (Clarke et al. 2001). Calmodulin is inhibited by TAM with an IC_{50} of 9 μM (Clarke et al. 2001). TAM can also induce intracellular oxidative stress (Clarke et al. 2001).

TAM and 4-OHT are high affinity ligands for the orphan nuclear receptor ERRγ, but not ERRα or ERRβ (Coward et al. 2001). 4-OHT was an ERRγ-selective antagonist (Coward et al. 2001). However, another group reported that 4-OHT was an antagonist for both ERRγ and ERRβ, but not ERRα (Tremblay et al. 2001). One recent report demonstrated that ERRα and ERRγ were overexpressed at the mRNA level in breast tumors compared to normal mammary epithelial cells and that ERRα overexpression correlated with hormonal insensitivity (Ariazi et al. 2002). The role of ERRs in SERM action remain to be defined.

13
Summary

SERMs are effective in the treatment of estrogen-dependent or responsive breast cancer (Fig. 7). SERMs bind ERα and ERβ and competitively block E_2 binding to the LBD. SERMs result in a conformation of the LBD that is distinct from that of E_2 which in vitro–and on certain genes in cells in which SERMs have antiestrogen activity–precludes ER interaction with coactivators. Because coactivator activity is necessary for the nucleosomal remodeling that is essential for upregulation of target gene transcription, the inhibition of ER–coactivator interaction by SERMs decreases ER-mediated activation of gene transcription. At target genes in cells in which SERMs have antiestrogenic activity, SERM-occupied ER interacts with corepressors NCoR and SMRT and their associated HDACs resulting in repressed chromatin structure and blockade of ER-activated gene transcription. In contrast, in endometrial cancer cells in which the SERM TAM is an agonist, ERα was recently shown to recruit coactivators and upregulate target genes regulated by the tethering mechanism of ER action in which ER interacts with AP-1 bound to its response element (Shang and Brown 2002). Understanding how SERMs impact ER interaction with target gene promoters, coactivators, corepressors, and other proteins will lead to a greater understanding of the role of estrogens and ER in the etiology and progression of breast cancer and should provide new therapeutic targets for disease prevention and control.

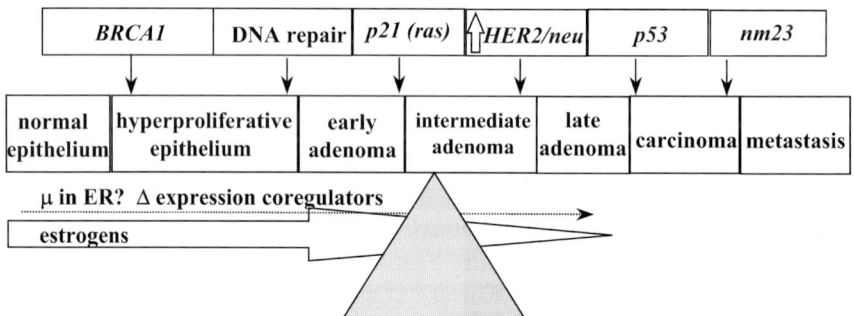

Fig. 7 SERMs block estrogen-activated ER and inhibit the progression of estrogen-dependent breast cancer. Shown is the proposed model of progression in multistage breast tumorigenesis based on Wolff and Weston (1997). Mammary epithelial cells require estrogen for replication, but as genetic changes occur, such as: mutations in BRCA1, p21, ERα, or coregulators; decreased DNA repair; loss of p53 expression; and ultimately, the loss of metastatic suppressor nm23 expression (Ouatas et al. 2003), cells become less responsive to estrogen and also SERM inhibition. The mechanisms for tamoxifen-resistance remain to be fully elucidated but overexpression of HER2/neu/ErbB2 is associated with estrogen-independence (Clarke et al. 2001)

References

Aalfs JD, Kingston RE (2000) What does 'chromatin remodeling' mean? Trends Biochem Sci 25:548–555

Aldous WK, Marean AJ, DeHart MJ, Matej LA, Moore KH (1999) Effects of tamoxifen on telomerase activity in breast carcinoma cell lines. Cancer 85:1523–1529

Allred DC, Baum M, Buzdar AU, Carlson RW, Dowsett M, Elledge RM, Gradishar WJ, Grana G, Howell A, Mamounas EP (2003) A Roundtable discussion of aromatase inhibitors as therapy for breast cancer. Breast J 9:213–222

Ariazi EA, Clark GM, Mertz JE (2002) Estrogen-related receptor alpha and estrogen-related receptor gamma associate with unfavorable and favorable biomarkers, respectively, in human breast cancer. Cancer Res 62:6510–6518

Bai Y, Giguere V (2003) Isoform-selective interactions between estrogen receptors and steroid receptor coactivators promoted by estradiol and ErbB-2 signaling in living cells. Mol Endocrinol 17:589–599

Beatson GT (1896) On the treatment of inoperable cases of carcinoma of the mamma: suggestions for a new method of treatment with illustrative cases. Lancet 2:104–107, 162–167

Belandia B, Orford RL, Hurst HC, Parker MG (2002) Targeting of SWI/SNF chromatin remodelling complexes to estrogen-responsive genes. EMBO J 21:4094–4103

Brzozowski AM, Pike AC, Dauter Z, Hubbard RE, Bonn T, Engstrom O, Ohma L, Greene GL, Gustafsson JA, Carlquist M (1997) Molecular basis of agonism and antagonism in the oestrogen receptor. Nature 389:753–758

Buzdar A (2003) Anastrozole as adjuvant therapy for early-stage breast cancer: implications of the ATAC Trial. Clin Breast Cancer 4 Suppl 1: S42–S48

Carroll JS, Prall OW, Musgrove EA, Sutherland RL (2000) A pure estrogen antagonist Inhibits cyclin E-Cdk2 activity in MCF-7 breast cancer cells and induces accumulation of p130-E2F4 complexes characteristic of quiescence. J Biol Chem 275:38221–38229

Chen H, Lin RJ, Xie W, Wilpitz D, Evans RM (1999) Regulation of hormone-induced histone hyperacetylation and gene activation via acetylation of an acetylase. Cell 98:675–686

Chen X, Danes C, Lowe M, Herliczek TW, Keyomarsi K (2000) Activation of the estrogen-signaling pathway by p21(WAF1/CIP1) in estrogen receptor-negative breast cancer cells. J Natl Cancer Inst 92:1403–1413

Clarke R, Skaar TC, Bouker KB, Davis N, Lee YR, Welch JN, Leonessa F (2001) Molecular and pharmacological aspects of antiestrogen resistance. J Steroid Biochem Mol Biol 76:71–84

Couse JF, Korach KS (1999) Estrogen receptor null mice: what have we learned and where will they lead us? Endocr Rev 20:358–417

Coward P, Lee D, Hull MV, Lehmann JM (2001) 4-Hydroxytamoxifen binds to and deactivates the estrogen-related receptor gamma. Proc Natl Acad Sci USA 98:8880–8884

Cowley SM, Hoare S, Mosselman S, Parker MG (1997) Estrogen receptors alpha and beta form heterodimers on DNA. J. Biol. Chem. 272:19858–19862

Cowley SM, Parker MG (1999) A comparison of transcriptional activation by ER alpha and ER beta. J Steroid Biochem Mol Biol 69:165–175

Crewe HK, Ellis SW, Lennard MS, Tucker GT (1997) Variable contribution of cytochromes P450 2D6, 2C9 and 3A4 to the 4-hydroxylation of tamoxifen by human liver microsomes. Biochem Pharmacol 53:171–178

Cuzick J, Powles T, Veronesi U, Forbes J, Edwards R, Ashley S, Boyle P, Baum M, Cawthorn S, Coates A, Hamed A, Howell A (2003) Overview of the main outcomes in breast-cancer prevention trials: First results from the International Breast Cancer Intervention Study (IBIS-I): a randomised prevention trial. Lancet 361:296–300

De Leo V, la Marca A, Morgante G, Lanzetta D, Setacci C, Petraglia F (2001) Randomized control study of the effects of raloxifene on serum lipids and homocysteine in older women. Am J Obstet Gynecol 184:350–353

Delaunay F, Pettersson K, Tujague M, Gustafsson JA (2000) Functional differences between the amino-terminal domains of estrogen receptors alpha and beta. Mol Pharmacol 58:584–590

Devlin-Leclerc J, Meng X, F. D, Leclerc P, Baulieu EE, Catelli MG (1998) Interaction and dissociation by ligands of estrogen receptor and Hsp90: the antiestrogen RU 58668 induces a protein synthesis-dependent clustering of the receptor in the cytoplasm. Mol Endocrinol 12:842–854

Dowsett M, Harper-Wynne C, Boeddinghaus I, Salter J, Hills M, Dixon M, Ebbs S, Gui G, Sacks N, Smith I (2001) HER-2 amplification impedes the antiproliferative effects of hormone therapy in estrogen receptor-positive primary breast cancer. Cancer Res 61:8452–8458

Dudley MW, Sheeler CQ, Wang H, Khan S (2000) Activation of the human estrogen receptor by the antiestrogens ICI 182,780 and tamoxifen in yeast genetic systems: Implications for their mechanism of action. Proc Natl Acad Sci USA 97:3696–3701

Eisen SF, Brown HA (2002) Selective estrogen receptor (ER) modulators differentially regulate phospholipase D catalytic activity in ER-negative breast cancer cells. Mol Pharmacol 62:911–920

Ellis PA, Saccani-Jotti G, Clarke R, Johnston SR, Anderson E, Howell A, A'Hern R, Salter J, Detre S, Nicholson R, Robertson J, Smith IE, Dowsett M (1997) Induction of apoptosis by tamoxifen and ICI 182780 in primary breast cancer. Int J Cancer 72:608–613

Elsby R, Ashby J, Sumpter JP, Brooks AN, Pennie WD, Maggs JL, Lefevre PA, Odum J, Beresford N, Paton D, Park BK (2000) Obstacles to the prediction of estrogenicity from chemical structure: assay-mediated metabolic transformation and the apparent promiscuous nature of the estrogen receptor. Biochem Pharmacol 60:1519–1530

Endoh H, Maruyama K, Masuhiro Y, Kobayashi Y, Goto M, Tai H, Yanagisawa J, Metzger D, Hashimoto S, Kato S (1999) Purification and identification of p68 RNA helicase acting as a transcriptional coactivator specific for the activation function 1 of human estrogen receptor alpha. Mol Cell Biol 19:5363–5372

Fan S, Ma YX, Wang C, Yuan RQ, Meng Q, Wang JA, Erdos M, Goldberg ID, Webb P, Kushner PJ, Pestell RG, Rosen EM (2001) Role of direct interaction in BRCA1 inhibition of estrogen receptor activity. Oncogene 20:77–87

Fisher B, Dignam J, Bryant J, DeCillis A, Wickerham DL, Wolmark N, Costantino J, Redmond C, Fisher ER, Bowman DM, Deschenes L, Dimitrov NV, Margolese RG, Robidoux A, Shibata H, Terz J, Paterson AH, Feldman MI, Farrar W, Evans J, Lickley HL (1996) Five versus more than five years of tamoxifen therapy for breast cancer patients with negative lymph nodes and estrogen receptor-positive tumors. J Natl Cancer Inst 88:1529–1542

Flouriot G, Brand H, Denger S, Metivier R, Kos M, Reid G, Sonntag-Buck V, Gannon F (2000) Identification of a new isoform of the human estrogen receptor-alpha (hER-alpha) that is encoded by distinct transcripts and that is able to repress hER-alpha activation function 1. EMBO J 19:4688–4700

Frasor J, Barnett DH, Danes JM, Hess R, Parlow AF, Katzenellenbogen BS (2003) Response-specific and ligand dose-dependent modulation of estrogen receptor (ER) alpha activity by ERbeta in the uterus. Endocrinology 144:3159–3166

Freedman AN, Graubard BI, Rao SR, McCaskill-Stevens W, Ballard-Barbash R, Gail MH (2003) Estimates of the Number of U.S. Women Who Could Benefit From Tamoxifen for Breast Cancer Chemoprevention. J Natl Cancer Inst 95:526–532

Fuchs-Young R, Glasebrook AL, Short LL, Draper MW, Rippy MK, Cole HW, Magee DE, Termine JD, Bryant HU (1995) Raloxifene is a tissue-selective agonist/antagonist that functions through the estrogen receptor. Ann NY Acad Sci 761:355–360

Fuqua SA, Schiff R, Parra I, Friedrichs WE, Su JL, McKee DD, Slentz-Kesler K, Moore LB, Willson TM, Moore JT (1999) Expression of wild-type estrogen receptor beta and variant isoforms in human breast cancer. Cancer Res 59:5425–5428

Green S, Chambon P (1987) Oestradiol induction of a glucocorticoid-responsive gene by a chimaeric receptor. Nature 325:75–78

Gustafsson J, Warner M (2000) Estrogen receptor beta in the breast: role in estrogen responsiveness and development of breast cancer. J Steroid Biochem Mol Biol 74:245–248

Hager GL, Smith CL, Fragoso G, Wolford R, Walker D, Barsony J, Hunt H (1998) Intranuclear trafficking and gene targeting by members of the steroid/nuclear receptor superfamily. J. Steroid Biochem Mol Biol 65:125–132

Hall JM, Chang CY, McDonnell DP (2000) Development of peptide antagonists that target estrogen receptor beta-coactivator interactions. Mol Endocrinol 14:2010–2023

Hall JM, Korach KS (2003) Stromal cell-derived factor 1, a novel target of estrogen receptor action, mediates the mitogenic effects of estradiol in ovarian and breast cancer cells. Mol Endocrinol 17:792–803

Hall JM, McDonnell DP (1999) The estrogen receptor beta-isoform (ERbeta) of the human estrogen receptor modulates ERalpha transcriptional activity and is a key regulator of the cellular response to estrogens and antiestrogens. Endocrinology 140:5566–5578

Hall JM, McDonnell DP, Korach KS (2002) Allosteric regulation of estrogen receptor structure, function, and coactivator recruitment by different estrogen response elements. Mol Endocrinol 16:469–486

Hodges LC, Cook JD, Lobenhofer EK, Li L, Bennett L, Bushel PR, Aldaz CM, Afshari CA, Walker CL (2003) Tamoxifen functions as a molecular agonist inducing cell cycle-associated genes in breast cancer cells. Mol Cancer Res 1:300–311

Horwitz KB (1994) How do breast cancers become hormone resistant? J Steroid Biochem Mol Biol 49:295–302

Huang ZQ, Li J, Sachs LM, Cole PA, Wong J (2003) A role for cofactor-cofactor and cofactor-histone interactions in targeting p300, SWI/SNF and mediator for transcription. EMBO J 22:2146–155

Iwao K, Miyoshi Y, Egawa C, Ikeda N, Noguchi S (2000) Quantitative analysis of estrogen receptor-beta mRNA and its variants in human breast cancers. Int J Cancer 88:733–736

Jackson TA, Richer JK, Bain DL, Takimoto GS, Tung l, Horwitz KB (1997) The partial agonist activity of antagonist-occupied steroid receptors is controlled by a novel hinge domain-binding coactivator L7/SPA and the corepressors N-CoR or SMRT. Mol. Endocrinol. 11:693–705

Jensen EV (1996) Steroid hormones, receptors, and antagonists. Ann NY Acad Sci 784:1–17

Jordan VC (1997) Tamoxifen treatment for breast cancer: concept to gold standard. Oncology: 7–13

Jordan VC (2001) Selective estrogen receptor modulation: a personal perspective. Cancer Res 61:5683–5687

Jordan VC (2002) A new day dawns: women without oestrogen or is a balance best? Breast Cancer Res 4:218-221

Jordan VC, Pappas SG (2003) Is tamoxifen the Rosetta stone for breast cancer? Tamoxifen: a most unlikely pioneering medicine: Chemoprevention of breast cancer: current and future prospects. J Natl Cancer Inst 95:338-340

Katzenellenbogen BS, Montano MM, Ediger TR, Sun J, Ekena K, Lazennec G, Martini PG, McInerney EM, Delage-Mourroux R, Weis K, Katzenellenbogen JA (2000) Estrogen receptors: selective ligands, partners, and distinctive pharmacology. Recent Prog Horm Res 55:163-193; discussion 194-195

Katzenellenbogen BS, Norman MJ, Eckert RL, Peltz SW, Mange WF (1984) Bioactivities, estrogen receptor interactions, and plasminogen activator- inducing activities of tamoxifen and hydroxytamoxifen isomers in MCF-7 human breast cancer cells. Cancer Res. 44:112-119

Klein-Hitpass L, Schorpp M, Wagner U, Ryffel GU (1986) An estrogen-responsive element derived from the 5' flanking region of the *Xenopus* vitellogenin A2 gene functions in transfected human cells. Cell 46:1053-1061

Klinge CM (2000) Estrogen receptor interaction with co-activators and co-repressors. Steroids 65:227-251

Klinge CM (2001) Estrogen receptor interaction with estrogen response elements. Nucleic Acids Res 29:2905-2919

Klinge CM, Bambara RA, Hilf R (1992) What differentiates antiestrogen-liganded *versus* estradiol-liganded estrogen receptor action? Oncology Res 4:1073-1081

Klinge CM, Jernigan SC, Smith SL, Tyulmenkov VV, Kulakosky PC (2001) Estrogen response element sequence impacts the conformation and transcriptional activity of estrogen receptor α. Mol Cell Endocrinol 174:151-166

Koh SS, Chen D, Lee YH, Stallcup MR (2001) Synergistic enhancement of nuclear receptor function by p160 coactivators and two coactivators with protein methyltransferase activities. J Biol Chem 276:1089-1098

Koszewski NJ, Notides AC (1991) Phosphate-sensitive binding of the estrogen receptor to its response elements. Mol Endocrinol 5:1129-1136

Kraichely DM, Sun J, Katzenellenbogen JA, Katzenellenbogen BS (2000) Conformational changes and coactivator recruitment by novel ligands for estrogen receptor-alpha and estrogen receptor-beta: correlations with biological character and distinct differences among SRC coactivator family members. Endocrinology 141:3534-3545

Kuiper GG, Carlsson B, Grandien J, Enmark E, Haggblad J, Nilsson S, Gustafsson J-A (1997) Comparison of the ligand binding specificity and transcript tissue distribution of estrogen receptors α and β. Endocrinology 138:863-870

Kuiper GG, Enmark E, Pelto-Huikko M, Nilsson S, Gustafsson J-A (1996) Cloning of a novel estrogen receptor expressed in rat prostate and ovary. Proc Natl Acad Sci USA 93:5925-5930

Kuiper GG, Shughrue PJ, Merchenthaler I, Gustafsson JA (1998) The estrogen receptor beta subtype: a novel mediator of estrogen action in neuroendocrine systems. Front Neuroendocrinol 19:253-286

Kulakosky PC, Jernigan SC, McCarty MA, Klinge CM (2002) Response element sequence regulates estrogen receptor alpha and beta affinity and activity. J Mol Endocrinol 29:137-152

Kushner PJ, Agard D, Feng WJ, Lopez G, Schiau A, Uht R, Webb P, Greene G (2000) Oestrogen receptor function at classical and alternative response elements. Novartis Found Symp 230:20-26; discussion 27-40

Lannigan DA (2003) Estrogen receptor phosphorylation. Steroids 68:1-9

Laudet V (1997) Evolution of the nuclear receptor superfamily: early diversification from an ancestral orphan receptor. J Mol Endocrinol 19:207–226

Lazennec G, Bresson D, Lucas A, Chauveau C, Vignon F (2001) ERbeta inhibits proliferation and invasion of breast cancer cells. Endocrinology 142:4120–4130

Lefstin JA, Yamamoto KR (1998) Allosteric effects of DNA on transcriptional regulators. Nature 392:885–888

Lerner LJ, Jordan VC (1990) Development of antiestrogens and their use in breast cancer: Eighth Cain memorial award lecture. Cancer Res 50:4177–4189

Leygue E, Dotzlaw H, Watson PH, Murphy LC (1999) Expression of estrogen receptor beta1, beta2, and beta5 messenger RNAs in human breast tissue. Cancer Res 59:1175–1179

Lieberman ME, Jordan VC, Fritsch M, Santos MA, Gorski J (1983) Direct and reversible inhibition of estradiol-stimulated prolactin synthesis by antiestrogens in vitro. J Biol Chem 258:4734–4740

Liu J, Knappenberger KS, Kack H, Andersson G, Nilsson E, Dartsch C, Scott CW (2003) A homogeneous in vitro functional assay for estrogen receptors: Coactivator recruitment. Mol Endocrinol 17:346–355

Llopis J, Westin S, Ricote M, Wang J, Cho CY, Kurokawa R, Mullen TM, Rose DW, Rosenfeld MG, Tsien RY, Glass CK (2000) Ligand-dependent interactions of coactivators steroid receptor coactivator-1 and peroxisome proliferator-activated receptor binding protein with nuclear hormone receptors can be imaged in live cells and are required for transcription. Proc Natl Acad Sci USA 97:4363–4368

Lonard DM, Nawaz Z, Smith CL, O'Malley BW (2000) The 26S proteasome is required for estrogen receptor-alpha and coactivator turnover and for efficient estrogen receptor-alpha transactivation. Mol Cell 5:939–948

Long BJ, Tilghman SL, Yue W, Thiantanawat A, Grigoryev DN, Brodie AM (1998) The steroidal antiestrogen ICI 182,780 is an inhibitor of cellular aromatase activity. J Steroid Biochem Mol Biol 67:293–304

Maglich JM, Sluder A, Guan X, Shi Y, McKee DD, Carrick K, Kamdar K, Willson TM, Moore JT (2001) Comparison of complete nuclear receptor sets from the human, Caenorhabditis elegans and Drosophila genomes. Genome Biol 2: RESEARCH0029

Martin AM, Weber BL (2000) Genetic and hormonal risk factors in breast cancer. J Natl Cancer Inst 92:1126–1135

Maruvada P, Baumann CT, Hager GL, Yen PM (2003) Dynamic shuttling and intranuclear mobility of nuclear hormone receptors. J Biol Chem 278:12425–12432

Matsuda K, Ochiai I, Nishi M, Kawata M (2002) Colocalization and ligand-dependent discrete distribution of the estrogen receptor (ER)alpha and ERbeta. Mol Endocrinol 16:2215–2230

McGuire WL, Clark GM, Dressler LG, Owens MA (1986) Role of steroid hormone receptors as prognostic factors in primary breast cancer. NCI Monogr: 19–23

McInerney EM, Tsai M-J, O'Malley BW, Katzenellenbogen BS (1996) Analysis of estrogen receptor transcriptional enhancement by a nuclear receptor coactivator. Proc Natl Acad Sci USA 93:10069–10073

McKenna NJ, O'Malley BW (2002) Minireview: nuclear receptor coactivators-an update. Endocrinology 143:2461–2465

Metivier R, Stark A, Flouriot G, Hubner MR, Brand H, Penot G, Manu D, Denger S, Reid G, Kos M, Russell RB, Kah O, Pakdel F, Gannon F (2002) A dynamic structural model for estrogen receptor-alpha activation by ligands, emphasizing the role of interactions between distant A and E domains. Mol Cell 10:1019–1032

Meyers MJ, Sun J, Carlson KE, Marriner GA, Katzenellenbogen BS, Katzenellenbogen JA (2001) Estrogen receptor-beta potency-selective ligands: structure-activity relationship studies of diarylpropionitriles and their acetylene and polar analogues. J Med Chem 44:4230–4251

Montano MM, Muller V, Trogaugh A, Katzenellenbogen BS (1995) The carboxy-terminal F domain of the human estrogen receptor: Role in the transcriptional activity of the receptor and the effectiveness of antiestrogens as estrogen antagonists. Mol. Endocrinol. 9:814–825

Morris C, Wakeling A, Robertson JF, Nicholson RI, Bundred NJ, Anderson E, Rayter Z, Dowsett M, Fox JN, Gee JM, Webster A, Wakeling AE, Dixon M, Howell A, Osborne CK (2002) Fulvestrant ('Faslodex')—a new treatment option for patients progressing on prior endocrine therapy. Comparison of the short-term biological effects of 7alpha-[9-(4,4,5,5,5-pentafluoropentylsulfinyl)-nonyl]estra-1,3,5, (10)-triene-3,17beta-diol (Faslodex) versus tamoxifen in postmenopausal women with primary breast cancer. ICI 182,780 (Faslodex): development of a novel, 'pure' antiestrogen. Endocr Relat Cancer 9:267–276

Naar AM, Beaurang PA, Robinson KM, Oliner JD, Avizonis D, Scheek S, Zwicker J, Kadonaga JT, Tjian R (1998) Chromatin, TAFs, and a novel multiprotein coactivator are required for synergistic activation by Sp1 and SREBP-1a *in vitro*. Genes Dev 12:3020–3031

Nardulli AM, Shapiro DJ (1992) Binding of the estrogen receptor DNA-binding domain to the estrogen response element induces DNA bending. Mol Cell Biol 12:2037–2042

Nilsson S, Makela S, Treuter E, Tujague M, Thomsen J, Andersson G, Enmark E, Pettersson K, Warner M, Gustafsson JA (2001) Mechanisms of estrogen action. Physiol Rev 81:1535–1565

Norris JD, Fan D, Sherk A, McDonnell DP (2002) A negative coregulator for the human ER. Mol Endocrinol 16:459–468

Norris JD, Paige LA, Christensen DJ, Chang CY, Huacani MR, Fan D, Hamilton PT, Fowlkes DM, McDonnell DP (1999) Peptide antagonists of the human estrogen receptor. Science 285:744–746

Oesterreich S, Zhang Q, Hopp T, Fuqua SA, Michaelis M, Zhao HH, Davie JR, Osborne CK, Lee AV (2000) Tamoxifen-bound estrogen receptor (ER) strongly interacts with the nuclear matrix protein HET/SAF-B, a novel inhibitor of ER-mediated transactivation. Mol Endocrinol 14:369–381

Omoto Y, Inoue S, Ogawa S, Toyama T, Yamashita H, Muramatsu M, Kobayashi S, Iwase H (2001) Clinical value of the wild-type estrogen receptor beta expression in breast cancer. Cancer Lett 163:207–212

Osborne CK, Coronado-Heinsohn EB, Hilsenbeck SG, McCue BL, Wakeling AE, McClelland RA, Manning DL, Nicholson RI (1995) Comparison of the effects of a pure steroidal antiestrogen with those of tamoxifen in a model of human breast cancer. J Natl Cancer Inst 87:746–750

Ouatas T, Salerno M, Palmieri D, Steeg PS (2003) Basic and translational advances in cancer metastasis: Nm23. J Bioenerg Biomembr 35:73–79

Pace P, Taylor J, Suntharalingam S, Coombes RC, Ali S (1997) Human estrogen receptor beta binds DNA in a manner similar to and dimerizes with estrogen receptor alpha. J Biol Chem 272:25832–25838

Pakdel F, Le Goff P, Katzenellenbogen BS (1993a) An assessment of the role of domain F and PEST sequences in estrogen receptor half-life and bioactivity. J Steroid Biochem Mol Biol 46:663–672

Pakdel F, Reese JC, Katzenellenbogen BS (1993b) Identification of charged residues in an N-terminal portion of the hormone-binding domain of human estrogen receptor important in transcriptional activity of the receptor. Mol Endocrinol 7:1408–1417

Pike AC, Brzozowski AM, Hubbard RE (2000) A structural biologist's view of the oestrogen receptor. J Steroid Biochem Mol Biol 74:261–268

Pike AC, Brzozowski AM, Hubbard RE, Bonn T, Thorsell AG, Engstrom O, Ljunggren J, Gustafsson JA, Carlquist M (1999) Structure of the ligand-binding domain of oestrogen receptor beta in the presence of a partial agonist and a full antagonist. EMBO J 18:4608–4618

Pike AC, Brzozowski AM, Walton J, Hubbard RE, Thorsell A, Li Y, Gustafsson J, Carlquist M (2001) Structural insights into the mode of action of a pure antiestrogen. Structure 9:145–153

Pink JJ, Jordan VC (1996) Models of estrogen receptor regulation by estrogens and antiestrogens in breast cancer cell lines. Cancer Res 56:2321–2330

Ramsey TL, Klinge CM (2001) Estrogen response element binding induces alterations in estrogen receptor a conformation as revealed by susceptibility to partial proteolysis. J. Mol. Endocrinol. 27:275–292

Reddel RR, Murphy LC, Hall RE, Sutherland RL (1985) Differential sensitivity of human breast cancer cell lines to the growth-inhibitory effects of tamoxifen. Cancer Res 45:1525–1531

Reid G, Hubner MR, Metivier R, Brand H, Denger S, Manu D, Beaudouin J, Ellenberg J, Gannon F (2003) Cyclic, proteasome-mediated turnover of unliganded and liganded ERalpha on responsive promoters is an integral feature of estrogen signaling. Mol Cell 11:695–707

Rich RL, Hoth LR, Geoghegan KF, Brown TA, LeMotte PK, Simons SP, Hensley P, Myszka DG (2002) Kinetic analysis of estrogen receptor/ligand interactions. Proc Natl Acad Sci USA 99:8562–8567

Romine LE, Wood JR, Lamia LA, Prendergast P, Edwards DP, Nardulli AM (1998) The high mobility group protein 1 enhances binding of the estrogen receptor DNA binding domain to the estrogen response element. Mol Endocrinol 12:664–674

Russo J, Hu YF, Tahin Q, Mihaila D, Slater C, Lareef MH, Russo IH (2001) Carcinogenicity of estrogens in human breast epithelial cells. Apmis 109:39–52

Safe S (2001) Transcriptional activation of genes by 17 beta-estradiol through estrogen receptor-Sp1 interactions. Vitam Horm 62:231–252

Samuni AM, Chuang EY, Krishna MC, Stein W, DeGraff W, Russo A, Mitchell JB (2003) Semiquinone radical intermediate in catecholic estrogen-mediated cytotoxicity and mutagenesis: chemoprevention strategies with antioxidants. Proc Natl Acad Sci USA 100:5390–5395

Schwabe JWR, Chapman L, Finch JT, Rhodes D (1993) The crystal structure of the estrogen receptor DNA-binding domain bound to DNA: How receptors discriminate between their response elements. Cell 75:567–578

Segnitz B, Gehring U (1995) Subunit structure of the nonactivated human estrogen receptor. Proc Natl Acad Sci USA 92:2179–2183

Shang Y, Brown M (2002) Molecular determinants for the tissue specificity of SERMs. Science 295:2465–2468

Shang Y, Hu X, DiRenzo J, Lazar MA, Brown M (2000) Cofactor dynamics and sufficiency in estrogen receptor-regulated transcription. Cell 103:843–852

Shapiro CL, Henderson IC (1994) Adjuvant therapy of breast cancer. Hematol Oncol Clinics of North America 8:213–231

Shiau AK, Barstad D, Loria PM, Cheng L, Kushner PJ, Agard DA, Greene GL (1998) The structural basis of estrogen receptor/coactivator recognition and the antagonism of this interaction by tamoxifen. Cell 95:927-937

Shiau AK, Barstad D, Radek JT, Meyers MJ, Nettles KW, Katzenellenbogen BS, Katzenellenbogen JA, Agard DA, Greene GL (2002) Structural characterization of a subtype-selective ligand reveals a novel mode of estrogen receptor antagonism. Nat Struct Biol 9:359-364

Shibutani S, Ravindernath A, Suzuki N, Terashima I, Sugarman SM, Grollman AP, Pearl ML (2000) Identification of tamoxifen-DNA adducts in the endometrium of women treated with tamoxifen. Carcinogenesis 21:1461-1467

Simoncini T, Varone G, Fornari L, Mannella P, Luisi M, Labrie F, Genazzani AR (2002) Genomic and nongenomic mechanisms of nitric oxide synthesis induction in human endothelial cells by a fourth-generation selective estrogen receptor modulator. Endocrinology 143:2052-2061

Smigel K (1998) Breast cancer prevention trial shows major benefit, some risk. J Natl Cancer Inst 90:647-648

Struhl K (1998) Histone acetylation and transcriptional regulatory mechanisms. Genes Dev 12:599-606

Sun J, Baudry J, Katzenellenbogen JA, Katzenellenbogen BS (2003) Molecular basis for the subtype discrimination of the estrogen receptor-beta-selective ligand, diarylpropionitrile. Mol Endocrinol 17:247-258

Tanenbaum DM, Wang Y, Williams SP, Sigler PB (1998) Crystallographic comparison of the estrogen and progesterone receptor's ligand binding domains. Proc Natl Acad Sci USA 95:5998-6003

Thornton JW (2001) Evolution of vertebrate steroid receptors from an ancestral estrogen receptor by ligand exploitation and serial genome expansions. Proc Natl Acad Sci USA 98:5671-5676

Toi M, Bando H, Saji S (2003) Decision tree and paradigms of primary breast cancer: changes elicited by preoperative therapy. Med Sci Monit 9: RA90-RA95

Tremblay GB, Bergeron D, Giguere V (2001) 4-Hydroxytamoxifen is an isoform-specific inhibitor of orphan estrogen-receptor-related (ERR) nuclear receptors beta and gamma. Endocrinology 142:4572-4575

Tsai KS, Yen ML, Pan HA, Wu MH, Cheng WC, Hsu SH, Yen BL, Huang KE, Anderson PW, Cox DA, Sashegyi A, Paul S, Silfen SL, Walsh BW, Godsland IF, De Leo V, la Marca A, Morgante G, Lanzetta D, Setacci C, Petraglia F, Johnston CC, Jr., Bjarnason NH, Cohen FJ, Shah A, Lindsay R, Mitlak BH, Huster W, Draper MW, Harper KD, Heath H, 3rd, Gennari C, Christiansen C, Arnaud CD, Delmas PD, Hozumi Y, Kawano M, Jordan VC (2001) Raloxifene versus continuous combined estrogen/progestin therapy: densitometric and biochemical effects in healthy postmenopausal Taiwanese women. Effects of raloxifene and hormone replacement therapy on markers of serum atherogenicity in healthy postmenopausal women. Effects of postmenopausal hormone replacement therapy on lipid, lipoprotein, and apolipoprotein (a) concentrations: analysis of studies published from 1974-2000. Randomized control study of the effects of raloxifene on serum lipids and homocysteine in older women. Long-term effects of raloxifene on bone mineral density, bone turnover, and serum lipid levels in early postmenopausal women: three-year data from 2 double-blind, randomized, placebo-controlled trials. In vitro study of the effect of raloxifene on lipid metabolism compared with tamoxifen. Osteoporos Int 12:1020-1025

Tzukerman MT, Esty A, Santiso-Mere D, Danielian P, Parker M, Stein R, B., Pike JW, McDonnell DP (1994) Human estrogen receptor transactivational capacity is determined by both cellular and promoter context and mediated by two functionally distinct intramolecular regions. Mol Endocrinol 8:21–30

Urnov FD, Wolffe AP (2001) A necessary good: nuclear hormone receptors and their chromatin templates. Mol Endocrinol 15:1–16

Wahli W, Martinez E (1991) Superfamily of steroid nuclear receptors: Positive and negative regulators of gene expression. FASEB J 5:2243–2249

Wakeling AE (1989) Comparative studies on the effects of steroidal and nonsteroidal oestrogen antagonists on the proliferation of human breast cancer cells. J Steroid Biochem 34:183–188

Wakeling AE (1993) Are breast tumours resistant to tamoxifen also resistant to pure antioestrogens? J Steroid Biochem Mol Biol 47:107–114

Wakeling AE, Bowler J (1992) ICI 182,780, a new antioestrogen with clinical potential. J Steroid Biochem Mol Biol 43:173–177

Wakeling AE, Dukes M, Bowler J (1991) A potent specific pure antiestrogen with clinical potential. Cancer Res 51:3867–3873

Wang C, Fu M, Angeletti RH, Siconolfi-Baez L, Reutens AT, Albanese C, Lisanti MP, Katzenellenbogen BS, Kato S, Hopp T, Fuqua SA, Lopez GN, Kushner PJ, Pestell RG (2001) Direct acetylation of the estrogen receptor alpha hinge region by p300 regulates transactivation and hormone sensitivity. J Biol Chem 276:18375–18383

Wang Z, Kyo S, Maida Y, Takakura M, Tanaka M, Yatabe N, Kanaya T, Nakamura M, Koike K, Hisamoto K, Ohmichi M, Inoue M (2002) Tamoxifen regulates human telomerase reverse transcriptase (hTERT) gene expression differently in breast and endometrial cancer cells. Oncogene 21:3517–3524

Weatherman RV, Scanlan TS (2001) Unique protein determinants of the subtype-selective ligand responses of the estrogen receptors (ERalpha and ERbeta) at AP-1 sites. J Biol Chem 276:3827–3832

Webb P, Nguyen P, Kushner PJ (2003) Differential SERM effects on corepressor binding dictate ERalpha activity in vivo. J Biol Chem 278:6912–6920

Webb P, Nguyen P, Shinsako J, Anderson C, Feng W, Nguyen MP, Chen D, Huang SM, Subramanian S, McKinerney E, Katzenellenbogen BS, Stallcup MR, Kushner PJ (1998) Estrogen receptor activation function 1 works by binding p160 coactivator proteins. Mol Endocrinol 12:1605–1618

Weihua Z, Andersson S, Cheng G, Simpson ER, Warner M, Gustafsson JA (2003) Update on estrogen signaling. FEBS Lett 546:17–24

Wijayaratne AL, McDonnell DP (2001) The human estrogen receptor-alpha is a ubiquitinated protein whose stability is affected differentially by agonists, antagonists, and selective estrogen receptor modulators. J Biol Chem 276:35684–35692

Wolff MS, Weston A (1997) Breast cancer risk and environmental exposures. Environ. Health Perspect. 105:891–896

Wood JR, Greene GL, Nardulli AM (1998) Estrogen response elements function as allosteric modulators of estrogen receptor conformation. Mol Cell Biol 18:1927–1934

Wood JR, Likhite VS, Loven MA, Nardulli AM (2001) Allosteric modulation of estrogen receptor conformation by different estrogen response elements. Mol Endocrinol 15:1114–1126

Wu L, Wu Y, Gathings B, Wan M, Li X, Grizzle W, Liu Z, Lu C, Mao Z, Cao X (2003) Smad4 as a transcription corepressor for estrogen receptor alpha. J Biol Chem 278:15192–15200

Xing W, Archer TK (1998) Upstream stimulatory factors mediate estrogen receptor activation of the cathepsin D promoter. Mol Endocrinol 12:1310–1321

Yamamoto Y, Wada O, Suzawa M, Yogiashi Y, Yano T, Kato S, Yanagisawa J (2001) The tamoxifen-responsive estrogen receptor alpha mutant D351Y shows reduced tamoxifen-dependent interaction with corepressor complexes. J Biol Chem 276:42684–42691

Zheng L, Annab LA, Afshari CA, Lee WH, Boyer TG (2001) BRCA1 mediates ligand-independent transcriptional repression of the estrogen receptor. Proc Natl Acad Sci USA 98:9587–9592

Zou A, Marschke KB, Arnold KE, Berger EM, Fitzgerald P, Mais DE, Allegretto EA (1999) Estrogen receptor beta activates the human retinoic acid receptor alpha- 1 promoter in response to tamoxifen and other estrogen receptor antagonists, but not in response to estrogen. Mol Endocrinol 13:418–430

Zwijsen RML, Buckle RS, Hijmans EM, Loomans CJM, Bernards R (1998) Ligand-independent recruitment of steroid receptor coactivators to estrogen receptor by cyclin D. Genes Dev 12:3488–3498

Artificial Zinc Finger Peptides: A Promising Tool in Biotechnology and Medicine

N. Corbi · V. Libri · C. Passananti (✉)

CNA, Istituto Biologia e Patologia Molecolari (IBPM), Viale Marx 43, 00137 Rome, Italy
claudio.passananti@ibpm.cnr.it

1	Zinc Finger DNA-Binding Domain	491
2	Zinc Finger Recognition Code	493
2.1	Zinc Finger Domains Selected and Assembled as Building Blocks	494
3	Control of Gene Expression by Artificial ZFP TFs: Advantages	496
3.1	Control of Gene Expression by Artificial ZFP TFs: Specificity and Effectiveness	498
4	Applications: A Promise in Target Validation, Biotechnology and Transcription Therapy	499
5	Approaches to Therapy: Perspectives	502
References		503

Abstract Since their first description about two decades ago, zinc finger DNA-binding proteins have constantly stimulated the creativity of scientists. In particular, the high versatility and modularity of the zinc finger domains make them optimal building blocks for constructing artificial transcription factors potentially able to control the expression of any desired gene in cells and organisms. Disease-related genes can be either turned off or enhanced for the purpose of treating conditions such as genetic disease, cancer or viral infection. Alternatively, the expression of beneficial genes can be boosted to generate animals or plants with advantageous characteristics. Moreover, zinc finger peptides can further expand the number of possible applications on the basis of their peculiar capability to specifically bind RNA. The opportunity to re-program the expression of specific genes at will represents a powerful tool in basic science, biotechnology and molecular medicine. In this chapter, we will focus on the different approaches followed to design and to deliver artificial zinc finger transcription factors with the final aim of approaching functional genomics, phenotypic engineering and human gene therapy.

Keywords Zinc finger · Recognition code · Artificial transcription factor · Biotechnology · Gene therapy

1
Zinc Finger DNA-Binding Domain

The zinc finger, defined as class Cys_2His_2, is small motif 28–30 amino acids long, folded into a compact globular module. The domain comprises an α

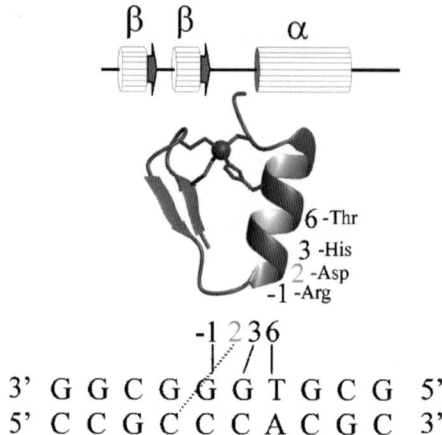

Fig. 1 Interactions between Zif268 middle finger and its DNA target. *Top*: Schematic diagram of a prototype Cys_2His_2 zinc finger motif. The domain comprises an α helix containing two invariant histidine residues coordinated through a zinc atom to two cysteines of an antiparallel β sheet. *Bottom*: Modular recognition between the central Zif268 finger and its triplet subsite DNA target. The individual finger domain binds essentially three base pairs of double-stranded DNA sequence (sense strand) with specific contacts through the amino-terminal part of the α helix in positions −1, 3, 6. The amino acid in position 2 (indicated in *gray*), usually an aspartate or a serine, makes an auxiliary contact on the antisense strand

helix containing two invariant histidine residues coordinated through a zinc atom to two cysteines of an antiparallel β sheet (Fig. 1). The zinc finger domain, originally discovered in the transcription factor TFIIIA, has been subsequently found in a huge variety of regulatory zinc finger proteins (ZFPs) (Miller et al. 1985; el Baradi and Pieler 1991; Schmiedeskamp and Klevit 1994; Passananti et al. 1995; Perez et al. 1996; Pengue and Lania 1996). It represents, so far, the most frequent DNA-binding domain in human transcription factors (Leon and Roth 2000). Cys_2His_2 domains occur in tandem arrays, composed usually of three or more zinc fingers. The X-ray crystal structures of the prototype Zif268 protein bound to its cognate DNA, revealed that an individual finger domain binds essentially three base pairs of double-stranded DNA sequence with specific contacts through the amino-terminal part of the α helix in position −1, 3, 6 (Fig. 1). Changes at these crucial residues alter the DNA-binding specificity of a finger, and this makes it possible to adapt fingers to recognize novel DNA sequences. Importantly, these peptides represent DNA-binding modules that can act as artificial transcription factors (ZFP ATFs) when fused to an appropriate regulatory domain.

2
Zinc Finger Recognition Code

The specificity of interactions observed between protein and DNA is frequently due to the complementarity of the surfaces of a protein α helix and the major groove of DNA (Pabo and Sauer 1992). In Cys_2His_2 domains, the physical interaction of α helix and major groove has indicated contacts between particular amino acids and specific DNA bases by a simple set of rules that can be combined to form a stereochemical recognition code. Even before the solution of Zif268 crystal structure the analysis of the large database of natural zinc finger domains combined with rational mutagenesis revealed some specific amino acid–base contacts (Nardelli et al. 1991, 1992; Desjarlais and Berg 1992a, 1992b). The determination of the Zif268 protein–DNA complex at 2.1 Å resolution was useful for understanding how Cys_2His_2 proteins recognize DNA (Pavletich and Pabo 1991). The three Zif268 fingers wrap around the DNA in an antiparallel mode and each finger has a similar relation to the DNA. As shown in Fig. 2, in each finger amino acids in positions −1, 3 and 6 of the α helix contact three adjacent bases (three-base subsite) on one strand of the DNA duplex. Contiguous zinc fingers recognize adjacent, but independent subsites. A one-to-one interaction between amino ac-

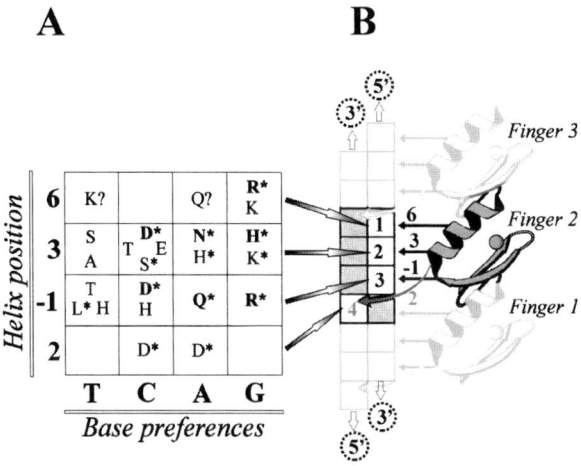

Fig. 2A, B Recognition code. A Amino acids at the crucial positions in the finger α helix (−1, 2, 3, 6) are listed in a matrix relating to the four bases at each position of a DNA subsite. Amino acid residues that arise recurrently from phage display selections are in *bold*, and *asterisks* indicate interactions observed in structural studies. Some correspondences are still uncertain (indicated by *question mark*) or poorly defined (left *blank*). B Schematic diagram of contacts between the middle finger of a prototype three-ZFP and its DNA-binding subsite. Contact between each finger and the antisense DNA strand is indicated by the *arrow*

ids and bases was the first simple description of the mode of zinc finger–DNA recognition. Dejarlais and Berg performed binding studies on variants of the central zinc finger of the transcription factor Sp1 and combined in a matrix the correlations between specific amino acids and bases (Desjarlais and Berg 1992c). They noted that the amino acid discrimination between different bases is rarely absolute, thus introducing the concept of the degeneracy of the code. A certain grade of degeneracy in binding specificity may be tolerated or perhaps even desired for designing synthetic ZFPs. A step forward in the elaboration of the code was obtained by the introduction of the phage display system. This method allows the selection of desirable mutants with new specificity from a large pool of randomized ZFPs (Choo and Klug 1995). The Pabo group and Jamieson et al. created a library of Zif268 variants randomizing the positions of the first finger (−1, 2, 3, 6) and selected them against the Zif268 target with a single modified triplet (Rebar and Pabo 1994; Jamieson et al. 1994). Choo and Klug (1994b), taking in account the importance of the context, performed more extensive randomization of Zif268 central finger α helix (−1, 1, 2, 3, 5, 6, 8). In addition, they confirmed their data in reverse by performing a SELEX *assay* (Systematic Evolution of Ligands by Exponential Enrichment) in which zinc fingers, with fixed α helix positions, yield specific DNA sequence from a randomized pool (Choo and Klug 1994a). Many research groups contributed to the construction of the code table (Fig. 2) (Choo and Klug 1997; Wolfe et al. 2000a; Lee et al., 2003). The design of ZFP, based on this code, had several applications (Desjarlais and Berg 1993; Choo et al. 1994; Corbi et al. 2000; McNamara and Ford 2000). Corbi and colleagues published two papers in which ZFPs were rationally designed to bind arbitrary DNA sequences (Corbi et al. 1997 1998). As a proof of specificity, using the SELEX assay the ZFPs selected strictly related target DNA sequences among randomized ones. These results validate some of the rules derived from the code. Therefore, design based on the code is a valid alternative to the selection approach, even considering further levels of complexity such as synergy between adjacent fingers (Isalan et al. 1997; Elrod-Erickson et al. 1998; Wolfe et al. 1999; Pabo and Nekludova 2000; Wolfe et al. 2001). Sera and Uranga (2002) proposed a rapid approach, based on a nondegenerate code to develop new ZFPs with satisfactory binding properties in a 'high-throughput manner.' This method allows manipulation of gene expression by screening multiple ZFP transcription factors (TFs) for multiple sites in a given promoter even in absence of information on chromatin structure.

2.1
Zinc Finger Domains Selected and Assembled as Building Blocks

The modularity in both structure and function of the zinc finger domain suggested that single domains of known specificity linked together could generate novel DNA-binding proteins. Three successful construction methods have been described: the parallel selection, the sequential selection, and the bipartite selection (Choo and Isalan 2000; Segal and Barbas 2000; Beerli and Barbas 2002).

The parallel selection approach combines three individual ZFPs previously selected by phage display. In particular, two Zif268 fixed fingers (anchor fingers), the first and the third one, ensure the correct positioning on target site, while the middle finger is randomized in the crucial positions (Choo and Klug 1995; Segal et al. 1999; Dreier et al. 2001). The selected 'monomers' provide prefabricated building blocks that can be assembled together (Segal 2002, 2003a; Liu et al. 2002). This method had several applications in the construction of artificial transcription factors that modulate the expression of diverse target genes even at the endogenous chromosomal site.

The sequential approach is based on a selection strategy that takes into account context-dependent interactions between fingers and subsites by serially selecting fingers, adding one finger at a time (Greisman and Pabo 1997). Thus, a new three-finger protein is created in which all the fingers have been selected and optimized in the context of the finger next to it.

The bipartite selection represents a variation of the two preceding methods with the intent to overcome the incompatibilities between adjacent fingers. Selection is carried out in parallel from two pre-made 'half' libraries, in which one-and-a-half fingers of the three-finger Zif268 are randomized. After selections the two zinc finger portions are assembled to form a novel three-ZFP (Isalan et al. 2001).

Other strategies for selecting ZFPs rely on yeast one-hybrid and bacterial two-hybrid systems (Bartsevich and Juliano 2000; Joung et al. 2000). These systems show two useful features: the selection is performed in the context of living cells and the best proteins can be obtained through a one-step selection. In addition, a novel selection system using mammalian cultured cells has been recently described (Blancafort et al. 2003).

Efforts in synthetic ZFPs design have also been focused on the assembly of polydactyl ZFPs with more than three fingers (Imanishi et al. 2001; Beerli and Barbas 2002). To this end two strategies have been used. The first is based on the connection of sets of fingers using linkers and assembling sets of fingers engineering the appropriate dimerization domain (Pabo et al. 2001; Nomura et al. 2003). Computer graphics modeling, with Zif268/DNA complex, suggested the possibility to connect two three-finger proteins using the conserved five-residue (TGEKP) linker that joins the fingers in many naturally occurring multi-finger domains (Liu et al. 1997; Kamiuchi et al. 1998). The second approach involves the use of dimerization domains to bring together two separate ZFPs. ZFPs fused either to the Gal4 dimerization domain or to a leucine zipper can dimerize and bind a longer DNA target (Pomerantz et al. 1998; Wolfe et al. 2000b). Dimeric complexes can also form when ZFPs are fused to steroid hormone receptor ligand-binding domains (LBDs) (Beerli et al. 2000b).

3
Control of Gene Expression by Artificial ZFP TFs: Advantages

Several strategies can be followed to alter gene expression in vivo. Methods such as homologous recombination, the use of antisense reagents, and RNA interference, can only knock down or downregulate a target gene, while ZFP ATFs, thanks to their plasticity and versatility, can either activate or repress transcription (Pabo et al. 2001; Beerli and Barbas 2002; Urnov and Rebar 2002). In fact, the ZFP DNA-binding domains can be fused to different functional domains such as VP16 from herpes simplex virus (HSV), p65 from

Fig. 3 Model structure of a designed ZFP ATF. A specific antibody epitope (*TAG*) precedes a nuclear localization signal (*NLS*). A designed zinc finger DNA-binding domain (*ZFP*) is followed by a desired regulatory domain (*RD*)

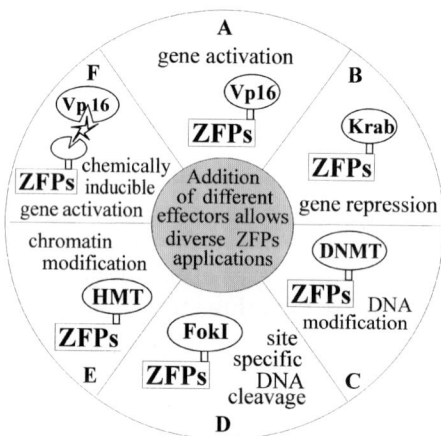

Fig. 4 ZFPs fused to different functional domains. An engineered ZFP can be fused to different functional domains for many different applications. (*A*) A ZFP fused to an activation domain, such as the VP16 domain (from HSV) for activating transcription. (*B*) A ZFP fused to a repression domain, such as the KRAB domain (from KOX protein), for repressing transcription. (*C*) A ZFP fused to a DNA methyltransferase (*DNMT*) domain for DNA modification. (*D*) A ZFP fused to a nuclease domain from a restriction enzyme, such as FokI, for site-specific DNA cleavage. (*E*) A ZFP fused to a histone methyltransferase (*HMT*) domain for chromatin modification. (*F*) A complete ZFP ATF that assembles only in presence of a small-molecule drug

Table 1 Genes regulated by ZFP ATFs

Target gene	ZFP-DBD	Regulatory domain	Disease related	References
BCR-ABL	Three-finger	None, VP16	Acute lymphoblastic leukemia	Choo et al. 1994
erB-2	Three-finger and six-finger	KRAB,ERD, SID,VP64 (VP16×4)	Several human tumors	Beerli et al. 1998; Beerli et al .2000
erB-3	Three-finger and six-finger	VP64 (VP16×4)	Several human tumors	Beerli et al. 2000; Dreier et al.2001
Utrophin	Three-finger	GAL4 and Sp1	Duchenne muscular dystrophy	Corbi et al. 2000
MDR1	Five-finger	KRAB (two copies)	Several human tumors	Bartsevich et al. 2000
Erythropoietin (EPO)	Three-finger	VP64 (VP16×4)	Hematological disorders	Zhang et al. 2000
VEGF-A	Three-finger and six-finger	VP16 and NF-kBp65	Several vascular diseases	Liu et al. 2001; Rebar et al.2002
PPARγ1,2	Six-finger	KRAB	Metabolic diseases	Ren et al. 2002
bax	Three-finger and five-finger	VP16	Several human tumors	Falke et al. 2003
IE HSV-1	Six-finger	KRAB	HSV infections	Papworth et al. 2003
HIV-1 Tat	Three-fingers and six-finger	KRAB	HIV infections	Reynolds et al.2003
IGF2	Three-finger	VP16 and NF-kBp65	Beckwith–Wiedemann syndrome	Jouvenot et al. 2003
H19	Three-finger	VP16 and NF-kBp65	Beckwith–Wiedemann syndrome	Jouvenot et al. 2003
CHK2	Six-finger	KRAB	Several human tumors	Tan et al. 2003

ZFP, zinc finger proteins; ATF, artificial transcription factors; DBD, DNA-binding domain.

the cellular transcription factor nuclear factor κB (NF-κB), for transcriptional activation, KRAB (Kruppel associated box) from natural zinc finger transcription factors, and Sid (Sin3A interaction domain), to get transcriptional repression (Figs. 3, 4 and Table 1). The modular feature of ZFP/DNA recognition makes possible the 'mix and match' approach, in which both natural fingers, with known site preferences, and synthetic fingers, with selected binding sites, are linked together to generate multifinger proteins able to target the desired DNA site (Beerli et al. 2000a; Kang and Kim 2000). At present, a huge collection of zinc finger modules is available and a computer program can easily search, within a specific promoter region, for binding sites that can be targeted by mixing ZFPs from the archive (Jamieson et al. 2003).

3.1
Control of Gene Expression by Artificial ZFP TFs: Specificity and Effectiveness

In order to make an artificial transcription factor acting as a 'natural transcription factor' many aspects need to be empirically tested, including: (a) the accessibility of the factor to the target site in the context of endogenous chromatin; (b) the position of the target site with respect to the target promoter; (c) the nature of the fused effector/regulatory domain (Yaghmai and Cutting 2002; Levine and Tjian 2003).

Once an ATF binds its DNA target and exerts the regulatory effect on transcription, its activity must be specific and should permit a desired level of gene expression in a given cell type, especially for human therapeutic use. In this regard one of the crucial issues in designing synthetic ZFP ATFs is the calibration of correct binding affinity/specificity. The specificity in terms of binding to a unique locus in the genome should be ensured by using polydactyl ZFPs containing six fingers and capable of targeting an 18-bp DNA sequence. In fact, statistically, a target sequence to be unique in the human genome must be at least 16 bp long. The increased length of the target sequence in principle should determine a proportional increase of the DNA-binding affinity. Nonetheless, on the basis of certain 'mysterious' properties of ZFPs, in some cases the affinities of three- and six-finger proteins toward their respective binding sites are almost equivalent. This could be due to an adverse cooperativity between certain fingers in binding long sequences (Ansari 2003). Moreover, in some cases exceeding a certain level of target site affinity is unfavorable because of increased binding to nonspecific DNA that could result in undesired pleiotropic effects (Beerli et al. 2000a). On the other hand, dose and timing of ZFP ATF expression must be finely calibrated for gene therapy use, allowing for the possibility to turn off the system promptly if any serious side effects occur (Pabo et al. 2001). These requisites can be accomplished by an inducible expression system based on a small molecule ligand able to trigger a pathway to activate a dormant transcription factor containing a signal responsive element. The resulting activated transcription factor could then regulate specifically the expression of a target gene. Gossen and co-workers were the pioneers in the field of inducible expression systems. They used functional domains derived from prokaryotes to develop chimeric transcription factors that bind to DNA only by removal (Tet-off system), or adding (Tet-on system,) doxycycline (Gossen and Bujard 1992; Gossen et al. 1995). Barbas' group reported the use of a tetracycline/doxycicline-dependent expression vector to regulate the expression of a ZFP ATF to obtain inducible regulation on the endougenous *erb-2* gene (Beerli et al. 2000a). However, this strategy is based on the delivery of two genes: one encoding the zinc finger protein under the control of a regulatable promoter containing the tetracycline response element and the other encoding the regulatory protein (tetracycline-controlled transactivator). Delivery of multiple vectors poses difficulties in a gene therapy setting. As an alternative approach, Barbas' group proposed the use of fusion proteins between designed zinc finger proteins and the eukaryotic nuclear hormone steroid receptor ligand-binding domains (LBDs) for the inducible control of gene expression (Beerli et al. 2000b). The LBDs are inactive until ligand is added externally. The naturally occurring nuclear hormone receptors bind to the ligand and translocate to the nucleus, where they bind to a nuclear response element, stimulating transcription. Steroid receptors bind DNA as dimers and typically recognize palindromic sequences. In this regard Barbas' group engineered a single-chain steroid hormone receptor LBD as result of an intramolecular rearrangement of two serially connected LBDs fused to a six-finger DNA-binding domain. This smart strategy allows targeting a desired 18-base pair long asymmetric target sequence by using a monomeric ligand-dependent transcriptional regulator, with the further advantage that only one gene needs to be delivered.

Another strategy includes the use of rapamycin based 'dimerizer system' to regulate the expression of a specific gene. In this case administration of the drug brings together the DNA-binding domain (ZPF based) and the activation domain, since each of them binds to separate halves of the small molecular drug (Pollock et al. 2002) (Fig. 4F).

4
Applications: A Promise in Target Validation, Biotechnology and Transcription Therapy

The conclusion of the human genome project had identified more than 30,000 genes (Venter et al. 2001). Now to systemically characterize the function of novel genes is of major importance. The ability of ZFP ATFs to perturb both up- and down-regulation of the expression of a specific gene provides a precious tool to deduce the function of a target gene product (Urnov et al. 2002; Lee et al. 2003). Moreover, considering that an aberrant transcription profile often represents a marker for a specific disease, the possibility to correct the abnormalities in gene expression by ZFP ATFs has the potentiality to reverse the disease process by the so called 'transcription therapy' (Pandolfi 2001). Choo and Klug were the first to realize a functional artificial ZFP (Choo et al. 1994). In particular, they engineered a three-ZFP able to bind specifically the unique nine-base pair region of the BCR-ABL fusion oncogene, generated by chromosomal translocation in acute lymphoblastic leukemia. In transfected cell lines the ZFP, able to target the BCR-ABL sequence, led to BCR-ABL mRNA reduction. After those promising results several studies have shown that designed ZFPs can regulate transcription of a target gene (Table 1). Passananti and his group engineered several synthetic three-ZFP ATFs on the basis of the recognition code. One of these artificial genes, named Jazz, was realized to target the promoter region of the utrophin, a dystrophin-related gene. Jazz was constructed with the aim of upregulating the expression level of utrophin in Duchenne muscular dystrophy (Corbi et al. 2000). The upregulation of utrophin gene expression, recognized as a plausible therapeutic approach in the treatment of Duchenne muscular dystrophy (Tinsley et al. 1996), could be achieved by ZFP ATFs as a valid alternative to pharmacological approaches. The use of ZFPs in this field could also provide a model system for screening compounds and for drug target validation.

Bartsevich and Juliano (2000) selected artificial ZFP ATFs able to regulate the transcription of the MDR1 multidrug resistance gene in the chromosomal context. Barbas' group selected six-finger Sp1-based ZFPs able to target and to regulate the endogenous *erB-2* and *erB-3* proto-oncogenes (Beerli et al. 1998, 2000a; Segal et al. 1999), frequently overexpressed in human tumors. Significantly, regulation of the two genes was highly specific, since ZFP ATFs can discern between the two genes even though their binding sites share 15 out of 18 nucleotides. Moreover, this research group obtained inducible expression of the artificial ZFP ATFs, particularly useful in view of gene therapy protocols. In another study Zhang et al. (2000) obtained, in transfected cell lines, the activation of the endogenous erythropoietin gene by a series of ZFP ATFs, whose activity was influenced by the capability to gain access to their recognition sequences within the chromatin infrastructure.

Similarly, Liu et al. (2001) used a panel of ZFP ATFs to target the endogenous locus of vascular endothelial growth factor (VEGF-A), a specific inducer of new blood vessel growth, involved in a variety of medical conditions. They showed that combinations of three-ZFP ATFs targeted to distinct adjacent chromosomal sites were functioning synergistically on VEGF-A activation. Moreover, comparing the behavior of ZFP ATFs targeted to accessible regions (identified by DNAase I hypersensitivity assay) with ZFP ATFs targeted to inaccessible regions, a clear difference was shown, underlining once again the importance of chromatin accessibility.

Peroxisome proliferator activated receptor-γ (PPARγ) (Ren et al. 2002) represents an interesting example of knockdown expression. PPARγ is a nuclear hormonal receptor and exists as two isoforms: γ1 and γ2. When a ZFP ATF repressed the expression of both isoforms, adipogenesis was restored selectively following PPARγ2–but not PPARγ1–reintroduction. The use of this ZFP ATF made it possible to distinguish the role of each isoform in adipogenesis. In addition, it could be functional for the design of specific drugs able to affect only the desired isoform (Jamieson et al. 2003). Customized ZFP ATFs were also able to inhibit transcription of viral genes and therefore viral replication. Papworth and colleagues (2003) designed ZFPs targeted to the viral immediate-early promoter of HSV-1 that were able to reduce the viral titer by about 90%. Similarly, they obtained repression of the HIV-1 long terminal repeat promoter and inhibition of HIV-1 replication (Reynolds et al. 2003). These results are promising to counteract viral infections in gene therapy. Falke and coworkers reported data on an artificial transcription factor able to stimulate the expression of the endogenous pro-apoptotic gene *bax*, one of the p53 targets (Falke et al. 2003). The p53 gene product is inactivated in over 50% of cancers. p53 negative cells fail to induce *bax* and the apoptotic pathway, and this seems to make them more refractory to chemotherapic agents. Therefore they propose the induction of the *bax* gene as a valuable tool in cancer chemotherapy, by diminishing survival of p53-deficient tumor cells. A recent study demonstrates the ability of a ZFP ATF to reactivate the transcriptionally silent IGF2 and H19 alleles, overriding the imprinted transcriptional status and highlighting the potential of synthetic transcription factors in the treatment of epigenetic lesions (Jouvenot et al. 2003). Tan and coworkers (2003) created a six-ZFP repressor that is able to abolish the function of the cell cycle checkpoint kinase CHK2 in different cell types. As proof of specificity, a microarray analysis demonstrated that CHK2 was the only gene in the genome that was repressed by the ZFP. This remarkable specificity is promising in view of both target validation and therapeutic interventions in vivo.

All of these data, obtained in cell lines expressing ZFP ATFs, either by transfection (stable or inducible) or infection, indicate stable, long-term target gene regulation. This is a fundamental starting point for the effective use of ZFP ATFs in a whole organism model. Indeed, expression of a ZFP ATF (delivered by adenovirus-based vectors) in a mouse ear model led to induction of the VEGF-A that stimulates the growth of blood vessels. These data demonstrated, for the first time, the feasibility of regulating, in vivo, crucial biological processes such as angiogenesis (Rebar et al. 2002). Interestingly, the induced activation of the *VEGF-A* gene led to the expression of its naturally occurring splice variants required for a correct angiogenesis. Of course, it is easier to deliver a single ZFP ATF than multiple splice variants. Therefore, a single intervention can be very advantageous in regulating complex biological processes in which coordination of multiple gene expression is required. In the field of large-scale ZFP ATF engineering, Kim's group proposed the elegant 'GeneGrip' method. This method is based on screening of 'natural' zinc fingers from the human genome that can be combined in the generation of modular novel ZFPs (Bae et al. 2003). Importantly, these factors by evading immune surveillance, may be an optimal tool for the development of gene therapy. In addition, ZFP ATFs could be selected randomly by ZFP libraries on the basis of an induced improved phenotype (Park et al. 2003). Such ZFP ATFs could also be used to regulate the expression of uncharacterized genes leading to the identification of the role of new genes (Blancafort et al. 2003).

Moreover, effector domains, other than activation or repression domains can be fused to ZFPs to generate novel proteins with novel functions (Fig.

4). For example ZFPs fused to restriction nucleases can recognize and cut specifically novel DNA sequences (Chandrasegaran and Smith 1999), ZFPs linked to a minimal histone methyltransferase catalytic domains can affect local methylation of histone H3K9 and consequently repress the expression of a target gene through chromatin modification (Snowden et al. 2002), ZFPs fused to C5-cytosine DNA methyltransferases has been proposed as a potential tool for promoter gene silencing through a DNA methylation-mediated cellular response (McNamara et al. 2002). Nonetheless, ZFPs by themselves can exert a phenotype by competing with endogenous factors for specific binding sites in vivo (Segal 2002).

ZFPs bind also RNA in a specific manner, thus becoming a tool with which to develop antiviral agents and correct defects of RNA metabolism associated with a variety of diseases (Friesen and Darby 2001; Lee et al. 2003). Finally, in the field of agricultural biotechnology, designed ZFP ATFs represent a novel and promising tool to alter the expression of specific target genes, in order to get healthier or more nutritious crop plants. In this regard genes related to disease/resistance could be turned on, while genes related to pathologic status or to the production of antinutritive proteins could be turned off (Stege et al. 2002; Sanchez et al. 2002; Segal et al. 2003b; Frommer and Beachy 2003).

5
Approaches to Therapy: Perspectives

ZFP ATFs designed for a therapeutic use must be efficiently delivered to the nuclei of cells within the tissue targets (Verma and Somia 1997). This is a crucial issue because viral vectors (adenovirus and retrovirus) provide efficient expression of ZFP ATFs in animal models, but in human gene therapy they can create problems in terms of safety. A strictly regulated targeted integration of a retrovirus could protect recipients from dangerous recombinations that can lead to events such as oncogenic activation (Somia and Verma 2000). A safer nascent approach consists in using ATFs directly as therapeutic agents after their expression and purification in systems such as bacteria or baculovirus (Falke and Juliano 2003). The realization of artificial transcription factors is still in its early stages, but preclinical and clinical trials could take place in the near future. The apparent low intrinsic toxicity of ZFP ATFs verified in transgenic organisms (Guan et al. 2002; Pasqualini et al. 2002) further confirms their potential for clinical applications. Another possible application is in engineering cell lines expressing ZFP capable of regulating transcription of pharmaceutically valuable proteins. ZFP ATFs can be also useful for validating genes as drug targets, therefore representing a valid alternative to the use of certain drugs that cause undesirable genome-wide effects. Nonetheless, the use of ZFP ATFs could also be recommended for regulation of disease-related genes that appear not to be controllable by drugs (Jamieson et al. 2003).

In conclusion, the hope of combating a vast array of diseases by gene therapy protocols could become a reality. Many branches of biology, from genetics to virology and immunology will need to contribute generously in order to achieve this goal.

Acknowledgements This work is dedicated to the memory of Prof. G. Thomas Passananti. We thank Dr. A.G. Ruscitti for precious assistance. This work was supported by Telethon A160 and Firb 'Epigenetica e Cromatina.' N. Corbi is recipient of a FIRC fellowship.

References

Ansari AZ (2003) Fingers reach for the genome. Nat Biotechnol 21:242–243

Bae KH, Do KY, Shin HC, Hwang MS, Ryu EH, Park KS, Yang HY, Lee DK, Lee Y, Park J, Sun KH, Kim HW, Yeh BI, Lee HW, Hyung SS, Yoon J, Seol W, Kim JS (2003) Human zinc fingers as building blocks in the construction of artificial transcription factors. Nat Biotechnol 21:275–280

Bartsevich VV, Juliano RL (2000) Regulation of the MDR1 gene by transcriptional repressors selected using peptide combinatorial libraries. Mol Pharmacol 58:1–10

Beerli RR, Segal DJ, Dreier B, Barbas CF3 (1998) Toward controlling gene expression at will: specific regulation of the erbB-2/HER-2 promoter by using polydactyl zinc finger proteins constructed from modular building blocks. Proc Natl Acad Sci USA 95:14628–14633

Beerli RR, Dreier B, Barbas CF, III (2000a) Positive and negative regulation of endogenous genes by designed transcription factors. Proc Natl Acad Sci USA 97:1495–1500

Beerli RR, Schopfer U, Dreier B, Barbas CF, III (2000b) Chemically regulated zinc finger transcription factors. J Biol Chem 275:32617–32627

Beerli RR, Barbas CF, III (2002) Engineering polydactyl zinc-finger transcription factors. Nat Biotechnol 20:135–141

Blancafort P, Magnenat L, Barbas CF (2003) Scanning the human genome with combinatorial transcription factor libraries. Nat Biotechnol 21:269–274

Chandrasegaran S, Smith J (1999) Chimeric restriction enzymes: what is next? Biol Chem 380:841–848

Choo Y, Klug A (1994a) Selection of DNA binding sites for zinc fingers using rationally randomized DNA reveals coded interactions. Proc Natl Acad Sci USA 91:11168–11172

Choo Y, Klug A (1994b) Toward a code for the interactions of zinc fingers with DNA: selection of randomized fingers displayed on phage. Proc Natl Acad Sci USA 91:11163–11167

Choo Y, Sanchez-Garcia I, Klug A (1994) In vivo repression by a site-specific DNA-binding protein designed against an oncogenic sequence. Nature 372:642–645

Choo Y, Klug A (1995) Designing DNA-binding proteins on the surface of filamentous phage. Curr Opin Biotechnol 6:431–436

Choo Y, Klug A (1997) Physical basis of a protein-DNA recognition code. Curr Opin Struct Biol 7:117–125

Choo Y, Isalan M (2000) Advances in zinc finger engineering. Curr Opin Struct Biol 10:411–416

Corbi N, Perez M, Maione R, Passananti C (1997) Synthesis of a new zinc finger peptide; comparison of its 'code' deduced and 'CASTing' derived binding sites. FEBS Lett 417:71–74

Corbi N, Libri V, Fanciulli M, Passananti C (1998) Binding properties of the artificial zinc fingers coding gene Sint1. Biochem Biophys Res Commun 253:686–692

Corbi N, Libri V, Fanciulli M, Tinsley JM, Davies KE, Passananti C (2000) The artificial zinc finger coding gene 'Jazz' binds the utrophin promoter and activates transcription. Gene Ther 7:1076–1083

Desjarlais, Berg JM (1992a) Redesigning the DNA-binding specificity of a zinc finger protein: a data base-guided approach. Proteins 13:272

Desjarlais JR, Berg JM (1992b) Redesigning the DNA-binding specificity of a zinc finger protein: a data base-guided approach. Proteins 12:101–104

Desjarlais JR, Berg JM (1992c) Toward rules relating zinc finger protein sequences and DNA binding site preferences. Proc Natl Acad Sci USA 89:7345–7349

Desjarlais JR, Berg JM (1993) Use of a zinc-finger consensus sequence framework and specificity rules to design specific DNA binding proteins. Proc Natl Acad Sci USA 90:2256–2260

Dreier B, Beerli RR, Segal DJ, Flippin JD, Barbas CF3 (2001) Development of zinc finger domains for recognition of the 5'-ANN-3' family of DNA sequences and their use in the construction of artificial transcription factors. J Biol Chem 276:29466–29478

el Baradi T, Pieler T (1991) Zinc finger proteins: what we know and what we would like to know. Mech Dev 35:155–169

Elrod-Erickson M, Benson TE, Pabo CO (1998) High-resolution structures of variant Zif268-DNA complexes: implications for understanding zinc finger-DNA recognition. Structure 6:451–464

Falke D, Fisher M, Ye D, Juliano RL (2003) Design of artificial transcription factors to selectively regulate the pro-apoptotic bax gene. Nucleic Acids Res 31: e10

Falke D, Juliano RL (2003) Selective gene regulation with designed transcription factors: implications for therapy. Curr Opin Mol Ther 5:161–166

Friesen WJ, Darby MK (2001) Specific RNA binding by a single C2H2 zinc finger. J Biol Chem 276:1968–1973

Frommer WB, Beachy R (2003) Plant biotechnology. A future for plant biotechnology? Naturally! Curr Opin Plant Biol. 6:147–149

Gossen M, Bujard H (1992) Tight control of gene expression in mammalian cells by tetracycline-responsive promoters. Proc Natl Acad Sci USA 89:5547–5551

Gossen M, Freundlieb S, Bender G, Muller G, Hillen W, Bujard H (1995) Transcriptional activation by tetracyclines in mammalian cells. Science 268:1766–1769

Greisman HA, Pabo CO (1997) A general strategy for selecting high-affinity zinc finger proteins for diverse DNA target sites. Science 275:657–661

Guan X, Stege J, Kim M, Dahmani Z, Fan N, Heifetz P, Barbas CF, III, Briggs SP (2002) Heritable endogenous gene regulation in plants with designed polydactyl zinc finger transcription factors. Proc Natl Acad Sci USA 99:13296–13301

Imanishi M, Hori Y, Nagaoka M, Sugiura Y (2001) Design of novel zinc finger proteins: towards artificial control of specific gene expression. Eur J Pharm Sci 13:91–97

Isalan M, Choo Y, Klug A (1997) Synergy between adjacent zinc fingers in sequence-specific DNA recognition. Proc Natl Acad Sci USA 94:5617–5621

Isalan M, Klug A, Choo Y (2001) A rapid, generally applicable method to engineer zinc fingers illustrated by targeting the HIV-1 promoter. Nat Biotechnol 19:656–660

Jamieson AC, Kim SH, Wells JA (1994) In vitro selection of zinc fingers with altered DNA-binding specificity. Biochemistry 33:5689–5695

Jamieson AC, Miller JC, Pabo CO (2003) Drug discovery with engineered zinc-finger proteins. Nat Rev Drug Discov 2:361–368

Joung JK, Ramm EI, Pabo CO (2000) A bacterial two-hybrid selection system for studying protein-DNA and protein-protein interactions. Proc Natl Acad Sci USA 97:7382–7387

Jouvenot Y, Ginjala V, Zhang L, Liu PQ, Oshimura M, Feinberg AP, Wolffe AP, Ohlsson R, Gregory PD (2003) Targeted regulation of imprinted genes by synthetic zinc-finger transcription factors. Gene Ther 10:513–522

Kamiuchi T, Abe E, Imanishi M, Kaji T, Nagaoka M, Sugiura Y (1998) Artificial nine zinc-finger peptide with 30 base pair binding sites. Biochemistry 37:13827–13834

Kang JS, Kim JS (2000) Zinc finger proteins as designer transcription factors. J Biol Chem 275:8742–8748

Lee DK, Seol W, Kim JS (2003) Custom DNA-binding proteins and artificial transcription factors. Curr Top Med Chem 3:645–657

Leon O, Roth M (2000) Zinc fingers: DNA binding and protein-protein interactions. Biol Res 33:21–30

Levine M, Tjian R (2003) Transcription regulation and animal diversity. Nature 424:147–151

Liu PQ, Rebar EJ, Zhang L, Liu Q, Jamieson AC, Liang Y, Qi H, Li PX, Chen B, Mendel MC, Zhong X, Lee YL, Eisenberg SP, Spratt SK, Case CC, Wolffe AP (2001) Regulation of an endogenous locus using a panel of designed zinc finger proteins targeted to accessible chromatin regions. Activation of vascular endothelial growth factor A. J Biol Chem 276:11323–11334

Liu Q, Segal DJ, Ghiara JB, Barbas CF3 (1997) Design of polydactyl zinc-finger proteins for unique addressing within complex genomes. Proc Natl Acad Sci USA 94:5525–5530

Liu Q, Xia Z, Zhong X, Case CC (2002) Validated zinc finger protein designs for all 16 GNN DNA triplet targets. J Biol Chem 277:3850–3856

McNamara AR, Ford KG (2000) A novel four zinc-finger protein targeted against p190(BcrAbl) fusion oncogene cDNA: utilisation of zinc-finger recognition codes. Nucleic Acids Res 28:4865–4872

McNamara AR, Hurd PJ, Smith AE, Ford KG (2002) Characterisation of site-biased DNA methyltransferases: specificity, affinity and subsite relationships. Nucl Acids Res 30:3818–3830

Miller J, McLachlan AD, Klug A (1985) Repetitive zinc-binding domains in the protein transcription factor IIIA from Xenopus oocytes. EMBO J 4:1609–1614

Nardelli J, Gibson TJ, Vesque C, Charnay P (1991) Base sequence discrimination by zinc-finger DNA-binding domains. Nature 349:175–178

Nardelli J, Gibson T, Charnay P (1992) Zinc finger-DNA recognition: analysis of base specificity by site-directed mutagenesis. Nucl Acids Res 20:4137–4144

Nomura W, Nagaoka M, Shiraishi Y, Sugiura Y (2003) Influence of TFIIIA-type linker at the N- or C-terminal of nine-zinc finger protein on DNA-binding site. Biochem Biophys Res Commun 300:87–92

Pabo CO, Sauer RT (1992) Transcription factors: structural families and principles of DNA recognition. Annu Rev Biochem 61:1053–1095

Pabo CO, Nekludova L (2000) Geometric analysis and comparison of protein-DNA interfaces: why is there no simple code for recognition? J Mol Biol 301:597–624

Pabo CO, Peisach E, Grant RA (2001) Design and selection of novel Cys2His2 zinc finger proteins. Annu Rev Biochem 70:313–340

Pandolfi PP (2001) Transcription therapy for cancer. Oncogene 20:3116–3127

Papworth M, Moore M, Isalan M, Minczuk M, Choo Y, Klug A (2003) Inhibition of herpes simplex virus 1 gene expression by designer zinc-finger transcription factors. Proc Natl Acad Sci USA 100:1621–1626

Park KS, Lee DK, Lee H, Lee Y, Jang YS, Kim YH, Yang HY, Lee SI, Seol W, Kim JS (2003) Phenotypic alteration of eukaryotic cells using randomized libraries of artificial transcription factors. Nat Biotechnol 21:1208–1214

Pasqualini R, Barbas CF, III, Arap W (2002) Vessel maneuvers: zinc fingers promote angiogenesis. Nat Med 8:1353–1354

Passananti C, Corbi N, Paggi MG, Russo MA, Perez M, Cotelli F, Stefanini M, Amati P (1995) The product of Zfp59 (Mfg2), a mouse gene expressed at the spermatid stage of spermatogenesis, accumulates in spermatozoa nuclei. Cell Growth Differ 6:1037–1044

Pavletich NP, Pabo CO (1991) Zinc finger-DNA recognition: crystal structure of a Zif268-DNA complex at 2.1 A. Science 252:809–817

Pengue G, Lania L (1996) Kruppel-associated box-mediated repression of RNA polymerase II promoters is influenced by the arrangement of basal promoter elements. Proc Natl Acad Sci USA 93:1015–1020

Perez M, Rompato G, Corbi N, De Gregorio L, Dragani TA, Passananti C (1996) Zfp60, a mouse zinc finger gene expressed transiently during in vitro muscle differentiation. FEBS Lett 387:117–121

Pollock R, Giel M, Linher K, Clackson T (2002) Regulation of endogenous gene expression with a small-molecule dimerizer. Nat Biotechnol 20:729–733

Pomerantz JL, Wolfe SA, Pabo CO (1998) Structure-based design of a dimeric zinc finger protein. Biochemistry 37:965–970

Rebar EJ, Pabo CO (1994) Zinc finger phage: affinity selection of fingers with new DNA-binding specificities. Science 263:671–673

Rebar EJ, Huang Y, Hickey R, Nath AK, Meoli D, Nath S, Chen B, Xu L, Liang Y, Jamieson AC, Zhang L, Spratt SK, Case CC, Wolffe A, Giordano FJ (2002) Induction of angiogenesis in a mouse model using engineered transcription factors. Nat Med 8:1427–1432

Ren D, Collingwood TN, Rebar EJ, Wolffe AP, Camp HS (2002) PPARgamma knockdown by engineered transcription factors: exogenous PPARgamma2 but not PPARgamma1 reactivates adipogenesis. Genes Dev 16:27–32

Reynolds L, Ullman C, Moore M, Isalan M, West MJ, Clapham P, Klug A, Choo Y (2003) Repression of the HIV-1 5' LTR promoter and inhibition of HIV-1 replication by using engineered zinc-finger transcription factors. Proc Natl Acad Sci USA 100:1615–1620

Sanchez JP, Ullman C, Moore M, Choo Y, Chua NH (2002) Regulation of gene expression in Arabidopsis thaliana by artificial zinc finger chimeras. Plant Cell Physiol 43:1465–1472

Segal DJ, Dreier B, Beerli RR, Barbas CF3 (1999) Toward controlling gene expression at will: selection and design of zinc finger domains recognizing each of the 5'-GNN-3' DNA target sequences. Proc Natl Acad Sci USA 96:2758–2763

Segal DJ, Barbas CF3 (2000) Design of novel sequence-specific DNA-binding proteins. Curr Opin Chem Biol 4:34–39

Segal DJ (2002) The use of zinc finger peptides to study the role of specific factor binding sites in the chromatin environment. Methods 26:76–83

Segal DJ, Beerli RR, Blancafort P, Dreier B, Effertz K, Huber A, Koksch B, Lund CV, Magnenat L, Valente D, Barbas CF3 (2003a) Evaluation of a modular strategy for the construction of novel polydactyl zinc finger dna-binding proteins. Biochemistry 42:2137–2148

Segal DJ, Stege JT, Barbas CF (2003b) Zinc fingers and a green thumb: manipulating gene expression in plants. Curr Opin Plant Biol 6:163–168

Sera T, Uranga C (2002) Rational design of artificial zinc-finger proteins using a nondegenerate recognition code table. Biochemistry 41:7074–7081

Snowden AW, Gregory PD, Case CC, Pabo CO (2002) Gene-specific targeting of H3K9 methylation is sufficient for initiating repression in vivo. Curr Biol 12:2159–2166

Somia N, Verma IM (2000) Gene therapy: trials and tribulations. Nat Rev Genet 1:91–99

Schmiedeskamp M and Klevit RE (1994) Zinc finger diversity. Curr Opin Struct Biol 4:28/35

Stege JT, Guan X, Ho T, Beachy RN, Barbas CF, III (2002) Controlling gene expression in plants using synthetic zinc finger transcription factors. Plant J 32:1077–1086

Tan S, Guschin D, Davalos A, Lee YL, Snowden AW, Jouvenot Y, Zhang HS, Howes K, McNamara AR, Lai A, Ullman C, Reynolds L, Moore M, Isalan M, Berg LP, Campos B, Qi H, Spratt SK, Case CC, Pabo CO, Campisi J, Gregory PD (2003) Zinc-finger protein-targeted gene regulation: genomewide single-gene specificity. Proc Natl Acad Sci USA 100:11997–12002

Tinsley JM, Potter AC, Phelps SR, Fisher R, Trickett JI, Davies KE (1996) Amelioration of the dystrophic phenotype of mdx mice using a truncated utrophin transgene. Nature 384:349–353

Urnov FD, Rebar EJ (2002) Designed transcription factors as tools for therapeutics and functional genomics. Biochem Pharmacol 64:919–923

Urnov FD, Rebar EJ, Reik A, Pandolfi PP (2002) Designed transcription factors as structural, functional and therapeutic probes of chromatin in vivo. Fourth in review series on chromatin dynamics. EMBO Rep 3:610–615

Venter JC, Adams MD, Myers EW, Li PW, Mural RJ, Sutton GG et al (2001) The sequence of the human genome. Science 291:1304–1351

Verma IM, Somia N (1997) Gene therapy—promises, problems and prospects. Nature 389:239–242

Wolfe SA, Greisman HA, Ramm EI, Pabo CO (1999) Analysis of zinc fingers optimized via phage display: evaluating the utility of a recognition code. J Mol Biol 285:1917–1934

Wolfe SA, Nekludova L, Pabo CO (2000a) DNA recognition by Cys2His2 zinc finger proteins. Annu Rev Biophys Biomol Struct 29:183–212

Wolfe SA, Ramm EI, Pabo CO (2000b) Combining structure-based design with phage display to create new Cys(2)His(2) zinc finger dimers. Struct Fold Des 8:739–750

Wolfe SA, Grant RA, Elrod-Erickson M, Pabo CO (2001) Beyond the 'recognition code': structures of two Cys2His2 zinc finger/TATA box complexes. Structure (Camb) 9:717–723

Yaghmai R, Cutting GR (2002) Optimized regulation of gene expression using artificial transcription factors. Mol Ther 5:685–694

Zhang L, Spratt SK, Liu Q, Johnstone B, Qi H, Raschke EE, Jamieson AC, Rebar EJ, Wolffe AP, Case CC (2000) Synthetic zinc finger transcription factor action at an endogenous chromosomal site. Activation of the human erythropoietin gene. J Biol Chem 275:33850–33860

Tetracycline-Controlled Transactivators and Their Potential Use in Gene Therapy Applications

D. Bohl · J.-M. Heard (✉)

Laboratoire Rétrovirus et Transfert Génétique, Institut Pasteur,
28 rue du Docteur Roux, 75724 Paris Cedex 15, France
jmheard@pasteur.fr

1	Introduction	510
2	The Tet-Off and Tet-On Systems	511
2.1	The Tet-Off System	511
2.2	The Tet-On System	513
2.3	Properties of the Tet Systems for In Vivo Applications	513
3	Gene Therapy Applications	514
3.1	Development of Tet-Regulatable Viral Vectors and Proof of Efficacy	515
3.1.1	Retroviral Vectors	515
3.1.1.1	Gamma-Retrovirus-Derived Vectors	515
3.1.1.2	Lentiviral Vectors	516
3.1.2	AAV Vectors	517
3.1.3	Ad Vectors	518
3.1.4	Improving Tet Systems	518
3.2	Gene Therapy Applications	520
3.2.1	The Systemic Delivery of Epo	520
3.2.2	Parkinson Disease	522
4	Modeling of Human CNS Pathologies	523
4.1	Description	523
4.2	Applications	524
4.2.1	Huntington Disease	525
4.2.2	Prion Diseases	526
4.3	Current Progress for Future Applications	526
5	Conclusions	527
References		527

Abstract The control of gene expression in higher eukaryotes has developed into an invaluable tool for both the design of gene therapy strategies and the study of gene function. Application of gene therapy to a number of diseases will require that therapeutic protein expression level is adjusted to prevent toxicity and to ensure biological efficacy. Amongst currently available chimeric regulatable systems, those derived from the tetracycline operon of *Escherichia coli* are by far the best documented and the most frequently used. This review will describe the various tetracycline systems. Advantages and drawbacks for in vitro studies and in vivo applications will be presented, as well as the development of viral vectors and gene therapy applications. Tetracycline regulatable expression systems progressively turned out to be effective tools for a broad range of investigations. These include studies of gene function in transgenic animals, analysis of complex and multi-staged biological processes including embryogenesis and cancer, and the generation of animal models of human disorders aimed at understanding the cause and progression of diseases. Strategies developed for these various purposes will be discussed and examples will be given, preferentially with respect to the central nervous system.

Keywords Regulation · Tetracycline · Gene transfer · Vectors · Transgenic mice

1
Introduction

The control of gene expression in higher eukaryotes has developed into an invaluable tool for the design of gene therapy strategies. Application of gene therapy to a number of diseases will require that therapeutic protein expression level is adjusted to prevent toxicity and to ensure biological efficacy. Safe and efficient expression levels of a therapeutic protein may vary during the course of a treatment as well as between patients. Though protein expression level can be controlled at various levels, including mRNA processing and transport, protein translation, and the formation of functional supramolecular assemblies, most of the available regulation systems operate on transcriptional promoter activity.

A first-generation of regulatable systems was based on the use of inducible promoters responding to physiological stimuli, including the mouse metalloprotein receptor (Mayo et al. 1982; Searl et al. 1985), steroid hormone inducible promoters (Israel and Kaufman 1989; Ko et al. 1989; Mader and White 1993; Picard et al. 1988; Webster et al. 1988), interferon responding promoters (Hug et al. 1988; Staeheli et al. 1986), and heat-shock inducible promoters (Schweinfest et al. 1988; Wurm et al. 1986). These systems generally suffer from high basal expression and weak induction. Advantages come from the use of an endogenous transactivator. This avoids the need of an expression system for exogenous transactivator and the risk that immune response would be mounted against a nonself protein. The major drawback is the absence of selectivity. Upon induction of the physiological stimulus, for example the administration of steroid hormones, all promoters containing responsive elements recognized by the transactivator are activated, leading to pleiotropic effects. These undesired consequences often unmask the expected biological effects of an increased expression of the protein encoded by the transgenic inducible system. Practically, this limitation impairs the use of regulatory systems based on physiological stimuli for research or gene therapy applications and emphasized the need to search for non-natural systems that provide selective control.

The second generation of systems relies on artificial transactivating proteins and responsive elements. Expression of the transactivator is induced in the cells of interest. Responsive elements acting on the inducible promoter are DNA sequences not found in the target cell genome. Activating stimuli are ideally chemical compounds devoid of biological effects Although these artificial systems may often achieve low background expression, high inducibility and accurate selectivity, in practice their use encounters two major difficulties. Since the artificial transactivator is, by definition, not present in the target cell, expression must be induced by introducing the corresponding gene under the control of a transcriptional promoter. The introduction of two transcriptional units in the same target cell is therefore mandatory for effectiveness. As a nonself protein, the artificial transactivator is potentially immunogenic and its expression exposes target cells to destruction by cytotoxic T cells. The feasibility of introducing two transcriptional units in every target cells efficiently, and the likelihood that transgenic tissues escape immune attack limit the in vivo applicability of these systems.

Amongst currently available chimeric regulatable systems, those derived from the tetracycline resistance operon of *Escherichia coli* are by far the best documented and the most frequently used. They provide valuable tools for potential therapeutic approaches to many of the diseases that may be amenable to gene therapy. This review will consider the various tetracycline systems. Advantages and drawbacks for in vitro studies and in vivo applications will be presented, as well as the development of viral vectors and gene therapy applications relying on these systems. Tetracycline regulatable expression systems progressively turned out to be effective tools for a broad range of investigations. These include studies of gene function in transgenic animals, analysis of complex and multi-staged biological processes including embryogenesis and cancer, and the generation of animal models of human disorders aimed at understanding the cause and progression of diseases. Strategies developed for these various purposes will be discussed and examples will be given, preferentially with respect to the central nervous system (CNS).

2
The Tet-Off and Tet-On Systems

2.1
The Tet-Off System

A system based on elements derived from the tetracycline resistance operon of the *E. coli* transposon Tn10 was developed in 1992 by Bujard's group (Fig.1A) (Gossen and Bujard 1992). A chimeric tetracycline-controlled transactivator (tTA) was constructed by fusing the C-terminal 129 amino acids of the acidic domain of the VP16 protein of the herpes simplex virus type-1 to the C terminus of the bacterial tetracycline repressor protein (tetR). In the

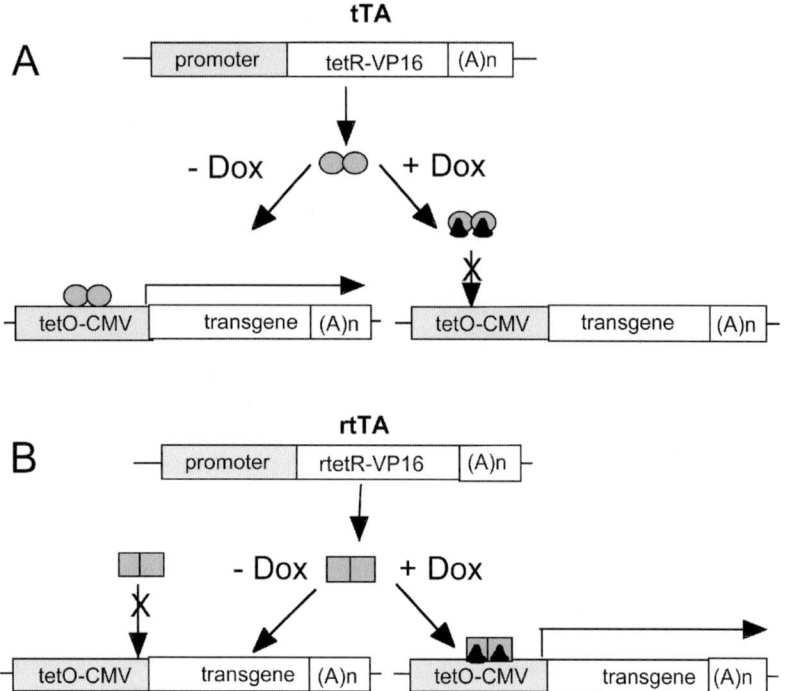

Fig. 1A, B Schematic of the tet-on and tet-off systems. **A** Schematic of the tet-off system. The tetracycline-responsive transactivator (*tTA*) gene is transcribed from a constitutive or tissue-specific promoter. In the presence of doxycycline, the transcription factor tTA is shifted off tetO binding sites, the transcription complex is not assembled and gene expression is silenced. In the absence of doxycycline, tTA binds to tetO sequences, transcription complexes are assembled on the minimal CMV promoter and gene expression is activated. **B** Schematic of the tet-on system. The tetracycline-responsive transactivator (*rtTA*) gene is transcribed from a constitutive or tissue-specific promoter. In the presence of doxycycline, rtTA binds to tetO sequences, transcription complexes are assembled on the minimal CMV promoter and gene expression is activated. In the absence of doxycycline, rtTA is shifted off tetO binding sites, the transcription complex is not assembled and gene expression is silenced

absence of tetracycline, the tetR moiety of tTA binds a regulatory region encompassing seven repeats of the 19-base pair tetO bacterial operator with high affinity and specificity. This element was placed immediately upstream of a minimal human cytomegalovirus promoter (tetO-CMV promoter). When bound to the promoter, the VP16 moiety of tTA induces the assembly of a transcriptional initiation complex (Ingles et al. 1991; Lin et al. 1991). In the presence of tetracycline or tetracycline derivatives such as anhydrotetracycline (Gossen and Bujard 1993), tTA preferentially binds the antibiotic, leading to conformational modifications that decrease affinity for the tetO operator. The system was referred to as the 'tet-off' system.

Bujard et al. reported that in clones of the human HeLa cell line stably expressing tTA, the expression of a luciferase gene could be modulated over a range of five orders of magnitude depending on the absence or the presence of 1 μM tetracycline in the culture medium (Gossen and Bujard 1992). The kinetics of the regulation was fast: gene expression decreased tenfold within 8 h and 50-fold within 12 h of antibiotic addition.

2.2
The Tet-On System

A second system was developed by the same group in 1995, allowing induction of gene expression in response to tetracycline derivatives (Gossen et al. 1995). As opposed to tet-off, this new system was referred to as 'tet-on.' The transactivator protein, called reverse transactivator (rtTA), requires tetracycline derivatives, preferentially doxycycline, for binding to the tetO operator (Fig. 1B). Differential activity of the two systems results from four amino acids changes in the tetR DNA-binding domain of the transactivator. Tetracycline derivatives inhibit tTA binding to tetO in the tet-off system, whereas they promote that of rtTA in the tet-on system.

In HeLa cells constitutively synthesizing rtTA, the expression of an integrated luciferase reporter gene inserted downstream of a tetO-CMV promoter was stimulated more than 1000-fold by the addition of doxycycline to the culture medium. Maximal induction occurred less than 24 h later. Addition of doxycycline would even induce transgene expression within 1 h (Hasan et al. 2001). Doxycycline-regulation was obtained in double transgenic mice, with induction values ranging from 1 to 7,000-fold depending on tissues (Kistner et al. 1996).

2.3
Properties of the Tet Systems for In Vivo Applications

The tet systems offer advantages that makes them attractive for in vivo regulation of gene expression and gene therapy. A characteristic feature is that both transactivator tetR moieties and cognate DNA-binding sites are of prokaryotic origin, thus ensuring restricted transactivation of transgenic sequences without transcriptional side effects. The price to pay for selectivity is the potential immunogenicity of the chimeric transactivator, which has unfortunately been documented in few large animals models (Favre et al. 2002; Latta-Mahieu et al. 2002). Modifications of the transactivation domains have been proposed to reduce immunogenicity. Baron et al. (1997) shortened the VP16 moiety to 12 amino acids without affecting the properties of the activation domain. A fusion between tetR and the KRAB repressor domain of the human zinc finger protein Kox1 was constructed (Deuschle et al. 1995). Binding of this chimeric protein to tetO blocked transcription efficiently and a 50-fold induction occurred after 3 days of tetracycline treatment. The activation domain located in the C terminal region of the human transcription factor nuclear factor (NF)-κB p65 protein was also proposed as an alternative to VP16 sequences (Rivera et al. 1996). The transcriptional activation domain of human E2F4, a protein ubiquitously expressed in vivo, was similarly fused to tetR with success (Akagi et al. 2001). The immunogenicity of these various chimeric transactivators is presently unknown.

Tetracycline and tetracycline derivatives are commonly used antibiotics in humans. Pharmacokinetics, pharmacodynamics and side effects are well documented. At low dosage, side effects are usually absent and tolerance is excellent. Because tetracycline and derivatives bind tetR with high affinity (10^{-9} M (Degenkolb et al. 1991), low dosages are effective for gene expression control, and side effects are rarely observed in mammalian cells and transgenic animals (Furth et al. 1994). Derivatives like doxycycline cross the blood–brain barrier, the placenta barrier, and are secreted in the milk, which may be appropriate for certain in vivo applications. Doxycycline can be administered either by oral route in sucrose solution at concentrations ranging from 200 µg/ml to 2 mg/ml, or as a time-release implanted capsule. Concentration is adjusted according to desired gene expression level.

The main difference between tTA and rtTA is the kinetics of activation and deactivation of gene expression. Using tTA as a transcriptional activator, the gene is kept silent by continuous doxycycline administration. Expression implies that doxycycline is totally eliminated. As the drug is lipophilic, long-term treatment may induce depots in fat tissues, the depletion of which is slow, delaying the induction of gene expression. Turning on gene expression by supplying doxycycline to the tet-on system may be preferable in that respect. System saturation is intrinsically more rapid and easier to control than depletion. For example, doxycycline depletion to activate gene expression via tTA can take 15–20 days in the mouse brain, whereas supplying doxycycline to rtTA takes only 3–6 days (Mansuy and Bujard 2000).

3
Gene Therapy Applications

Gene expression regulatory systems must conform with certain prerequisites if they are to be considered for gene therapy applications: (a) the two components of the regulation system must be efficiently introduced into a large proportion of targets, which are usually primary cells; (b) gene expression level must be negligible in the absence of inducer; (c) induction must be intense, modulable and reversible; (d) Regulation must be sustained over long periods of time in recipient animals.

Efficient introduction of the two components of the regulatory system in poorly transfectable cells necessitates the use of viral vectors. A number of viral vectors carrying tet-regulatable expression system has been developed for gene transfer strategies. Most are based on retroviruses, adenoviruses (Ad), or adeno-associated viruses (AAV).

3.1
Development of Tet-Regulatable Viral Vectors and Proof of Efficacy

3.1.1
Retroviral Vectors

Retrovirus-based gene transfer vectors include vectors derived from gamma-retroviruses, most often from mouse origin, or from lentiviruses. These vectors are efficient tools for gene delivery, integration, and expression in mammalian cells, both in vitro and in vivo (Verma and Somia 1997).

Gamma-retrovirus-derived vectors were the first gene transfer vectors developed with the purpose of gene therapy application. A majority of the patients enrolled in gene therapy trials have received retroviral vectors. Vectors are derived from the Moloney murine leukemia virus, the genome of which consists of two 9-kb RNA strands encased inside an enveloped capsid. Retroviral vectors efficiently penetrate and randomly integrate their genetic material into the chromosome of rapidly dividing cells, but they are ineffective in resting or slow dividing cells. This restriction is a major limitation to their use in gene therapy applications, which practically imposes ex vivo gene transfer to cultured cells. Such restriction does not exist for lentiviruses, which penetrate and integrate their genome into target cells independently of the cell division cycle.

3.1.1.1
Gamma-Retrovirus-Derived Vectors

The transfer of the two elements of the tet-off system using gamma-retrovirus vector has been achieved by various strategies. Authors have reported success using a single vector with two transcriptional units. Either internal promoters or internal ribosome entry sites were used (Hofman et al. 1996; Hoshimaru et al. 1996; Hwang et al. 1996; Iida et al. 1996; Paulus et al. 1996). Others used two vectors, one coding for the transactivator, the other one containing the gene of interest under the control of the tetracycline inducible promoter tetO-CMV (Bohl and Heard 1997; Bohl et al. 1997). Toxic squelching effects of the VP16 domain (Gossen and Bujard 1992) often complicated the isolation of packaging cells producing high vector titers.

Single vector systems were validated in immortalized cell lines (Fishman et al. 1994; Hoshimaru et al. 1996; Hwang et al. 1996; Iida et al. 1996; Paulus et al. 1996; Schultze et al. 1996) and primary muscle cell cultures (Hofman et al. 1996). Gene expression was modulated in the range of 300-fold in populations of NIH-3T3 cells (Hwang et al. 1996; Paulus et al. 1996) and 600-fold in some clones of primary myogenic cells (Hofman et al. 1996). Background activity of the minimal tetO-CMV promoter depended on the target cell type (Ackland-Berglund and Leib 1995). It increased with the presence of several promoters in the same vector (Iida et al. 1996; Paulus et al. 1996). The use of a single vector with adjacent or bicistronic transcription units exposes the recipient cells to the generation of undesired recombinant genomes. Difficulties were encountered in steadily maintaining high vector-producing cell lines. Promoter interference sometimes led to suboptimal expression and/or regulation (Emerman and Temin 1984). With the aim of eliminating promoter interference, single vectors systems were deigned in which expression of both tTA and the reporter gene were placed under the control of the tetO-CMV promoter (Agha-Mohammadi et al. 1997; Hofman et al. 1996). A drawback of this system was the slower kinetics of induction due to the retroactive control loop.

The efficiency of the two-vector tet-on system was documented in mice transplanted with primary transduced myoblasts (Bohl et al. 1997). Iterative supply or removal of doxycycline in the drinking water led to concommittant on/off switching of erythropoietin (Epo) secretion for 5 months in the absence of significant background. The absence of an immune response against transgenic muscle fibers expressing the rtTA chimera was an important observation in this study. Long-lasting regulation of Epo secretion was also observed in mice implanted with immunoprotective capsules containing allogeneic transduced fibroblasts (Serguera et al. 1999).

3.1.1.2
Lentiviral Vectors

Lentiviral vectors such as HIV-1-derived vectors are promising tools for gene therapy applications demanding long-term expression. However, they have not been used in humans so far. The lentivirus genome consists of two 10-kb RNA strands encased inside enveloped capsids. Long-term gene expression after lentivirus-mediated gene transfer has been documented in many tissues, including the CNS, the hematopoietic system, the skeletal muscles and the liver. Capacity to incorporate a tet-regulated expression cassette and to transfer it into target cells was reported, using either a single or two independent vectors (Kafri et al. 2000; Reiser et al. 2000). However, both in vitro and in vivo studies with either tTA or rtTA showed to significant leakiness of the tetracycline regulatory systems when introduced into lentivirus vectors. Putative causes of leakiness are a selective pressure against cells producing high levels of tTA, the residual affinity of rtTA to its operator, and/or some specific feature of the vector structure. Search for improved vector design is needed to generate efficient tet-controlled lentivirus vectors devoid of significant basal activity.

3.1.2
AAV Vectors

AAV vectors have broad utility in both experimental and applied gene therapy; AAV vectors providing regulatable expression are thus regarded as important tools. The tetracycline system is the most often used in these vectors.

AAV vectors are recognized as safe and effective gene transfer vehicles for gene therapy. Because of the absence of pathogenicity of AAV in humans, the capacity to transduce nondividing cells, the long-term gene expression in vivo and the availability of manufacturing procedures suitable for clinical grade material, AAV vectors fulfill the most relevant criteria for gene therapy applications. They are especially attractive for treating disorders of the CNS. The most significant limitation of AAV vectors is the small size of the single-stranded DNA AAV genome (4.7 kb), which cannot accommodate more than 4.5 kb of foreign DNA.

The tetracycline system has been incorporated into AAV vectors. Various configurations were proposed and have proved effective, using either a single genome or two independent vectors (Bohl et al. 1998; Fitzsimons et al. 2001; Haberman et al. 1998; Rendhal et al. 1998). The single-vector configuration has the advantage of ensuring delivery of both components in every targeted cell, although the size limitation complicates vector design. Efficient delivery of the mouse Epo cDNA expressed under the control of the tet-off or the tet-on system was reported into mouse skeletal muscles. This was performed either using a single (Bohl et al. 1998) or two separate AAV vectors (Rendhal et al. 1998). Long-term regulation of hematocrit levels and Epo serum concentrations was observed in both settings. Epo secretion was switched on and off iteratively over several months without decrease of induction levels. Background expression was negligible (Bohl et al. 1998). The same system was, however, much less efficient when implemented in nonhuman primates (Favre et al. 2002). Similar tet-on AAV vectors encoding the cynomolgus Epo cDNA were prepared and injected intramuscularly in macaques. Induction levels appeared very low, suggesting that the nuclear concentration of the transactivator was a limiting factor in this context (Smith-Arica et al. 2000; Urlinger et al. 2000). Improvements were obtained by driving rtTA expression through a stronger promoter. Though induction was more efficient, background Epo secretion level was high.

AAV vectors hold promises for treating neurological disorders with respect to their ability to induce long-term gene expression without toxicity or immune response. Fitzsimmons et al. (2001) reported AAV–enhanced green fluorescent protein (eGFP) transduction and tight tet-off regulation in the rat striatum. Whereas hundreds of eGFP-positive neurons were detected upon activation, there were only a few positive cells when doxycycline was provided (induction 10–49-fold) (Fitzsimons et al. 2001; Haberman et al. 1998). Authors presumed that the presence of both promoter and enhancer elements in the AAV genome terminal repeats could account for the background expression in the 'off' state. Leakiness was actually reduced and induction increased from 49- to 204-fold by the insertion of insulator sequences (Haberman et al. 1998).

3.1.3
Ad Vectors

Adenoviruses contain a single, double-stranded 39-kb linear DNA genome encased within a nonenveloped capsid. Vectors derived from adenoviruses allow remarkably efficient in vivo gene transfer to a wide spectrum of cell types and organs. In immunocompetent hosts, Ad vector-mediated gene transfer is transient, a consequence of the induction of robust innate and antigen-specific immune responses against transgenic cells. Both virus and transgene encoded peptides elicit this response, the viral component probably being responsible for boosting the antigenicity of transgene products that could otherwise be tolerated. There are nevertheless broad gene therapy applications for Ad vectors, which have been administrated to thousands of patients without significant side effects.

Tetracycline regulatable Ad vectors used either tTA or rtTA to regulate expression of marker or toxic proteins and consisted of either one or two vectors that allowed in vitro and in vivo regulation (Corti et al. 1999; Harding et al. 1998; Hu et al. 1997; Massie et al. 1998; Molin et al. 1998; Neering et al. 1996; Yoshida and Hamada 1997).

Several studies highlighted that in the context of a single Ad vector the tet-off system was more effective than the tet-on system for regulating transgene expression (Mizuguchi and Hayakawa 2002; Xu et al. 2003). The vector containing the tet-off system provided negative control of gene expression ranging from 20- to 500-fold, depending on cell type and condition. By contrast, gene expression was increased by only two- to 28-fold with the tet-on system and effective doxycycline concentration was about 100-fold higher than for the tet-off system (Mizuguchi and Hayakawa 2002).

3.1.4
Improving Tet Systems

Leakiness related to rtTA residual transactivating affinity in the absence of tetracycline, or related to the operator in the absence of transactivator, was often reported as a potential concern for accurate gene expression regulation. Attempts at improving the tet-on system were recently described. Second-generation rtTAs have been developed by Urlinger et al. (2000). New rtTA variants were selected after sequential mutagenesis in *Saccharomyces cerevisiae* for low or undetectable doxycycline-independent gene expression and high inducibility. A variant called rtTA2S-M2 functioned at 10-fold lower doxycycline concentration than rtTA and showed no background expression in the absence of doxycycline. The VP16 activation domain was replaced with a minimal activation domain consisting of 12 amino acids (Baron et al. 1995) and the coding sequence was optimized for expression in human cells. Stringent control of gene expression using these news variants was reported in vitro and in mice (Lamartina et al. 2002). In in vitro transient transfection assays, rtTA variants displayed basal activity markedly lower than that of rtTA and the magnitude of induction was 250- to 630-fold higher. Mouse skeletal muscles were electroinjected with plasmids expressing Epo under the control of rtTA or the rtTA variants. Variants displayed a considerably lower background activity and larger window of induction than rtTA in vivo. Stringent repeated on and off switch of Epo secretion and hematocrit levels were observed in the absence of background activity over a 10-month period.

A tetracycline-responsive transcriptional silencer (tTS) was also assessed as a means of reducing the basal expression of tet-on system (Freunlieb et al. 1999). tTS is a chimera between tetR and the silencing domain of the human protein Kid1. tTS binds rtTA responsive promoters in the absence of doxycycline, while addition of doxycycline prevents binding, relieves repression, and promotes the association with rtTA. A switch between an actively repressed to an actively stimulated form of the promoter is thus obtained. A ten- to 200-fold repression was measured in transient expression assays, whereas inducibility was increased by three orders of magnitude in stably transfected HeLa cells. Plasmids co-expressing tTS and the rtTA variant rtTA2S-M2 were electrotransferred into mouse skeletal muscles (Lamartina et al. 2003) and a 1,000-fold induction of serum alkaline phosphatase gene expression was obtained. Regulation of Epo gene expression over the long term was stringent, strictly depending on doxycycline dosage and without leakiness.

New rtTA variants will probably improve tet-regulated AAV vectors. However, the limited capacity of AAV vectors prevents accommodation of both tTS and rtTA in the same genome. This may be done in Ad vectors, whose capacity is larger. Improved performances of the tet-on system are expected. Reduced basal expression and increased induction level were actually reported in vitro as well as in skeletal muscles (Salucci et al. 2002). Using two lentivirus vectors, one coding for rtTA2S-M2 and the other one containing *lac*Z with the minimal *tet*O promoter, Koponen et al. (2003) showed that cotransduction into rat brain led to a tight control of lacZ expression over a short period of time. Vigna et al. (Vigna et al. 2002) transduced $CD34^+$ cells with a tet-off lentivirus vector containing the tTA2S transactivator, which has a codon-optimized sequence and a minimal transactivation domain. On/off switching of the reporter gene was documented over a period of 20 weeks in human hematopoietic cells transplantated in NOD/SCID mice.

3.2
Gene Therapy Applications

Gene therapy applications for which tet-regulated gene expression has been considered are listed in Table 1. Issues raised by the systemic delivery of hormones, or by the delivery of therapeutic gene to CNS cells are discussed below, taking as examples Epo for a treatment of β-thalassemia and tyrosine hydroxylase (TH) for that of Parkinson disease.

Table 1 Examples of tetracycline-regulatable systems used in animal models of gene therapy applications

Disease/pathology	Transfer protocol	Gene of interest	Reference
β-Thalassemia	AAV vector	Erythropoietin	Samakoglu et al. 2002
Chronic arthritis	DNA electroinjection	Interleukin 10	Perez et al. 2002
Parkinson disease	Adenoviral vector	Tyrosine hydroxylase	Corti et al. 1999
Fat loss	AAV vector	Leptin	Wilsey et al. 2002
Axonal regrowth	Retroviral vector	Nerve growth factor	Blesh et al. 2001
Retinopathies	AAV vector	Green fluorescent protein	Folliot et al. 2003

3.2.1
The Systemic Delivery of Epo

Erythropoietin is a glycoprotein hormone secreted by the kidney and in some situations by the liver. Its main role is to induce terminal differentiation of erythroid precursors into red blood cells. Recombinant human Epo (rHuEpo) was the first hematopoietic growth factor produced by genetic engineering. rHuEpo is administrated to patients with chronic renal failure. It stimulates erythropoietic activity, leading to increased reticulocyte cell counts and hemoglobin concentrations in the blood. rHuEpo may also be useful in the treatment of Epo-responsive anemia associated with AIDS and cancer, self-transfusions, and hemoglobinopathies of inflammatory origin. However, for most countries, treating all candidate patients would imply a reduction of rHuEpo costs or the development of alternative delivery methods. Epo secretion from genetically modified tissues could play a role in this context. Ideally, the delivery of adequate amounts of Epo would reproduce physiological regulation. Though hypoxia-inducible gene expression systems have been reported (Payen et al. 2001; Rinsch et al. 1997) these systems are not yet available for gene therapy application. Regulation by an exogenous inducer such as tetracycline would presumably be acceptable for severe disorders, at least during initial phases of human application.

The systemic delivery of high and controlled amounts of Epo might be beneficial in β-thalassemia patients. β-thalassemia is the most prevalent monogenic disorder in the world, with millions of patients in South-East Asia. It results from the deficient synthesis of β-globin and the subsequent accumulation of unpaired α globin chains. Re-stimulation in the adult of the expression of the fetal γ-globin chain would circumvent the genetic impairment, and the correction of the globin chain imbalance would likely improve clinical manifestations. Evidence was provided that Epo could play this role (for a review see (Bohl and Heard 2000). Proposing gene therapy to developing countries supposes a simple, low cost, low risk procedure. Long-term Epo delivery at controlled levels must ensure biological efficacy in the absence of polycythemia.

Using an AAV vector expressing the mouse Epo cDNA under the control of the tet-on system, Epo delivery could be adjusted to stimulate fetal erythropoiesis at levels consistent with the maintenance of β-thalassemic mice in steady-state normocythemia (Samakoglu et al. 2002). Doxycycline dosage was adapted to hematocrit. Serum Epo concentrations were maintained at a level more than 14.4-fold higher than in normal mice and 2.8-fold higher than in untreated β-thalassemic mice. Circulating red blood cells essentially synthesized β-minor globin, the mouse equivalent to human fetal γ-globin.

Long-term control of systemic Epo delivery was further explored in non-human primates, with the perspective of human applications (Favre et al. 2002). The cynomolgus macaque Epo cDNA was expressed under the control of the tet-on system in a single AAV vector. The vector was injected in skeletal muscles. Five-day induction pulses with doxycycline were performed monthly for 13 months. Favre et al. (2002) showed successful initial inductions in all macaques with a tight regulation and a rapid deinduction pattern upon doxycycline withdrawal. Maximum Epo concentrations were reached 2–3 days after the first doxycycline administration and returned to background levels less than 5 days after the treatment was stopped. However, sustained regulation over the entire experimental period was observed in one out of six macaques only. Epo secretion in response to doxycycline was progressively lost in other animals. Humoral and cellular immune responses against the rtTA transactivator protein and a loss of transgenic cells was demonstrated in these animals. Humoral and cellular immune responses against rtTA2S-M2 was also documented after intramuscular plasmid or Ad vector injection in primates (Latta-Mahieu et al. 2002). These results contrast with previous studies in mice, in which rtTA or tTA expression in skeletal muscle did not induce immune reaction (Bohl et al. 1998; Rendhal et al. 1998). Major histocompatibility complex haplotype and/or T-cell repertoire variation between outbred individuals may account for different reactions amongst individuals in macaques. These observations underscore the importance of preclinical studies in nonhuman primates as opposed to inbred murine models. Strategies considered with the aim of reducing transgene product immunogenicity include the use of tissue-specific promoters that could prevent expression of potentially immunogenic proteins in antigen-presenting cells (Ralph et al. 2000; Smith-Arica et al. 2000) and the administration of immunosuppressive agents at the time of transgene delivery.

3.2.2
Parkinson Disease

Parkinson disease is a severe neurodegenerative disorder. A loss of dopaminergic neurons in the substantia nigra and of the neurotransmitter dopamine in striatal cells is responsible for tremor, rigidity, bradykinesia and cognitive abnormalities. L-Dopa therapy, which controls clinical manifestations at the onset, loses efficacy with time (Marsden and Parkes 1976). Increasing difficulty in adjusting dosage leads to highly disabling fluctuations of the motor response (Obeso et al. 1994). A control of dopamine delivery to the striatum that would mimic the complex mechanism of dopaminergic stimulation might circumvent this difficulty. TH is the rate-limiting enzyme in the synthesis of dopamine and studies have shown that the transplantation of cells expressing TH in the dopamine-depleted striatum can supply dopamine locally and partially restore the lesion-induced deficits (Fisher et al. 1991). A gene therapy approach could thus consist of transferring the TH cDNA into striatal cells and controlling its expression level. Alternatively, delivering controlled amounts of neuron survival factors to the substantia nigra might prevent cell death.

Ad vectors expressing either the human TH (Corti et al. 1999; Corti et al. 1999) or the human glial cell line-derived neurotrophic factor (Kojima et al. 1997) under the control of a tet-off regulatable system have been constructed. TH vectors were either introduced in vitro into human neural progenitor cells that were further transplanted in the striatum, or directly injected in the striatum (Corti et al. 1999a, 1999b). Recipient rats were unilaterally lesioned with 6-hydroxydopamine to deplete dopamine (6-OHDA) (Corti et al. 1999b). TH immunolabeling of striatal cells could be modulated depending on the administration of doxycycline in the drinking water (1 mg/ml). A 20-fold lower dose of doxycycline was sufficient to control TH transcription. Apomorphine-induced rotational asymmetry in 6-OHDA lesioned rats is a behavioral indicator of the extent of dopamine restoration in the denervated striatum. A significant reduction of the number of apomorphine-induced rotations was observed from week 1 to week 4 after vector injection. However, the rotation scores were only transiently reduced and were similar to preinjection values 6 weeks after vector injection, a time at which TH immunolabeling was still present. The authors hypothesized that this phenomenon was related to the residual concentration of the co-factor BH4 in lesioned rats, as the hydroxylation of tyrosine mediated by TH requires this cofactor. This was supported by the finding that a subcutaneous injection of BH4 induced a substantial increase of tissue Dopa levels in treated rats.

4
Modeling of Human CNS Pathologies

4.1
Description

Classical transgenic and gene-targeted mouse mutants are powerful model systems in which to study the pathogenesis of neurodegenerative diseases. However, questions of fundamental importance for the development of therapy cannot be addressed using classical techniques, such as the identification of events upstream of that induced by the gene of interest or the determination of whether a process is reversible or irreversible. In vivo temporally and regionally specified expression is required to address those questions. Tetracycline-regulated systems are useful tools in this respect.

Adapting the tet system to mice necessitates crossing of two transgenic lines: the first line expresses tTA under the control of a strong promoter; and the second line carries the transgene of interest downstream of a tetO-CMV promoter. Using β-galactosidase or luciferase markers and the tet-off system, Furth et al. (1994) showed repression of gene expression within 7 days of tetracycline treatment in transgenic mice. More rapid kinetics were obtained with doxycycline, reducing the mRNA levels by 50% within 2 days. Repression was fully reversible within 7 days (Tremblay 1998).

The combination of two transgenic lines often leads to unexpected patterns of expression. Different lines may yield different expression patterns when bred with the same tTA line (Furth et al. 1994; Ghavami et al. 1999; Yamamoto et al. 2000). The mechanisms responsible for this variability are not understood: they could be related to the expression of tTA or to the accessibility of the tetO promoter. Gene replacement (knockin) was proposed to overcome variability of expression—the strategy involves the insertion of a single cassette containing both tTA and tetO-transgene sequences downstream of promoters of interest by homologous recombination (Bond et al. 2000). The transgene is then expressed as the intact promoter. Nevertheless, gene expression levels differ from that of the wild-type promoter due to amplification by the tet system.

The tet-off system is exposed to the consequences of a long-term or prenatal exposure to tetracycline. As the drug is stored in bones and cleared slowly, delayed transgene expression is usual upon removal. The tet-on system might be preferable in this context. However, delays before induction depends on the tissue of interest, showing variations from 1 day (Kistner et al. 1996) to 6 days (Mansuy et al. 1998) days in adult transgenic mice. Moreover the doxycycline dosages required for rtTA induction may be 100-fold higher than for tTA suppression. Because of these features, tet system are not suitable for investigating acute phenomena. Studies in transgenic mice may be facilitated by the rtTA2s-M2 transactivator (Urlinger et al. 2000) that is more sensitive, stable, and has negligible background expression.

Another system that has potential applications for transgenic studies is the bidirectional tetO promoter, which allows coexpression and coregulation of two genes in stochiometric amounts (Baron et al. 1995; Chtarto et al. 2003; Krestel et al. 2001). This is of particular interest when the expression of the gene of interest is difficult to monitor. The second gene can be an easily detectable marker allowing identification of transgene positive cells in tissue. For example, Yamamoto et al. (2000) developed a transgenic mouse model expressing a mutant huntingtin gene and the lacZ marker simultaneously. Gene expression was then analyzed by X-gal staining.

4.2
Applications

More than 20 mouse lines expressing tTA or rtTA in various cell types or tissues and over 25 lines containing a variety of target genes under the control of the tetO promoter have been described to date (Table 2). Examples of their use for investigations in the CNS are given below.

Table 2 Examples of mouse strains transgenic for tTA or rtTA, in which tissue-specific promoters control transactivator gene expression

Tissue	Transactivator	Target gene	Reference
Liver	tTA	Luciferase	Kistner et al. 1996
Liver	tTA	TGF-β1	Ueberham et al. 2003
Heart	tTA	Diphtheria toxin A	Lee et al. 1998
CNS: hippocampus, striatum	tTA	Mutated calcineurin	Mansuy et al. 1998
CNS: hippocampus, cortex, striatum, olfactory bulb, cerebellum	rtTA	Calcineurin inhibitor	Malleret et al. 2001
CNS: hippocampus, striatum, cortex	tTA	Huntingtin	Yamamoto et al. 2000
Embryonic neural crest	tTA/rtTA	Ednrb	Shin et al. 1999
CNS	tTA	Prion	Tremblay et al. 1998
CNS	tTA	LacZ	Mayford et al. 1996
Hematopoietic cells	tTA	Myc	Felsher and Bishop 1999
Hematopoietic cells	tTA	Bcr-Abl1	Huettner et al. 2000
Retina	tTA	BDNF	Okoye et al. 2003
Muscle	tTA	Utrophin	Squire et al. 2002

Bcr-Abl1, breakpoint cluster region Abelson protein 1; Ednrb, endothelin receptor B; TGFβ1, transforming growth factor 1; BDNF, brain-derived neurotrophic factor.

4.2.1
Huntington Disease

One of the most recent breakthroughs in the use of the tet regulatory systems was the creation of a conditional animal model for Huntington disease (HD) (Yamamoto et al. 2000). This inherited and neurodegenerative disorder is caused by a CAG expansion mutation in exon 1 of the huntingtin gene. The mutation results in a polyglutamine (polyQ) stretch at the N terminus of the huntingtin protein, which can lead to novel protein–protein interactions (Petersen et al. 1999) and form insoluble aggregates. Transgenic mice that express exon 1 of huntingtin with a pathogenic stretch of CAG develop intranuclear protein aggregates, striatal atrophy, changes in dopamine receptor levels and motor deficits. These findings demonstrate that expression of huntingtin with a polyQ stretch leads to HD-like pathology. Two questions relevant to pathophysiology could not be addressed in the classical transgenic models: do pathological events require continued expression of the mutant protein that initiated the process? are behavioral and/or biochemical phenotypes reversible? To answer these questions, Yamamoto et al. constructed a transgenic mouse model in which the expression of huntingtin gene exon 1 carrying an expansion of 94 CAG repeats was regulated by tetracycline (Yamamoto et al. 2000). The transgene and lacZ were placed under the control of the bi-directional promoter (Baron et al. 1995) in a first transgenic line. Expression was restricted to the forebrain by breeding this line with animals expressing tTA under the control of the CamKIIa promoter (Mayford et al. 1996). As the bi-directional promoter allows control of the mutant huntingtin gene and the lacZ marker gene simultaneously, gene expression induced by tTA was monitored by X-gal staining. Expression of the abnormal huntingtin progressively induced a behavioral and pathological phenotype reminiscent of HD. Mice developed cytoplasmic and nuclear aggregates, striatal atrophy, reactive astrocytes, and a motor phenotype. When the phenotype was well established (at 18 weeks of age), doxycycline was given to mice, leading to arrest the production of mutant huntingtin. Sixteen weeks later, aggregates had disappeared from the striatum and were greatly reduced in the cortex. The motor phenotype reversed to control. These results provided evidence for a role of huntingtin in the occurrence of HD, suggesting that the progressive nature of HD is dependent upon the continuous production of huntingtin carrying a polyQ stretch. Reversion of clinical manifestation upon removal of the mutated protein suggests that brain tissues are plastic enough to recover from the insult.

4.2.2
Prion Diseases

Studies of transmissible spongiform encephalopathy, or prion disease, have demonstrated that these neurodegenerative illnesses result from the conversion of the cellular prion protein PrPc into a pathogenic protease-resistant isoform PrPsc. The function of PrPc is unknown. Inactivation of the PrPc gene by gene targeting in mice does not induce deleterious effect, but protects animals from developing the disease after inoculation of the infectious agent, suggesting that PrPc might be involved in the accumulation of the prion. To study the function of PrPc and to examine whether it is required for prion formation, a mouse model was developed in which PrPc expression is regulated by tTA on a PrPc knockout background (Tremblay et al. 1998). tTA expression was driven from the prion promoter. Induction of high levels of PrPc during embryogenesis and postnatal development caused lethality. The phenomenon could be rescued by suppression of the transgene expression. In contrast, high PrPc expression levels were not detrimental in adulthood. Injection of PrPsc in animals expressing the protein resulted in the accumulation of PrPsc and to neurodegenerative symptoms. In mice receiving doxycycline, PrPc expression was minimal (less than 15% of normal levels) and no neurological dysfunction was observed. These results suggest that though PrPc may not play an essential role in the adult mouse brain, it is a prerequisite for acquiring the prion disease.

4.3
Current Progress for Future Applications

Recently a combination of both tTA and rtTA variants of the tetracycline-regulatable system and mutated tetO sequences were developed and used to achieve exclusive control of two gene activities within the same cells at different concentrations of doxycycline (Baron et al. 1999). This approach is based on the differences of affinities of tTA and rtTA for doxyxycline. The tTA2-1 transactivator activates the minimal promoter (tetO4) in the absence of doxycycline, whereas the rtTA2-1 transactivator is inactive. In the presence of doxycycline concentrations in the range of 100–3,000 ng/ml, rtTA2-1 stimulates another tetO sequence (tetO6), whereas tTA2-1 is switched off. At intermediate doxycycline levels (e.g., 30 ng/ml) none of the promoters is active.

This system could be used for the generation of novel conditional mutants in transgenic mice. For example, it could allow switching from a wild-type to a mutant allele at a defined developmental state, then reversing to the expression of either the wild-type or the null allele, depending on doxycycline stimulation level.

5
Conclusions

Tetracycline regulatable systems offer numerous advantages: strict on/off regulation, high inducibility, short responses times, specificity, no interference with cellular pathways, bioavailability of a nontoxic inducer, and dose dependence. Induction/decay kinetics operate on a time scale of days, reflecting the early point of regulation and the many potentially rate-limiting steps between transcription and protein production. These kinetics are well appropriate for controlling secretion levels of proteins that regulate cell growth and differentiation events. Many applications of the tet systems have been described in this field. However, there are proteins of which secretion levels are tightly controlled that are secreted with much faster kinetics. For example, insulin is secreted in a brief and precise pulse of 4 h. Even faster kinetics operate for neurotransmitters. An artificial system allowing regulation of protein secretion in a much faster range than the tet system was described by Rivera et al. (2000). In this system, proteins are made constitutively, but they are engineered to accumulate in the endoplasmic reticulum until the addition of a small molecule ligand, which releases the stored pool. Pulsed delivery of insulin with this system corrected hyperglycemia in mice.

Applications of the tetracycline systems have been proposed in much broader domains than that of the control of protein secretion level. They were applied to the analysis of complex and multi-staged biological processes, such as embryogenesis and cancer, and to model human neuropathologies. Transgenic mouse models have been invaluable in the effort to unravel the pathogenic processes underlying neurodegenerative diseases. Inducible expression systems, although still at an early stage of development, offer a more powerful means of approaching these complex questions by allowing investigators to express pathogenic proteins at specified times and discrete tissue regions. However, adjusting expression at the desired level in transgenic animals still remains a challenging issue.

References

Ackland-Berglund CE, Leib DA (1995). Efficacy of tetracycline-controlled gene expression is influenced by cell type. BioTechniques 18:196–200

Agha-Mohammadi S, Alvarez-Vallina L, Ashworth LJ, Hawkins RE (1997) Delay in resumption of the activity of tetracycline-regulatable promoter following removal of tetracycline analogues. Gene Ther 4:993–997

Akagi K, Kanai M, Saya H, Kozu T, Berns A (2001) A novel tetracycline-dependent transactivator with E2F4 transcriptional activation domain. Nucl Acids Res 29:E23

Baron U, Freundlieb S, Gossen M, Bujard H (1995) Co-regulation of two gene activities by tetracycline via bidirectional promoter. Nucl Acids Res 23:3605–3606

Baron U, Schnappinger D, Helbl V, Gossen M, Hillen W, Bujard H (1999) Generation of conditional mutants in higher eukaryotes by switching between the expression of two genes. Proc Natl Acad Sci USA 96:1013-1018

Blesh A, Conner J, Tuszynski M (2001) Modulation of neuronal survival and axonal growth in vivo by tetracycline-regulated neurotrophin expression. Gene Ther 8:954-960

Bohl D, Heard J (2000) Delivering erythropoietin through genetically engineered cells. J Am Soc Nephrol 11:S159-S162

Bohl D, Heard J (1997) Modulation of erythropoietin delivery from engineered muscles in mice. Human Gene Ther 8:195-204

Bohl D, Naffakh N, Heard JM (1997). Long term control of erythropoietin secretion levels by tetracycline in mice transplanted with engineered primary myoblasts. Nat Med 3:299-312

Bohl D, Salvetti A, Moullier P, Heard JM (1998). Control of erythropoietin delivery by doxycycline in mice after intramuscular injection of adeno-associated vector. Blood 92:1512-1517

Bond C, Sprengel R, Bissonnette J, Kaufmann W, Probnow D, Neelands T, Storck T, Baetscher M, Jerenic J, Maylie J, Knaus H, Seeburg P, Adelman J (2000) Respiration and parturition affected by conditional overexpression of the Ca2+-activated K+ channel subunit, SK3. Science 289:1942-1946

Chtarto A, Bender H, Hanemann C, Kemp T, Lehtonen E, Levivier M, Brotchi J, Velu T, Tenenbaum L (2003) Tetracycline-inducible transgene expression mediated by a single AAV vector. Gene Ther 10:84-94

Corti O, Sabate O, Horellou P, Colin P, Dumas S, Buchet D, Buc-Caron M-H, Mallet J (1999a) A single adenovirus vector mediates doxycycline-controlled expression of tyrosine hydroxylase in brain grafts of human neural progenitors. Nat Biotechnol 17:349-354

Corti O, Sanchez-Capelo A, Colin P, Hanoun N, Hamon M, Mallet J (1999b) Long-term doxycycline-controlled expression of human tyrosine hydroxylase after direct adenovirus-mediated gene transfer to a rat model of Parkinson's disease. Proc Natl Acad Sci USA 96:12120-12125

Degenkolb J, Takahashi M, Ellestad GA, Hillen W (1991) Structural requirements of tetracycline-tet repressor interaction: determination of equilibrium binding constants for tetracycline analogs with the tet repressor. Antimicrob Agents Chemother 35:1591-1595

Deuschle U, Meyer WK-H, Thiesen HJ (1995) Tetracycline-reversible silencing of eukaryotic promoters. Mol Cell Biol 15:1907-1914

Emerman M, Temin HM (1984) Genes with promoters in retrovirus vectors can be independently suppressed by an epigenetic mechanism. Cell 39:459-467

Favre D, Blouin V, Provost N, Spisek R, Porrot F, Bohl D, Marmé F, Chérel Y, Salvetti A, Hurtrel B, Heard J, Rivière Y, Moullier P (2002) Lack of immune response against the tetracycline-dependent transactivator correlates with long-term doxycycline-regulated transgene expression in nonhuman primates after intramuscular injection of recombinant adeno-associated virus. J Virol 76:11605-11611

Felsher D, Bishop J (1999) Reversible tumorigenesis by Myc in hematopoietic lineages. Mol Cell 4:199-207

Fisher LJ, Jinnah HA, Kale LC, Higgins GA, Gage FH (1991) Survival and function of intrastriatally grafted primary fibroblasts genetically modified to produce L-dopa. Neuron 6:371-380

Fishman GI, Kaplan ML, Buttrick PM (1994) Tetracycline-regulated cardiac gene expression in vivo. J Clin Invest 93:1864–1868
Fitzsimons H, McKenzie J, During M (2001) Insulators coupled to a minimal bidirectional tet cassette for tight regulation of rAAV-mediated gene transfer in the mammalian brain. Gene Ther 8:1675–1681
Folliot S, Briot D, Conrath H, Provost N, Cherel Y, Moullier P, Rolling F (2003) Sustained tetracycline-regulated transgene expression in vivo in rat retinal ganglion cells using a single type 2 adeno-associated vector. J Gene Med 5:493–501
Freunlieb S, Chirra-Müller C, Bujard H (1999) A tetracycline controlled activation/repression system with increased potential for gene transfer into mammalian cells. J Gene Med 1;4–12
Furth PA, St. Onge L, Böger H, Gruss P, Gossens M, Kistner A, Bujard H, Hennighausen L (1994) Temporal control of gene expression in transgenic mice by a tetracycline-responsive promoter. Proc Natl Acad Sci USA 91:9302–9306
Ghavami A, Stark K, Jareb M, Ramboz S, Segu L, Hen R (1999) Differential addressing of 5-HT1A and 5-HT1B receptors in epithelial cells and neurons. J Cell Sci 112:967–976
Gossen M, Bujard H (1993) Anhydrotetracyclin, a novel effector for tetracycline controlled gene expression in eukaryotic cells. Nucl Acids Res 21:4411–4412
Gossen M, Bujard H (1992) Tight control of gene expression in mammalian cells by tetracycline-responsive promoters. Proc Natl Acad Sci USA 89:5547–5551
Gossen M, Freundlieb S, Bender G, Müller G, Hillen W, Bujard H (1995) Transcriptional activation by tetracyclines in mammalian cells. Science 268:1766–1769
Haberman RP, McCown TJ, Samulski RJ (1998) Inducible long-term gene expression in brain with adeno-associated virus gene transfer. Gene Ther 5:1604–1611
Harding TC, Geddes BJ, Murphy D, Knight D, Uney JB (1998) Switching transgene expression in the brain using adenoviral tetracycline-regulatable system. Nat Biotechnol 16:553–555
Hasan M, Schonig K, Berger S, Graewe W, Bujard H (2001) Long-term, noninvasive imaging of regulated gene expression in living mice. Genesis 29:116–122
Hofman A, Nolan G, Blau HB (1996) Rapid retroviral delivery of tetracycline-inducible genes in a single autoregulatory cassette. Proc Natl Acad Sci USA 93:5185–5190
Hoshimaru M, Ray J, Sah DWY, Gage FH (1996) Differentiation of the immortalized adult neuronal progenitor cell line HC2S2 into neurons by regulatable suppression of the v-myc oncogene. Proc Natl Acad Sci USA 93:1518–1523
Hu SX, Ji W, Zhou Y, Logothetis C, Xu H-J (1997) Development of an adenovirus vector with tetracycline-regulatable human tumor necrosis factor alpha gene expression. Cancer Res 57:3339–3343
Huettner C, Zhang P, van Etten R, Tenen D (2000) Reversibility of acute B cell leukemia induced by BCR-ABL1. Nat Genet 24:57–60
Hug H, Costas M, Staeheli P, Aebi M, Weissmann C (1988) Organization of the murine Mx gene and characterization of its interferon- and virus- inducible promoter. Mol Cell Biol 8:3065–3079
Hwang JJ, Scuric Z, Anderson WF (1996) Novel retroviral vector transferring a suicide gene and a selectable marker gene with enhanced gene expression by using a tetracycline-responsive expression system. J Virol 70:8138–8141
Iida A, Chen ST, Friedmann T, Yee JK (1996) Inducible gene expression by retrovirus-mediated transfer of a modified tetracycline-regulated system. J Virol 70:6054–6059
Ingles CJ, Shales M, Cress WD, Triezenberg SJ, Greenblatt J (1991) Reduced binding of TFIID to transcriptionally compromised mutants of VP16. Nature 351:588–590

Israel D, Kaufman RJ (1989) Highly inducible expression from vectors containing multiple GRE's in CHO cells overexpressing the glucocorticoid receptor. Nucl Acids Res 17:4589–4604

Kafri T, van Praag H, Gage F, Verma I (2000) Lentiviral vectors: regulated gene expression. Mol Ther 1:516–521

Kistner A, Gossen M, Zimmermann F, Jerecic J, Ullmer C, Lubbert H, Bujard H (1996) Doxycycline-mediated quantitative and tissue-specific control of gene expression in transgenic mice. Proc Natl Acad Sci USA 93:10933–10938

Ko MSH, Takahashi N, Sugiyama N, Takano T (1989) An auto-inducible vector conferring high glucocorticoid inducibility upon stable transformant cells. Gene 84:383–389

Kojima H, Abiru Y, Sakajiri K, Watabe K, Ihishi N, Takamori M, Hatanaka H, Yagi K (1997) Adenovirus-mediated transduction with human glial cell line-derived neurotrophic factor gene prevents 1-methyl-4-phenyl-1,2,3,6-tetrahydropyridine-induced dopamine depletion in striatum of mouse brain. Biochem Biophys Res Commun 238:569–573

Krestel H, Mayford M, Seeburg P, Sprengel R (2001) A GFP-equipped bidirectional expression module well suited for monitoring tetracycline-regulated gene expression in mouse. Nucl Acids Res 29:e39

Lamartina S, Roscilli G, Rinaudo C, Sporeno E, Silvi L, Hillen W, Bujard H, CorteseR, Ciliberto G, Toniatti C (2002) Stringent control of gene expression in vivo by using novel doxycycline-dependent trans-activators. Hum Gene Ther 13:199–210

Lamartina S, Silvi L, Roscilli G, Casimiro D, Simon A, Davies M, Shiver J, Rinaudo C, Zampaglione I, Fattori E, Colloca S, Gonzalez Paz O, Laufer R, Bujard H, Cortese R, Ciliberto G, Toniatti, C. (2003) Construction of an rtTA2S-M2/tTSkid-based transcription regulatory switch that displays no basal activity, good inducibility, and high responsiveness to doxycycline in mice and non-human primates. Mol Ther 7:271–280

Latta-Mahieu M, Rolland M, Caillet C, Wang M, Kennel P, Mahfouz I, Loquet I, Dedieu J, Mahfoudi A, Trannoy E, Thuiller V (2002) Gene transfer of a chimeric trans-activator is immunogenic and results in short-lived transgene expression. Hum Gene Ther 13:1611–1620

Lee P, Morley G, Huang Q, Fischer A, Seiler S, Horner J, FactorS, Baidya D, Jalife J, Fishman G. (1998) Conditional lineage ablation to model human diseases. Proc Natl Acad Sci USA 95:11371–11376

Lin Y-S, Ha I, Maldonado E, Reinberg D, Green MR (1991) Binding of general transcription factors TFIIB to an acidic activating region. Nature 353:569–571

Mader S, White JS (1993) A steroid-inducible promoter for the controlled overexpression of cloned genes in eukaryotic cells. Proc Natl Acad Sci USA 90:5603–5607

Malleret G, Haditsch U, Genoux D, Jones M, Bliss T, Vanhoose A, Weitlauf C, Kandel E, Winder D, Mansuy I (2001) Inducible and reversible enhancement of learning, memory, and long-term potentiation by genetic inhibition of calcineurin. Cell 104:675–686

Mansuy I, Bujard H (2000) Tetracycline-regulated gene expression in the brain. Curr Opin Neurobiol 10:593–596

Mansuy I, Mayford M, Jacob B, Kandel E, Bach M (1998) Restricted and regulated overexpression reveals calcineurin as a key component in the transition from short-term to long-term memory. Cell 92:39–49

Marsden CD, Parkes, JD (1976) "On-off" effects in patients with Parkinson's disease on chronic levodopa therapy. Lancet 1:292–296

Massie B, Couture F, Lamoureux L, Mosser DD, Guibault C, Jolicoeur P, Belanger F, Langelier Y (1998) Inducible overexpression of a toxic protein by an adenovirus vector with a tetracycline-regulatable expression cassette. J Virol 72:2289–2296

Mayford M, Bach ME, Huang YY, Wang L, Hawkins RD, Kandel ER (1996) Control of memory formation through regulated expression of a CaMKII transgene. Science 274:1678–1683

Mayo KE, Warren R, Palmiter RD (1982) The mouse metallothionein-I gene is transcriptionally regulated by cadmium following transfaction into human or mouse cells. Cell 29:99–108

Mizuguchi H, Hayakawa T (2002) The tet-off system is more effective than the tet-on system for regulating transgene expression in a single adenovirus vector. J Gene Med 4:240–247

Molin M, Shoshan M-C, Ohman-Forslund K, Linder S, Akusjarvi G (1998) Two novel adenovirus vector systems permitting regulated protein expression in gene transfer experiments. J Virol 72:8358–8361

Neering SJ, Hardy SF, Minamoto D, Spratt SK, Jordan CT (1996) Transduction of primitive human hematopoietic cells with recombinant adenovirus vectors. Blood 88:1147–1155

Obeso JA, Grandas F, Herrero MT, Horowski R (1994) The role of pulsatile versus continuous dopamine receptor stimulation for functional recovery in Parkinson's disease. Eur J Neurosci 6:889–897

Okoye G, Zimmer J, Sung J, Gehlbach P, Deering T, Nambu H, Hackett S, Melia M, Esumi N, Zack D, Campochiaro P (2003) Increased expression of brain-derived neurotrophic factor preserves retinal funtion and slows cell death from rhodopsin mutation or oxidative damage. J Neurosci 23:4164–4172

Paulus W, Baur I, Boyce FM, Breakefield XO, Reeves SA (1996) Self-contained, tetracycline-regulated retroviral system for gene delivery to mammalian cells. J Virol 70:62–67

Payen E, Bettan M, Henri A, Tomkiewitcz E, Houque A, Kuzniak I, Zuber J, Scherman D, Beuzard Y (2001) Oxygen tension and a pharmacological switch in the regulation of transgene expression for gene therapy. J Gene Med 3:498–504

Perez N, Plence P, Millet V, Greuet D, Minot C, Noel D, Danos O, Jorgensen C, Apparailly F (2002) Tetracycline transcriptional silencer tightly controls transgene expression after in vivo intramuscular electro-transfer: application to interleukin 10 therapy in experimental arthritis. Hum Gene Ther 13:2161–2172

Petersen A, Mani K, Brundin P (1999) Recent advances on the pathogenesis of Huntington's disease. Exp Neurol 157:1–18

Picard D, Salser SJ, Yamamoto KR (1988) A movable and regulable inactivation function within the steroid binding domain of the glucocorticoid Receptor. Cell 54:1073–1080

Reiser J, Lai Z, Zhang X, Brady R (2000) Development of multigene and regulated lentivirus vectors. J Virol 74:10589–10599

Rendhal KG, Leff SE, Otten GR, Spratt SK, Bohl D, Van Roey M, Donahue BA, Cohen LK, Mandel RJ, Danos O, Snyder RO (1998) Regulation of gene expression in vivo following transduction by two separate rAAV vectors. Nat Biotechnol 16:757–761

Rinsch C, Régulier E, Déglon N, Dalle B, Beuzard Y, Aebischer P (1997) A gene therapy approach to regulated delivery of erythropoietin as a function of oxygen tension. Hum Gene Ther 8:1881–1889

Rivera VM, Clackson T, Natesan S, Pollock R, Amara JF, Keenan T, Magari SR, Phillips T, Courage NL, Cerasoli F, Dennis AH, Gilman M (1996) A humanized system for pharmacologic control of gene expression. Nat Med 2:1028–1032

Salucci V, Scarito A, Aurisicchio L, Lamartina S, Nicolaus G, Giampaoli S, Gonzalez-Paz O, Toniatti C, Bujard H, Hillen W, Ciliberto G, Palombo F (2002) Tight control of gene expression by a helper-dependent adenovirus vector carrying the rtTA2S-M2 tetracycline transactivator and repressor system. Gene Ther 9:1415–1421

Samakoglu S, Bohl D, Heard JM (2002) Mechanisms leading to sustained reversion of β-thalassemia in mice by doxycycline-controlled Epo delivery from muscles. Mol Ther 6:793–803

Schultze N, Burki Y, Lang Y, Certa U, Bluethmann H (1996) Efficient control of gene expression by single step integration of the tetracycline system in transgenic mice. Nat Biotech 14:499–505

Schweinfest CW, Jorcyk CL, Fujiwara S, Papas TS (1988) A heat-shock inductible eukaryotic expression vector. Gene 71:207–210

Searl PF, Stuart GW, Palmiter RD (1985) Building a metal-responsive promoter with synthetic regulatory elements. Mol Cell Biol 5:1480–1489

Serguera C, Bohl D, Rolland E, Prevost P, Heard JM (1999) Control of erythropoietin secretion by doxycycline or mifepristone in mice bearing polymer encapsulated engineered cells. Hum Gene Ther 10:375–383

Shin M, Levorse J, Ingram R, Tilgham S (1999) The temporal requirement for endothelin receptor-B signalling during neural crest development. Nature 402:496–501

Smith-Arica J, Morelli A, Larregina A, Smith J, Lowenstein P, Castro M (2000) Cell-type specific and regulatable transgenesis in the adult brain: adenovirus-encoded combined transcroptional targeting and inducible transgene expression. Mol Ther 2:

Squire S, Raymackers J, Vandebrouck C, Potter A, Tinsley J, Fisher R, Gillis J, Davies K (2002) Prevention of pathology in mdx mice by expression of utrophin: analysis using an inducible transgenic expression system. Hum Mol Genet11:3333–3344

Staeheli P, Danielson P, Haller O, Sutcliffe JG (1986) Transcriptional activation of the mouse Mx gene by type 1 interferon. Mol Cell Biol 6:4770–4774

Tremblay P, Meiner Z, Galou M, Heinrich C, Petromilli C, Lisse T, Cayetano J, Torchia M, Mobley W, Bujard H, DeArmond S, Prusiner S (1998) Doxycycline control of prion protein transgene expression modulates prion disease in mice. Proc Natl Acad Sci USA 95:12580–12585

Ueberham E, Low R, Ueberman U, Schönig K, Bujard H, Gebhardt R (2003) Conditional tetracycline-regulated expression of TGF-beta1 in liver of transgenic mice leads to reversible intermediary fibrosis. Hepatology 37:1067–1078

Urlinger S, Baron U, Thellman M, Hasan M, Bujard H, Hillen W (2000) Exploring the sequence space for tetracycline-dependent transcriptional activators: novel mutations yield expanded range and sensitivity. Proc Natl Acad Sci USA 97:7963–7968

Verma IM, Somia N (1997) Gene therapy promises, problems and prospects. Nature 389:239–242

Vigna E, Cavalieri S, Ailles L, Geuna M, Loew R, Bujard H, Naldini L (2002) Robust and efficient regulation of transgene expression in vivo by improved tetracycline-dependent lentiviral vectors. Mol Ther 5:252–261

Webster NJG, Green S, Jin JR, Chambon P (1988) The hormone-binding domains of the estrogen and glucocorticoid receptors contain an inducible transcription activation function. Cell 54:199–207

Wilsey J, Zolotukhin S, Prima V, Shek E, Matheny M, Scarpace P (2002) Hypothalamic delivery of doxycycline-inducible leptin gene allows for reversible transgene expression and physiological responses. Gene Ther 9:1492–1499

Wurm FM, Gwinn KA, Kingston RE (1986) Inducible overexpression of a mouse c-myc protein in mammalian cells. Proc Natl Acad Sci USA 83:5414–5418

Xu Z, Mizuguchi H, Mayumi T, Hayakawa T (2003) Regulated gene expression from adenovirus vectors: a systematic comparison of various inducible systems. Gene 309: 145–151

Yamamoto A, Lucas J, Hen R (2000) Reversal of neuropathology and motor dysfunction in a conditional model of Huntington's disease. Cell 101:57–66

Yoshida Y, Hamada H (1997) Adenovirus-mediated inducible gene expression through tetracycline-controllable transactivator with nuclear localization signal. Biochem Biophys Res Commun 230:426–430

Modulating Transcription with Artificial Regulators

A. K. Mapp[1] (✉) · A. Z. Ansari[2] · Z. Wu[1] · Z. Lu[3]

[1] Department of Chemistry and Department of Medicinal Chemistry,
University of Michigan, Ann Arbor, MI 48109-1055, USA
amapp@umich.edu; zqw@umich.edu
[2] Department of Biochemistry and the Genome Center,
University of Wisconsin, Madison, WI 53706, USA
[3] Memorial Sloan-Kettering Cancer Center, New York, NY 10021, USA

1	Introduction	536
2	ATF Design: Lessons from Nature	538
3	ATFs: A Modular Design Strategy	540
3.1	ATF Design Strategy	540
3.2	Polyamide-Based ATFs	541
3.3	Triplex-Forming Oligonucleotide-Based ATFs	546
3.4	Peptide Nucleic Acids	550
4	Artificial Activation Domains	552
4.1	B42 and AH	552
4.2	P201: The Most Potent Artificial Activation Domain	554
4.3	Peptides That Target Specific Proteins	556
4.3.1	Activation Domains That Target CBP/p300	556
4.3.2	Activation Domains That Target Gal80	557
4.3.3	Activation Domains That Target Gal11	558
4.4	RNA Activators: Nonpeptidic Activating Domains	560
5	Future Directions in the Field	562
5.1	Repression of Transcription	562
5.2	Cell Permeability	563
5.3	Future Applications of ATFs: Transcription-Based Therapeutics	564
6	Conclusions	564
	References	565

Abstract Many human diseases are characterized by altered gene expression patterns caused by malfunctioning transcriptional regulators. This has inspired intense interest in the development of artificial transcription factors that regulate the expression of specific genes either positively or negatively. Particular success has been achieved through the use of DNA-binding molecules such as triplex-forming oligonucleotides, peptide nucleic acids, and polyamides that upregulate transcription in vitro and, in some cases, in cell culture. Despite impressive advances in the last decade, however, artificial transcription factors that reconstitute all of the functions of natural regulators are not yet a reality. In particular, it has proven difficult to design artificial transcription factors composed entirely of non-natural components. Such factors will be powerful chemical tools for unraveling the mechanisms of transcriptional regulation and in the long term offer considerable therapeutic potential.

Keywords Artificial transcription factor · Small molecule · Polyamide · Triplex-forming oligonucleotide · Peptide nucleic acid

1
Introduction

Although almost all cells within an organism contain the same genetic information, the differing patterns of gene expression in each cell type gives rise to tissues with diverse morphology and function. The intimate relationship between specific gene expression patterns and the ultimate fate and function of a given cell has been reinforced by information gleaned from the recent decoding of the human genome along with genomewide expression profiling experiments (Lockhart and Winzeler 2000). Moreover, it has also been demonstrated that gene expression patterns in diseased cells are often substantially altered relative to those in normal cells, indicating a fault or disruption in the regulatory network even when the disease cannot be morphologically differentiated by other methods (Alizadeh et al. 2000; Duncan et al. 1998; Pandolfi 2001; Perou et al. 2000).

Gene expression is initiated by a variety of intra- and extracellular signals that mobilize transcriptional regulatory factors to the requisite genes. (Levy and Darnell 2002; Ptashne and Gann 2001; Shi and Massague 2003) These regulators subsequently directly govern the transcription of the specified genes, prescribing both the time scale and level of up- or downregulation. As gatekeepers of regulated gene expression, it is not surprising that in many human diseases it is often malfunctioning transcriptional regulators that produce the altered patterns of gene expression at the heart of the ailment (Levy and Darnell 2002; Pandolfi 2001). As a result, molecules that can target specific genes and control their expression are attractive subjects for study (Ansari 2001; Ansari and Mapp 2002a; Denison and Kodadek 1998; Dervan 2001; Gottesfeld et al. 2000; Mapp 2003) In principle, artificial transcriptional factors (ATF) could be designed to up- or downregulate any gene or a subset of targeted genes in a given cell without influencing the expression of any other gene in the genome. Such factors would be powerful tools for unraveling the key transcriptional events that govern cell fate and, in the long term, have significant therapeutic potential.

This is a nascent field at the intersection of chemistry, biology, and medicine and there are—as yet—few examples of nonproteinaceous ATFs for us to discuss. There are several examples of protein-based ATFs, particularly those based upon the zinc finger scaffold, but these are outside the scope of this chapter. In addition to the discussion of zinc finger proteins in the

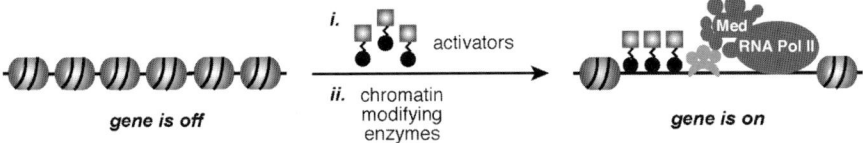

Fig. 1 A model of transcriptional activation. Activators localize at specific sites on DNA and participate in the recruitment of chromatin remodeling enzymes that alter the chromatin structure along with the RNA polymerase II holoenzyme and general transcription factors to initiate transcription

chapter by Corbi et al. in this volume, several recent reviews summarize advances in this arena (Beerli and Barbas 2002; Pabo et al. 2001; Yaghmai and Cutting 2002). As we describe these key first steps in the field herein, we will point out the general principles currently used for ATF design and describe a few of the challenges on the horizon before fully functional small molecule ATFs will become a reality.

We define an artificial transcription factor as a functional entity that can regulate transcription either as a single molecule or as an ensemble of interacting molecules (Fig. 1). To achieve this, an ATF directly influences the transcription of a specifically targeted gene or a set of genes in either a positive or negative fashion. This parallels the function of endogenous transcriptional regulators; the yeast transcriptional activator Gal4, for example, activates only the subset of yeast genes (7 of ~6,000) necessary for galactose (and melibiose) metabolism through binding to specific sites on DNA and interacting with the proteins that constitute the transcriptional machinery (Ptashne and Gann 2001). In addition, similar to natural transcription factors, ATFs may not be limited to functioning only in transcriptional initiation but may also regulate subsequent steps such as elongation or termination (Bentley 1999; Blau et al. 1996; Brown et al. 1998).

Within this definition, a variety of molecules that influence transcriptional responses are not considered to be ATFs. Molecules that trigger signal transduction pathways that lead to activation of specific transcriptional cascades, or that affect cellular localization of transcription factors are not ATFs (Levy and Darnell 2002; Ptashne and Gann 2001; Shi and Massague 2003). In addition, molecules that affect transcription factor or transcript stability or the translational efficiency of transcripts are also not ATFs as such molecules participate in post-transcriptional events (Liu et al. 2001). Thus, double-stranded, short interfering RNA molecules (siRNA) that lead to transcript degradation and RNA antisense are not artificial transcription factors (Dykxhoorn et al. 2003). This definition also excludes molecules such as histone deacetylase inhibitors that do not affect transcription directly but can lead to global changes in gene expression patterns by acting on general enzymes that play a role in transcription (Marks et al. 2001). Moieties that interfere with the binding of an inhibitor to an activating region, such as inhibitors of the p53-MDM2 interaction, or the binding of a repressor to its cofactor, while causing a long-range effect on gene expression, do not directly target specific genes or gene clusters and so are also not ATFs (Duncan et al. 2001; Stoll et al. 2001). Finally, it is also possible for a molecule to be considered an ATF in one functional context but be excluded in another. For example, siRNA not only leads to transcript instability, but it can also direct promoter methylation in plants, directly silencing transcriptional initiation (Bender 2001). Thus, in the latter functional context, siRNA would be an ATF. This is an important distinction as virtually all molecules that affect a biological process will affect transcription patterns but few will do so by directly and specifically targeting a gene or set of genes.

2
ATF Design: Lessons from Nature

The current design approach for all ATFs is inspired by the modular construction of endogenous transcriptional regulators and this necessitates a brief overview of natural regulator composition. Virtually all natural regulators contain two essential modules, a DNA-binding domain (DBD) and a regulatory domain (Ptashne 1997; Ptashne and Gann 1997). The DBD is responsible for much of the gene-targeting specificity of a transcriptional regulator as it localizes the regulator adjacent to the gene under its control. A wide variety of protein DBDs have been characterized structurally and biochemically (Garvie and Wolberger 2001), and such studies have lent support to the model that the primary role of the DBD within transcriptional regulators is to target specific sites within genomic DNA. In contrast, the regulatory modules typically play a less critical role in which gene is selected for regulation, but rather mediate their effects, either up- or downregulation, directly on the gene to which they are targeted. Much less is known about the structure and function of regulatory domains. For transcriptional activators, the activation domains (ADs) are typically categorized by their amino acid sequence content: acid-rich (Ptashne and Gann 2001), glutamine-rich (Courey et al. 1989; Courey and Tjian 1988; Gerber et al. 1994), and proline-rich (Mermod et al. 1989) regions are common. Many ADs are thought to adopt recognizable secondary structures only when bound to their various partner proteins. An excellent example of this is the transcriptional activation domain of CBP which forms a helix upon binding to its target, CREB (Radhakrishnan et al. 1997, 1999) These types of studies have been difficult both to carry out and to interpret, however, as the physiologically relevant targets of most ADs are not unequivocally resolved. The structural requirements for repressor domain function are also fairly poorly defined and, as

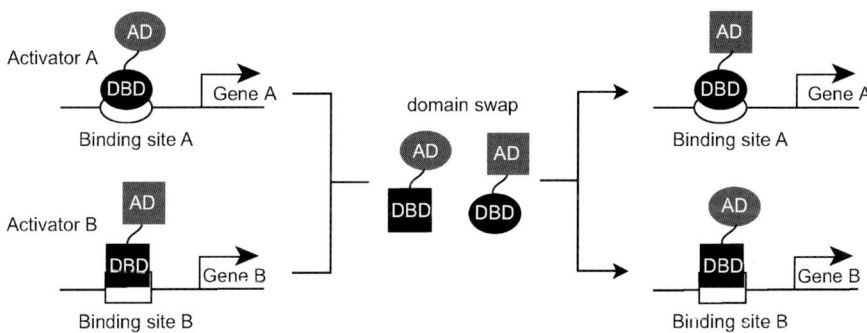

Fig. 2 Domain swapping experiments demonstrate that transcriptional activators are modular and that the domains are exchangeable. An activating domain (*AD*) taken from transcription factor A and attached to the DNA binding domain (*DBD*) from transcription factor B will stimulate transcription of the genes that are bound by the DBD of transcription factor B, and it will not specifically affect expression of any other gene. Similarly a repression module will inhibit the expression of genes dependent on where it is targeted

with ADs, repressor domains are often described in accordance with their amino acid content (Ansari and Mapp 2002; Ptashne and Gann 2001).

Not only are endogenous transcriptional regulators modular in construction, but the functional modules can in most cases also be interchanged (Ansari 2001; Brent and Ptashne 1985; Hope and Struhl 1986; Ptashne and Gann 2001; Sadowski et al. 1988). In other words, an AD taken from transcription factor A and attached to the DBD from transcription factor B will upregulate the transcription of all the genes that are bound by the DNA-binding module of transcription factor B and will not specifically affect expression of any other gene (Fig. 2). By the same token, a repressing module will repress the expression of all genes to which it is targeted. It is not necessary for the two functional domains to residue within the same protein; they can also exist in two or more separate but interacting proteins. The latter case is more often seen with repressors; in those cases, repressor module-bearing proteins are targeted to specific genes/promoters by specific interactions with other DNA-binding proteins (Mannervik et al. 1999; Ptashne and Gann 2001; Smith and Johnson 2000).

The modular construction of transcriptional regulators does not reflect the myriad functions in which they participate (Fig. 3). Transcriptional activators, for example, do much more than simply bind to a specific site on DNA and recruit the transcriptional machinery; they also respond to external stimuli, traffic to the nucleus, bind to DNA as part of a multi-protein ensemble, interact with many additional proteins in a cooperative manner, and are subject to various covalent modifications that regulate overall activity and lifetime (Bouwman and Philipsen 2002; Gonzalez and Montminy 1989;

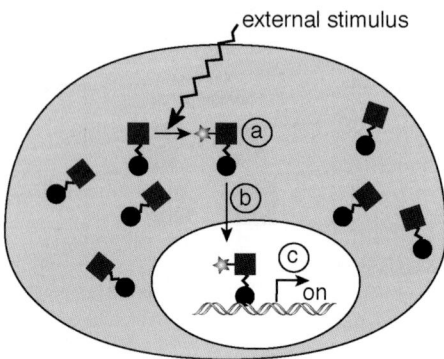

Fig. 3 Transcriptional regulators participate in many recognition events as part of their function. An activator responds to an external signal relayed by a cell surface receptor that then leads to covalent modification of the activator, (*a*). The activator then traffics to the nucleus (*b*), binds to DNA, and interacts with various components of the transcriptional machinery in order to turn the gene on (*c*)

Jackson and Tjian 1988; Lamarre-Vincent and Hsieh-Wilson 2003; Lamph et al. 1990; Levy and Darnell 2002; Mayr and Montminy 2001; Sadowski et al. 1991; Salghetti et al. 2001; Shi and Massague 2003; Stad et al. 2001; Wells et al. 2003). The central question that has emerged in field of ATF design is whether designed regulators can reproduce the specificity and complexity of function of the endogenous regulators. As will be observed in the ensuing discussion, this remains a worthy challenge.

3
ATFs: A Modular Design Strategy

3.1
ATF Design Strategy

Given the modular construction of endogenous transcriptional regulators, it is not surprising that the primary approach for generating ATFs has relied upon the modular replacement of the functional domains of endogenous transcriptional regulators with artificial counterparts (Ansari and Mapp 2002a). As will be seen, this has been most successful with generating artificial transcriptional activators containing novel DBDs. For this reason, we have divided the discussion into sections based upon the DBD used. We will also focus largely upon artificial transcription activators because artificial repressors that do not use protein DBDs remain to be developed (see Sect. 5.2 for a discussion of inhibition strategies). Unlike protein-based ATFs (Beerli and Barbas 2002) in which the two functional modules function fairly independently, one of the concerns in small molecule-based ATFs is that attachment of an activation domain or similar functionality will adversely affect the specificity and/or thermodynamics of DNA binding. In addition, an inherent challenge for these types of constructs is synthetic; since the cellular machinery cannot, for the most part, be used to construct such chimeras; synthetic strategies for incorporating very different types of functionality must be devised.

3.2
Polyamide-Based ATFs

Collaborative efforts by the groups of M. Ptashne and P. Dervan led to the construction of ATFs based upon small molecule DBDs related to the natural product distamycin (Mapp et al. 2000). Several different motifs have since been reported and function well in an in vitro system (Ansari et al. 2001; Arora et al. 2002). Although not yet tested in vivo, studies of this class of ATFs have provided insights into the design of future variants.

The natural product distamycin is a crescent-shaped molecule that binds preferentially to A,T-rich sequences in the minor groove of DNA with dissociation constants in the low micromolar range (Fig. 4; Arcamone et al. 1964). Distamycin binds to DNA via a network of hydrogen bonds with the functional groups present in the minor groove; also contributing energetically are electrostatic and van der Waals' interactions. Although distamycin itself is of somewhat limited use as a DBD given its moderate affinity and relatively poor specificity for individual DNA sequences, it has served as a model for the development of a class of DNA-binding molecules called 'polyamides' that are vastly superior in terms of DNA-binding affinity and specificity (Dervan and Edelson 2003). Polyamides are composed not only of pyrrole amino acids but also imidazole and hydroxypyrrole amino acids and incorporation of heterocycles with altered hydrogen bond donor and acceptor properties enables the recognition of A·T, T·A, G·C, and C·G base pairs (Fig. 4). Although there is a wide variety of polyamide structural motifs that has been developed for the recognition of specific DNA sequences, the most common is the so-called hairpin polyamide in which two polyamide arms are covalently linked by a flexible amino acid tether. The hairpin polyamides target specific sequences with dissociation constants comparable to endogenous DNA-binding proteins in vitro and in some cases, have been found to penetrate cells and traffic to the nucleus (Dickinson et al. 1998; Gottesfeld et al. 1997; Janssen et al. 2000a, 2000b). Perhaps more important for future ATF design, polyamides are readily prepared by solid phase synthesis methods (Baird and Dervan 1996; Belitsky et al. 2002b; Wurtz et al. 2001), making them attractive starting points for the design of ATFs and other DNA-targeting conjugates.

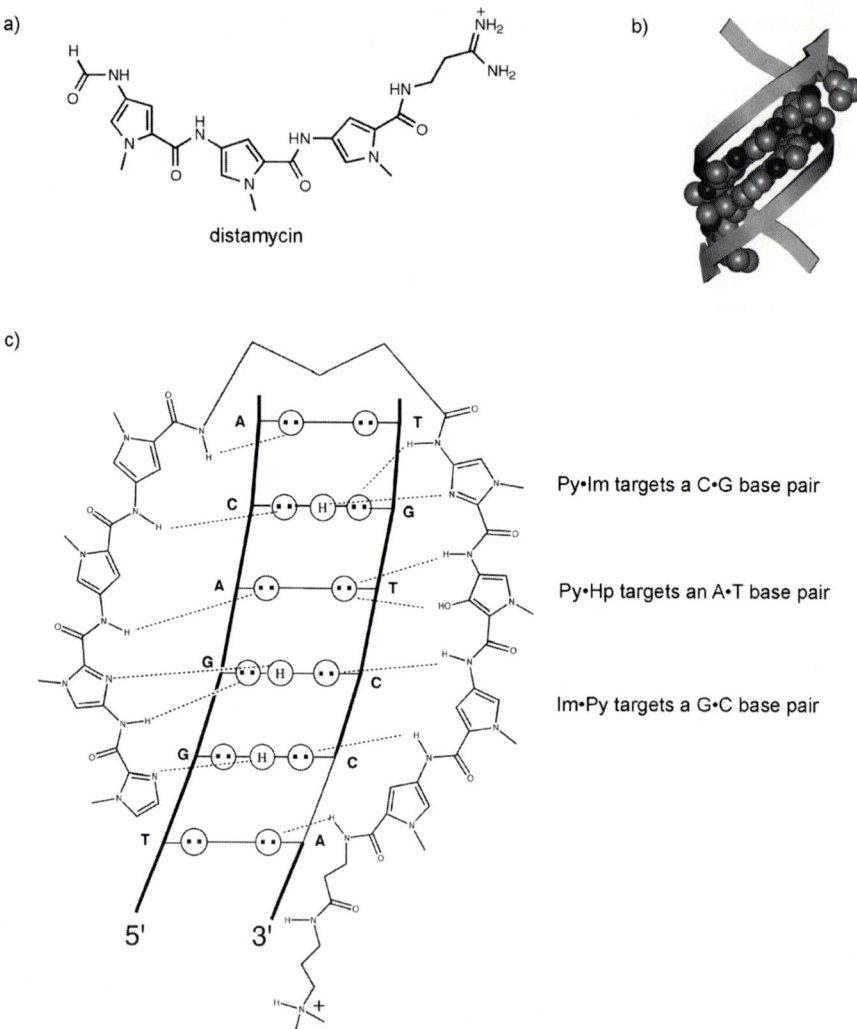

Fig. 4a–c Model of DNA minor groove recognition by polyamides. **a** Polyamides are derived from the natural product distamycin, composed of three pyrrole amino acids. **b** Polyamides, including distamycin, can bind to DNA in a 2:1 polyamide:DNA motif as shown. **c** When two polyamides are linked as shown, the resulting structure, a 'hairpin polyamide,' is able to discriminate all four base pairs in the DNA minor groove through the formation of specific hydrogen bonds and through shape-specific recognition of certain pairings

The first polyamide-based ATFs incorporated a dimerization domain as well as a DBD and an activating module (Fig. 5). As mentioned in Sect. 2, the DBDs and ADs of most endogenous activators are present within the same polypeptide chain and are often connected by a dimerization domain. For example, the potent yeast activators Gcn4 and Gal4 contain a dimeriza-

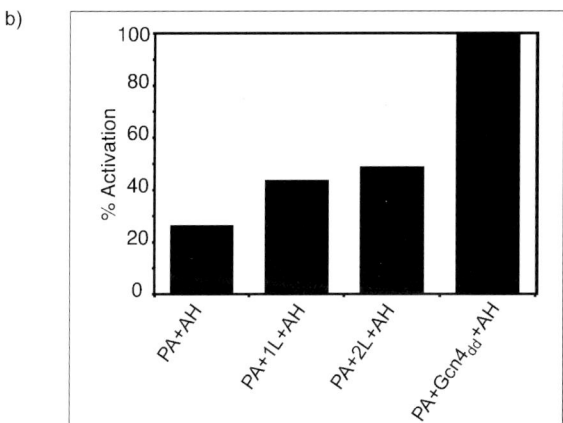

Fig. 5a, b Hairpin polyamide–peptide conjugates activate transcription in vitro. a The first generation polyamide–peptide ATFs consisted of a hairpin polyamide as a DBD covalently linked to the AH activation domain. The two functional domains were connected either via a dimerization domain [Gcn4(251-281)] (*Gcn4dd*) or through a flexible polyethylene glycol-derived linker as shown. b The polyamide–peptide ATFs activate transcription in vitro and the extent of activation is related to the length and composition of the linking domain. The most potent of the ATFs is PA+Gcn4dd+AH but replacement of this semi-rigid linker with one or two flexible tethers as in PA+1L+AH and PA+2L+AH leads to only a 50% decrease in potency

tion domain between the DBD and the dimerization unit. Since it was not clear at that time if the dimerization domain played some functional role in transcriptional activation, this unit was included in the design (Mapp et al. 2000). The dimerization module from Gcn4 was chosen for these constructs as it has been well characterized both structurally and functionally and can be readily prepared through solid phase peptide synthesis (Agre et al. 1989; Ellenberger et al. 1992; Oshea et al. 1991; Weiss et al. 1990). The polyamide selected as the DBD was a hairpin designed to bind to the sequence 5'-TGT-TAT-3' with a dissociation constant of 1.1 nM when unmodified (Trauger et al. 1996). The AD was AH, a peptide designed by Ptashne and co-workers to mimic the characteristics of many endogenous activating domains in a minimal motif (see Sect. 4.1) (Giniger and Ptashne 1987). The construct was designed to bind to a pseudo-palindromic sequence 5'-ATAACA(N)7TGTTAT-3'. The synthetic strategy used to prepare this construct has proven general for the preparation of polyamide-based ATF constructs (Mapp and Dervan 2000), and is derived from the native ligation strategy originally developed by Kent and co-workers for preparing synthetic proteins (Dawson et al. 1994).

Although the polyamide–peptide conjugate bound to the target pseudo-palindromic site in which the half-sites were separated by seven base pairs with good specificity, the affinity was approximately an order of magnitude decreased (11 nM) relative to that of the parent polyamide. The binding data were, however, fit with a cooperative binding isotherm, indicating that a positive interaction between adjacent polyamide–peptide conjugates was occurring. However, fusion of the peptide at the carboxy terminus of the polyamide proved to have an overall deleterious effect upon binding.

The activation potential of the constructs was assessed using standard in vitro transcription assays with yeast nuclear extracts (Lue et al. 1989, 1991). In these experiments, each construct was incubated with a DNA template consisting of five cognate palindromic binding sites upstream of a minimal viral promoter and PA+Gcn4$_{dd}$+AH upregulated transcription 13-fold compared to transcription reactions containing no construct (basal). The ability of PA+Gcn4$_{dd}$+AH to upregulate transcription was dependent upon the presence of the cognate binding site, as replacement of the cognate site with a binding site containing a single base pair mismatch resulted in a significant decrease in transcription levels. Similar to endogenous activators, the addition of free activating region to the reactions inhibited activated transcription with no effect on basal (activator independent) transcription. Taken together, these experiments and others pieces of evidence are consistent with conjugate PA+Gcn4+AH functioning in an analogous fashion to endogenous regulators; in other words, the construct binds to DNA and subsequently recruits the transcriptional machinery to the nearby promoter region in order to activate transcription.

Although these results were promising in that a wholly un-natural DBD could replace a protein (in other words, no biopolymer), the resulting construct was still rather large (MW=7,465). Thus, a series of compounds was prepared in which the Gcn4 dimerization peptide was replaced by a flexible linker (Fig. 5). In this case, the linker was a polyethylene glycol-derived link-

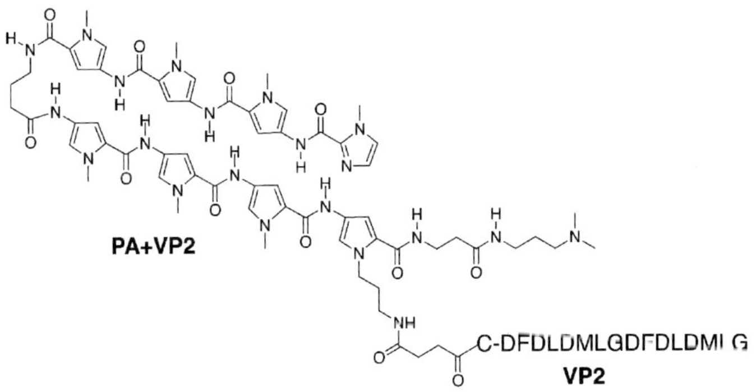

Fig. 6 A more potent polyamide-based ATF. Replacement of the activation domain with a more potent variant, the VP16-derived activation domain VP2, and altering the position of attachment led to a more potent, smaller ATF

er that provided water solubility and was easily accessed synthetically. This replacement led to a smaller construct (PA+1L+AH, MW=4,165) that functioned with only a twofold decrease in activity relative to the Gcn4$_{dd}$-containing construct. The decrease in activity was attributed to the greater flexibility of the polyether tether that could not project the activating region away from the DNA as effectively as the Gcn4$_{dd}$ in order for it to interact with the transcriptional machinery.

The potency of this class of artificial activators was further increased by replacing the activating region with a more active variant, the VP16-derived activating region VP2 (Fig. 6) (Ansari et al. 2001). In addition, when the activating region is moved to a different location within the polyamide, the affinity for the cognate DNA sequence is improved, the size of the construct is decreased, and the conjugate remains quite active despite the lack of a polar linker region. This streamlined design provides an attractive starting point for the design of the next generation of polyamide-based artificial transcription factors.

The role of effective projection of the activating region away from the DNA was further investigated by the design and testing of polyamide–peptide conjugates containing polyproline helices as the linker region (Fig. 7) (Arora et al. 2002). As polyproline containing 6–15 residues forms rigid helices of increasing length, the linker region served as a molecular ruler with which to determine the effect of projection on activity. The constructs were designed to introduce a spacing of 18, 27, 36, and 45 Å in between the polyamide DBD and the activating domain, the VP2 activator peptide as well as AH. The activity of the conjugates was determined by in vitro transcription assays. In this case, conjugate PA+(Pro)$_9$+VP2 with a 36-Å spacer was the most active of the group, suggesting that there may be an optimal linker length for artificial transcription factors. The optimal projection distance

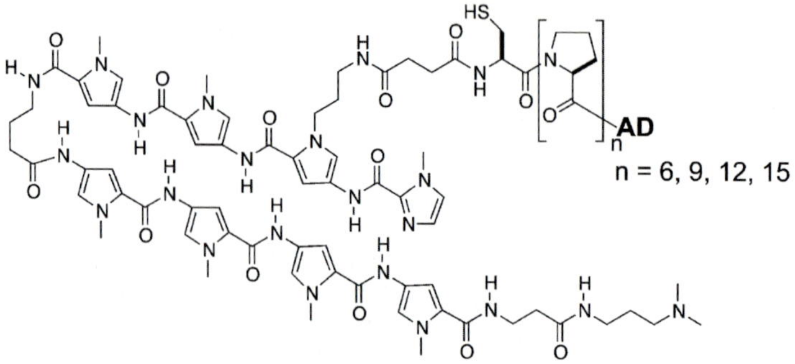

AD = PEFPGIELQELQELQALLQQ
AH
DFDLDMLGDFDLDMLG
VP2

Fig. 7 Polyamide ATFs with polyproline linker regions. The flexible tether of the earlier designs was replaced with polyproline regions of incremental lengths in order to probe the role of projection in transcriptional potency

will likely be dependent upon a variety of factors—promoter structure, activating domain composition and size, and position of the cognate binding site relative to the transcription start site, for example—but is nevertheless an important design concept that must be considered for future ATFs. In addition, this study provides an example of how ATFs are useful not just as prototypes for transcription-based therapeutics but also as tools for studying transcription mechanism.

The next frontier for these constructs is a demonstration of function in vivo. Recent studies by Dervan indicate that this should be possible given proper functionalization of the polyamide (Belitsky et al. 2002a).

3.3
Triplex-Forming Oligonucleotide-Based ATFs

The first example of an ATF using a nonprotein DBD was reported by Svinarchuk and coworkers in 1999 (Kuznetsova et al. 1999). Their design included an AD derived from the viral co-activator VP16 and a triplex-forming oligonucleotide (TFO) as a DBD.

TFOs are oligonucleotides that bind to the major groove of duplex DNA through Hoogsteen and reverse-Hoogsteen interactions with purine-rich tracts (Fig. 8) (Faria and Giovannangeli 2001; Giovannangeli and Helene 2000; Ledoan et al. 1987; Miller 1996; Moser and Dervan 1987). The requirement for consecutive purines in the same strand is a strict one as only A and G bases have additional functionality in the major groove enabling the formation of two hydrogen bonds with the third strand. The various common

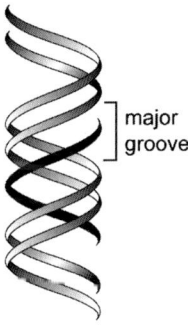

Fig. 8 Triplex-forming oligonucleotides bind to the major groove of DNA as a third strand

pairings are indicated in Fig. 9. This places a limitation on the variety of sequences that TFOs can target and despite a great deal of effort expended by a number of researchers (Giovannangeli and Helene 2000) it has proven difficult to expand the motif past this requirement. Despite this inherent limitation, TFOs have a number of advantageous characteristics making them attractive choices for artificial DBDs. It is, for example, straightforward to design a TFO that binds to a unique site within a genome as sequences >15 base pairs in length are routinely targeted. In addition, TFOs are easily synthesized using automated synthesis techniques and can also be readily modified postsynthetically by the incorporation of reactive functional groups into the automated synthesis protocol. Consistent with their structure and composition, cell permeability and stability of TFOs can be problematic (Sedelnikova et al. 1999). A number of successful approaches have been developed to deal with the former problem, however, such as the use of cationic complexing agents and covalent attachment of nuclear transport sequences (Zanta et al. 1999) and TFOs have been demonstrated to bind to cognate DNA sites in a cellular context (Faria et al. 2001). With regard to stability, replacement of the phosphodiester backbone with groups such as phosphonothioesters has proven in some cases to provide superior characteristics (Faria and Giovannangeli 2001).

The TFO used as a DBD by Svinarchuk and co-workers in their construction of the first nonproteinaceous artificial activator was designed to target the *vpx* gene of HIV-2 and was chosen due to its known propensity to form extremely stable triple helices (Fig. 10) (Debin et al. 1997; Kuznetsova et al. 1999). Covalently attached to this oligonucleotide were repeats of a core sequence taken from the potent viral co-activator VP16 (Tanaka 1996) via the formation of a phosphoramidite bond; constructs with either two or three repeats of this sequence were prepared. In addition, the peptide ADs were attached at either the 5′ or the 3′ end such that effects of attachment position could be investigated. Gel electrophoretic mobility shift assays were used to

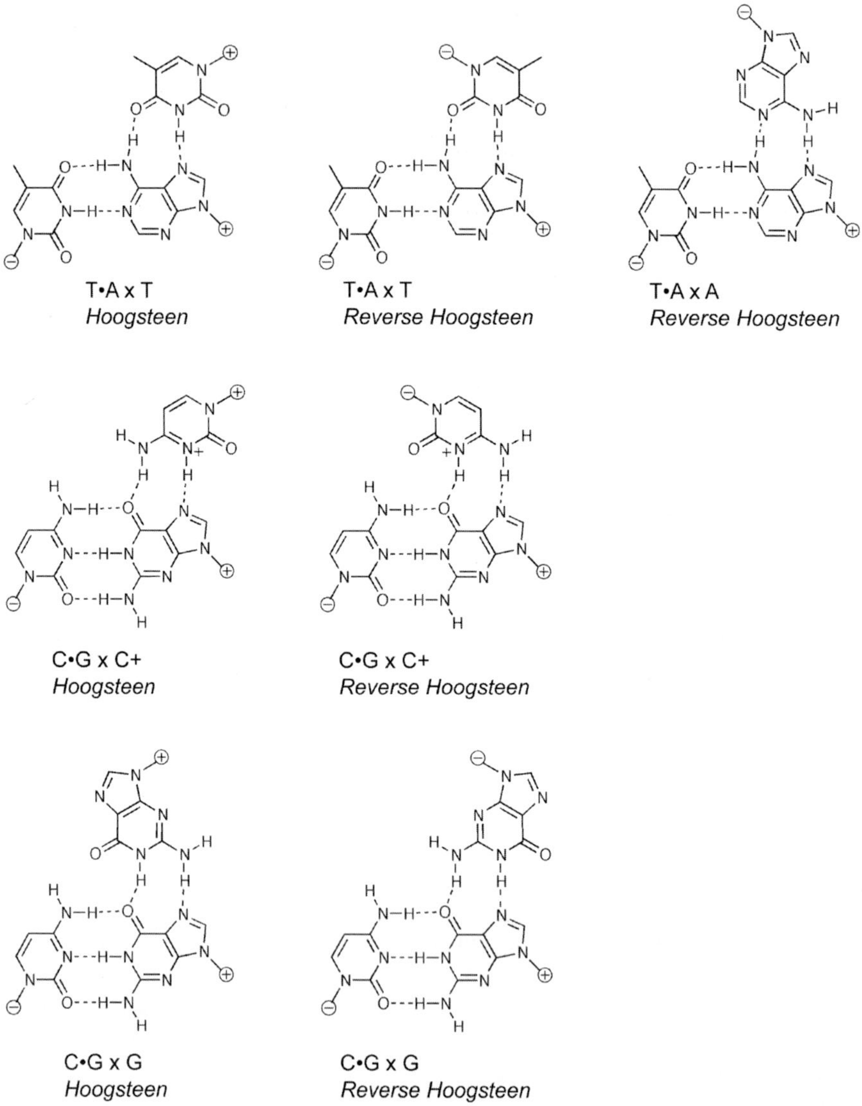

T·A x T
Hoogsteen

T·A x T
Reverse Hoogsteen

T·A x A
Reverse Hoogsteen

C·G x C+
Hoogsteen

C·G x C+
Reverse Hoogsteen

C·G x G
Hoogsteen

C·G x G
Reverse Hoogsteen

Fig. 9 Common TFO pairings. The triplex strand forms specific hydrogen bonds with the purine-rich strand either through Hoogsteen (T·AxT, C·GxC+. C·GxG) or reverse-Hoogsteen (T·AxT, T·AxA, C·GxC+, C·GxG) interactions. Two of these motifs (the Hoogsteen and non-Hoogsteen C′GxC pairings) require protonated C to occur and arc thus pH-sensitive

TFO-3'-P₃
5'-GGAGGGGGAGGAGGAGGAGG-3'-(Gly)₃Pro(AlaAspAlaLeuAspAspPheAspLeuAspMetLeuPro)₃

TFO-5'-P₃
3'-GGAGGAGGAGGAGGGGGAGG-5'-(Gly)₃Pro(AlaAspAlaLeuAspAspPheAspLeuAspMetLeuPro)₃

Fig. 10 Triplex-forming oligonucleotide + activation domain constructs prepared by Svinarchuk and coworkers (Kuznetsova et al. 1999). In addition to those shown, constructs with only two repeats of the core AD sequence were prepared also but showed little or no activator function in cell culture

verify that attachment of the peptides did not negatively impact triplex-forming propensity and the stability of the resulting triple-helical complexes.

Once the constructs were prepared and characterized, a transient transfection assay was used to monitor their effectiveness as transcriptional activators. The TFO–peptide conjugates were precomplexed with a template consisting of binding sites for the TFOs as part of a c-*fos* promoter with a luciferase reporter. Electroporation was then used to deliver the complex into 3T3 cells and luciferase activity was compared against an internal β-galactoside standard. Only one of the four constructs proved to be active in the functional assay, TFO-5'-P3, consisting of three repeats of the core VP16 sequence attached at the 5' position of the oligonucleotide, and the observed activity was low (fourfold over background). As outlined by the authors of the study, there is a variety of factors that could influence the low activity. Only one promoter was examined, for example, and the spacing between the TFO-binding sites and the transcription start site was not optimized. Based upon subsequent experiments reported in the literature (see Sect. 3.2 and an additional TFO example shown below), it is also likely that the lack of a linker of any significant length between the TFO and the AD may also have negatively impacted function. Nonetheless, this report represented an important step forward in the design and function of artificial transcriptional activators as it extended the concept of modular construction beyond protein-based ATFs.

Some 3 years after the first TFO-based ATF was described, Stanojevic and Young reported an updated design that increased the overall activity elicited by such constructs (Fig. 11) (Stanojevic and Young 2002). A slightly longer TFO was used in this case but the most significant departure was the inclusion of a long polyethyleneglycol tether linking the AD to the TFO. A similar AD derived from VP16 was used as the AD. A modified thymidine residue was incorporated that contained a free amine functional group. The amine was then used to covalently attach the peptide to the oligonucleotide. Two different ADs were attached, a 14-amino acid residue unit and a 29-amino acid residue version of VP16. Both of these activated transcription as measured by in vitro transcription reactions and TFO-ATF29 showed activity comparable to a Gal4-VP16 positive control (30–40-fold relative to basal).

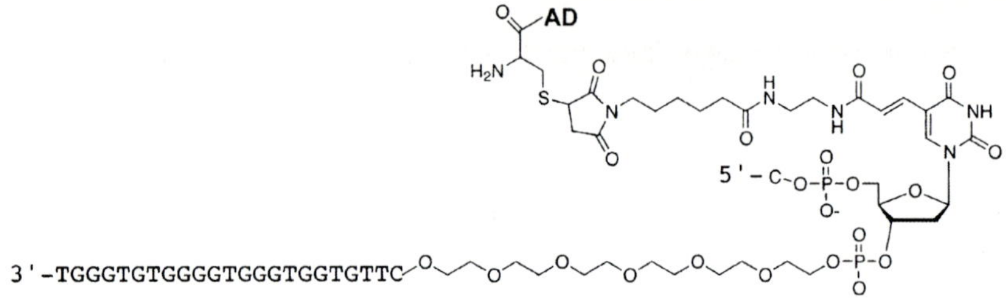

3'-TGGGTGTGGGGTGGGTGGTGTTC-O~O~O~O~O~O~O-

AD = GSDALDDFDLDMLGSDALDDFDLDMLGS
ATF29
GSDALDDFDLDML
ATF14

Fig. 11 Updated TFO-based ATF design. The incorporation of a flexible linker and a more potent activation domain led to a dramatic increase in potency relative to the design shown in Fig. 10. (Stanojevic and Young 2002)

Thus, incorporation of a flexible tether into the design dramatically increased the potency of TFO-based ATFs; this is analogous to the effect originally reported by Dervan and Ptashne with polyamide-based ATFs (Ansari et al. 2001; Mapp et al. 2000). Transient transfection assays in HeLa or BHK-21 cells with reporter constructs containing five TFO-binding sites showed a surprising reversal in the order of activity; the construct with the smaller AD actually increases in activity, 30-fold activation for TFO-ATF14 and five-fold activation for TFO-ATF29. Both constructs appear to traffic to the nucleus to similar degrees and it is thus unclear what is the origin of the differences between the in vitro and the in vivo activities.

As exemplified by the above experiments, TFOs have a promising future in ATF design. The key limitation to this strategy is the relatively small group of DNA sequences that can be effectively targeted by TFOs given the requirement for purine-rich strands. In addition, delivery into cells continues to be a general challenge as it is for all other DNA-binding molecules.

3.4
Peptide Nucleic Acids

Although peptide nucleic acids (PNAs) have not yet been used to construct an artificial transcriptional activator, a recent study of a PNA–AD conjugate indicates the considerable potential this class of DBD holds (Liu et al. 2002).

PNA is structurally related to TFO except that the phosphodiester backbone is replaced by a charge-neutral amide bond (Fig. 12) (Br7aasch and Corey 2001; Faria and Giovannangeli 2001; Nielsen 2001). This confers a number of advantages over TFOs, the most apparent of which is the increased stability in a cellular environment. As PNAs do not have a phosphodiester backbone, they are not cleaved by nucleases and the non-natural ar-

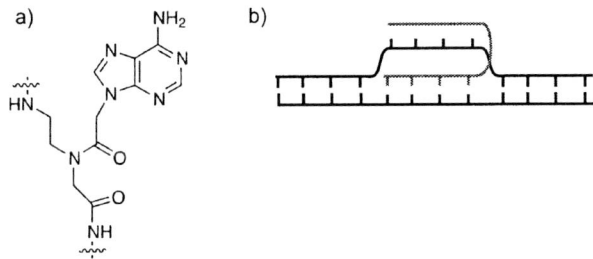

Fig. 12a, b Peptide nucleic acid (*PNA*) composition and DNA binding. **a** In PNA the phosphodiester backbone of an oligonucleotide is replaced with an amide backbone. **b** Unlike TFOs, bis-PNAs strand-invade duplex DNA in order to bind, causing a loop-like formation

rangement of the peptide backbone renders them resistant to cleavage by endogenous proteases. In addition, PNA is straightforward to synthesize by solid-phase synthetic protocols and can be readily conjugated to a variety of different molecules without significant impact upon DNA-binding specificity and affinity.

A major difference between PNA and triplex-forming oligonucleotides is the DNA-binding motif. Unlike TFOs, PNA typically does not bind via a simple third-strand interaction but rather through the formation of an invasion complex. In the case of a simple PNA designed to target a homopurine stretch of DNA, one PNA invades and interacts with the homopurine stretch by the formation of Watson–Crick-like base pairs. The two PNAs can be connected with a covalent tether to create a so-called clamp PNA or bis-PNA and this provides DNA-binding molecules with exceptional stability.

Recently, Kodadek and co-workers described the synthesis and preliminary characterization of a clamp PNA conjugated to a peptide AD, Gal80BP-A (see Sect. 4.3.2 for details of this AD) (Fig. 13) (Liu et al. 2002). This was synthesized in a straightforward fashion with the PNA portion of the conjugate prepared by automated solid phase synthesis and the AD was then added manually. A short, hydrophilic amino acid similar to that used in the polyamide examples was used as a tether. This construct was found to bind

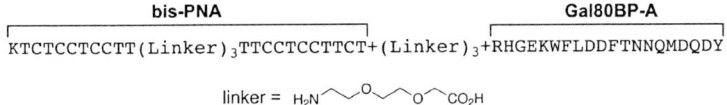

Fig. 13 First-generation design of a PNA-based ATF. When the AD Gal80BP-A is covalently linked to a bis- or clamp PNA via a hydrophilic tether, the resulting construct recruits a key coactivator to DNA, indicating that this may function as an ATF. Note that the Gal80BP-A sequence is written such that the carboxyl-terminal residue is on the *left side* and the amino-terminal residue is on the *right side*

tightly and specifically to its cognate DNA sequence 5′-AAGGAGGAGA-3′ through a strand-invasion mechanism as ascertained by gel mobility shift assays. In addition, pull-down experiments were used to demonstrate that the construct recruits the co-activator Gal11 to DNA. This indicates that the PNA+AD chimera has the necessary functionality to activate transcription and future results for this class of compounds will no doubt be important for the field.

4
Artificial Activation Domains

In all of the examples of ATFs discussed in earlier sections, the activation domain used is a peptide either derived from an endogenous activating region or closely related in sequence to an endogenous AD. The development of nonbiopolymer-based ADs remains one of the outstanding challenges in ATF design and such molecules will undoubtedly prove to be powerful tools for understanding transcriptional regulator function and mechanism. In this section, we summarize the evolution of artificial ADs and point towards future directions.

4.1
B42 and AH

One of the earliest artificial ADs is B42, first reported in 1987 (Ma and Ptashne 1987). B42 was isolated from a screen for activation potential of random fragments of the *Escherichia coli* genome fused to the DBD of Gal4. The screen was carried out in yeast and a number of activating regions, ranging from 12 to 81 amino acid residues in size were isolated (Fig. 14). The activation domains thus isolated bore a strong resemblance to endogenous amphipathic or acidic ADs such as Gal4, containing a preponderance of negatively charged residues interspersed with hydrophobic ones. All of the ADs functioned as activators with the extent of activation generally correlating with

	Sequence	Activity
B42	INKDIEECNA IIEQFIDYLR TGQEMPMEMA DQAINVVPGM TPKTILHAGP PIQPDWLKSN GFHEIEADVN DTSLLLSGD	756
B7	ISEPDYGALL DDMFFHDGSD IPTDLIEQEQ RVFHAHFRHF LD	678
B3	IAALEDNDDL FGHGLFLV	359

Fig. 14 The first artificial activation domain. Shown are three activation domains isolated from fusing segments of the *E. coli* genome to a Gal4 DNA binding domain and screening for activators. In general, the potency of the ADs (expressed in the figure as β-galactosidase units) was related to the size of the activating regions (Ma and Ptashne 1987)

	Sequence	Activity
AH	PEFPGI**ELQELQELQ**ALLQQQ	420
Scrambled AH	PEFPGI**ELELELQQQ**ALLQQQ	7

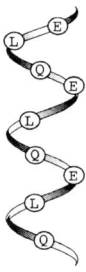

Fig. 15 A designed artificial activation domain. As many eukaryotic activating regions are composed of a combination of polar and hydrophobic residues and are thought to form amphipathic helices, a peptide designed to mimic such a structure, AH, was prepared and found to function well as an activator when fused to a Gal4 DNA binding domain (residues 1–147). The activity is expressed in β-galactosidase units. When the core sequence was scrambled (scrambled AH) the function was lost almost completely (Giniger and Ptashne 1987)

the size of the AD and the net negative charge of the sequence. Of this group, only B42 is commonly used in ATF construction but these experiments revealed that it is fairly straightforward to generate artificial ADs, at least those which bear resemblance to endogenous versions.

Since many eukaryotic ADs contained a combination of hydrophobic and polar amino acid residues and were, at the time, hypothesized to form an amphipathic helix, it was proposed in the late 1980s that a peptide designed to form an amphipathic helix should function as a transcriptional activator when tethered to a DBD. Thus the peptide AH, for amphipathic helix, was designed and tested in yeast (Giniger and Ptashne 1987). As seen in Fig. 15, it is a relatively small peptide, consisting of 20 residues and a core repeat of glutamic acid, leucine, and glutamine as the amphipathic sequence. When fused to Gal4(1–147) and tested for activation in a yeast strain bearing a β-galactosidase reporter, this artificial activator construct upregulated approximately 20% as efficiently as full-length Gal4, impressive activity for a much smaller protein construct. Although there is no direct structural evidence indicating that AH forms an amphipathic helix, the activation potential is dependent upon the amino acid sequence, consistent with a requirement for secondary structure. When the amino acids were scrambled such that an amphipathic helix would be unlikely to be formed, no activation was observed. Further, when the polar glutamine residues were replaced with valine to provide an additional hydrophobic face, the resulting AD functioned only modestly, activating transcription to 10% of the levels provided by Gal4(1–147)+AH. This AD also functions well in metazoan systems (Horikoshi et al. 1988; Lin et al. 1988), reinforcing the notion that a number of key mechanistic aspects of activator function are conserved. In addition, AH does not need a protein DBD in order to function; as illustrated in

Sect. 3.2, a hairpin polyamide suffices as a DBD to provide a designer activator with no direct counterparts in nature.

4.2
P201: The Most Potent Artificial Activation Domain

With early success in isolating activating regions from random sequences of the *E. coli* genome, a more recent attempt utilized the improved methods of oligonucleotide synthesis and cloning to isolate smaller peptides that function as potent activating modules (Lu et al. 2000, 2002) A library of eight random residues fused to the Gal4 DBD [Gal4(1–100)] was generated and screened for strong activators (Fig. 16). A nonbiased library of eight residues generates a library of 20^8 molecules; as yeast transformation efficiency is limited to $\sim 10^7$ only a subset of the possible sequences were sampled. Early concerns about identifying peptides that would function robustly were ameliorated by the surprising frequency with which such clones were obtained. Nearly one in a thousand peptides (0.1% frequency) activate as well as B42 or AH and of those several are as potent as Gal4 (one of the most potent activators in yeast). As a class, however, none of the strong activators shared the two key features of natural activating regions in that these peptides did not bear acidic residues and did not show any propensity to form amphipathic helices. One similarity, however, is that they do contain a high percentage (65.4%) of hydrophobic residues as do most natural ADs.

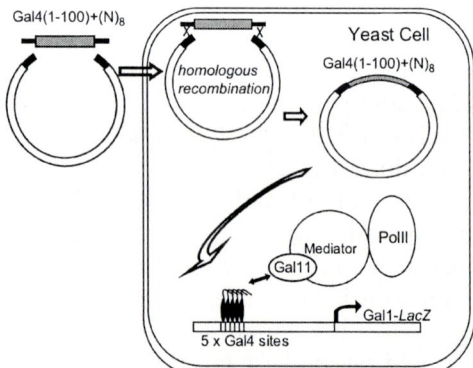

Fig. 16 The most potent artificial activator. The potent AD P201 was isolated from a screen of random libraries of eight-residue peptides fused to Gal4(1–100). This was accomplished by co-transformation of a linearized plasmid encoding Gal4(1–100) and libraries of oligonucleotides encoding eight-residue peptides of varying sequence composition. The short oligonucleotides and the linearized plasmid had identical sequences at the end and thus underwent homologous recombination once inside the yeast. A reporter gene, β-galactosidase, under control of the new Gal4-containing constructs was used to identify the strongest activators from the library (Lu et al. 2000, 2002)

Among several candidates one octameric peptide, P201 (YLLPTCIP), that activated better than Gal4 was chosen for further analysis. P201 functioned as an activator when fused to two different types of natural DBDs. However, the 'domain-swap' experiments suggested that in addition to the eight amino acid peptide, a few residues of the original Gal4 DBD were required for full activity of P201. These residues of Gal4 were inert on their own (or when fused to numerous other octamer peptides in the screen) yet a subsequent alanine scanning experiment identified two key hydrophobic residues of the Gal4 that were adjacent to the P201 peptide as being important for overall function.

Given the ability of strong yeast activators to function across species, it was surprising that P201 was incapable of activating transcription in mammalian cell lines.(Lu et al. 2000) This immediately suggested that P201 was able to interact with a yeast-specific component of the transcriptional machinery. Ptashne and co-workers used a clever selection strategy to identify the target of this species-specific activator as Gal11, a nonessential component of the 'Mediator' complex in yeast (Lee and Young 2000; Myers and Kornberg 2000). A common target emerging from studies of transcriptional activators is Mediator, a multi-protein complex proposed to act as a bridge between DNA-bound activators and RNA polymerase II (Myers and Kornberg 2000). Gal11 is a Mediator component unique to yeast that has been implicated as a target of endogenous activators (Jeong et al. 2001; Koh et al. 1998; Myers and Kornberg 2000; Park et al. 2000). In addition, when Gal11 is fused to a protein DBD it activates transcription robustly (Barberis et al. 1995a). P201 seems to be exquisitely dependent on Gal11 in order to function as an activator. This contrasts with natural eukaryotic activators, which normally have several putative targets in the transcriptional machinery (Koh et al. 1998; Ptashne and Gann 2001). Gal11 is particularly interesting because it is one of the few components of the machinery that can elicit strong transcription of a nearby gene when tethered upstream by a fused DBD (Barberis et al. 1995b; Gaudreau et al. 1999). Moreover, when it bears a point mutation (N342V) it is able to fortuitously interact with the dimerization domain of Gal4 and this interaction suffices to stimulate robust transcription of a nearby gene (Hidalgo et al. 2001; Himmelfarb et al. 1990; Barberis et al. 1995b). Thus, Gal11 appears to be a privileged target and could very well serve as an important drugable target that is yeast-specific with homologs present in pathogenic yeast *Candida albicans* and no clear homolog in humans. The identification of P201 as a robust yeast-specific activator that functions primarily through its interactions with Gal11 strongly suggests that small molecules that target Gal11 may serve as a starting point for the generation of a new class of fungicides.

4.3
Peptides That Target Specific Proteins

Given the advent of a multitude of efficient screening and selection technologies, it is perhaps not surprising that in the last few years, attention has turned to generating novel ADs by identifying molecules that interact with specific targets in the transcriptional machinery. Interestingly, however, such experiments have in general yielded ADs that bear strong resemblance to endogenous, amphipathic ADs as outlined below.

4.3.1
Activation Domains That Target CBP/p300

The KIX domain of p300/CBP interacts with a variety of activation domains including those from CREB, the SREBPs, c-Myb, and c-Jun as part of its coactivator function (Lamph et al. 1990; Zor et al. 2002). This domain has been well characterized both biochemically and structurally and presents a hydrophobic groove formed from two parallel helices for binding to ADs (Radhakrishnan et al. 1997, 1999) Montminy and coworkers used a phage display library selection to generate ligands for the KIX domain of p300/CBP with the goal of identifying new minimal activator domains with increased potency that would function in an analogous fashion to endogenous ADs (Frangioni et al. 2000). The library consisted of eight completely randomized residues flanked by a polar (GlySer)4 linker region. In the first round of the selection, an apparent consensus sequence was derived based upon the peptides that showed the greatest apparent affinity for the target and a biased phage display library was constructed based upon this consensus (Fig. 17). From a second round of selection with this library, a more detailed consensus sequence was arrived at (WWVYE/DLLF) and the ligand identified as the optimal binder had a K_d for KIX (16 mM) within the range of an endogenous KIX ligand, the minimal c-Myb AD (2 mM). When fused to a protein DBD, Gal4(1–147), the optimal KIX-binding peptide was approximately 40-fold more active than the DBD alone in human 293 cells.

	Sequence	K_d for CBP/p300	Fold activation
KBP 1.66	SSLVDLILF(GS)$_4$	55.1 µM	10
KBP 2.20	SWAVYELLF(GS)$_4$	16.1 µM	40

Fig. 17 p300/CBP binding peptides that function as activation domains. The optimal first-round selection peptide (*KBP 1.66*) bound the target protein more weakly than the optimal second-round peptide (*KBP 2.20*) and was also a weaker activation domain. Both ADs were tested as fusions to Gal4(1–147) (Frangioni et al. 2000)

The activation effected by the KIX-binding peptide isolated in the above experiments appears to occur through a specific p300/CBP-binding interaction. When the peptide is coupled to agarose, for example, it is able to precipitate p300/CBP from cell lysate selectively whereas an unrelated peptide of the same length does not. In addition, when a p300/CBP sequestering protein, E1A, was expressed in cells containing the Gal4(1–147)+KIX-binding peptide construct, transcription of the reporter gene under control of the KIX-binding peptide was abrogated. Competition binding experiments with a fluorophore-labeled version of the KIX-binding peptide and endogenous KIX-binding ligands were carried out in order to investigate possible binding modes. These experiments revealed that the KIX-binding peptide targets an overlapping but distinct binding site within the KIX domain.

Although the new ADs isolated in the selection experiments are closely related to the endogenous KIX ligands in terms of both composition and mechanism, these experiments provided an additional demonstration of the ease with which ADs can be generated. One might question why sequences of endogenous KIX ligands were not isolated from this set of experiments but the answer probably lies in the many roles that the endogenous KIX ligands play in addition to KIX binding. Thus, these results provide more insight into the complex role of ADs and hint at the difficulties that lie ahead in designing nonbiopolymer counterparts.

4.3.2
Activation Domains That Target Gal80

Gal80 is a yeast inhibitor that binds specifically to the activating region of Gal4, preventing Gal4 from stimulating the expression of the *GAL* family of genes in the absence of galactose or in the presence of glucose in the media. This Gal80-mediated inhibition is relieved in the presence of galactose. Kodadek and Han used phage display selections to identify 20-amino acid peptides that bind to Gal80 and also function effectively as activating regions when fused to the Gal4 DBD (Fig. 18) (Han and Kodadek 2000). The researchers used a phage library composed of approximately 10^9 unique sequences and six rounds of selection against a His6-tagged version of Gal80

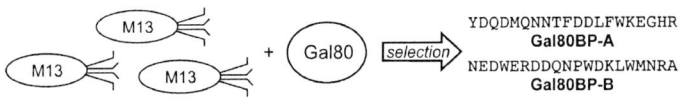

Fig. 18 Gal80-binding peptides function as activation domains. Phage display screening against the transcriptional inhibitor Gal80, both a His$_6$-tagged and a GST-tagged variant, was used to identify two new peptides, Gal80BP-A and Gal80BP-B, that function well as activation domains when fused to a DNA binding domain such as Gal4(1–147) (Han and Kodadek 2000)

as well as a glutathione S-transferase (GST)-tagged version in order to minimize the isolation of GST-binding or His6-binding peptides. Two 20-mers were identified from this and although the sequences differ from that of Gal4, they are similar in terms of ratio of hydrophobic, polar, and negatively charged amino acid residues. One of the two, Gal80BP-A, was chosen for further study and was found to bind to Gal80 with a K_d of approximately 300 nM, significantly less tightly than the native Gal4 AD with an estimated K_d of less than 25 nM. Nonetheless, Gal4 AD and Gal80BP-A appear to target the same binding site as competition experiments demonstrated that Gal4 AD was able to displace Gal80BP-A, although an allosteric mechanism cannot be ruled out. As would be expected, Gal80BP-A also functions as an AD when fused to a protein DBD. Also consistent with other activating regions, natural or artificial, this peptide interacts with the amino-terminal region of Gal11. Similar to other examples in this chapter, the linker between the DBD and the AD proved to be important. Including a 20-residue linker, either via an unrelated peptide sequence or tandem copies of the AD, significantly increased activity such that a construct containing Gal4(1–147) and two copies of Gal80BP-A activated transcription at approximately 33% of the rate of a chimera containing Gal4(1–147) and the Gal4 AD as measured in *Saccharomyces cerevisiae*. This novel strategy of targeting repressors to identify activating regions may prove useful for generating ADs that can mimic aspects of endogenous activator function beyond simple contact with the transcriptional machinery. In addition, these experiments provided additional evidence for the superimposition of several functions upon AD sequence.

4.3.3
Activation Domains That Target Gal11

Mapp and co-workers recently reported that ligands for Gal11 function as ADs when localized to DNA, presumably via a Gal11 interaction (Fig. 19) (Wu et al. 2003). The ligands were identified from a screen of two eight-residue synthetic peptide libraries, each with four randomized positions for a total of 320,000 total unique sequences. The carboxy-terminal residues of Library 1 are similar to sequences often found in eukaryotic ADs, thought to form amphipathic helices upon interaction with the transcriptional machinery. Library 2 has a proline at position 3 to bias the screen against the isolation of peptides similar to endogenous activators. From the screens, 37 peptide sequences were isolated. Unlike the examples described in earlier sections, at least the three different types of sequences were isolated, amphipathic sequences similar to endogenous ADs and the artificial ADs described earlier, hydrophobic sequences, and sequences bearing an excess of positively charged residues. This probably reflects the use of a screen rather than a selection.

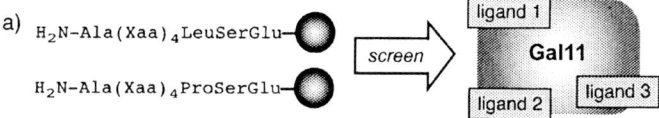
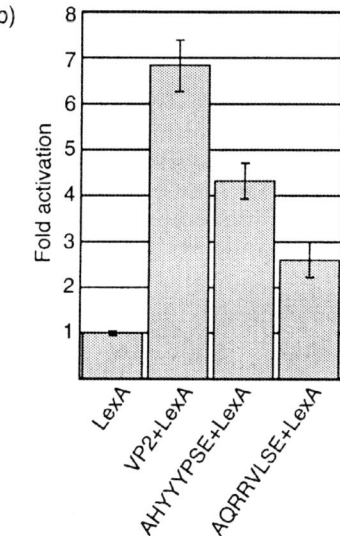

Fig. 19a, b Gal11-binding peptides isolated from a screen of synthetic peptide libraries. Two new artificial activation domains were isolated from a screen of synthetic peptide libraries (libraries 1 and 2) based upon their ability to bind to the transcriptional machinery component Gal11 (**a**). These ADs have unique sequence composition compared to previously isolated activation regions and function nearly as well as a VP16-derived activator in yeast (**b**) (Wu et al. 2003)

To test if the Gal11 binding interaction would lead to function, the ligands were fused to a protein DBD (LexA) and tested for activator function in yeast. While a number of the ligands functioned as weak activators, two were selected for further study as their sequences were unlike endogenous ADs. These two peptides, AQRRVLSE and AHYYYPSE, proved to be 38% and 65% as active as the positive control, a VP2+LexA construct, despite being only eight residues in length. Through binding affinity measurements and competition experiments, it was determined that the two new ADs target unique binding surfaces within Gal11 and that the levels of functional potency are probably related to the binding site locations. The ADs bind to Gal11 with affinities similar to those of endogenous activators such as VP2; the overall moderate levels of activity elicited by the new ADs may be a reflection of the difficulty with identifying an optimal DBD for an AD selected from a synthetic library. This can be overcome by the use of a synthetic DBD such as a TFO or polyamide that will more closely reflect the peptide presentation in the screen. This same screening strategy can also be applied to small molecule ADs, perhaps paving the way for the identification of small molecule activation domains.

4.4
RNA Activators: Nonpeptidic Activating Domains

Unlike DBDs, efforts to generate novel ADs have largely been limited to peptides. In principle, there is no reason why a nonpeptidic ligand that interacts with certain affinity with one or more components of the transcriptional machinery should not be able to function as an activating module. To test this notion, Ptashne and colleagues studied randomized RNA molecules that were tethered to DNA upstream of a reporter gene (Saha et al. 2003). The decision to test the ability of RNA to serve as an AD was based on several observations. First, short RNA molecules ('aptamers') that bind with high affinity and specificity to a variety of targets ranging from large proteins to small molecules can be readily isolated. Second, RNA molecules bear two key physical characteristics that are analogous to natural activators—an excess of negative charge and the presence of hydrophobic functional groups. Third, unlike DNA, RNA molecules can adopt a variety of three-dimensional conformations with relative ease; this property may facilitate the identification of molecules that target very specific components of the machinery. Finally, using a rapid genetic selection it is possible to scan a focused library of 1–10 million molecules to identify those that could activate transcription of reporter genes.

In the initial study, 10 residues of a short RNA hairpin were randomized and this provided a combinatorial library of approximately 10^6 molecules. To display the residues in an extended conformation, the randomized sequence was positioned in a loop region between a well-characterized palindromic sequence that forms a stable RNA duplex in yeast. The hairpins themselves are incapable of binding DNA and an additional segment of RNA that binds specifically to the MS2 bacteriophage coat protein was thus included in the construct. The coat protein itself was fused to a DBD; thus a DBD–MS2 coat protein could be targeted to a reporter gene that bears the binding sites of the DBD. The RNA library would be tethered to the MS2 coat proteins and thus be delivered to the reporter gene. RNA sequences that could stimulate reporter gene activity were found with reasonable frequency (Fig. 20). Of these, the most active eight molecules were isolated and found to have a consensus sequence motif. This came as a surprise as similar selection with peptidic activators often do not provide a readily discernable consensus (Lu et al. 2000). The hairpin scaffold within which these RNA activators (rioboactivators) were presented may have limited the choice of activa-

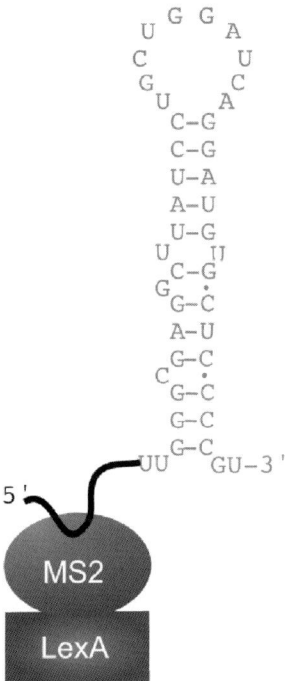

Fig. 20 Riboactivators, a new class of artificial activators. RNA hairpins such as the one shown can function as activation domains when localized to DNA through interaction with a RNA binding protein MS2 fused to a DNA binding domain (Saha et al. 2003; Buskirk et al. 2003)

tors. Moreover a limited number of residues were randomized and the authors suggested that a longer variable region could in principle yield stronger and more diverse activators. Indeed, a subsequent study by Liu and coworkers using longer randomized RNA molecules in the identical system did yield stronger activators, although these are still not as robust as the endogenous activator Gal4 or the artificial peptidic activator P201 described in the previous section (Buskirk et al. 2003).

While these studies make it clear that RNA molecules can serve as activating modules the mechanism by which they do so remains to be determined. Efforts to identify targets of the hairpin activator showed that TBP would bind robustly to these sequences under very stringent ionic conditions (1 M NaCl) and in the presence of detergents (0.5% NP40). In general, the hairpin riboactivator turned out to be very 'sticky' and was able to bind a variety of proteins (A.Z. Ansari, S. Saha, M. Ptashne, unpublished results). This may be physiologically relevant and may be the cause for the moderate activation levels observed in vivo. Thus, targets of the riboactivators remain to found and questions remain whether these activators interact with the transcriptional machinery or serve to recruit with some low affinity the natural bona fide activators which then stimulate gene expression. It should be noted however, that targets of natural activators, especially the acidic activators, have been elusive (Ptashne and Gann 2001). Thus, despite the ability of riboactivators to contact components of the transcriptional machinery, determining the physiological relevance of such contacts at the exclusion of all other possible contacts in vivo is a nontrivial task.

One approach to circumvent the problem of identifying the targets of riboactivators would be to use RNA aptamers that target specific components of the transcriptional machinery. In fact, RNA aptamers that bind RNA polymerase II and TBP have been identified (Thomas et al. 1997). In the former case these aptamers have been shown to interfere with the function of the polymerase. Tethering the RNA pol II-binding aptamers upstream of promoters may provide nonpeptidic repressor modules whereas tethering the TBP-binding aptamer may stimulate transcription. Similarly, aptamers that bind Gal11 or other components of the machinery could be isolated. The value of the studies comes not only from the demonstration that RNA molecules can function as transcriptional activators but that they may usher in the era of nonpeptidic regulatory modules. Coupled with nonpeptidic DNA-binding modules one can now reasonably imagine the generation of a new class of artificial regulators that probably will contribute to the field of transcription-based therapeutics.

5
Future Directions in the Field

5.1
Repression of Transcription

Regulatory domains that repress transcription remain relatively poorly understood, and there are only a few examples of short peptides that can function as repressors when fused to protein DBDs (Andrulis et al. 1998; Fisher and Caudy 1998; Saha et al. 1993). In general, these function only modestly and none have been tested when fused to a nonproteinaceous DBD such as TFO or a hairpin polyamide. As a result, the most common strategy has been to design DBDs to compete for the binding site of an endogenous transcription factor. This has been successfully accomplished with PNA, TFO, and polyamides with some impressive results in vitro and in vivo (Besch et al. 2002; Cutrona et al. 2000; Dickinson et al. 1998; Faria et al. 2000; Janssen et al. 2000a, 2000b). However, there are disadvantages to this strategy. It demands that one have a very precise knowledge of which binding sites are occupied in vivo and it also requires that the ATF outcompete the endogenous factor for binding to its cognate site, a requirement that is hard to achieve in cases where a large number of transcription factors bind cooperatively to adjacent sites (Maniatis et al. 1998; Merika and Thanos 2001). Thus, it may be more desirable in the long term to generate modular ATFs that have a repression module attached to a DBD, and one that can repress over varying distances from the promoter.

An interesting alternative to the above strategy has been developed by Hurley and co-workers who discovered that a cationic porphyrin, TMPyP4, can stabilize alternate DNA structures present upstream of promoters (Siddiqui-Jain et al. 2002). TMPyP4 does not bear a repression domain but by stabilizing a G-quadruplex formation upstream of a promoter, the necessary regulatory proteins are unable to bind and expression is repressed. One such promoter that has successfully been targeted is that of *c-myc*, a proto-oncogene. The inhibition of *myc* expression is correlated with inhibition of tumor growth in vivo (Grand et al. 2002). Not all porphyrins stabilize G-quadruplex structures, and a slightly different derivative failed to elicit a similar response, providing evidence for the specificity of their active compound. While this approach may not be effective for targeting any desired promoter, it is a powerful strategy for silencing genes that bear G-quadruplexes in the promoter vicinity, and molecules that stabilize other unusual DNA conformations may prove to be equally interesting as inhibitors of transcription for other classes of genes.

5.2
Cell Permeability

Delivery of ATFs into cells and organisms continues to be a considerable obstacle. For example, the two TFO-based regulators discussed in Sect. 3.3 were delivered into cells either by electroporation or with cationic liposomes (Kuznetsova et al. 1999; Stanojevic and Young 2002). This is a nonsustainable way to modulate transcriptional regulation and it severely narrows the scope of the tissues that can be targeted. In contrast, PNA can be delivered into the nuclei of cells via a short peptide, the nuclear localization signal (NLS), or the steroid dihydroxytestosterone (Boffa et al. 2000; Cutrona et al. 2000; Koppelhus and Nielsen 2003). Thus, one can envision future versions of ATFs containing a delivery module in addition to the DBD and the AD. The NLS, certain hydrophobic peptides, an arginine-rich peptide derived from the HIV transcriptional regulator Tat, as well as peptidomimetic derivatives of Tat can impart potent cell membrane transducing properties on the molecules to which they are coupled (Hawiger 1999; Rothbard et al. 2002; Schwarze and Dowdy 2000; Torgerson et al. 1998). Attaching these molecules to PNA, TFO, or polyamide ATFs could improve cellular uptake and perhaps nuclear trafficking properties. Such use of a delivery module may make ATFs of all varieties more viable in cells and organisms.

5.3
Future Applications of ATFs: Transcription-Based Therapeutics

A chapter about nonproteinaceous artificial transcriptional activators would not be complete without at least one example of how these molecules could be used in a disease model. We end, then, with a discussion about a potential long-term therapeutic target for ATFs, acute promyelocytic leukemia (APL). APL occurs due to a block in hemopoietic development of myeloid cells at the promyelocytic stage (He et al. 1999; Melnick and Licht 1999). APL is associated with chromosomal translocations that lead to a fusion of the retinoic acid receptor (RARa) DNA and ligand-binding domains to either the promyelocytic leukemia (PML) gene product, promyelocytic leukemia zinc finger gene product or nucleophosmin gene (He et al. 1999; Melnick and Licht 1999; Slack and Gallagher 1999). In healthy cells, RARa activates transcription in response to its ligand, retinoic acid (RA). However, the chromosomal translocations described above generate chimeric X-RARa proteins that function as transcriptional repressors (Slack and Gallagher 1999). Repression occurs due to interactions mediated by the fusion partners with histone deactelyases (Grignani et al. 1998; He et al. 1998; Lin et al. 1998). Consequently genes that must be activated are now actively repressed at physiological levels of RA. Moreover the X-RARa fusion is still capable of dimerizing with a partner retinoid x-receptor (RXR) (Perez et al. 1993), and it is proposed that the fusion proteins behave as dominant-negative mutants by competing with the wild-type alleles of RARa for DNA binding and RXR binding. While patients suffering from APL-associated with the PML–RARa fusion-go into remission upon treatment with high doses of RA, they inevitably suffer a relapse (He et al. 1999; Jurcic et al. 1995). Artificial transcriptional activators that target the promoter regions of genes under RARa control would represent an exciting new paradigm for the treatment of this disease. The mouse models for this form of myeloid leukemia are well established (He et al. 1999) and present an ideal system in which to test synthetic transcriptional activators. It is hoped that future generations of ATFs can be developed to target currently intractable diseases such as APL.

6
Conclusions

While there have been great strides in the development of ATFs in the past few years, many challenges remain to be addressed. Specific delivery of ATFs, and other bioactive molecules, to cells and tissues is an immediate challenge that has fueled much recent research. Perhaps the most apparent omission in the field is the lack of small molecule activation or repression domains. While it has proven surprisingly easy to generate peptidic or nucleic acid activating regions, this has not held true for other types of molecules. Part of the difficulty may arise from the fact that transcriptional regulation is largely governed by protein–protein interactions and there are few organic molecules, either natural or unnatural products, with such capabilities. This remains one of the outstanding challenges in bio-organic chemistry (Cochran 2001; Toogood 2002). Success in this area will thus have impact far beyond the development of artificial transcriptional regulators. Given the multitude of challenges and the many exciting applications for ATFs, this will remain a fruitful area of research for many years to come.

Acknowledgements AKM is grateful for financial support from the NIH (GM65330), the Burroughs Wellcome Fund (New Investigator in the Toxicological Sciences), and the March of Dimes (Basil O'Connor Starter Scholar). AZA thanks the University of Wisconsin for a Steenbock Career Development award.

References

Agre P, Johnson PF, McKnight SL (1989) Cognate DNA-binding specificity retained after leucine zipper exchange between Gcn4 and C/Ebp. Science 246:922–926

Alizadeh AA, Eisen MB, Davis RE, Ma C, Lossos IS, Rosenwald A, Boldrick JG, Sabet H, Tran T, Yu X, Powell JI, Yang LM, Marti GE, Moore T, Hudson J, Lu LS, Lewis DB, Tibshirani R, Sherlock G, Chan WC, Greiner TC, Weisenburger DD, Armitage JO, Warnke R, Levy R, Wilson W, Grever MR, Byrd JC, Botstein D, Brown PO, Staudt LM (2000) Distinct types of diffuse large B-cell lymphoma identified by gene expression profiling. Nature 403: 503–511

Andrulis ED, Neiman AM, appulla DC, Sternglanz R (1998) Perinuclear localization of chromatin facilitates transcriptional silencing. Nature 394:592–595

Ansari AZ (2001) Regulating gene expression: The design of synthetic transcriptional regulators. Curr Org Chemy 5:903–921

Ansari AZ, Mapp AK (2002a) Modular design of artificial transcription factors. Curr Opin Chem Biol 6:765–772

Ansari AZ, Mapp AK, Nguyen DH, Dervan PB, Ptashne M (2001) Towards a minimal motif for artificial transcriptional activators. Chem Biol 8:583–592

Arcamone F, Nicolell.V, Penco S, Orezzi P, Pirelli A (1964) Structure + synthesis of distamycin A. Nature 203:1064–1067

Arora PS, Ansari AZ, Best TP, Ptashne M, Dervan PB (2002) Design of artificial transcriptional activators with rigid poly-L-proline linkers. J Am Chem Soc 124:13067–13071

Baird EE, Dervan PB (1996) Solid phase synthesis of polyamides containing imidazole and pyrrole amino acids. J Am Chem Soc 118:6141–6146

Barberis A, Pearlberg J, Simkovich N, Farrell S, Reinagel P, Bamdad C, Sigal G, Ptashne M (1995a) Contact with a component of the polymerase II holoenzyme suffices for gene activation. Cell 81:359–368

Beerli RR, Barbas CF (2002) Engineering polydactyl zinc-finger transcription factors. Nature Biotechnol 20:135–141

Belitsky JM, Leslie SJ, Arora PS, Beerman TA, Dervan PB (2002a) Cellular uptake of N-methylpyrrole/N-methylimidazole polyamide-dye conjugates. Bioorg Medic Chem 10:3313–3318

Belitsky JM, Nguyen DH, Wurtz NR, Dervan PB (2002b) Solid-phase synthesis of DNA binding polyamides on oxime resin. Bioorg Medic Chem 10:2767–2774

Bender J (2001) A vicious cycle: RNA silencing and DNA methylation in plants. Cell 106:129–132

Bentley D (1999) Coupling RNA polymerase II transcription with pre-mRNA processing. Curr Opin Cell Biol 11:347–351

Besch R, Giovannangeli C, Kammerbauer C, Degitz K (2002) Specific inhibition of ICAM-1 expression mediated by gene targeting with triplex-forming oligonucleotides. J Biol Chem 277:32473–32479

Blau J, Xiao H, McCracken S, Ohare P, Greenblatt J, Bentley D (1996) Three functional classes of transcriptional activation domains. Mol Cell Biol 16:2044–2055

Boffa LC, Scarfi S, Mariani MR, Damonte G, Allfrey VG, Benatti U, Morris PL (2000) Dihydrotestosterone as a selective cellular/nuclear localization vector for anti-gene peptide nucleic acid in prostatic carcinoma cells. Cancer Res 60:2258–2262

Bouwman P, Philipsen S (2002) Regulation of the activity of Sp1-related transcription factors. Mol Cell Endocrinol 195:27–38

Braasch DA, Corey DR (2001) Synthesis, analysis, purification, and intracellular delivery of peptide nucleic acids. Methods 23: 97–107

Brent R, Ptashne M (1985) A eukaryotic transcriptional activator bearing the DNA specificity of a prokaryotic repressor. Cell 43:729–736

Brown SA, Weirich CS, Newton EM, Kingston RE (1998) Transcriptional activation domains stimulate initiation and elongation at different times and via different residues. EMBO J 17:3146–3154

Buskirk AR, Kehayova PD, Landrigan A, Liu DR (2003) In vivo evolution of an RNA-based transcriptional activator. Chem Biol 10:533–540

Cochran AG (2001) Protein-protein interfaces: mimics and inhibitors. Curr Opin Chem Biol 5:654–659

Courey AJ, Holtzman DA, Jackson SP, Tjian R (1989) Synergistic activation by the glutamine-rich domains of human transcription factor Sp1. Cell 59:827–836

Courey AJ, Tjian R (1988) Analysis of Sp1 in vivo reveals multiple transcriptional domains, including a novel glutamine-rich activation motif. Cell 55:887–898

Cutrona G, Carpaneto EM, Ulivi M, Roncella S, Landt O, Ferrarini M, Boffa LC (2000) Effects in live cells of a c-myc anti-gene PNA linked to a nuclear localization signal. Nat Biotechnol 18:300–303

Dawson PE, Muir TW, Clarklewis I, Kent SBH (1994) Synthesis of proteins by native chemical ligation. Science 266:776–779

Debin A, Malvy C, Svinarchuk F (1997) Investigation of the formation and intracellular stability of purine (puuine/pyrimidine) triplexes. Nucl Acids Res 25:1965–1974

Denison C, Kodadek T (1998) Small-molecule-based strategies for controlling gene expression. Chem Biol 5:R129–R145

Dervan PB (2001) Molecular recognition of DNA by small molecules. Bioorg Medic Chem 9:2215–2235

Dervan PB, Edelson BS (2003) Recognition of the DNA minor groove by pyrrole-imidazole polyamides. Curr Opin Struct Biol 13:284–299

Dickinson LA, Gulizia RJ, Trauger JW, Baird EE, Mosier DE, Gottesfeld JM, Dervan PB (1998) Inhibition of RNA polymerase II transcription in human cells by synthetic DNA-binding ligands. Proc Natl Acad Sci USA 95:12890–12895

Duncan SA, Navas MA, Dufort D, Rossant J, Stoffel M (1998) Regulation of a transcription factor network required for differentiation and metabolism. Science 281:692–695

Duncan SJ, Gruschow S, Williams DH, McNicholas C, Purewal R, Hajek M, Gerlitz M, Martin S, Wrigley SK, Moore M (2001) Isolation and structure elucidation of chlorofusin, a novel p53-MDM2 antagonist from a Fusarium sp. J Am Chem Soc 123:554–560

Dykxhoorn DM, Novina CD, Sharp PA (2003) Killing the messenger: Short RNAs that silence gene expression. Nat Rev Mol Cell Biol 4:457–467

Ellenberger TE, Brandl CJ, Struhl K, Harrison SC (1992) The Gcn4 basic region leucine zipper binds DNA as a dimer of uninterrupted alpha-helices—crystal-structure of the protein DNA complex. Cell 71:1223–1237

Faria M, Giovannangeli C (2001) Triplex-forming molecules: from concepts to applications. J Gene Med 3:299–310

Faria M, Wood CD, Perrouault L, Nelson JS, Winter A, White MRH, Helene C, Giovannangeli C (2000) Targeted inhibition of transcription elongation in cells mediated by triplex-forming oligonucleotides. Proc Natl Acad Sci USA 97:3862–3867

Faria M, Wood CD, White MRH, Helene C, Giovannangeli C (2001) Transcription inhibition induced by modified triple helix-forming oligonucleotides: A quantitative assay for evaluation in cells. J Mol Biol 306:15–24

Fisher AL, Caudy M (1998) Groucho proteins: transcriptional corepressors for specific subsets of DNA-binding transcription factors in vertebrates and invertebrates. Genes Dev 12:1931–1940

Frangioni JV, LaRiccia LM, Cantley LC, Montminy MR (2000) Minimal activators that bind to the KIX domain of p300/CBP identified by phage display screening. Nat Biotechnol 18:1080–1085

Garvie CW, Wolberger C (2001) Recognition of specific DNA sequences. Mol Cell 8:937–946

Gaudreau L, Keaveney M, Nevado J, Zaman Z, Bryant GO, Struhl K, Ptashne M (1999) Transcriptional activation by artificial recruitment in yeast is influenced by promoter architecture and downstream sequences. Proc Natl Acad Sci USA 96:2668–2673

Gerber HP, Seipel K, Georgiev O, Hofferer M, Hug M, Rusconi S, Schaffner W (1994) Transcriptional activation modulated by homopolymeric glutamine and proline stretches. Science 263:808–811

Giniger E, Ptashne M (1987) Transcription in yeast activated by a putative amphipathic alpha-helix linked to a DNA-binding unit. Nature 330:670–672

Giovannangeli C, Helene C (2000) Triplex-forming molecules for modulation of DNA information processing. Curr Opin Molec Therapeutics 2:288–296

Gonzalez GA, Montminy MR (1989) Cyclic-Amp stimulates somatostatin gene-transcription by phosphorylation of Creb at serine-133. Cell 59:675–680

Gottesfeld JM, Neely L, Trauger JW, Baird EE, Dervan PB (1997) Regulation of gene expression by small molecules. Nature 387:202–205

Gottesfeld JM, Turner JM, Dervan PB (2000) Chemical approaches to control gene expression. Gene Exp 9:77–91

Grand CL, Han HY, Munoz RM, Weitman S, Von Hoff DD, Hurley LH, Bearss DJ (2002) The cationic porphyrin TMPyP4 down-regulates c-MYC and human telomerase reverse transcriptase expression and inhibits tumor growth in vivo. Mol Cancer Ther 1:565–573

Grignani F, De Matteis S, Nervi C, Tomassoni L, Gelmetti V, Cioce M, Fanelli M, Ruthardt M, Ferrara FF, Zamir I, Seiser C, Lazar MA, Minucci S, Pelicci PG (1998) Fusion proteins of the retinoic acid receptor-alpha recruit histone deacetylase in promyelocytic leukaemia. Nature 391:815–818

Han Y, Kodadek T (2000) Peptides selected to bind the Gal80 repressor are potent transcriptional activation domains in yeast. J Biol Chem 275:14979–14984

Hawiger J (1999) Noninvasive intracellular delivery of functional peptides and proteins. Curr Opin Chem Biol 3:89–94

He LZ, Guidez F, Tribioli C, Peruzzi D, Ruthardt M, Zelent A, Pandolfi PP (1998) Distinct interactions of PMLRAR alpha and PLZF-RAR alpha with co-repressors determine differential responses to RA in APL. Nature Genet 18:126–135

He LZ, Merghoub T, Pandolfi PP (1999) In vivo analysis of the molecular pathogenesis of acute promyelocytic leukemia in the mouse and its therapeutic implications. Oncogene 18:5278–5292

Hidalgo P, Ansari AZ, Schmidt P, Hare B, Simkovich N, Farrell S, Shin EJ, Ptashne M, Wagner G (2001) Recruitment of the transcriptional machinery through GAL11P: structure and interactions of the GAL4 dimerization domain. Genes Dev 15:1007–1020

Himmelfarb HJ, Pearlberg J, Last DH, Ptashne M (1990) Gal11p—a yeast mutation that potentiates the effect of weak Gal4-derived activators. Cell 63:1299–1309

Hope IA, Struhl K (1986) Functional dissection of a eukaryotic transcriptional activator protein, Gcn4 of Yeast. Cell 46:885–894

Horikoshi M, Carey MF, Kakidani H, Roeder RG (1988) Mechanism of action of a yeast activator—direct effect of Gal4 derivatives on mammalian TfIId–promoter interactions. Cell 54:665–669

Jackson SP, Tjian R (1988) O-Glycosylation of eukaryotic transcription factors—implications for mechanisms of transcriptional regulation. Cell 55:125–133

Janssen S, Cuvier O, Muller M, Laemmli UK (2000a) Specific gain- and loss-of-function phenotypes induced by satellite-specific DNA-binding drugs fed to *Drosophila* melanogaster. Mol Cell 6:1013–1024

Janssen S, Durussel T, Laemmli UK (2000b) Chromatin opening of DNA satellites by targeted sequence-specific drugs. Mol Cell 6:999–1011

Jeong CJ, Yang SH, Xie YQ, Zhang L, Johnston SA, Kodadek T (2001) Evidence that Gal11 protein is a target of the Gal4 activation domain in the mediator. Biochemistry 40:9421–9427

Jurcic JG, Caron PC, Miller WH, Yao TJ, Maslak P, Finn RD, Larson SM, Warrel RP, Scheinberg DA (1995) Sequential targeted therapy for relapsed acute promyelocytic leukemia with all-trans-retinoic acid and anti-Cd33 monoclonal-antibody M195. Leukemia 9:244–248

Koh SS, Ansari AZ, Ptashne M, Young RA (1998) An activator target in the RNA polymerase II holoenzyme. Mol Cell 1:895–904

Koppelhus U, Nielsen PE (2003) Cellular delivery of peptide nucleic acid (PNA). Adv Drug Del Rev 55:267–280

Kuznetsova S, Ait-Si-Ali S, Nagibneva I, Troalen F, Le Villain JP, Harel-Bellan A, Svinarchuk F (1999) Gene activation by triplex-forming oligonucleotide coupled to the activating domain of protein VP16. Nucl Acids Res 27:3995–4000

Lamarre-Vincent N, Hsieh-Wilson LC (2003) Dynamic glycosylation of the transcription factor CREB: A potential role in gene regulation. J Am Chem Soc 125:6612–6613

Lamph WW, Dwarki VJ, Ofir R, Montminy M, Verma IM (1990) Negative and positive regulation by transcription factor camp response element-binding protein is modulated by phosphorylation. Proc Natl Acad Sci USA 87:4320–4324

Ledoan T, Perrouault L, Chassignol M, Thuong NT, Helene C (1987) Sequence-targeted chemical modifications of nucleic-acids by complementary oligonucleotides covalently linked to porphyrins. Nucl Acids Res 15:8643–8659

Lee TI, Young RA (2000) Transcription of eukaryotic protein-coding genes. Annu Rev Genet 34:77–137

Levy DE, Darnell JE (2002) STATs: Transcriptional control and biological impact. Nat Rev Mol Cell Biol 3:651–662

Lin RJ, Nagy L, Inoue S, Shao WL, Miller WH, Evans RM (1998) Role of the histone deacetylase complex in acute promyelocytic leukaemia. Nature 391:811–814

Lin YS, Carey MF, Ptashne M, Green MR (1988) Gal4 derivatives function alone and synergistically with mammalian activators in vitro. Cell 54:659–664

Liu B, Han Y, Corey DR, Kodadek T (2002) Toward synthetic transcription activators: Recruitment of transcription factors to DNA by a PNA-peptide chimera. J Am Chem Soc 124:1838–1839

Liu DY, Liu XY, Robinson D, Seele L, Veach RA, Timmons S, Collins RD, Ballard DW, Hawiger J (2001) Transgene- and peptide-directed inhibition of transcription factor NF-kappa B protects mice from superantigen-induced lethal shock. Blood 98:37B–37B

Lockhart DJ, Winzeler EA (2000) Genomics, gene expression and DNA arrays. Nature 405:827–836

Lu XY, Ansari AZ, Ptashne M (2000) An artificial transcriptional activating region with unusual properties. Proc Natl Acad Sci USA 97:1988–1992

Lu Z, Ansari AZ, Lu XY, Ogirala A, Ptashne M (2002) A target essential for the activity of a nonacidic yeast transcriptional activator. Proc Natl Acad Sci USA 99:8591–8596

Lue NF, Flanagan PM, Kelleher RJ, Edwards AM, Kornberg RD (1991) RNA polymerase-II transcription in vitro. Methods Enzymol 194:545–550

Lue NF, Flanagan PM, Sugimoto K, Kornberg RD (1989) Initiation by yeast RNA polymerase-II at the adenoviral major late promoter in vitro. Science 246:661–664

Ma J, Ptashne M (1987) A new class of yeast transcriptional activators. Cell 51:113–119

Maniatis T, Falvo JV, Kim TH, Kim TK, Lin CH, Parekh BS, Wathelet MG (1998) Structure and function of the interferon-beta enhanceosome. Cold Spring Harbor Symp Quant Biol 63:609–620

Mannervik M, Nibu Y, Zhang H, Levine M (1999) Transcriptional coregulators in development. Science 284:606–609

Mapp AK (2003) Regulating transcription: a chemical perspective. Org Biomol Chem 1:2217–2220

Mapp AK, Ansari AZ, Ptashne M, Dervan PB (2000) Activation of gene expression by small molecule transcription factors. Proc Natl Acad Sci USA 97:3930–3935

Mapp AK, Dervan PB (2000) Preparation of thioesters for the ligation of peptides with non-native substrates. Tetrahedron Lett 41:9451–9454

Marks PA, Rifkind RA, Richon VM, Breslow R, Miller T, Kelly WK (2001) Histone deacetylases and cancer: Causes and therapies. Nat Rev Cancer 1:194–202

Mayr B, Montminy M (2001) Transcriptional regulation by the phosphorylation-dependent factor CREB. Nat Rev Mol Cell Biol 2:599–609

Melnick A, Licht JD (1999) Deconstructing a disease: RAR alpha, its fusion partners, and their roles in the pathogenesis of acute promyelocytic leukemia. Blood 93:3167–3215

Merika M, Thanos D (2001) Enhanceosomes. Curr Opin Genet Dev 11:205–208

Mermod N, Oneill EA, Kelly TJ, Tjian R (1989) The proline-rich transcriptional activator of Ctf/Nf-I Is distinct from the replication and DNA-binding domain. Cell 58:741–753

Miller PS (1996) Development of antisense and antigene oligonucleotide analogs. Progr Nucl Acid Res Mol Biol 52:261–291

Moser HE, Dervan PB (1987) Sequence-specific cleavage of double helical DNA by triple helix formation. Science 238: 645–650

Myers LC, Kornberg RD (2000) Mediator of transcriptional regulation. Annu Rev Biochem 69:729–749

Nielsen PE (2001) Peptide nucleic acid targeting of double-stranded DNA. Drug–Nucl Acid Interact 340:329–340

Oshea EK, Klemm JD, Kim PS, Alber T (1991) X-Ray structure of the Gcn4 leucine zipper, a 2-stranded, parallel coiled coil. Science 254:539–544

Pabo CO, Peisach E, Grant RA (2001) Design and selection of novel Cys(2)His(2) zinc finger proteins. Annu Rev Biochem 70:313–340

Pandolfi PP (2001) Transcription therapy for cancer. Oncogene 20:3116–3127

Park JM, Kim HS, Han SJ, Hwang MS, Lee YC, Kim YJ (2000) In vivo requirement of activator-specific binding targets of mediator. Mol Cell Biol 20: 8709–8719

Perez A, Kastner P, Sethi S, Lutz Y, Reibel C, Chambon P (1993) PML RAR homodimers—distinct DNA-binding properties and heteromeric interactions with Rxr. EMBO J 12:3171–3182

Perou CM, Sorlie T, Eisen MB, van de Rijn M, Jeffrey SS, Rees CA, Pollack JR, Ross DT, Johnsen H, Aksien LA, Fluge O, Pergamenschikov A, Williams C, Zhu SX, Lonning PE, Borresen-Dale AL, Brown PO, Botstein D (2000) Molecular portraits of human breast tumours. Nature 406:747–752

Ptashne M (1997) Control of gene transcription: An outline. Nat Med 3:1069–1072

Ptashne M, Gann A (1997) Transcriptional activation by recruitment. Nature 386:569–577

Ptashne M, Gann A (2001) Genes & Signals. Cold Spring Harbor Laboratory, New York

Radhakrishnan I, Perez-Alvarado GC, Parker D, Dyson HJ, Montminy MR, Wright PE (1999) Structural analyses of CREB-CBP transcriptional activator-coactivator complexes by NMR spectroscopy: Implications for mapping the boundaries of structural domains. J Mol Biol 287:859–865

Radhakrishnan I, PerezAlvarado GC, Parker D, Dyson HJ, Montminy MR, Wright PE (1997) Solution structure of the KIX domain of CBP bound to the transactivation domain of CREB: A model for activator:coactivator interactions. Cell 91:741–752

Rothbard JB, Kreider E, VanDeusen CL, Wright L, Wylie BL, Wender PA (2002) Arginine-rich molecular transporters for drug delivery: role of backbone spacing in cellular uptake. J Med Chem 45:3612–3618

Sadowski I, Ma J, Triezenberg S, Ptashne M (1988) Gal4-Vp16 is an unusually potent transcriptional activator. Nature 335:563–564

Sadowski I, Niedbala D, Wood K, Ptashne M (1991) Gal4 is phosphorylated as a consequence of transcriptional activation. Proc Natl Acad Sci USA 88:10510–10514

Saha S, Ansari AZ, Jarrell KA, Ptashne M (2003) RNA sequences that work as transcriptional activating regions. Nucl Acids Res 31:1565–1570

Saha S, Brickman JM, Lehming N, Ptashne M (1993) New eukaryotic transcriptional repressors. Nature 363:648–652

Salghetti SE, Caudy AA, Chenoweth JG, Tansey WP (2001) Regulation of transcriptional activation domain function by ubiquitin. Science 293:1651–1653

Schwarze SR, Dowdy SF (2000) In vivo protein transduction: intracellular delivery of biologically active proteins, compounds and DNA. Trends Pharmacol Sci 21:45–48

Sedelnikova OA, Panyutin IG, Luu AN, Neumann RD (1999) The stability of DNA triplexes inside cells as studied by iodine-125 radioprinting. Nucl Acids Res 27:3844–3850

Shi YG, Massague J (2003) Mechanisms of TGF-beta signaling from cell membrane to the nucleus. Cell 113:685–700

Siddiqui-Jain A, Grand CL, Bearss DJ, Hurley LH (2002) Direct evidence for a G-quadruplex in a promoter region and its targeting with a small molecule to repress c-MYC transcription. Proc Natl Acad Sci USA 99:11593–11598

Slack JE, Gallagher RE (1999) The molecular biology of acute promyelocytic leukemia. Cancer Treat Res 99:75–124

Smith RL, Johnson AD (2000) Turning genes off by Ssn6-Tup1: a conserved system of transcriptional repression in eukaryotes. Trends Biochem Sci 25:325–330

Stad R, Little NA, Xirodimas DP, Frenk R, van der Eb AJ, Lane DP, Saville MK, Jochemsen AG (2001) Mdmx stabilizes p53 and Mdm2 via two distinct mechanisms. EMBO Rep 2:1029–1034

Stanojevic D, Young RA (2002) A highly potent artificial transcription factor. Biochemistry 41:7209–7216

Stoll R, Renner C, Hansen S, Palme S, Klein C, Belling A, Zeslawski W, Kamionka M, Rehm T, Muhlhahn P, Schumacher R, Hesse F, Kaluza B, Voelter W, Engh RA, Holak TA (2001) Chalcone derivatives antagonize interactions between the human oncoprotein MDM2 and p53. Biochemistry 40:336–344

Tanaka M (1996) Modulation of promoter occupancy by cooperative DNA binding and activation-domain function is a major determinant of transcriptional regulation in vivo. P Natl Acad Sci USA 93:4311–4315

Thomas M, Chedin S, Carles C, Riva M, Famulok M, Sentenac A (1997) Selective targeting and inhibition of yeast RNA polymerase II by RNA aptamers. J Biol Chem 272:27980–27986

Toogood PL (2002) Inhibition of protein-protein association by small molecules: Approaches and progress. J Medic Chem 45:1543–1558

Torgerson TR, Colosia AD, Donahue JP, Lin YZ, Hawiger J (1998) Regulation of NF-kappa B, AP-1, NFAT, and STAT1 nuclear import in T lymphocytes by noninvasive delivery of peptide carrying the nuclear localization sequence of NF-kappa B p50. J Immunol 161:6084–6092

Trauger JW, Baird EE, Dervan PB (1996) Recognition of DNA by designed ligands at subnanomolar concentrations. Nature 382:559–561

Weiss MA, Ellenberger T, Wobbe CR, Lee JP, Harrison SC, Struhl K (1990) Folding transition in the DNA-binding domain of Gcn4 on specific binding to DNA. Nature 347:575–578

Wells L, Whalen SA, Hart GW (2003) O-GlcNAc: a regulatory post-translational modification. Biochem Biophys Res Commun 302:435–441

Wu Z, Belanger G, Brennan BB, Lum JK, Minter AR, Rowe SP, Plachetka A, Majmudar CY, Mapp AK (2003) Targeting the transcriptional machinery with unique artificial transcriptional activators. J Am Chem Soc 125:12390–12391

Wurtz NR, Turner JM, Baird EE, Dervan PB (2001) Fmoc solid phase synthesis of polyamides containing pyrrole and imidazole amino acids. Org Lett 3:1201–1203

Yaghmai R, Cutting GR (2002) Optimized regulation of gene expression using artificial transcription factors. Mol Ther 5:685–694

Zanta MA, Belguise-Valladier P, Behr JP (1999) Gene delivery: A single nuclear localization signal peptide is sufficient to carry DNA to the cell nucleus. Proc Natl Acad Sci USA 96:91–96

Zor T, Mayr BM, Dyson HJ, Montminy MR, Wright PE (2002) Roles of phosphorylation and helix propensity in the binding of the KIX domain of CREB-binding protein by constitutive (c-Myb) and inducible (CREB) activators. J Biol Chem 277:42241–42248

Subject Index

AAV vector   517, 519, 521
acetylation   7, 71, 142, 475
activation domain (AD)   538, 552
activin   172
acute lymphoblastic leukemia (ALL)   266
- myeloid leukemia (AML)   266
- promyelocytic leukemia (APL)   564
Ad vector   518, 522
adaptor protein disabled-2 (Dab-2)   177
adenovirus   502
- E2 gene promoter   279
AF-1/2   463
affinity   473
Affymetrix GeneChips   72, 79, 80
aflatoxin B1   224
aging   230
Akt   177
Alk 1   192
Alk 5   192
ALL (acute lymphoblastic leukemia)   266
allosteric effector   474
alpha fetoprotein   395
alternative splicing   145
ALY   152
amino-terminal transregulatory domain   310
aminothiol   237
AML (acute myeloid leukemia)   175, 191, 266
amphipathic helix   553
anchor finger   495
angiogenesis   192, 218
ankyrin   343
ANOVA   83

antibiotics   514
anticancer drug   239
antiestrogen   458
antisense reagent   496
AP-1   184, 471
APC   131, 140, 153, 156, 220
APL (acute promyelocytic leukemia)   564
apoptosis   179, 214, 288, 318
- transactivation-independent   216
architectural transcription factor   128, 151
ARF   225, 289
armadillo   131, 142
artificial transcriptional factor (ATF)   536
ASPP family members   217
ATF (artificial transcriptional factor)   498, 536
ATF-2   184
- polyamid-based   541
atherosclerosis   239, 347
ATM   219
- mutations   225
atopic dermatitis   447
ATP-chromatin remodeling complex   476
axin   131, 133, 136, 153, 155

B42   552
basal transcriptional machinery   48
bax   501
Bcl 2   229
Bcl 3   329
bHLHZIP protein   174
bidirectional tetO promoter   524
bilirubin   425
BMP   172

bone metastasis 186, 187
– morphogenetic protein 4 (BMP4) 147
breast cancer 227, 456
– risk factors 456
Brg1 51, 150
Brinker 58
Brm 51
bromodomain 15

cancer 315, 320
– colorectal 156
– of the breast 227, 456
– therapy 167, 231
CAR (constitutive androstane receptor) 412, 414, 419
– signal 422
carcinogen 229
carcinoma 228
CARD (caspase recruitment domain) 334
catenin
– β-catenin 124, 31, 141, 147, 150, 151, 154, 220
– γ-catenin 131
CBF-1 48
CBP 6, 135, 142, 190
– coactivator 16
CDB3 236
CDK (cyclin-dependent kinase) 278
– inhibitor 181, 238
Cdx1 130
cell competence 172
cell cycle 278, 280, 293, 300, 390
– – arrest 181, 214
– – regulation 293
– motility 186
– permeability 563
– transformation 392
cellular
– network 301
– proliferation 282
– senescence 218
– transformation 262, 264
c-fos promoter 549
chemoprevention 458
chemosensitivity 233
chemotherapeutic agent 232
chicken ovalbumin upstream promoter transcription factor (COUP-TFI/II) 47
ChIP (chromatin immunoprecipitation) 43, 471

– assay 294
Chk2 220
chromatin 15, 173
– immunoprecipitation (ChIP) 10, 43, 71, 87
– – assay 76
– remodeling 18, 51, 467
– structure 150
chromodomain 15
chronic inflammatory disease 344
ciliary neurotrophic factor (CNTF) 111
cis-regulatory element 37
c-jun 153
CK Iε 140
CK II 141
c-*myc* 6, 133, 153, 156, 181, 289, 310, 563
– targets 314
coactivator 126, 150, 173, 475
cofactor recruitment 130
colchicine 427
colorectal cancer 156
competition 47
conductin 133
conformation 474
conformational difference 467
constitutive androstane receptor (CAR) 414
core antigen 378
corepressor 39, 126, 150, 173, 477
CoREST 52
corticosteroid 448
Coup-TF (chicken ovalbumin upstream promoter transcription factor) 47
CP31398 237
CRARF domain 129, 135
Cre recombinase 107
Cre/loxP system 227
CREB 6, 12, 75, 538, 556
CRM1 296
CSL 48
CtBP (C-terminal-binding protein) 52, 58, 133, 142, 150, 154, 155, 191
CTGF 187
Cushing syndrome 352
cyclin D1 133, 153
cyclin-dependent kinase (CDK) 278, 315
cyclophosphamide 232
CYP (cytochrome P450) 410
– 3A 413
– gene promoters 411

cytochrome C release   216
– P450 (CYP)   410
cytokine   167, 341, 424
– expression   338
– signaling pathways   99

Dab-2 (adaptor protein disabled-2)   177
DBD (DNA-binding domain)   463, 547
death receptor pathway   215
decoy   440, 444
dermatitis   343
destruction complex   125
dexamethasone   413, 419, 427
DHFR (dihydrofolate reductase)   283
dihydrofolate reductase (DHFR)   283
dihydroxytestosterone   563
dimerization domain   542
dimethylbenzanthracene   230
dimethylnitrosamine   229
distamycin   541
DNA
– adducts   458
– bending   128, 151, 152
– binding   128, 263
– – activity   170
– – cofactors   174
– – domain (DBD)   5, 267, 538
– – motif   551
– contact mutant   223
– damage   317, 319
– double-strand break repair   211
– excision repair   218
– methylation   54
– methyltransferase   54
– microarrays   10
– mismatch repair   154
– synthesis   291, 292
dominant-negative caspase 9
   (C9DN)   229
dopachrome tautomerase   136
Dorsal   56
downstream promoter element (DPE)   8
doxycycline   498, 513
DPE (downstream promoter element)   8
DPN   466
Duchenne muscular dystrophy   500
dye swap   81

E(z)   55
E1A   142

E1AF   262
E2F   49, 81, 181, 286, 443
– family   278
E2F1   220
– apoptosis   288
– gene knockouts   284
E2F2   289
E2F3   290
E3 ubiquitin ligase   172
E-cadherin   183
Ectromelia virus (EV)   105
Elf/Ese subfamily   262
ellipiticine   237
Emμ-Myc
– lymphoma   232
– mice   229
EMSA   468
encephalomyelitis   343
ENCODE Project   87
endogenous estrogen   456
endoglin   193
endometrial cancer   478
enhancer   9, 37
enterocolitis   108
epigenetic event   187
epithelial plasticity   183
epithelial-to-mesenchymal transition
   182
Epo   104
ER (estrogen receptor)   57, 457
ER α 459, 479
– degradation   469
ER β 460, 479
ERKO   461
erythropoietin   520
E-selectin   349
17β-estradiol   81, 456
estrogen
– receptor (ER)   57, 457
– response element (ERE)   84, 465
Ets
– factor   263
– – overexpression   264
– mediated transformation-specific
   signaling   269
– target genes   264
– transcription factor   260
Ets2DBD   268
ETV4   262
even-skipped gene

Evi-1   191
Ewing sarcoma tumor cell   268
Ewing's tumor   265

F domain   467
Faslodex (Fulvestrant)   461
FAST-1   174
Fli1   265
forkhead transcription factor FKHRL1   180
Friend leukemia virus   226
5-FU   233
FXR ligand   417

GADD45   214
gain-of-function mutant   228
GAL4   310
– dimerization   495
Gal4   4, 50, 552, 557
Gal11   555, 558
Gal80   50, 557
galactose metabolism   537
gamma irridation   232
gastric epithelial cell   180
Gcn5   15
G-CSF   108
gel mobility shift assay   444
gene network   87
– therapy   234, 441, 443, 448, 499, 509, 510, 514
genomic heterogeneity   317
– instability   316
genomics   72
genotoxic agent   224
glomerulonephritis   445
glucocorticoid   352
– hormones   419
– receptor (GR)   49, 352, 411, 418
– response element (GRE)   418
Grg   132, 150
Groucho   52, 57
ground squirrel hepatitis virus (GSHV)   378
growth factor   153
GS domain   169
GSK3β 140, 154

hairpin activator   561
HBV
– genome   393
– polymerase   383
HBx   378, 379
– homo-dimerization   380
Hda1   52
HDAC (histone deacetylase)   51, 52, 151, 297, 477
HDAC 1   143
helix-loop-helix motif (HLH motif)   329
hepadnavirus   378
hepatocarcinogenesis   393
hepatocellular carcinoma (HCC)   381
hepatocyte   392, 414, 422
hepatoma cell   391
HER-2/Neu   189
hereditary hemorrhagic telangiectasia   193
heterochromatin   282
heterodimer   467
heterodimerization domain   310
hierarchical cluster (HC)   83
HIF1α 85, 221
high-mobility group (HMG)   126
hinge region   465
histone   15
– acetyltransferase (HAT)   16, 425
– deacetylase (HDAC)   132, 297
– methyltransferase   51, 53
HIV (human immunodeficiency virus)   152, 441
HLH (helix-loop-helix) motif   330
HMEC (human mammary epithelial cell)   318, 319
HMG (high-mobility group) domain   128, 134, 138
holoenzyme   13
homeostasis   73, 167, 391
hormone replacement therapy   456
HP1   55
hsp 70   469
hsp 90   468
human immunodeficiency virus (HIV)   152
– mammary epithelial cell (HMEC)   318
– T lymphotrophic virus (HTLV)   383
Huntington disease   525
hydrocephalus   298
4-hydroxytamoxifen   459
hyperbilirubinemia   425
hypercholesterolemia   354
hyperglycaemia   349
hypertension   349

ICAM (intercellular adhesion molecule) 445
ICAM 1   339
ICI164,384   460
ICI182,780   460
IGF-BP 3   215
IGF-BP 5   109
IKK 2   355
– dimer   330
– signalosome   329, 333
immunogenicity   513, 522
immunoprecipitation assay   295
inactivating mutation   187
incontinentia pigmenti   343
inflammation   239, 332, 338, 424
– acute   344
– chronic   344, 351
– phases   339
inhibitor of κB (IκB)   327
initiator element   8
INK4a   225
innate immunity   338
intercellular adhesion molecule (ICAM)   445
interferon (IFN)
– IFN-α 101, 102
– IFN-β 101, 102, 333
– – enhanceosome   9
– IFN-γ 102
– IFN-4   112
interleukin (IL)
– IL-1   331, 335
– – receptor antagonist   355
– IL-1β 334, 336
– IL-11   187
– IL-12   103
– – p40   108
– IL-2   104
– IL-4   112
involution   178
IRAK1   336
ISWI   19

JAK (Janus kinase)   98, 100

keratin 5   110
keratinocyte   109
Kip1   320
KIX ligand   557
Knirps   47, 58
knockout mice   100

LDL (low density lipoprotein)   347
LEF1   135, 136, 138, 139, 143–145
lentiviral vector   516, 519
leptomycin B   238
leukocyte   339
– apoptosis   342
Li-Fraumeni syndrome   227
ligand-binding pocket   466
linearized plasmid   554
linker domain   172
Lit-1   138, 139
LMP-2   48
loop
– design   78
– L3 loop   171
– L45 loop   171
low density lipoprotein (LDL)   347
LOWESS   82
LXXLL   468
lymphotoxin β (LTβ)   338
lysozyme M   107

malignant progression   185
mammalian cell cycle   278
mammary epithelium   167
– gland   109
MAPK pathway   344
matrix metalloproteinase (MMP)   340
Max   300, 311
MBI/II   311
MCF7   74
MDM2   219
Mdm2   6, 50
mediator   14
MEF (mouse embryonic fibroblast)   290, 292, 298, 330
metastasis   185, 218
methylation   7, 16
methylphosphonation   449
methyltransferase   17, 475
Mga   300
MH1 domain   170
MIAME standard   79
microarray   77
– normalization   82
microsatellite instability   154, 188
microtubule   426
microvessel density   193
migration   178
MIP-1α/β 102

missense mutation   188, 222
MITF   136, 152
mitochondrial pathway   215
mitogen-activated protein (MAP) kinase   382, 389
– phosphorylation   172
mLEF1   139
MMP-1   346
motoneuron   110
mouse embryonic fibroblast (MEF)   330
– model   226
mutant p53 reactivation   236
mutator phenotype   316
Myc   221, 300
myocardial infarction   446

NCoR   48, 52, 472, 477
NEMO   329, 337, 354
neointimal hyperplasia   444
NES (nuclear export signal)   328
neural induction   172
neurodegenerative
– disease   523
– disorder   239
neutral sphingomyelinase   336
NFAT   12
NIH3T3   317
NLK   138, 139
NLS (nuclear localizations signal)   296, 311
NSAID (non-steroidal anti-inflammatory drug)   352
nuclear export signal (NES)   328
– factor-κB (NF-κB)   7, 326, 425, 445
– hormone receptor   6
– import   171
– localizations signal (NLS)   296
nullizygosity   287

OAZ   174
oligodeoxynucleotide (ODN)   440, 446
oligonucleotide   79, 549
– microarray   72
ONYX015   235
organ site tropism   186
osteocalcin   134

p15^{Ink4b} 181
p19ARF   220

p21
– gene   214
– p21^{Cip1} 181
p27   320
– p27^{Kip1} 191
p38 MAP kinase   184
p53   6, 50, 174, 286, 391
– consensus binding site   211
– dependent repression   216
– heterozygous mice   226
– mutations   222
– negative mouse embryos   288
– null mice   225
– protein   210, 213
– status   231, 233
– target genes   212
– tumor-derive mutants   216
p63   217
p70(s6 k)   182
p73   217, 288
p130   297
p300   16, 135, 142, 150, 190, 217
palindromic
– binding site   544
– consensus ERE sequence   473
PAM   84
Pangolin   142
papilloma   186
Parkinson disease   522
pathogen-associated molecular pattern (PAMP)   332
PcG   55
PEA3   262
peptide growth factors   168
– nucleic acid (PNA)   550
peroxisome proliferator activated receptor-γ (PPAR-γ)   500
phosphorothioation   449
phosphorylation   7, 71, 169, 280, 464
Pi3 kinase   177
PI3K/Akt   222
PIASy   143, 144
pifithrin-α 236
Pit-1   56
PKC (protein kinase C)   337
plakoglobin   131
PML body   144
PNA (peptide nucleic acid)   550
pocket protein   302
polyamid   541, 542, 545, 560

Subject Index

polycomb 55
polymerase 378
polymorphism 188
polyproline 545
polyubiquination 327
Pontin52 150
Pop1 139, 142
post-translational modification 221
POU domain 71
PPAR (peroxisome proliferator activated receptor) 353
PPAR γ 500
PPT 466
pRb (product of the retinoblastoma tumor suppressor gene) 279, 286, 294
- protein family members 300
- wild-type 281
PRC-1 55
pregnane X
- receptor (PXR) 412
- response element (PXRE) 420
preinitiation complex 11
PRIMA-1 237
prion disease 526
proapoptotic activity 392
product of the retinoblastoma tumor suppressor gene (pRb) 279
progesterone receptor (PR) 456
promoter 8, 37, 181
- interference 516
protease activity 390
proteasome 7, 354, 390, 469
protein inhibitor of the activated STAT (PIAS) 113
- kinase C (PKC) 337, 382
- phosphatase-2A (PP2A) 177
protein-protein interaction 171
proto-oncogene 125, 154, 283, 310
P-selectin 349
PTEN 215
PTHrP 186
PU.1 265
PXR (pregnane X receptor) 412, 419
- signal 422

raloxifene (RAL) 458
RANK (receptor activator of NF-κB) 346
rapamycin 499
RAR (retinoic acid receptor) 564

Ras 183
- oncogenic 189
- signaling pathways 266, 269
Ras-Raf-MAP kinase 388
rat embryonic cell (REC) 319
Rat1A 317
Rb (retinoblastoma) 49, 52
RBP-Jκ 48
receptor
- serine/threonine kinase 169
- tyrosine kinase 172
Rel-homology domain (RHD) 326
respiratory syncytical virus (RSV) 105
restenosis 443
retinoblastoma (Rb) 49, 315, 443
- phosphorylation 315
retinoic acid receptor (RAR) 564
retroviral vector 515
retrovirus 502
rheumatoid arthritis (RA) 345
RhoA 177
riboactivator 560
ribozyme 235
RIF 423
RNA
- interference (RNAi) 71, 73
- metabolism 502
- molecule 560
- polymerase 4, 441, 475, 555
roscovitine 238
Rpd3 52
RSV (respiratory syncytical virus) 105
Runx 134, 175, 180

SAGE (serial analysis gene expression) 73
Saos-2 280
sarcoma 228
SCF (stem cell factor) 105
SELEX assay 494
serial analysis gene expression (SAGE) 73
SERM 460, 480
- ER-independent activities 479
SET domain 53
short heterodimer partner (SHP) 416
short interfering RNA (siRNA) 75
14-3-3 sigma 214
signal transduction 124, 380, 387, 388
Sin3 52
Sin3 a 217

SIP1   185
siRNA (short interfering RNA)   75, 235
skeletal bone   186
Ski   190
Slug   185
SMAD   169
– Co-SMADs   170
– inhibitory   170
– receptor-activated   170
Smad   129, 136, 152
Smad 5   193
Smad 7   179
small molecule inhibitor   131
SMRT   48, 52, 472, 477
Smurf1/2   190
Snail   185
SNF2   18
SnoN   190
SOCS   111
Sp1   12
spindle cell carcinoma   186
SRC (steroid receptor coactivator)   20
SREBP (sterol regulatory element-binding protein)   7
SSXF motif   171
STAT   98, 100
– knockout mice   101
stem cell   152
– factor (SCF)   105
steroid receptor coactivator (SRC)   20
steroid/nuclear receptor (SN/NR) superfamily   462
sterol regulatory element-binding protein (SREBP)   7
structural mutant   223
Su(H)   48
SU(var)3-9   53
SUMO   137, 143
sumoylation   138
surface antigen   378
survival   179
SUV39H1   53
SWI/SNF   18
synoviocyte   346, 347
synthetic peptide   236

T cell   107
– apoptosis   285
– factor   124
TAD (transcriptional activation domain)   5

TAF9   6
tamoxifen (TAM)   57, 457
– agonist activity   478
– metabolites   458, 459
TAR decoy   441
target gene   194
TATA box   8, 70
TATA-binding protein   6
TBP-associated factor   11
TCF 1   137, 147, 157
TCF 3   135, 145
TCF 4   139, 143–145, 155, 157
TCF 4E   129
TCF protein   129
TCRα   130, 134, 137, 152
Tel   265
TEL2   260
– gene family   260
tethering mechanism   471
tet-off system   513
tet-on system   513
tetracycline   498, 509, 511, 514
– repressor   4
12-O-tetradecanoyl-phorbol-13-acetat (TPA)   230
TF signaling network   86
TFII A   11, 12
TFII B   11, 13
TFII D   11, 70
TFII E   11
TFII F   11
TFII H   11, 13
TFO (triplex-forming oligonucleotide)   546
TGF-β   136, 155, 167
– antagonist   186
β-thalassemia   521
thiazoledinedione (TZD)   353
thymic epithelium   110
thymidin kinase (TK)   283
thyroid hormone receptor   57
TLE   52, 132, 150
TLR (Toll-like receptor)   332
TNF-α   331, 335, 341, 350
Toll-like receptor (TLR)   332
total gene expression analysis (TOGA)   72
Tpo   104
TRADD   335
TRAM   334

transactivation   381, 383, 388
transactivator   510, 511
transcription factor   211, 457, 492
- therapy   499
transcriptional
- activation   130, 216, 476
- - domain (TAD)   5
- activator protein   4
- apparatus   384
- control   126
- corepressor   132
- repression   130
- - mechanisms   45
- silencer   519
transducin-like-enhancer-of-split
   in humans   132
transphosphorylation   382
trichostatin A (TSA)   52
TRIF   334
tripartite model   472
triple knockout mouse   299
triplex-forming oligonucleotide
   (TFO)   546, 547, 551
TSA (trichostatin A)   52
tumor
- latency   186
- profiling   233
- progression   155
- promoter   167
- suppression   210
- suppressor   125, 156, 167
tumorigenesis   155, 210, 286, 315
Tup1   49

ubiquitin   7
ubiquitination   313

ubiquitinylation   7
upstream activating sequence   9
UV irridation   221
- radiation   223

vascular
- cellular adhesion molecule
   (VCAM)   339, 445
- endothelial growth factor (VEGF)   500
- smooth muscle cell (VSMC)   443
VDR (vitamine D receptor)   415
- signal   422
VEGF (vascular endothelial growth factor)
   500
v-*ets*   260
viral vector   509, 515
virus life cycle   393
vitamine D   423
- receptor (VDR)   415
- response elements (VDRE)   416
v-*myc*   309
VP16 protein   4

Wig-1 protein   219
Wnt   147, 149
- growth factor   124, 130
- signaling   135, 142, 147, 150
woodchuck hepatitis virus (WHV)   378

X ORF   378

ZEB protein   175
ZFP   494
zinc finger   5, 464, 491
- nuclear   219
- protein   536

Printing: Saladruck, Berlin
Binding: Stein+Lehmann, Berlin